TETRAHEDRON ORGANIC CHEMISTRY SERIES
Series Editors: J E Baldwin, FRS & P D Magnus, FRS

VOLUME 15

Expanded, Contracted

& Isomeric Porphyrins

Related Pergamon Titles of Interest

BOOKS

Tetrahedron Organic Chemistry Series:
CARRUTHERS: Cycloaddition Reactions in Organic Synthesis
DEROME: Modern NMR Techniques for Chemistry Research
GAWLEY & AUBÉ: Principles of Asymmetric Synthesis
HASSNER & STUMER: Organic Syntheses based on Name Reactions and Unnamed Reactions
LEVY & TANG: The Chemistry of *C*-Glycosides
PAULMIER: Selenium Reagents & Intermediates in Organic Synthesis
PERLMUTTER: Conjugate Addition Reactions in Organic Synthesis
SIMPKINS: Sulphones in Organic Synthesis
WILLIAMS: Synthesis of Optically Active Alpha-Amino Acids 2nd Edn*
WONG & WHITESIDES: Enzymes in Synthetic Organic Chemistry

JOURNALS

BIOORGANIC & MEDICINAL CHEMISTRY
BIOORGANIC & MEDICINAL CHEMISTRY LETTERS
TETRAHEDRON
TETRAHEDRON: ASYMMETRY
TETRAHEDRON LETTERS

Full details of all Elsevier Science publications/free specimen copy of any Elsevier Science journal are available on request from your nearest Elsevier Science office

*In Preparation

Expanded, Contracted & Isomeric Porphyrins

JONATHAN L. SESSLER

and

STEVEN J. WEGHORN

The University of Texas at Austin

PERGAMON

U.K.	Elsevier Science Ltd., The Boulevard, Langford Lane, Kidlington, Oxford OX5 1GB, U.K.
U.S.A.	Elsevier Science Inc., 665 Avenue of the Americas, New York, NY 10010, U.S.A.
JAPAN	Elsevier Science Japan, Tsunashima Building Annex, 3-20-12 Yushima, Bunkyo-ku, Tokyo 113, Japan

First Edition 1997

Library of Congress Cataloging in Publication Data

A catalog record for this book is available from the Library of Congress

British Library Cataloguing in Publication Data

A catalogue record for this book is available from the British Library

ISBN 0 08 0420923 Hardcover
ISBN 0 08 0420931 Flexicover

Printed and Bound in Great Britain by
Redwood Books, Trowbridge, Wiltshire

CONTENTS

Preface   xi

Chapter 1. Introduction   1

  1   Background   1
    1.1 Reasons for Interest   2
    1.2 Scope   3
    1.3 Nomenclature   3
    1.4 Graphical Overview of the Field   6
    1.5 References   7

Chapter 2. Contracted Porphyrins   11

  2   Introduction   11
    2.1 Corrole   13
        2.1.1 Synthesis of the Corrole Macrocycle   17
        2.1.2 Metallocorroles   31
        2.1.3 Reduction and Oxidation of Metallocorroles   62
        2.1.4 N-Substituted Corroles   64
        2.1.5 "Corrologen"   84
        2.1.6 Corrole–Corrole and Corrole–Porphyrin Dimers   85
    2.2 Isocorrole   88
        2.2.1 Synthesis   88
        2.2.2 Metalation Chemistry   93
    2.3 Contracted Porphyrins Containing Fewer than Four Pyrrole-like Subunits   95
        2.3.1 Subphthalocyanines   95
        2.3.2 Subtriazaporphyrins   107
        2.3.3 Heteroatom-bridged [18]Annulenes   107
        2.3.4 Other Systems of Interest   116
    2.4 References   120

Chapter 3. Isomeric Porphyrins   127

  3   Introduction   127
    3.1 [18]Porphyrin-(2.0.2.0) ("Porphycene")   128
        3.1.1 Synthesis and Structure   128
        3.1.2 Metalloporphycenes   141

	3.1.3	Porphycene Transformations	148
3.2		[18]Porphyrin-(2.1.0.1) ("Corrphycene" or "Porphycerin")	162
	3.2.1	Synthesis	162
	3.2.2	Metal Complexes of Corrphycene	164
3.3		[18]Porphyrin-(2.1.1.0) ("Hemiporphycene")	167
3.4		[18]Porphyrin-(3.0.1.0) ("Isoporphycene")	170
3.5		"Mutant" or "N-Confused" Porphyrins	170
3.6		Future Directions in Porphyrin Isomer Chemistry	177
3.7		References	181

Chapter 4. Expanded Porphyrins Containing Four Pyrroles 185

4	Introduction		185
4.1	Homoporphyrins		185
4.2	Vinylogous Porphyrins (Odd–Odd Systems)		201
	4.2.1	Historical Overview	202
	4.2.2	Bisvinylogous Porphyrins	203
	4.2.3	Tetravinylogous Porphyrins	209
	4.2.4	Octavinylogous Porphyrins	215
4.3	Stretched Porphycenes (Even–Even Systems)		216
4.4	Heteroatom-containing Stretched Porphycenes		230
4.5	Odd–Even Systems		237
4.6	Miscellaneous Systems of Interest		240
4.7	References		249

Chapter 5. Contracted Expanded Porphyrins: Sapphyrins and Smaragdyrins 253

5	Introduction		253
5.1	Sapphyrins		253
5.2	*meso*-Aryl-substituted Sapphyrins		259
5.3	Heterosapphyrins		265
5.4	Optical Properties of Sapphyrins and Heterosapphyrins		267
	5.4.1	Sapphyrins and *meso*-Aryl-substituted Sapphyrins	267
	5.4.2	Heterosapphyrins	270
5.5	Metal Complexes of Sapphyrins		272
	5.5.1	Complexes with First-row Transition Elements	272
	5.5.2	Sitting-a-top Complexes	273
	5.5.3	Uranyl Cation Coordination by Modified Sapphyrins	277

	5.5.4	Metal Complexes of Heterosapphyrins	280
5.6		Three-dimensional Sapphyrins and Sapphyrin Conjugates	283
	5.6.1	Sapphyrin Conjugates	284
	5.6.2	"Capped" Sapphyrins	289
	5.6.3	Sapphyrin Oligomers	290
5.7		Smaragdyrins and Heterosmaragdyrins (Norsapphyrins)	294
	5.7.1	Contributions from Grigg and Johnson	294
	5.7.2	Contributions from Woodward and coworkers	297
	5.7.3	Contributions from Sessler	298
5.8		References	299

Chapter 6. Other Carbon-bridged Pentapyrrolic Systems   303

6		Introduction	303
6.1		Orangarin: Two Bridging Carbons	303
6.2		Isosmaragdyrin: Three Bridging Carbons	305
6.3		Sapphyrin "Isomers": Four Bridging Carbons	308
	6.3.1	Ozaphyrins	308
	6.3.2	[22]Dehydropentaphyrin-(2.1.0.0.1) and [22]Pentaphyrin-(2.1.0.0.1)	311
6.4		Pentaphyrins	312
	6.4.1	Pentaazapentaphyrins	313
	6.4.2	Heteropentaphyrins	318
6.5		Other Systems	323
	6.5.1	Pentaoxa[30]pentaphyrin-(2.2.2.2.2)	323
	6.5.2	Pentathia[30]pentaphyrin-(2.2.2.2.2)	324
	6.5.3	Decavinylogous Pentaphyrinogen	326
	6.5.4	"Inverted" (Non-conjugated) Pentaphyrin	326
6.6		References	328

Chapter 7. Carbon-linked Hexapyrrolic Systems and Heteroatom Analogs   329

7		Introduction	329
7.1		Hexaphyrins	329
	7.1.1	Synthesis	329
	7.1.2	Metalation Chemistry	331
	7.1.3	Hexathiahexaphyrinogen	334
7.2		Rubyrins and Heterorubyrins	335
	7.2.1	Rubyrins	335
	7.2.2	Heterorubyrins	338
7.3		Rosarins	339

	7.3.1	Synthesis	339
	7.3.2	Metalation Chemistry	341
7.4	Rosarinogens and Heterorosarinogens		343
7.5	Bronzaphyrins		344
7.6	Amethyrins		348
	7.6.1	Synthesis	348
	7.6.2	Metalation Chemistry	352
7.7	Other Systems of Interest		356
	7.7.1	Hexaoxa- and Hexathia[36]hexaphyrin-(2.2.2.2.2.2)	356
	7.7.2	Cram's Cavitand	362
	7.7.3	Hexathia[30]hexaphyrin-(2.0.2.0.2.0)	363
	7.7.4	Pyrazole-containing Hexaphyrin-like Systems	364
7.8	References		366

Chapter 8. Higher Order Systems   369

8	Introduction		369
	8.1	Turcasarin: Decaphyrin-(1.0.1.0.0.1.0.1.0.0)	369
	8.2	Cyclooctapyrroles	371
	8.2.1	Tetrahydrooctaphyrin-(2.1.0.1.2.1.0.1) and Octaphyrin-(2.1.0.1.2.1.0.1)	371
	8.2.2	Octaphyrin-(1.1.1.0.1.1.1.0)	376
	8.2.3	Octaphyrin-(1.0.1.0.1.0.1.0)	379
	8.3	Pyrrole–Thiophene Decamers: Hexathiatetraaza[44]decaphyrins-(2.0.0.0.0.2.0.0.0.0)	380
	8.4	References	383

Chapter 9. Nitrogen-bridged Expanded Porphyrins   385

9	Introduction		385
	9.1	Schiff base-derived Expanded Porphyrins	386
	9.1.1	The "[2 + 2]" Approach	386
	9.1.2	The "[1 + 1]" Approach	392
	9.2	Other Nitrogen-bridged Expanded Porphyrins	409
	9.2.1	Superphthalocyanines	409
	9.2.2	Porphocyanines	415
	9.2.3	14-Aza[26]porphyrin-(5.1.5.1)	419
	9.3	Cryptand-like Expanded Porphyrins	421
	9.4	References	424

Chapter 10. Applications   429

  10   Introduction   429
      10.1   Magnetic Resonance Imaging   429
            10.1.1   Background: Utility of Contrast Agents   429
            10.1.2   Gadolinium(III) Texaphyrins   432
      10.2   Photodynamic Therapy and Photodynamic Viral   433
            Inactivation
            10.2.1   Introduction: Need for Photosensitizers   433
            10.2.2   Cadmium(II) and Lutetium(III) Texaphyrins   436
            10.2.3   Porphycenes and Expanded Porphycenes   440
            10.2.4   Sapphyrins and Vinylogous Porphyrins   442
      10.3   X-Ray Radiation Therapy Enhancement   442
            10.3.1   General Overview: Potential Benefit of   442
                    Sensitizers
            10.3.2   Gadolinium(III) Texaphyrins   442
      10.4   Antisense Applications: RNA Hydrolysis and DNA   444
            Photolysis
            10.4.1   Introduction   444
            10.4.2   Photolytic Strategies   447
            10.4.3   RNA Hydrolysis: Lanthanide(III)   449
                    Texaphyrins
            10.4.4   Combined Strategies: Yttrium(III)   453
                    Texaphyrins
      10.5   Expanded Porphyrins as Anion-binding Agents   453
            10.5.1   Introduction   453
            10.5.2   Anion Binding in the Solid State   454
            10.5.3   Anion Binding in Solution: Sapphyrins,   478
                    Rubyrins, Anthraphyrins
            10.5.4   Sapphyrin–Oligonucleotide Interactions   485
      10.6   References   490

Index   505

PREFACE

For a people, a book can precipitate an important beginning, or mark a critical milestone; the same is true for molecules. A timely monograph can widely advertise that a class of compounds has truly arrived on the chemical scene, can pay homage to the past, render accurately the present, and highlight the excitement of the future. Thus a monograph can accomplish something that is simply not possible within the confines of briefer reviews or even shorter original research contributions.

Of course, for a book to function in this way, the field must be ready. Enough must have been accomplished to allow for thorough review and enough must remain undone that readers, particularly those from allied disciplines, can sense easily the challenges and opportunities presented by the area. In this way a "mere" book can help bring to life a new field of chemistry.

It is the opinion of the authors that the chemistry of porphyrin analogues, porphyrin-like systems that are "expanded, contracted, or isomeric", as compared to their well-studied tetrapyrrolic "parents", now deserves a life of its own. Over the last decades, particularly in the last few years, so much has been accomplished that, on the one hand, the sheer mass of information precludes writing an all-encompassing journal-type review. On the other hand, this same progress has served to make clear that the chemistry of porphyrin analogues is prescient in terms of a number of current applications and future research possibilities. Thus, a synopsis of all that is known appears fully warranted.

The task of summarizing all that is known is, needless to say, a daunting, near-impossible one. Accordingly, the authors would like to apologize in advance for any errors of omission or commission they may have made. Conversely, the authors would like to thank all those who have contributed to the positive elements of this book. First, we would like to salute two colleagues, Prof. Burchard Franck of the University of Münster and Prof. Emanuel Vogel of the University of Cologne, whose recent retirements helped propel the authors into writing this book *now* as opposed to "next year". Their friendship and inspiration is deeply appreciated. They have helped the authors and the field grow in ways that cannot be honored with words. Their further contributions, limited now by retirement, are eagerly awaited. Second, the authors would like to thank those colleagues, the above individuals along with Prof. David Dolphin, who were so kind as to share information with the authors in advance of publication. Third, a debt is due to Dr. Vincent Lynch of this department. His direct assistance, as well as general tutelage with regard to the manipulation of X-ray diffraction data files, allowed many of the requisite structural figures to be generated. Fourth, many coworkers and collaborators, past and present, need to be thanked. Without their assistance the authors' own contributions would have been far more limited. This is particularly true of colleagues at Pharmacyclics Inc. whose work contributed to many of the texaphyrin-related advances discussed in Chapter 10. Special thanks is also due to Dr. Tarak D. Mody, Dr. Darren J. Magda, Dr. Philip A. Gale, Mr. John Genge, Mrs. Petra I. Sansom, and Mr. Andrei Andrievsky for their technical help in putting this chapter together. Fifth, the National Science Foundation, the National Institutes of Health, Pharmacyclics Inc., and the Coordinating Board of the State of Texas, are gratefully acknowledged; without continued financial support from these sources, many of the University of Texas contributions reviewed in this monograph would not have been possible and the time needed to generate the text itself would never have been found. Finally, thanks are given to all those who were so kind as to look over the manuscript and whose comments helped make a better work; the authors are indebted to you all.

Jonathan L. Sessler
Steven J. Weghorn

1   Background

The porphyrins (e.g., **1.1**; Figure 1.0.1) are a class of naturally occurring macrocycles that are ubiquitous in our world, and they have been called the "pigments of life".[1,2] This auspicious designation reflects their importance in numerous biological functions. Indeed, life as we understand it relies on the full range of biological processes that are either performed by or catalyzed by porphyrin-containing proteins. Chlorophyll-containing photosynthetic reaction centers in plants, for instance, convert light energy into chemical energy while producing oxygen along the way. It is this oxygen, evolved from photosynthesis, which is transported, stored, and reduced by heme-containing proteins in many organisms, including mammals. Not surprisingly, therefore, these molecules remain of fundamental interest to chemists and biochemists. Indeed, they continue to be intensely investigated by researchers world-wide.

Figure 1.0.1   Structure **1.1**

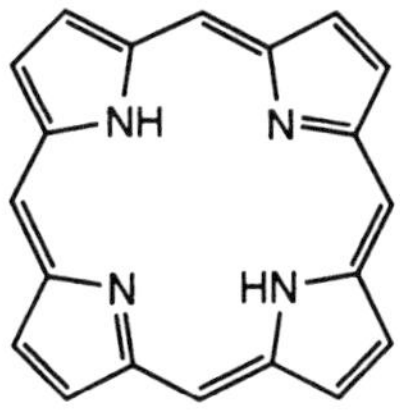

1.1
porphyrin

Inspired by the importance of the porphyrins, a new research direction has emerged in recent years that is devoted to the preparation and study of non-porphyrin polypyrrole macrocycles. Here, the principal objectives have been to generate completely synthetic systems that bear some structural resemblance to naturally occurring porphyrin derivatives while being quite different in their specific chemical makeup. Within this context, three different research directions have evolved, namely those involving the syntheses of contracted, isomeric, and expanded porphyrins, respectively. It is the chemistry of these systems that is the subject of this book. Specifically, this book is intended to provide the first review of contracted and iso- meric porphyrins, while at the same time updating the review on expanded porphyr- ins that appeared in *Topics in Current Chemistry* in 1991.[3] Owing to the newness of the field, the emphasis of this book will be on synthesis and characterization (all work on porphyrin isomers and much of that associated with expanded porphyrins has only appeared in the last 10 years). One chapter on applications has, however, been included. Also, in the context of the preparative portions of the text, some efforts have been made to explain why various porphyrin analog targets are of interest.

1.1   Reasons for Interest

Much of the interest associated with porphyrin analogs stems from the fact that they do resemble the porphyrins. On one level, this has made them of interest as potential ligands. Indeed, one of the main stimuli driving research in the area of contracted, isomeric, and expanded porphyrin chemistry has been a desire to see if the rich coordination chemistry of porphyrins can be modified, extended, or refined by resorting to ligand systems that are smaller, larger, or "reorganized" as compared to those of the porphyrins. Such desires have led *inter alia* to the discovery of new coordination modes and/or oxidation states for classic "porphyrin-type" metals, such as iron and nickel in the case of contracted and isomeric porphyrins. They have also led to the discovery that certain expanded porphyrin systems function as ligands *par excellence* for cations of the lanthanide and actinide series. Finally, related studies have shown that certain porphyrin analogs are capable of functioning as binucleating ligands. Taken in concert, these various findings have spawned inter- est in using expanded porphyrins as metal sequestering agents, as contrast agents in magnetic resonance imaging (MRI), as RNA-cleaving catalysts for antisense appli- cations, and as enzyme models in bioinorganic chemistry. Several of these applica- tions (but not all) are highlighted in Chapter 10.

On another level, the electronic similarities (or lack thereof) between the naturally occurring porphyrins and the various contracted, isomeric, and expanded analogs has made these latter systems of both theoretical and practical interest. Porphyrins and many of the wide range of analogs described in this book may be considered as being heteroatom-bridged annulenes. They have thus attracted atten- tion as potentially congeneric systems wherein the limits of classic Hückel $(4n + 2)$ π- electron aromaticity may be explored. This has led to the synthesis of ever larger macrocycles with greatly increased π-conjugation pathways. Such systems, as it has transpired, often display absorbance bands that are considerably red-shifted com- pared to those present in porphyrins. This has led to the consideration that certain expanded porphyrins could be useful as photodynamic therapy (PDT) photosensit- izers (cf. Chapter 10). An appreciation of electronic structure differences *vis à vis* the porphyrins has also led at least one expanded porphyrin to be developed as a poten- tial X-ray radiation therapy (XRT) sensitizer for cancer therapy (cf. Chapter 10).

One area, unprecedented in the chemistry of porphyrins, where porphyrin analogs have made their mark is in anion binding. This application, which has so far been demonstrated rigorously in but a few protonated expanded porphyrin systems (and really only developed in the case of sapphyrin—a prototypic expanded porphyrin described in Chapter 5), is making the chemistry of synthetic polypyrrole macrocycles of potential interest in a variety of areas wherein anion recognition and transport are important (cf. Chapter 10). These include antiviral drug delivery, anion sensing, and nucleotide recognition and transport, as well as chromatography-based purifications. Thus, while this aspect of porphyrin analog chemistry departs significantly from its antecedents in naturally occurring tetrapyrrolic systems, it is serving to illustrate in a very dramatic way how the synthesis of new materials can lead the science of pyrrole-containing macrocycles into completely unexpected yet inherently interesting directions.

1.2   Scope

To define the scope of the presentation, the following definitions will be employed throughout this book. First, the term "contracted porphyrin" will be used to refer to any macrocycle containing at least three pyrrole or pyrrole-like subunits, arranged in a conjugated (or at least in an all-sp^2-hybridized) framework. The term "isomeric porphyrin" will be used to describe structural variants of the parent $C_{20}H_{14}N_4$ porphyrin macrocycle, including macrocycles derived from simply "scrambling" the four pyrrolic subunits and four bridging carbon atoms. Finally, an "expanded porphyrin" will be defined as any macrocycle containing at least 17 atoms in a cyclic, conjugated framework that contains at least three pyrrole or pyrrole-like heterocyclic subunits. This definition is somewhat more limited than that used in the 1991 *Topics in Current Chemistry* review.[3] However, this restriction in scope was deemed necessary so as to: (1) avoid unnecessary repetition of areas detailed in the previous review for which no new major developments have been reported; and (2) exclude from coverage topics that might be better treated in the context of other more generalized reviews of macrocyclic or heterocyclic chemistry. Also excluded from this review are corrins,[4–12] dehydrocorrins,[13,14] porphyrins[15,16] and their heteroatom analogs,[17–20] porphyrinogens,[21] phthalocyanines,[22–25] and porphyrazines.[26–28] Key references to the original literature are, however, included at the end of this introduction for the benefit of the reader interested in these related topics.

1.3   Nomenclature

For the sake of simplicity, the term "pyrrole-derived" will be used to refer to any macrocycle that contains any combination of five-membered, one-heteroatom-containing cycles, including pyrrole, furan, thiophene, and related systems. According to common convention, the 2- and 5-positions of these five-membered heterocycles will typically be referred to as the alpha (α) positions, while the 3- and 4-

positions will be called the beta (β) positions (Figure 1.3.1). This same nomenclature will be used when these heterocycles are incorporated into a macrocyclic structure even though this usage is less traditional. In the macrocycles themselves, the term "*meso*" or "*meso*-like" will be used to designate the bridging atoms that separate the heterocyclic constituents. This latter nomenclature derives from that of the porphyrins, wherein the 5-, 10-, 15-, and 20-positions are termed *meso*-positions.

Figure 1.3.1   Structures of pyrrole and porphyrin showing their general naming/numbering schemes

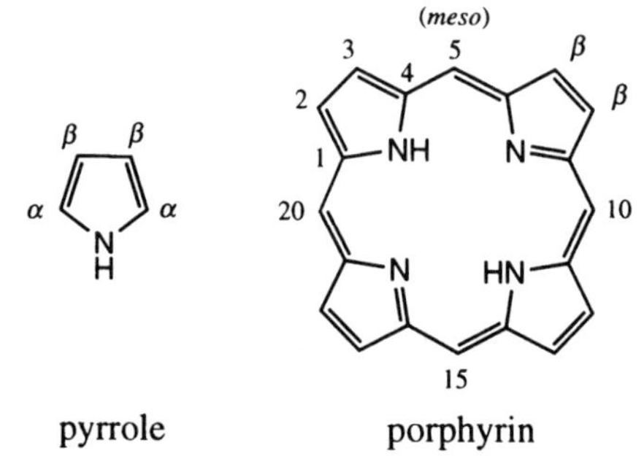

pyrrole   porphyrin

      Owing to the complexity of the applicable IUPAC nomenclature rules, many of the porphyrin analogs discussed in this book were assigned trivial names by their original discoverers. Unfortunately, these assigned trivial names have not been based upon one unified approach to naming. These trivial names are often derived from a word stem based on the color or other feature of the macrocycle, followed by the suffix "phyrin" or "rin" taken from por*phyrin*. R. B. Woodward began this trend when he assigned the trivial name "sapphyrin" to a pentapyrrolic macrocycle (e.g., **1.2**) that he found crystallizes as a dark blue solid (Figure 1.3.2).[29–31] Sessler followed suit when he named the bright red, six-pyrrole-containing macrocycle **1.3** "rubyrin"[32] and the large, "Texas-sized" system **1.4** "texaphyrin".[33,34] In a similar vein, LeGoff termed his vinylogous porphyrin **1.5** "platyrin" from the Greek "*platys*" meaning "wide or broad".[35] The less colorful trivial names "pentaphyrin" (e.g., **1.6**) and "hexaphyrin" (e.g., **1.7**) were introduced by Gossauer and were originally expected to refer to two classes of five- and six-pyrrole-containing expanded porphyrins he developed.[36–38] These names have now been generalized so as to fit within the context of Franck's systematic approach to expanded porphyrin nomenclature.[39] According to this nomenclature, which has now been modified and extended for the purposes of this book, hexaphyrin **1.7** would be referred to as [26]hexaphyrin-(1.1.1.1.1.1) indicating a 26 π-electron hexapyrrolic macrocycle bridged by single sp^2-hybridized centers. Similarly, "platyrin" **1.5** would be termed [22]porphyrin-(3.1.3.1), since it is a 22 π-electron macrocycle that contains four pyrrolic subunits linked in alternative fashion by first three and then one carbon atom spacers.

Figure 1.3.2   Structures **1.2–1.7**

The solid and dashed lines shown in the texaphyrin metal complex **1.4** are used to indicate anionic/ covalent vs. neutral/coordinative ligand–metal bonds, respectively. Such a convention is used throughout this book, primarily in an effort to allow the reader to appreciate the formal oxidation state of a given cation in a given complex.

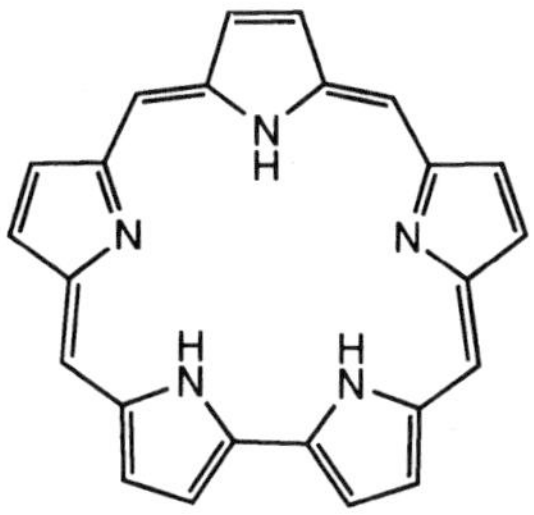

1.2	**1.3**	**1.4**
"sapphyrin"	"rubyrin"	"texaphyrin"
[22]pentaphyrin-(1.1.1.1.0)	[26]hexaphyrin-(1.1.0.1.1.0)	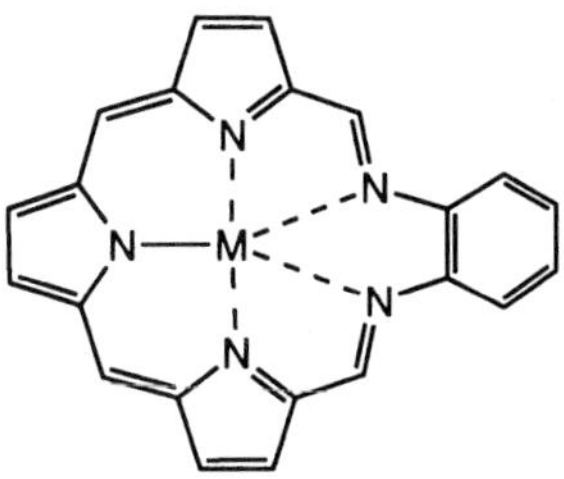

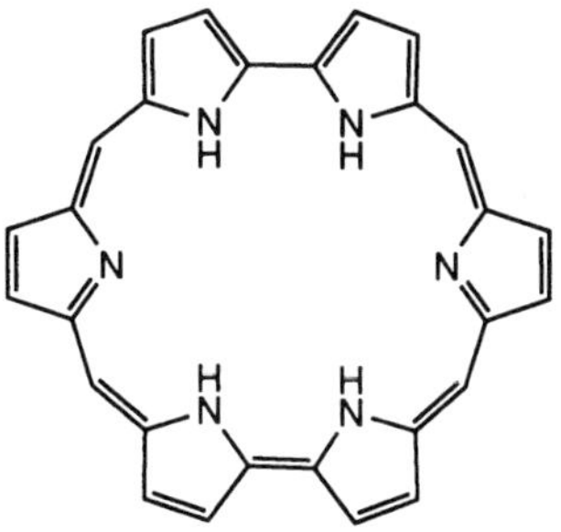

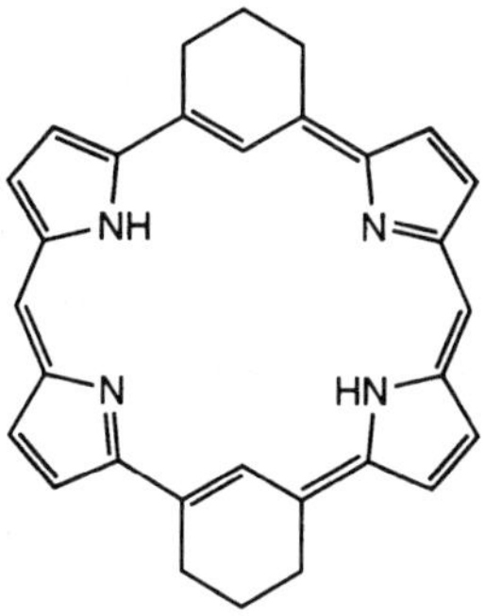

1.5	**1.6**	**1.7**
"platyrin"	"pentaphyrin"	"hexaphyrin"
"bisvinylogous porphyrin"	[22]pentaphyrin-(1.1.1.1.1)	[26]hexaphyrin-(1.1.1.1.1.1)
[22]porphyrin-(3.1.3.1)		

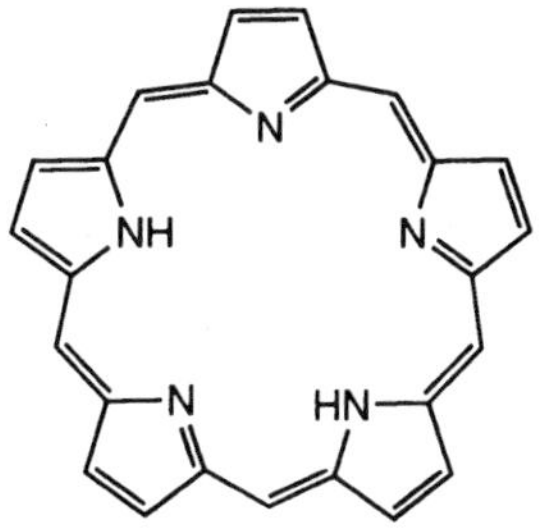

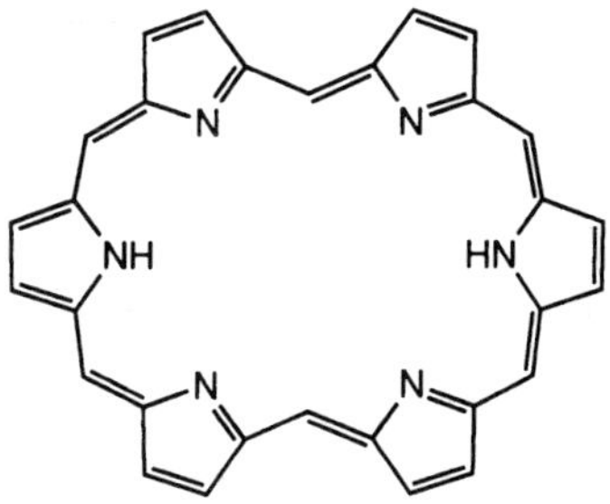

Throughout this book an effort has been made to use both the original authors' trivial names as well as designations based on a Franck-type approach. Since the rationale for using this latter systematic nomenclature is to allow for easy inter-comparisons between classes of molecules, a formal definition is appropriate here. For a given macrocycle, the number of pyrrole or pyrrole-like substituents is first determined and the system designated as being a porphyrin (four subunits), pentaphyrin (five subunits), hexaphyrin (six subunits), etc., as warranted. This compound name is then preceded by a number (in brackets) that indicates the (shortest) π-conjugation pathway available to the molecule. It is also followed by a sequence (in parentheses), which, starting with the largest bridging spacer and continuing around the ring, serves to define the number of carbon (or other) atoms between the heterocyclic subunits. (In the original nomenclature put forward by Franck, the numbering does not begin with the largest bridge; his naming system is thus somewhat different from that used in this book.) As needed, the whole name is prefaced by designations, such as mono- or dioxa, to indicate the number of heteroatoms in the system. Thus, the porphyrin isomer "porphycene" (e.g., **1.8**) is [18]porphyrin-(2.0.2.0) and the heterosapphyrin analog **1.9** is dioxa-[22]pentaphyrin-(1.1.1.1.0) (Figure 1.3.3).

Figure 1.3.3   Structures **1.8** and **1.9**

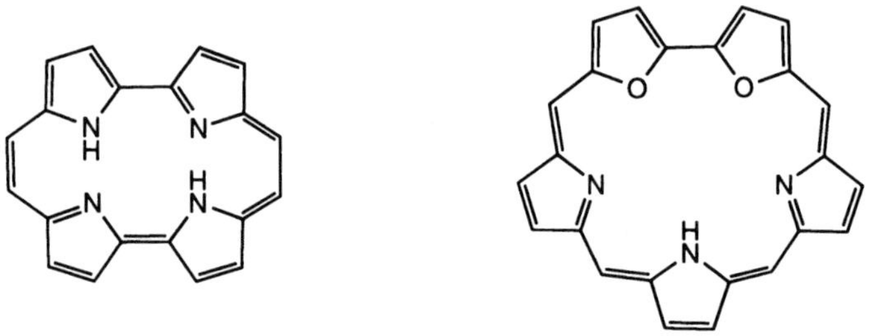

1.8
"porphycene"
[18]porphyrin-(2.0.2.0)

1.9
"dioxasapphyrin"
dioxa[22]pentaphyrin-(1.1.1.0)

1.4   Graphical Overview of the Field

This book is divided into three sections, which are meant to detail the current status of the generalized contracted (Chapter 2), isomeric (Chapter 3), and expanded porphyrin (Chapters 4–9) fields. Each of these sections, and, indeed, the individual chapters themselves, are designed so that they can be read as separate units. Still, for the sake of organization and, it is hoped, the benefit of the reader, an organizational "flow chart of porphyrin analogs" has been constructed (Figure 1.4.1). This chart is intended to provide a broad, "at-one-glance" overview of the field. Specifically, it is meant to illustrate what has been accomplished to date and to show how one particular system relates, at least conceptually, to any other given one. An ancillary feature of this chart is that it also makes clear that which has not yet been done. Thus, while serving as a graphical introduction to this book, written with the intention of saluting accomplishments made to date, this chart could also serve the salubrious role of inspiring future researchers to "fill in the holes". To the extent it fulfills this mission, this chart will have helped to advance further the emerging discipline of porphyrin analog research.

Figure 1.4.1   Flow chart of porphyrin analogs.
Names in parentheses denote species containing one or more two-atom *meso* bridges. Items in quotes correspond to hypothetical (i.e., still unknown) structures. The designations "4P-6BC" and "6P-5BC" are meant to refer to two hypothetical systems containing four pyrrole-like subunits and six *meso* bridges, and six pyrrole-like subunits and five *meso* bridges, respectively. This chart does not specify certain connectivity or configurational (e.g., *cis* vs. *trans*) arrangements of pyrrole and *meso* groups. It also does not include all the possible (i.e., still-hypothetical) "nP-mBC" systems, nor does it include higher order (e.g., octapyrrolic and decapyrrolic) systems.

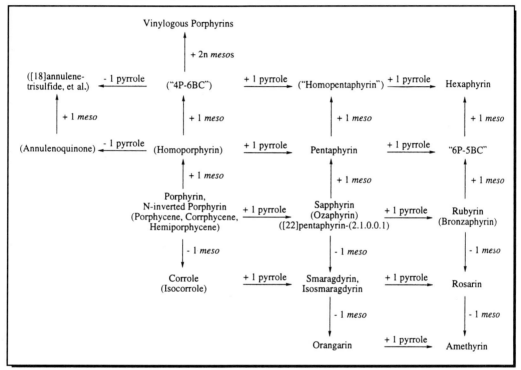

Note in Proof

Professor David Dolphin of the University of British Columbia has informed the authors of this monograph that he and co-author Ayub Jasat have produced a manuscript for *Chemical Reviews* that also summarizes the expanded porphyrin field. The interested reader is urged to make reference to this work.

1.5   References

1.   Battersby, A. R.; Fookes, C. J. R.; Matcham, G. W. J.; McDonald, E. *Nature (London)* **1980**, *285*, 17–21.
2.   Kräutler, B. *Chimia* **1987**, *41*, 277–292.
3.   Sessler, J. L.; Burrell, A. K. *Top. Curr. Chem.* **1991**, *161*, 177–273.

4. Jackson, A. H. In *The Porphyrins*, Vol. 1; Dolphin, D., Ed.; Academic Press: New York, 1978; pp 341–363.

5. Grigg, R. In *The Porphyrins*, Vol. 2; Dolphin, D., Ed.; Academic Press: New York, 1978; pp 328–351.

6. Dolphin, D.; Harris, R. L. N.; Huppatz, J. L.; Johnson, A. W.; Kay, I. T. *J. Chem. Soc. (C)* **1966**, 30–40.

7. Grigg, R.; Johnson, A. W.; Shelton, K. W. *J. Chem. Soc. (C)* **1968**, 1291–1296.

8. Grigg, R.; Johnson, A. W.; Kenyon, R.; Math, V. B.; Richardson, K. *J. Chem. Soc. (C)* **1969**, 176–182.

9. Dicker, I. D.; Grigg, R.; Johnson, A. W.; Pinnock, H.; Richardson, K.; van den Broek, P. *J. Chem. Soc. (C)* **1971**, 536–547.

10. Rasetti, V.; Pfaltz, A.; Kratky, C.; Eschenmoser, A. *Proc. Natl Acad. Sci. USA* **1981**, *78*, 16–19.

11. Kräutler, B.; Konrat, R.; Stupperich, E.; Färber, G.; Gruber, K.; Kratky, C. *Inorg. Chem.* **1994**, *33*, 4128–4139.

12. Blanche, F.; Cameron, B.; Crouzet, J.; Debussche, L.; Thibaut, D.; Vuilhorgne, M.; Leeper, F. J.; Battersby, A. R. *Angew. Chem. Int. Ed. Eng.* **1995**, *34*, 383–411.

13. Genokhova, N. S.; Melent'eva, T. A.; Berezovskii, V. M. *Russ. Chem. Rev.* **1980**, *49*, 1056–1067.

14. Melent'eva, T. A. *Russ. Chem. Rev.* **1983**, *52*, 641–661. See also reference 3.

15. See, e.g., *Porphyrins and Metalloporphyrins*; Smith, K. M., Ed.; Elsevier: Amsterdam, 1975.

16. See, e.g., *The Porphyrins*, Vols 1–7; Dolphin, D., Ed.; Academic Press: New York, 1978.

17. Johnson, A. W. *Pure Appl. Chem.* **1971**, *28*, 195–217.

18. Pandian, R. P.; Chandrashekar, T. K. *Inorg. Chem.* **1994**, *33*, 3317–3324.

19. Vogel, E.; Haas, W.; Knipp, B.; Lex, J.; Schmickler, H. *Angew. Chem. Int. Ed. Eng.* **1988**, *27*, 406–409.

20. Vogel, E.; Pohl, M.; Hermann, A.; Wiß, T.; König, C.; Lex, J.; Gross, M.; Gisselbrecht, J. P. *Angew. Chem. Int. Ed. Eng.* **1996**, *35*, 1520–1524.

21. See, e.g., Crescenzi, R.; Solari, E.; Floriani, C.; Chiesi-Villa, A.; Rizzoli, C. *Inorg. Chem.* **1996**, *35*, 2413–2414, and references therein.

22. See, e.g., *Phthalocyanine Compounds*, Moser, F. H. and Thomas, A. L., Eds; Reinhold Publishing: New York, 1963.

23. See, e.g., Lever, A. B. P. In *Advances in Inorganic Chemistry and Radiochemistry*, Vol. 7; Emeléus, H. J. and Sharp, A. G., Eds; Academic Press: New York, 1965; pp 27–114.

24. See, e.g., *The Phthalocyanines*, Vols I, II; Moser, F. H. and Thomas, A. L., Eds; CRC Press, Inc.: Boca Raton, 1983.

25. See, e.g., *Phthalocyanines*, Vols 1–3; Leznoff, C. C. and Lever, A. B. P., Eds; VCH: New York, 1989.

26. See, e.g., Ghosh, A.; Gossman, P. G.; Almlöf, J. *J. Am. Chem. Soc.* **1994**, *116*, 1932–1940.

27. See, e.g., Ghosh, A.; Fitzgerald, J.; Gassman, P. G.; Almlöf, J. *Inorg. Chem.* **1994**, *33*, 6057–6061.

28. See, e.g., van Nostrum, C. F.; Benneker, F. B. G.; Brussaard, H.; Kooijman, H.; Veldman, N.; Spek, A. L.; Schoonman, J.; Feiters, M. C.; Nolte, R. J. M. *Inorg. Chem.* **1996**, *35*, 959–969.

29. Bauer, V. J.; Clive, D. L. J.; Dolphin, D.; Paine III, J. B.; Harris, F. L.; King, M. M.; Loder, J.; Wang, S.-W. C.; Woodward, R. B. *J. Am. Chem. Soc.* **1983**, *105*, 6429–6436.

30. Sessler, J. L.; Cyr, M. J.; Burrell, A. K. *Synlett* **1991**, 127–134.

31. Král, V.; Furuta, H.; Shreder, K.; Lynch, V.; Sessler, J. L. *J. Am. Chem. Soc.* **1996**, *118*, 1595–1607. See also Chapter 5.

32. Sessler, J. L.; Morishima, T.; Lynch, V. *Angew. Chem. Int. Ed. Eng.* **1991**, *30*, 977–980. See also Chapter 7.

33. Sessler, J. L.; Hemmi, G.; Mody, T.; Murai, T.; Burrell, A.; Young, S. W. *Acc. Chem. Res.* **1994**, *27*, 43–50.

34. Sessler, J. L.; Král, V.; Hoehner, M.; Chin, K. O. A.; Dávila, R. M. *Pure Appl. Chem.* **1996**, *68*, 1291–1295. See also Chapter 9.

35. Berger, R. A.; LeGoff, E. *Tetrahedron Lett.* **1973**, *44*, 4225. See also Chapter 4.

36. Rexhausen, H.; Gossauer, A. *J. Chem. Soc., Chem. Commun.* **1983**, 275.

37. Burrell, A. K.; Hemmi, G.; Lynch, V.; Sessler, J. L. *J. Am. Chem. Soc.* **1991**, *113*, 4690–4692. See also Chapter 6.

38. Charrière, R.; Jenny, T. A.; Rexhausen, H.; Gossauer, A. *Heterocycles*, **1993**, *36*, 1561–1575. See also Chapter 7.

39. Gosmann, M.; Franck, B. *Angew. Chem. Int. Ed. Eng.* **1986**, *25*, 1100–1101.

Chapter Two   Contracted Porphyrins

2   Introduction

With the elucidation of the structure of vitamin B_{12} (**2.1**) in 1955[1-4] (see also some recent reviews concerning vitamin B_{12} and its coenzymes[5-8]) came the first evidence for the existence of a class of naturally occurring tetrapyrrolic macrocycles that were similar in structure to the porphyrins, but missing one of the *meso* carbon bridges (Figure 2.0.1). Although important for many other reasons, this so-called "corrin" chromophore of B_{12} (e.g., **2.2**, shown in its simplified, unsubstituted form below) is of interest in the context of this book in that it is formally a contracted porphyrin. This fact is not of arbitrary pedagogical interest for, indeed, it is well known that it is a larger porphyrin-like macrocycle that is the direct precursor to the contracted, electronically reduced corrin core of vitamin B_{12} and its biologically active coenzyme form.

Since the initial disclosure of the basic corrin structure, there has been a considerable body of effort devoted toward the synthesis of macrocycles related to this chromophore that may be considered as being intermediates between porphyrin and corrin. These macrocycles, namely the "dehydrocorrins" (e.g., tetradehydrocorrin **2.3**) and the "corroles" (e.g., **2.4**), represent interesting classes of contracted porphyrins that warrant specific mention here. The interest in these molecules derives in part from the fact that they could represent "milestones" along the biosynthetic pathway leading to vitamin B_{12}. They are, however, also of interest from a non-biological perspective. Simply stated, this is because corrole-type macrocycles possess unique electronic and chemical characteristics, the study of which can help one to understand better the chemistry of all porphyrin analogs.

In 1978 two reviews on the chemistry of the abovementioned contracted porphyrins (i.e., corrole, dehydrocorrins, and corrins) appeared.[9,10] Two more reviews pertaining to corroles and octadehydrocorrins appeared in 1980 and 1983, respectively.[11,12] Thus, this chapter is intended to bring together into one source much of the information contained in these reviews while at the same time bringing the treatment of this subject up to date. However, upon delving into the original literature, it became apparent that the earlier reviews did not offer a complete overview of all relevant literature to that time. Further, it became apparent quite early on in our investigations of the literature that a disappointingly large number of early claims in the corrole literature were later shown to be incorrect, or at the very least incomplete. In light of the often confusing early reports, this review has been extended "back in time" to include much of what was previously covered. This has been done with an

11

eye toward providing both a more complete description of all the reported corrole analogs and alerting the reader to mistakes made in the early days of this field. With regard to this latter point, we acknowledge, of course, that mistakes are easily made, especially in a field as complex as this one. Nonetheless, the hope is that we will help steer the reader away from confusing "traps" that could arise as the result of browsing through the early corrole-related literature. Along the way we will necessarily provide the reader with some perspective on the challenges faced by pioneering workers in the field.

Figure 2.0.1   Structures **2.1–2.5**

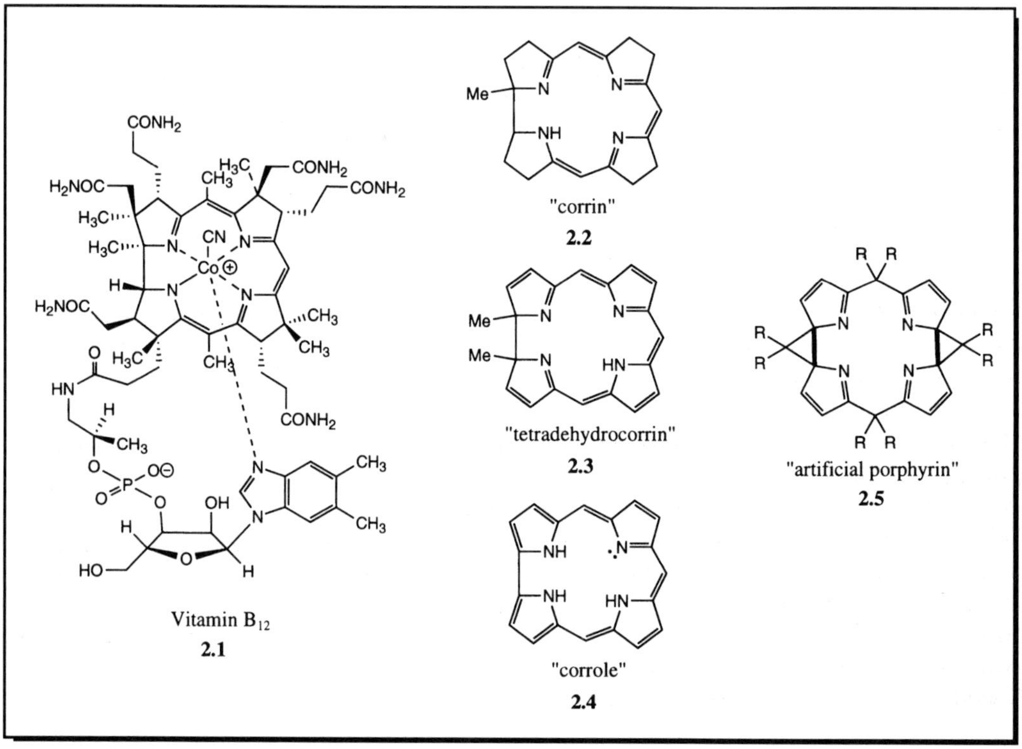

The review presented here has a more synthetic focus than its predecessors. Detailed information about the physical properties of the corrin-related macrocycles is, therefore, not included here. Instead, the reader is referred to the earlier reviews, as well as to a number of relevant papers, for detailed descriptions of the physical properties of corrin-type systems.[13–18] Still, in this chapter, a complete, up-to-date discussion of corrole and heterocorrole synthesis and metalation properties will be presented. Also, two sections will be devoted to other synthetic contracted porphyrins, including isocorroles, and several systems that contain fewer than four pyrrole-like subunits in their macrocyclic framework.

Because of its immense scope, a detailed description of corrins (and vitamin B_{12}) will not be presented here. The reader is instead referred to reviews of B_{12} chemistry and its biosynthesis that have appeared recently.[5–8] Further, because they are more directly related to the corrins than are the corroles, the chemistry of the dehydrocorrins will not be discussed here. Also not included in this review are the so-called artificial porphyrins of Floriani, *et al.* (e.g., **2.5**), since it is deemed by these authors in their review that these macrocycles are more dehydrocorrin-like than corrole-like in their nature.[19] Other systems omitted here include the "spiro" porphyrins of Battersby and coworkers,[20] the tetraphosphole macrocycles of Mathey and coworkers[21] and the tetrapyrrolic systems of Bartczak[22] and Smith and co-workers.[23] Thus, the emphasis will be on those contracted porphyrins that are most removed, in structural and chemical terms, from the macrocyclic unit found in coenzyme B_{12} and its analogs.

2.1   Corrole

As mentioned above, corrole may be considered as being a formal intermediate in nature between porphyrin and corrin. Corrole, like the parent corrin, has a direct link between the A and D heterocyclic rings. The numbering of the corroles is consistent with that used for the porphyrins and is shown below (Figure 2.1.1). Note that the numbering begins on the alpha (α) carbon of ring A that is adjacent to ring D (position 1) and continues around the macrocycle to the alpha carbon of ring D (position 19). In order to maintain a numbering analogy with that used for the porphyrins, early corrole researchers numbered the nitrogen atoms from 21 to 24. As a result, position 20 ended up being eliminated in this nomenclature scheme. While this is not too confusing, it should be noted that on occasion a different numbering system has also appeared in the literature.[24–26] In this scheme, the number 20 is not omitted, the nitrogen atoms are thus numbered 20–23. For the sake of consistency, the first of these two numbering schemes will be used here exclusively. Additional nomenclature has been borrowed from that of the porphyrins. Specifically, the bridging carbon atoms will be referred to as *meso*-positions.

Figure 2.1.1   Structure of corrole showing the atom numbering scheme

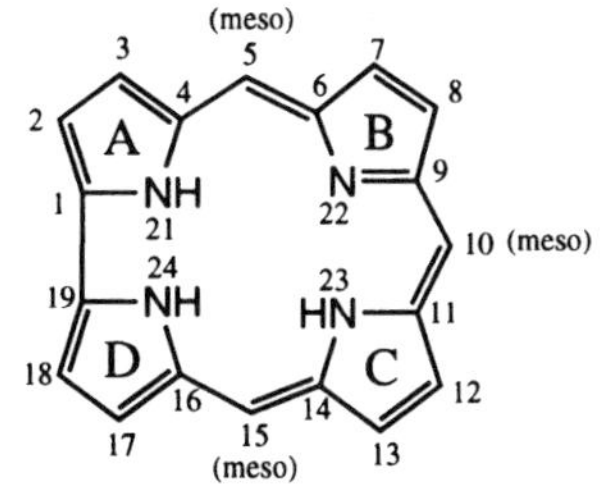

Free-base corrole (e.g., **2.6**), like porphyrin, contains an 18 π-electron pathway and, as such, can sustain a diamagnetic, aromatic ring current (Scheme 2.1.1). Corroles are much stronger acids (and weaker bases) than porphyrins. Thus, unlike porphyrin, corrole forms a stable anion when treated with aqueous alkali. This anion, best represented by structure **2.7**, is also an aromatic 18 π-electron system, which may account in part for its special stability. In any event, the neutral form of the macrocycle may be regenerated on acidification. Thus, the corrole-to-porphyrin relationship in terms of acidity allows for comparison to be made between these two systems and their smaller six π-electron counterparts benzene and cyclopentadienyl anion (Figure 2.1.2). In other words, cyclopentadienyl anion may be regarded as a "contracted benzene" just as corrole anion may be considered as being a "contracted porphyrin".

Scheme 2.1.1

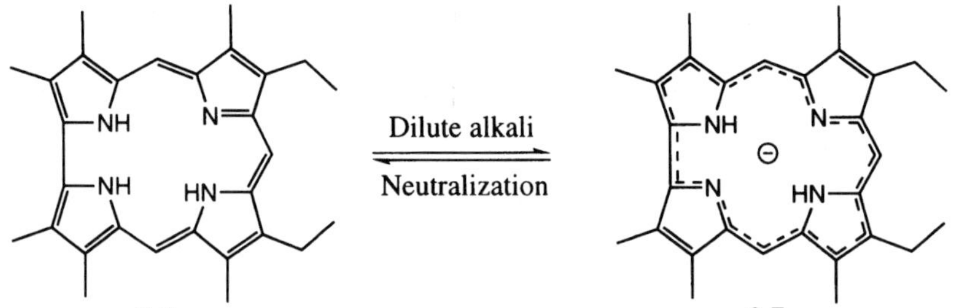

Figure 2.1.2   Structural Comparison between Porphyrin vs. Corrole Anion and Benzene vs. Cyclopentadienyl Anion

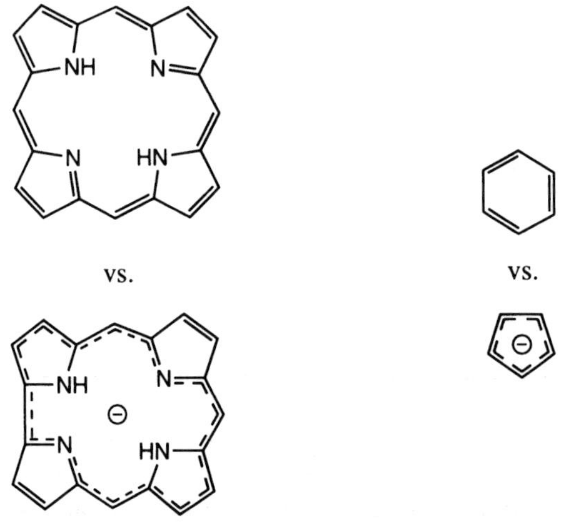

Free-base corroles as well as their anionic and monoprotonated adducts are generally characterized by several strong ultraviolet (UV)-visible absorption bands. As is true for the porphyrins, the position and intensity of these bands presumably reflects the extended aromatic conjugation present within the molecule. Indeed the spectra of corroles resemble those of the porphyrins in that there is typically a strong Soret-like transition near 400 nm, as well as three weaker, long-wavelength Q-type transitions in the 500–600 nm spectral region.[27]

A single crystal X-ray structure of a free-base corrole was obtained in 1971, and is shown below in Figure 2.1.3.[28] Based upon this structure and upon molecular calculations, free-base corrole was determined to bear a "proton hole" at N(22) (as drawn for **2.4**), with the other three nitrogen atoms protonated.[25] Where appropriate throughout this discussion, the corrole framework will be shown in this manner. This does not mean, however, that a specific location is being designated for what are, presumably, highly delocalized double bonds.

Figure 2.1.3   Single Crystal X-ray Diffraction Structure of Free-base 8,12-Diethyl-2,3,7,13,17,18-Hexamethylcorrole **2.6**.
This figure was generated using information down-loaded from the Cambridge Crystallographic Data Centre and corresponds to a structure originally reported in reference 28. Atom labeling scheme: carbon: ○; nitrogen: ●. Hydrogen atoms have been omitted for clarity

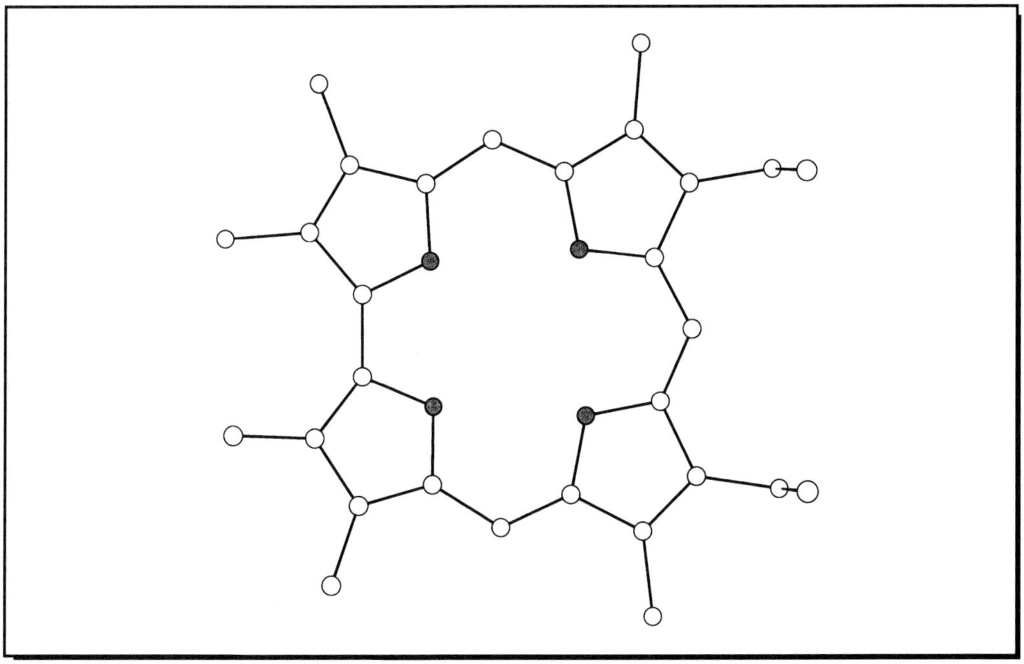

Corrole also forms stable protonated species. For example, corrole can be monoprotonated by treatment with weak acid to afford adducts such as **2.8**, which, like corrole, display strong visible absorption bands (*vide supra*) (Scheme

2.1.2). A single crystal X-ray diffraction analysis carried out on the mono-HBr salt of 8,12-diethyl-2,3,7,13,17,18-hexamethylcorrole **2.8** (X = Br) helped confirm the fact that it is the monoprotonated adduct that is formed under non-forcing conditions (Figure 2.1.4).[29] When corrole is treated with strong acid (e.g., sulfuric acid), however, the macrocycle loses its intense visible absorptions. These findings were rationalized by suggesting that it is a dicationic species (e.g., **2.9**) that is formed under these conditions. While this putative dication was originally formulated with protonation occurring at C(10),[27] it was later determined that the site of this second protonation is more likely to be at C(5) (i.e., as shown).[30] Later studies on the protonation of *N*-methyl corroles actually revealed that corrole can be sequentially protonated *three* times (this is discussed in detail later on in this chapter). In addition, nuclear magnetic resonance (NMR) experiments showed that rapid exchange of all three *meso* protons occurs when corrole is treated with deuterotrifluoroacetic acid.[27] In spite of this, and the loss of conjugation it implies on protonation with strong acid, the multiply protonated forms of corrole remain quite stable.

Scheme 2.1.2

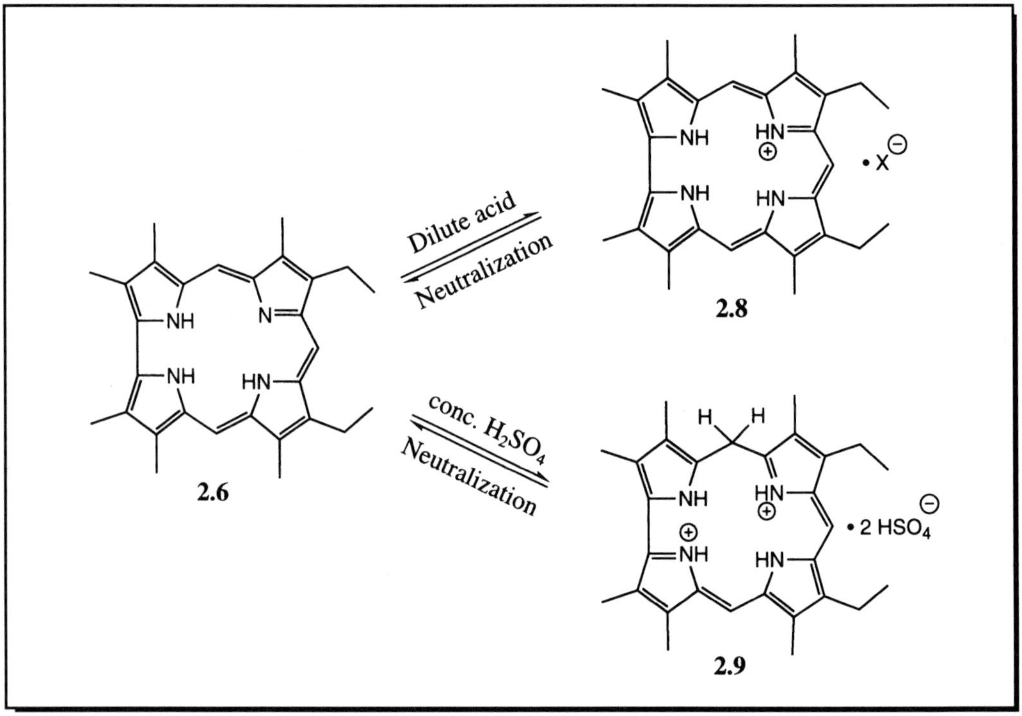

The internal core of corrole resembles that of the porphyrins in that there are four inward-pointing nitrogen atoms. In the case of the porphyrins, these heteroatoms serve to ligate a remarkably diverse range of metal cations.[31,32] This is also true for corroles. However, unlike porphyrin, which is a dianionic ligand, the fully

Figure 2.1.4   Single Crystal X-Ray Diffraction Structure of 8,12-Diethyl-2,3,7,13,17,18-Hexamethylcorrole **2.8** in the Form of its Hydrobromide Salt.
This figure was generated using information down-loaded from the Cambridge Crystallographic Data Centre and corresponds to a structure originally reported in reference 29. Atom labeling scheme: carbon: ○; nitrogen: ●; bromine: ◉. Hydrogen atoms have been omitted for clarity

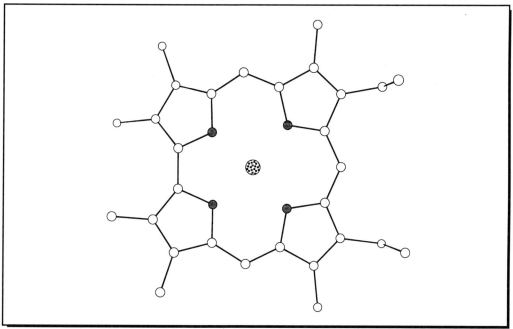

deprotonated form of corrole is a trianionic ligand. Indeed, several neutral metal(III) complexes of corrole have been prepared. Corrole has even been shown to form stable complexes with tetravalent metals. The details of this metalation chemistry, however, are complicated. Indeed, many of the original structural assignments put forward by Johnson and others have recently been called into question by Vogel and coworkers. Nonetheless, since this chapter is presented in a semi-chronological sequence, these original structures are given throughout. Only during a discussion of the contributions by Vogel and his group in Section 2.1.2.2 will the proposed alternate structural interpretations of certain metallocorroles be presented. This is something about which the reader should be cognizant of in advance.

2.1.1   Synthesis of the Corrole Macrocycle

2.1.1.1   Heterocorrole Syntheses

The first efforts to prepare corrole were presented by A. W. Johnson in 1960.[33] They involved an attempt to effect ring closure of a metallo–dibromo

5,5′-bi(dipyrrylmethane) (e.g., **2.10**–**2.12**) in the presence of formaldehyde. The compounds formed under these conditions were initially formulated as being the metallo–corroles **2.13**–**2.15** (i.e., species that contain a metal–hydroxide bond) (Scheme 2.1.3). In 1961, Johnson and coworkers published an amendment stating that while the species formed in this reaction do contain oxygen, the oxygen was actually part of the ring system (e.g., **2.17**), and that formaldehyde played no part in the reaction.[34] Also in 1961, and later in 1963, he rescinded his claim to have prepared the Cu and Co derivatives of oxacorrole in this manner.[34,35] Rather, he suggested that a mono–amide-containing linear tetrapyrrole (e.g., **2.16**) is actually formed in these cases, and that this species can, in another reaction, be subsequently cyclized to the metallo–oxacorroles **2.17** and **2.18** (Scheme 2.1.4). Even later (in 1965), Johnson amended his initial report of corrole formation from the Pd-tetrapyrrole **2.10** and actually went so far as to state that any prior claims to having made corrole in this manner were specious.[36] Thus, the synthetic chemistry of the corroles got off to a rocky start, with the first 5 years (1960–1965) being plagued by this series of unfortunate characterization errors.

Scheme 2.1.3

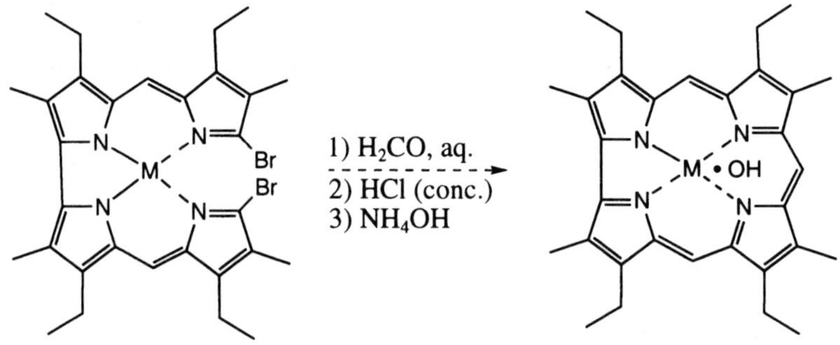

1) H_2CO, aq.
2) HCl (conc.)
3) NH_4OH

2.10. M = Pd
2.11. M = Co
2.12. M = Cu

2.13. M = Pd
2.14. M = Co
2.15. M = Cu

In spite of the above, and in spite of what might have been inferred from the early papers, the Cu and Co oxacorroles **2.17** and **2.18** were apparently successfully made by the Johnson group. These, in turn, were later demetalated successfully using sulfuric acid to give the metal-free species **2.19**.[35] This metal-free corrole was reported to react further with various metal salts to afford the corresponding metalated derivatives. Unfortunately, however, specific details as to which metals were used and the properties of the resulting metallocorroles were never presented in the open literature. They are thus not known to the present authors.

At about the same time that the oxacorroles were being studied, Johnson, *et al.* successfully prepared the metallo–derivatives of the imino- and *N*-methylimino-corrole derivatives **2.21–2.24** (Scheme 2.1.5). This he did starting from the corresponding metallo–linear tetrapyrroles **2.10**, **2.20**, and **2.12**.[35] Interestingly, in the case of the Cu(II)imino-corrole **2.24**, treatment with sulfuric acid effected demetalation and liberated a metal-free corrole. By contrast, the Pd(II)imino-corroles **2.21–2.23** reportedly could not be demetalated.

Scheme 2.1.4

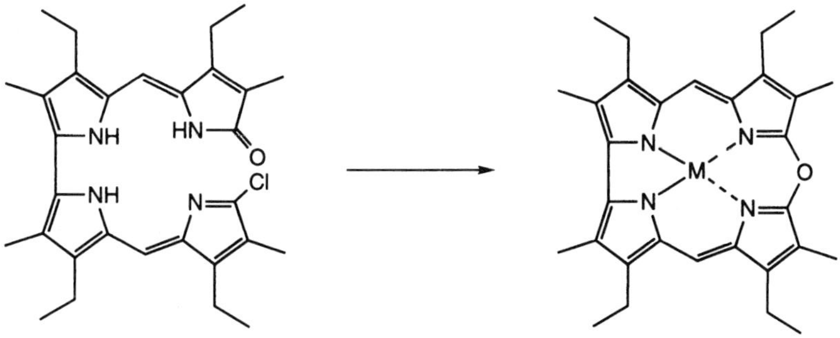

2.16

2.17. M = Co
2.18. M = Cu
2.19. M = H$_2$ $\Big\}$ conc. H$_2$SO$_4$

Scheme 2.1.5

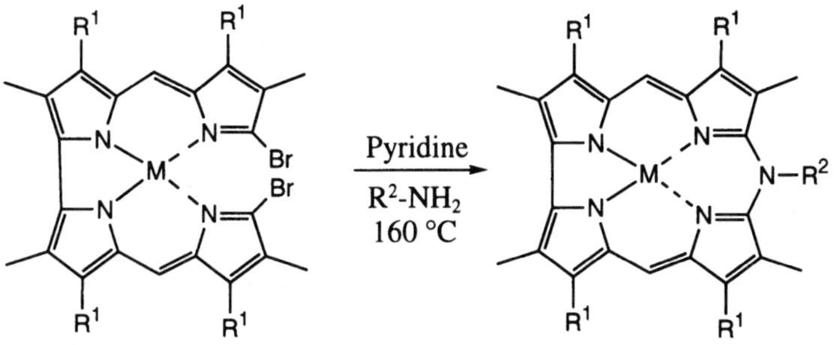

Pyridine
R^2-NH$_2$
160 °C

2.10. M = Pd, R^1 = Et
2.20. M = Pd, R^1 = CO$_2$Et
2.12. M = Cu, R^1 = Et

2.21. M = Pd, R^1 = Et, R^2 = H
2.22. M = Pd, R^1 = CO$_2$Et, R^2 = H
2.23. M = Pd, R^1 = Et, R^2 = Me
2.24. M = Cu, R^1 = Et, R^2 = H

Johnson, *et al.* also reported the synthesis of the Pd(II) thiacorrole **2.25**.[35] This macrocycle could be prepared from the corresponding tetrapyrrole **2.10** by treatment with sodium sulfide in pyridine (Scheme 2.1.6). In contrast to the imino-corroles referred to above, the same transformation could not be effected using the Cu(II)tetrapyrrole **2.12**.

Scheme 2.1.6

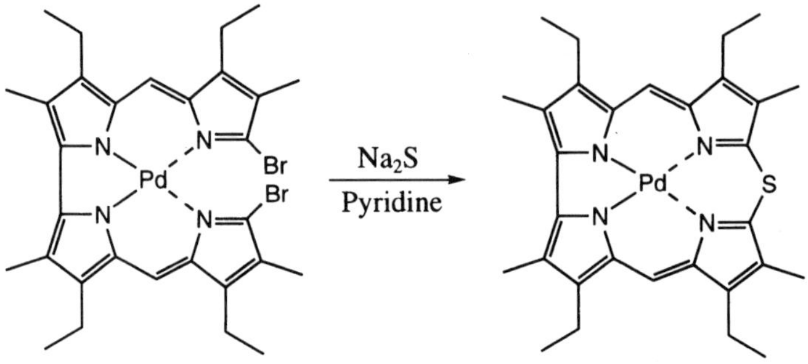

An alternative, "2 + 2" approach to *meso*-thiacorroles was reported in 1972 by Broadhurst, *et al.* It involved the condensation between the bis(formylpyrrolyl) sulfide **2.26** and the bis(carboxypyrrolyl)sulfide **2.27**.[37,38] The *meso*-dithiaphlorin **2.28** so obtained was then subject to sulfur extrusion to afford the *meso*-thiacorroles **2.29** and **2.30** in a nearly 1:1 ratio, and in 42% combined yield (Scheme 2.1.7). The metalated *meso*-thiacorroles **2.32** and **2.33** were prepared in a similar fashion from the corresponding metalated *meso*-dithiaphlorin **2.31**. Incorporation of metal ions into the dithia-macrocycle enhanced the rate of sulfur extrusion, but did not affect the product distribution ratio.[37,38]

Not long after reporting the synthesis of *meso*-heterocorroles, Broadhurst, *et al.* reported two "2 + 2" approaches to the synthesis heterocorroles in which one or more of the pyrrole rings is formally exchanged for furan.[38–39] The first of these involved the acid-catalyzed condensation between diformyl bifuran (**2.34**) and a diacid dipyrrylmethane (e.g., **2.35–2.37**) (Scheme 2.1.8). This type of reaction afforded low yields (*c.* 7%) of the desired furan-containing corroles **2.38–2.40**. Surprisingly, in addition to the expected corroles, small yields (*c.* 9%) of a macro-cycle containing two furan rings and three pyrrole rings (arising from cleavage–recombination processes during the reaction) were isolated. These were determined to be hetero-analogs of the sapphyrins (**2.41** and **2.42**) (see Chapter 5 for a full discussion of sapphyrin and its analogs).

An alternative route to these bifuran-containing corroles was also pursued; it involved carrying out an acid-catalyzed condensation between the diformyldifuryl sulfide **2.43** and an appropriate diacid dipyrrylmethane species (e.g., **2.35–2.37**) (Scheme 2.1.9).[38–39] The reaction proceeds presumably *via* sulfur-containing macro-cyclic intermediates such as **2.44–2.46**. However, none of these putative intermediates could ever be detected. Rather, spontaneous sulfur extrusion was presumed to take

Scheme 2.1.7

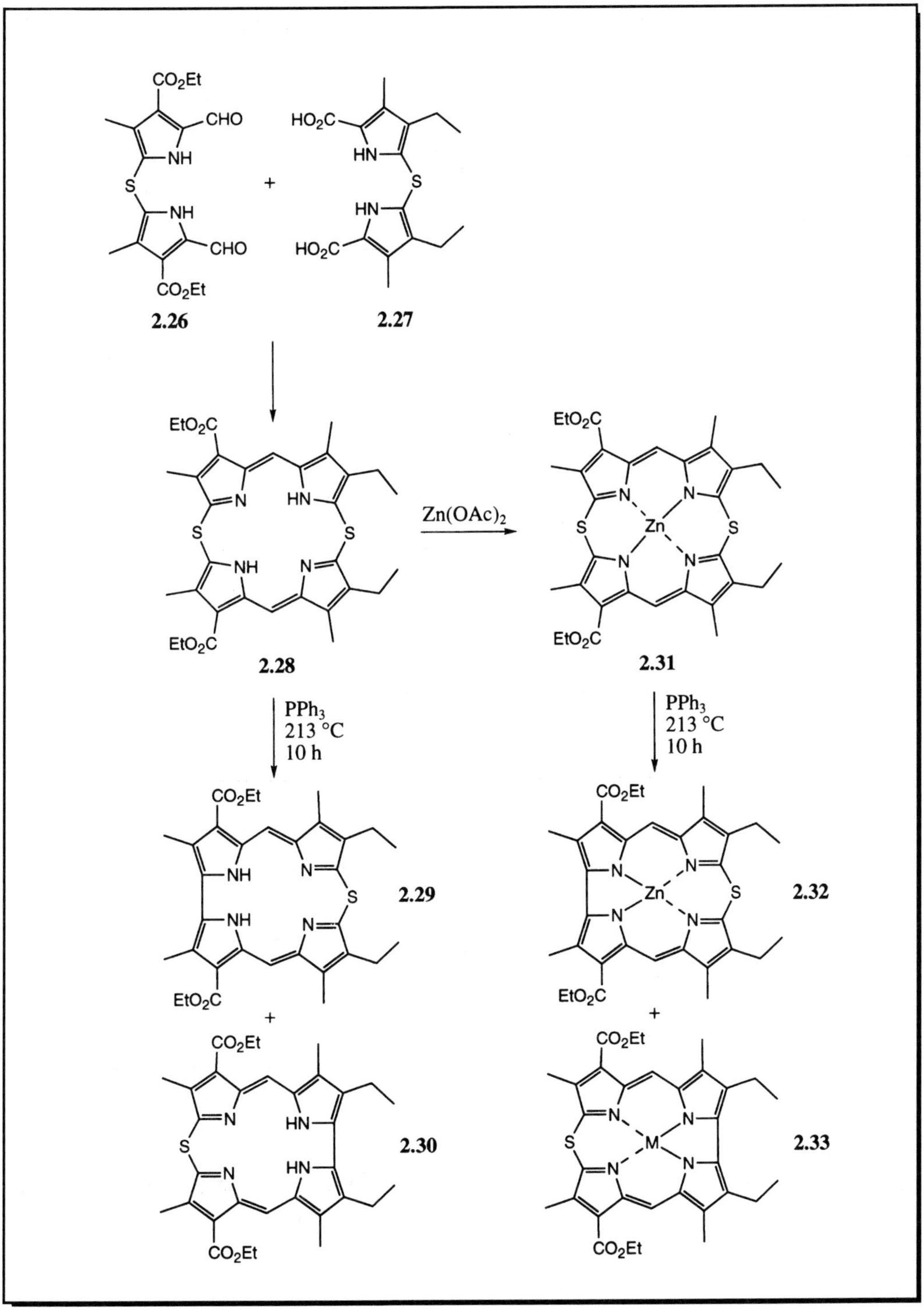

place (so as to afford the corresponding corrole derivatives directly). Nevertheless, this method afforded higher yields (27–30%) of the desired macrocycles **2.38**–**2.40**, along with another corrole (derived from **2.43** and dipyrrylmethanes **2.36** or **2.37**) of presumed structure **2.47** and **2.48**, which contained only one furan heterocycle. While Broadhurst, *et al.* demonstrated that the bifuran-containing corroles **2.38**–**2.40** did not form stable complexes with divalent metals (e.g., Co, Ni, Cu, Zn), the mono-furan compounds **2.47** and **2.48** did form metal complexes when treated with appropriate salts of Ni(II), Co(II), and Cu(II).[37,38] Unfortunately, owing to limited quantities of these mono-furan-containing corroles, no further details as to the exact nature of the metal complexes were made available. Thus, in the opinion of the authors, these systems remain as targets for further characterization studies.

Scheme 2.1.8

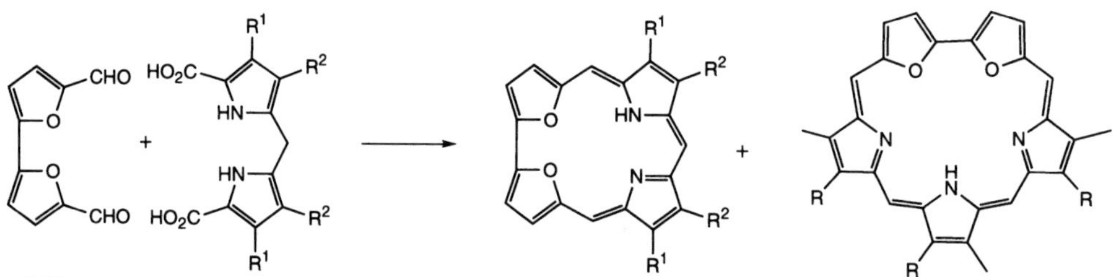

2.34	**2.35**. $R^1 = R^2 = Me$	**2.38**. $R^1 = R^2 = Me$	**2.41**. R = Me
	2.36. $R^1 = Me, R^2 = Et$	**2.39**. $R^1 = Me, R^2 = Et$	**2.42**. R = Et
	2.37. $R^1 = R^2 = Et$	**2.40**. $R^1 = R^2 = Et$	

The bifuran-derived corrole derivative **2.39** was later subjected to metalation conditions using tetracarbonyldi-μ-chlorodirhodium(I) in chloroform.[40] Proton NMR spectroscopic analyses of the resulting metal complex revealed a highly symmetrical structure. Based on this inference, the complex was formulated as being the rhodium(I) dicarbonylcorrole **2.49**, wherein nitrogen atoms N(22) and N(23) provide coordination to the rhodium center (Scheme 2.1.10). On the basis of these findings, it appears as if bifuran-containing corrole derivatives could be well suited for the coordination of monovalent metals in general. However, no further reports of such monovalent complexes have appeared in the literature.

Using each of the "2 + 2" synthetic pathways discussed above, Broadhurst, *et al.* also attempted to prepare bithiophene-containing corroles (cf. Schemes 2.1.8 and 2.1.9).[37,38] However, in no case could any corrole-like products be obtained. This failure likely reflects the increased steric bulk of the sulfur atoms; these presumably hinder effective macrocyclization.

Heterocorroles in which all four of the inward-oriented donor atoms are oxygen have recently been described by the Vogel group.[41,42] The first of these were prepared *via* a "2 + 2" strategy that involved the condensation of the diformyldifuryl sulfide **2.43** with a difurylmethane. In the specific case of **2.43** and **2.50**, treat-

Scheme 2.1.9

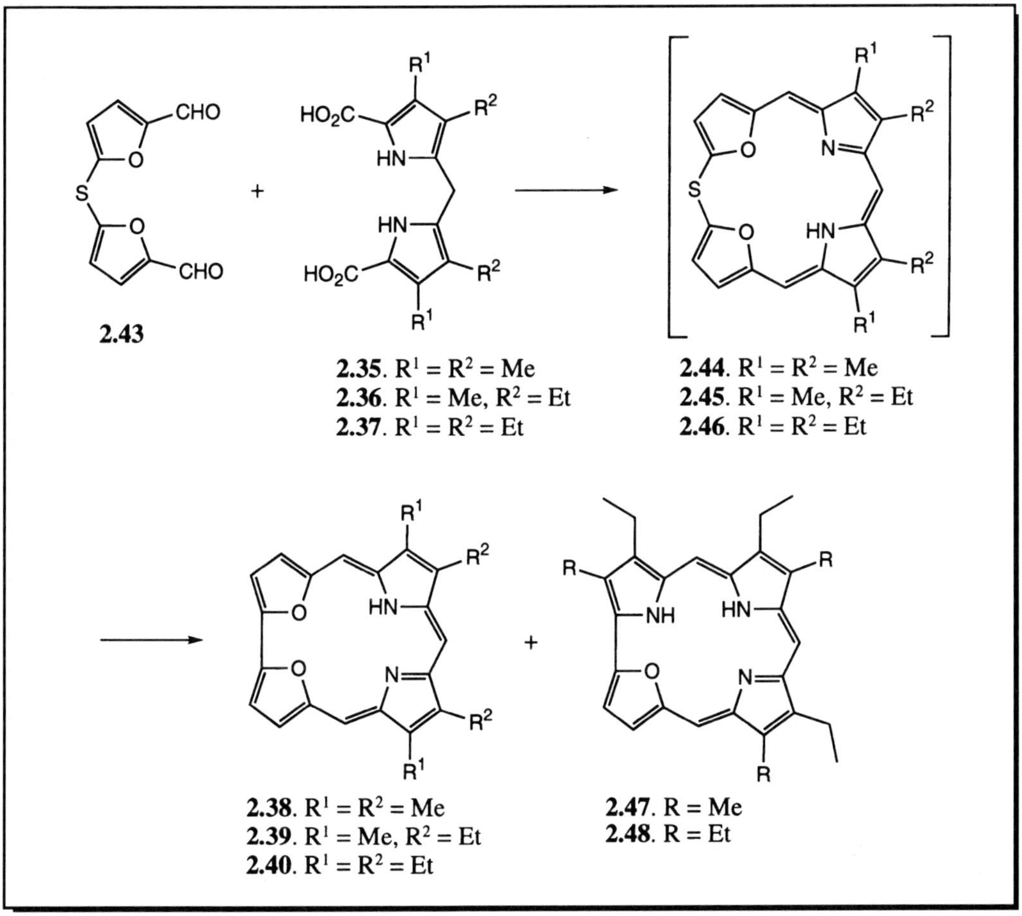

2.43

2.35. R^1 = R^2 = Me
2.36. R^1 = Me, R^2 = Et
2.37. R^1 = R^2 = Et

2.44. R^1 = R^2 = Me
2.45. R^1 = Me, R^2 = Et
2.46. R^1 = R^2 = Et

2.38. R^1 = R^2 = Me
2.39. R^1 = Me, R^2 = Et
2.40. R^1 = R^2 = Et

2.47. R = Me
2.48. R = Et

Scheme 2.1.10

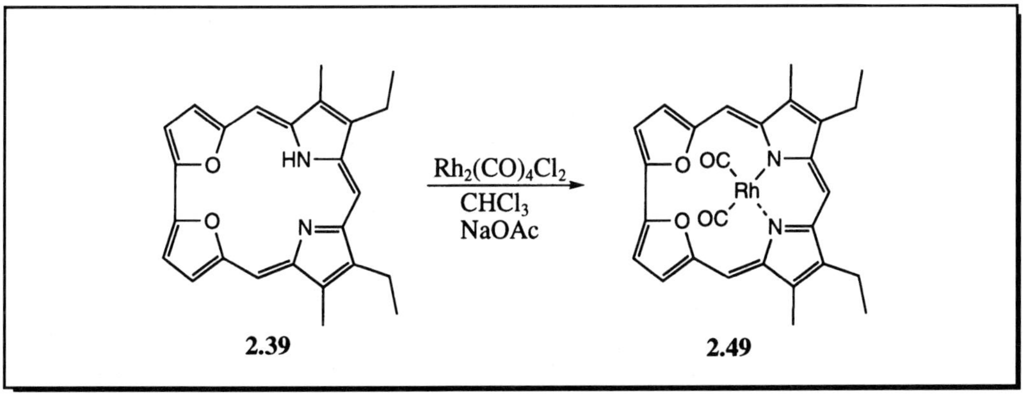

2.39 $\xrightarrow[\text{NaOAc}]{\text{Rh}_2(\text{CO})_4\text{Cl}_2 \atop \text{CHCl}_3}$ **2.49**

ment with HBr followed by 70% HClO$_4$ afforded the substituent-free tetraoxacorrole **2.55a** in 6.4% yield.[41] These same conditions were also used to prepare the 10-alkyl corroles **2.56–2.58** and the 10-phenyl corrole **2.9** (in 0.7%, 3.5%, 0.5%, and 13.4% yield, respectively) starting from **2.43** and the corresponding methylene-substituted difurylmethane derivative **2.51–2.54** (Scheme 2.1.11). As was presumably true in the case of the dioxacorroles **2.38–2.40** (presented in Scheme 2.1.9), the putative sulfur-bridged thiaphlorin intermediate present in Vogel's reactions could not be isolated.

Scheme 2.1.11

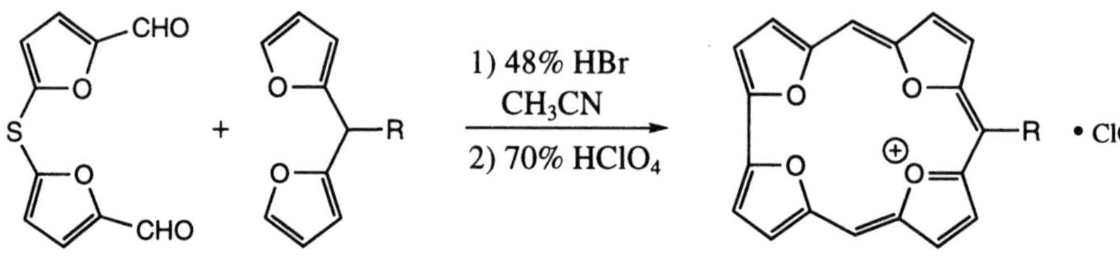

2.43	**2.50**. R = H	**2.55a**. R = H
	2.51. R = Me	**2.56**. R = Me
	2.52. R = Et	**2.57**. R = Et
	2.53. R = *n*-Pr	**2.58**. R = *n*-Pr
	2.54. R = phenyl	**2.59**. R = phenyl

The octa-alkyl derivative of tetraoxacorrole has also been prepared by the group of Vogel.[42] In this case, however, tetraoxacorrole cation **2.62** was prepared from the hydroxymethyl-substituted mono-furan precursor **2.60**. In this way, **2.60** was cyclocondensed using ρ-toluene sulfonic acid catalyst to afford the "tetraoxacorrologen" intermediate **2.61**. This intermediate was then treated with CuCl$_2$ followed by perchloric acid to afford the 18 π-electron, "ring-contracted" monocation **2.62** as its perchlorate salt (Scheme 2.1.12). It is worth noting that tetraoxaporphyrinogen (**2.63**) was also isolated from this reaction in 16% yield (Figure 2.1.5). It appears, therefore, that formation of corrologen **2.61** or corrole **2.62** does not arise as the result of expulsion of a methylene bridge from tetraoxaporphyrinogen. In spite of this critical observation, it is important to appreciate that an exact mechanism for the formation of **2.62** has yet to be worked out.

Tetraoxacorrole cation **2.55a** has been subjected to preliminary investigations in terms of its reactions with nucleophiles.[41] Interestingly, nucleophilic attack appears to occur predominantly at the 5-position of the corrole ring. For instance, treatment of **2.55a** with NaOH in dimethylsulfoxide (DMSO) followed by neutralization with HCl afforded 5-oxotetraoxacorrole **2.64** (Scheme 2.1.13). When treated with sodium borohydride, reduction of **2.55a** was found to produce the 5-hydrotetraoxacorrole **2.65** in 86% yield (Scheme 2.1.14). Corrologen **2.65** proved to be quite stable. Nevertheless, it could be smoothly reoxidized to the aromatic TFA salt **2.55b** by treatment with bis(trifluoroacetoxy)iodobenzene (Scheme 2.1.15).

Scheme 2.1.12

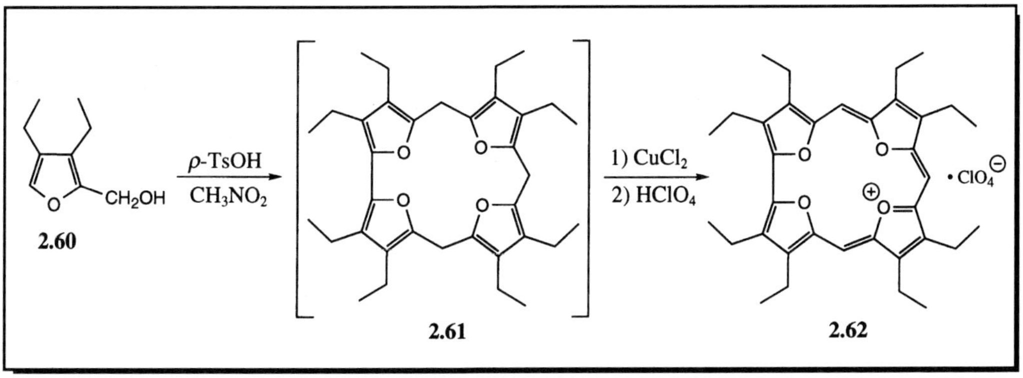

Figure 2.1.5   Structure **2.63**

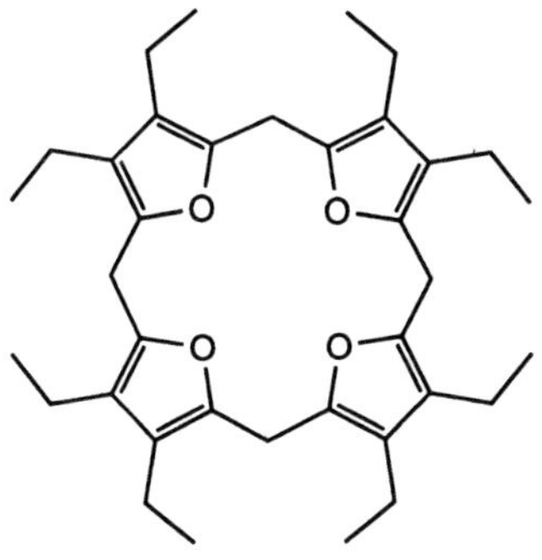

2.63

Scheme 2.1.13

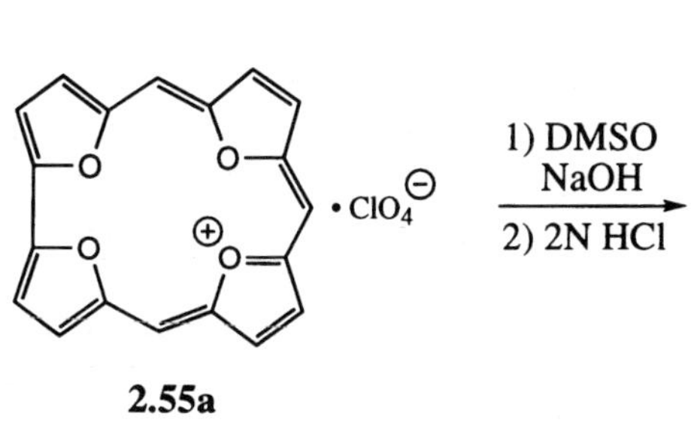

Scheme 2.1.14

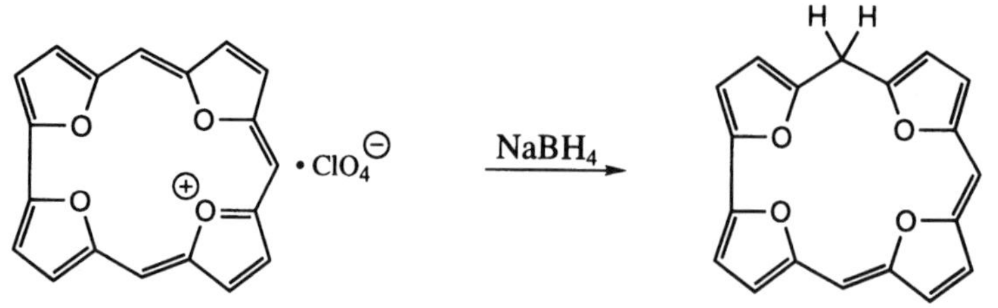

2.55a   **2.65**

Scheme 2.1.15

2.55b

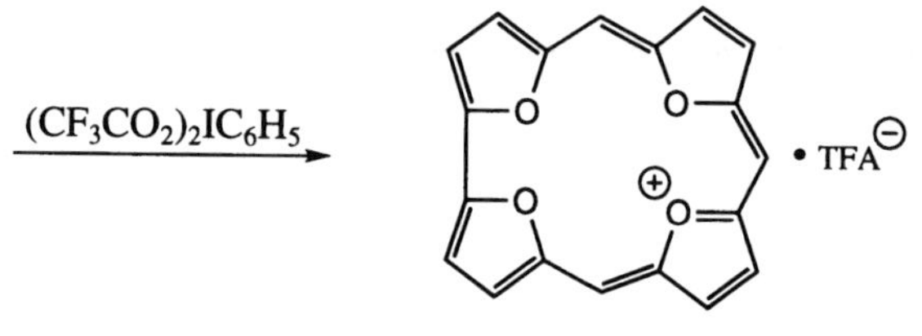

2.1.1.2   Corroles from Tetrapyrrolic Precursors

In 1964, 4 years after their initial incorrect claim, A. W. Johnson and co-workers[27] succeeded in preparing a derivative of the parent tetra*pyrrolic*, all-carbon-bridged corrole. The key step in this synthetic strategy involved the light-promoted cyclization of a free-base dideoxybiladiene-ac, **2.66**. This afforded the corresponding corrole **2.6** in *ca.* 60% yield (Scheme 2.1.16).[27] Variations on this same approach, using the hydrobromide salts of dideoxybiladienes-ac (**2.67–2.75**), were later used to prepare corrole **2.6** as well as a variety of other substituted corroles (**2.76–2.83**) (Scheme 2.1.17). Here, the yields were generally on the order of 20–60%.[43–45]

Typical conditions for these cyclizations involved using a 200 W tungsten lamp to irradiate a methanol solution of the dideoxybiladiene-ac in the presence of dilute aqueous ammonia,[43,45] although other methods not involving irradiation have also been reported.[44,46,47] It is interesting to note that irradiation of the dideoxybiladiene-ac salt **2.72** in the presence of hydroquinone improved the yield of corrole but slightly (36% vs. 33%). Addition of ρ-*t*-butylcatechol also had no effect on the yield (it decreased from 33% to 31%). The apparent lack of an effect of these additives, which might be expected to intercept any free-radical intermediates during the course of the reaction, seems to indicate that this particular cyclization does not involve a free-radical mechanism.[45]

Scheme 2.1.16

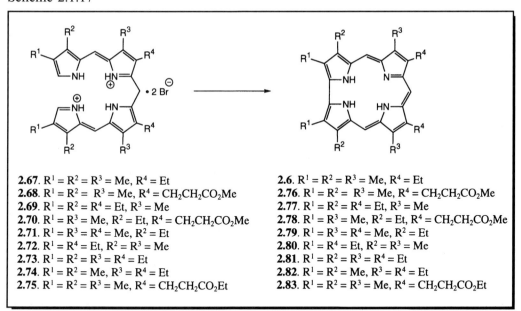

2.66   **2.6**

Scheme 2.1.17

2.67. $R^1 = R^2 = R^3 = Me$, $R^4 = Et$	**2.6**. $R^1 = R^2 = R^3 = Me$, $R^4 = Et$
2.68. $R^1 = R^2 = R^3 = Me$, $R^4 = CH_2CH_2CO_2Me$	**2.76**. $R^1 = R^2 = R^3 = Me$, $R^4 = CH_2CH_2CO_2Me$
2.69. $R^1 = R^2 = R^4 = Et$, $R^3 = Me$	**2.77**. $R^1 = R^2 = R^4 = Et$, $R^3 = Me$
2.70. $R^1 = R^3 = Me$, $R^2 = Et$, $R^4 = CH_2CH_2CO_2Me$	**2.78**. $R^1 = R^3 = Me$, $R^2 = Et$, $R^4 = CH_2CH_2CO_2Me$
2.71. $R^1 = R^3 = R^4 = Me$, $R^2 = Et$	**2.79**. $R^1 = R^3 = R^4 = Me$, $R^2 = Et$
2.72. $R^1 = R^4 = Et$, $R^2 = R^3 = Me$	**2.80**. $R^1 = R^4 = Et$, $R^2 = R^3 = Me$
2.73. $R^1 = R^2 = R^3 = R^4 = Et$	**2.81**. $R^1 = R^2 = R^3 = R^4 = Et$
2.74. $R^1 = R^2 = Me$, $R^3 = R^4 = Et$	**2.82**. $R^1 = R^2 = Me$, $R^3 = R^4 = Et$
2.75. $R^1 = R^2 = R^3 = Me$, $R^4 = CH_2CH_2CO_2Et$	**2.83**. $R^1 = R^2 = R^3 = Me$, $R^4 = CH_2CH_2CO_2Et$

An extensive investigation into the cyclization of free-base dideoxybiladiene-ac was carried out in 1966.[48] In this report, Johnson and coworkers demonstrated that the cyclization of dideoxybiladiene-ac **2.66** to the corresponding metal-free corrole **2.6** could, in fact, be catalyzed by the addition of any of the following: light, $K_2Fe(CN)_6$, $FeCl_3$, ceric sulfate, H_2O_2, benzoyl peroxide, and di-*t*-butylperoxide. All of these reagents effected efficient cyclization (yields on the order of 68–84%)

in the presence of ammonia, and in the presence or absence of molecular oxygen. These results led to the suggestion that the formation of these metal-free corroles proceeds *via* a free-radical mechanism. This conclusion is, of course, in contradiction with that drawn about the mechanism of the light-induced cyclization of dideoxybiladiene-ac HBr-salts, as discussed immediately above.

The synthesis of 10-substituted corroles from dideoxybiladiene-ac dihydrobromides has recently been reported by Paolesse, *et al.*[46] Here, base-catalyzed ring closure of **2.84** was found to afford the 10-carboxymethyl-substituted corroles **2.87**. The 10-substituted free-base corroles **2.88** and **2.89** were also prepared in this way (Scheme 2.1.18). Using a similar strategy, but starting from dideoxybiladiene salt **2.90**, Paolesse and coworkers synthesized the 5-methoxycarbonyl-substituted corrole **2.91** (Scheme 2.1.19).

Scheme 2.1.18

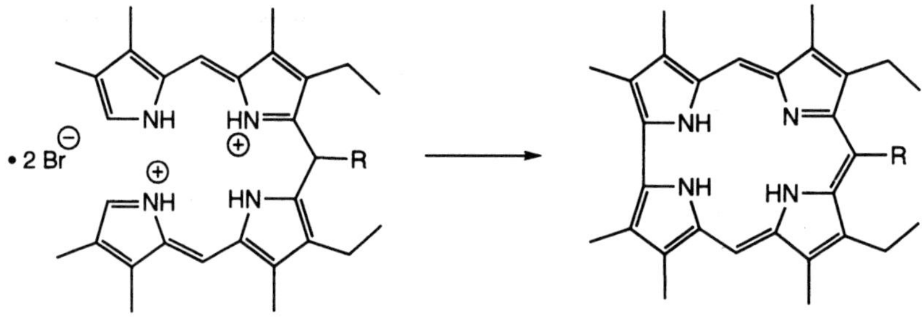

2.84. R = CO$_2$Me
2.85. R = CH$_2$CO$_2$Me
2.86. R = CH$_2$CH$_2$Cl

2.87. R = CO$_2$Me
2.88. R = CH$_2$CO$_2$Me
2.89. R = CH$_2$CH$_2$Cl

Scheme 2.1.19

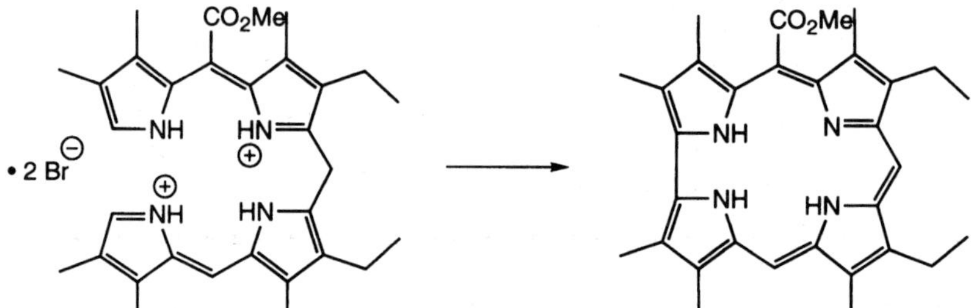

2.90

2.91

The formation of metal-free corroles from 1,19-dihalodideoxybiladienes-ac has also been demonstrated. The first example of this came in 1975 when Engel and Gossauer reported the preparation of corroles bearing substitution patterns corresponding to those of uroporphyrin III and 12-decarboxyuroporphyrin III (i.e., corroles **2.94** and **2.95**); these were made from the corresponding diiodobiladienes-ac (**2.92** and **2.93**) (Scheme 2.1.20).[49] More recently (1992), Smith and coworkers demonstrated that corrole **2.76** could be easily prepared in yields > 25% by heating the 1,19-dibromo derivative **2.96** in methanol at reflux (Scheme 2.1.21).[50] Importantly, Smith and coworkers also showed that this method allowed for the ready preparation of unsymmetrically substituted corroles (e.g., **2.99** and **2.100**). However, this approach, which relies on the use of halo-derivatives, has found little general use. This, presumably, reflects the tedious nature of the syntheses required to obtain the needed α-halo dideoxybiladienes-ac.

Scheme 2.1.20

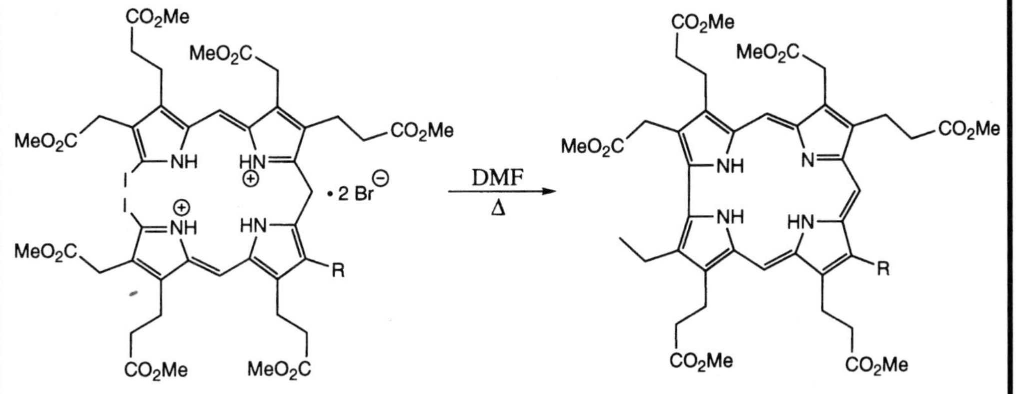

2.92. R = Me
2.93. R = CH₂CO₂Me

2.94. R = Me
2.95. R = CH₂CO₂Me

2.1.1.3   Corroles from Bipyrrolic Precursors

One example of a synthetic approach to free-base corroles is known that does not involve the cyclization of dideoxybiladienes-ac and related species. This involves the acid catalyzed "2 + 2" condensation between the bis(formylpyrrolyl) sulfide **2.26** and the diacid dipyrrylmethane species **2.36**. This afforded the *meso*-thiaphlorin macrocycle **2.101** in 52% yield (Scheme 2.1.22).[37,38] Heating this product at reflux in *o*-dichlorobenzene effects sulfur extrusion and generates corrole **2.102** in 35–40% yield. Interestingly, when the extrusion process is performed in the presence of triphenylphosphine (PPh₃), the yield of corrole increases to 60%. However, a mechanistic rationale for this latter interesting finding has yet to be put forth.

Scheme 2.1.21

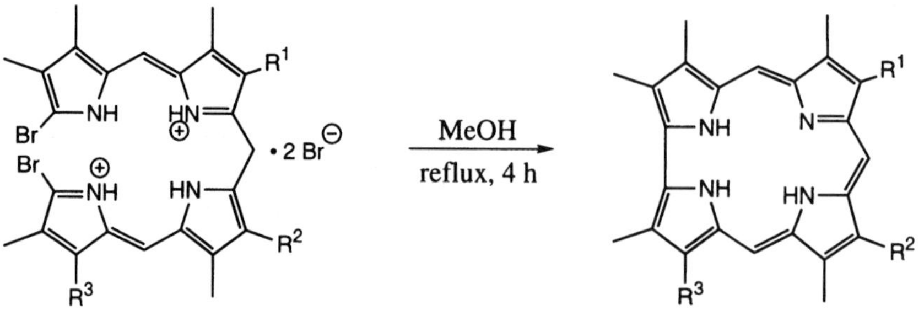

2.96. $R^1 = R^2 = CH_2CH_2CO_2Me$, $R^3 = Me$
2.97. $R^1 = CH_2CH_2Cl$, $R^2 = Et$, $R^3 = Me$
2.98. $R^1 = Me$, $R^2 = R^3 = CH_2CH_2CO_2Me$

2.76. $R^1 = R^2 = CH_2CH_2CO_2Me$, $R^3 = Me$
2.99. $R^1 = CH_2CH_2Cl$, $R^2 = Et$, $R^3 = Me$
2.100. $R^1 = Me$, $R^2 = R^3 = CH_2CH_2CO_2Me$

Scheme 2.1.22

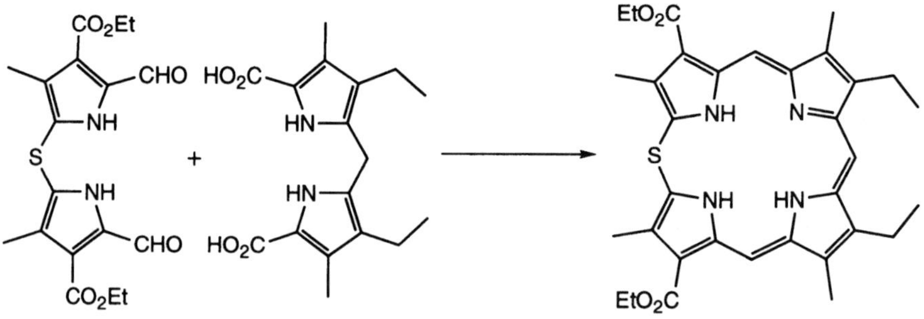

2.26   **2.36**   **2.101**

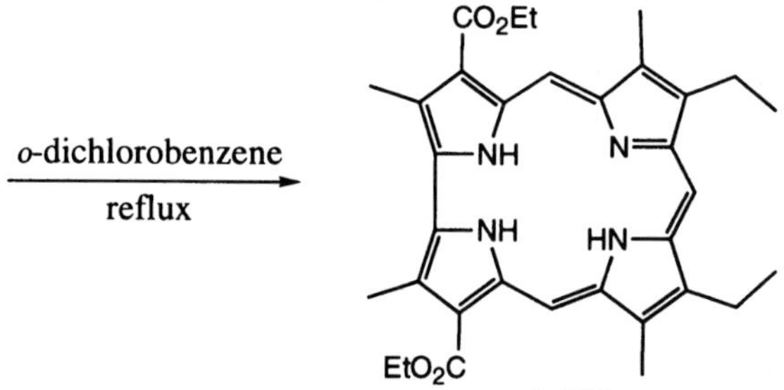

2.102

2.1.2   *Metallocorroles*

2.1.2.1   **Original work from Johnson and Others**

2.1.2.1.1   *Metallocorroles from tetrapyrrolic precursors*

The dideoxybiladiene-ac approach to corroles was also applied to the preparation of metallocorroles. The first report of this sort of reaction appeared in 1964. It described the preparation of the nickel(II) corrole **2.105** from the corresponding metal(II) dideoxybiladiene-ac **2.103** (Scheme 2.1.23).[27] Interestingly, this approach did not work in the case of zinc(II) dideoxybiladiene-ac (**2.104**). This latter result was rationalized in terms of the known coordination chemistry of Zn(II). In particular, it was postulated that the presumed tetrahedral coordination of the zinc(II) center served to lock the reactive termini of the biladiene into positions that were unsuited for cyclization.

In the case of the successful Ni(II) cyclization, the initial report (1964) indicated that the reaction could be carried out by irradiating with a 200 W tungsten lamp. It was later determined (in 1966) that, in contrast to the case of the metal-free dideoxybiladienes, light did not serve to activate this particular metal-dependent cyclization. Conversely, the presence of a base was found to be essential.[48] As part of this 1966 study, it was also found that cyclization of the metal(II)dideoxybiladienes-ac in the presence of di-*t*-butyl peroxide afforded only small yields of a mixture of macrocycles.[48] The above results led to the suggestion that the cyclization of metal(II)dideoxybiladienes-ac does not proceed *via* a free-radical mechanism. It was thus proposed that the initial reaction involved abstraction of a proton from the central (C(10)) methylene group of the dideoxybiladiene-ac.

Scheme 2.1.23

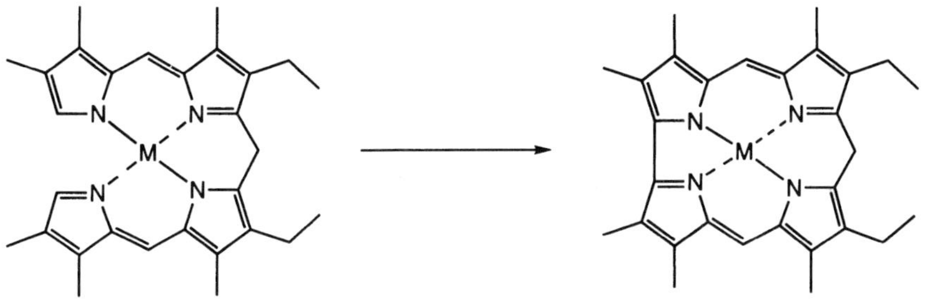

2.103. M = Ni   **2.105**. M = Ni
2.104. M = Zn

Johnson, *et al.* showed in 1965 that the hydrobromide salts of dideoxybiladienes-ac (e.g., **2.67–2.69**) could be converted directly to the corresponding nickel(II) or copper(II) corroles **2.105**, **2.107–2.110**) by treatment with the appropriate metal salt (Scheme 2.1.24).[43] This same approach was later employed by Murakami, *et al.* in 1981 to obtain the metal(II) corroles **2.111–2.114** from dideoxybiladienes-ac **2.74** and **2.75**.[51] It was also used by Boschi, *et al.* in 1990 to generate the manganese(II) corrole **2.115** from dideoxybiladiene-ac **2.106**.[52]

Scheme 2.1.24

2.67. $R^1 = R^2 = R^3 = Me$, $R^4 = Et$
2.68. $R^1 = R^2 = R^3 = Me$, $R^4 = CH_2CH_2CO_2Me$
2.69. $R^1 = R^2 = R^4 = Et$, $R^3 = Me$
2.74. $R^1 = R^2 = Me$, $R^3 = R^4 = Et$
2.75. $R^1 = R^2 = R^3 = Me$, $R^4 = CH_2CH_2CO_2Et$
2.106. $R^1 = R^2 = R^3 = R^4 = Me$

2.105. $R^1 = R^2 = R^3 = Me$, $R^4 = Et$, M = Ni
2.107. $R^1 = R^2 = R^3 = Me$, $R^4 = Et$, M = Cu
2.108. $R^1 = R^2 = R^3 = Me$, $R^4 = CH_2CH_2CO_2Me$, M = Cu
2.109. $R^1 = R^2 = R^4 = Et$, $R^3 = Me$, M = Ni
2.110. $R^1 = R^2 = R^4 = Et$, $R^3 = Me$, M = Cu
2.111. $R^1 = R^2 = Me$, $R^3 = R^4 = Et$, M = Ni
2.112. $R^1 = R^2 = Me$, $R^3 = R^4 = Et$, M = Cu
2.113. $R^1 = R^2 = R^3 = Me$, $R^4 = CH_2CH_2CO_2Et$, M = Ni
2.114. $R^1 = R^2 = R^3 = Me$, $R^4 = CH_2CH_2CO_2Et$, M = Cu
2.115. $R^1 = R^2 = R^3 = R^4 = Me$, M = Mn

Based on chemical and spectroscopic evidence, the generally accepted structures of the above metallocorroles are ones in which the metals are divalent and the corrole is acting as a dianionic ligand.[43,51,53–56] That is to say, in each complex, only two of the three pyrrolic hydrogen atoms have been replaced by the metal ion, and the complexes are, therefore, neutral (i.e., no other ligands or counteranions are required). In the case of nickel(II) complexes, spectroscopic arguments have been put forth[43,51,56] in support of these systems being best represented by the tautomeric form given in generalized structure **2.116**. As part of this formulation, the remaining pyrrolic hydrogen atom was arbitrarily assigned to *meso* carbon C-10. This results in two hydrogen atoms being bound at this *meso* position and hence *inter alia* to a disruption of the macrocyclic π-electron system. In the case of copper, a structure similar to that of the nickel complexes has also been proposed in the literature (i.e., **2.117**).[43,51] Interestingly, for this complex, an alternate interpretation in which the remaining hydrogen atom is assigned to the original pyrrolic nitrogen atom (i.e., as depicted by structure **2.118**), has also been proposed (Figure 2.1.6).[45] However, as will be noted in Section 2.1.2.2, structural assignments alternative to the above (i.e., **2.116–2.118**), have recently been suggested, at least for the copper and nickel complexes.[47]

Figure 2.1.6   Structures **2.116–2.118**

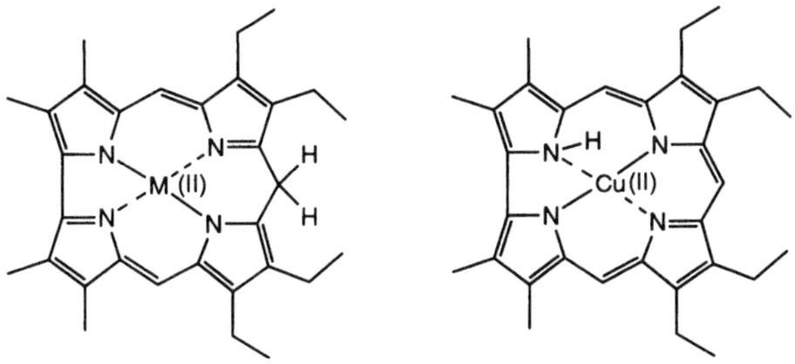

2.116. M = Ni
2.117. M = Cu

2.118

It was found that addition of hydroxide anion in dimethylformamide or dimethylsulfoxide to metal(II) corrole complexes results in the appearance of much sharper absorption bands relative to the starting compounds. These findings were considered consistent with the idea that an anionic, 18 π-electron aromatic corrole complex (e.g., **2.119**) is formed as the result of what appears to be a formal deprotonation process (Scheme 2.1.25).[43,51,56] That deprotonation actually occurs was inferred from acid–base titrations involving nickel(II) and copper(II) corroles.[56] The conclusion that these species are anionic aromatic compounds came from an appreciation that their electronic spectra resemble those recorded for divalent metallo–porphyrins. In any event, the anion that results was found to be quenched upon acidification, regenerating the corresponding non-aromatic metallocorroles.[51]

Scheme 2.1.25

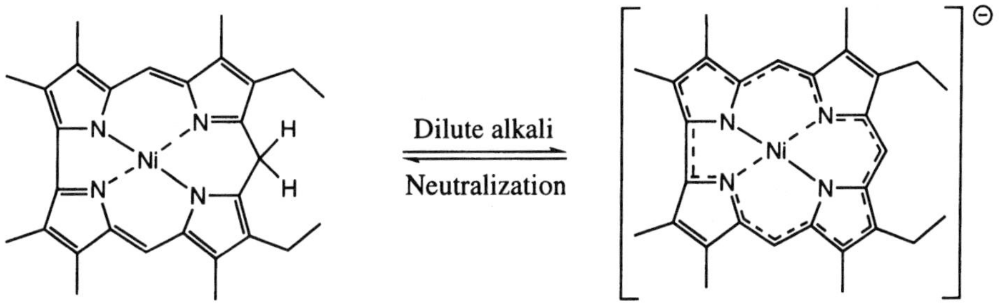

2.105   Dilute alkali
                          Neutralization   **2.119**

Trivalent cobalt corroles may also be prepared from dideoxybiladienes-ac as shown in Scheme 2.1.26.[57–59] Towards this end, Co(OAc)$_2$ is reacted with the dihydrobromide salt of a dideoxybiladiene-ac (e.g., **2.120**) in the presence of PPh$_3$ or a pyridine base. This affords the corresponding cobalt(III) corrole (**2.122**)[†] with one axial ligand.[57] Using analogous conditions, the cobalt(III) corroles **2.123**–**2.126** could be prepared from the corresponding dideoxybiladiene-ac in the presence of a coordinating base. (In one instance, the axial ligand was exchanged from PPh_3 to pyridine by heating complex **2.124a** in pyridine to afford complex **2.124b**[57].) The *meso*-substituted cobalt corroles **2.127**–**2.130** were also prepared from the corresponding dideoxybiladienes-ac **2.84**–**2.86** and **2.90** by treatment with $Co(OAc)_2$ in the presence of triphenylphosphine (Schemes 2.1.27 and 2.1.28).[46] In all of these cobalt insertion reactions, the metal spontaneously oxidizes from cobalt(II) to cobalt(III). This affords the *aromatic*, conjugated corrole derivatives.

Scheme 2.1.26

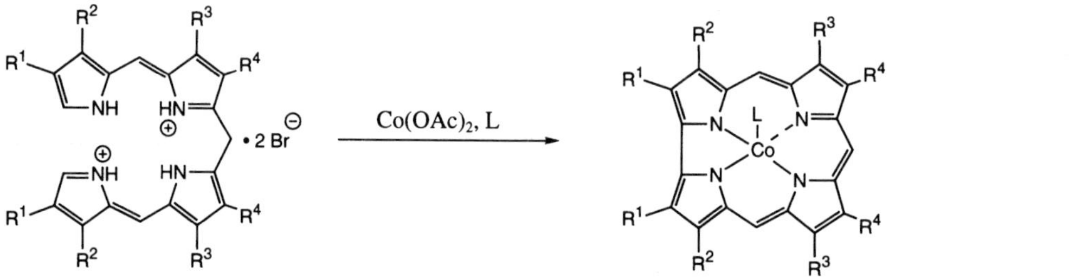

2.120. $R^1 = R^2 = R^3 = R^4 = H$
2.106. $R^1 = R^2 = R^3 = R^4 = Me$
2.121. $R^1 = R^3 = R^4 = Et, R^2 = Me$
2.72. $R^1 = R^4 = Et, R^2 = R^3 = Me$
2.74. $R^1 = R^2 = Me, R^3 = R^4 = Et$

pyridine, Δ

2.122. $R^1 = R^2 = R^3 = R^4 = H, L = PPh_3$
2.123. $R^1 = R^2 = R^3 = R^4 = Me, L = PPh_3$
2.124a. $R^1 = R^3 = R^4 = Et, R^2 = Me, L = PPh_3$
2.124b. $R^1 = R^3 = R^4 = Et, R^2 = Me, L = Py$
2.125a. $R^1 = R^4 = Et, R^2 = R^3 = Me, L = Py$
2.125b. $R^1 = R^4 = Et, R^2 = R^3 = Me, L = PPh_3$
2.126a. $R^1 = R^2 = Me, R^3 = R^4 = Et, L = Py$
2.126b. $R^1 = R^2 = Me, R^3 = R^4 = Et, L = 4\text{-Me-Py}$
2.126c. $R^1 = R^2 = Me, R^3 = R^4 = Et, L = 4\text{-Cn-Py}$
2.126d. $R^1 = R^2 = Me, R^3 = R^4 = Et, L = 3\text{-Ac-Py}$
2.126e. $R^1 = R^2 = Me, R^3 = R^4 = Et, L = 3\text{-OH-Py}$
2.126f. $R^1 = R^2 = Me, R^3 = R^4 = Et, L = 3\text{-Me-Py}$
2.126g. $R^1 = R^2 = Me, R^3 = R^4 = Et, L = 2\text{-Me-Py}$

    A single crystal X-ray structure of the $Co(PPh_3)$ corrole **2.122** confirmed the oxidation state of the metal and the nearly in-plane coordination of the cobalt atom.[58] In this particular case, the cobalt(III) center sits 0.38 Å above the ring plane and lies in a distorted square pyramidal coordination geometry (Figure 2.1.7). Unfortunately, owing to disorder in the orientation of the corrole ligand more specific information regarding the geometry of the corrole ligand could not be obtained.

[†]In the case of the completely unsubstituted dideoxybiladiene **2.120**, cobalt salts are the only ones known to catalyze corrole formation.[57]

Scheme 2.1.27

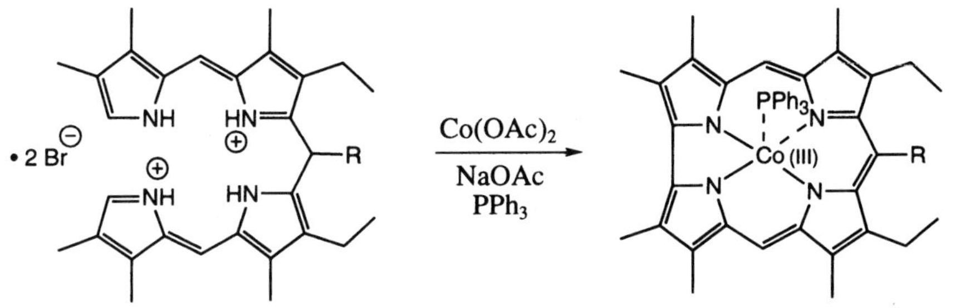

2.84. R = CO$_2$Me
2.85. R = CH$_2$CO$_2$Me
2.86. R = CH$_2$CH$_2$Cl

2.127. R = CO$_2$Me
2.128. R = CH$_2$CO$_2$Me
2.129. R = CH$_2$CH$_2$Cl

Scheme 2.1.28

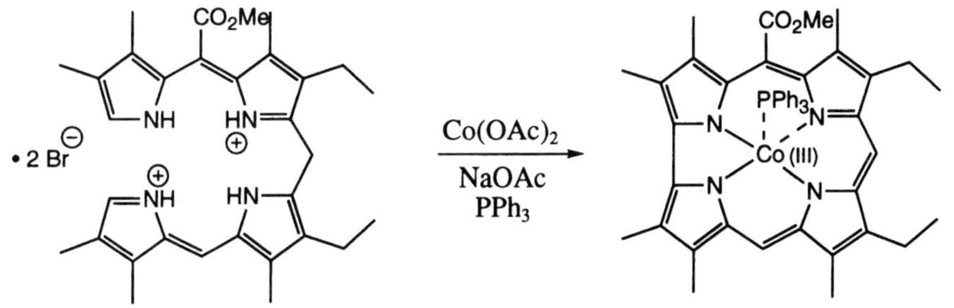

2.90   **2.130**

When heated in dimethylformamide (DMF) in the absence of a coordinating base, the octaethyldideoxybiladiene-ac **2.73** reacts with Co(OAc)$_2$ to afford the square planar cobalt(III) corrole **2.131** (Scheme 2.1.29).[59] In the absence of axially ligating bases, it was determined that this complex actually exists in a monomer–dimer–polymer equilibrium at higher concentrations (represented schematically in Scheme 2.1.30). In fact, as inferred from NMR spectroscopic analyses, at concentration ranges above *ca.* 5 × 10^{-3} M in chloroform, the cobalt(III) corrole **2.131** exists predominantly as a polymeric species. When low concentrations of pyridine are added, the aggregation is disrupted in favor of formation of monomeric pentacoordinate cobalt(III) corrole. At high pyridine concentration, it appears that hexacoordinate cobalt(III) species are formed.[59]

Trivalent metallocorrole complexes containing metals other than cobalt have also been prepared from hydrobromide salts of dideoxybiladienes-ac. For instance, in 1988, Boschi, *et al.* reported two approaches to the formation of in-plane trivalent rhodium complexes. The first involved reacting octamethylbiladiene-ac **2.106** with

RhCl$_3$ in the presence of a coordinating base such as PPh$_3$, methyldiphenylphosphine (PPh$_2$Me), tri-phenylarsine (AsPh$_3$), or triethylamine (Et$_3$N).[60] The resulting trivalent rhodium corroles (**2.132a–e**) were shown to have one axially coordinating ligand that stabilized the complex significantly (Scheme 2.1.31). In fact, all attempts to isolate axially unsubstituted Rh-corroles failed, even though it appears that such complexes appear to form in the absence of ligand, L. Interestingly, when this reaction is carried out in DMF in the absence of ligand L, the dimethylamine Rh-corrole **2.132e** is isolated. The axial ligand in this case arises presumably from thermal decomposition of the DMF solvent.[60]

Figure 2.1.7   Single Crystal X-Ray Diffraction Structure of Cobalt(III) Corrole Triphenylphosphine **2.122**.

This figure was generated using information down-loaded from the Cambridge Crystallographic Data Centre and corresponds to a structure originally reported in reference 58. The cobalt atom sits 0.38 Å above the mean macrocyclic plane and lies in a distorted square pyramidal coordination geometry. A high degree of disorder is present in the crystal and leads to a structural refinement in which an "extra" *meso*-like carbon atom is present in the structure. Atom labeling scheme: cobalt: ⊕; carbon: ○; nitrogen: ◉; phosphorus: ⦶. Hydrogen atoms have been omitted for clarity

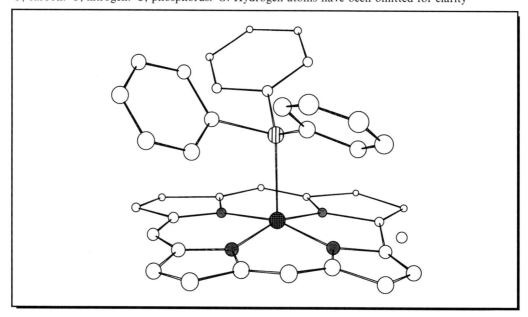

When Rh(III) corrole **2.132a** (L = PPh$_3$) is treated with excess *t*-butylisocyanide, the corrole complex **2.133a** is obtained, which contains a hexacoordinate rhodium center (Scheme 2.1.32).[60] A similar product (**2.133b**) can be obtained from **2.132a** by treatment with benzylisocyanide.

The second approach to the generation of trivalent rhodiumcorrole complexes involved reacting dideoxybiladiene-ac **2.106** in methanol with tetracarbonyldi-μ-chlororhodium(I) (Rh$_2$(CO)$_4$Cl$_2$).[60] After addition of PPh$_3$, a Rh(III) corrole **2.134** is obtained, which contains one PPh_3 and one CO coordinated as axial ligands to the metal center (Scheme 2.1.33). It is presumed that first an out-of-plane Rh(I) corrole complex is formed and that this undergoes metal-centered oxidation during the course of the reaction to afford the final, isolated Rh(III) corrole.[52] In 1990, Boschi, *et al.* prepared the manganese(III) and iron(III) octamethylcorroles **2.135** and **2.136**.[‡] They did this by treating the corresponding dideoxybiladiene-ac (**2.106**) with $Mn(OAc)_3$ and $FeCl_3$ (Scheme 2.1.34).[52] Interestingly, under analogous reaction conditions, attempts to prepare the corresponding chromium(III) and ruthenium(III) corroles from $Cr(OAc)_3$ and $RuCl_3$, respectively, afforded only the metal-free corroles.

Scheme 2.1.29

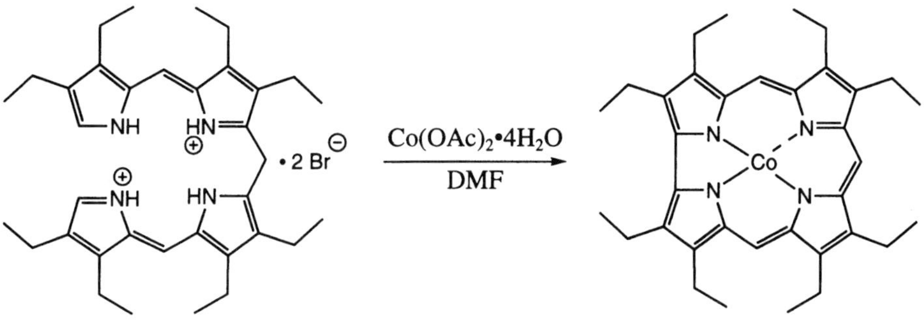

2.73   **2.131**

Scheme 2.1.30

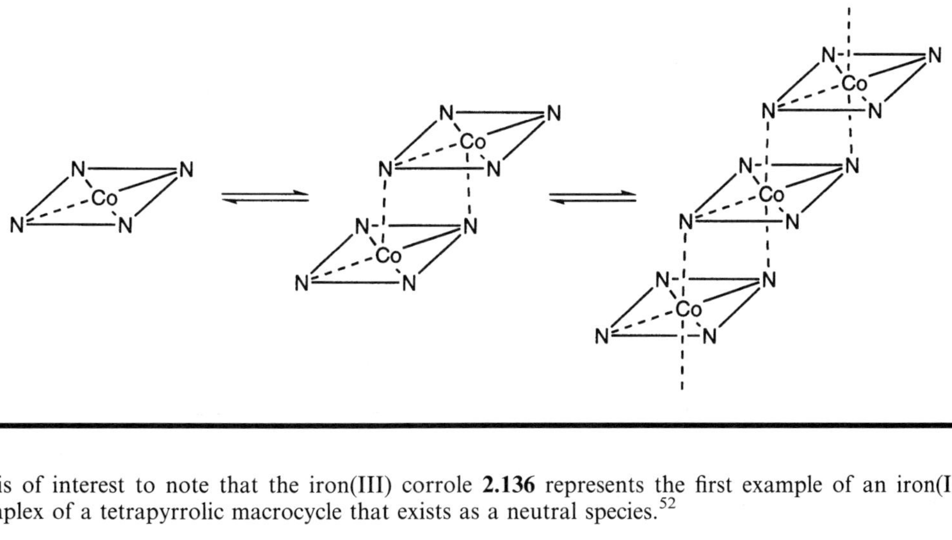

[‡] It is of interest to note that the iron(III) corrole **2.136** represents the first example of an iron(III) complex of a tetrapyrrolic macrocycle that exists as a neutral species.[52]

The iron(III) octamethylcorrole **2.136** was later studied in terms of its reactivity towards axial ligands.[61] In this work, Licoccia, *et al.* determined that, like the ligand-free cobalt(III) corroles, aggregation of the macrocycles occurs in the absence of ligating base. In the presence of pyridine, on the other hand, there is evidence to support the coordination of one axial pyridine ligand. In this instance, the sixth coordination site at iron may be occupied by water. With axially coordinated pyridine, magnetic moment measurements were consistent with the iron center being in a mixed-spin state ($S = 1/2, 5/2$). This behavior is similar to that seen for synthetic 5-coordinate iron(III) porphyrins wherein the metal atom is often found to be in a mixed-spin state.[62]

Scheme 2.1.31

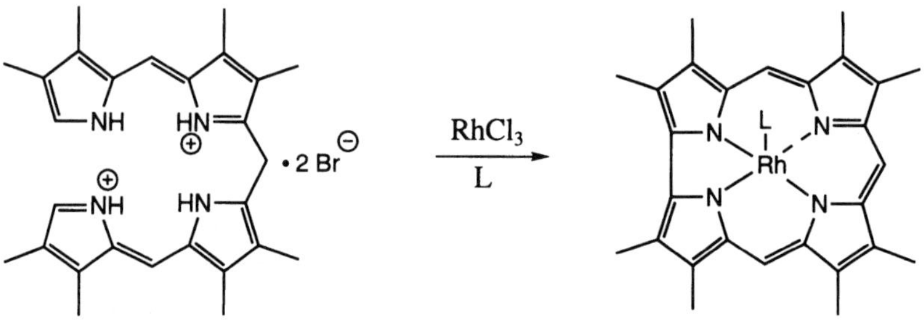

2.106

2.132a. L = PPh$_3$
2.132b. L = PPh$_2$Me
2.132c. L = AsPh$_3$
2.132d. L = NEt$_3$
2.132e. L = NHMe$_2$

Scheme 2.1.32

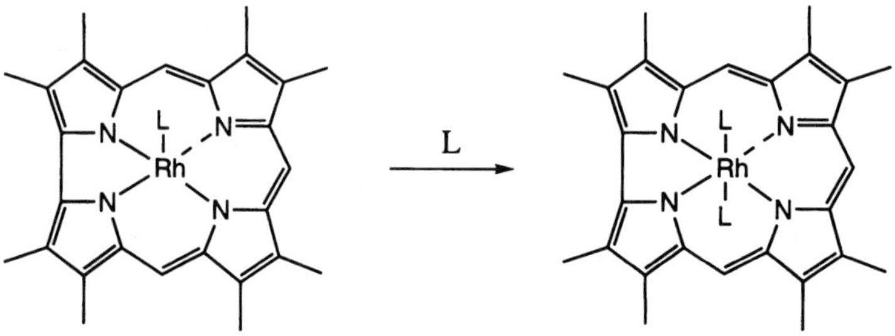

2.132

2.133a. L = *t*-butylisocyanide
2.133b. L = benzylisocyanide

Addition of chloride anion to the ligand-free (i.e., four coordinate) iron(III) corrole leads to a complex with chloride coordinated apically to the iron center. To verify the occurrence of this presumed axial ligation, *t*-butyl isocyanide, a chloride anion surrogate, was added to a chloroform solution of ligand-free iron(III) corrole. In this instance, NMR, as well as infrared (IR) spectroscopic data were found to be consistent with the proposed axial ligation.[62]

Scheme 2.1.33

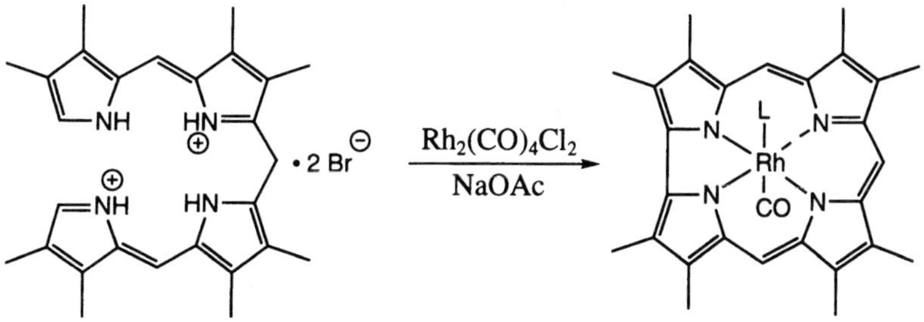

2.106   **2.134.** L = PPh$_3$

Scheme 2.1.34

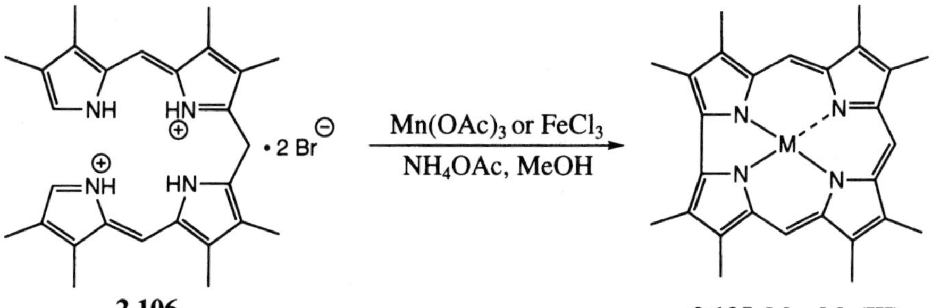

2.106   **2.135.** M = Mn(III)
                                             2.136. M = Fe(III)

Main-group metallocorroles have been prepared from dideoxybiladienes-ac as shown in Schemes 2.1.35 and 2.1.36. For instance, treating dideoxybiladiene-ac **2.106** with SnCl$_2$ in methanol in the presence of sodium acetate affords the aromatic Sn(IV) corrole salt **2.137a**. Here, as in some of the examples described earlier, the initially low-valent tin(II) center is thought to undergo spontaneous oxidation to tin(IV) during the course of the reaction.[63] If sodium acetate is replaced by triethyl amine in these metal insertions, the tin(IV) corrole **2.137b** is obtained in the form of its corresponding chloride salt. The acetate and chloride anions in these tin(IV) complexes can also be exchanged. This can be done, for instance, by treating complex **2.137a** with NaCl and HCl, or by treating complex **2.137b** with AgNO$_3$ and NaOAc.

Also, the tin(IV) corrole complex **2.137c**, in which hydroxide is the counter anion, may be obtained by subjecting complex **2.137a** to chromatography on basic alumina.

Scheme 2.1.35

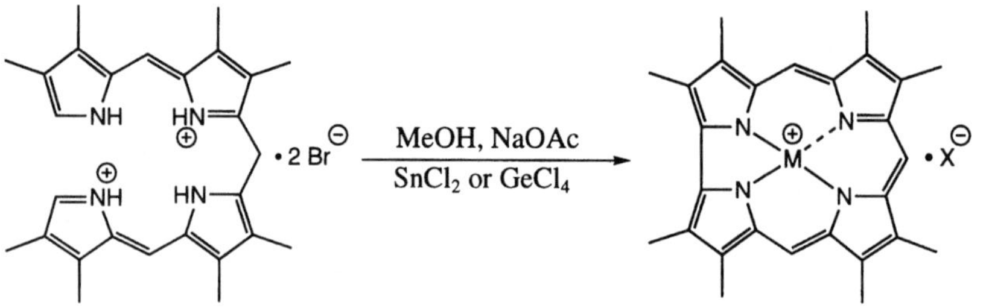

2.106

NaCl, HCl (**2.137a**. M = Sn, X = OAc) $AgNO_3$, NaOAc
2.137b. M = Sn, X = Cl
2.137c. M = Sn, X = OH

NaCl, HCl (**2.138a**. M = Ge, X = OAc) $AgNO_3$, NaOAc
2.138b. M = Ge, X = Cl

Similar complexes of germanium (i.e., **2.138a,b**) may be obtained by reacting dideoxybiladiene **2.106** with $GeCl_4$.[63] Likewise, indium(III) corroles may be prepared *via* methods similar to those used to generate the tin(IV) complexes.[63] Specifically, treating dideoxybiladiene-ac **2.106** with $InCl_3$ affords a 12% yield of the neutral aromatic indium(III) corrole **2.139**, along with variable amounts of metal-free octamethylcorrole. However, this particular metal complex proved to be somewhat unstable, decomposing extensively during chromatography.

Scheme 2.1.36

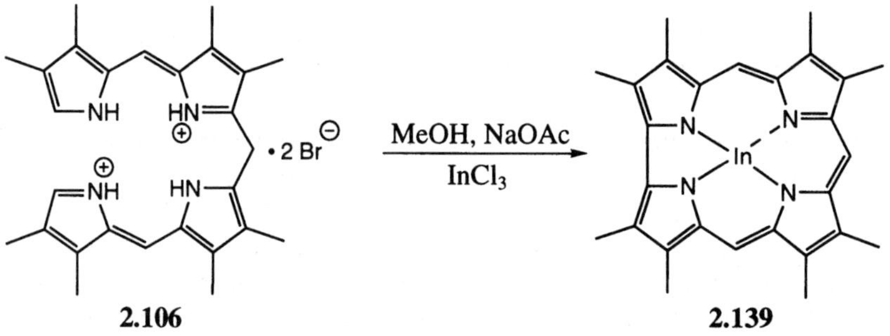

2.106 **2.139**

2.1.2.1.2   *Metallocorroles from bipyrrolic precursors*

A more direct, "2 + 2" approach to the synthesis of cobalt(III) corroles has been described.[57] It involves condensing a diformyl bipyrrole such as **2.140** with a diacid dipyrrylmethane such as **2.36**. This approach is thus similar to the one used to obtain the bifuran-containing corroles **2.38–2.40** described earlier. In the present instance, the diacid bipyrrole **2.141** may also be reacted with a diformyl dipyrrylmethane such as **2.142**. This affords corrole **2.125b** (Scheme 2.1.37). In either case, the reaction must be carried out in the presence of Co(II) and PPh₃.§ This requirement for a presumably coordinating metal cation is in stark contrast to what is seen in the case of the bifuran analog; there, no metal is needed to template the reaction.[37–39]

Scheme 2.1.37

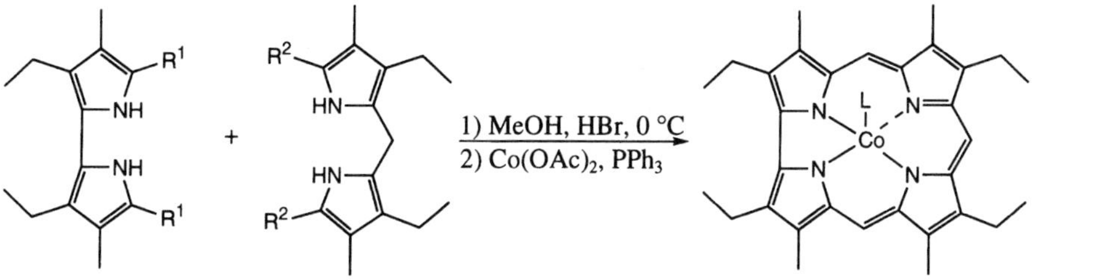

| 2.140. R¹ = CHO | 2.36. R² = CO₂H |
| 2.141. R¹ = CO₂H | 2.142. R² = CHO |

2.125b. L = PPh₃
2.125a. L = pyridine
2.125c. L = p-tolylisocyanide

The Co(III) corrole **2.125b** that results from the above reactions contains an axially coordinating PPh₃ ligand. This may be exchanged, in an equilibrium sense, with either pyridine or p-tolylisocyanide.[57] This is achieved by heating the Co(III) corrole(L) in the presence of the desired coordinating base. The pyridine complex **2.125a** has been shown to form a stable protonation product on addition of HBr. While the exact nature of this protonated complex has not yet been elucidated, it may be transformed back to the starting Co(III) corrole(Py) species **2.125a** by the addition of pyridine. This pyridine complex and the corresponding p-tolylisocyanide one may also, if desired, be treated with PPh₃ to regenerate the original Co(III) corrole(PPh₃) complex **2.125b**.

---

§Vogel, *et al.* have recently reinvestigated the acid-catalyzed condensation between diformyl bipyrrole and dipyrrylmethane, in the absence of a metal cation. Interestingly, these researchers reported the isolation of cyclic octapyrroles from these reactions.[127] For a complete discussion of this and other related findings, see Chapter 8.

2.1.2.1.3　　*Meso-phenyl-metallocorroles*

In 1993, Paolesse, *et al.* described the synthesis of the first *meso*-phenyl substituted corroles.[64,65] These were obtained from the mono-phenyl substituted dideoxybiladiene-ac **2.145**, a species obtained, in turn, from the acid-catalyzed condensation of the monoformyl pyrrole **2.143** and the diacid dipyrrylmethane derivative **2.144**.[64] Once the hydrobromide salt of this phenyl-substituted dideoxybiladiene-ac **2.145** was in hand, cobalt(II)-mediated cyclization afforded the mono-*meso*-phenyl-cobalt(III) corrole **2.146** in 40% yield (Scheme 2.1.38). As in the case of the *meso*-unsubstituted corroles, addition of PPh$_3$ proved necessary for the isolation of the product.

Scheme 2.1.38

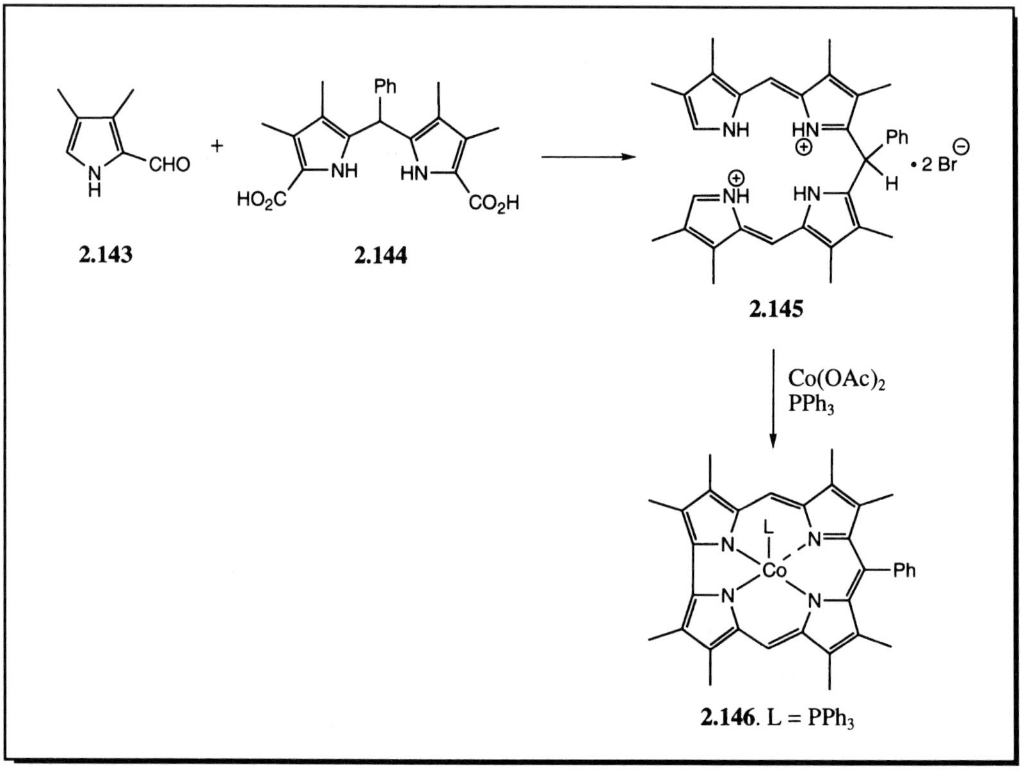

The synthesis of bis-(*meso*-phenyl)cobalt(III) corroles **2.150** and **2.151** was achieved by Licoccia and coworkers in 1994.[65] The procedure used involved first carrying out an acid-catalyzed condensation between the diacid dipyrrylmethane **2.35** and the α-hydroxybenzyl pyrrole **2.147**. This afforded a mixture of two dideoxybiladienes-ac (i.e., **2.148** and **2.149**), which were not isolated. Rather, as the next and final step in the synthesis, these were treated directly with Co(OAc)$_2$ in the presence of PPh$_3$. This yielded a 1:2 ratio of the two cobalt(III) corrole products **2.150** and **2.151** (total yield: 20%) (Scheme 2.1.39).

Scheme 2.1.39

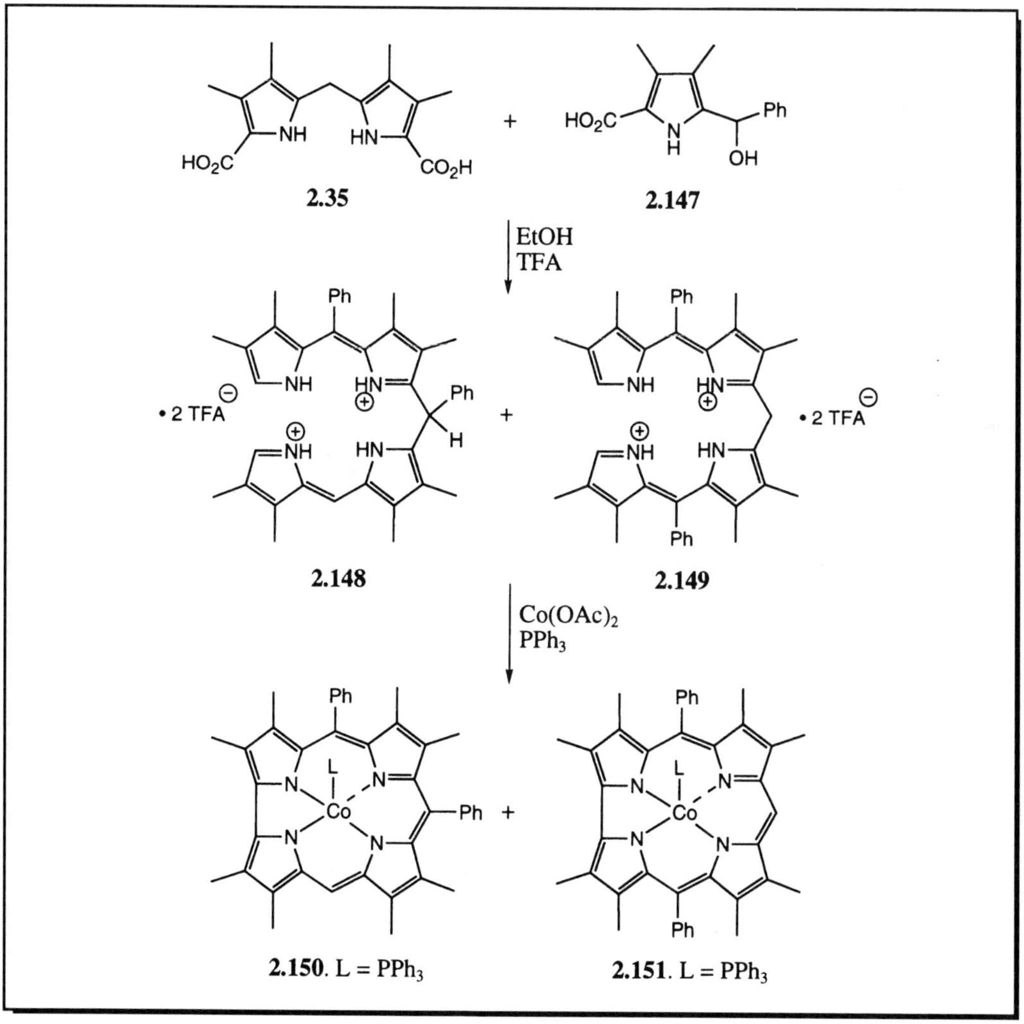

2.35   **2.147**

EtOH
TFA

2.148   **2.149**

Co(OAc)₂
PPh₃

2.150. L = PPh₃   **2.151.** L = PPh₃

To date, four synthetic approaches have been used to produce tris-(*meso*-phenyl)cobalt(III) corroles successfully. Three of these were reported by Paolesse, *et al.* in 1993[64] and are shown in Scheme 2.1.40. Of these, one involved as its first step the POCl₃-catalyzed reaction between the 2-benzoylpyrrole-5-carboxylic acid **2.152** and the phenyl-substituted dipyrrylmethane **2.144**. The resulting biladiene-ac was then treated without isolation with Co(OAc)₂ and PPh₃ to give the desired cobalt(III) corrole **2.153** in 2% yield.¶

¶In the presence of high PPh₃ concentrations, a second axially coordinating PPh₃ will add to the metal center (to afford a hexacoordinate cobalt(III) corrole).

Scheme 2.1.40

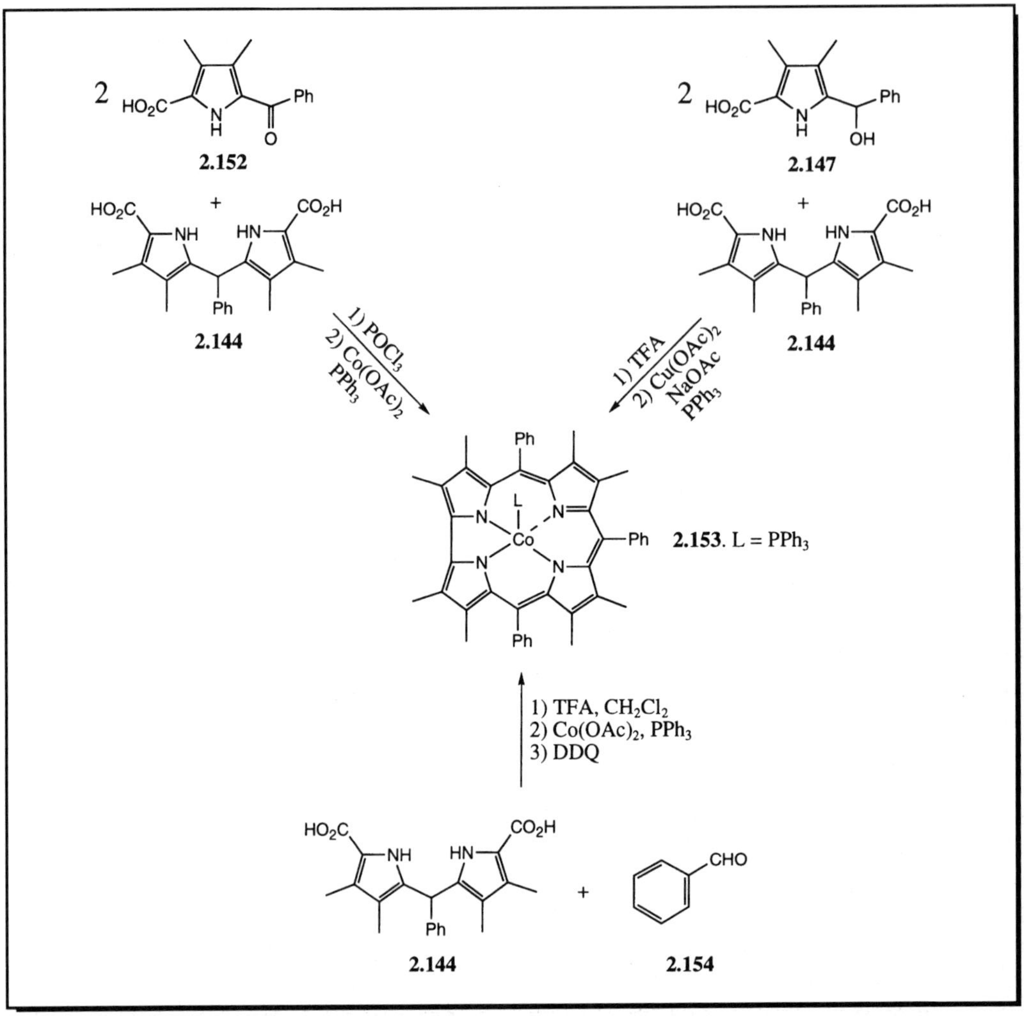

It was presumed that the major factor causing the low yields in the above reaction was the low reactivity of the benzoylpyrrole **2.152**. This low reactivity required the use of drastic conditions (POCl₃) to effect the requisite condensation. To overcome this, a second strategy was tested by Paolesse, *et al.* In it, the benzoyl-pyrrole **2.152** was replaced with the reduced 2-hydroxybenzylpyrrole derivative **2.147**.[64] The conditions required to effect condensation between this pyrrole and the dipyrrylmethane **2.144** were much more mild (trifluoroacetic acid (TFA) instead of POCl₃). Again, the tetrapyrrolic condensation product of this reaction could not be isolated. Therefore, the reaction mixture was simply buffered with sodium acetate, and the cyclization to the corrole **2.153** was then carried out *in situ*. This procedure allowed for a significant increase in the yield (to 20%) of the cobalt(III) corrole **2.153**.

The tedious preparation of the monopyrrolic precursor **2.147** necessary for the tris-(phenyl)cobalt(III) corrole **2.153** led Paolesse, *et al.* to consider a third approach to preparing tris-(phenyl)cobalt(III) corroles.[64] It centers around an initial acid-catalyzed condensation between the diacid dipyrrylmethane **2.144** with benzaldehyde (**2.154**) that is followed by a subsequent cyclization step carried out in the presence of Co(OAc)$_2$ and PPh$_3$. In this case, it was found that 2,3-dichloro-5,6-dicyano-1,4-benzoquinone (DDQ) helped mediate the requisite final oxidation step. While this approach did give a slightly lower yield of corrole (18%), it had the advantage of being quite streamlined.

An even more streamlined synthesis of cobalt(III) corroles was reported in 1994 by Licoccia and coworkers.[65] Here, the success of the approach rests on the fact that the hydroxybenzylpyrrole **2.147** undergoes acid-catalyzed condensation to give the presumed intermediate tetraphenyl porphyrinogen species **2.155**. This latter material is then treated directly with Co(OAc)$_2$ and PPh$_3$ to afford the tris-(phenyl)cobalt(III) corrole **2.156** in 25% yield (Scheme 2.1.41). Here, coordinative binding to the cobalt atom most likely catalyzes the critical ring contraction step. To the extent this is true, this ring contraction step represents a model for the uroporphyrinogen-corrin transformation occurring during the biosynthesis of coenzyme B$_{12}$. In any event, it is of interest to note that no other metal has been found that will engender it. Manganese, iron, and rhodium cations, for instance, specifically fail to catalyze this ring contraction. Their presence leads instead to the production of metalated and free-base porphyrins. Nickel(II) also fails to catalyze the contraction, even though it is known that nickel(II), as well as cobalt(II), are known to promote the formation of contracted macrocycles in natural systems (see Eschenmoser[5] and references therein).

A single crystal X-ray structure of cobalt(III) tris-(*meso*-phenyl)corrole **2.156** has been obtained (Figure 2.1.8).[65] In this structure, as in that of the *meso*-unsubstituted cobalt(III) corrole complex **2.122**, the cobalt atom lies just above the mean plane of the macrocycle (0.28 Å in **2.156** vs. 0.38 Å in **2.122**).[58] Interestingly, when the structure of **2.125** is compared to other metallocorrole structures (discussed later),[52,66] it is apparent that neither *meso*- nor β-substituents together or independently significantly alter the general corrole structure. This is in contrast to what is found in the case of porphyrins; here, significant distortions can arise from steric interactions between *meso*-aryl groups and flanking β-alkyl substituents.

Scheme 2.1.41

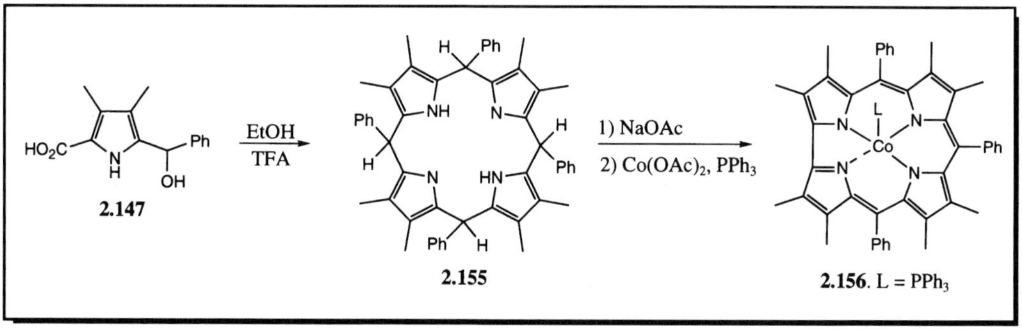

Figure 2.1.8 Single Crystal X-Ray Diffraction Structure of Cobalt(III) 5,10,15-
   Triphenylcorrole Triphenylphosphine **2.156**.

This figure was generated using information down-loaded from the Cambridge Crystallographic
Data Centre and corresponds to a structure originally reported in reference 65. The cobalt atom sits
0.28 Å above the mean macrocyclic plane (average displacement from the plane: 0.14 Å) and lies in
an approximately square pyramidal coordination environment. Atom labeling scheme: cobalt: ⬤;
carbon: ○; nitrogen: ◉; phosphorus: ⬤. Hydrogen atoms have been omitted for clarity

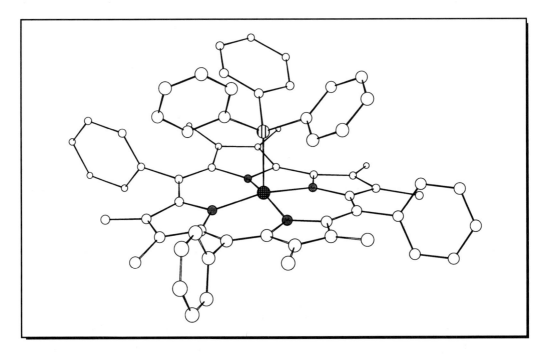

2.1.2.1.4   Metallocorroles from preformed corroles

It is perhaps not surprising that metallocorroles may be prepared from pre-
formed metal-free corroles as well as from linear pyrrolic precursors. In fact, the
former metal insertion approach has allowed a considerable number of metallocorroles to be prepared, including complexes containing mono-, di-, tri-, and tetravalent
metal cations (as discussed above in Sections 2.1.2.1.1–2.1.2.1.3). The following section will describe examples of the latter approach to metallocorroles, that is, *via*
insertion of a metal center into a pre-formed corrole ring.

To date, only one example of a monovalent metallocorrole has been reported.
It was reported in 1976 by Grigg, *et al.* and involves a rhodium corrole, which was
obtained in 36% yield as the result of reacting free-base diethyl-hexamethyl corrole
2.6 with $Rh_2(CO)_4Cl_2$.[67] Unlike the trivalent corrole complex **2.134**, obtained earlier
by the treatment of a dideoxybiladiene-ac with this same metal salt, the complex
isolated in this instance analyzed as being the monovalent $Rh(CO)_2$corrole, **2.157**
(Scheme 2.1.42). This complex was later prepared in 72% yield,[68] although it was

also reported that the complex undergoes extensive decomposition upon recrystallization. Nevertheless, sufficient pure material could be obtained such that binding of the metal center to rings A and B of the corrole skeleton, in accord with the structure represented in Scheme 2.1.42, could be established *via* NMR spectroscopic analyses.[67]

Scheme 2.1.42

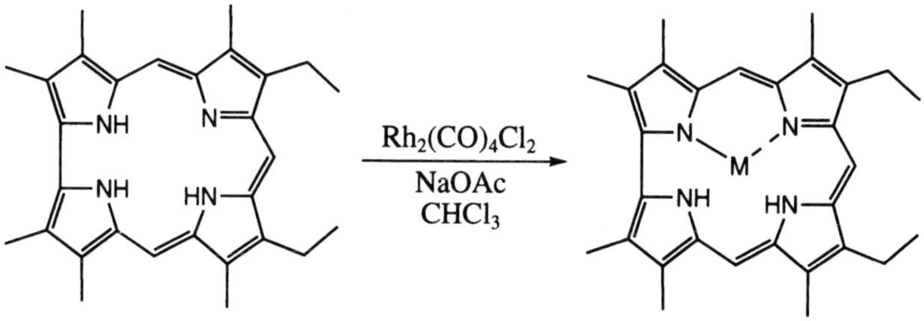

2.6   Rh$_2$(CO)$_4$Cl$_2$ / NaOAc / CHCl$_3$ **2.157**. M = Rh(CO)$_2$

As implied above, free-base corrole can also be used as the starting material for the formation of divalent transition metal complexes. For instance, the diethyl-hexamethyl free-base corrole **2.6** has been reported to form stable, non-aromatic complexes with Ni(II), Cu(II), and Zn(II) (i.e., **2.105**, **2.107**, and **2.158**).[43] These complexes are identical to those prepared from ring-closure reactions involving the precursor dideoxybiladiene-ac, as described earlier. Similarly, the Ni(II) and Cu(II) corrole complexes **2.109** and **2.110** have been prepared from the corresponding free-base corrole **2.77**,[43] just as **2.159** has been prepared from **2.80**.[45] It should be noted, however, that the structural assignments of the metallocorroles of Scheme 2.1.43 are to be considered tentative in light of the recent findings of Vogel and coworkers (cf. Section 2.1.2.2). Also, the copper corrole structures **2.107**, **2.110**, and **2.159** (Scheme 2.1.43) are derived from those presented by Johnson and coworkers in 1971.[45] This representation thus differs from that given in 1965 (and used in Scheme 2.1.24) in the context of an analogous metallocorrole complex.[43]

Treatment of the free-base corroles **2.79** or **2.80** with Pd(OAc)$_2$ in pyridine (Scheme 2.1.44) affords the respective aromatic pyridinium salts of the anionic palladium complexes **2.160** and **2.161** (in 92% and 89% yield, respectively).[45] A similar charged zinc(II) complex (**2.163**) has also been prepared by this method in 20% yield (Scheme 2.1.45).[63] It should be noted here that in the 1971 report of the anionic palladium complex from Johnson and coworkers the palladium(II) center was represented as bearing the negative charge.[45] Later, similar localized anionic representations of metal(II) corroles were given by others.[26,51,56] However, this assignment may or may not be correct. For this reason, and because it is the overall system that bears what is most likely a delocalized negative charge, we suggest that it may be less confusing to represent such systems as we have for complexes **2.160**, **2.161**, and

2.163. Such a representation of metal(II) corrole anions has been used previously.[11,12,55]

Scheme 2.1.43

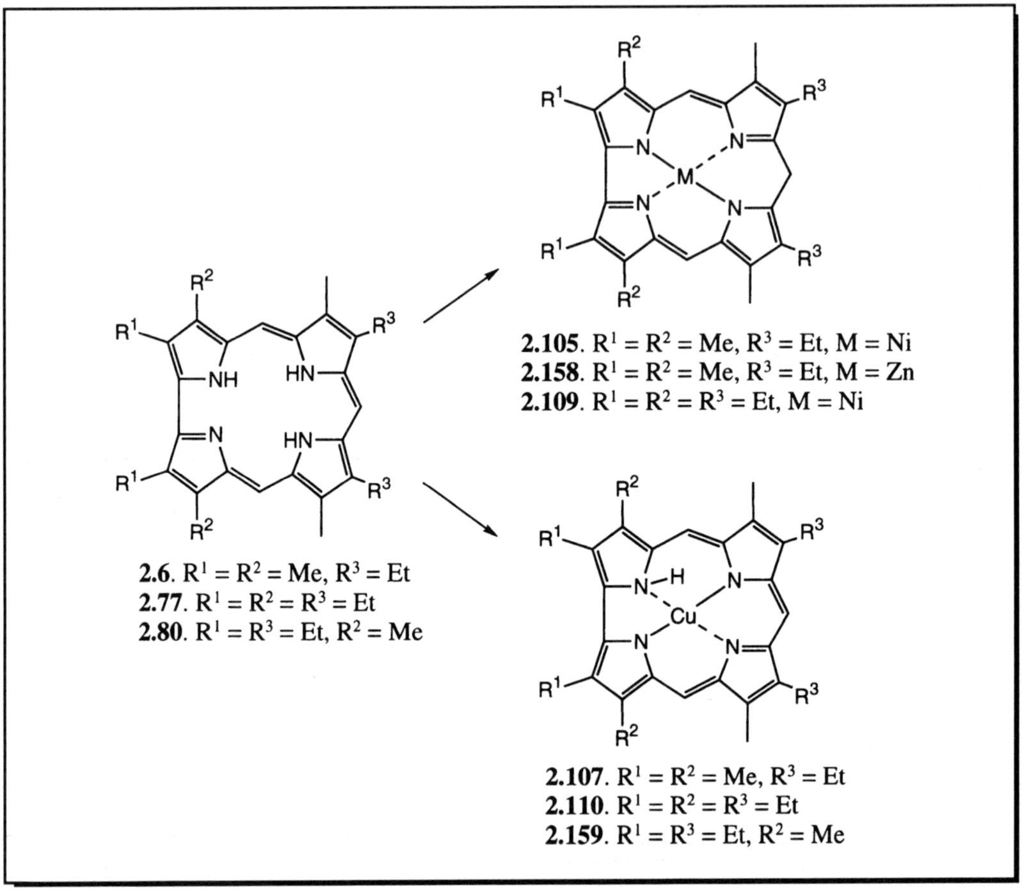

2.105. $R^1 = R^2 = Me$, $R^3 = Et$, $M = Ni$
2.158. $R^1 = R^2 = Me$, $R^3 = Et$, $M = Zn$
2.109. $R^1 = R^2 = R^3 = Et$, $M = Ni$

2.6. $R^1 = R^2 = Me$, $R^3 = Et$
2.77. $R^1 = R^2 = R^3 = Et$
2.80. $R^1 = R^3 = Et$, $R^2 = Me$

2.107. $R^1 = R^2 = Me$, $R^3 = Et$
2.110. $R^1 = R^2 = R^3 = Et$
2.159. $R^1 = R^3 = Et$, $R^2 = Me$

Cobalt(III) corroles can also be prepared from the free-base macrocycle. The
first reported example of this came from Johnson and Kay in 1965.[43] In this report,
8,12-diethyl-2,3,7,13,17,18-hexamethylcorrole **2.6** was treated with Co(OAc)$_2$ in pyridine. The cobalt complex reportedly isolated was formulated as the "cobaltic"
complex represented by structure **2.164** (Scheme 2.1.46). Johnson and Kay further
reported in 1965 that heating this complex in methanol afforded a somewhat
unstable pyridine-free "cobaltous" corrole (e.g., **2.165**) for which no structure was
offered. Treating this supposed "cobaltous" corrole with pyridine regenerated the
"cobaltic" corrole **2.164**.

A diaxial, bis-pyridine ligation mode was also put forward for the cobalt
corrole **2.166** (Figure 2.1.9). This complex, which was reported in 1971, was obtained

by treating an *N*-methylcorrole (discussed later in this chapter) with Co(ClO$_4$)$_2$·6EtOH in pyridine.[45] Confusingly, Johnson and coworkers state in a footnote to this same 1971 paper, that the dipyridine cobalt complex **2.166** presented in an earlier 1965 report was actually incorrectly reported as being a cobalt(II) derivative.[43] The compound in question, however, is the material that is referred to as the cobaltic (cobalt(III)) derivative. Unfortunately, it was not until the publication of a subsequent 1973 paper that the confusion this footnote engendered was finally laid to rest.[57] At that time, Johnson, *et al.* determined that the octahedral dipyridine complexes **2.164** and **2.166** were actually the monopyridine complexes **2.167** and **2.125b**, and that only in solutions containing excess pyridine were these octahedral complexes stable (as judged by electronic absorption spectral studies).[57] Subsequent to this initial work, three other cobalt(III) corroles (i.e., **2.128**, **2.168**, and **2.169**) bearing only one axial pyridine were reported.[59] These complexes were prepared from the corresponding free-base corroles and are represented in Scheme 2.1.47.

Scheme 2.1.44

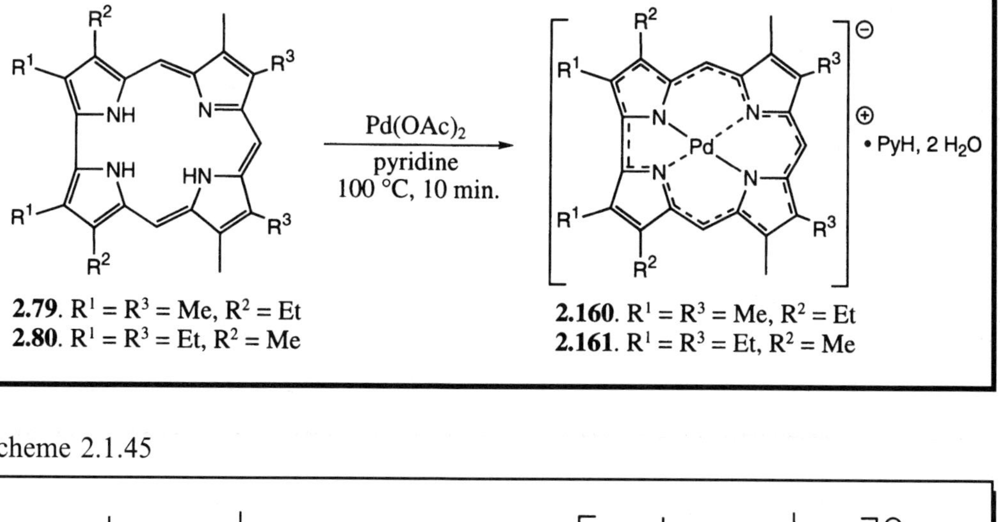

2.79. R^1 = R^3 = Me, R^2 = Et
2.80. R^1 = R^3 = Et, R^2 = Me

Pd(OAc)$_2$
pyridine
100 °C, 10 min.

2.160. R^1 = R^3 = Me, R^2 = Et
2.161. R^1 = R^3 = Et, R^2 = Me

• PyH, 2 H$_2$O

Scheme 2.1.45

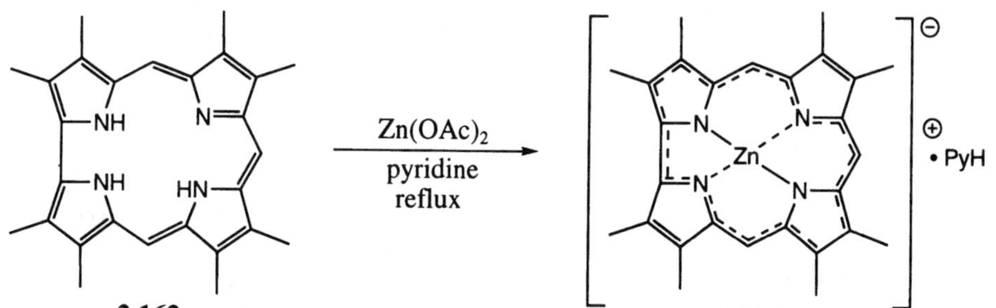

2.162

Zn(OAc)$_2$
pyridine
reflux

2.163

• PyH

Scheme 2.1.46

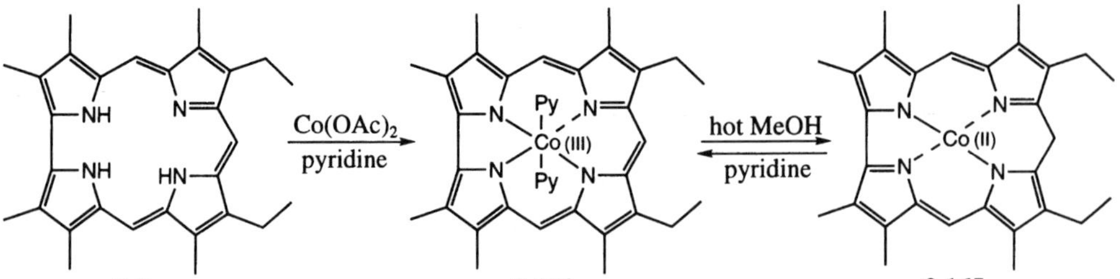

2.6 **2.164** "Cobaltic" **2.165** "Cobaltous"

Scheme 2.1.47

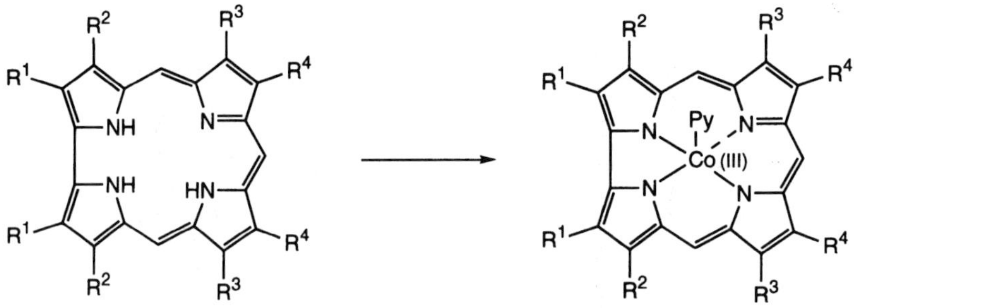

2.6. $R^1 = R^2 = R^3 = Me$, $R^4 = Et$
2.80. $R^1 = R^4 = Et$, $R^2 = R^3 = Me$
2.82. $R^1 = R^2 = Me$, $R^3 = R^4 = Et$
2.81. $R^1 = R^2 = R^3 = R^4 = Et$
2.83. $R^1 = R^2 = R^3 = Me$, $R^4 = CH_2CH_2CO_2Et$

2.167. $R^1 = R^2 = R^3 = Me$, $R^4 = Et$
2.125a. $R^1 = R^4 = Et$, $R^2 = R^3 = Me$
2.128. $R^1 = R^2 = Me$, $R^3 = R^4 = Et$
2.168. $R^1 = R^2 = R^3 = R^4 = Et$
2.169. $R^1 = R^2 = R^3 = Me$, $R^4 = CH_2CH_2CO_2Et$

Returning to the original work of Johnson, *et al.*, it was also determined later that the compound considered by Johnson and Kay[43] in 1965 to be the pyridine-free cobalt(II) ("cobaltous") corrole complex (e.g., **2.165**) was in fact a square planar axial-ligand-free, cobalt(III) complex (**2.170**). This assignment was made possible in 1973 when a *bona fide* paramagnetic, square planar cobalt(III) corrole complex **2.170** (i.e., corresponding to Johnson's presumed cobaltous species **2.165**) was prepared (Scheme 2.1.48). It was obtained by boiling the Co(III) corrole complex **2.167** in chloroform.[57] Further support for this axial-ligand-free formulation (**2.170**) was obtained in 1978 with the observation that the square planar Co(III) corrole derivative **2.171**, prepared using the direct insertion approach described earlier in this chapter, has a strong tendency to form oligomers in solution.[59]

Figure 2.1.9 Structure **2.166**

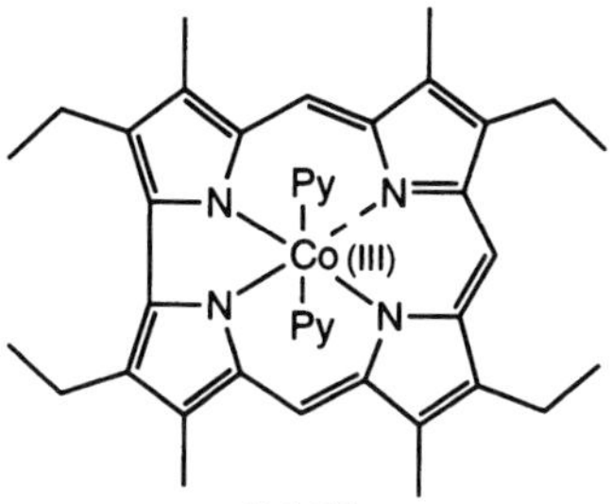

2.166

Scheme 2.1.48

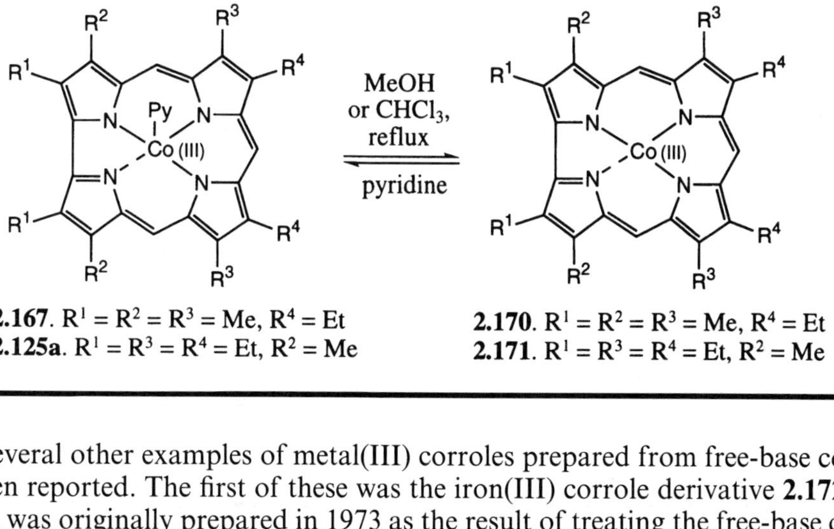

2.167. R^1 = R^2 = R^3 = Me, R^4 = Et
2.125a. R^1 = R^3 = R^4 = Et, R^2 = Me

2.170. R^1 = R^2 = R^3 = Me, R^4 = Et
2.171. R^1 = R^3 = R^4 = Et, R^2 = Me

Several other examples of metal(III) corroles prepared from free-base corroles have been reported. The first of these was the iron(III) corrole derivative **2.172**. This complex was originally prepared in 1973 as the result of treating the free-base corrole **2.6** with phenyllithium in THF, followed by FeCl$_2$ (Scheme 2.1.49).[57] This same complex was later prepared from corrole **2.6** by reacting it with FeCl$_2$ in the presence of pyridine.[69] In both cases, the initially bound iron(II) center undergoes spontaneous oxidation (to iron(III)) during the reaction and/or workup.

A similar iron(III) corrole **2.135** (Scheme 2.1.50) was also prepared in 1990 by Boschi, *et al.*[52] In this instance, the metalation conditions involved treating octamethylcorrole **2.162** with FeCl$_3$ in DMF. Using an analogous metalation approach, the manganese(III) corrole **2.136** could also be prepared from free-base corrole **2.162** by reacting it with Mn(OAc)$_3$ in DMF. Interestingly, however, treating **2.162** with Cr(OAc)$_3$ in DMF afforded only decomposition products. Likewise, when RuCl$_3$ was used, only octamethylporphyrin was obtained. Presumably, the octamethylporphyrin so obtained is the result of metal-catalyzed cleavage-recombination reactions involving the corrole ring.

Scheme 2.1.49

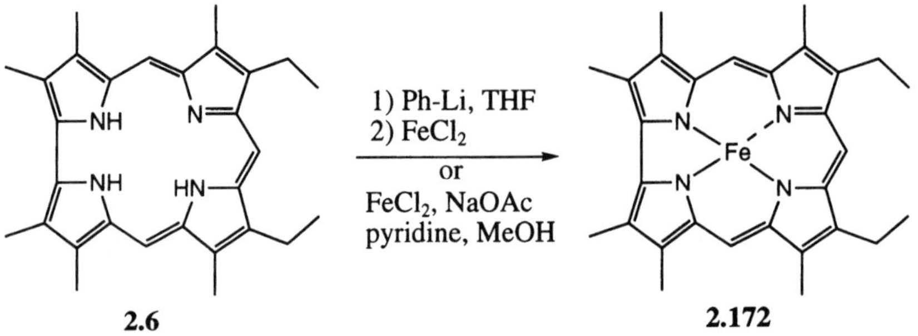

When octamethylcorrole **2.162** was treated with $Cr(CO)_6$ or $Mn_2(CO)_{10}$ in toluene, the respective Cr(III) and Mn(III) corroles **2.173** and **2.136** were obtained (Scheme 2.1.51).[52] When $Fe(CO)_5$ was used instead, a small amount of I_2 had to be added to catalyze the formation of the corresponding iron(III) corrole, **2.135**. With $Ru_3(CO)_{12}$, the analogous reaction produced only octamethylporphyrin, obtained in the form of both a ruthenium complex and free-base ligand. As immediately above, this may be rationalized in terms of metal-catalyzed reactions wherein the corrole undergoes ring opening to form linear oligopyrrolic species that then cyclize to give porphyrinic products.

Scheme 2.1.50

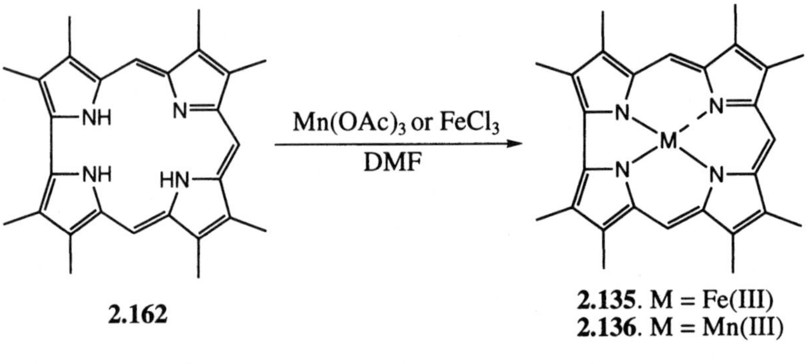

In benzene, octamethylcorrole **2.162** reacts with $Rh_2(CO)_4Cl_2$ in the presence of $AsPh_3$ to afford both the rhodium(I) dicarbonyl complex **2.174** and the Rh(III) corrole derivative **2.132c** (in 10% and 75% yield, respectively) (Scheme 2.1.52).[52]

After isolation, complex **2.132c** (Figure 2.1.10) proved to be identical to the product prepared from the corresponding dideoxybiladiene-ac precursor (*vide supra*). Further, a single crystal X-ray diffraction structural analysis of this complex confirmed the nearly in-plane location of the rhodium atom. The rhodium atom sits, in fact, 0.26 Å above the plane formed by the nitrogen atoms, which are themselves coplanar. The coordination geometry about the rhodium center is thus distorted square pyramidal, with all four of the Rh-N bonds being non-equivalent. The slight distortion in the macrocycle is similar to that observed in other corrole structures,[52,58,65,66] indicating that the basic macrocyclic ligand in this particular complex is relatively unaffected by rhodium coordination.

Scheme 2.1.51

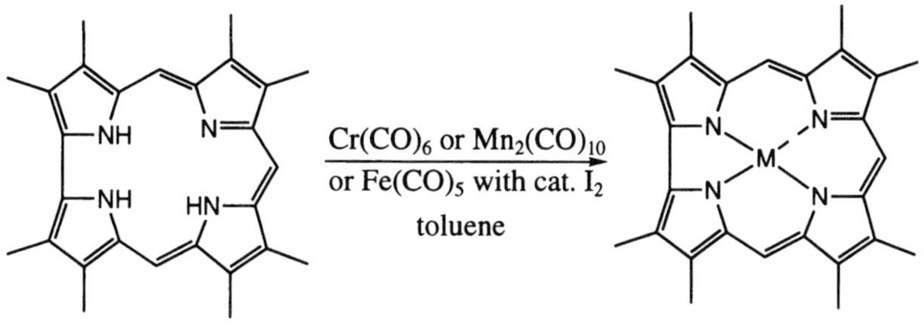

2.162   Cr(CO)$_6$ or Mn$_2$(CO)$_{10}$ or Fe(CO)$_5$ with cat. I$_2$ toluene

2.173. M = Cr
2.136. M = Mn
2.135. M = Fe

Scheme 2.1.52

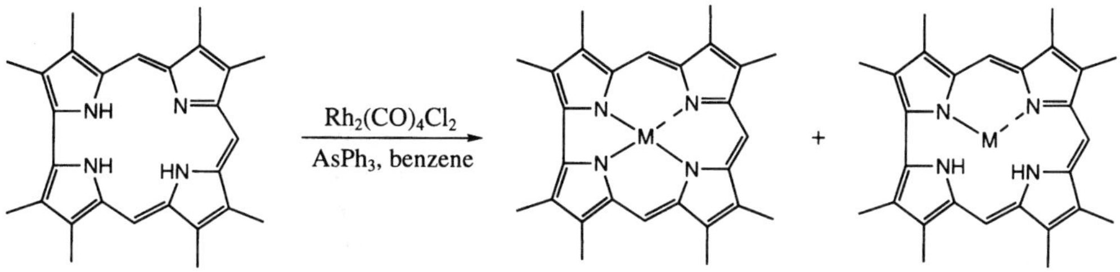

2.162   Rh$_2$(CO)$_4$Cl$_2$ AsPh$_3$, benzene   **2.132c**. M = RhAsPh$_3$ +   **2.174**. M = Rh(CO)$_2$

To date, there has only been one report of a main-group element, indium(III), which has been successfully inserted into a pre-formed corrole.[63] This insertion was achieved by treating octamethylcorrole **2.162** with InCl$_3$ in boiling DMF. What was obtained in 15% yield was a complex that proved identical to the In(III) corrole

2.139 discussed earlier (Scheme 2.1.53).[63] Although singular, this example serves to highlight the fact that this approach to corrole main-group chemistry could be generalized further. That this has not happened could simply reflect the fact that the alternative approach to metallocorroles involving direct metal insertion, concurrent with cyclization of a dideoxybiladiene-ac, is so easy.

Figure 2.1.10   Single Crystal X-Ray Diffraction Structure of Rhodium(III) 2,3,7,8,12,13,17,18-Octamethylcorrole Triphenylarsine **2.132c**.

This figure was generated using information down-loaded from the Cambridge Crystallographic Data Centre and corresponds to a structure originally reported in reference 52. The rhodium atom sits 0.26 Å above the plane defined by the four nitrogen atoms (0.31 Å above the mean macrocyclic plane) and lies in an approximately square pyramidal coordination environment. Atom labeling scheme: rhodium: ◉; carbon: ○; nitrogen: ●; arsenic: ◑. Hydrogen atoms have been omitted for clarity

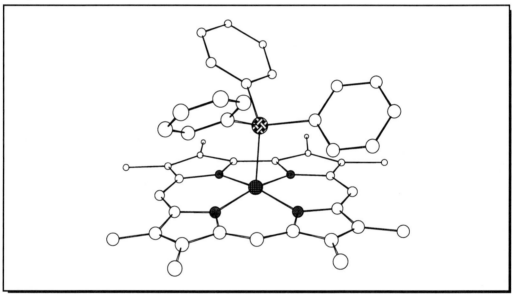

Recently, Vogel, *et al.* have described the synthesis of what may formally be considered as being an iron(IV) corrole.[66] It was obtained as the result of a sequence of reactions that involved first treating octaethylcorrole **2.81** with $Fe_2(CO)_9$, followed by air. The intermediate μ-oxo-iron(IV) dimer (**2.175**) that results was then, as the second step in the sequence, treated with HCl. This cleaves the dimer to form the iron(IV) corrole(Cl) monomer **2.176**. This product can be treated with PhMgBr to afford the iron(IV) corrole complex **2.177** containing an axial phenyl substituent (Scheme 2.1.54).

Also described by Vogel, *et al.* in 1994 was the preparation of an iron(III) corrole containing an axially coordinating pyridine ligand.[66] This was prepared *via* two different routes. First, free-base octaethylcorrole **2.81** was treated as above with $Fe_2(CO)_9$. However, here, instead of oxidizing to the μ-oxo dimer (i.e., **2.175**), the resulting intermediate product was exposed to excess pyridine to generate the iron(III) corrole species **2.178** (Scheme 2.1.55). Alternatively, **2.178** could be prepared by treating the μ-oxo iron(IV) dimeric species **2.175** with hydrazine in the presence of pyridine.

Scheme 2.1.53

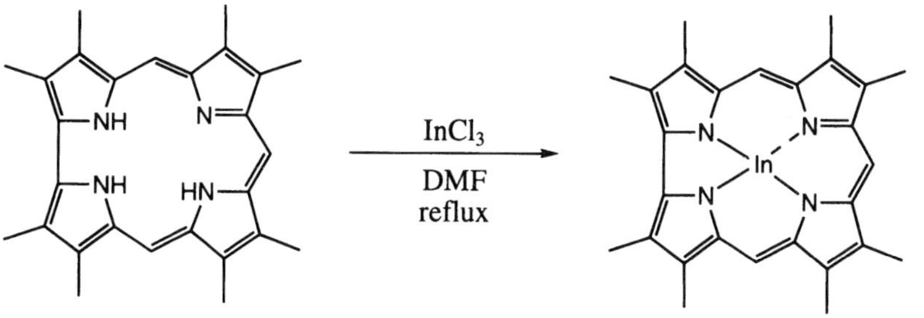

2.162   **2.139**

Single crystal X-ray diffraction studies of each of the above iron(III) and iron(IV) corrole complexes (**2.175–2.178**) showed that in each case the corrole macrocycles are nearly planar.[66] In the case of both the iron(III) corrole **2.178** (Figure 2.1.11) and the σ-phenyl iron(IV) corrole **2.177**, the iron atom was found to lie 0.27 Å above the mean plane of the macrocycle. The structures of the iron(IV) corrole species **2.175** (Figure 2.1.12) and **2.176**, on the other hand, revealed the iron atom as being *c.* 0.40 Å above the mean macrocycle plane.

Formally, pentavalent neutral metallocorroles have been prepared by Murakami and coworkers.[70–72] The first of these was the oxomolybdenum(V) corrole derivative **2.179**.[70] This complex was prepared by heating free-base corrole **2.82** with molybdenum pentachloride in oxygen-free decalin (Scheme 2.1.56). Alternatively, molybdenum hexacarbonyl ($Mo(CO)_6$) could be used as the metal source. In both cases, oxidation to the oxomolybdenum complex **2.179** was believed to occur during workup (involving chromatography on neutral alumina followed by recrystallization). In this way, complex **2.179** was isolated in *c.* 40% yield. Similar yields of the oxochromium(V) complex **2.180** could be achieved *via* the reaction of **2.82** with anhydrous chromium(II) chloride in DMF.[71] Here too, spontaneous oxidation during workup was used to afford the formally pentavalent oxo-complex **2.180**.

2.1.2.2   Recent Work from Vogel: A New Interpretation

As hinted at previously, Vogel and coworkers have recently offered an alternative interpretation of what occurs when corroles are reacted with divalent metals.[47] This work, which has been largely focused on nickel and copper, has served to call into question the structural and metal oxidation state assignments not only of the complexes derived from these cations, but also other metallocorrole complexes reported in the literature.

Scheme 2.1.54

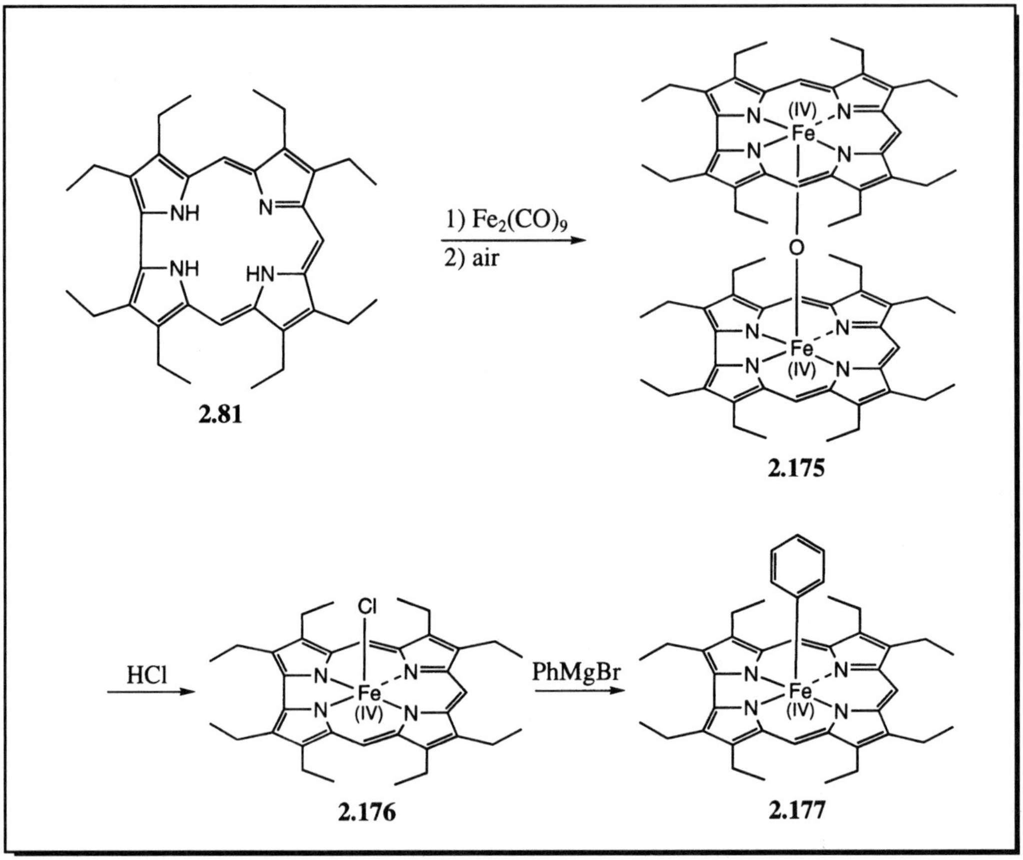

In terms of specifics, when Vogel and coworkers treated tetraethyl-tetra-methylcorrole **2.82** with nickel acetate tetrahydrate in DMF in the presence of air, a metal complex was isolated, which is spectroscopically identical to the complex previously studied by Murakami (Scheme 2.1.57).[51] However, based on various findings detailed below, the complex isolated by Vogel was formulated as being the π-radical nickel(II) complex represented by structure **2.111b**, and not that given previously (i.e., **2.111a** in Figure 2.1.13).[47]

The reaction of corrole **2.82** with copper(II) acetate was also found to produce a neutral complex that is spectroscopically identical to that previously described.[51] In this case, however, the complex that results was assigned structure **2.181**, wherein oxidation of the metal center from +2 to +3 appears to have occurred (Scheme 2.1.58).[47] This is in marked contrast to the two previous structural assignments (i.e.,

2.112a and **2.112b**; Figure 2.1.13) in which a neutral *divalent* metal complex was assumed to have been formed.

Scheme 2.1.55

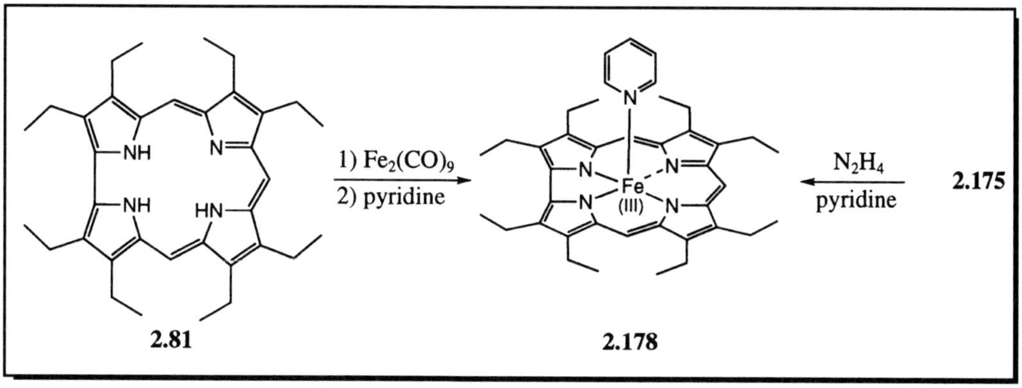

Figure 2.1.11   Single Crystal X-Ray Diffraction Structure of the Mono-pyridine Adduct of Iron(III) 2,3,7,8,12,13,17,18-Octaethylcorrole **2.178**.

This figure was generated using information down-loaded from the Cambridge Crystallographic Data Centre and corresponds to a structure originally reported in reference 66. The iron atom sits 0.27 Å above the mean macrocyclic plane (the maximum deviation of C and N atoms from the mean ring plane is 0.22 Å). Atom labeling scheme: iron: ◉; carbon: ○; nitrogen: ●. Hydrogen atoms have been omitted for clarity

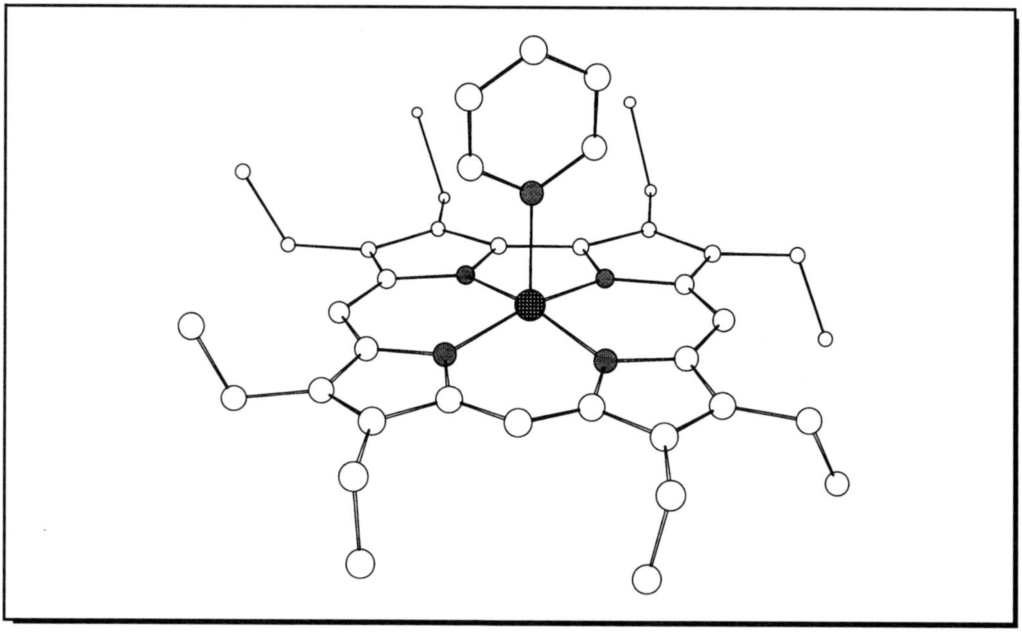

Figure 2.1.12   Single Crystal X-Ray Diffraction Structure of μ-oxo-iron(IV)
2,3,7,8,12,13,17,18-Octaethylcorrole **2.177**.
This figure was generated using information down-loaded from the Cambridge Crystallographic
Data Centre and corresponds to a structure originally reported in reference 66. The iron atoms are
situated *c.* 0.40 Å above their respective mean macrocyclic planes and reside in what are
approximately square pyramidal coordination environments. The Fe-O-Fe angle was found to be
170°. Atom labeling scheme: iron: ◉; carbon: ○; nitrogen: ●; oxygen: ◉. Hydrogen atoms have
been omitted for clarity

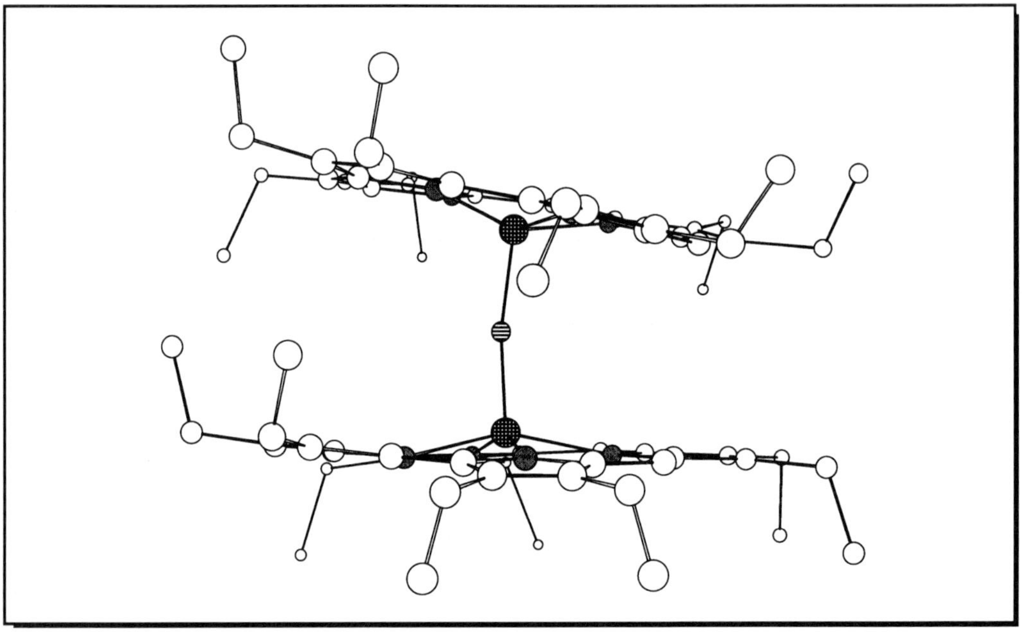

Scheme 2.1.56

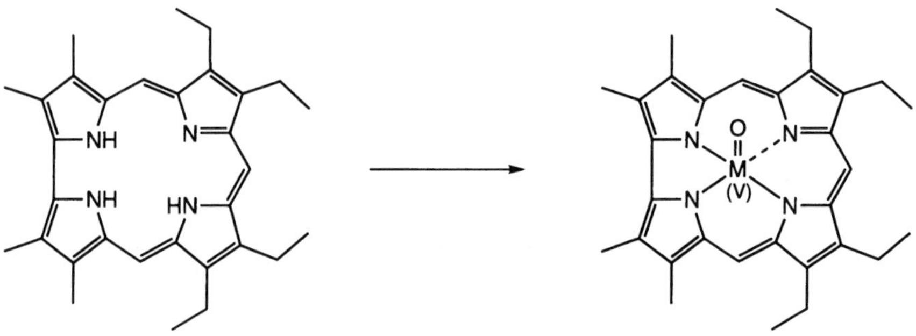

2.82   **2.179**. M = Mo
                                 2.180. M = Cr

Scheme 2.1.57

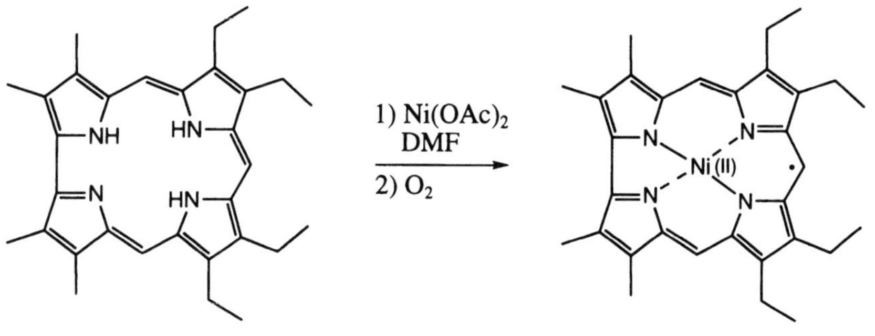

2.82   2.111b

Figure 2.1.13   Structures **2.111a**, **2.112a**, and **2.112b**

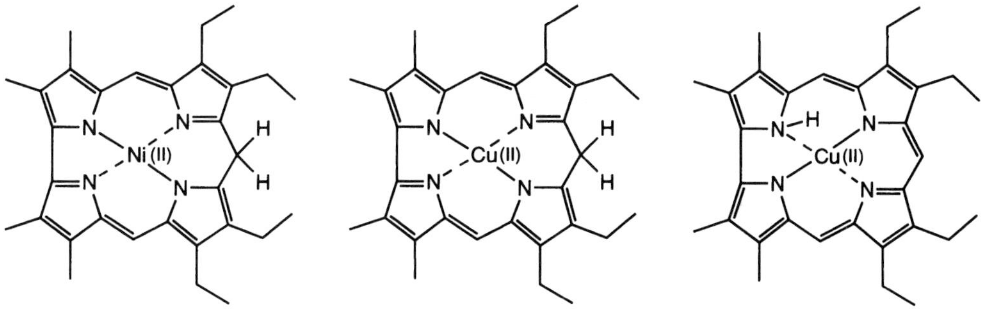

2.111a   2.112a   2.112b

Scheme 2.1.58

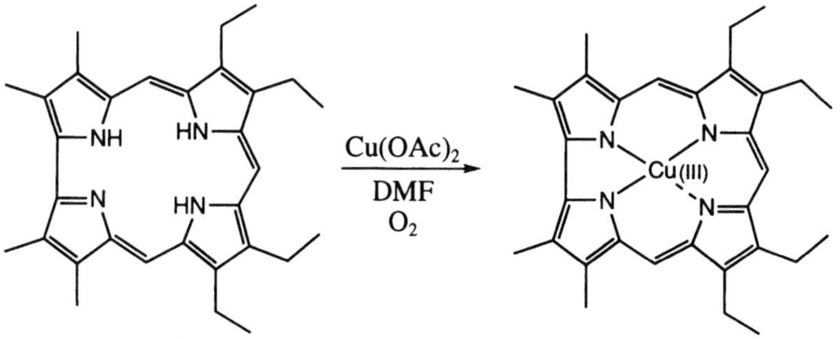

2.82   2.181

The conclusion that the above nickel and copper complexes (**2.111** and **2.181**) are formally trivalent (overall) was made after determining the effective magnetic moment of these complexes.[47] In the case of the copper complex **2.181**, these experiments revealed the presence of an effectively diamagnetic metal center. This fact led to the assignment of **2.181** as having a Cu(III) center. On the other hand, similar magnetic susceptibility experiments carried out on nickel complex **2.111** revealed that it was effectively paramagnetic. This is a situation that could be explained either by a complex that contains a trivalent nickel center, or one that has an unpaired electron in a π-orbital of the corrole ligand. This latter alternative was chosen on the basis of a UV-vis spectroscopic analysis. Specifically, the spectrum of **2.111** revealed the presence of a long-wavelength (i.e., at *ca.* 650 nm) Q-band transition. Since similar bathochromically shifted Q-bands have been reported for certain ligand-oxidized metallo–porphyrins, the conclusion that the corrole ring had also been oxidized appeared reasonable.[73–76]

It is also worth mentioning that, in the case of the copper complex **2.181**, variable temperature ¹H NMR experiments revealed the presence of a highly temperature-dependent electronic system.[47] That is to say, the complex appears to be diamagnetic at lower temperatures and paramagnetic at higher temperatures. This was interpreted as being indicative of an equilibrium that exists between a diamagnetic form such as **2.181a** (containing a copper(III) center), and a higher energy triplet form such as **2.181b** (Figure 2.1.14).

Figure 2.1.14   Structures **2.181a** and **2.181b**

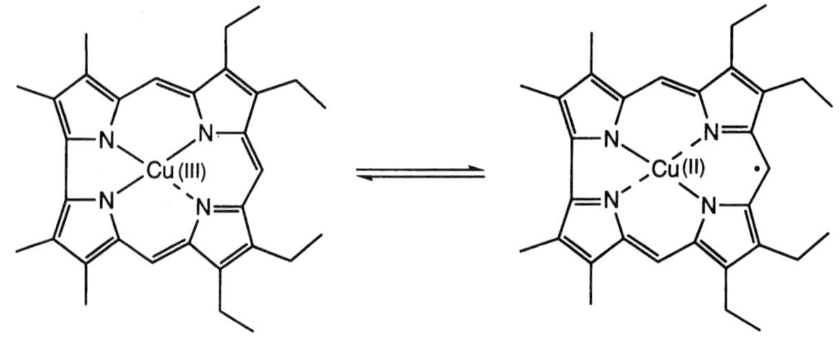

| 2.181a | 2.181b |

Interestingly, Vogel and coworkers have found that other metallocorroles containing a π-delocalized radical core can also be easily formed.[47] This has been explicitly demonstrated in the case of the phenyl-substituted cobalt(III) corrole **2.182** and the phenyl-substituted cationic iron(IV) corrole **2.183**. The cobalt(III) π-radical complex **2.182** was prepared from the "starting" cobalt(III) corrole **2.131** by reacting it with phenylmagnesiumbromide followed by air (Scheme 2.1.59). The corresponding iron complex **2.183** was isolated by treating the σ-phenyl iron(IV) corrole **2.177** with saturated aqueous FeCl₃, followed by washing with 1 M perchloric acid (Scheme 2.1.60)

Scheme 2.1.59

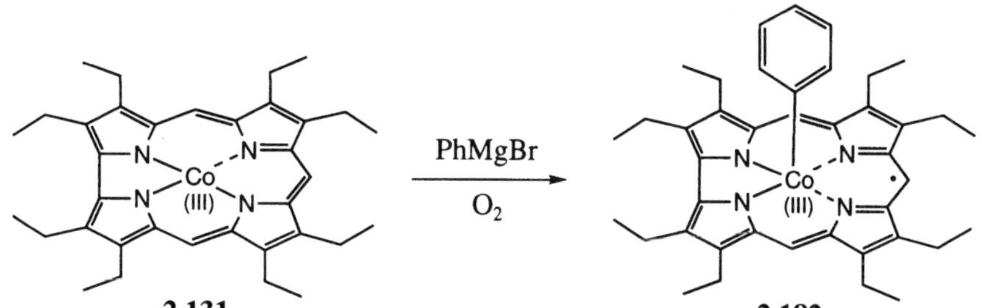

2.131   PhMgBr / O₂   **2.182**

Scheme 2.1.60

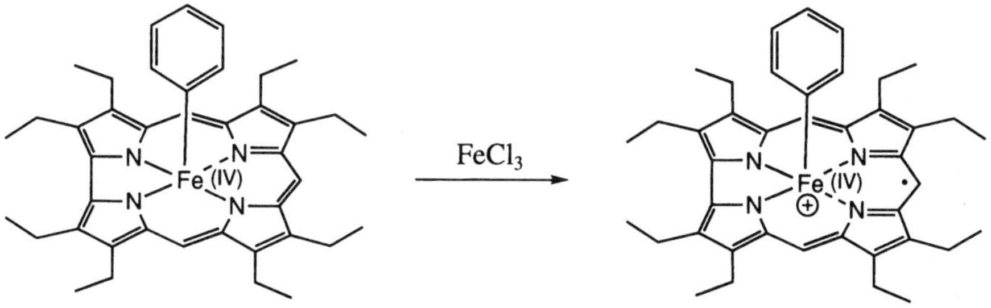

2.177   FeCl₃   **2.183**

The conclusion that the cobalt and iron complexes **2.182** and **2.183** are formally π-radical species is supported by a wealth of spectroscopic evidence. For instance, the ¹H NMR spectrum of the cobalt complex **2.182** indicated the presence of a paramagnetic system with resonances that are consistent with the proposed cobalt(III) formulation (as opposed to a low-spin, paramagnetic cobalt(IV) corrole). Further, the UV-vis absorption spectrum recorded for complex **2.182** was found to be remarkably similar to those of porphyrin π-radicals.[73–76] In the case of the iron complex **2.183**, Mössbauer spectroscopy was used to confirm the assignment of the complex as having a formally tetravalent metal and a π-radical carbon skeleton. Here, measurements at 120 K revealed that the formal removal of one electron from the neutral species **2.177** had very little effect on the Mössbauer spectrum. This was interpreted as an indication that oxidation had occurred at the corrole ligand, and not at the metal center. Had metal oxidation occurred, more dramatic differences in the Mössbauer spectrum would have been observed.

2.1.3   Reduction and Oxidation of Metallocorroles

As discussed in Section 2.1.2.1 above, there have been several attempts to prepare cobalt(II) corroles. The first of these involved the reacting a biladiene-ac or metal-free corrole with a cobalt(II) salt. Unfortunately, as is now well established, this approach actually affords the corresponding Co(III) corrole. On the other hand, Co(II) corroles may be prepared *via* the reduction of the easily obtained Co(III) precursors. The first success enjoyed along these lines was by Hush and Woolsey. These workers reduced 8,12-diethyl-2,3,7,13,17,18-hexamethyl-cobalt(III) corrole **2.184** *in vacuo* in tetrahydrofuran (THF) using a sodium film. This afforded a Co(II) corrole anion, of indeterminate structure.[26] These same workers also reported the use of tetra-*n*-butyl ammonium hydroxide and NaBH$_4$ in DMF as reductants, although these latter conditions give reaction rates that are reportedly much less favorable than those obtained using sodium film reduction. Also, with these reductants, the reactions do not proceed to completion.

The cobalt(II) corrole anion prepared as above was characterized primarily by electron spin resonance (esr) and absorption spectroscopy.[26] When prepared *via* sodium film reduction, the cobalt(II) corrole oxidizes rapidly to the corresponding Co(III) corrole on exposure to air. When prepared by the other methods, it is moderately stable in air in the presence of a reducing agent. Attempts to prepare the neutral form of the initial Co(II) corrole anion, by protonation with perchloric acid, resulted in formal oxidation to the Co(III) derivative. Interestingly, further protonation of the Co(III) corrole with perchloric acid led to what appeared to be a protonated Co(III) corrole. Certainly, the absorption spectrum of this species is similar to that of the corresponding neutral nickel(II) corrole complex. However, the exact nature of this protonated material has not been fully elucidated.

In 1974, Hush and Woolsey reported that treating the cobalt(II) corrole species discussed above with pyridine induced little in the way of spectral changes. These authors took this as an indication that the divalent cobalt corrole complex in question shows little or no tendency towards axial coordination.[26] In 1978, however, Murakami, *et al.* reported a different result, stating that one pyridine molecule does in fact coordinate axially to the cobalt center of an anionic Co(II) corrole.[59]

In work along somewhat related lines, Murakami, *et al.* investigated the hydroxide-mediated reduction of 8,12-diethyl-2,3,7,13,17,18-hexamethyl-cobalt(III) corrole **2.184**.[69] This involved treating **2.184** with *n*-butyl ammonium hydroxide in the presence of ethyl vinyl ether in an aprotic solvent such as dichloromethane, benzene, or DMF. Using a similar approach, these researchers also found that iron(III) corrole **2.185** could be reduced to give the corresponding iron(II) corrole (Scheme 2.1.61). In this work, it was determined that olefins, on oxygenation with hydroxide ion, can act as the requisite electron donors for reduction.

The one-electron reduction of cobalt(III) corroles to cobalt(II) corroles can also be effected electrochemically.[57,69,77] As expected, increasing the donating ability of the axially coordinating ligand decreases the ease of reduction. It also affects the extent to which the reaction can be made to be reversible under normal laboratory conditions. For instance, reductions involving pyridine cobalt(III) corroles are generally reversible, whereas those of the corresponding PPh_3 adducts are not. On the other hand, interestingly enough, electrochemical *oxidations* of cobalt(III) corroles are often quasi-reversible, with three one-electron oxidation waves commonly being observed.[77,78]

Scheme 2.1.61

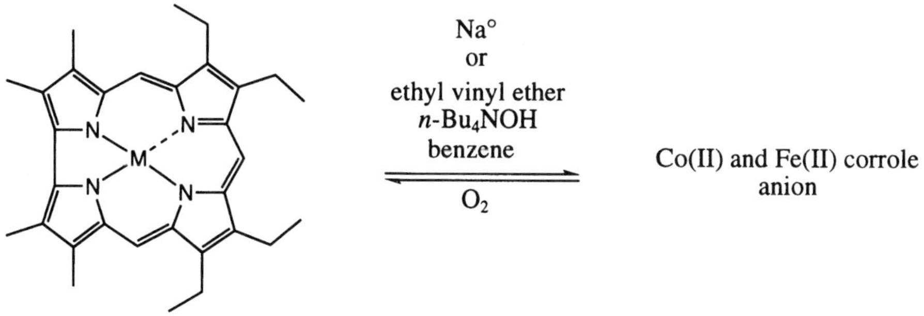

2.184. M = Co(III)
2.185. M = Fe(III)

In addition to the above, direct chemical oxidations of cobalt(III) corroles have also been attempted. For instance, in early work by Johnson and coworkers, triphenylphosphinecobalt(III)-2,8,12,18-tetraethyl-3,7,13,17-tetramethyl-corrole **2.125b** was treated with an excess of DDQ at room temperature.[57] Surprisingly, the resulting aromatic product was determined to be the 3,17-diformyl cobalt(III) corrole derivative **2.186** (Scheme 2.1.62). The regioselectivity of the oxidation of the methyl groups was determined by exposing the triphenylphosphine complex of cobalt(III)-3,17-dimethyl-2,7,8,12,13,18-hexaethylcorrole **2.124a** to DDQ. The resulting oxidized product (**2.187**) was found to contain formyl groups at the same positions as did the product in the initial tetraethyl-tetramethyl case. This was taken as evidence that the oxidation had indeed taken place at the 3- and 17-positions of the corrole ring.[||]

Vogel and coworkers have recently described the reduction of some metallo-corroles using a variety of chemical agents to effect reduction.[47] In the first case, reduction of nickel complex **2.111** to the anionic complex **2.188** was effected by treating the starting complex with hydrazine hydrate. Analogous reductions of the copper(III) corrole **2.181** and cobalt(III) corrole **2.131** to the corresponding anionic complexes **2.189** and **2.190** could also be carried out by treating the starting oxidized species with hydrazine hydrate and sodium borohydride, respectively (Scheme

---

[||]The structural representation of the diformylcorrole derivatives **2.186** and **2.187** are apparently incorrect in A. W. Johnson's and coworkers' 1973 paper. Indeed, these representations depict a tetraanionic ligand with a presumed tetravalent cobalt rather than the neutral cobalt(III) corrole species presented here and, it would appear, supported by the actual experimental details given in the very paper in question.[57]

2.1.63). Interestingly, in the cases of the Ni(II) and Cu(II) reduction products (**2.188** and **2.189**), the UV-vis spectra recorded proved nearly identical to those of previously described systems, prepared *via* treatment with hydroxide anion in DMF or with other agents such as potassium metal or NaBH₄.[43,45,51,53,55,56] The original researchers in this area ascribed the spectral changes that occurred upon treatment with such agents to deprotonation of originally neutral metal(II) corrole ligands. Vogel and coworkers suggest a different explanation. They propose that these reactions are better described as being true reduction processes involving what they consider to be the "starting" π-radical **2.111b** and copper(III) complex **2.181** (see Section 2.1.2.2).

Scheme 2.1.62

2.125b. R = Me. L = PPh₃
2.124a. R = Et. L = PPh₃

2.186. R = Me. L = PPh₃
2.187. R = Et. L = PPh₃

Oxidations of metallocorroles have also been investigated by Vogel and coworkers.[47,79] For instance, these researchers found that treatment of σ-phenyl-cobalt(III) corrole **2.182** with Fe(ClO₄)₃ resulted in oxidation to cation **2.191** (Scheme 2.1.64). They also found that iron(III) salts could be used to effect oxidation of nitrosyl iron(III) corrole **2.192**. In this case, however, it was a π-cation radical species (i.e., **2.193**), which was obtained upon by treatment with, for instance, iron(III) chloride (Scheme 2.1.65).[47,79] Similar oxidations of tetravalent complexes have also been carried out by Vogel and coworkers (see Scheme 2.1.60).[47,79]

2.1.4   *N-Substituted Corroles*

2.1.4.1   Preparation of N-Alkyl and N-Acetyl Corroles

Considerable effort has been directed toward the preparation of N-substituted corroles. This research was inspired, presumably, by the fact that similar N-substituted species can be obtained in the porphyrin series.[9] In any event, objectives associated with this line of investigation include assessing the extent to which a change in internal steric crowding can affect the electronic, basicity, and metal coordination properties of a given type of tetraaza macrocycle.

Scheme 2.1.63

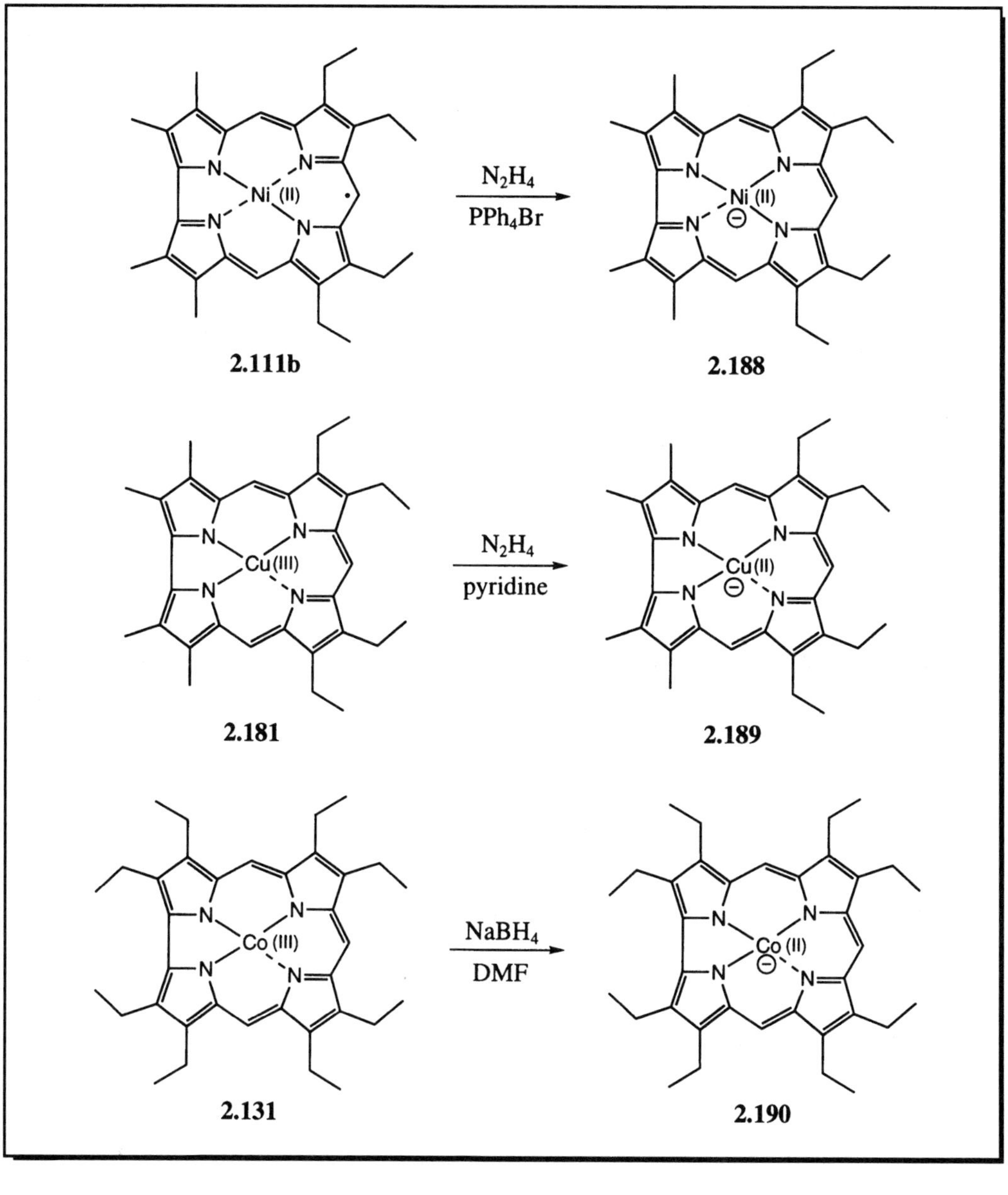

Ni (II)
N₂H₄
PPh₄Br
2.111b
2.188
Cu(III)
Cu(II)
N₂H₄
pyridine
2.181
2.189
Co (III)
Co (II)
NaBH₄
DMF
2.131
2.190

Scheme 2.1.64

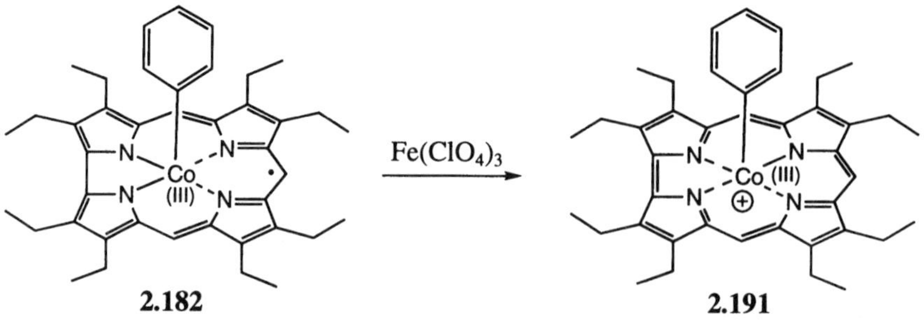

2.182   Fe(ClO$_4$)$_3$ **2.191**

Scheme 2.1.65

2.192   FeCl$_3$ **2.193**

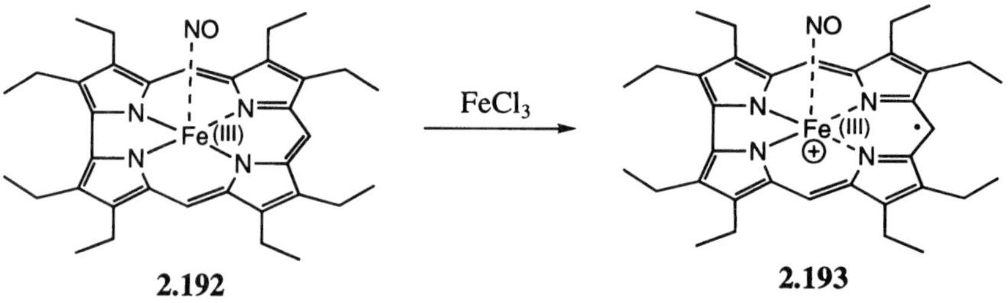

The first preparation of an N-substituted corrole was reported by Johnson and Kay in 1965.[36,43] These workers reported that reaction of 8,12-diethyl-2,3,7,13,17,18-hexamethylcorrole **2.6** with potassium carbonate and methyliodide in refluxing acetone gives rise to a roughly 3 : 1 mixture of two compounds. These were subsequently determined to be the isomeric N(21)- and N(22)-methyl corrole species **2.194** and **2.195**, respectively (Scheme 2.1.66). Here, the N(21)-methyl corrole was assigned as being the dominant product. However, it was also found that this product, and its N(22)-methyl isomer are stable under the methylation conditions. This supports the contention that, under these reaction conditions, the key methylation process is under kinetic control.

In terms of spectroscopic characteristics, it was determined that the UV-vis absorption spectrum of the N(21)-methyl corrole **2.194** is very similar to that of the normal, N-unsubstituted corrole.[43] On the other hand, the UV-vis spectrum of the N(22)-methyl corrole **2.195** was found to be markedly different from that of either the N(21)-methyl-substituted species **2.194** or the N-unsubstituted corrole **2.6**. Furthermore, in the relevant ^{1}H NMR spectra, the signals of the internal and external protons of the N(22)-methyl corrole **2.195** were found to be spread out compared to those of the N(21)-methyl or N-unsubstituted species. This latter finding was deemed consistent with there being a greater ring current in the N(22)-substituted material than in these two comparison compounds.[36,43] However, it could also reflect a variety of other factors, including different degrees of out-of-plane macrocycle distortion, that were apparently not considered by the original authors.

Scheme 2.1.66

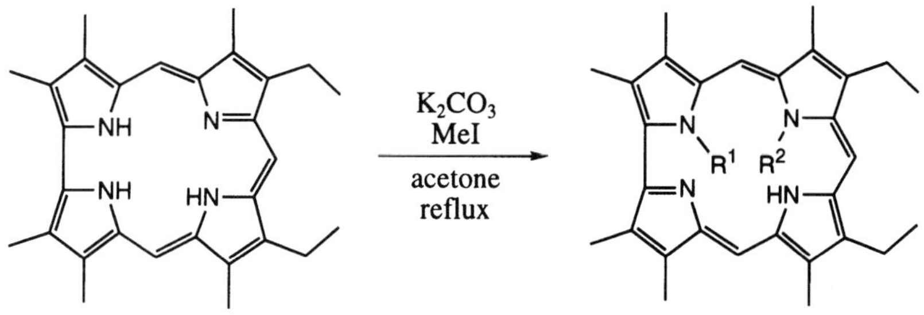

K$_2$CO$_3$
MeI

acetone
reflux

2.6

2.194. R^1 = Me, R^2 = H
+
2.195. R^1 = H, R^2 = Me

The preparation of the N(21)- and N(22)-substituted methyl corroles **2.196** and **2.197** has also been described,[43] as have the N(21)- and N(22)-ethyl- and allyl-substituted species **2.198**–**2.203** (Scheme 2.1.67).[45,80] Additionally, methylation reactions involving 3,17-carbethoxy corrole **2.102** and methyl iodide have been carried out using diisopropylethyl amine as the base.[37,38] In this instance, in contrast to what happens in the all-β-alkyl series, the N(22)-methyl corrole **2.205** forms as the major product (by a nearly 2:1 margin over the N(21)-substituted alternative, **2.204**) (Scheme 2.1.68). Interestingly, the same paper that describes this latter chemistry also reports an alternative synthesis of N(21)-methyl-3,17-carbethoxy corrole **2.204**.[37,38] In this approach, the thiaphlorin **2.101** was methylated using methyl iodide in diisopropylethyl amine to afford the N-methylated product **2.206** (Scheme 2.1.69). This latter species was then subject to a heat-induced sulfur extrusion. This gave the N(21)-methyl corrole **2.204** in 85% yield.** A clear advantage of this stepwise procedure was that it allowed the N(21)-methyl-3,17-dicarbethoxy corrole **2.204** to be prepared selectively.

In other work involving substitution at a single nitrogen, Johnson and co-workers[30] reported that heating 8,12-diethyl-2,3,7,13,17,18-hexamethyl corrole **2.6** with acetic anhydride gives the N(21)-acyl-substituted corrole **2.207** in 11% yield, free of any N(22)-substituted contaminant (Scheme 2.1.70). Applying this same procedure to 3,8,12,17-tetraethyl-2,7,13,18-tetramethyl corrole **2.79**, it was reported, gave

---

It proved much more difficult to extrude sulfur from the N-methylated derivative **2.206. A higher temperature was required than was in the case of the N-unsubstituted counterpart **2.101** (described earlier in this chapter).

the corresponding N(21)-acylated species **2.208**, in 14.5% yield. The available NMR data for **2.207** and **2.208** were considered consistent with the assigned structures (i.e., acylation on N(21)), although the electronic spectra of the N(21)-acylated species **2.207** and **2.208** more closely resemble those of N(22)-*alkyl*-substituted species.

Scheme 2.1.67

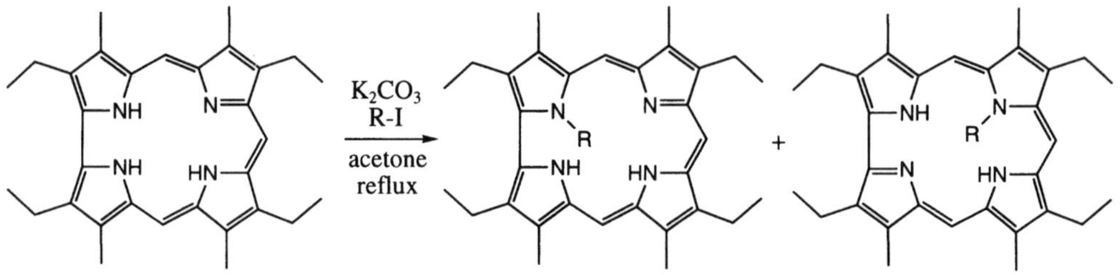

2.80	
	2.196. R = Me
	2.198. R = Et
	2.200. R = CH₂CHCH₂
	2.202. R = CH₂CHCMe₂

2.196. R = Me   **2.197**. R = Me
2.198. R = Et   **2.199**. R = Et
2.200. R = CH₂CHCH₂   **2.201**. R = CH₂CHCH₂
2.202. R = CH₂CHCMe₂   **2.203**. R = CH₂CHCMe₂

K₂CO₃, R-I, acetone reflux

Scheme 2.1.68

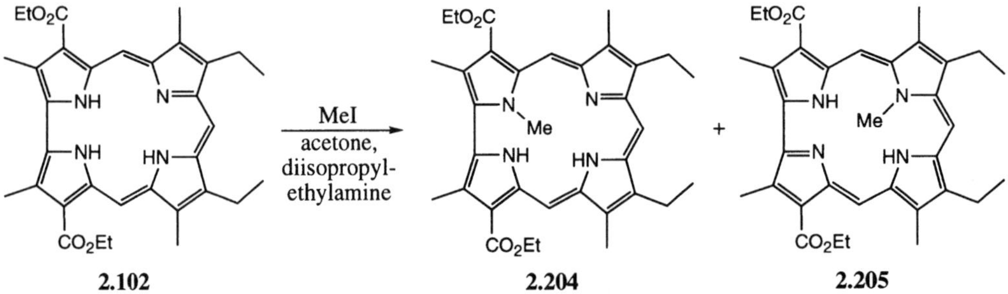

2.102   MeI, acetone, diisopropyl-ethylamine   **2.204**   +   **2.205**

2.1.4.2   *N,N'-Dimethylcorrole*

Under more forcing conditions than those initially employed to prepare mono *N*-methyl corroles (see above), Johnson and coworkers were able to prepare *N,N'*-dimethyl corroles as shown in Scheme 2.1.71.[80] In the original report on this work, appearing as a communication in 1970, no attempt was made to determine the exact regioselectivity of the second alkylation. In fact, treatment of either corrole **2.80**, or *N*(21)- or *N*(22)-methyl corroles **2.196** or **2.197** with methyl iodide in a sealed tube was reported to afford the same species in all cases. This product was proposed to be either **2.209** *or* **2.210** (as the hydroiodic salt). It was later suggested that, in each of these reactions, it was actually the *N*(21), *N*(22)-dimethyl corrole iodide **2.209** that is most likely formed under the reaction conditions.[30] It was also found that when heated to 180 °C in *o*-dichlorobenzene, this putative *N*(21),*N*(22)-dimethyl corrole iodide (i.e., **2.209**) undergoes thermal cleavage so as to afford the *N*(21)-methyl corrole **2.196**.[30]

Scheme 2.1.69

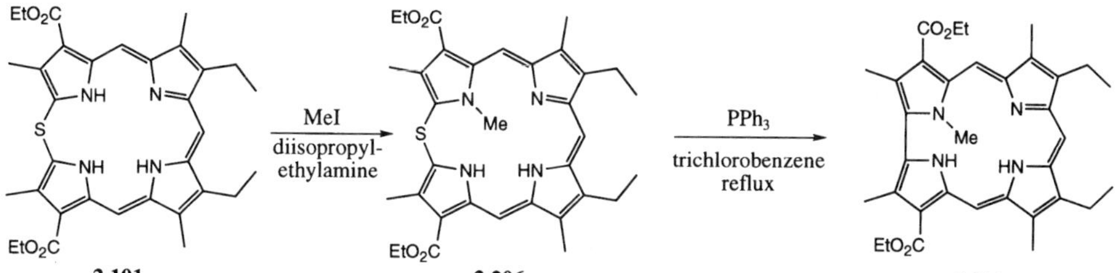

| **2.101** | **2.206** | **2.204** |

Scheme 2.1.70

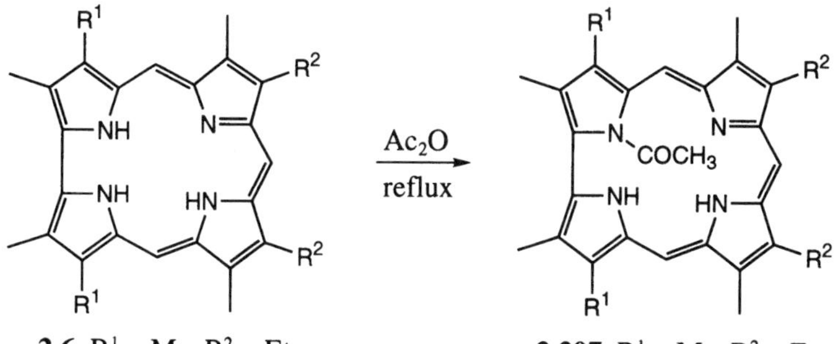

2.6. R¹ = Me, R² = Et   **2.207.** R¹ = Me, R² = Et
2.79. R¹ = Et, R² = Me   **2.208.** R¹ = Et, R² = Me

Interestingly, the second methyl group introduced into **2.196** or **2.197** was found to be incorporated at a greater rate than the first. This led to the suggestion that both the nucleophilicity and basicity of the mono-*N*-methyl corroles were greater that of the parent, N-unsubstituted corrole system. In fact, *N*-alkyl corroles do appear to be quite basic in that, unlike simple corroles, they do not form anions when treated with base. Moreover, the *N*(21),*N*(22)-dimethyl corrole iodide **2.209** is reportedly so basic that it cannot be isolated in its free-base form.[30]

Scheme 2.1.71

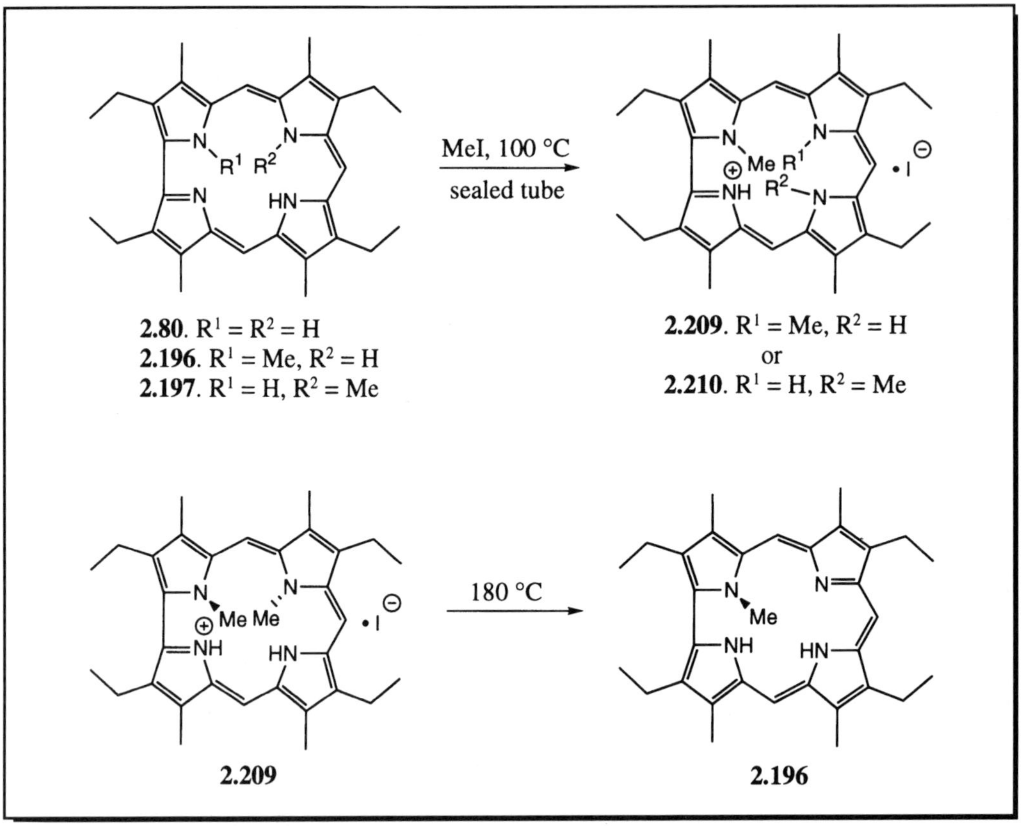

2.80. R^1 = R^2 = H
2.196. R^1 = Me, R^2 = H
2.197. R^1 = H, R^2 = Me

2.209. R^1 = Me, R^2 = H
or
2.210. R^1 = H, R^2 = Me

MeI, 100 °C
sealed tube

180 °C

2.209

2.196

2.1.4.3   N-Substitution of Metallocorroles

In 1968, Johnson and coworkers reported the reaction of the nickel(II) corrole **2.105** with methyl iodide in acetone using sodium hydroxide as the base.[53] The product obtained was originally considered to be species **2.211** in which methylation had occurred at the metal center (Scheme 2.1.72). However, this proposed structure was later discarded in favor of the correct formulation **2.212**, an N(21)-methylated nickel(II) corrole derivative.[81] This same type of alkylation reaction could be used to prepare the N(21)-alkyl nickel(II) corroles **2.113** and **2.114**, as well as the N(21)-alkyl copper(II) corroles **2.215** and **2.216** starting from **2.107** and **2.159**, respectively (Scheme 2.1.73).[45,81]

A single crystal X-ray diffraction analysis of this abovementioned N(21)-methyl copper(II) corrole **2.215** confirmed the site of methylation to be at N(21).[81] This same structure also revealed a rather distorted molecule with the N-methylated pyrrole ring being twisted out of the mean plane of the remaining pyrrolic portions of the macrocycle. Thus, this X-ray structure was seminal in its time. It provided both a key proof of structure (and a key correction of an earlier structural misassignment) and gave important insight into how such substituted corrole systems might appear in three dimensions.

Scheme 2.1.72

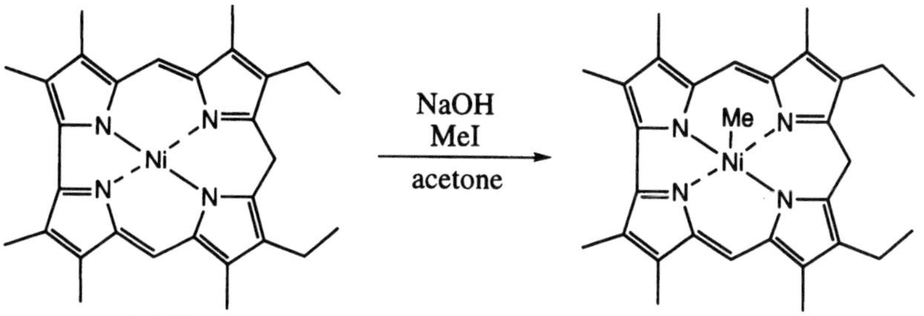

2.105   **2.211**

Scheme 2.1.73

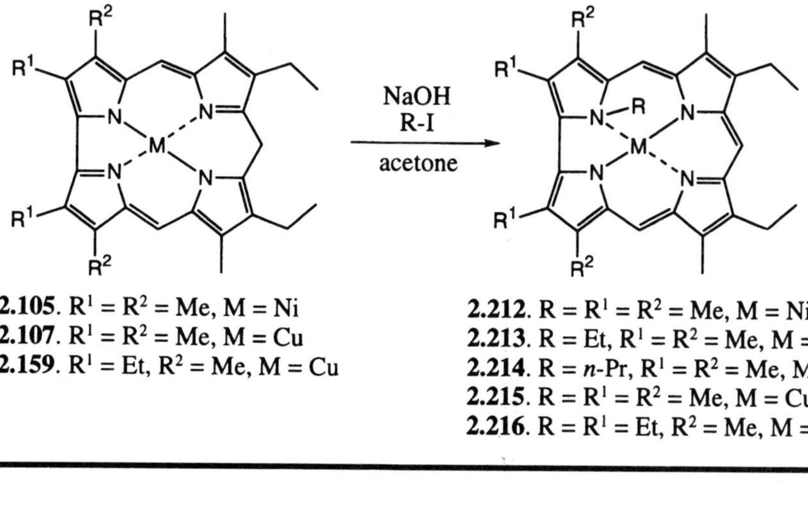

2.105. $R^1 = R^2 = $ Me, M = Ni
2.107. $R^1 = R^2 = $ Me, M = Cu
2.159. $R^1 = $ Et, $R^2 = $ Me, M = Cu

2.212. R = $R^1 = R^2 = $ Me, M = Ni
2.213. R = Et, $R^1 = R^2 = $ Me, M = Ni
2.214. R = *n*-Pr, $R^1 = R^2 = $ Me, M = Ni
2.215. R = $R^1 = R^2 = $ Me, M = Cu
2.216. R = $R^1 = $ Et, $R^2 = $ Me, M = Cu

Alkylation reactions involving other electrophiles are known. For instance, treatment of the anion of nickel(II) corrole **2.105** with allyl bromide was found to afford a mixture of two compounds that proved to be C-allyl metallocorroles.[54] The major product was found to be the 3-allyl-3-methyl-nickel(II) corrole **2.218**, and the minor component the *meso*-diallyl derivative **2.219** (Scheme 2.1.74). In the first instance, the regioselectivity of the addition (i.e., at the 3-position) was specifically confirmed by treating the 3,17-diethylcorrole **2.217** with allyl bromide; this, as expected, afforded the 3-allyl-3-ethyl derivative **2.220**.

Scheme 2.1.74

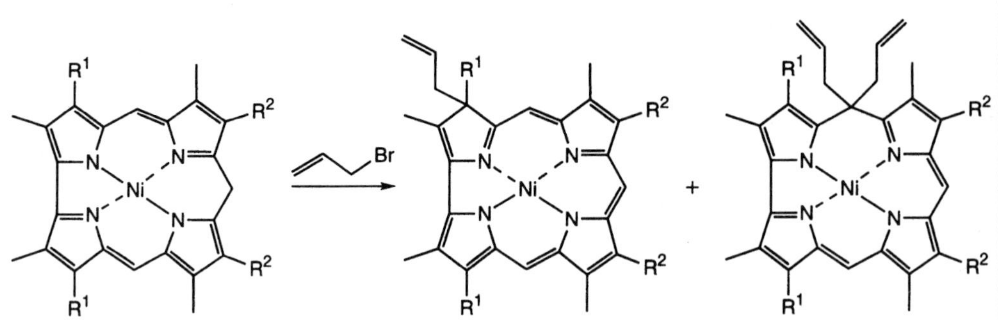

2.105. R^1 = Me, R^2 = Et   **2.218.** R^1 = Me, R^2 = Et   **2.219.** R^1 = Me, R^2 = Et
2.217. R^1 = Et, R^2 = Me   **2.220.** R^1 = Et, R^2 = Me

In the case of the pyridinium salt of the palladium(II) corrole derivative **2.161**, methylation gave a mixture of two products, namely $N(21)$-methyl corrole **2.221** and 3,3-dimethyl corrole **2.222** (Scheme 2.1.75).[45] As above, the regioselectivity of the C-methylation was elucidated by examining the methylation products of the 3,17-diethylcorrole derivative **2.160**. Further, in this specific instance, it was determined that under the methylating conditions, compound **2.223** could not be obtained from **2.224**, nor could this latter be made from **2.223**. This was taken as an indication that these methylation reactions were under kinetic control. Interestingly, however, both of the methylated products were found to undergo reversible protonation in the presence of TFA, in both cases, the site of protonation was determined (by NMR spectroscopic analysis) to be at the C(5) *meso*-position.

Scheme 2.1.75

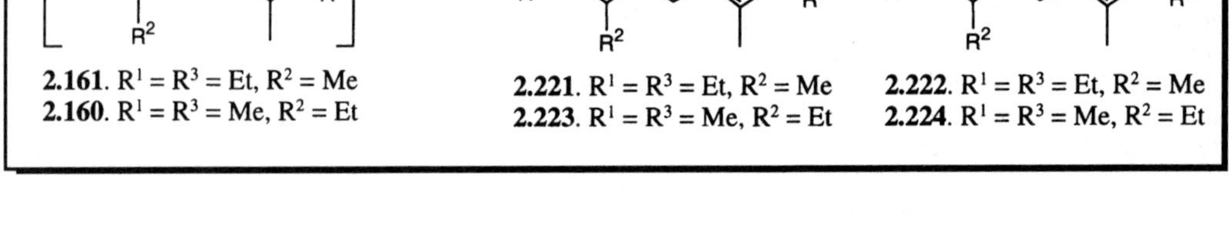

2.161. R^1 = R^3 = Et, R^2 = Me   **2.221.** R^1 = R^3 = Et, R^2 = Me   **2.222.** R^1 = R^3 = Et, R^2 = Me
2.160. R^1 = R^3 = Me, R^2 = Et   **2.223.** R^1 = R^3 = Me, R^2 = Et   **2.224.** R^1 = R^3 = Me, R^2 = Et

Cobalt(III) corroles have also been subject to N-substitution. In 1973, for instance, Johnson and coworkers reported the synthesis of the N(21)-aryl-substituted cobalt(II) corroles **2.225–2.227**.[57] These were obtained by treating the corresponding cobalt(III) corrole(pyridine) derivatives **2.167** and **2.125a** with aryl lithium or phenyl magnesium bromide (Scheme 2.1.76).[††] These products exhibited no significant spectral changes upon the addition of pyridine. This is consistent with these complexes having but little or no affinity for axial ligands. Interestingly, however, when either of the *N*-aryl corroles **2.225** or **2.226** was heated in the presence of triphenylphosphine, only the corresponding dealkylated product **2.167b** was obtained. Under these conditions, no thermal rearrangements were observed.

Scheme 2.1.76

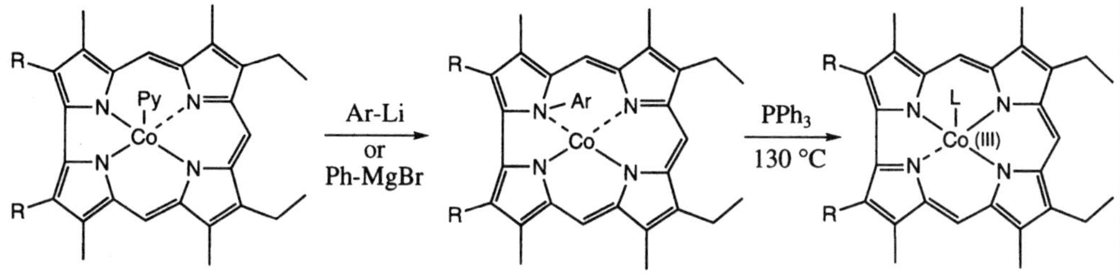

2.167. R = Me	**2.225**. R = Me, Ar = Ph	**2.167b**. R = Me. L = PPh₃
2.125a. R = Et	**2.226**. R = Me, Ar = *p*-tolyl	
	2.227. R = Et, Ar = Ph	

An alternative approach to preparing N-substituted cobalt(II) corroles was also tested in 1973 by Johnson and coworkers.[57] Here, cobalt(III) 2,8,12,18-tetra-ethyl-3,7,13,17-tetramethylcorrole **2.125a** was subject to reduction with NaBH₄ to give what was presumed to be a cobalt(II) corrole. This latter species was then treated with methyl iodide followed by PPh₃. Unfortunately, this sequence of reactions afforded only the non-alkylated cobalt(III) corrole derivative, **2.125b** (Scheme 2.1.77).

[††]This type of reaction could not be effected using alkyl-lithium or alkyl-Grignard reagents. Also, the products represented by **2.225** and **2.227** with presumed divalent cobalt centers as offered by Johnson and coworkers should be considered as being tentative only; no conclusive data were actually given in support of the proposed formulations.[57] There is also a discrepancy between the main textual body and the experimental section of the 1973 paper by Johnson and coworkers as to which Co(III) derivative was actually used in these reactions.[57] In the main body, it is stated that the square planar (implying pyridine-free) Co(III) corrole is used, whereas in the experimental section, it is stated that the pyridine derivatives were used.

Scheme 2.1.77

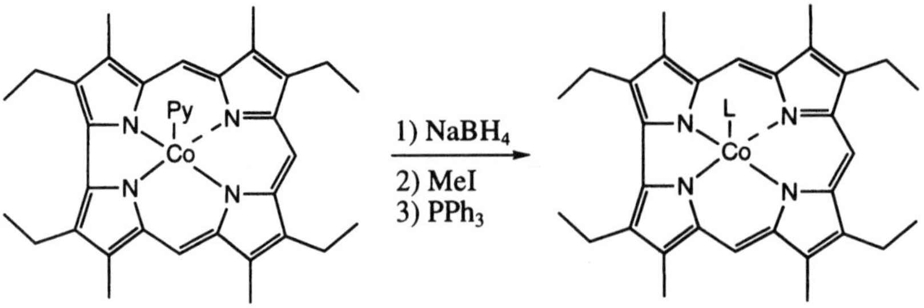

2.125a   1) NaBH$_4$ **2.125b**. L = PPh$_3$
      2) MeI
      3) PPh$_3$

2.1.4.4   **Metalation of *N*-Alkyl Corroles**

Attempts to insert metal cations into "pre-formed" N-substituted corroles have also been made. Some of these efforts have met with success. In 1976, for instance, Grigg and coworkers demonstrated the ability of *N*(21)- and *N*(22)-methyl corroles **2.194** and **2.195** to form stable complexes with monovalent rhodium.[40,82] Specifically, *N*(21)-methyl corrole **2.194** was treated with tetracarbonyldi-μ-chlororhodium(I). The product of this reaction was found to be a mono-rhodium complex (**2.228**) (Scheme 2.1.78), in which the rhodium atom sits well above the plane of the macrocycle, as determined by single crystal X-ray diffraction analysis (Figure 2.1.15).[40,82] In this complex, the C and D rings of the corrole were found to be involved in coordination to the rhodium center, with all four pyrrole rings of the corrole being essentially coplanar. Still, the two pyrrole rings involved in coordination to rhodium were observed to be tilted slightly toward the metal center, while the other two pyrroles were found to be tilted below the mean plane of the macrocycle.

Scheme 2.1.78

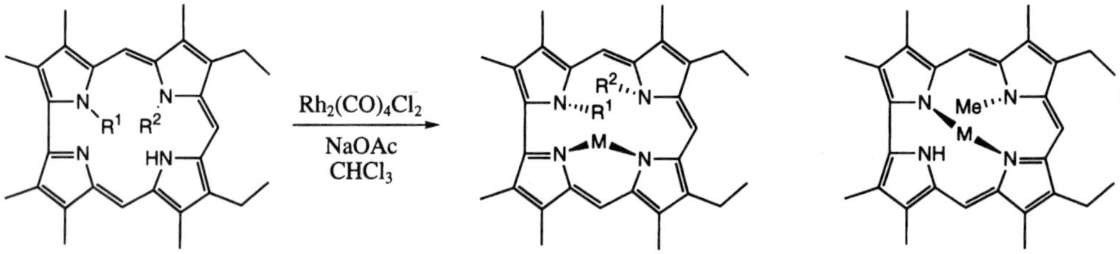

2.194. R^1 = Me, R^2 = H   Rh$_2$(CO)$_4$Cl$_2$ **2.228**. R^1 = Me, R^2 = H, M = Rh(CO)$_2$ **2.230**
2.195. R^1 = H, R^2 = Me   NaOAc   **2.229**. R^1 = H, R^2 = Me, M = Rh(CO)$_2$
                          CHCl$_3$

Figure 2.1.15   Single Crystal X-Ray Diffraction Structure of
Dicarbonylrhodium(I) *N*(21)-Methyl-8,12-diethyl-2,3,7,13,17,18-hexamethylcorrole
2.228.
This figure was generated using information down-loaded from the Cambridge Crystallographic
Data Centre and corresponds to a structure originally reported in reference 82. The *N*(21)-methyl
group and Rh(CO)$_2$ moiety are *trans* to one another, and the maximum deviation from the
macrocyclic plane is 0.05 Å. Atom labeling scheme: rhodium: ◉; carbon: ○; nitrogen: ●; oxygen:
⊜. Hydrogen atoms have been omitted for clarity

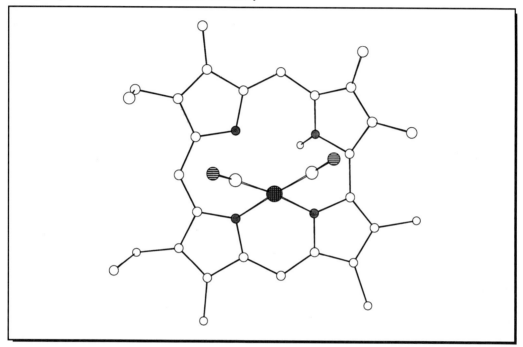

When *N*(22)-methyl corrole **2.195** was treated with tetracarbonyldi-μ-
chlororhodium(I), two isomeric products were obtained in a 7:1 ratio. As was
observed in the case of the corresponding N(21)-substituted derivative, the major
product was determined to have the C and D rings of the corrole involved in metal
coordination (i.e., **2.229**). This structural assignment was confirmed *via* a single
crystal X-ray structural analysis (Figure 2.1.16). By contrast, the minor isomer was
found to have the A and C corrole rings involved in metal ligation (i.e., **2.230**).[40]
Interestingly, it was determined that an equilibrium exists between the two isomers
with the free-energy difference between the two being 0.13 kcal mol^{-1}. Thus, longer
initial reaction times or warming of the original product mixture provided a final
13:16 distribution of products **2.229** and **2.230**.[40]

It is interesting to note that although *N*-alkyl Ni(II) corroles can be prepared
via the alkylation of N-unsubstituted nickel(II) corroles,[53,81] the corresponding
metal-free *N*-alkyl corroles appear reluctant to form stable Ni(II) complexes when
treated with nickel(II) ions.[80] It is possible, however, to prepare strong palladium(II)

and copper(II) *N*-alkyl corroles from metal-free *N*-alkyl corroles. For instance, treating the *N*(21)-methyl corrole **2.196** with Pd(OAc)$_2$ in acetic acid afforded the *N*(21)-methyl Pd(II) corrole **2.221** in 73% yield (Scheme 2.1.79).[80] Similarly, *N*(21)-ethyl corrole **2.198** can be converted into its Pd(II) and Cu(II) adducts, **2.231** and **2.216** by treating with the appropriate salts.[45] The *N*(22)-ethyl corrole **2.199** has also been metalated using Cu(OAc)$_2$ to afford *N*(22)-ethyl copper(II) corrole **2.232** in 69% yield.[45] Both this particular metallo–corrole and its *N*(21)-ethyl isomer **2.216** have been shown to form stable, but reversible protonation products on addition of TFA.[45] However, the exact site of this protonation has yet to be elucidated.

Figure 2.1.16   Single Crystal X-Ray Diffraction Structure of Dicarbonylrhodium(I) *N*(22)-Methyl-8,12-diethyl-2,3,7,13,17,18-hexamethylcorrole **2.229**.

This figure was generated using information down-loaded from the Cambridge Crystallographic Data Centre and corresponds to a structure originally reported in reference 40. The *N*(22)-methyl group and Rh(CO)$_2$ moiety are *trans* to one another, and the maximum deviation from the macrocyclic plane is 0.16 Å. Atom labeling scheme: rhodium: ●; carbon: ○; nitrogen: ◉; oxygen: ◒. Hydrogen atoms have been omitted for clarity

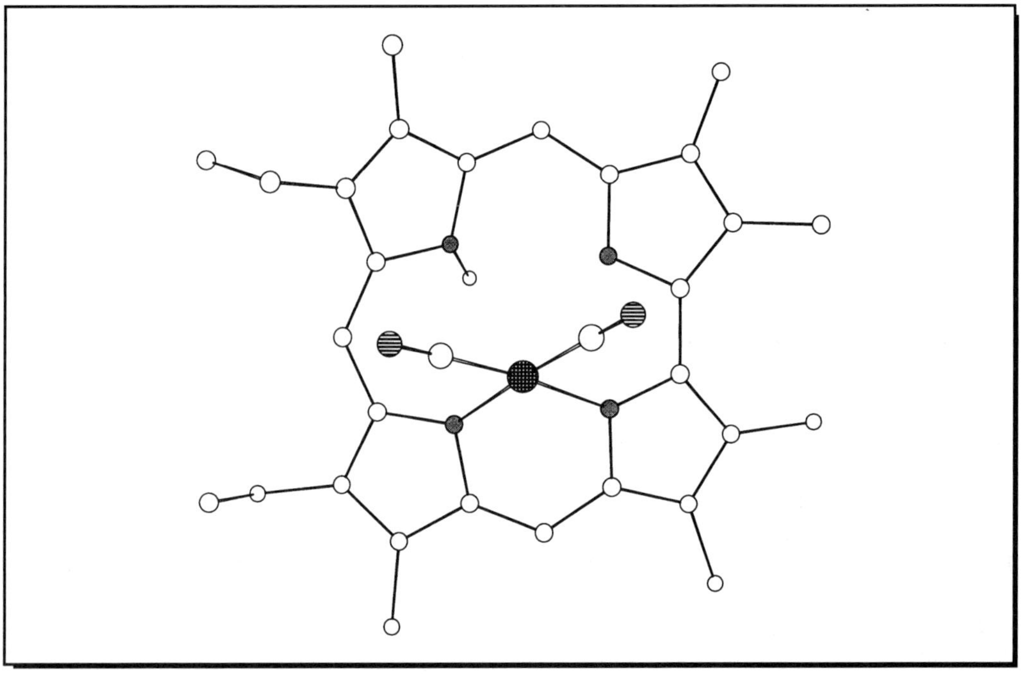

Attempts to insert copper(II) into either *N*(21)-allyl or *N*(22)-allyl corroles using Cu(OAc)$_2$ as the metal source led only to the formation of N-unsubstituted copper(II) corroles. On the other hand, heating *N*(21)-allyl or *N*(22)-allyl corroles with Ni(OAc)$_2$ led to no reaction whatsoever. (The starting materials were isolated unchanged.) These results were interpreted as meaning that, in the case of copper, metal insertion precedes cleavage of the *N*-allyl bonds.[45]

Scheme 2.1.79

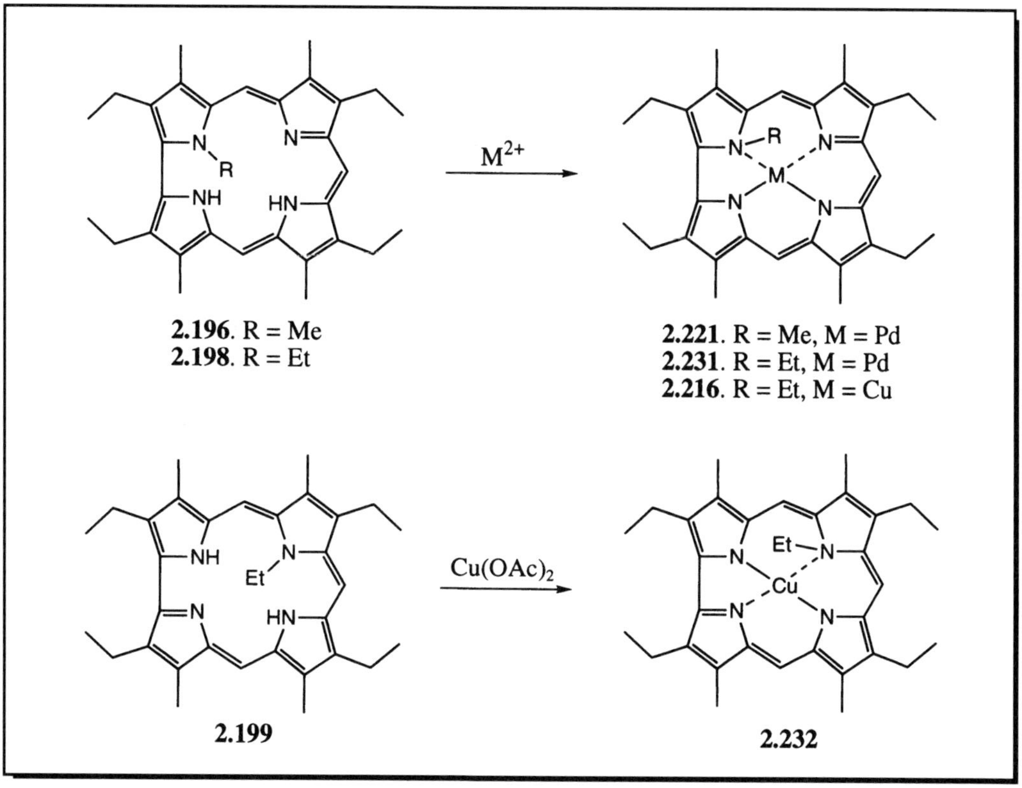

2.196. R = Me
2.198. R = Et

M^{2+}

2.221. R = Me, M = Pd
2.231. R = Et, M = Pd
2.216. R = Et, M = Cu

2.199

Cu(OAc)$_2$

2.232

Attempts to insert cobalt(II) into *N*(21)-methyl corrole **2.196** have also been made. In 1971, for instance, Johnson, *et al.* tried treating corrole **2.196** with Co(ClO$_4$)$_2$·6EtOH in the presence of pyridine.[45] However, only the demethylated cobalt(III) corrole(Py) species **2.125a** was obtained (Scheme 2.1.80).‡‡

One other approach to metalated *N*(21)-methyl corroles was reported by Johnson and coworkers in the early 1970s.[37,38] In this instance, the *N*-methyl thiaphlorin **2.206** was used as the starting material. It was heated in the presence of Pd(OAc)$_2$ in acetic acid to give the metalated corrole derivative **2.233** directly (see Scheme 2.1.81). The rate of sulfur extrusion in the presence of the palladium(II) ions was found to be dramatically increased in comparison to that of the analogous metal-free system **2.206** or the N-unsubstituted thiaphlorins **2.102** (Scheme 2.1.22). It was proposed that this rate enhancement reflected the "extra" stabilization that

---

‡‡As mentioned earlier in the text, the cobalt(III) corrole **2.125a** was formulated incorrectly in the original literature (i.e., as the bis-pyridine adduct). The correct formulation is shown here.

would accrue as the incoming Pd(II) center began to coordinate the four ligand nitrogen atoms.[37,38]

Scheme 2.1.80

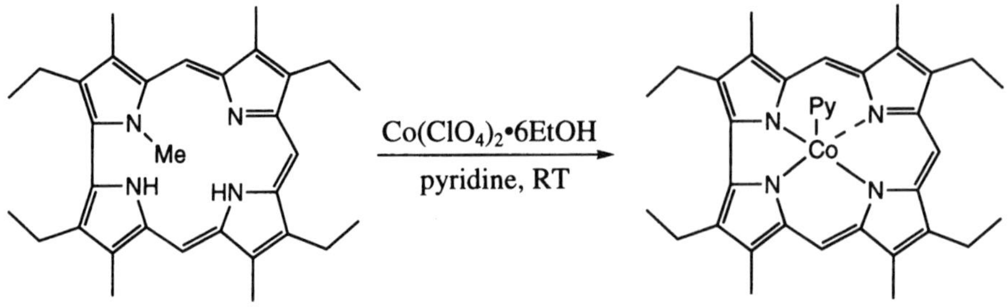

2.196   2.125a

Scheme 2.1.81

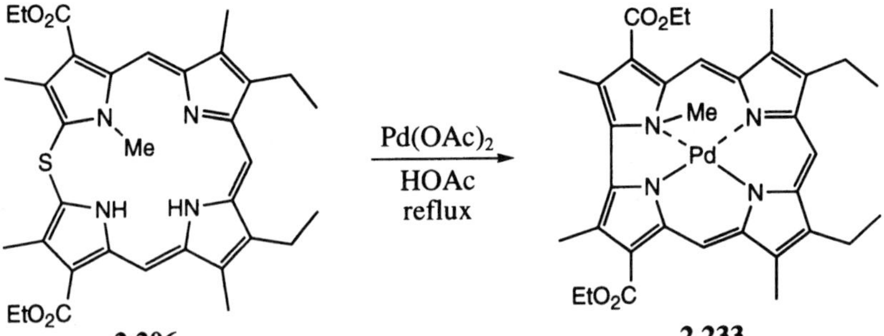

2.206   2.233

2.1.4.5   Thermal Stability of N-substituted Corroles and Metallocorroles

It has been demonstrated[80] that mono-*N*-alkyl corroles do not thermally iso-merize even after 12 hours at 256 °C. The *N*(21)-allyl corrole **2.200** was also found to be resistant to thermal isomerization; it remained unchanged even after being heated for 8 hours at 110 °C. However, the *N*(22)-allyl corroles are less thermally robust. For instance, heating *N*(22)-allyl corrole **2.201** for eight hours at 110 °C affords a 29% yield of the parent, unsubstituted corrole **2.80**, along with a 24% yield of *N*(21)-allyl corrole **2.200** as an isomerization product (Scheme 2.1.82).[80] Similarly, the *N*(22)-dimethyl allyl corrole **2.203** rearranges to give the *N*(21)-dimethyl allyl corrole species **2.202** in 16% yield. It also gives the unsubstituted corrole **2.80** in 41% yield. In the case of this latter isomerization, there is no observed inversion of the dimethyl allyl group during the course of reaction. This finding led to the suggestion that the rearrangement involves a free-radical process, rather than a sigmatropic one.[80]

The two *N*(21)-alkyl nickel(II) corroles (namely, **2.212** and **2.213**) have been shown to undergo facile thermal rearrangements, yielding the corresponding 3,3-dialkyl corrole species **2.234** and **2.235**, respectively (Scheme 2.1.83).[53] In a similar fashion, the related *N*(21)-methyl palladium(II) species **2.221** and **2.223** were also found to undergo thermal isomerization, again giving 3,3-dimethyl derivatives (**2.222** and **2.224**, respectively).[45,80] By contrast, subjecting the palladium(II) *N*(22)-methyl derivative **2.236** to thermolysis conditions (180 °C) was found to give rise to substantial decomposition as well as a small amount of the demetalated *N*(22)-methyl corrole **2.197** (Scheme 2.1.84).[§§]

When 3,3-dimethyl Pd(II) corrole **2.222** was heated to reflux in allyl bromide and then treated with sodium perchlorate, the main product obtained was the C(17)-allyl derivative **2.237** (isolated in 73% yield as the perchlorate salt) (Scheme 2.1.85).[45] The site of addition of the allyl fragment is also the site at which protonation of the 3,3-dialkyl metallo–corroles appears to occur. For instance, treating the nickel(II) 3,3-dimethyl- or 3-methyl-3-ethyl corroles **2.234** and **2.235** with TFA affords the corresponding C(17) protonated trifluoroacetate salts **2.238** and **2.239**, respectively (Scheme 2.1.86).[53] Similarly, protonation of palladium(II) 2,8,12,18-tetraethyl-3,3,7,13,17-pentamethyl corrole **2.222** with TFA affords the C(17)-protonation product **2.240**.[80]

In contrast to the nickel and palladium corrole complexes discussed above, the copper(II) *N*(21)-alkyl corroles **2.215** and **2.216** do not appear to undergo thermal isomerization. In fact, subjecting these species to thermolysis conditions leads only to loss of the *N*-alkyl group, giving rise to the corresponding N-unsubstituted copper(II) corroles **2.107** and **2.159**, respectively (Scheme 2.1.87).[45,83] A similar resistance to rearrangement was seen in the case of the copper(II) *N*(22)-ethyl corrole **2.232**. Here, however, rather than dealkylation, heating gives rise only to decomposition products.

The exact origin of the disparate thermal stability exhibited by the geometrically similar nickel(II) and copper(II) corroles is not yet fully known. It has been suggested, however, that the differences in the electronic structures of the two metal centers might be critical.[45] In particular, the unpaired electron present in the d^9 copper(II) complexes could play a role in stabilizing a decomposition-inducing homolytic cleavage pathway involving the *N*-alkyl groups. Such a decomposition would be much less likely in the case of divalent nickel.

[§§]A claim to have both prepared the palladium(II) *N*(22)-alkyl corrole **2.236** and studied its thermal properties was made by Johnson and coworkers in 1970.[80] In the body of a follow-up 1971 paper, however, these same workers state that these palladium(II) *N*(22)-alkyl corroles were too unstable to isolate.[45]

Scheme 2.1.82

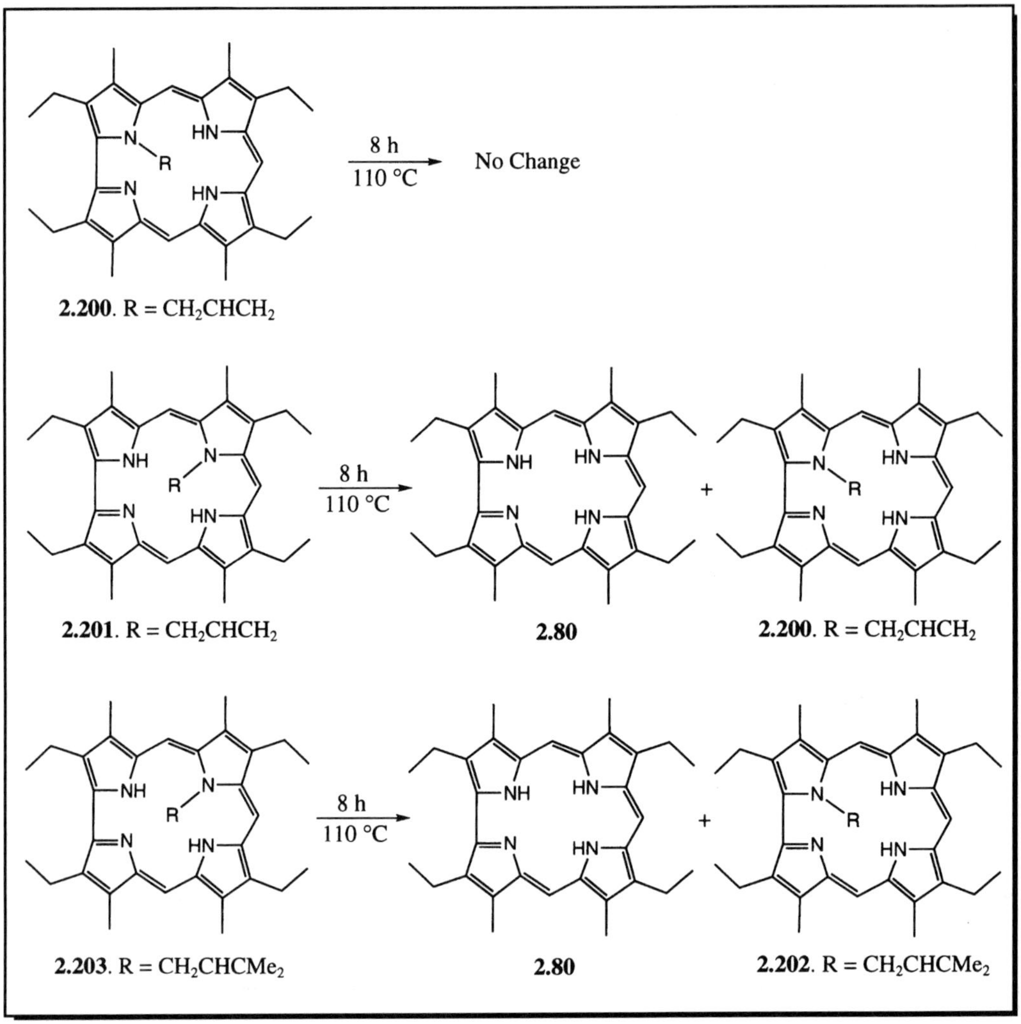
8 h
110 °C
No Change
2.200. R = CH_2CHCH_2
8 h
110 °C
+
2.201. R = CH_2CHCH_2
2.80
2.200. R = CH_2CHCH_2
8 h
110 °C
+
2.203. R = CH_2CHCMe_2
2.80
2.202. R = CH_2CHCMe_2

2.1.4.6 Protonation of *N*(21)- and *N*(22)-Methyl Corroles

Initial reports concerning the protonation behavior of corrole involved claims of having observed both a mono- and a diprotonated species. In the case of the monoprotonated macrocycle, the UV-vis absorption bands remained relatively unchanged compared to those of the free-base species. In the case of diprotonated corroles, on the other hand, the second protonation event was presumed to occur at one of the *meso* carbons, resulting in a disruption of the π-electron conjugation. This latter conclusion was based upon the dramatic changes observed to occur in the UV-vis absorption spectrum.

Scheme 2.1.83

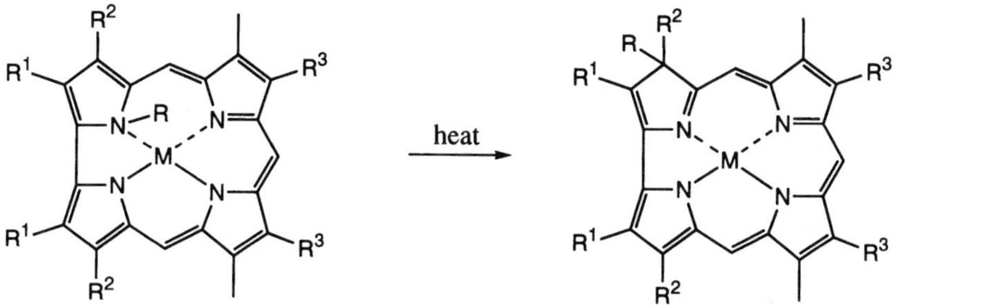

2.212. R = R^1 = R^2 = Me, R^3 = Et, M = Ni
2.213. R = R^3 = Et, R^1 = R^2 = Me, M = Ni
2.221. R = R^2 = Me, R^1 = R^3 = Et, M = Pd
2.223. R = R^1 = R^3 = Me, R^2 = Et, M = Pd

2.234. R = R^1 = R^2 = Me, R^3 = Et, M = Ni
2.235. R = R^3 = Et, R^1 = R^2 = Me, M = Ni
2.222. R = R^2 = Me, R^1 = R^3 = Et, M = Pd
2.224. R = R^1 = R^3 = Me, R^2 = Et, M = Pd

Scheme 2.1.84

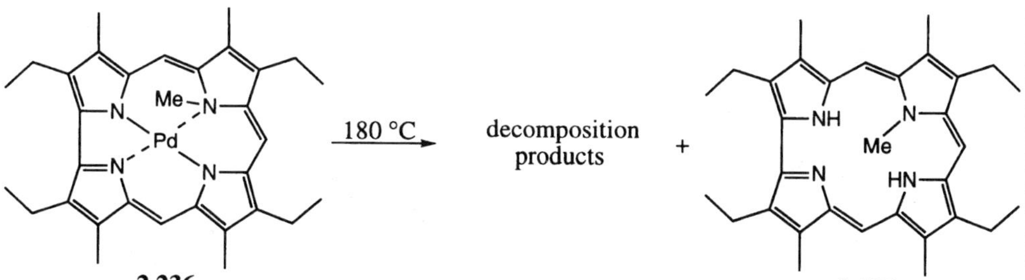

2.236 **2.197**

Scheme 2.1.85

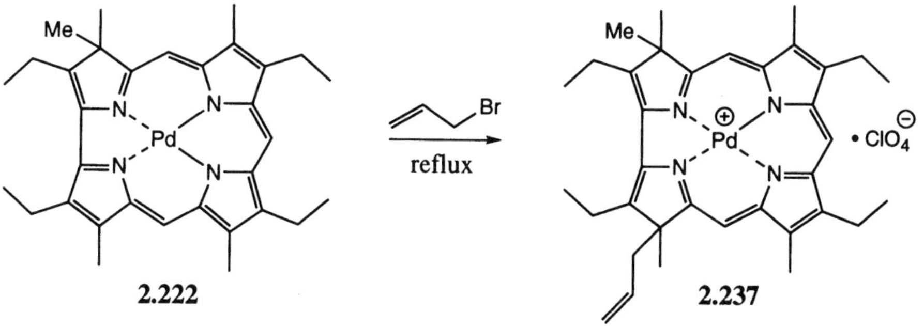

2.222 **2.237**

Scheme 2.1.86

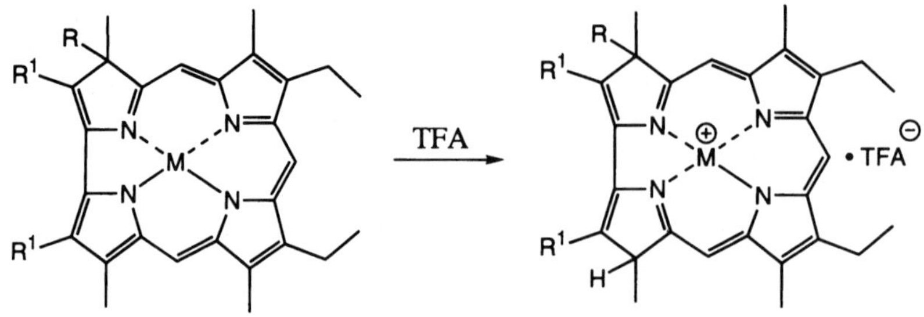

2.234. R = R¹ = Me, M = Ni
2.235. R = Et, R¹ = Me, M = Ni
2.222. R = Me, R¹ = Et, M = Pd

2.238. R = R¹ = Me, M = Ni
2.239. R = Et, R¹ = Me, M = Ni
2.240. R = Me, R¹ = Et, M = Pd

Scheme 2.1.87

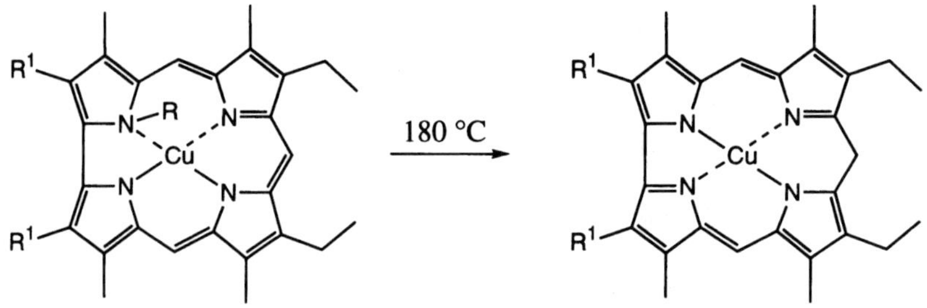

2.215. R = R¹ = Me
2.216. R = R¹ = Et

2.107. R¹ = Me
2.159. R¹ = Et

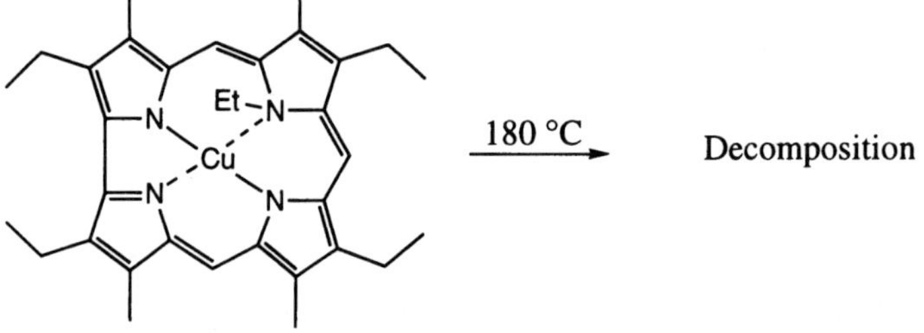

2.232

Decomposition

In order to gain a more complete understanding of the protonation chemistry of corroles, a study of the behavior of *N*-methyl corroles in acidic media was carried out.[84] In this study it was found, for instance, that *N*(21)-methyl corrole (**2.194**) exists

in the form of a mono cation at all concentrations of acetic acid in toluene. The
N(22)-isomer **2.195** behaves similarly in pure acetic acid, and in acetic acid/toluene
mixtures. However, in the case of this N(22)-isomer, there are second and third
protonation events that occur to 50% completion at 0.75 M and 1.0 M H_2SO_4 in
HOAc, respectively. The second protonation occurs with retention of a Soret absorbance in the UV-vis spectral region. The site at which this second protonation occurs
remains uncertain, however. The third protonation, on the other hand, results in a
loss of any type of Soret band absorbance. It is thus considered to involve reaction at
the C(5) *meso*-position.[30] Unfortunately, further studies into the protonation chemistry of this and other corroles have not been carried out. It, therefore, remains
uncertain as to whether the results obtained for the protonation of N-methyl corrole
2.194 and **2.195** can be generalized to include other N-substituted or even N-unsubstituted corroles.

2.1.4.7   N-Substituted Heterocorroles

Only very few examples of experiments devoted to the alkylation of heterocorroles have appeared in the literature. These have involved alkylation of the
bifuran-containing corrole **2.39** with methyl iodide or ethyl iodide.[80] For instance,
the mono-N-methyl corrole **2.241** was prepared in this way from corrole **2.39** in 56%
yield (Scheme 2.1.88). In an analogous manner, the mono-N-ethyl corrole **2.242** was
prepared in 45% yield starting from **2.39**. In both cases, the mono-N-alkyl corroles
were isolated as their hydroiodide salts. In fact, these corroles were found to be so
basic that they could not be isolated in their free-base forms.[30]

Scheme 2.1.88

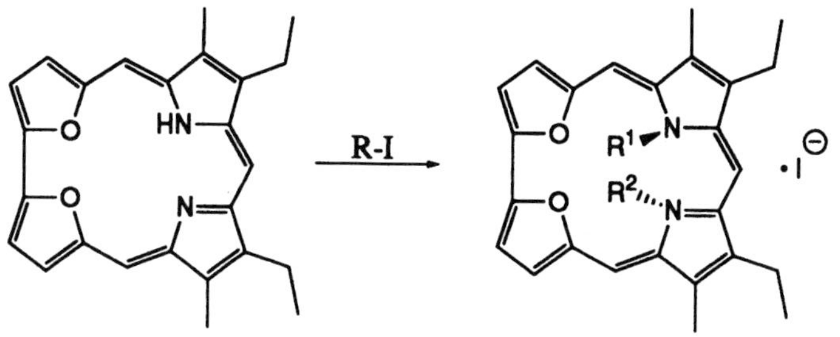

2.39

2.241. R^1 = Me, R^2 = H
2.242. R^1 = Et, R^2 = H
2.243. R^1 = R^2 = Me
2.244. R^1 = R^2 = Et

Under more forcing conditions, the *N,N'*-dialkyl corroles **2.243** and **2.244** were prepared in high yield from the starting corrole **2.39**. The N-substituents were concluded to be in a *trans* disposition with respect to one another (i.e., one is found above the mean macrocyclic plane and the other below), thereby making the molecule achiral. This was determined by partial resolving of the D-camphorsulphonate derivative of corrole **2.244**.[80]

The bifuran-containing corrole **2.39** was also subject to acylating conditions. In this instance, treatment of corrole **2.39** with acetyl chloride using aluminum trichloride as the Lewis acid afforded the C(5)-acetyl-substituted corrole derivative **2.245** in 31% yield (Scheme 2.1.89).[30] In addition to this product, a di-substituted corrole was also obtained. This was presumed to be either a 2,18- or a 3,17-diacetyl-substituted corrole. However, no evidence was put forward in support of either of these assignments.

Scheme 2.1.89

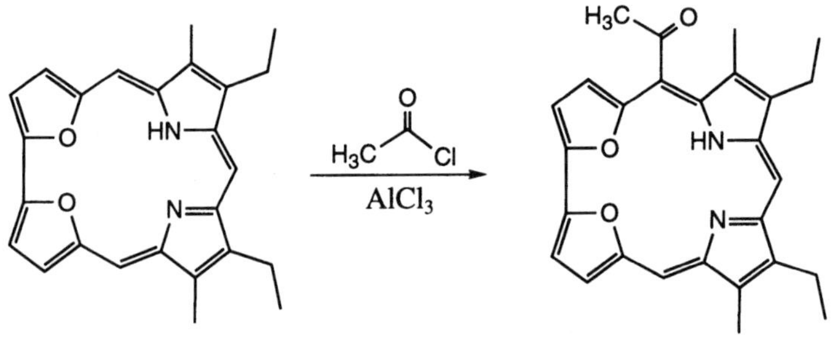

2.39   **2.245**

2.1.5   "Corrologen"

Free-base corrole may be reduced in the presence of hydrogen gas and a platinum catalyst to afford the corresponding "corrologen". For instance, in 1965 Johnson reported the reduction of 8,12-diethyl-2,3,7,13,17,18-hexamethylcorrole **2.6** to give the colorless, somewhat unstable corrologen **2.246** (Scheme 2.1.90).[36] This type of reduction was later performed on the uroporphyrin III-like substituted corroles **2.94** and **2.95**. In this instance, corrologens **2.247** and **2.248** were obtained.[85] These corrologens undergo ready autoxidation in the presence of air, reverting back to the corresponding starting conjugated corroles. Alternatively, these corrologens can be oxidized by treatment with electron-deficient agents such as DDQ.[36,85] Interestingly, however, it was observed that at least in the case of corrologen **2.246**, oxidation in the presence of acid leads to the formation of porphyrins. Under these conditions, no trace of corrole was detected.[43]

Scheme 2.1.90

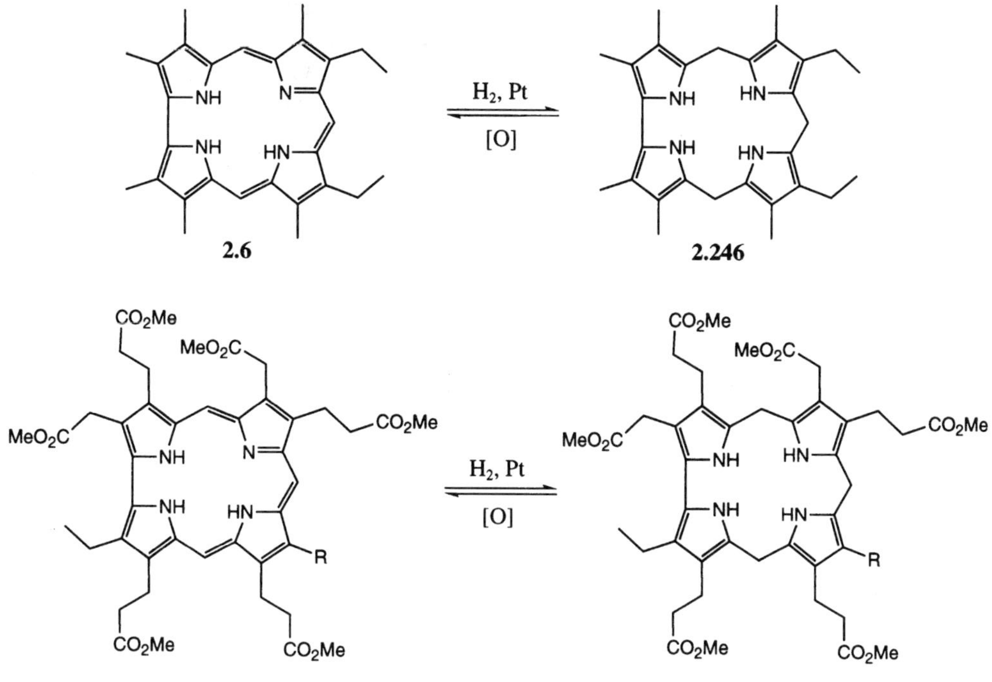

2.6   2.246

2.94. R = Me
2.95. R = CH₂CO₂Me

2.247. R = Me
2.248. R = CH₂CO₂Me

2.1.6   Corrole–Corrole and Corrole–Porphyrin Dimers

Quite recently, Paolesse and coworkers have reported the synthesis of some novel corrole–corrole dimer systems.[46] The first of these, system **2.250**, was prepared from the phenyl-linked bis(dideoxybiladiene-ac) tetrabromide **2.249** using chemistry akin to that used to prepare simple monocorroles (Scheme 2.1.91). Specifically, tetrabromide **2.249a** was dissolved in NaHCO₃-saturated methanol and treated with chloranil. After working up with hydrazine followed by chromatography, the corrole dimer **2.250a** was isolated in 46% yield. In this same way, the ethyl-substituted system **2.250b** was prepared from bis(dideoxybiladiene-ac) tetrabromide **2.249b** in 49% yield. The biscorrole system **2.252** with a *meta*-phenyl linkage was similarly prepared starting from **2.251** in 39% yield (Scheme 2.1.92).

The bis-cobalt(III) complexes **2.255** and **2.256** were also synthesized by Paolesse, *et al.*[46] Here, in analogy to what was done to produce the corresponding metallocorrole monomers, ring closure of the dimeric bis(dideoxybiladiene-ac) tetrabromides **2.249a** and **2.51** was effected using cobalt(II) acetate as catalyst. In the case of **2.249a**, cobalt(II) acetate-mediated ring closure in the presence of triphenylphosphine afforded the bis-metalated species **2.253** in 52% yield (Scheme 2.1.93). The cobalt(III) *m*-phenylbiscorrole **2.254** was prepared in a like manner starting from **2.251** (Scheme 2.1.94). However, in this instance, the ring closure yield was significantly lower (i.e., 28%). This lower yield was presumed to reflect the greater instability of this complex under the purification conditions employed.[46]

Scheme 2.1.91

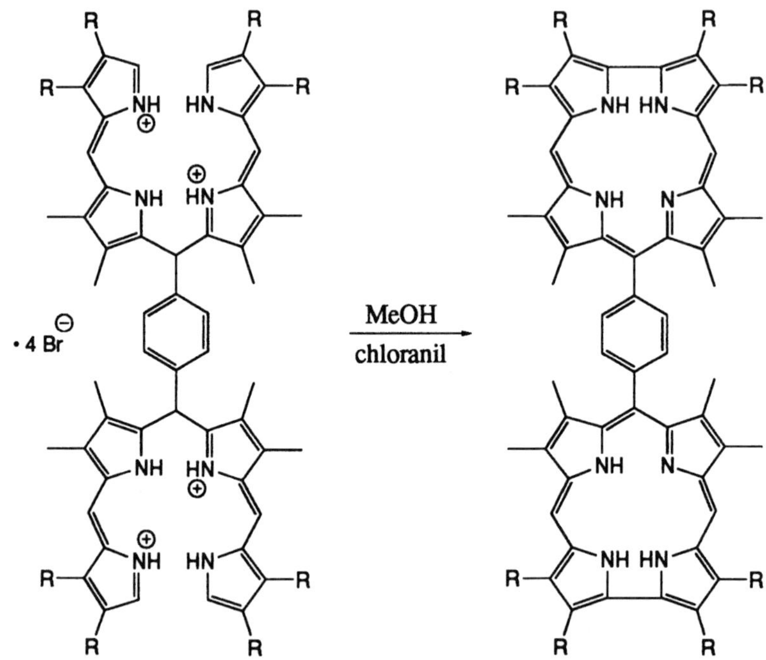

2.249a. R = Me   2.250a. R = Me
2.249b. R = Et   2.250b. R = Et

The corrole–corrole dimers of Schemes 2.1.91 and 2.1.92 each exhibit optical spectra that are similar to those of an octaalkylcorrole, with no significant modifications in either the Soret or Q-band regions.[46] Thus, the corrole dimer system **2.250a**, for instance, exhibits a Soret band at 400/410 nm ($\varepsilon = 112\,000$ and $82\,000$ $M^{-1}\,cm^{-1}$, respectively) and Q-bands at 542 nm and 597 nm ($\varepsilon = 24\,000$ and $28\,000$ $M^{-1}cm^{-1}$, respectively). Similarly, the electronic absorption spectra of complexes **2.253** and **2.254** do not differ significantly from spectra of analogous monomeric cobalt(III) corrole complexes. This leads the authors of this monograph to conclude that little electronic interaction exists between the two subunits.

The synthesis of the corrole–porphyrin pseudo-dimers **2.256a** and **2.256b** was also described by Paolesse and coworkers.[46] These chimerical systems were prepared from the corresponding dideoxybiladiene-ac–porphyrin units **2.255a** and **2.255b** *via* an oxidative ring-closing procedure similar to that used to prepare the homo-corrole dimers **2.250** and **2.252** (Scheme 2.1.95). The heterobimetallic corrole–porphyrin pseudo-dimer **2.259** was also prepared by Paolesse, *et al.*[46] In this case, a final cobalt(II)-catalyzed ring closure, involving the dideoxybiladiene-ac–porphyrinato nickel complex **2.250**, was employed to afford complex **2.259** in 41% yield (Scheme 2.1.96).

Scheme 2.1.92

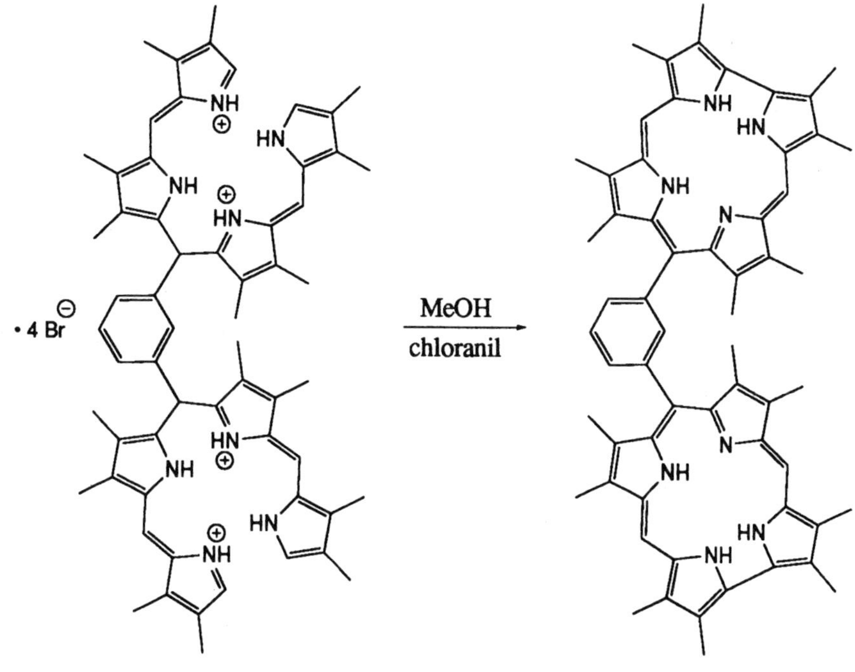

2.251   **2.252**

In the case of the metal-free species **2.256**, the electronic absorption spectrum shows characteristic Q-like absorption bands ascribable to both the porphyrin and corrole subunits. In the Soret region, however, the intensity of the porphyrin absorbance overwhelms that of the corrole absorbance. In complex **2.259**, only the absorbance bands of the porphyrin are visible, both in the Soret and in the Q-band region. Again, this was rationalised in terms of the much greater absorption intensity of the metalloporphyrin chromophore compared to that of the metallocorrole.[46]

Scheme 2.1.93

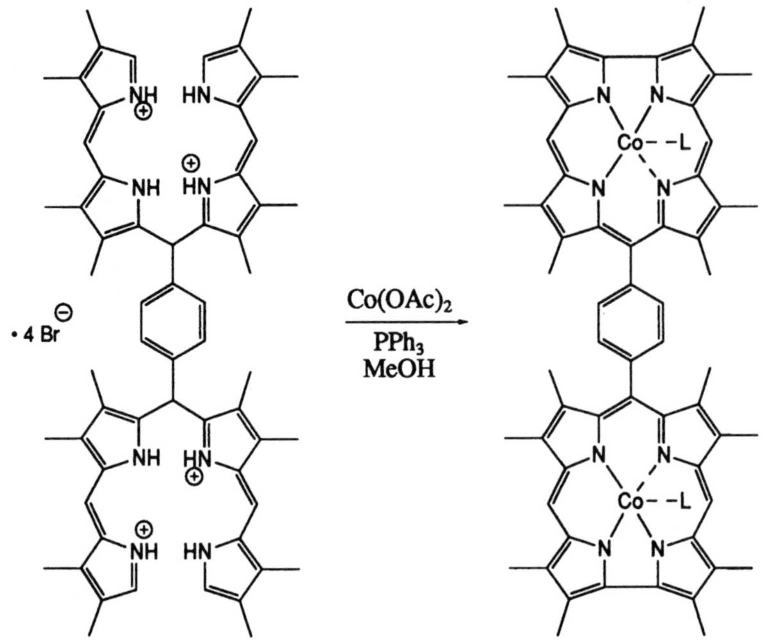

$Co(OAc)_2$
$\dfrac{}{\substack{PPh_3 \\ MeOH}}$

· 4 Br$^{\ominus}$

2.249a

2.253. L = PPh$_3$

Scheme 2.1.94

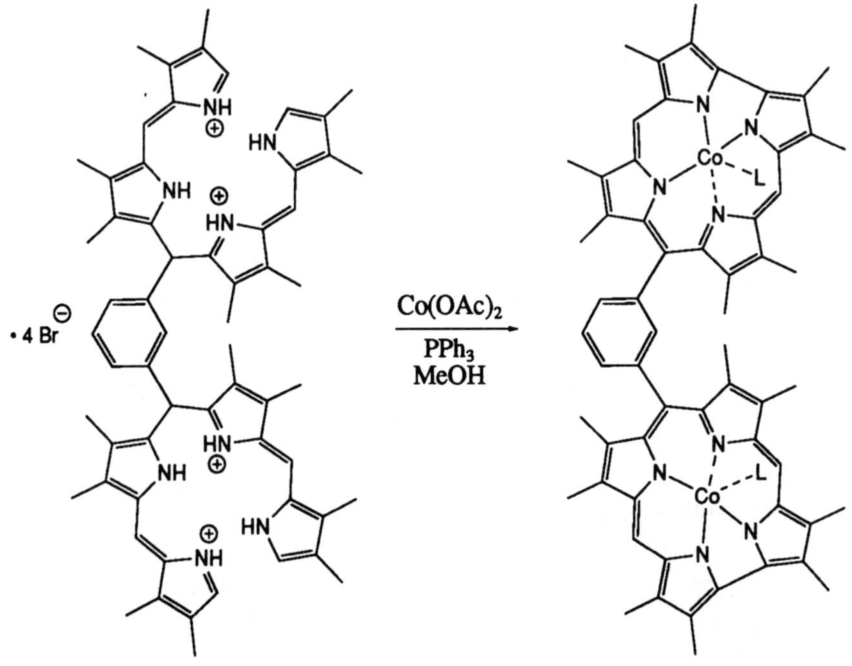

$Co(OAc)_2$
$\dfrac{}{\substack{PPh_3 \\ MeOH}}$

· 4 Br$^{\ominus}$

2.251

2.254. L = PPh$_3$

Scheme 2.1.95

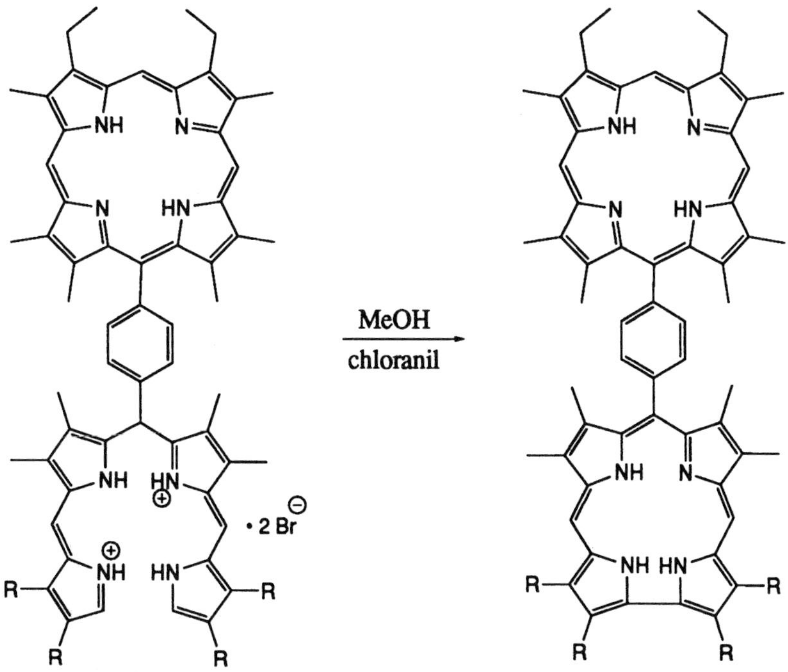

2.255a. R = Me
2.255b. R = Et

MeOH
chloranil

2.256a. R = Me
2.256b. R = Et

Scheme 2.1.96

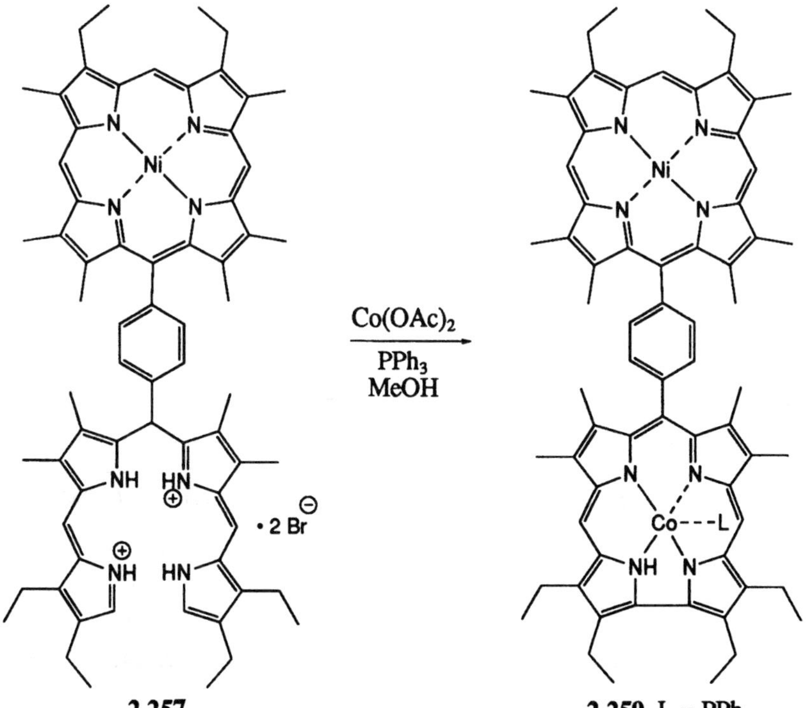

2.257

Co(OAc)$_2$
PPh$_3$
MeOH

2.259. L = PPh$_3$

2.2   Isocorrole

2.2.1   Synthesis

The preparation of a corrole isomer, termed isocorrole, has recently been reported by Vogel and coworkers.[83,86–89] The general, unsubstituted form of this compound is shown in Figure 2.2.1 and has the structure **2.261**. Isocorrole may be regarded as being derived from the porphyrin isomer porphycene (e.g., **2.260**) by formal removal of one of its four *meso* carbon atoms. Considered in this light, isocorrole **2.261** bears the same relationship to porphycene as the parent corrole does to porphyrin.

Figure 2.2.1   Generalized Structures of Porphycene (**2.260**) and Isocorrole (**2.261**)

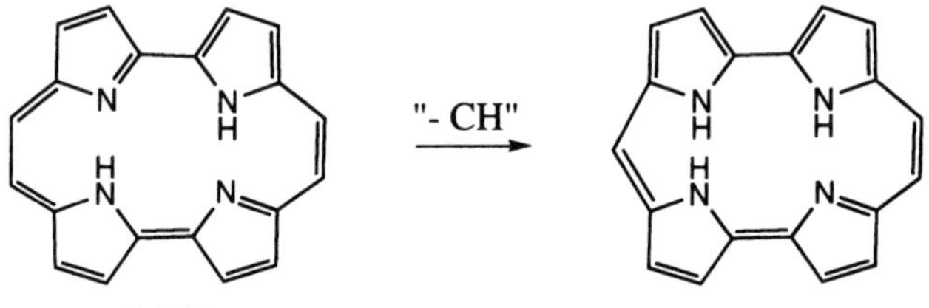

2.260	**2.261**
"porphycene"	"isocorrole"

The first isocorroles to be prepared were, in fact, isolated as a minor by-product of the reductive carbonyl coupling reaction between two diformyl bipyrroles used to prepare porphycene (discussed in Chapter 3). For instance, octa-alkyl isocorroles **2.265** and **2.267** were isolated in *ca.* 0.4% yield in the course of trying to prepare porphycenes **2.264** and **2.266** (Scheme 2.2.1).[86] While the mechanistic details of this low-yielding, isocorrole-forming process remain to be unraveled, follow-up studies revealed that isocorroles did not arise as the result of a ring-contraction process occurring during the course of the reductive carbonyl coupling reaction.[88,89] It was further determined that these isocorrole by-products were only obtained from porphycene-forming reactions in which the bipyrrole precursor was fully substituted in its β-positions. The determinants of this apparent steric requirement remain unidentified, however.

It was later found that isocorroles could in fact be prepared from porphycenes *via* a direct ring-contraction process. For instance, it was found that treating porphycene **2.268** with aqueous potassium carbonate in DMF afforded isocorrole **2.269** in 34% yield (Scheme 2.2.2).[86,88] Under slightly different conditions (i.e., KOH in DMF), octaethylisocorrole **2.267** was prepared from the corresponding octaethylporphycene (**2.266**).[87–89] Interestingly, this type of ring contraction (i.e., to generate what are formyl-substituted isocorroles), could not be achieved in the case of etioporphycene **2.264**.[88,89] At present, an adequate explanation for this apparent anomaly is still unavailable.

Scheme 2.2.1

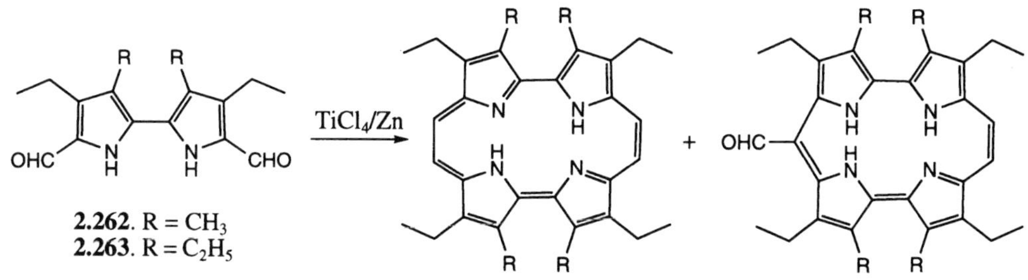

2.262. R = CH₃
2.263. R = C₂H₅

TiCl₄/Zn

2.264. R = CH₃
2.266. R = C₂H₅

+ OHC

2.265. R = CH₃
2.267. R = C₂H₅

Scheme 2.2.2

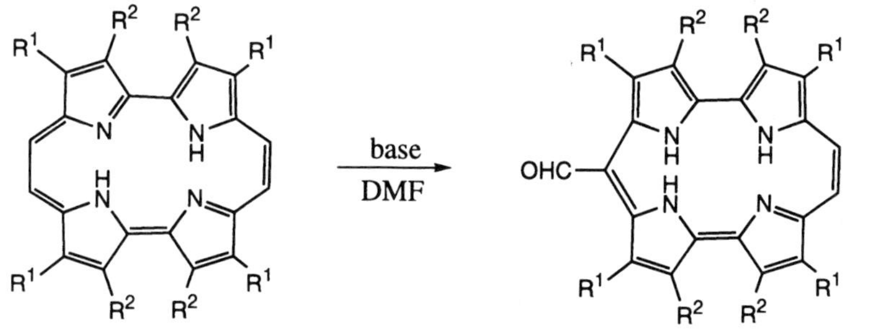

2.268. R¹ = n-C₃H₇, R² = Br
2.266. R¹ = R² = C₂H₅

base
DMF

2.269. R¹ = n-C₃H₇, R² = Br
2.267. R¹ = R² = C₂H₅

    A single crystal X-ray diffraction analysis has been carried out on the bromo-substituted isocorrole derivative **2.269**.[86] This analysis revealed a somewhat non-planar macrocyclic structure (Figure 2.2.2). The non-planarity in **2.269** was presumed to arise primarily as the result of NH steric interactions present within the core of the macrocycle. As a consequence of these interactions, the two pyrroles of the dipyrrylmethene-like end of the macrocycle each twist by 23° in opposite directions out of the mean macrocyclic plane. This out-of-plane rotation of two pyrrole rings is in contrast to that observed in a typical corrole structure. In this latter case, three of the pyrrole rings are nearly coplanar, and one of the pyrroles is rotated out of plane but only by 8–10°.[28]

Figure 2.2.2   Single Crystal X-Ray Diffraction Structure of Isocorrole Aldehyde
2.269.
This figure was generated using information down-loaded from the Cambridge Crystallographic
Data Centre and corresponds to a structure originally reported in reference 86. The average
deviation from macrocyclic planarity is 0.01 Å. However, the A and D rings (those adjacent to the
meso-carbon bearing the formyl group) are each rotated out of the macrocyclic plane by 23°. Atom
labeling scheme: carbon: ○; nitrogen: ●; oxygen: ⊜; bromine: ⊗. Hydrogen atoms have been
omitted for clarity

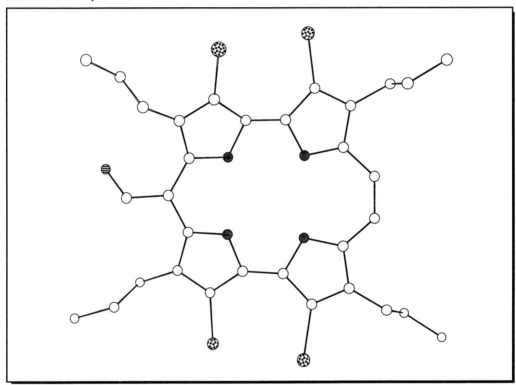

Fully *meso*-unsubstituted isocorroles (i.e., ones without a *meso*-formyl group)
have also been prepared by Vogel and coworkers.[88,89] This was first achieved using
a variation of the ring-contraction reaction used to prepare isocorroles **2.265** and
2.267 (*vide supra*). Here, it was determined that treating etioporphycene **2.264** with
aqueous KOH in a 2:1 mixture of DMSO and diglyme afforded the *meso*-unsub-
stituted etioisocorrole **2.270** in *ca.* 1% yield (Scheme 2.2.3). These same conditions
were used to prepare the *meso*-unsubstituted octaethylisocorrole **2.271** from the
corresponding octaethylporphycene **2.266** in *ca.* 10% yield. It is of interest to
note that in the case of octaethylporphycene **2.266** a change in the solvent ratio
from 2:1 DMSO:diglyme to 1:2 DMSO:diglyme resulted in a change in the
ring-contraction product. That is, simply changing the DMSO-to-diglyme solvent
ratio in Scheme 2.2.3 from 2:1 to 1:2 resulted in the formation of the formyl-

substituted isocorrole (i.e., **2.267** in Scheme 2.2.1) and not the *meso*-unsubstituted product **2.271**.[88,89] In the case of etioporphycene **2.264**, this same solvent ratio change also had an effect on the outcome of the reaction. However, in this instance, when a 1 : 2 solvent ratio is used, no ring contraction was observed to occur at all.[88,89]

Scheme 2.2.3

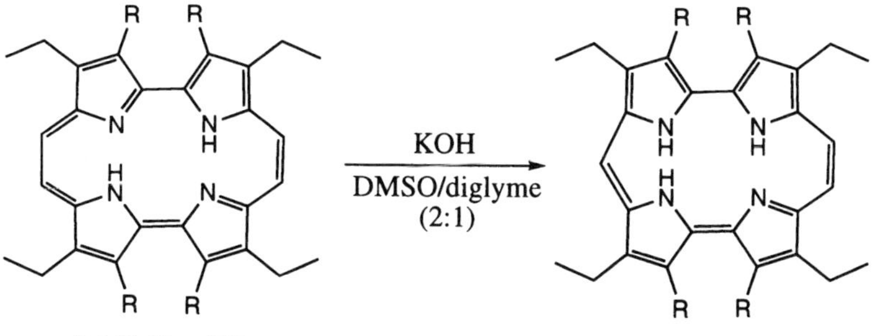

2.264. R = CH₃
2.266. R = C₂H₅

KOH
DMSO/diglyme
(2:1)

2.270. R = CH₃
2.271. R = C₂H₅

Vogel and coworkers have also explored an alternative approach to the synthesis of isocorroles.[83,88,89] This more rational approach involves the low-valent titanium-mediated intramolecular reductive coupling of an α,ω-formyl-substituted linear tetrapyrrolic species. Specifically, ring closure of the linear species **2.272**, carried out under these McMurry-type conditions, afforded the etioisocorrole **2.270** in 10% yield (Scheme 2.2.4). The putative "corrologen", thought to occur as an intermediate in this reaction, was not isolated. Rather, oxidation by molecular oxygen or FeCl₃ during workup was presumed to occur spontaneously.

Scheme 2.2.4

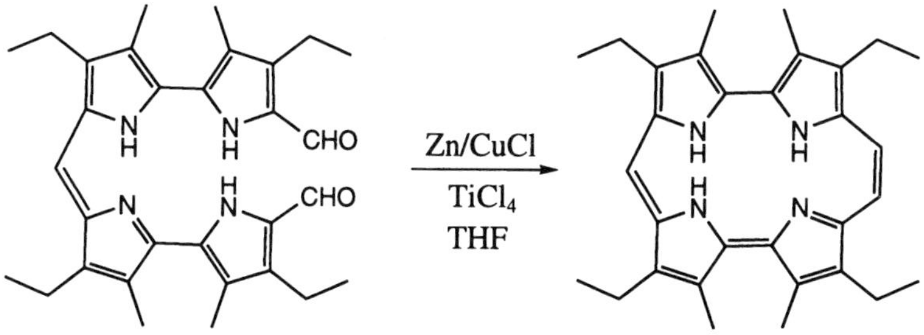

2.272

Zn/CuCl
TiCl₄
THF

2.270

2.2.2   *Metalation Chemistry*

The obvious structural relationship between corrole and isocorrole led Vogel and coworkers to explore the coordination chemistry of these latter isomeric systems.[83,86–89] For instance, in preliminary studies involving the reaction of *meso*-formyl isocorrole **2.269** with di- and trivalent metal ions, it was found that complexes do indeed form. However, these complexes proved to be highly unstable and decomposed too quickly to allow for characterization. One notable exception to this apparent kinetic instability has been discovered, however. It was found, for instance, that the reaction of tetrabromo-substituted isocorrole **2.269** with cobalt(II) salts in the presence of pyridine resulted in the formation of a stable product.[86] Similarly, octaethylisocorrole **2.267** forms a stable complex under these conditions.[87] On the basis of ^{1}H NMR spectroscopic analyses, these complexes were assigned as being diamagnetic cobalt(III) complexes (i.e., **2.273** and **2.274**) bearing two axially coordinating pyridine ligands (Scheme 2.2.5). An analogous bis-pyridyl cobalt(III) product (i.e., **2.275a**) was also obtained from the *meso*-unsubstituted etioisocorrole **2.270** as the result of carrying out an analogous metalation procedure (Scheme 2.2.6).[83,88,89] Here, repeated crystallization of complex **2.275a** from *t*-butyl pyridine and methanol afforded the *t*-butyl pyridyl cobalt(III) complex **2.275b**.

Scheme 2.2.5

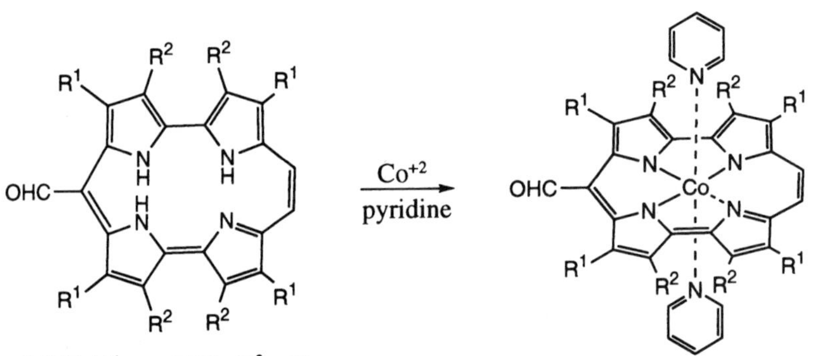

2.269. R^1 = *n*-C$_3$H$_7$, R^2 = Br
2.267. R^1 = R^2 = C$_2$H$_5$

2.273. R^1 = *n*-C$_3$H$_7$, R^2 = Br
2.274. R^1 = R^2 = C$_2$H$_5$

Scheme 2.2.6

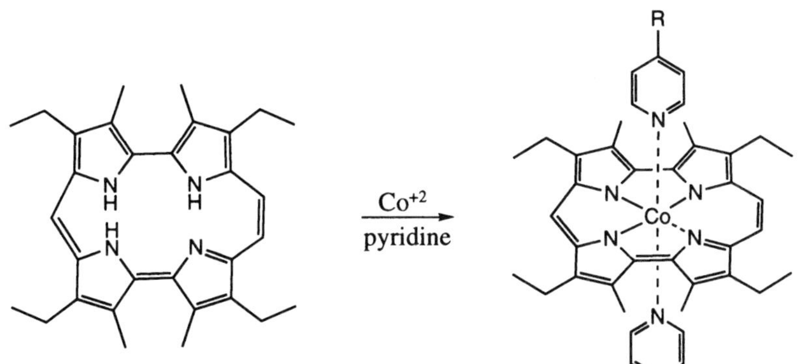

2.270

$\xrightarrow{\text{Co}^{+2}}$
pyridine

R

$\begin{cases} \textbf{2.275a}. \ R = H \\ \textbf{2.275b}. \ R = t\text{-}C_4H_9 \end{cases}$

2.3   Contracted Porphyrins Containing Fewer than Four Pyrrole-like Subunits

2.3.1   Subphthalocyanines

2.3.1.1   Synthesis

The phthalocyanines (e.g., **2.277**) are well known for their ready availability and their tendency to form stable complexes with a wide variety of metals.[90-93] They are typically prepared *via* the metal-templated macrocyclization of a phthalonitrile derivative (e.g., **2.276**) as shown in Scheme 2.3.1. The ease of preparation as well as the remarkable coordinative ability of the phthalocyanines led Meller and Ossko in 1972 to attempt to prepare boron-containing phthalocyanine derivatives. This they tried *via* the reaction of phthalonitrile with haloboranes.[94] What they in fact obtained, however, was not the desired tetrameric phthalocyanine derivative, but rather a "contracted" trimeric phthalocyanine **2.284**, a species that has since come to be known as "subphthalocyanine" (Scheme 2.3.2). A similar fluorine-containing subphthalocyanine **2.285** was also prepared using this procedure.

Since the initial disclosure of subphthalocyanines **2.284** and **2.285**, several other examples of subphthalocyanines have been reported. The first of these, the bromo-derivative of the *t*-butyl-substituted macrocycle **2.286** was prepared in 1990 by Kobayashi, *et al.*[95] The synthesis of unsubstituted and *t*-butyl-substituted sub-phthalocyanines **2.287** and **2.288** (Scheme 2.3.2), in which the axial substituent at the boron center is a phenyl group, has also been reported.[96,97] (In the case of the *t*-butyl-substituted compound, the two possible structural isomers **2.288a** and **2.288b** (Figure 2.3.1) were separated by high-performance liquid chromatography (HPLC).[96])

Finally, a synthesis of the chloro-derivative of the nitro-substituted subphthalocyanine **2.289** was recently reported by Torres and coworkers.[98]

Scheme 2.3.1

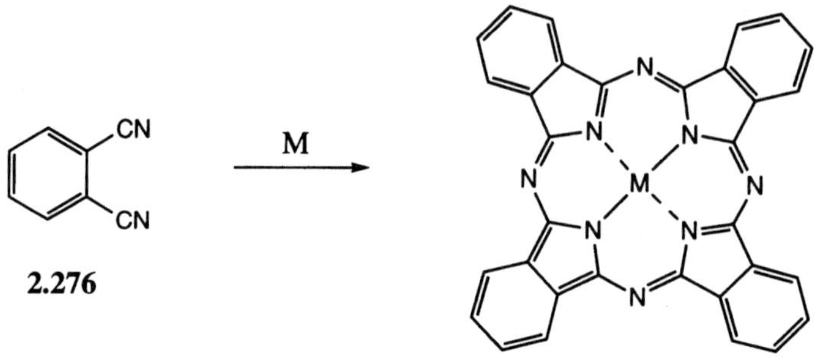

2.276

2.277
"phthalocyanine"

Scheme 2.3.2

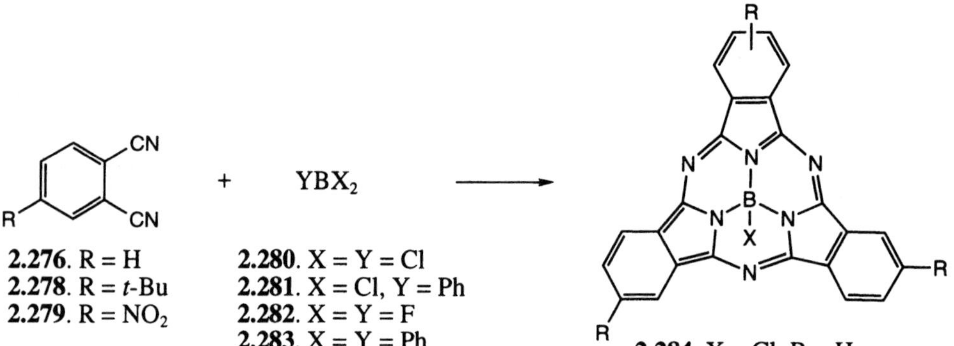

+   YBX₂

2.276. R = H   **2.280**. X = Y = Cl
2.278. R = *t*-Bu   **2.281**. X = Cl, Y = Ph
2.279. R = NO₂   **2.282**. X = Y = F
                         2.283. X = Y = Ph

2.284. X = Cl, R = H
2.285. X = F, R = H
2.286. X = Br, R = *t*-Bu
2.287. X = Ph, R = H
2.288. X = Ph, R = *t*-Bu
2.289. X = Cl, R = NO₂

In addition to the above subphthalocyanines, the hexakis(alkylthio)-substi-
tuted system **2.291** was prepared by Bekaroglu and coworkers from the bis(alkylthio)phthalonitrile derivative **2.290** (Scheme 2.3.3).[99] Also, in what may be
considered an intellectual extension on the synthesis of subphthalocyanines,

Rauschnabel and Hanack have reported the synthesis of the related sub*n*aphthalocyanines **2.294** and **2.295**.[97] The synthesis of **2.294** involved heating the 2,3-dicyanonaphthalene **2.292** in naphthalene in the presence of PhBCl₂ (Scheme 2.3.4). This afforded a 25% crude yield of **2.294**, a product that proved somewhat light-sensitive. In a similar way, the *t*-butyl derivative **2.295** was also prepared from the *t*-butyl-substituted 2,3-dicyanonaphthalene **2.293**. As expected, this latter compound exhibited better solubility than its unsubstituted "parent". Unfortunately, it too proved rather unstable in the presence of light.

Figure 2.3.1   Structures **2.288a** and **2.288b**

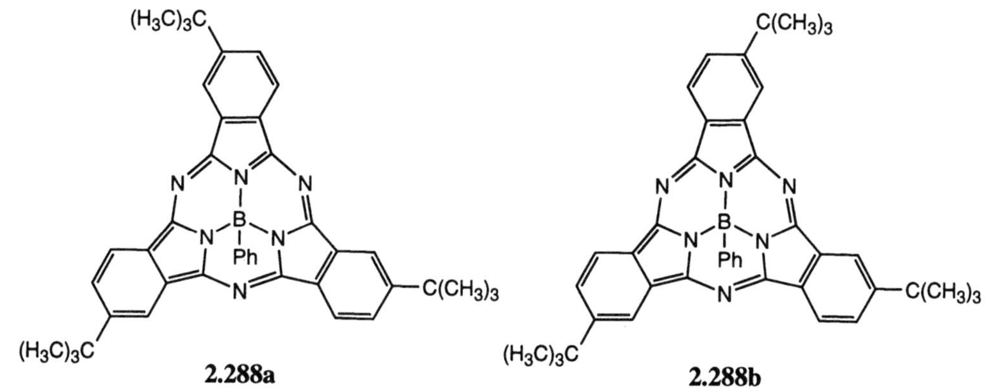

Scheme 2.3.3

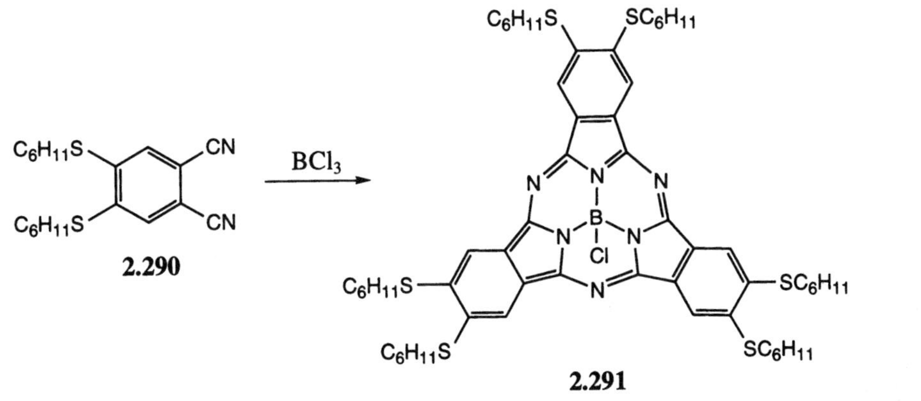

Scheme 2.3.4

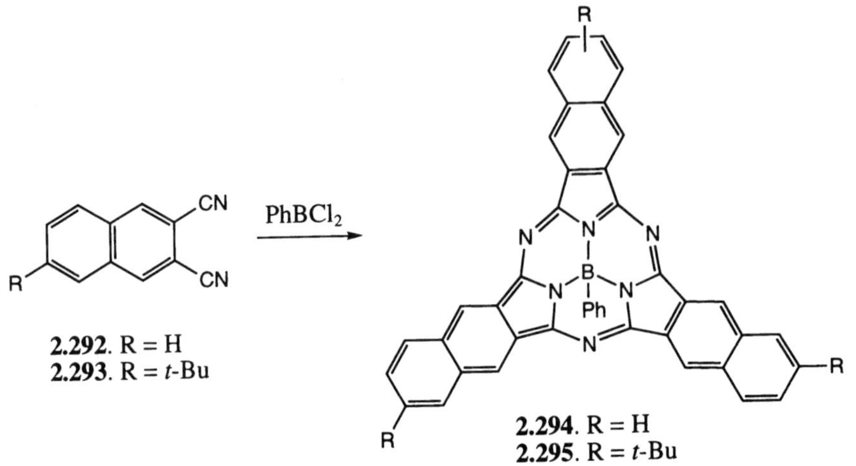

2.292. R = H
2.293. R = *t*-Bu

2.294. R = H
2.295. R = *t*-Bu

Subphthalocyanines contain a delocalized 14 π-electron conjugated pathway, and are brightly colored compounds, both in the solid state and in organic solution. As such, they exhibit fairly strong absorption bands in their visible electronic spectrum. For instance, subphthalocyanine **2.284** exhibits a Soret-like absorbance band at *ca.* 305 nm and a more intense Q-like transition at 565 nm ($\varepsilon = 50\,100\ M^{-1}\,cm^{-1}$ and $89\,100\ M^{-1}\,cm^{-1}$, respectively, in $CHCl_3$).[100] These bands are blue-shifted relative to those of the metallophthalocyanines (by *c.* 20–30 nm and 120–130 nm in the Soret and Q-band regions, respectively), and exhibit absorption coefficients that typically are smaller than those of the metallophthalocyanines.[90–93,101]

A single crystal X-ray diffraction analysis has been carried out on both the chloro- and the phenyl-substituted subphthalocyanines **2.284** and **2.287** (Figures 2.3.2 and 2.3.3).[97,102] These analyses served to show that subphthalocyanines lie in a bowl-shaped conformation. This is, of course, very different from the near-planar conformation of the parent phthalocyanines. Presumably, this bowl-shaped structure accounts, in part, for the decreased molar absorptivities of the subphthalocyanines relative to their phthalocyanine "parents". Nevertheless, despite the non-planar nature of these macrocycles, the subphthalocyanines are capable of supporting an induced diamagnetic ring current (as judged by NMR spectroscopy). Thus, they may appropriately be considered as being aromatic.

2.3.1.2   Ring Expansion to Phthalocyanines

While the subphthalocyanines are interesting in their own right, much of the recent impetus for preparing these macrocycles derives from their use as precursors in the synthesis of unsymmetrically substituted phthalocyanines, which are otherwise difficult to prepare. The phthalocyanines and their symmetrically substituted derivatives have been extensively studied since they often exhibit properties such as electrical conductivity, electrochromism, mesophase formation, and photosensitivity that make them of considerable interest in the preparation of novel materials.[90–93,103] Unsymmetrically substituted phthalocyanines are also considered useful in these as well as other applications such as nonlinear optics and thin-film formation where such systems offer certain advantages over their symmetrically substituted counterparts.[90–93,103] Because of this, an efficient synthesis of unsymmetrically substituted phthalocyanines would be very helpful. Such simplified syntheses, derived from ring expansions of subphthalocyanines, have been developed. Several examples of this approach are described below.

Figure 2.3.2   Single Crystal X-Ray Diffraction Structure of Subphthalocyanine(Cl) **2.284**.

This figure was generated using information down-loaded from the Cambridge Crystallographic Data Centre and corresponds to a structure originally reported in reference 102. This structure reveals the bowl-shaped form of the macrocycle and a tetrahedral coordination geometry about the boron atom. The average B–N distance is 1.46 Å and the B–Cl distance is 1.86 Å. Atom labeling scheme: boron: ◕; carbon: ○; nitrogen: ●; chlorine: ◔. Hydrogen atoms have been omitted for clarity

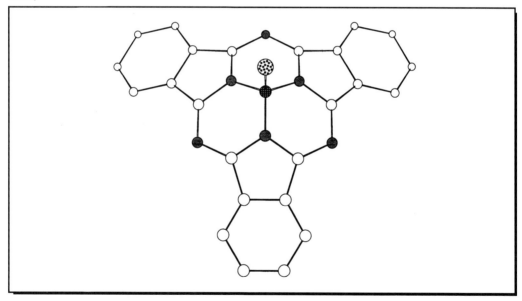

The first documented example of a ring-expansion reaction involving a subphthalocyanine came in 1990.[95] Here, Kobayashi and Osa and coworkers reported that by treating the *t*-butyl-substituted subphthalocyanine **2.286** with succinimide (**2.296**), one could obtain the unsymmetrically substituted phthalocyanine **2.297** in 13% yield (Scheme 2.3.5). Similarly, treating subphthalocyanine **2.286** with the diiminoisoindoline analogs **2.298–2.300**, afforded the unsymmetrically substituted phthalocyanines **2.301–2.303** in 8–20% yield (Scheme 2.3.6).[95] Later, independent work by the teams of Kasuga,[104] Hanack,[105] and Wöhrle[106] resulted in the preparation of the unsymmetrical phthalocyanines **2.310–2.315**. These compounds were obtained starting with the unsubstituted subphthalocyanine derivative **2.284** and the diiminoisoindolines **2.304–2.309**, respectively (Scheme 2.3.7). Similarly, ring expansion reactions involving the hexakis(alkylthio)-substituted subphthalocyanine **2.291**, using either **2.298** or the nitro-substituted derivative **2.305**, allowed Bekaroglu and coworkers to isolate systems **2.316** and **2.317** (Scheme 2.3.8).[99] Finally, Bekaroglu and coworkers have also described how phthalocyanines **2.320, 2.321,** and **2.323** containing one macrocyclic substituent may be prepared from **2.284** and the requisite diiminoisoindoline (i.e., **2.318, 2.319,** and **2.322**) (Schemes 2.3.9 and 2.3.10).[107]

Figure 2.3.3   Single Crystal X-Ray Diffraction Structure of Subphthalocyanine(phenyl) **2.287**.

This figure was generated using information down-loaded from the Cambridge Crystallographic Data Centre and corresponds to a structure originally reported in reference 97. This structure reveals the bowl-shaped form of the macrocycle and a tetrahedral coordination geometry about the boron atom. The average B–N distance is 1.50 Å and the B–C distance is 1.59 Å. Atom labeling scheme: boron: ◕; carbon: ○; nitrogen: ●. Hydrogen atoms have been omitted for clarity

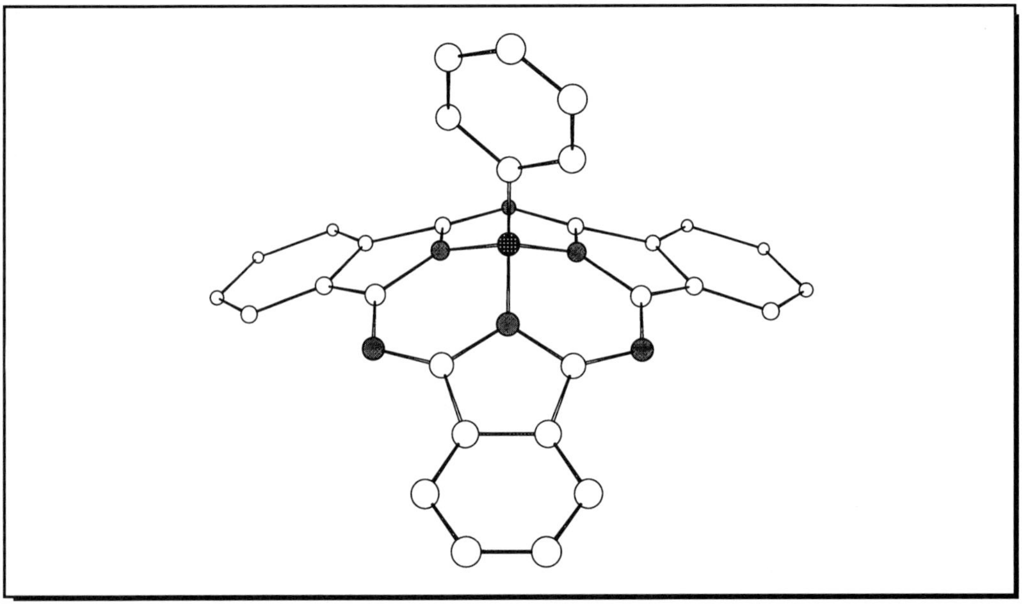

That subphthalocyanines undergo ring expansion so readily is presumed to reflect their strained nature as embodied in their cone-shaped conformation. This does not mean, however, that, in a preparative sense, these ring-expansion reactions will always yield the desired products. For instance, Wöhrle and coworkers found that the reaction between subphthalocyanine **2.284** and the diiminoisoindolines **2.308**

and **2.309** gave only low yields of the desired insertion products (i.e., **2.314** and **2.315**) while producing a mixture of several other products that included both unsubstituted and differently substituted phthalocyanines.[106] Also, while Hanack and coworkers obtained a reasonable (60%) yield of the amino-substituted phthalocyanine **2.312** from **2.184** and diiminoisoindoline **2.306**, they too obtained unsubstituted phthalocyanine as a by-product of their reaction[105].

Scheme 2.3.5

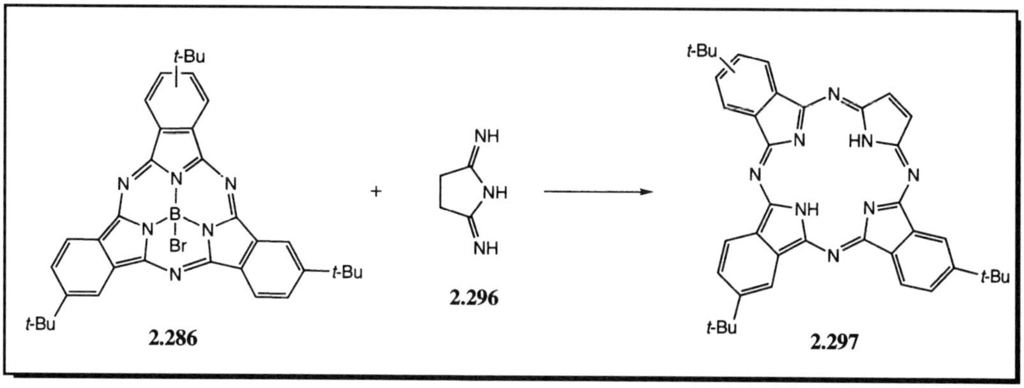

Scheme 2.3.6

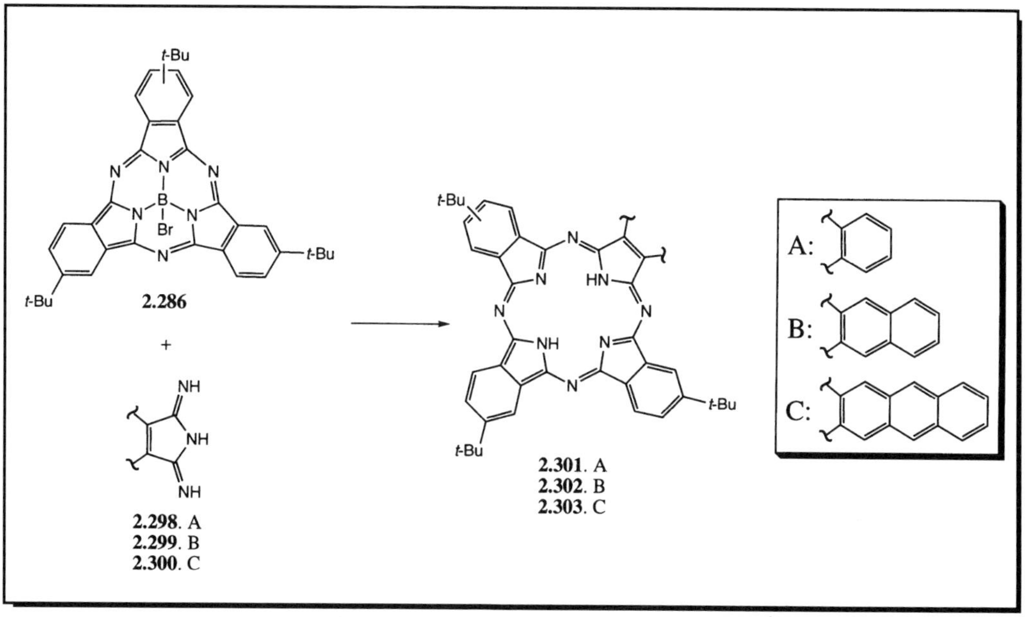

Scheme 2.3.7

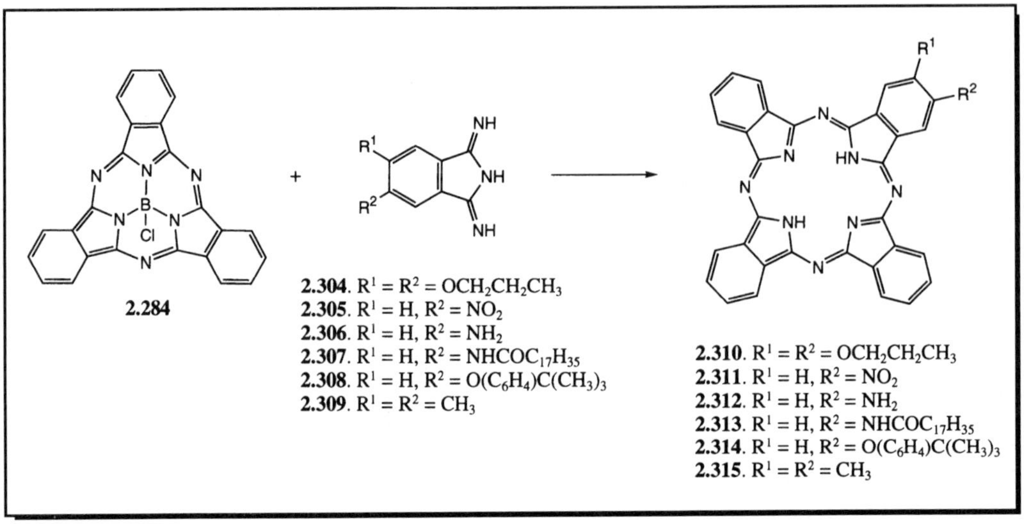

2.284

2.304. $R^1 = R^2 = OCH_2CH_2CH_3$
2.305. $R^1 = H, R^2 = NO_2$
2.306. $R^1 = H, R^2 = NH_2$
2.307. $R^1 = H, R^2 = NHCOC_{17}H_{35}$
2.308. $R^1 = H, R^2 = O(C_6H_4)C(CH_3)_3$
2.309. $R^1 = R^2 = CH_3$

2.310. $R^1 = R^2 = OCH_2CH_2CH_3$
2.311. $R^1 = H, R^2 = NO_2$
2.312. $R^1 = H, R^2 = NH_2$
2.313. $R^1 = H, R^2 = NHCOC_{17}H_{35}$
2.314. $R^1 = H, R^2 = O(C_6H_4)C(CH_3)_3$
2.315. $R^1 = R^2 = CH_3$

Scheme 2.3.8

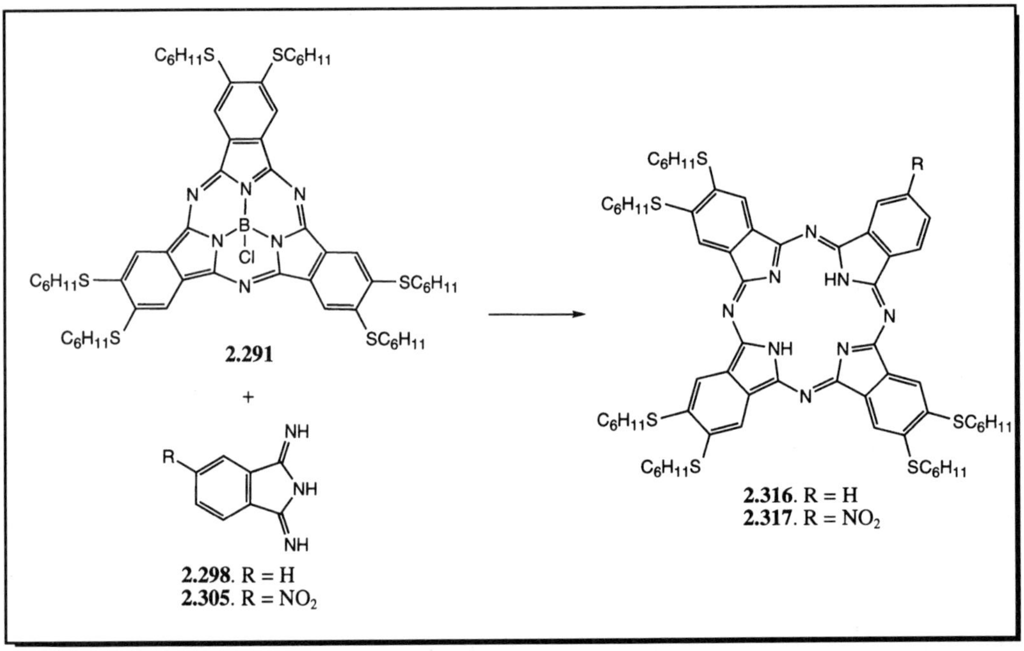

2.291

2.298. R = H
2.305. R = NO_2

2.316. R = H
2.317. R = NO_2

In an attempt to explain these findings, Wöhrle and coworkers found that subphthalocyanine **2.284** can be converted to unsubstituted phthalocyanine by treatment with the weak base, zinc(II) acetate dihydrate (Scheme 2.3.11).[106] On the other hand, no reaction occurred when $ZnCl_2 \cdot H_2O$ was used. This and other experiments led to the conclusion that subphthalocyanine does not undergo a concerted reaction when treated with diiminoisoindoline. Rather, a multistep reaction was proposed to occur. Such a reaction was presumed to involve first a base-catalyzed decomposition of the subphthalocyanine ring followed by reaction of the resulting activated fragments either with themselves, or with the diiminoisoindoline. A related mechanism was proposed by Hanack and coworkers.[105] In this case, an initial reaction of diiminoisoindoline **2.306** with macrocycle **2.284** was proposed as giving rise to a linear four-unit species such as **2.325**. This species, it was suggested, could then either cyclize to give the desired phthalocyanine **2.312** or disproportionate to give the two two-unit species **2.326** and **2.327**. These latter entities could then either react with each other (to give **2.312**), or self-condense to give both unsubstituted and diamino-substituted phthalocyanine (Scheme 2.3.12).

Scheme 2.3.9

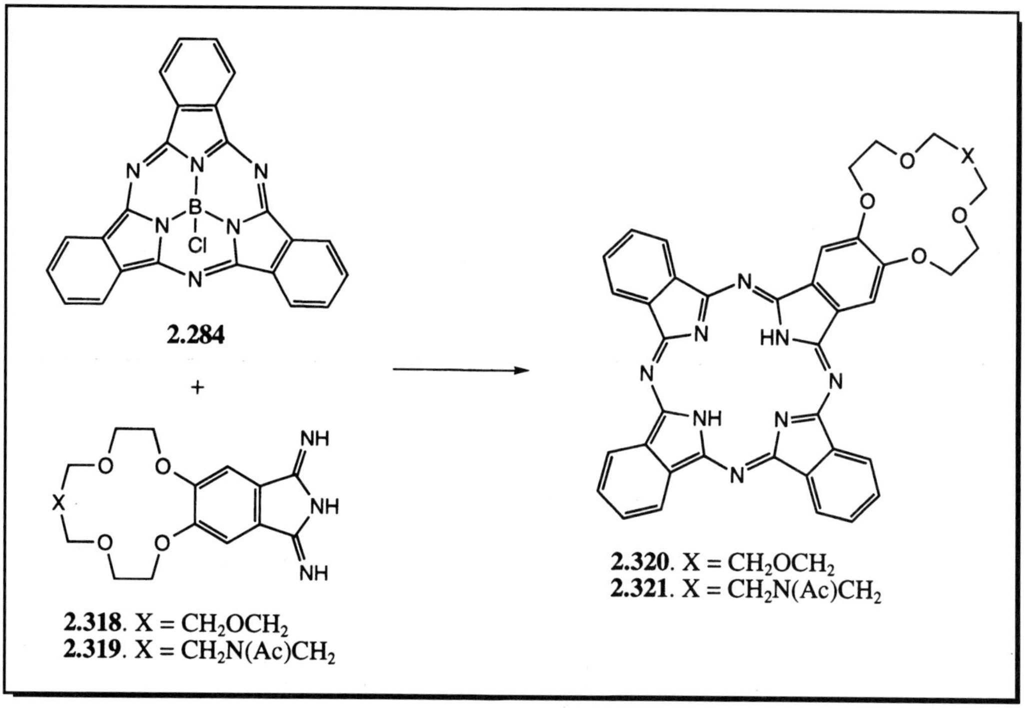

2.284

+

2.318. X = CH$_2$OCH$_2$
2.319. X = CH$_2$N(Ac)CH$_2$

2.320. X = CH$_2$OCH$_2$
2.321. X = CH$_2$N(Ac)CH$_2$

In an attempt to preclude the formation of the various side-products observed in the above reactions, Wöhrle and coworkers developed a modified means of effecting this type of ring expansion.[106] They found that reacting the subphthalocyanine **2.284** with a phthalonitrile derivative such as **2.328** or **2.329** in the presence of zinc(II) acetate led to reasonable yields of the zinc(II) mono-substituted phthalocyanine **2.308** or **2.330** (Scheme 2.3.13). Importantly, they also found that the amount of mono-substituted phthalocyanine product in these reactions was substantially increased relative to analogous reactions involving diiminoisoindolines.

Scheme 2.3.10

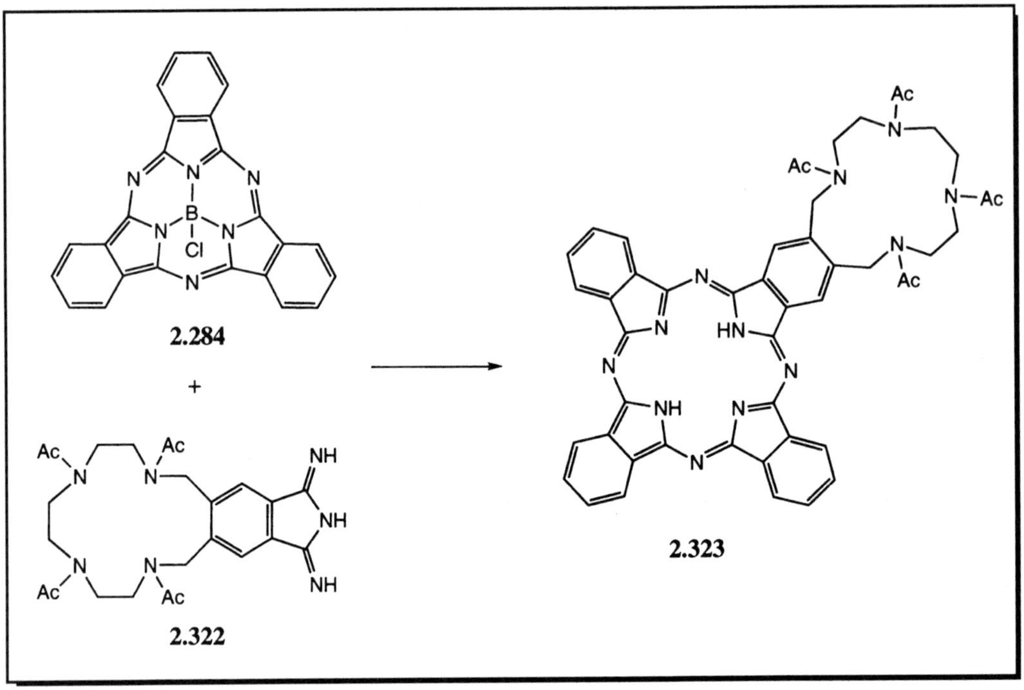

2.284

+

2.322

2.323

2.3.1.3   Subphthalocyanine Dimer

One example of a dimeric subphthalocyanine has recently been reported.[101] This macrocycle (2.332) was prepared by condensing of an excess of 4-*t*-butylphthalonitrile 2.278 with 1,2,4,5-tetracyanobenzene (2.331) in the presence of Ph$_2$BBr. The resulting laterally bridged system 2.332 is obtained in 2.8% yield (along with a 24% yield of the monomeric subphthalocyanine 2.286) (Scheme 2.3.14). Despite the presumed non-planar nature of the subphthalocyanine units, ^{1}H NMR data collected for this system indicate that it, like the monomeric systems discussed above, is aromatic. On the other hand, compound 2.332 displays spectral properties that differ dramatically from those of the monomeric macrocycle 2.286. For instance, compared to that of 2.286, the UV-vis absorption spectrum of 2.332 is rather featureless, although absorption bands are observed in nearly the same spectral region as in the case of 2.286. The emission spectrum of 2.332 also resembles that of the monomeric species **2.286**. The actual bands, however, are somewhat blue-shifted relative to this control.

Scheme 2.3.11

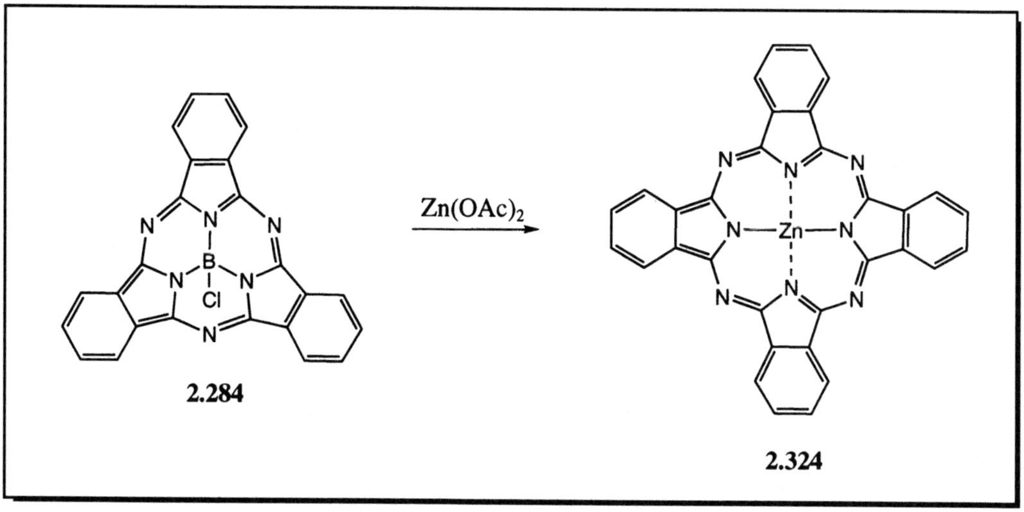

Scheme 2.3.12

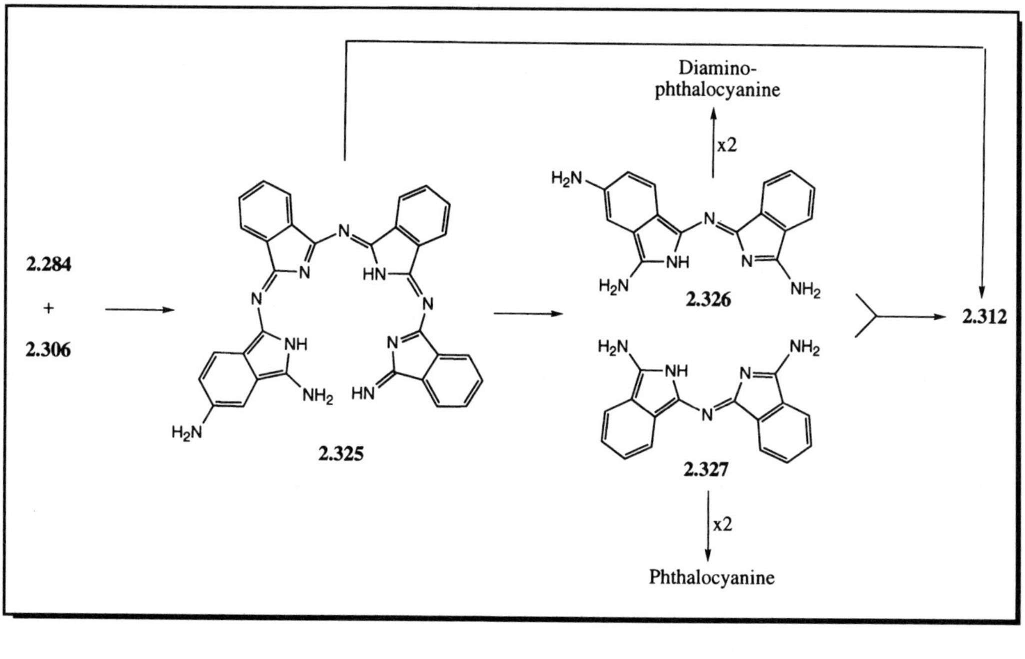

Scheme 2.3.13

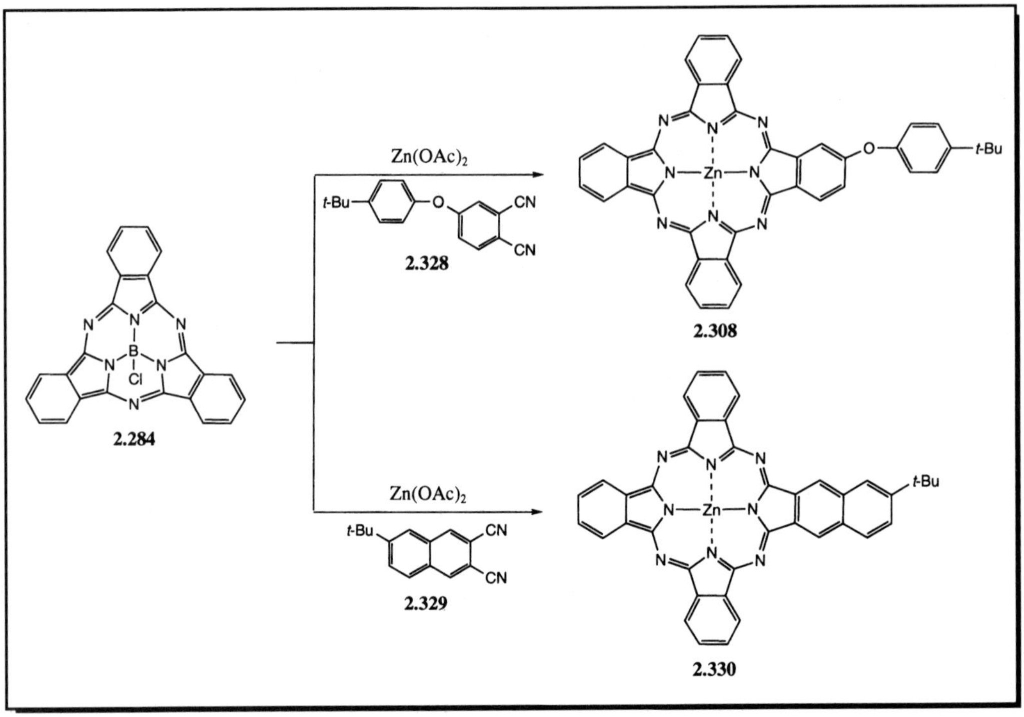

Scheme 2.3.14

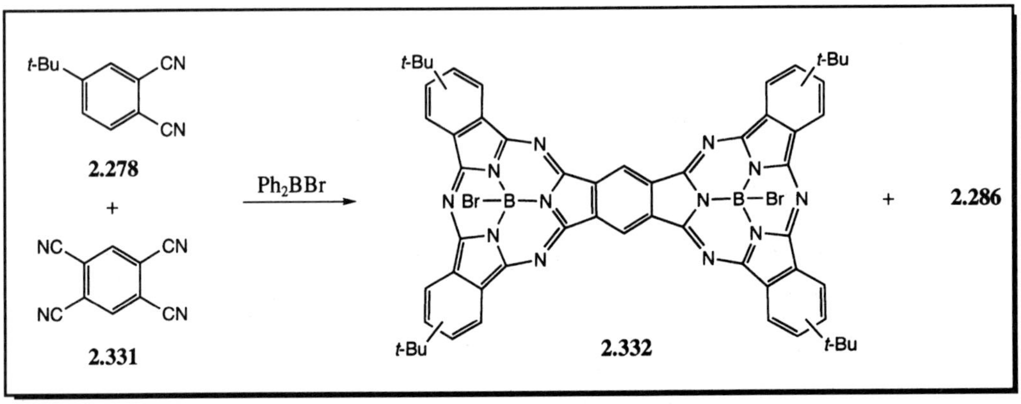

2.3.2   Subtriazaporphyrins

Rauschnabel and Hanack recently reported the first example of what they termed a "subtriaazaporphyrin".[97] This macrocycle was prepared in a fashion analogous to that used to prepare the subphthalocyanines discussed above. Specifically, when (E)-di(4-*tert*-butylphenyl)fumaronitrile **2.333** was heated in 1-chloronaphthalene in the presence of BCl$_3$, the subtriazaporphyrin **2.334** was obtained in *ca.* 15% yield (Scheme 2.3.15). Unfortunately, few details as to the exact nature of this molecule have so far appeared in the literature.

Scheme 2.3.15

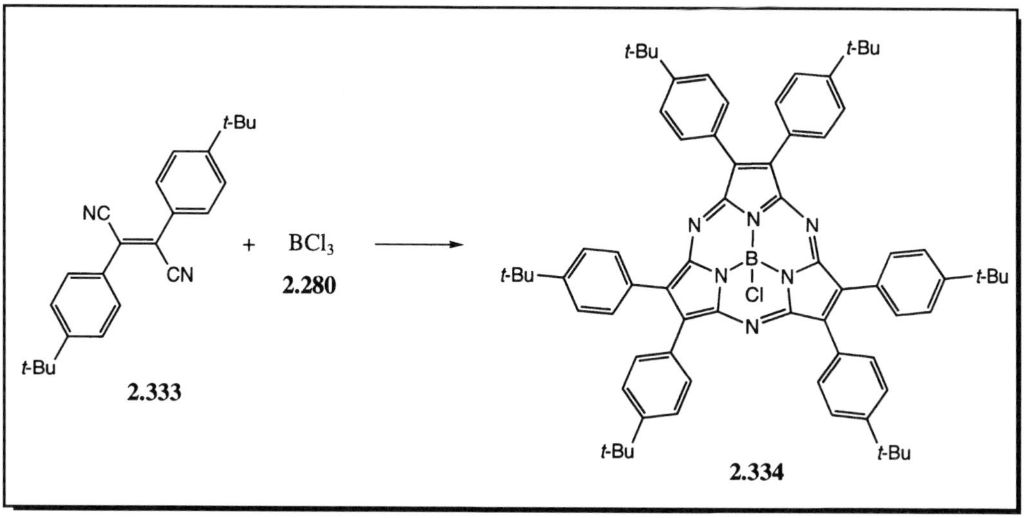

2.3.3   *Heteroatom-bridged [18]Annulenes*

2.3.3.1   Synthesis and General Aromaticity Concerns

In 1964, Badger and coworkers described what was the first representative of a new class of heteroatom-bridged [18]annulenes. This they did with their report on the synthesis of the so-called [18]annulene trisulfide **2.335** (Figure 2.3.4).[108] Macrocycle **2.335** represents just one of ten possible macrocycles (i.e., **2.335–2.344**) of this type that could in theory be derived by formally substituting one, two or all three of the thiophene subunits with pyrrole and/or furan. Indeed, since the initial report of Badger, *et al.*, many of these systems have, in fact, been prepared.[109–118] In this section, a review of the various synthetic strategies used to prepare these macrocycles is provided.

Like much of the work reviewed in this monograph, one of the major stimuli driving the preparation of the bridged [18]annulenes derived from a desire to study further the limits of, and requirements for, aromaticity. In particular, it was considered that these molecules would be analogs of the aromatic [18]annulene **2.345**. However, in the present instance, the structural difference would be the replacement of the six "inner" protons of **2.345** by three oxygen, sulfur, or NH groups, or some combination of the three.

Figure 2.3.4   Structures **2.335–2.345**

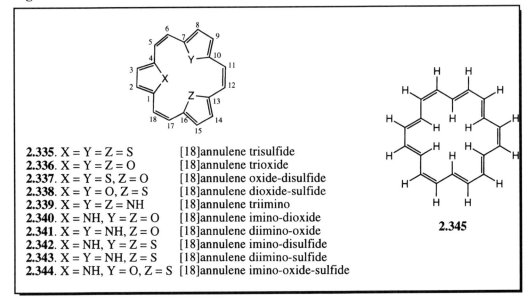

2.335. X = Y = Z = S	[18]annulene trisulfide
2.336. X = Y = Z = O	[18]annulene trioxide
2.337. X = Y = S, Z = O	[18]annulene oxide-disulfide
2.338. X = Y = O, Z = S	[18]annulene dioxide-sulfide
2.339. X = Y = Z = NH	[18]annulene triimino
2.340. X = NH, Y = Z = O	[18]annulene imino-dioxide
2.341. X = Y = NH, Z = O	[18]annulene diimino-oxide
2.342. X = NH, Y = Z = S	[18]annulene imino-disulfide
2.343. X = Y = NH, Z = S	[18]annulene diimino-sulfide
2.344. X = NH, Y = O, Z = S	[18]annulene imino-oxide-sulfide

2.345

Calculations carried out on all of the molecules of this set (**2.335–2.344**) allowed predictions to be made as to which should be aromatic.[119] The prediction for each macrocycle is listed in Table 2.3.1, along with the experimentally determined result (where available). This model, which was based on an index of aromatic stabilization per π-electron, did not take into account unfavorable effects owing to the larger size of NH and S relative to O.[119] In spite of this obvious oversimplification, in those cases where experimental data are available, these calculations have proved to be remarkably accurate.

Prior to this theoretical study, simple CPK modeling studies carried out by Badger, *et al.* on several of these macrocycles indicated that systems derived from two or more thiophene units would likely be non-planar, and thus most likely non-aromatic.[112] If, however, one or fewer thiophene units were present, it was predicted that a planar, aromatic macrocycle would result. It was Badger's intent, therefore, to see whether one could in fact "dial-in" planarity and hence aromaticity as a result of effecting known changes in the within-ring steric factors.

In terms of actual chemistry, the first example of this class of macrocycles to be prepared was, as mentioned above, the [18]annulene-trisulfide derivative **2.335**.[108,111] This macrocycle was obtained by reacting thiophene-2,5-diacetic acid **2.346** with

Table 2.3.1   Predicted vs. Experimental Determination of Aromaticity for [18]Annulene 2.345 and the Bridged [18]Annulenes 2.335–2.344

Macrocycle	Theoretical Prediction	Experimental Evidence	References
[18]Annulene **(2.345)**	Aromatic	Aromatic	128–131
[18]Annulene Trisulfide **(2.335)**	Non-aromatic	Polyenic	108, 111, 117, 118
[18]Annulene Trioxide **(2.336)**	Aromatic	Aromatic	109, 113
[18]Annulene Oxide-disulfide **(2.337)**	Non-aromatic	Polyenic	109, 112
[18]Annulene Dioxide-sulfide **(2.338)**	Aromatic	Aromatic	109, 112
[18]Annulene Triimino **(2.339)**	Non-aromatic	Unknown	—
[18]Annulene Imino-dioxide **(2.340)**	Aromatic	Aromatic	117
[18]Annulene Diimino-oxide **(2.341)**	Non-aromatic	Unknown	—
[18]Annulene Imino-disulfide **(2.342)**	Non-aromatic	Polyenic	115
[18]Annulene Diimino-sulfide **(2.343)**	Non-aromatic	Unknown	—
[18]Annulene Imino-oxide-sulfide **(2.344)**	Aromatic	Aromatic	117

methyl *cis*-α,β-di-(5-formyl-2-thienyl)acrylate **2.349**. This afforded the diacid deriva-tive **2.352a** as a red precipitate. Purification was generally carried out by converting this crude product to the corresponding triester **2.352b** (HCl in methanol, reflux). Saponification (KOH in aqueous ethanol) then yielded the triacid derivative **2.352c**. Decarboxylation, effected by heating to 210–220 °C in the presence of quinoline and copper chromite for 75 minutes then yielded the desired [18]annulene trisulfide **2.335**. The overall yield of this reaction sequence was 14%. Each of the other macrocycles (i.e., **2.353a–c** through **2.357a–d**) and the unsubstituted derivatives derived therefrom

(**2.336**–**2.338** and **2.343**) were prepared in an analogous fashion by Badger *et al.* Here, the relevant starting compounds were **2.346-2.351**[109,110,112–115] (Scheme 2.3.16).

Scheme 2.3.16

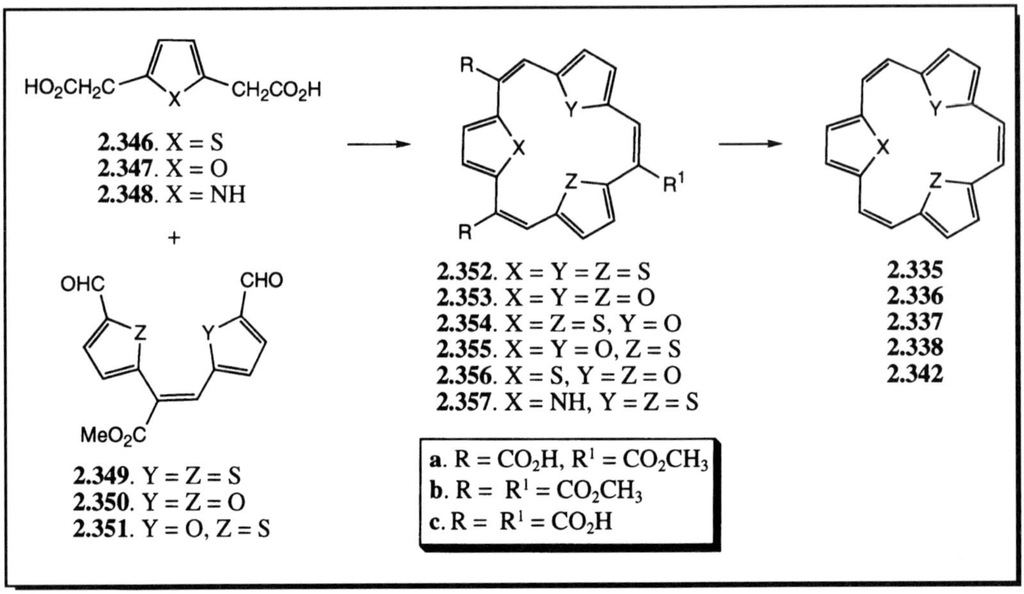

HO$_2$CH$_2$C X CH$_2$CO$_2$H

2.346. X = S
2.347. X = O
2.348. X = NH

+

OHC CHO
Z Y

MeO$_2$C

2.349. Y = Z = S
2.350. Y = Z = O
2.351. Y = O, Z = S

2.352. X = Y = Z = S
2.353. X = Y = Z = O
2.354. X = Z = S, Y = O
2.355. X = Y = O, Z = S
2.356. X = S, Y = Z = O
2.357. X = NH, Y = Z = S

a. R = CO$_2$H, R^1 = CO$_2$CH$_3$
b. R = R^1 = CO$_2$CH$_3$
c. R = R^1 = CO$_2$H

2.335
2.336
2.337
2.338
2.342

Since the first reports of Badger and coworkers, other groups have devised alternative methods to synthesize bridged [18]annulenes. The first of these, reported by Elix in 1969, was a low-yielding synthesis of [18]annulene trioxide **2.336**, prepared *via* a Wittig-type self-condensation of ylid **2.358** (Scheme 2.3.17).[116] Much more recently, an approach to these macrocycles based on a McMurry-type coupling has been reported. It represents a significant simplification in terms of generating this kind of symmetrical molecule.[117,118] In one example of this chemistry, Vogel and coworkers reported the synthesis (in 4% yield) of [18]annulene trioxide **2.336** obtained *via* the low-valent titanium-mediated cyclotrimerization of 2,5-diformyl furan **2.359** (Scheme 2.3.18).[117] Using this same methodology, the [18]annulene tri-sulfide **2.335**[117] was prepared in 5% yield from 2,5-diformyl thiophene **2.360**.¶¶ Surprisingly, Cava and coworkers have achieved yields as high as 38% (based on **2.360**) when preparing this same [18]annulene trisulfide (i.e., **2.335**) while employing similar McMurry-type conditions.)[118] This seeming dichotomy most likely reflects subtle but important differences in the reaction conditions. (For instance, Cava and coworkers add pyridine to the reaction mixture, and use Zn instead of a Zn/Cu couple to prepare the of low-valent titanium reagent.)[118]

¶¶In each of these trimerization reactions (Schemes 2.3.17 and 2.3.18), higher order systems were also isolated. These are discussed in Chapters 4, 6, and 7, where appropriate.

Scheme 2.3.217

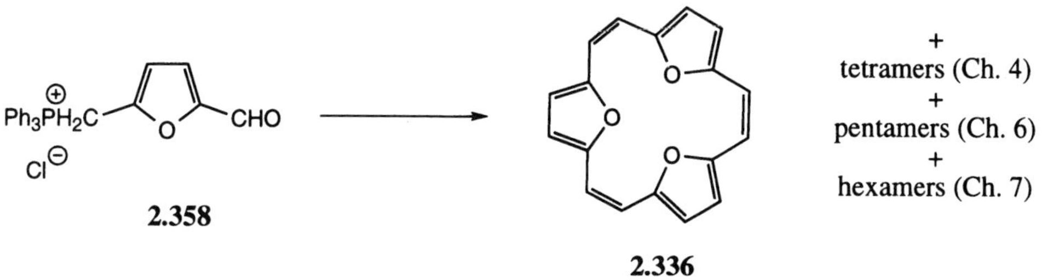

2.358

2.336

+
tetramers (Ch. 4)
+
pentamers (Ch. 6)
+
hexamers (Ch. 7)

Scheme 2.3.218

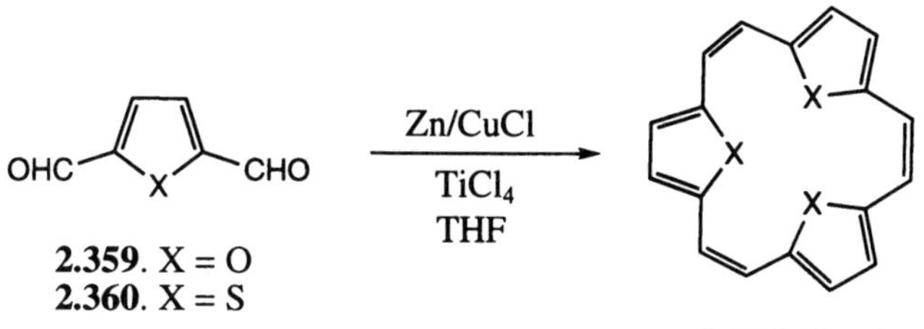

2.359. X = O
2.360. X = S

Zn/CuCl
TiCl$_4$
THF

2.336. X = O
2.335. X = S

As an intellectual extension of the above work, the group of Vogel used the McMurry coupling reaction to prepare the hexaethyl derivative of [18]annulene trisulfide (**2.362**) (Scheme 2.3.19).[117] These researchers also explored the synthesis of unsymmetrically bridged imino [18]annulenes using this same approach. In this context they found, for instance, that mixing 2,5-diformyl furan (**2.359**) with 2,5-diformyl pyrrole (**2.363**) in a 2:1 ratio under reductive cyclocondensation conditions gave rise to the [18]annulene imino-dioxide **2.340** in 4.5% yield (Scheme 2.3.20).[117] In the same way, Vogel and coworkers also succeeded in preparing the [18]annulene imino-disulfide **2.342** in 3.5% yield starting from a 2:1 mixture of **2.360** and **2.363**. Finally, McMurry coupling of a 1:1:1 mixture of 2,5-diformyl furan, 2,5-diformyl thiophene, and 2,5-diformyl pyrrole was found to produce the [18]annulene imino-oxide-sulfide **2.344** as well as the [18]annulene imino-dioxide **2.340** as the main cyclization products (Scheme 2.3.21).

Of the ten conceivable bridged [18]annulenes based on O, S, and NH (i.e., **2.335–2.344**), the three that have thus far eluded synthesis are those containing two or more pyrrole rings (see Table 2.3.1). Whether this reflects some sort of intrinsic instability or "mere" synthetic inaccessibility remains the subject of debate. What is known, however, is that in spite of significant effort,[117,120] a generalized McMurry-type approach fails to yield the [18]annulene triimino macrocycle **2.339**. This is true whether using simple 2,5-diformyl pyrrole (**2.363**) or 3,4-diethyl-2,5-diformyl pyrrole (**2.365**) is used as the starting material (Scheme 2.3.22). Interestingly, reactions of this type lead to low yields (*c*. 1%) of the corresponding cyclic tetramers **2.366** and **2.367** (cf. Chapter 4).

Scheme 2.3.19

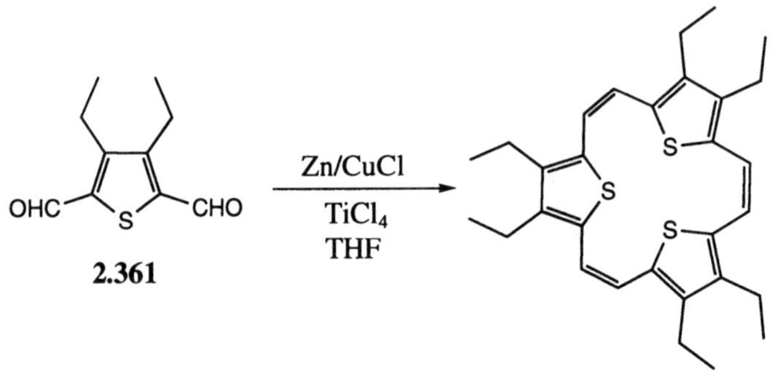

2.361 **2.362**

Scheme 2.3.20

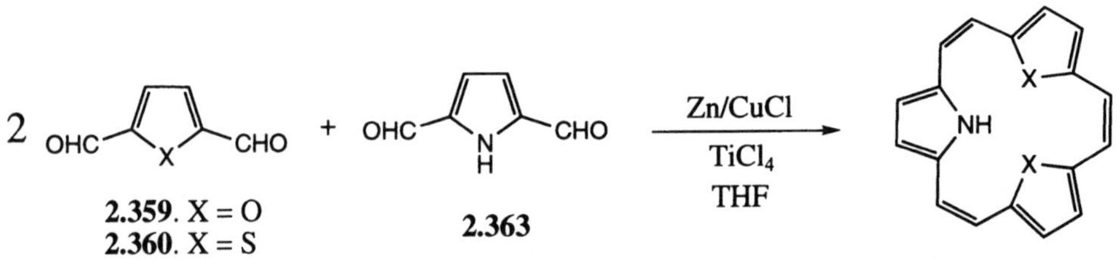

2.359. X = O
2.360. X = S
2.363
2.340. X = O
2.342. X = S

In marked contradistinction to the above, Vogel, *et al.* found that when *N*-methyl- or *N*-ethyl-2,5-diformyl pyrrole (**2.368** or **2.369**) are subjected to McMurry conditions, the cyclic trimers **2.370** and **2.371** could be isolated, albeit in yields of less than 1%.[83] Here, ^{1}H NMR spectroscopic experiments led to the conclusion that these systems were olefinic, rather than aromatic, in nature. These same experiments revealed, however, that **2.370** and **2.371** possess a predominant conformation in which one of the pyrrole rings is rotated outward (as shown in Scheme 2.3.23), presumably, to accommodate the steric bulk of the nitrogen protecting groups. In any event, because of these conformational effects, systems **2.370** and **2.371** have different three-dimensional shapes than the furan- and thiophene-containing [18]annulenes, therefore, direct comparisons cannot be made. Further unfortunate is the fact that all efforts to remove the nitrogen protecting groups (e.g., using LiI in collidine or AlCl₃ in benzene) have so far failed. Thus, the all-pyrrole-derived [18]annulene system **2.339** remains an elusive target and an as-yet unknown entity.

Scheme 2.3.21

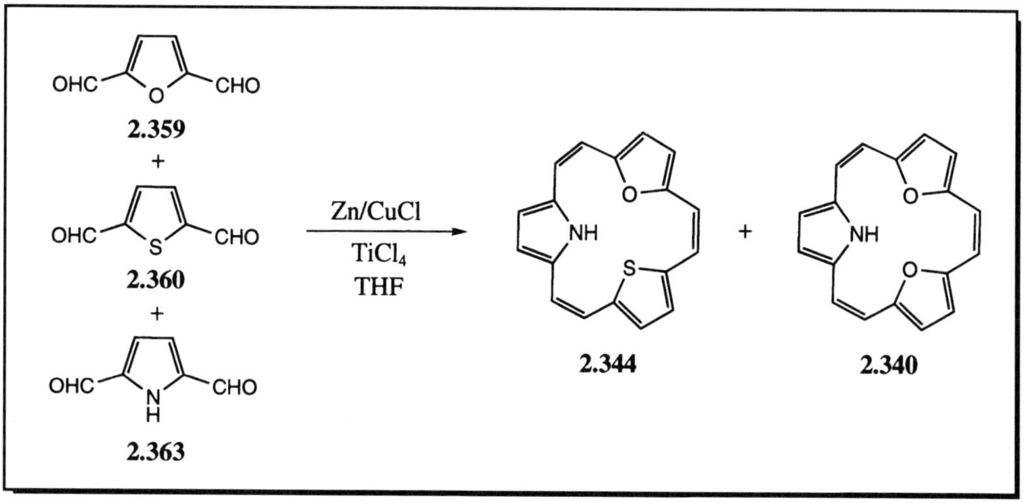

Scheme 2.3.22

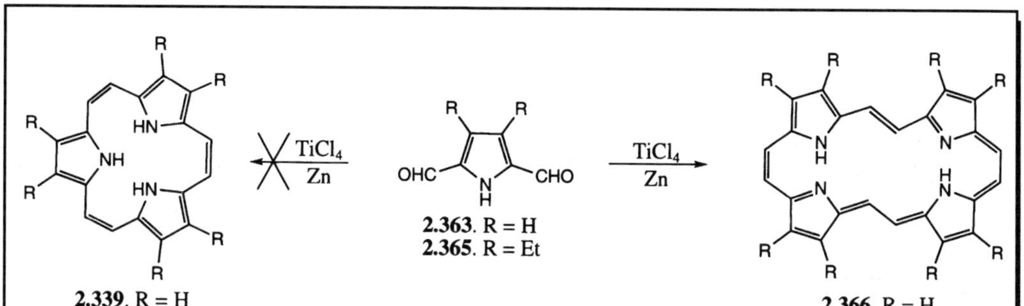

2.3.3.2   **UV-vis Spectral Properties of Bridged [18]Annulenes**

Generally speaking, the UV-vis spectral properties of the heteroatom-bridged [18]annulenes can be divided into two categories. Not surprisingly, the division line separating these categories has its origins in whether the system in question is aromatic or non-aromatic. Typically, the non-aromatic macrocycles all display similar spectral features characterized by relatively weak and very broad absorption bands in the 220–430 nm spectral regions. For instance, in the case of [18]annulene trisulfide **2.335**, the main absorption bands are found at 223 nm, 287 nm, and 376 nm (ε = 19 700, 20 900, and 5200 $M^{-1}cm^{-1}$, respectively, in methanol.)[117] The situation is a little bit different for the presumed-to-be aromatic systems. While these planar, π-electron-delocalized macrocycles exhibit absorption bands in the same general spectral region as the non-aromatic systems, the bands are sharper and significantly more intense. In the particular case of the [18]annulene trioxide **2.336**, for instance, the primary absorbance bands are found at 219 nm, 232 nm, 307 nm, 322 nm, 331 nm, and 404 nm (ε = 19 400, 14 800, 41 600, 146 500, 436 000, and 23 300 $M^{-1}cm^{-1}$, respectively, in methanol).[117] Thus, spectroscopically speaking, there is a real and discernible difference between these two subclasses of contracted porphyrin analog.

Scheme 2.3.23

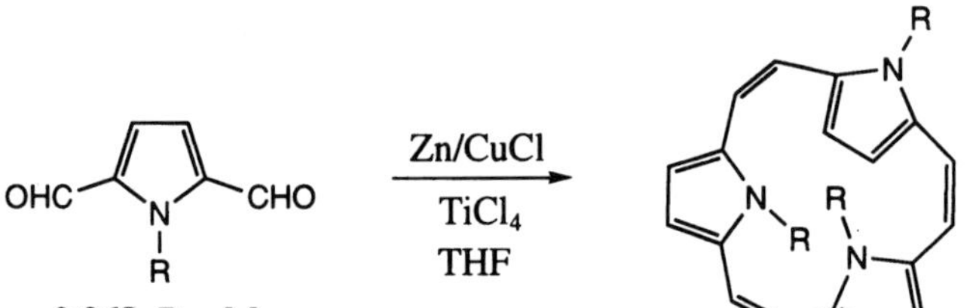

2.368. R = Me
2.369. R = Et

2.370. R = Me
2.371. R = Et

2.3.3.3   Calix [3]indoles

A class of macrocyclic system that bears some structural similarity to the still-hypothetical [18]annulene triimino species **2.339** has recently been described by Black, *et al.*[121–124] The first representative of this new family, the so-called calix [3]indole **2.376**, was prepared in 12% yield *via* the POCl₃-catalyzed condensation between indole **2.372** and benzaldehyde (**2.154**) (Scheme 2.3.24).[121] The related symmetrical *meso*-aryl systems **2.377–2.379** were prepared in analogous fashion in yields of 83%, 10%, and 81%, respectively.

Later, an alternative synthesis of similar symmetrical calix [3]indoles was reported.[122] In this work, also from Black's group, it was disclosed that macrocycles such as **2.386–2.391** could be prepared in *ca.* 60% yield *via* the acetic acid-catalyzed trimerization of the corresponding 7-hydroxymethyl indole derivatives **2.380–2.385** (Scheme 2.3.25). Calix [3]indole **2.390** can also be obtained by treatment of the 2-hydroxymethyl indole compound **2.392** with p-TsOH in dichloromethane. In 1996, Black and coworkers reported yet a third, but related, means of synthesizing symmetrical calix [3]indoles.[124] This approach centers around carrying out the acid-catalyzed cyclotrimerization of either 7-(α-hydroxyamide)indole **2.393** or 2-(α-hydroxyamide)indole **2.397**. In either case it is the calix [3]indole **2.395** that is produced (Scheme 2.3.26). Using similar chemistry, the *t*-butylamide-substituted macrocycle **2.396** could be prepared from either the functionalized hydroxymethyl-indole **2.394** or its isomer **2.398**.

Scheme 2.3.24

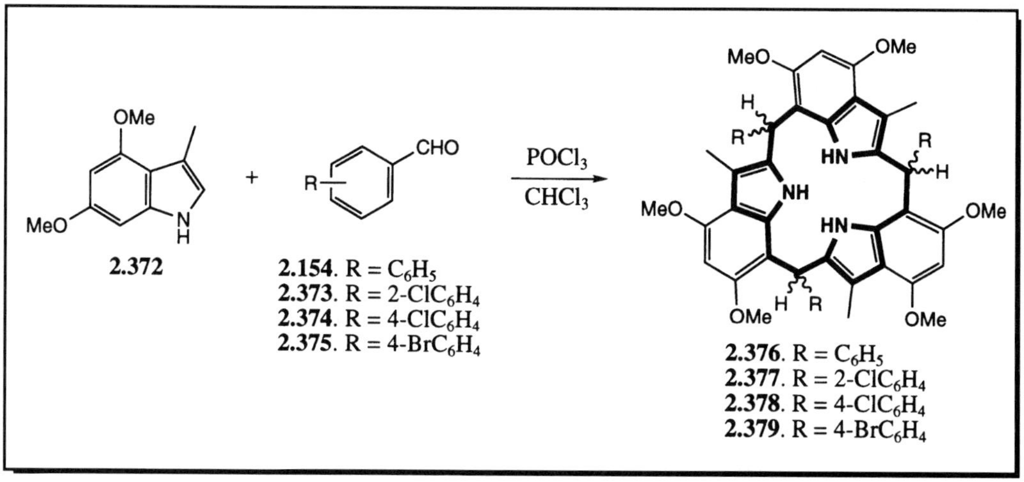
OMe
MeO
N
H
2.372
+
R
CHO
2.154. R = C_6H_5
2.373. R = 2-ClC_6H_4
2.374. R = 4-ClC_6H_4
2.375. R = 4-BrC_6H_4
$POCl_3$
$CHCl_3$
MeO
OMe
H
R
NH
HN
HN
OMe
OMe
H R
2.376. R = C_6H_5
2.377. R = 2-ClC_6H_4
2.378. R = 4-ClC_6H_4
2.379. R = 4-BrC_6H_4

Scheme 2.3.25

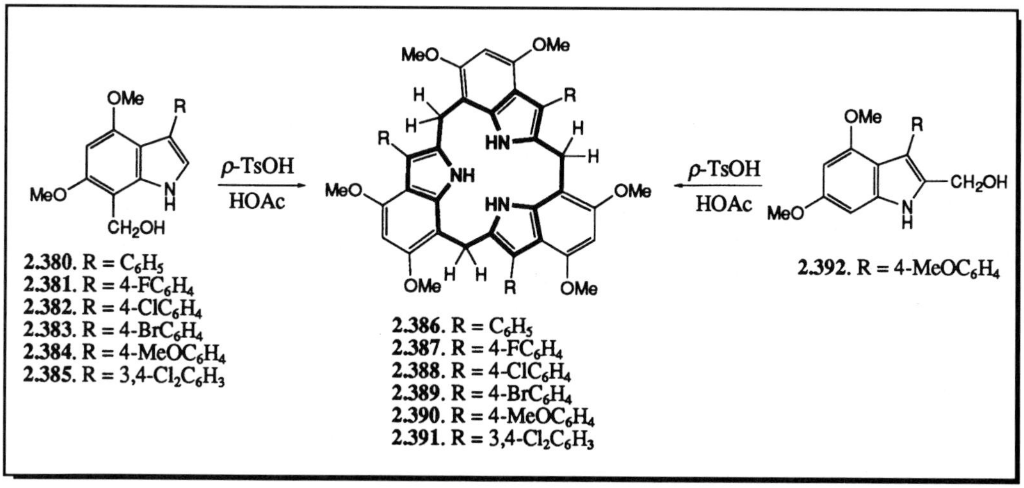
OMe
R
MeO
N
H
CH_2OH
2.380. R = C_6H_5
2.381. R = 4-FC_6H_4
2.382. R = 4-ClC_6H_4
2.383. R = 4-BrC_6H_4
2.384. R = 4-$MeOC_6H_4$
2.385. R = 3,4-$Cl_2C_6H_3$
p-TsOH
HOAc
MeO
OMe
H
R
NH
HN
HN
OMe
2.386. R = C_6H_5
2.387. R = 4-FC_6H_4
2.388. R = 4-ClC_6H_4
2.389. R = 4-BrC_6H_4
2.390. R = 4-$MeOC_6H_4$
2.391. R = 3,4-$Cl_2C_6H_3$
p-TsOH
HOAc
CH_2OH
2.392. R = 4-$MeOC_6H_4$

The above calix [3]indoles **2.376–2.379**, **2.386–2.391** and **2.395–2.396** all possess the basic substructure contained in the hypothetical [18]annulene triimino **2.339** (i.e., three "pyrrole" subunits and six *meso* carbons arranged in a (2.2.2) fashion).

However, these new systems contain three saturated *meso* carbon atoms in their macrocyclic framework. Thus, any kind of direct comparison between these systems and the bridged [18]annulenes **2.335–2.344** is not justified. Unfortunately, access to fully conjugated systems from any of these macrocycles appears rather unlikely because they contain indole subunits instead of pyrroles. Indeed, perhaps for this very reason, no conjugated systems of this type have been reported in the literature.

Scheme 2.3.26

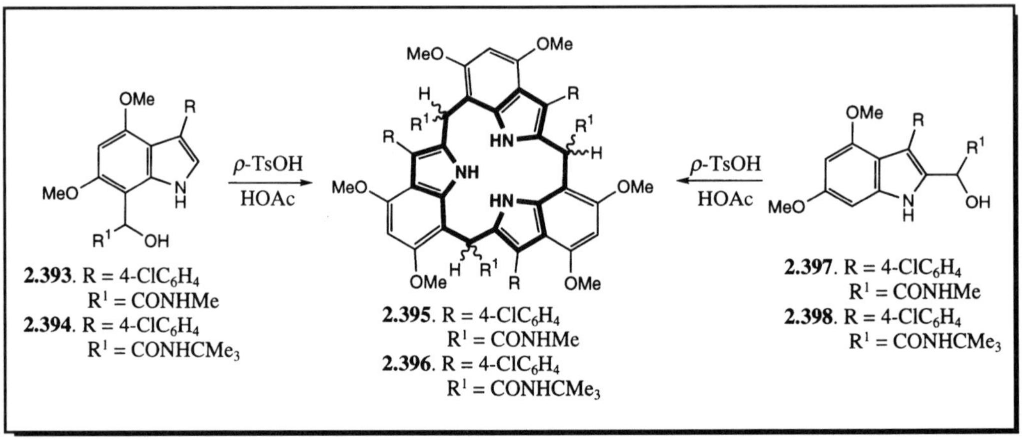
OMe
R
MeO
N
H
R1
OH
p-TsOH
HOAc
MeO
OMe
H
R1
R
R
R1
HN
NH
HN
H
MeO
OMe
OMe
H
R1
R
OMe
p-TsOH
HOAc
OMe
R
R1
MeO
N
H
OH
2.393. R = 4-ClC6H4
R1 = CONHMe
2.394. R = 4-ClC6H4
R1 = CONHCMe3
2.395. R = 4-ClC6H4
R1 = CONHMe
2.396. R = 4-ClC6H4
R1 = CONHCMe3
2.397. R = 4-ClC6H4
R1 = CONHMe
2.398. R = 4-ClC6H4
R1 = CONHCMe3

2.3.4 *Other Systems of Interest*

A handful of other heteroatom-bridged annulene-type contracted porphyrins have appeared in the literature. The first of these were reported in 1973 by Cresp and Sargent and consisted of the [17]annulenone derivatives **2.400a–c**.[125,126] These were prepared using a procedure similar to that used by Badger, *et al.*[108–115] to obtain several of the bridged [18]annulene derivatives discussed above. Specifically, the 2,2′-bis-5 formylfuryl ketone **2.399** was condensed with furan-2,5-diacetic acid **2.347**. Esterification using methanolic HCl afforded the methoxycarbonyl-substituted annulenone **2.400a** (Scheme 2.3.27). Saponification in the presence of KOH then yielded the diacid **2.400b**. This diacid, in turn, was decarboxylated by heating in quinoline in the presence of copper chromite so as to yield the unsubstituted annulenone **2.400c**.[126]

Cresp and Sargent also reported the preparation of the trifuran-containing annulenone **2.400c**, a compound they prepared *via* two different Wittig-type pathways. The first of these involved the LiOEt catalyzed reaction between the 2,2′-bis-5 formylfuryl ketone **2.399** and the furan-2,5-bis(triphenylphosphonium chloride) **2.401**. This produced the annulenone **2.400c** in 8% yield (Scheme 2.3.28).[126] The second approach involved simply switching the functionality present on the two precursors. Thus, the 2,5-bis(triphenylphosphonium chloride) species **2.405** was reacted with furan-2,5-dicarbaldehyde **2.359** yield to afford **2.400c** in 9.6% yield.[126]

Scheme 2.3.27

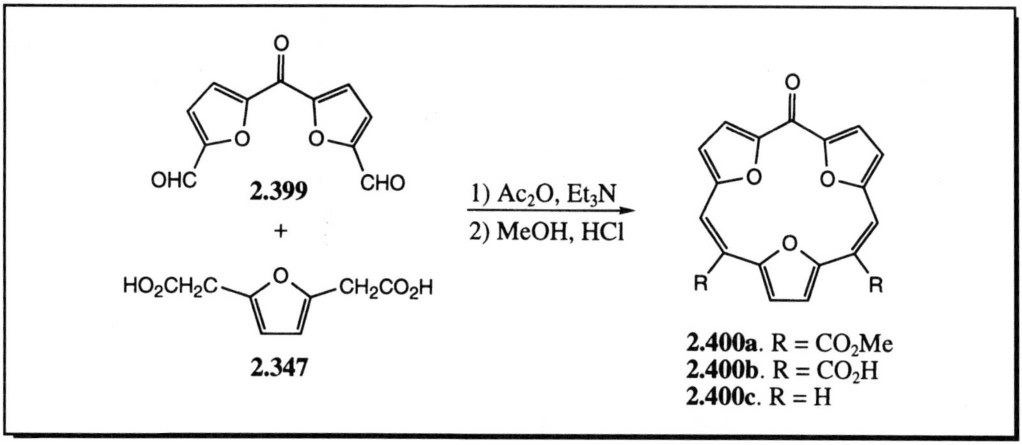

2.400a. R = CO$_2$Me
2.400b. R = CO$_2$H
2.400c. R = H

Scheme 2.3.28

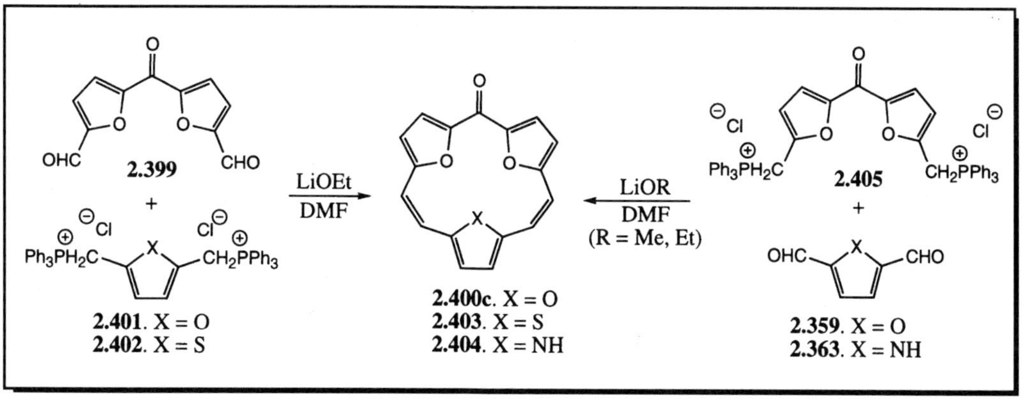

2.401. X = O
2.402. X = S

2.400c. X = O
2.403. X = S
2.404. X = NH

2.359. X = O
2.363. X = NH

Using chemistry analogous to the above, Cresp and Sargent also prepared the thiophene-containing and the pyrrole-containing annulenones **2.403** and **2.404**.[126] The first of these was obtained by reacting the dialdehyde **2.399** with thiophene-2,5-bis(triphenylphosphonium chloride) **2.402**; this afforded **2.403** in 8.5% yield (Scheme 2.3.41). The second product, the pyrrole-containing annulenone **2.404**, was prepared from the bis(phosphonium) salt **2.405** and 2,5-diformyl pyrrole **2.363** in 13.7% yield.[126] As part of this study, Cresp and Sargent found that the epimino-diepoxy [17]annulenone **2.404** is a planar paratropic 16 π-electron system, just as is the triepoxy [17]annulenone **2.400c**. On the other hand, the sulfur-containing epithio-diepoxy [17]annulenone **2.403** was found not only to be nonplanar, but also to exhibit no appreciable paratropicity.[126] These findings are thus consistent with the results Badger, *et al.* had found previously while studying their bridged [18]annulenes.[108-115]

In 1973 Sargent and Cresp also reported syntheses of the homoannulenes **2.406–2.408**. These were prepared in 70–90% yield *via* a reductive deoxygenation procedure that involves treating the corresponding annulenone (**2.400c**, **2.403**, and **2.404**) with $LiAlH_4$ and aluminum chloride (Scheme 2.3.29).[126] The methoxy-substituted homoannulenes **2.409** and **2.410** were also prepared *via* reduction of the corresponding annulenone (i.e., **2.400c** and **2.403**) by a sequence involving first reduction with either $LiAlH_4$ or $NaBH_4$, and then methylation with methyl iodide (Scheme 2.3.30).

Scheme 2.3.29

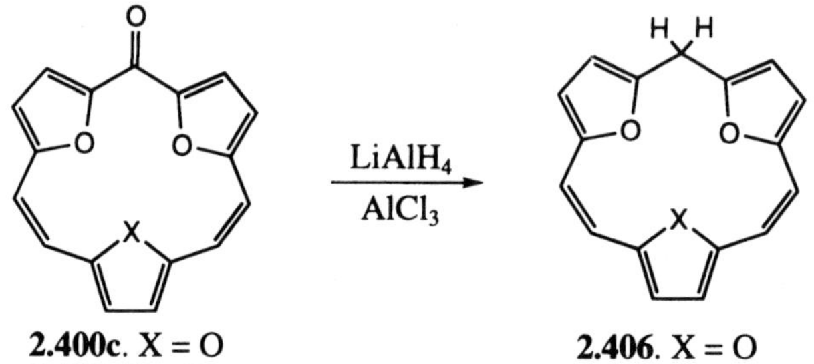

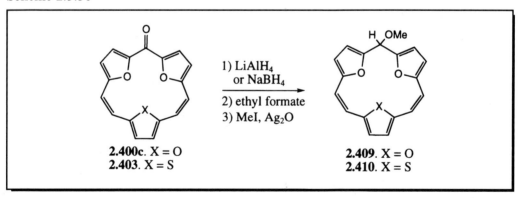

2.400c. X = O
2.403. X = S
2.404. X = NH

2.406. X = O
2.407. X = S
2.408. X = NH

Scheme 2.3.30

2.400c. X = O
2.403. X = S

2.409. X = O
2.410. X = S

Other systems containing three heteropentacycles include the sulfur-bridged macrocyclic system **2.411** (diepoxyepithiacycloheptadecin). This was prepared in 21.3% yield *via* a multistep synthesis that involved first reacting 5,5′-dithiodi-2-furaldehyde **2.43** with thiophene-2,5-diacetic acid **2.346**, followed by esterification with methanolic HCl (Scheme 2.3.31).[125] Saponification of the resulting diester (**2.411a**) using KOH then afforded the diacid derivative **2.411b**. Finally, this latter species was decarboxylated using copper chromite in quinoline to afford the unsubstituted macrocycle **2.411c** in 18.7% yield (based on **2.411a**). The triepoxy-thiacyclohepta-decin derivatives **2.412a–c** could also be prepared using a similar approach.[125] By contrast, the thiacycloheptadecin derivatives **2.411c** and **2.412c** proved easier to prepare *via* Wittig chemistry. Specifically, they were obtained by coupling the difuran derivative **2.43** to the appropriate bis(triphenyl phosphonium chloride) salt **2.401** or **2.402** in the presence of lithium ethoxide in DMF (Scheme 2.3.32).[125]

Scheme 2.3.31

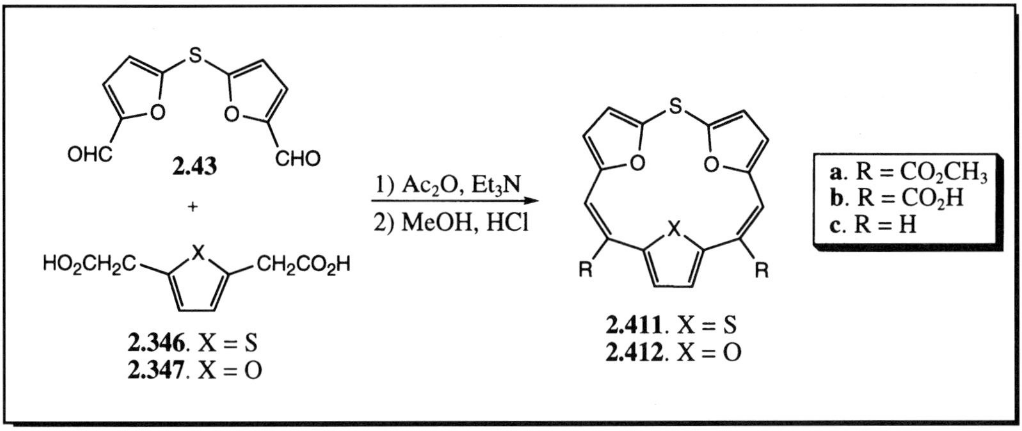

Scheme 2.3.32

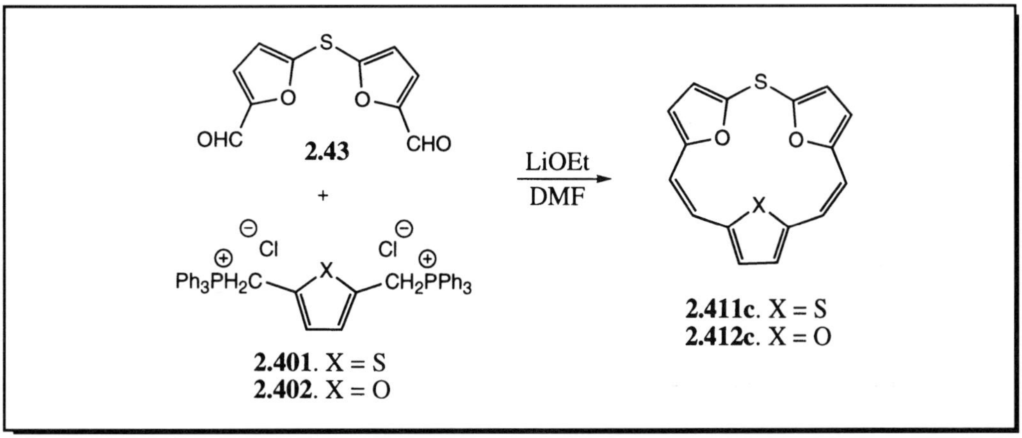

The sulfur-bridged species **2.411c** and **2.412c** may formally be considered as being 18 π-electron macrocycles. However, it was rationalized *via* Stuart–Briegleb model analysis that the triepoxy macrocycle **2.412c** could adopt a planar conforma-tion, whereas the thiophene-containing systems could not.[125] Thus it was expected that the former (only) would display characteristics consistent with aromaticity. To test these predictions, NMR spectroscopic data were collected for these compounds and then compared to those obtained for **2.406** and **2.407**.[125] Surprisingly, as compared to the homoannulene systems **2.406** and **2.407** (species that, by virtue of containing an sp^3-hybridized center, cannot sustain a diamagnetic ring current), it was determined that neither **2.411c** nor **2.412c** is aromatic. Since the 16 π-electron triepoxy [17]annulenone **2.400c** showed marked paratropicity, and the 18 π-electron [18]annulene trioxide **2.336** was clearly diatropic, it was concluded, on the basis of the above findings, that the ability of these thiaannulenes to sustain a ring current is limited. Certainly, it is considerably less than that of analogous annulenes (or annulenones) of comparable ring size.

Note in Proof

Subsequent to the submission of this manuscript, a communication detailing the interpretation of copper (II) and nickel (II) corrole as II radical cations was published by Vogel *et al.*[132] Another report in the calix[3]-indole area also appeared.[133]

2.4   References

1. Hodgkin, D. C.; Johnson, A. W.; Todd, A. R. *Chem. Soc.* (London) *Spec. Publ.* No. 3 **1955**, 109–123.
2. Hodgkin, D. C. *Bull. Soc. Franc. Mineral. Cryst.* **1955**, *78*, 106–115.
3. Hodgkin, D. C.; Kamper, J.; Lindsey, J.; Mackay, M.; Pickworth, J.; Robertson, H. J.; Shoemaker, C. B.; White, J. G.; Prosen, R. J.; Trueblood, K. N. *Proc. Royal Soc., Ser. A* **1957**, *242*, 228–263.
4. Bonnett, R. *Chem. Rev.* **1963**, *63*, 573–605.
5. Eschenmoser, A. *Angew. Chem. Int. Ed. Eng.* **1988**, *27*, 5–39.
6. Scott, A. I. *Angew. Chem. Int. Ed. Eng.* **1993**, *32*, 1223–1243.
7. Battersby, A. R. *Science*, **1994**, *264*, 1551–1557.
8. Blanche, F.; Cameron, B.; Crouzet, J.; Debussche, L.; Thibaut, D.; Vuilhorgne, M.; Leeper, F. J.; Battersby, A. R. *Angew. Chem. Int. Ed. Eng.* **1995**, *34*, 383–411.
9. Jackson, A. H. In *The Porphyrins*, Vol. 1; Dolphin, D., Ed.; Academic Press: New York, 1978; pp. 341–363.
10. Grigg, R. In *The Porphyrins*, Vol. 2; Dolphin, D., Ed.; Academic Press: New York, 1978; pp. 328–351.
11. Genokhova, N. S.; Melent'eva, T. A.; Berezovskii, V. M. *Russ. Chem. Rev.* **1980**, *49*, 1056–1067.
12. Melent'eva, T. A. *Russ. Chem. Rev.* **1983**, *52*, 641–661.
13. Dolphin, D.; Harris, R. L. N.; Huppatz, J. L.; Johnson, A. W.; Kay, I. T. *J. Chem. Soc. (C)* **1966**, 30–40.

14. Grigg, R.; Johnson, A. W.; Shelton, K. W. *J. Chem. Soc. (C)* **1968**, 1291–1296.

15. Grigg, R.; Johnson, A. W.; Kenyon, R.; Math, V. B.; Richardson, K. *J. Chem. Soc. (C)* **1969**, 176–182.

16. Dicker, I. D.; Grigg, R.; Johnson, A. W.; Pinnock, H.; Richardson, K.; van den Broek, P. *J. Chem. Soc. (C)* **1971**, 536–547.

17. Rasetti, V.; Pfaltz, A.; Kratky, C.; Eschenmoser, A. *Proc. Natl Acad. Sci. USA* **1981**, *78*, 16–19.

18. Kräutler, B.; Konrat, R.; Stupperich, E.; Färber, G.; Gruber, K.; Kratky, C. *Inorg. Chem.* **1994**, *33*, 4128–4139.

19. Floriani, C. *J. Chem. Soc., Chem. Commun.* **1996**, 1257–1263.

20. Spivey, A. C.; Capretta, A.; Frampton, C. S.; Leeper, F. J.; Battersby, A. R. *J., Chem. Soc. Chem. Commun.* **1995**, 1789–1790.

21. Laporte, F.; Mercier, F.; Ricard, L.; Mathey, F. *J. Am. Chem. Soc.* **1994**, *116*, 3306–3311.

22. Bartczak, T. J. *Acta Crystallogr., Sect. A* **1978**, *34*, S127.

23. Sturrock, E. D.; Bull, J. R.; Kirsch, R. E.; Pandey, R. E.; Senge, M. O.; Smith, K. M. *J. Chem. Soc., Chem. Commun.* **1993**, 872–874.

24. Hush, N. S.; Dyke, J. M.; Williams, M. L.; Woolsey, I. S. *Mol. Phys.* **1969**, *17*, 559–560.

25. Dyke, J. M.; Hush, N. S.; Williams, M. L.; Woolsey, I. S. *Mol. Phys.* **1971**, *20*, 1149–1152.

26. Hush, N. S.; Woolsey, I. S. *J. Chem. Soc., Dalton Trans.* **1974**, 24–34.

27. Johnson, A. W.; Kay, I. T. *Proc. Chem. Soc.* **1964**, 89–90.

28. Harrison, H. R.; Hodder, O. J. R.; Hodgkin, D. C. *J. Chem. Soc. (B)* **1971**, 640–645.

29. Anderson, B. F.; Bartczak, T. J.; Hodgkin, D. C. *J. Chem. Soc., Perkin Trans. II*, **1974**, 977–980.

30. Broadhurst, M. J.; Grigg, R.; Shelton, G.; Johnson, A. W. *J. Chem. Soc., Perkin Trans. I*, **1972**, 143–151.

31. Johnson, A. W. In *Porphyrins and Metalloporphyrins*; Smith, K. M., Ed.; Elsevier: Amsterdam, 1976; pp. 729–754.

32. Buchler, J. W. In *The Porphyrins*, Vol. 1; Dolphin, D., Ed.; Academic Press: New York, 1978; pp. 390–474.

33. Johnson, A. W.; Price, R. *J. Chem. Soc.* **1960**, 1649–1653.

34. Johnson, A. W.; Kay, I. T. *Proc. Chem. Soc.* **1961**, 168–169.

35. Johnson, A. W.; Kay, I. T.; Rodrigo, R. *J. Chem. Soc.* **1963**, 2336–2342.

36. Johnson, A. W.; Kay, I. T. *Proc. Roy. Soc. Ser. A* **1965**, *288*, 334–341.

37. Broadhurst, M. J.; Grigg, R.; Johnson, A. W. *J. Chem. Soc., Chem. Commun.* **1970**, 807–809.

38. Broadhurst, M. J.; Grigg, R.; Johnson, A. W. *J. Chem. Soc., Perkin Trans. I* **1972**, 1124–1135.

39. Broadhurst, M. J.; Grigg, R.; Johnson, A. W. *J. Chem. Soc., Chem. Commun.* **1969**, 23–24.

40. Abeysekera, A. M.; Grigg, R.; Trocha-Grimshaw, J.; King, T. J. *J. Chem. Soc., Perkin Trans. I* **1979**, 2184–2192.
41. Behrens, F., Ph.D. Dissertation; University of Cologne, Germany, **1994**.
42. Dörr, J., Ph.D. Dissertation; University of Cologne, Germany, **1994**.
43. Johnson, A. W.; Kay, I. T. *J. Chem. Soc.* **1965**, 1620–1629.
44. Harris, R. L. N.; Johnson, A. W.; Kay, I. T. *J. Chem. Soc. (C)* **1966**, 22–29.
45. Grigg, R.; Johnson, A. W.; Shelton, G. *J. Chem. Soc. (C)* **1971**, 2287–2294.
46. Paolesse, R.; Pandey, R. K.; Forsyth, T. P.; Jaquinod, L.; Gerzevske, K. R.; Nurco, D. J.; Senge, M. O.; Licoccia, S.; Boschi, T.; Smith, K. M. *J. Am. Chem. Soc.* **1996**, *118*, 3869–3882.
47. Will, S., Ph.D. Dissertation; University of Cologne, Germany, **1996**.
48. Dolphin, D.; Johnson, A. W.; Leng, J.; van den Broek, P. *J. Chem. Soc. (C)* **1966**, 880–884.
49. Engel, J.; Gossauer, A. *J. Chem. Soc., Chem. Commun.* **1975**, 713–714.
50. Pandey, R. K.; Zhou, H.; Gerzevske, K.; Smith, K. M. *J. Chem. Soc., Chem. Commun.* **1992**, 183–185.
51. Murakami, Y.; Matsuda, Y.; Sakata, K.; Yamada, S.; Tanaka, Y.; Aoyama, Y. *Bull. Chem. Soc. Jpn.* **1981**, *54*, 163–169.
52. Boschi, T.; Licoccia, S.; Paolesse, R.; Tagliatesta, P.; Azarnia, M. *J. Chem. Soc., Dalton Trans.* **1990**, 463–468.
53. Grigg, R.; Johnson, A. W.; Shelton, G. *J. Chem. Soc., Chem. Commun.* **1968**, 1151–1152.
54. Johnson, A. W. *Pure Appl. Chem.* **1970**, *23*, 375–389.
55. Grigg, R.; Johnson, A. W.; Shelton, G. *Liebigs Ann. Chem.* **1971**, *746*, 32–53.
56. Hush, N. S.; Dyke, J. M.; Williams, M. L.; Woolsey, I. S. *J. Chem. Soc. Dalton Trans.* **1974**, 395–399.
57. Conlon, M.; Johnson, A. W.; Overend, W. R.; Rajapaksa, D.; Elson, C. M. *J. Chem. Soc., Perkin Trans. I*, **1973**, 2281–2288.
58. Hitchcock, P. B.; McLaughlin, G. M. *J. Chem. Soc. Dalton Trans.* **1976**, 1927–1930.
59. Murakami, Y.; Yamada, S.; Matsuda, Y.; Sakata, K. *Bull. Chem. Soc. Jpn.* **1978**, *51*, 123–129.
60. Boschi, T.; Licoccia, S.; Paolesse, R.; Tagliatesta, P. *Inorg. Chim. Acta* **1988**, *141*, 169–171
61. Licoccia, S.; Paci, M.; Paolesse, R.; Boschi, T. *J. Chem. Soc. Dalton Trans.* **1991**, 461–466.
62. Shedbalkar, V. P.; Dugad, L. B.; Mazumdar, S.; Mitra, S. *Inorg. Chim. Acta* **1988**, *148*, 17–20.
63. Paolesse, R.; Licoccia, S.; Boschi, T. *Inorg. Chim. Acta* **1990**, *178*, 9–12.
64. Paolesse, R.; Licoccia, S.; Fanciullo, M.; Morgante, E.; Boschi, T. *Inorg. Chim. Acta* **1993**, *203*, 107–114.
65. Paolesse, R.; Licoccia, S.; Bandoli, G.; Dolmella, A.; Boschi, T. *Inorg. Chem.* **1994**, *33*, 1171–1176.
66. Vogel, E.; Will, S.; Tilling, A. S.; Neumann, L. Lex, J.; Bill, E.; Trautwein, A. X.; Wieghardt, K. *Angew. Chem. Int. Ed. Engl.* **1994**, *33*, 731–735.

67. Grigg, R.; Trocha-Grimshaw, J.; Viswanatha, V. *Tetrahedron Lett.* **1976**, 289–292.

68. Abeysekera, A. M.; Grigg, R.; Trocha-Grimshaw, J.; Viswanatha, V. *J. Chem. Soc. Perkin Trans. I* **1977**, 36–44.

69. Murakami, Y.; Aoyama, Y.; Hayashida, M. *J. Chem. Soc., Chem. Commun.* **1980**, 501–502.

70. Murakami, Y.; Matsuda, Y.; Yamada, S. *Chem. Lett.* **1977**, 689–692.

71. Matsuda, Y.; Yamada, S.; Murakami, Y. *Inorg. Chim. Acta* **1980**, *44*, L309–L311.

72. Murakami, Y.; Matsuda, Y.; Yamada, S. *J. Chem. Soc. Dalton Trans.* **1981**, 855–861.

73. Fajer, J.; Borg, D. C.; Forman, A.; Dolphin, D.; Felton, R. H. *J. Am. Chem. Soc.* **1970**, *92*, 3451–3459.

74. Fuhrhop, J.-H.; Mauzerall, D. *J. Am. Chem. Soc.* **1969**, *91*, 196–198.

75. Ozawa, S.; Watanabe, Y.; Morishima, I. *Inorg. Chem.* **1992**, *31*, 4042–4043.

76. Renner, M. W.; Barkigia, K. M.; Zhang, Y.; Medforth, C. J.; Smith, K. M.; Fajer, J. *J. Am. Chem. Soc.* **1994**, *116*, 8582–8592.

77. Kadish, K. M.; Koh, W.; Tagliatesta, P.; Sazou, D.; Paolesse, R.; Licoccia, S.; Boschi, T. *Inorg. Chem.* **1992**, *31*, 2305–2313.

78. Similar redox properties were found for Rh(III) octamethylcorrole. See reference 77.

79. Autret, M.; Will, S.; Caemelbecke, E. V.; Lex, J.; Gisselbrecht, J.-P.; Gross, M.; Vogel, E.; Kadish, K. M. *J. Am. Chem. Soc.* **1994**, *116*, 9141–9149.

80. Broadhurst, M. J.; Grigg, R.; Shelton, G.; Johnson, A. W. *J. Chem. Soc., Chem. Commun.* **1970**, 231–233.

81. Grigg, R.; King, T. J.; Shelton, G. *J. Chem. Soc. Chem. Commun.* **1970**, 56.

82. Abeysekera, A. M.; Grigg, R.; Trocha-Grimshaw, J.; Viswanatha, V. *Tetrahedron Lett.* **1976**, 3189–3192.

83. Euteneuer, P., Ph.D. Dissertation; University of Cologne, Germany, **1995**.

84. Grigg, R.; Hamilton, R. J.; Jozefowicz, M. L.; Rochester, C. H.; Terrell, R. J.; Wickwar, H. *J. Chem. Soc. Perkin Trans. II*, **1973**, 407–413.

85. Engel, J.; Gossauer, A.; Johnson, A. W. *J. Chem. Soc. Perkin Trans. I* **1978**, 871–875.

86. Will, S.; Rahbar, A.; Schmickler, H.; Lex, J.; Vogel, E. *Angew. Chem. Int. Ed. Engl.* **1990**, *29*, 1390–1393.

87. Vogel, E. *Pure Appl. Chem.* **1993**, *65*, 143–152.

88. Neumann, K. L., Ph.D. Dissertation; University of Cologne, Germany, **1994**.

89. Vogel, E. *J. Heterocyclic Chem.* **1996**, *33*, 1461–1487.

90. *Phthalocyanine Compounds*, Moser, F. H. and Thomas, A. L., Eds.; Reinhold Publishing: New York, 1963.

91. Lever, A. B. P. In *Advances in Inorganic Chemistry and Radiochemistry*, Vol. 7; Emeléus, H. J. and Sharp, A. G., Eds.; Academic Press: New York, 1965; pp. 27–114.

92. *The Phthalocyanines*, Vols I and II; Moser, F. H. and Thomas, A. L., Eds.; CRC Press, Inc: Boca Raton, 1983.

93. *Phthalocyanines*, Vols 1–3; Leznoff, C. C. and Lever, A. B. P., Eds.; VCH: New York, 1989.
94. Meller, A.; Ossko, A. *Monats. Chem.* **1972**, *103*, 150–155.
95. Kobayashi, N.; Kondo, R.; Nakajima, S.; Osa, T. *J. Am. Chem. Soc.* **1990**, *112*, 9640–9641.
96. Hanack, M.; Geyer, M. *J. Chem. Soc., Chem. Commun.* **1994**, 2253–2254.
97. Rauschnabel, J.; Hanack, M. *Tetrahedron Lett.* **1995**, *36*, 1629–1632.
98. Sastre, A.; Torres, T.; Díaz-García, M. A.; Agulló-López, F.; Dhenaut, C.; Brasselet, S.; Ledoux, I.; Zyss, J. *J. Am. Chem. Soc.* **1996**, *118*, 2746–2747.
99. Dabak, S.; Gül, A.; Bekaroglu, Ö. *Chem. Ber.* **1994**, *127*, 2009–2012.
100. Díaz-García, M. A.; Agulló-López, F.; Sastre, A.; Torres, T.; Torruellas, W. E.; Stegeman, G. I. *J. Phys. Chem.* **1995**, *99*, 14988–14991.
101. Kobayashi, N. *J. Chem. Soc. Chem. Commun.* **1991**, 1203–1205.
102. Kietaibl, H. *Monats. Chem.* **1974**, *105*, 405–418.
103. Kobayashi, N.; Togashi, M.; Osa, T.; Ishii, K.; Yamauchi, S.; Hino, H. *J. Am. Chem. Soc.* **1996**, *118*, 1073–1085.
104. Kasuga, K.; Idehara, T.; Handa, M. *Inorg. Chim. Acta* **1992**, *196*, 127–128.
105. Sastre, A.; Torres, T.; Hanack, M. *Tetrahedron Lett.* **1995**, *36*, 8501–8504.
106. Weitemeyer, A.; Kliesch, H.; Wöhrle, D. *J. Org. Chem.* **1995**, *60*, 4900–4904.
107. Musluoglu, E.; Gürek, A.; Ahsen, V.; Gül, A.; Bekaroglu, Ö. *Chem. Ber.* **1992**, *125*, 2337–2339.
108. Badger, G. M.; Elix, J. A.; Lewis, G. E. *Proc. Chem. Soc.*, **1964**, 82.
109. Badger, G. M.; Elix, J. A.; Lewis, G. E.; Singh, U. P.; Spotswood, T. M. *J. Chem. Soc. Chem. Commun.* **1965**, 269–270.
110. Badger, G. M.; Elix, J. A.; Lewis, G. E.; Singh, U. P.; Spotswood, T. M. *J. Chem. Soc. Chem. Commun.* **1965**, 492.
111. Badger, G. M.; Elix, J. A.; Lewis, G. E. *Aust. J. Chem.* **1965**, *18*, 70–89.
112. Badger, G. M.; Lewis, G. E.; Singh, U. P. *Aust. J. Chem.* **1966**, *19*, 257–268.
113. Badger, G. M.; Elix, J. A.; Lewis, G. E. *Aust. J. Chem.* **1966**, *19*, 1221–1241.
114. Badger, G. M.; Lewis, G. E.; Singh, U. P. *Aust. J. Chem.* **1966**, *19*, 1461–1476.
115. Badger, G. M.; Lewis, G. E.; Singh, U. P. *Aust. J. Chem.* **1967**, *20*, 1635–1642.
116. Elix, J. A. *Aust. J. Chem.* **1969**, *22*, 1951–1962.
117. Schall, R., Ph.D. Dissertation; University of Cologne, Germany, **1993**.
118. Hu, Z.; Atwood, J. L.; Cava, M. P. *J. Org. Chem.* **1994**, *59*, 8071–8075.
119. Knop, J. V.; Milun, M.; Trinajstic, N. *J. Heterocyclic Chem.* **1976**, *13*, 505–508.
120. Vogel, E.; Jux, N.; Rodriguez-Val, E.; Lex, J.; Schmickler, H. *Angew. Chem. Int. Ed. Eng.* **1990**, *29*, 1387–1390.
121. Black, D. St C.; Craig, D. C.; Kumar, N. *J. Chem. Soc., Chem. Commun.* **1989**, 425–426.
122. Black, D. St C.; Bowyer, M. C.; Kumar, N.; Mitchell, P. S. R. *J. Chem. Soc. Chem. Commun.* **1993**, 819-821.
123. Black, D. St C.; Craig, D. C.; Kumar, N. *Tetrahedron Lett.* **1995**, *36*, 8075–8078.

124. Black, D. St C.; Craig, D. C.; Kumar, N.; McConnell, D. B. *Tetrahedron Lett.* **1996**, *37*, 241–244.

125. Cresp, T. M.; Sargent, M. V. *J. Chem. Soc. Perkin Trans. I* **1973**, 1786–1790.

126. Cresp, T. M.; Sargent, M. V. *J. Chem. Soc. Perkin Trans. I* **1973**, 2961–2971.

127. Vogel, E.; Bröring, M.; Fink, J.; Rosen, D.; Schmickler, H.; Lex, J.; Chan, K. W. K.; Wu, Y.-D.; Plattner, D. A.; Nendel, M.; Houk, K. N. *Angew. Chem. Int. Ed. Eng.* **1995**, *34*, 2511–2514.

128. Sondheimer, F.; Wolovsky, R.; Amiel, Y. *J. Am. Chem. Soc.* **1962**, *84*, 274–284.

129. Jackman, L. M.; Sondheimer, F.; Amiel, Y.; Ben-Efraim, D. A.; Gaoni, Y.; Wolovsky, R.; Bothner-By, A. A. *J. Am. Chem. Soc.* **1962**, *84*, 4307–4312.

130. Sondheimer, F.; Calder, I. C.; Elix, J. A.; Gaoni, Y.; Garratt, P. J.; Grohman, K.; Di Mayo, G.; Mayer, J.; Sargent, M. V.; Wolovsky, R. *Chem. Soc. (London) Spec. Publ. No. 21* **1967**, 75–107.

131. Haddon, R. C.; Haddon, V. R.; Jackman, L. M. *Fortschr. Chem. Forsch.* **1971**, *16*, 103–220.

132. Will, S.; Lex, J.; Vogel, E.; Schmickler, H.; Gisselbrecht, J.-P.; Haubtmann, C.; Bernard, M.; Gross, M. *Angew. Chem. Int. Ed. Engl.* **1997**, *36*, 357–361.

133. Black, D. St L.; Craig, D. C.; Kumar, N. *Aust. J. Chem.* **1996**, *49*, 311–318.

3 Introduction

The abundance and relevance of porphyrins in nature has inspired voluminous work devoted to the preparation of molecules designed to understand and mimic the biological processes mediated by these naturally occurring compounds (see, for instance, Smith[1] and Dolphin[2]). In spite of the ever-increasing attention being dedicated to the porphyrins, surprisingly little research has had as its primary goal the preparation of structural variants of porphyrins ([18]porphyrin-(1.1.1.1) **3.1**), which contain the same $C_{20}H_{14}N_4$ constituents in a different configurational form (e.g., **3.2**–**3.6**) (Figure 3.0.1).[3–5] In fact, the first synthetic porphyrin isomer was only reported in 1986 by Vogel and coworkers.[6] This isomer, porphycene ([18]porphyrin-(2.0.2.0) **3.2**), has attracted considerable attention. Nonetheless, in spite of this attention, no new examples of porphyrin isomers had appeared in the literature through the end of 1993. The calendar year 1994, however, proved to be much more fruitful. In fact, during that single year, four independent papers appeared, which, taken together, reported the synthesis and characterization of three new isomeric porphyrins.

In this chapter, the chemistry of isomeric porphyrins will be reviewed with an emphasis being placed on synthesis. A general discussion of the preparative chemistry of porphycene **3.2** and its derivatives will, therefore, be presented first. This will be followed by a brief summary of three other recently isolated "nitrogen-in" isomers of porphyrin, namely [18]porphyrin-(2.1.0.1) (e.g., **3.3**), [18]porphyrin-(2.1.1.0) (e.g., **3.4**), and [18]porphyrin-(3.0.1.0) (e.g., **3.5**). These three isomers, like porphyrin and porphycene, contain four pyrrole and four sp^2-hybridized *meso*-like carbons, and maintain an aromatic π-conjugation pathway (indicated by the bold bonds in structures **3.1**–**3.5**). Also discussed in this chapter is another type of porphyrin isomer, namely one of structure **3.6**, which is derived from "flipping out" one of the internal pyrrolic nitrogen atoms. Not discussed, however, is the chemistry of other $C_{20}H_{14}N_4$ systems. In other words, for the purposes of this review, a porphyrin isomer is defined as being a macrocycle that contains four pyrroles and four *meso*-like bridging carbon atoms, arranged in some fashion other than the alternating one in which they are found in porphyrins.

Figure 3.0.1 Structures **3.1–3.6** Showing Atom Numbering Schemes and the Main 18 π-Electron Conjugation Pathways Present in Porphyrin and Some of its Isomers

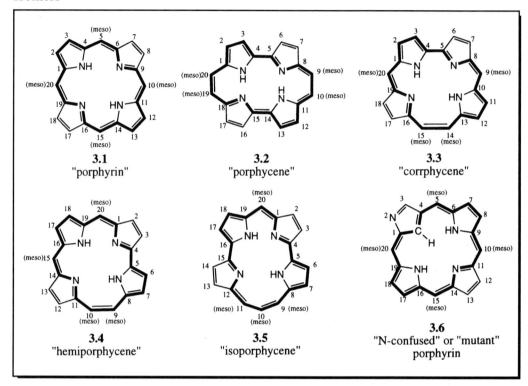

3.1
"porphyrin"

3.2
"porphycene"

3.3
"corrphycene"

3.4
"hemiporphycene"

3.5
"isoporphycene"

3.6
"N-confused" or "mutant"
porphyrin

3.1   [18]Porphyrin-(2.0.2.0) ("Porphycene")

3.1.1   Synthesis and Structure

3.1.1.1   Porphycene

As mentioned above, the first well-characterized example of a porphyrin isomer, compound **3.2**, was reported by Vogel and coworkers in 1986.[6] It was prepared in *ca.* 3% final, purified yield from two 5,5′-diformyl-2,2′-bipyrrole subunits **3.7**, *via* a reductive McMurry-type coupling as shown in Scheme 3.1.1. Interestingly, in the course of this reaction process, the initial condensation product (presumably the 20 π-electron species represented by **3.8**) was not isolated. Rather, in analogy to what is observed in the case of porphyrin, where the still unknown "isophlorin" (e.g., **3.9**; Figure 3.1.1) lies as a presumed intermediate along the reaction pathway, this presumed [20]annulene **3.8** was found to be highly susceptible to autoxidation, undergoing spontaneous two-electron oxidation under the reaction conditions. The isolated product is thus the 18-π-electron aromatic porphycene **3.2**. This latter species, prepared initially free of *meso-* or β-alkyl substituents, was observed to contain structural features reminiscent of those found in both the porphyrins and acenes (*vide infra*). This led Vogel to propose the trivial name "porphycene" for this new class of compound.[6]

Scheme 3.1.1

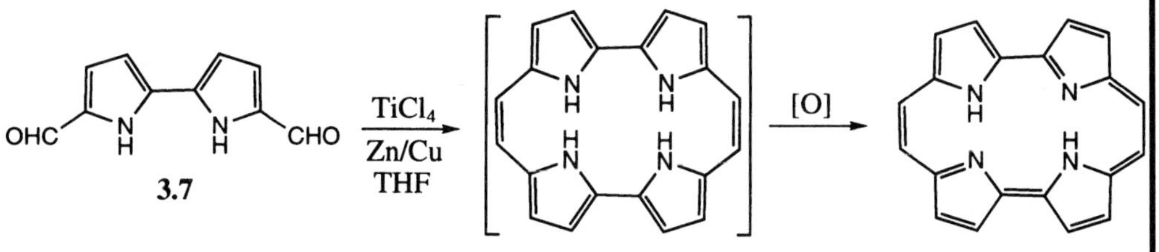

Figure 3.1.1   Structure **3.9**

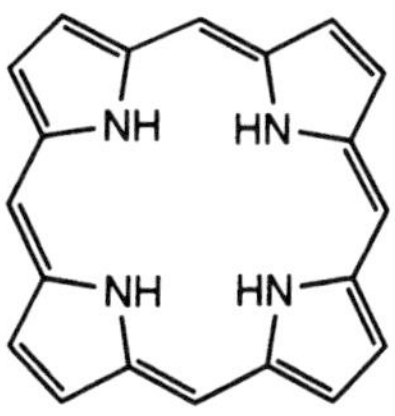

3.9
"isophlorin"

Proton NMR data collected for the prototypic species **3.2** supported the basic proposal that compounds of the porphycene class are best considered as being true aromatic porphyrin isomers.[6,7] For instance, the internal NH protons in **3.2** experience an upfield shift (to $\delta = 3.15$ ppm) relative to simple pyrrolic NH protons, and a similar but opposite (i.e., downfield) shift is observed for the β-pyrrolic protons. These shifts were taken as indicating the presence of a diamagnetic (aromatic) ring current in the macrocyclic system. Further support for the proposal that **3.2** is aromatic came from UV-vis spectral analyses.[6] Here, in particular, it was found that the absorption spectrum of **3.2** recorded in benzene was similar in many ways to that of unsubstituted porphyrin. A split Soret-like band ($\lambda_{max} = 358$ and 370 nm), and three longer wavelength Q-type bands ($\lambda_{max} = 558$, 596, and 630 nm) are observed, for instance, in the spectrum. Finally, a single crystal X-ray structural analysis of **3.2** revealed, as would be expected for a porphyrin-like molecule, a virtually planar macrocycle (the maximum displacement of the C and N atoms from the mean heavy atom plane is only ± 0.04 Å) (Figure 3.1.2). This same

X-ray diffraction analysis also revealed that the four nitrogen atoms of the central N₄ cavity of porphycene form a rectangle, the sides of which are 2.83 Å and 2.63 Å in length. The core of porphycene is thus smaller than that of porphyrin wherein the N–N distance between adjacent pyrroles is 2.89 Å.[6]

Figure 3.1.2   Single Crystal X-Ray Diffraction Structure of Porphycene **3.2**.
This figure was generated using information down-loaded from the Cambridge Crystallographic Data Centre and corresponds to a structure originally reported in reference 6. The porphycene molecule is centrosymmetric and virtually planar (the maximum deviation from the mean macrocyclic plane is 0.04 Å). The four nitrogen atoms form a rectangle, the sides of which are 2.83 Å and 2.63 Å in length. Top: top view; bottom: side view. Atom labeling scheme: carbon: ○; nitrogen: ●. Hydrogen atoms have been omitted for clarity

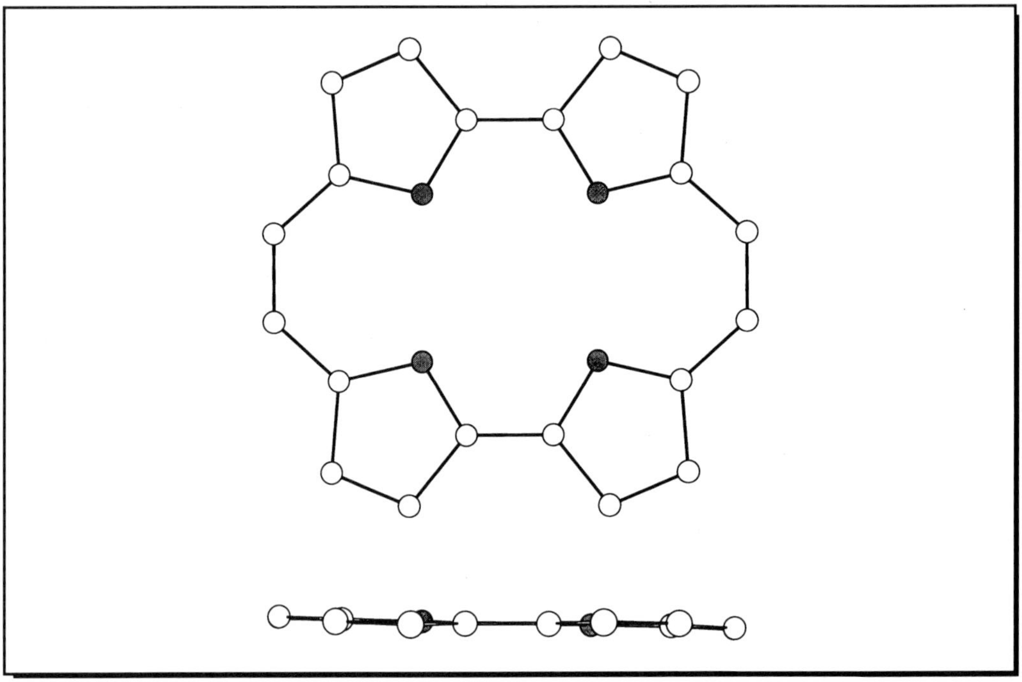

Because of the somewhat poor solubility and crystallinity of the porphyrin-like species **3.2** and its metal complexes (*vide infra*), the 2,7,12,17-tetrapropyl substituted analog **3.20** was prepared by Vogel and coworkers shortly after the original porphycene synthesis was complete (see Figure 3.0.1 for atom numbering scheme).[8] It, too, was made *via* a reductive carbonyl coupling procedure involving the appropriate diformyl bipyrrolic precursor (**3.10**), as have numerous other β-substituted porphycenes reported by the Vogel group. Examples of these latter include both the tetra-substituted alkyl and alkoxy systems **3.21**–**3.26**, the octa-alkyl derivatives **3.27**–**3.29** (Scheme 3.1.2), and the unsymmetrically functionalized porphycene **3.30** (Scheme 3.1.3).[8–14] Many of these were prepared in considerably higher yields (*ca.* 10–25%) than was the original unsubstituted parent **3.2**.

Scheme 3.1.2

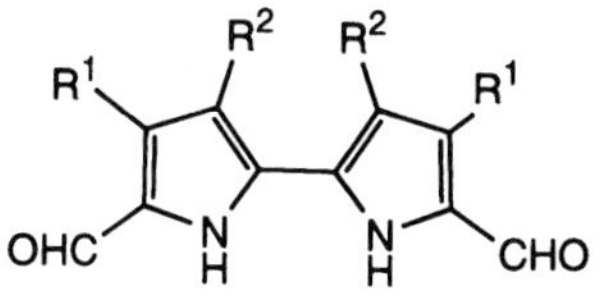

TiCl$_4$

Zn/Cu

THF

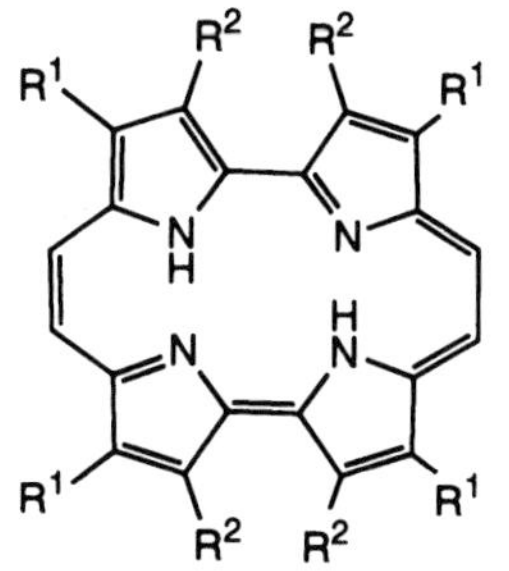

3.10. R^1 = *n*-Pr, R^2 = H
3.11. R^1 = Me, R^2 = H
3.12. R^1 = Et, R^2 = H
3.13. R^1 = *t*-Bu, R^2 = H
3.14. R^1 = Ph, R^2 = H
3.15. R^1 = $(CH_2)_2OMe$, R^2 = H
3.16. R^1 = $(CH_2)_2OEt$, R^2 = H
3.17. R^1 = R^2 = Me
3.18. R^1 = R^2 = Et
3.19. R^1 = Et, R^2 = Me

3.20. R^1 = *n*-Pr, R^2 = H
3.21. R^1 = Me, R^2 = H
3.22. R^1 = Et, R^2 = H
3.23. R^1 = *t*-Bu, R^2 = H
3.24. R^1 = Ph, R^2 = H
3.25. R^1 = $(CH_2)_2OMe$, R^2 = H
3.26. R^1 = $(CH_2)_2OEt$, R^2 = H
3.27. R^1 = R^2 = Me
3.28. R^1 = R^2 = Et
3.29. R^1 = Et, R^2 = Me

Scheme 3.1.3

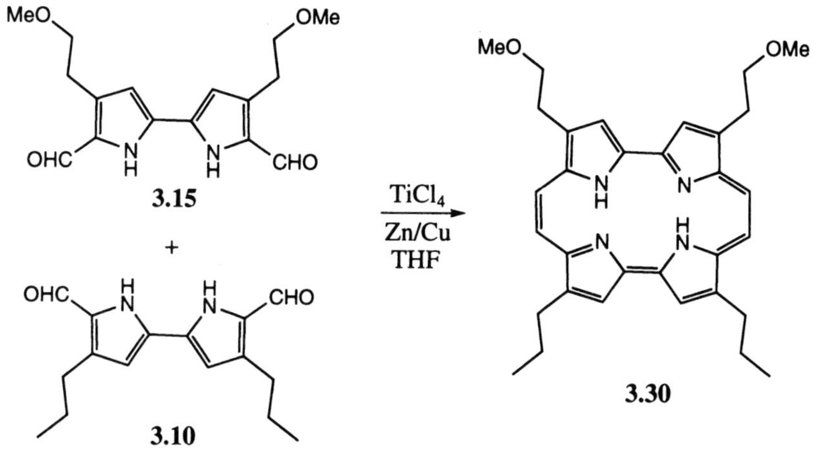

3.15 + **3.10** → (TiCl$_4$, Zn/Cu, THF) **3.30**

Many of the substituted porphycenes have been analyzed by X-ray diffraction methods. In the case of the tetrapropylporphycene **3.20** (Figure 3.1.3), such a solid-state structural analysis revealed very little change in the planarity and overall geometry of the macrocycle compared to the β-free parent **3.2**.[8] On the other hand, an X-ray structure of the octaethylporphycene **3.28** (Figure 3.1.4) showed a slight twist, presumably due to van der Waals repulsions between the ethyl groups in the 3,6- and 13,16-positions.[9] This distortion allows the N$_4$ coordination center to adopt an approximately square planar arrangement in the solid state. This is in contrast to the rectangular arrangement present in the more planar systems **3.2** and **3.20**.

Figure 3.1.3   Single Crystal X-Ray Diffraction Structure of 2,7,12,17-Tetra-*n*-propylporphycene **3.20**.
This figure was generated using information down-loaded from the Cambridge Crystallographic Data Centre and corresponds to a structure originally reported in reference 8. Porphycene **3.20** is a centrosymmetric molecule and has a virtually planar macrocyclic ring (the maximum deviation from the mean macrocyclic plane is 0.04 Å). Atom labeling scheme: carbon: ○; nitrogen: ●. Hydrogen atoms have been omitted for clarity

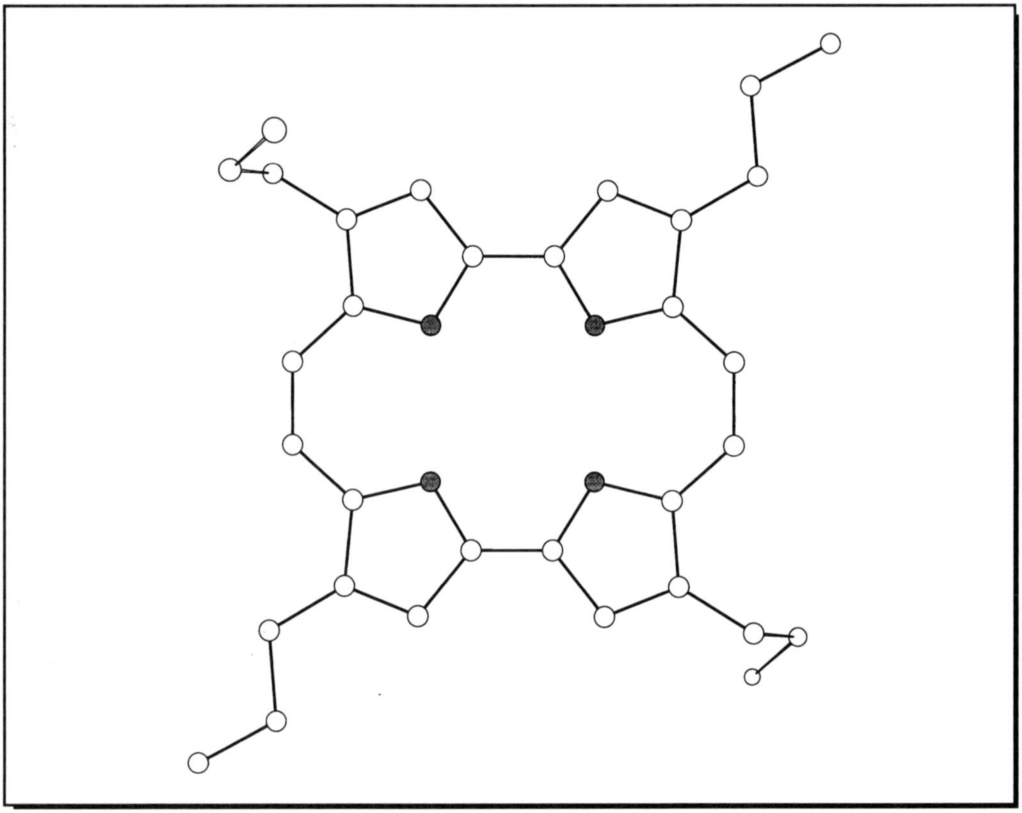

In addition to changing the β-substitution pattern of porphycene, Vogel and coworkers have introduced substitution at the *meso*-positions of the macrocycle. Specifically, the 9,10,19,20-tetraalkyl porphycenes **3.37–3.39** were prepared *via* the reductive carbonyl coupling of the corresponding 5,5'-diacyl-2,2'-bipyrroles **3.31–3.33** (Scheme 3.1.4).[15] Here, in contradistinction to what is true for the

Figure 3.1.4 Single Crystal X-Ray Diffraction Structure of 2,3,6,7,12,13,16,17-Octaethylporphycene **3.28**.
This figure was generated using information down-loaded from the Cambridge Crystallographic Data Centre and corresponds to a structure originally reported in reference 9. The central cavity of porphycene **3.28** adopts an approximately square arrangement. The molecule also experiences a slight twist about the bipyrrole linkages (the maximum deviation from the mean macrocyclic plane is 0.27 Å). Top: top view; bottom: side view. Atom labeling scheme: carbon: ○; nitrogen: ●. Hydrogen atoms have been omitted for clarity

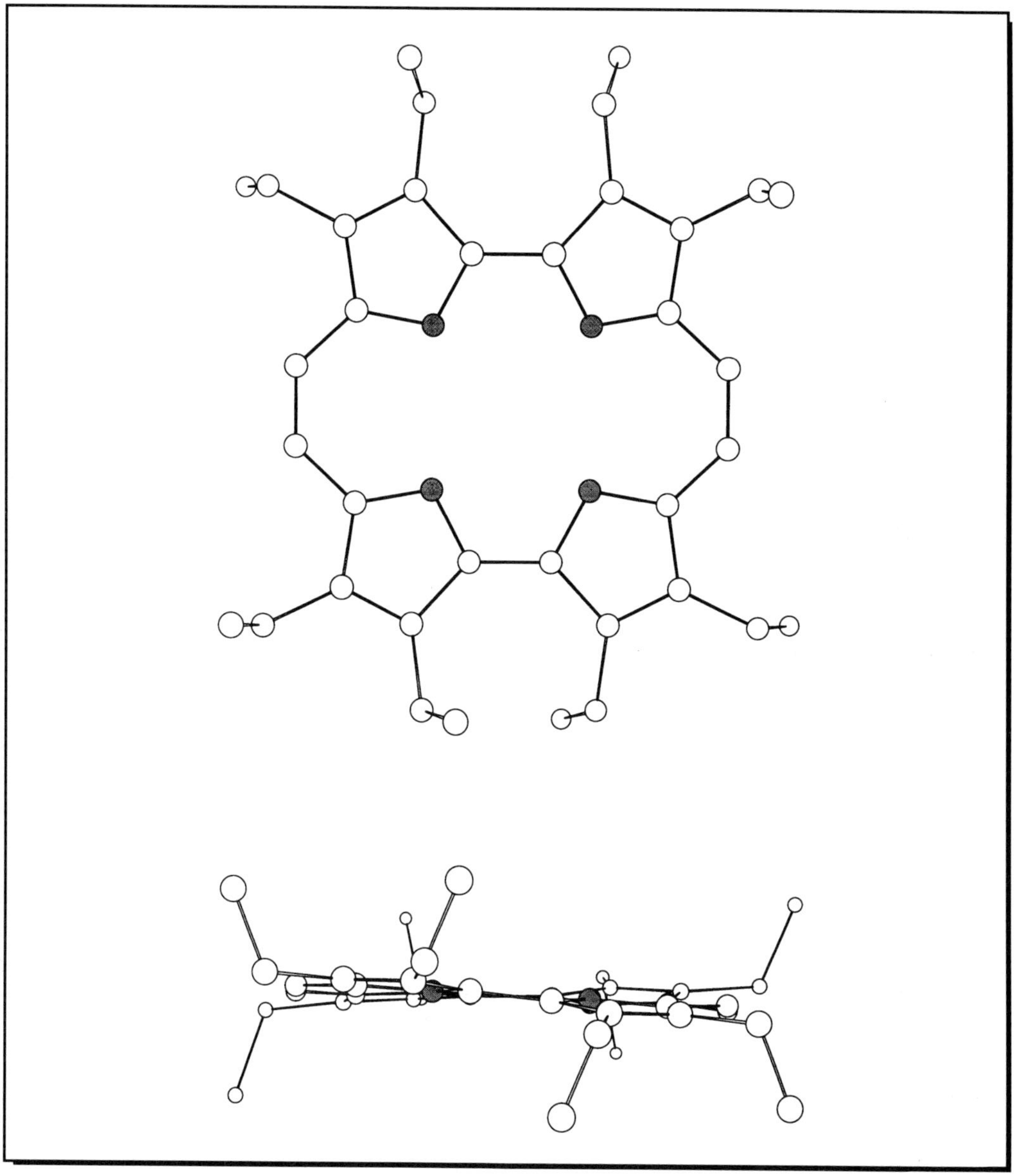

meso-unsubstituted porphycenes, the intermediate dihydroporphycenes **3.34–3.36** could be isolated as colorless compounds. In particular, the 9,10,19,20-tetrapropyl substituted dihydroporphycene **3.36** was isolated and characterized structurally *via* single crystal X-ray diffraction analysis.[16] The resulting structure revealed a cyclophane-type conformation for **3.36** (represented schematically in Figure 3.1.5) in which it was presumed that steric interactions between the neighboring alkyl chains served to slow dramatically the rate of pyrrole ring rotation. This, in turn, it was speculated, would provide a kinetic barrier against adopting the near-planar conformation required for oxidation.[17] On the other hand, once oxidized, the aromatic 9,10,19,20-alkyl substituted porphycenes exhibited only a slight conformational strain, as judged, again, by a solid state X-ray diffraction analysis of the tetrapropyl derivative **3.39** (the molecule, although still essentially planar, suffers a 6.5° twisting distortion around the C(9)–C(10) bond; Figure 3.1.6). Here, the slight strain that was inferred was considered to arise from unfavorable steric interactions between the alkyl substituents.[16]

Scheme 3.1.4

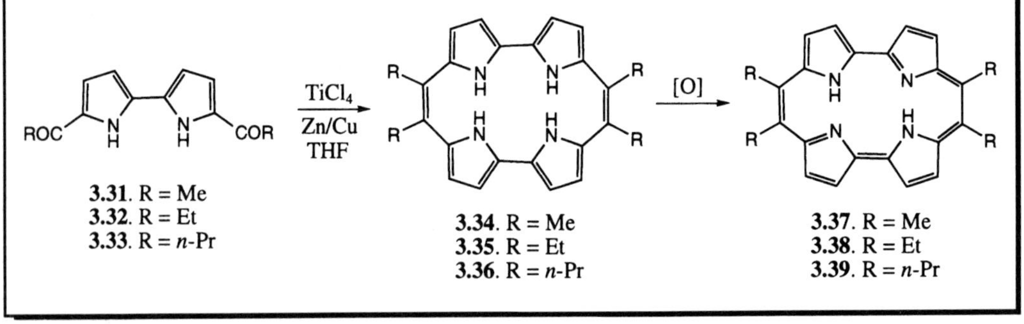

3.31. R = Me
3.32. R = Et
3.33. R = *n*-Pr

3.34. R = Me
3.35. R = Et
3.36. R = *n*-Pr

3.37. R = Me
3.38. R = Et
3.39. R = *n*-Pr

Figure 3.1.5   Schematic Representation of the Cyclophane-type Structure of **3.36**

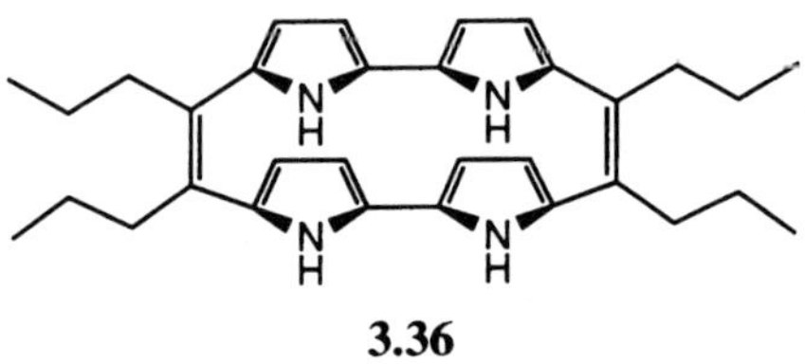

3.36

Another structural variant of the original porphycene skeleton was recently mentioned in the context of a personal review by Vogel.[16] This is the fused benzoporphycene derivative **3.42**, a species that is obtained from the reductive carbonyl coupling of the diformyl β-fused bipyrrole **3.40** (Scheme 3.1.5). Like the 9,10,19,20-substituted porphycenes, the intermediate dihydroporphycene obtained in this instance (**3.41**) could also be isolated. In fact, this species proved significantly more resistant to oxidation than one would perhaps expect; DDQ (2,3-dichloro-5,6-dicyano-1,4-benzoquinone) was required to effect an oxidation that normally occurs spontaneously in the presence of air. In any event, the resulting benzoporphycene was found to be stable and to display unusual electronic properties. Unfortunately, a detailed report describing the chemistry of this system has yet to appear. Thus, detailed information is currently lacking.

Figure 3.1.6   Single Crystal X-Ray Diffraction Structure of 9,10,19,20-tetra-*n*-propylporphycene **3.39**.
This figure was generated using information down-loaded from the Cambridge Crystallographic Data Centre and corresponds to a structure originally reported in reference 16. Porphycene **3.39** is present in the crystal as a centrosymmetric molecule that is nearly planar (the maximum deviation from the mean macrocyclic plane is 0.04 Å). Top: top view; bottom: side view. Atom labeling scheme: carbon: ○; nitrogen: ●. Hydrogen atoms have been omitted for clarity

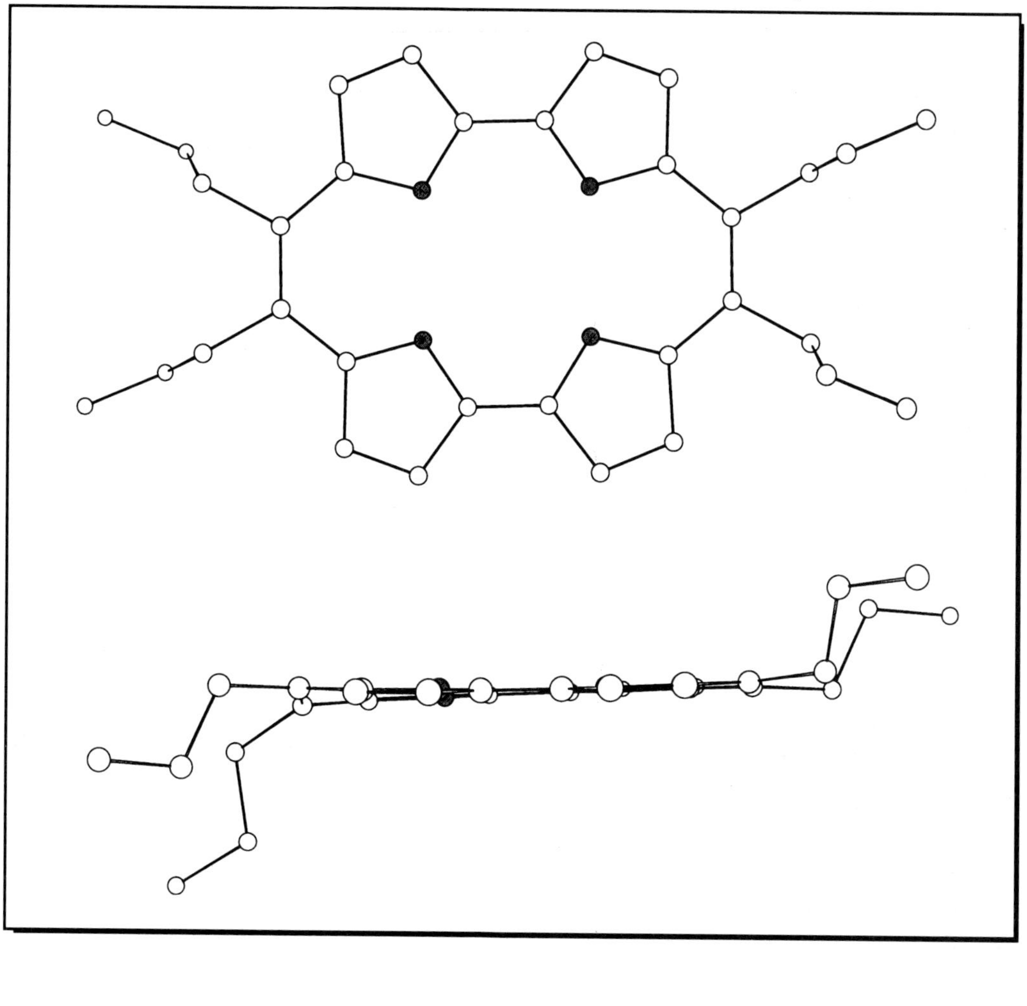

Scheme 3.1.5

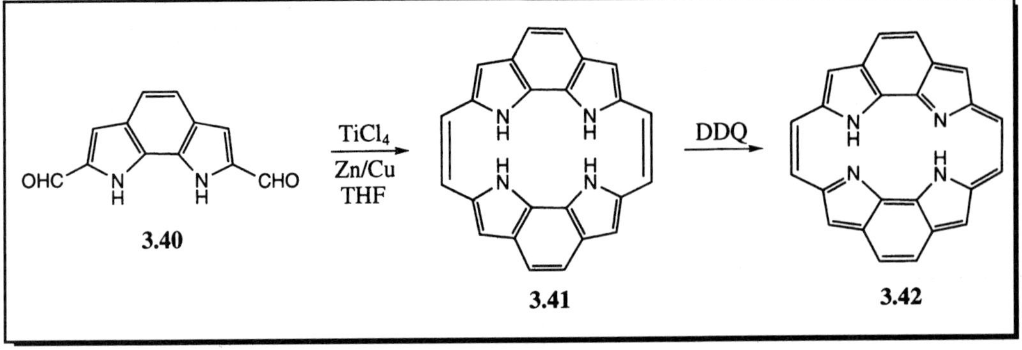

3.1.1.2   Heteroporphycene

Several examples of porphycene-like macrocycles in which the pyrrolic nitrogen atoms have been formally exchanged for differing heteroatoms have appeared in the literature.[18–20] The first of these, the tetraoxaporphycene **3.45**, was reported by Vogel in 1990.[18] The synthesis of this molecule involved the low-valent titanium-mediated reductive coupling between two molecules of 5,5'-diformyl-2,2'-bifuran **3.43**. The resulting 20 π-electron macrocycle **3.44** was readily isolated (Scheme 3.1.6), and was shown to sustain a pronounced *paramagnetic* ring current in the ^{1}H NMR spectrum (Figure 3.1.7). In marked contrast to the cyclophane-type structure observed in the case of the isoelectronic *N,N'*-dihydroporphycenes, however, this particular species was found to be quite planar in the solid state, as deduced from a single crystal X-ray diffraction analysis.

Scheme 3.1.6

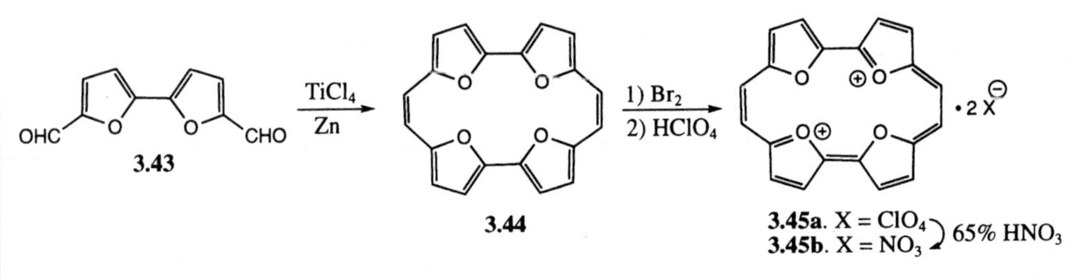

The 20 π-electron macrocycle **3.44** was found to undergo ready two-electron oxidation when treated with Br_2 and perchloric acid.[18] This afforded the 18 π-electron aromatic dication **3.45**. This species was originally isolated in the form of its bisperchlorate salt (**3.45a**). It was found, however, that this latter could be converted into the corresponding nitrate salt (**3.45b**) by treating with 65% nitric acid. Both salts proved to be aromatic as evidenced by: (1) the observation of downfield-shifted signals in the relevant ^{1}H NMR spectra; and (2) the presence of UV-vis absorbances in the electronic spectra, which resembled very closely those of the N_4-porphycenes.[18] The X-ray structure of the nitrate salt (**3.45b**) also revealed a nearly perfectly planar ring system (the maximal deviations of the C and O atoms from the mean plane are ±0.02 Å; Figure 3.1.8)). Like its N_4 congeners, the inward-pointing heteroatoms of the O_4-porphycene species **3.45b** define a well-formed rectangle. This core shape thus stands in contrast to what is observed in the case of the 20 π-electron O_4-dihydro-porphycene, **3.44**. Here, the four oxygen atoms form nearly a perfect square.

Figure 3.1.7   Single Crystal X-Ray Diffraction Structure of "Tetraoxaporphycenogen" **3.44**.

This figure was generated using information down-loaded from the Cambridge Crystallographic Data Centre and corresponds to a structure originally reported in reference 18. Macrocycle **3.44** is centrosymmetric in the solid state and has a slightly puckered ring framework (maximum deviation from the mean macrocyclic plane is 0.13 Å). The four oxygen atoms adopt an approximately square conformation. Atom labeling scheme: carbon: ○; oxygen: ◒. Hydrogen atoms have been omitted for clarity

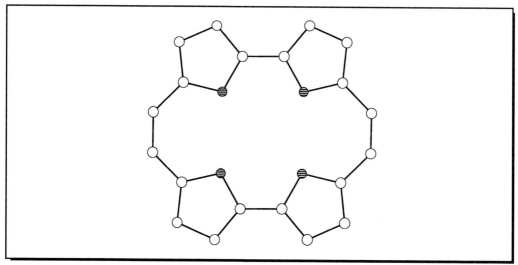

The synthesis of the dioxaporphycene **3.49** has also been reported by Vogel and coworkers.[20] To prepare this new derivative, the mixed furan-pyrrole dialdehyde **3.46** was subject to reductive McMurry-type dimerization. This afforded the air-stable intermediate **3.47** free of isomeric species such as **3.48**. Accordingly, oxidation, effected by treating with concentrated nitric acid and 70% perchloric acid, afforded the dioxaporphycene salt **3.49** in 94% yield (based on **3.47**) (Scheme 3.1.7). As in the case of the tetraoxaporphycene **3.45** above, ^{1}H NMR and UV-vis spectroscopic analyses supported the contention that this new dioxaporphycene derivative (shown in its diprotonated form as structure **3.49**) is aromatic.

Figure 3.1.8 Single Crystal X-Ray Diffraction Structure of Tetraoxaporphycene Salt **3.45b**.

This figure was generated using information down-loaded from the Cambridge Crystallographic Data Centre and corresponds to a structure originally reported in reference 18. Macrocycle **3.45b** is centrosymmetric in the solid state and is rather planar (maximum deviation from the mean macrocyclic plane is 0.02 Å). The four oxygen atoms define a rectangular core. Atom labeling scheme: carbon: ○; oxygen: ⊖; nitrogen: ●. Hydrogen atoms have been omitted for clarity

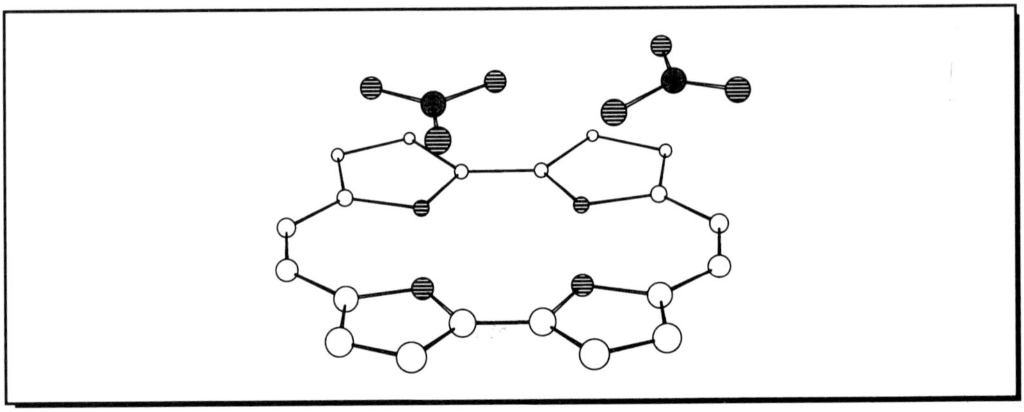

Scheme 3.1.7

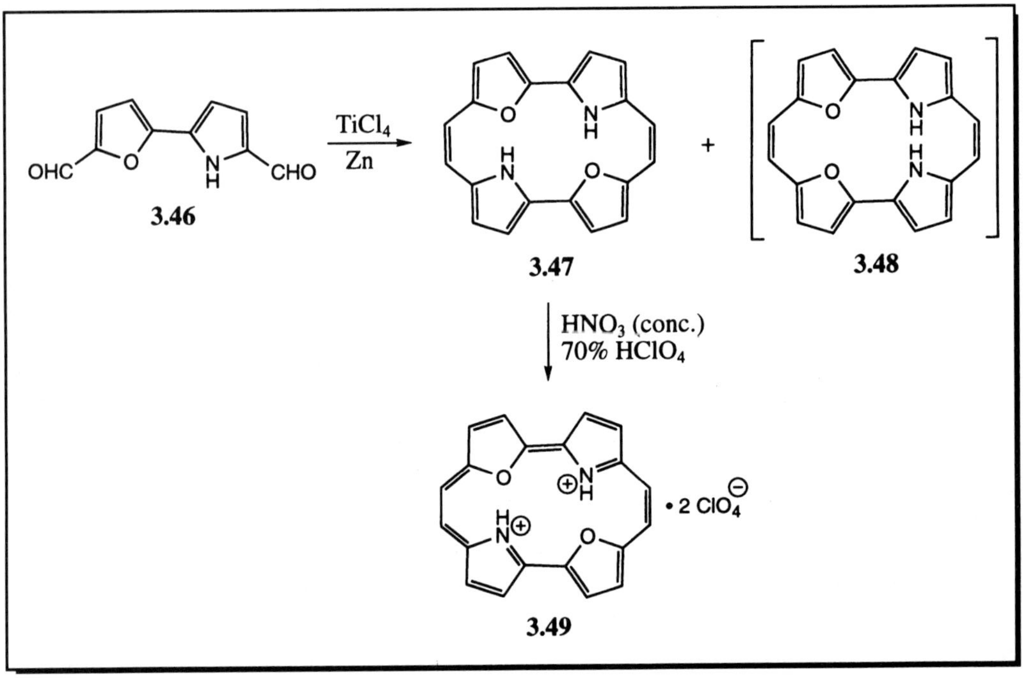

In addition to the above, two air-stable, crystalline 20-π-electron dithia-*N,N'*-dihydroporphycene derivatives, specifically **3.51** and **3.52**, have been prepared recently by Neidlein and coworkers (Scheme 3.1.8).[19] As could perhaps be imagined, these species were also prepared *via* a low-valent titanium-mediated reductive coupling procedure, involving the diformyl thiophene-pyrrole **3.50**, carried out in the presence of TiCl₄ and Zn.[19] Interestingly, it is not only **3.51** but also **3.52** that is isolated under these conditions. This stands in direct contrast to what is observed in the case of the mixed furan-pyrrole coupling discussed above (Scheme 3.1.7); there, only one of the two possible isomers was isolated.

Scheme 3.1.8

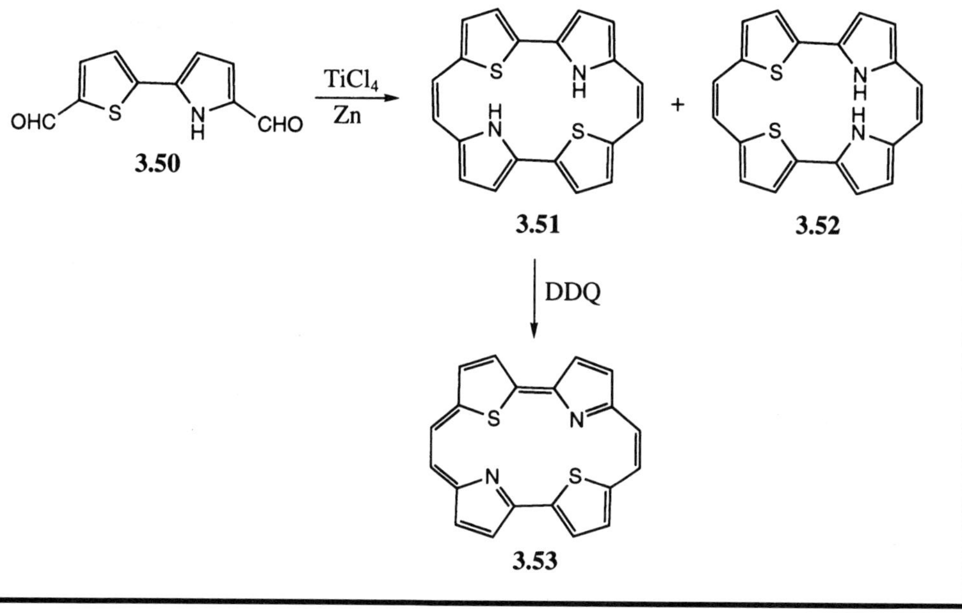

As was true in the case of the dioxa derivative **3.47**, the dithia species **3.51** could be oxidized to its corresponding aromatic 18 π-electron dicationic species **3.53**. The oxidant used in this case was DDQ. It gave **3.53** in quantitative yield.[19]

Given the ready manner in which **3.51** can be oxidized to **3.53**, it is clear that porphycene, like porphyrin before it,[21–23] can accommodate two thiophene rings in place of two pyrroles, while maintaining a fully conjugated 18 π-electron aromatic structure. In the specific case of **3.53**, this particular conclusion was supported by a single crystal X-ray diffraction analysis (Figure 3.1.9); here, the structure that emerged was found to be essentially planar. This finding, in turn, served to confirm that the two sulfur atoms, despite their large size (compared to nitrogen), could be easily accommodated within the core of the macrocycle without disrupting the π-conjugation of the macrocycle.

Figure 3.1.9   Single Crystal X-Ray Diffraction Structure of Dithiaporphycene **3.53**.
This figure was generated using information down-loaded from the Cambridge Crystallographic Data Centre and corresponds to a structure originally reported in reference 19. The macrocycle is very nearly planar (the maximum deviation from the mean macrocyclic plane is 0.08 Å). Atom labeling scheme: carbon: ○; nitrogen: ●; sulfur: ⦚. Hydrogen atoms have been omitted for clarity

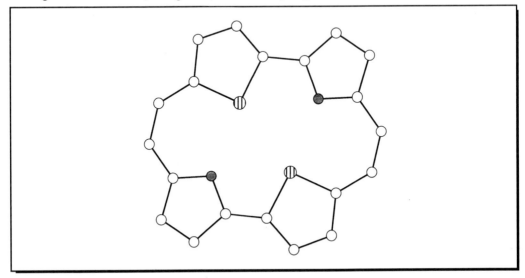

The synthesis of the tetrathia analog of the porphycenes has also been attempted. This was reported independently in 1993 by Merz and collaborators[24] and in 1994 by Cava and coworkers.[25] In both instances, the reductive carbonyl coupling of the 2,2'-bithiophene-5,5'-dicarbaldehyde **3.54**, effected using low valent titanium, afforded moderate yields (8.4% and 10.5%, respectively) of the [20]annulene system **3.55** (Scheme 3.1.9). The UV-vis absorption spectrum of this compound (in CH_3CN) revealed only two broad, relatively weak transitions at 272 nm and 333 nm; a finding that led to the suggestion that this molecule was highly distorted. The expected non-planar structure of **3.55** was then subsequently confirmed *via* a single crystal X-ray structural analysis (see reference 24). In this case, it was the relatively large size of the sulfur atoms that presumably precluded the formation of a planar structure.

Compound **3.55** was studied both in terms of its reduction and its oxidation properties. Perhaps not surprisingly, compound **3.55** was found to be resistant to both chemical and electrochemical oxidation, presumably owing to the steric demands of the sulfur atoms mentioned above.[24,25] On the other hand, **3.55**, when studied electrochemically, displayed a single reversible two-electron reduction wave. The annulene **3.55** could also be reduced using sodium metal. In both cases (chemical and electrochemical), the species obtained was presumed to be the diatropic 22 π-electron dianion **3.56**. The assignment of **3.56** as being a diatropic system was based primarily on ^{1}H NMR spectroscopic data of the chemically reduced species. Specifically, the observed diatropic shifts for the dianion **3.56**, on the order of 1.75 ppm (relative to **3.55**), were taken as an indication that **3.56** possesses a weak macrocyclic ring current.[24]

Scheme 3.1.9

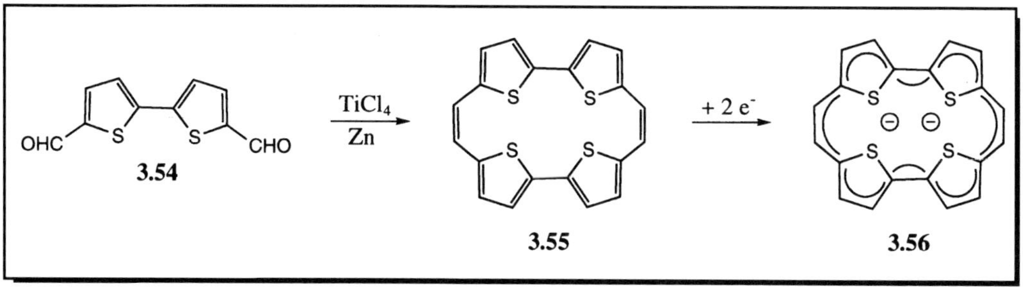

3.1.2   *Metalloporphycenes*

Like porphyrin, porphycene in its neutral form can be considered as being a doubly protonated version of a dianionic ligand. Thus, it was hoped that porphycene would act as an effective cheland, mimicking the rich metalation chemistry of the porphyrins. Thanks to the efforts of Vogel and others, much of this promise has now been transformed into experimental reality. Thus, in spite of the fact that the porphycenes possess a central cavity that is smaller than that of the porphyrins, porphycene has indeed been found to form stable, neutral complexes with a wide range of divalent metal cations.[3–6,8,9,11,15,16,26,27] Additionally, one example of a porphycene complex formed from a monovalent metal has been reported,[28] as have several examples of complexes of porphycene with higher-valent metals.[3–5,12,16,29–34]

The rich metalation chemistry of the porphycenes was alluded to in Vogel's seminal publication in the area.[6] In it, it was noted that the all-unsubstituted porphycene **3.2** forms stable cobalt(II), copper(II), nickel(II), and zinc(II) complexes (i.e., **3.57–3.60**), when treated with the appropriate divalent metal salt (Scheme 3.1.10).[6] Later it was shown that porphycene **3.2** also forms stable complexes with divalent palladium and platinum (**3.61** and **3.62**).[16] Unfortunately, at present no X-ray structural information is available for any of these latter complexes.

As mentioned earlier, the poor solubility and crystallinity characteristics of the metal complexes derived from porphycene **3.2** inspired the synthesis of a number of β-alkyl-substituted analogs. Not surprisingly, these β-substituted systems showed metal coordination properties similar to those of their unsubstituted counterpart. For instance, the 2,7,12,17-tetrapropylporphycene **3.20** has been shown to form stable complexes with Co(II), Ni(II), Cu(II), Pd(II), and Pt(II) (i.e., **3.63–3.67**).[3–5,8,26,27]

A single crystal X-ray diffraction structure of the Ni(II) 2,7,12,17-tetrapropyl-porphycene **3.65** revealed that, like the free ligand, the molecule is of centrosymmetric symmetry.[8] These same analyses revealed that the complex, like the metal-free ligand, is nearly planar. However, complexation of this metal cation was found to have a dramatic effect on the internal core geometry of the ligand. For instance, in the free ligand (**3.20**), the distance between N(1) and N(2) (see Figure 3.1.10 for numbering) was found to be 0.22 Å greater than the distance between N(1) and N(3). By contrast, in complex **3.65**, the distance between N(1) and N(2) was found to be 0.25 Å *shorter* than that between N(1) and N(3) (Figure 3.1.11). This distortion of the core geometry was presumed to have its genesis in a need to establish a more regular square-planar coordination environment about the metal center.

Scheme 3.1.10

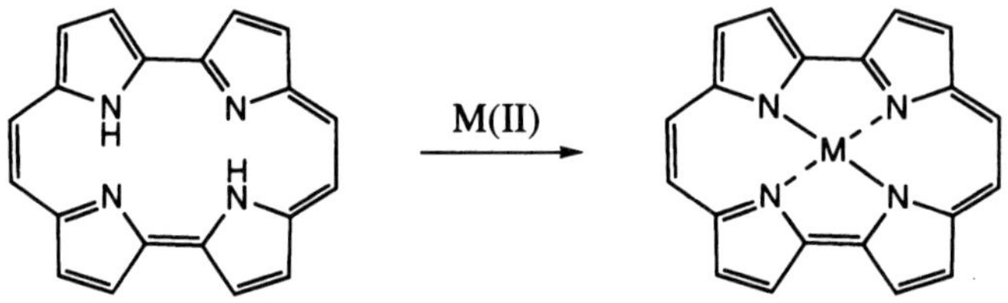

3.2

M(II) →

3.57. M = Co
3.58. M = Cu
3.59. M = Ni
3.60. M = Zn
3.61. M = Pd
3.62. M = Pt

Figure 3.1.10   Structures **3.63–3.71**

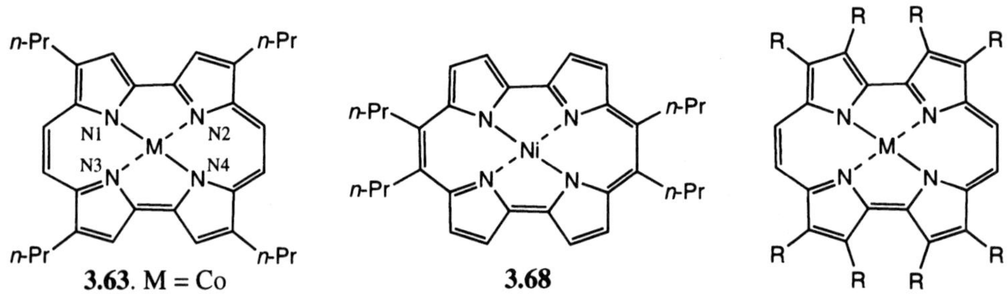

3.63. M = Co
3.64. M = Cu
3.65. M = Ni
3.66. M = Pd
3.67. M = Pt

3.68

3.69. R = Me, M = Ni
3.70. R = Et, M = Ni
3.71. R = Et, M = Zn

The isomeric species, 9,10,19,20-tetrapropylporphycene **3.68**, has been demonstrated as being capable of forming a stable complex with divalent nickel.[15] As evident from a single crystal X-ray structural analysis of this complex (Figure 3.1.12), it behaves in many ways like the nickel(II) complex of 2,7,12,17-tetrapropylporphycene (**3.65**). In particular, the ligand distorts significantly so as to provide a nearly square-planar arrangement of the coordinating nitrogen atoms. As a result of this internal core distortion, the interaction between the peripheral *n*-propyl chains is presumably enhanced. This serves to twist the macrocycle further around the C(9)–C(10) bonds. As a result, the dihedral angles about these bonds increase from 6.5° in the case of the free-base ligand (**3.39**) to 13.4° in the case of the metal complex.

Figure 3.1.11   Single Crystal X-Ray Diffraction Structure of Nickel(II) 2,7,12,17-Tetrapropylporphycene **3.65**.
This figure was generated using information down-loaded from the Cambridge Crystallographic Data Centre and corresponds to a structure originally reported in reference 8. Nickel(II) porphycene **3.65** is present in the crystal as a centrosymmetric molecule that is nearly planar (the maximum deviation from the mean macrocyclic plane for atoms in the porphycene core is 0.04 Å). The average Ni–N bond length is 1.90 Å. Atom labeling scheme: nickel: ◉; carbon: ○; nitrogen: ●. Hydrogen atoms have been omitted for clarity

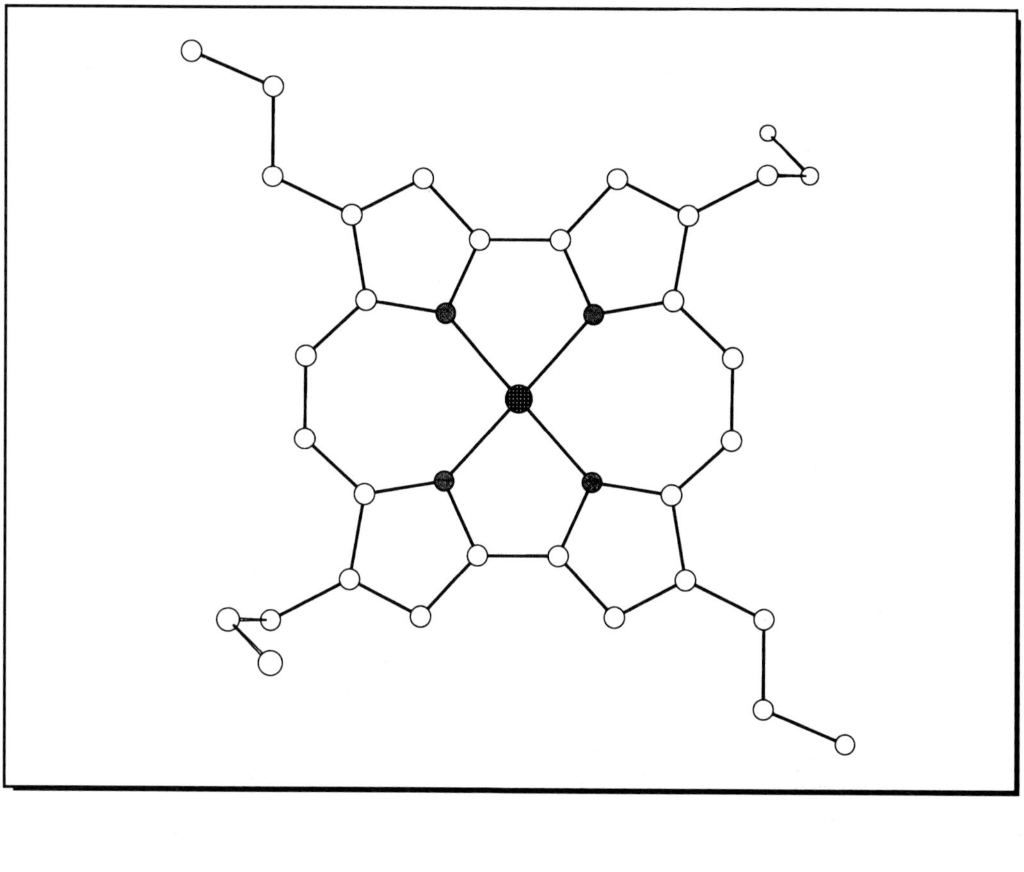

Figure 3.1.12   Single Crystal X-Ray Diffraction Structure of Nickel(II) 9,10,19,20-Tetra-*n*-propylporphycene **3.68**.

This figure was generated using information down-loaded from the Cambridge Crystallographic Data Centre and corresponds to a structure originally reported in reference 15. Nickel(II) porphycene **3.68** is asymmetrically oriented in the crystal and deviates somewhat from planarity (the maximum deviation of the porphycene atoms from the mean macrocyclic plane is 0.45 Å). The average Ni–N bond length is 1.89 Å. Atom labeling scheme: nickel: ●; carbon: ○; nitrogen: ●. Hydrogen atoms have been omitted for clarity

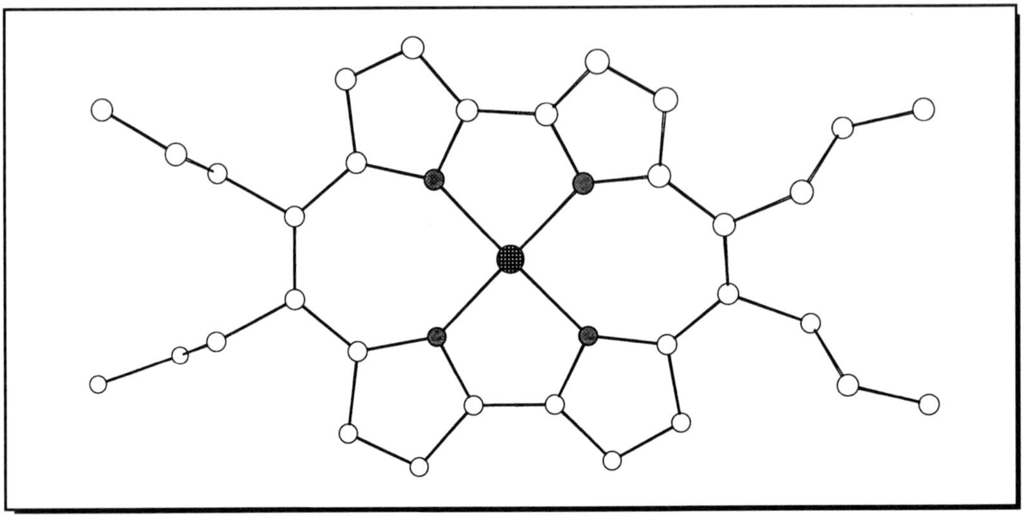

The octa-alkyl porphycenes are also able to support the formation of metal(II) complexes.[9,10] Examples of these include the nickel(II) complex of octamethylporphycene (**3.69**) and the nickel(II) and zinc(II) complexes of octaethylporphycene (**3.70** and **3.71**). As shown by single crystal X-ray diffraction data, incorporation of divalent zinc into octaethylporphycene results in the formation of a nearly planar structure (Figure 3.1.13). Considering the somewhat non-planar nature of the starting metal-free system **3.28**, the insertion of this cation apparently serves to enforce a marked increase in ligand planarity.

It is interesting to note that 9,10,19,20-tetrapropylporphycene **3.39** does not form a stable complex with zinc(II). Similarly, 2,7,12,17-tetrapropylporphycene **3.20** forms only a moderately stable Zn(II) complex. By contrast, 2,3,6,7,12,13,16,17-octaethylporphycene **3.28** forms highly stable complexes with Zn(II). These results indicate that the potential exists to adjust considerably the metal coordination properties of the porphycenes simply by varying the peripheral substituents.[15]

In addition to forming complexes with divalent cations, the porphycenes act as ligands for certain trivalent metal cations. In many cases this chemistry parallels that seen in the porphyrin series. For instance, the μ-oxo-diiron(III)porphycenates **3.72** and **3.73**[12] were prepared from the corresponding free-base porphycene **3.20** or **3.23** using a procedure analogous to that used to prepare the corresponding μ-oxo-diiron(III)porphyrinates.[35] Thus, the reaction of **3.20** or **3.23** with Fe(acac)₃ in

Figure 3.1.13   Single Crystal X-Ray Diffraction Structure of Zinc(II) 2,3,6,7,12,13,16,17-Octaethylporphycene **3.71**.
This figure was generated using information down-loaded from the Cambridge Crystallographic Data Centre and corresponds to a structure originally reported in reference 9. Zinc(II) porphycene **3.71** is present in the crystal as a centrosymmetric molecule, which is nearly planar (the maximum deviation of the porphycene atoms from the mean macrocyclic plane is 0.05 Å). The N_4 core is rectangular in shape and the average Zn–N bond length is 2.01 Å. Top: top view; bottom: side view. Atom labeling scheme: nickel: ●; carbon: ○; nitrogen: ●. Hydrogen atoms have been omitted for clarity

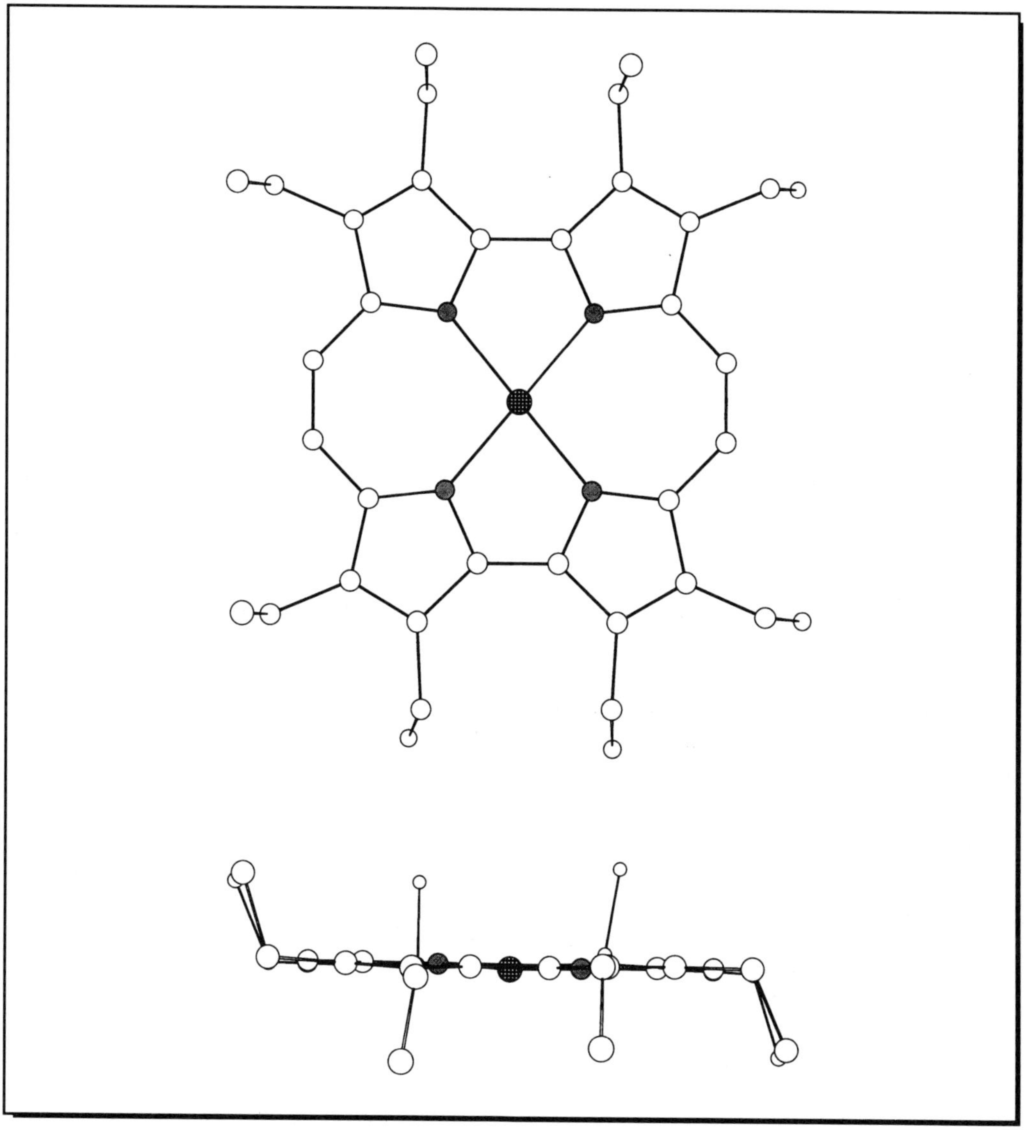

phenol, followed by treatment with NaOH, was found to produce the μ-oxo-diiron(III)porphycenates **3.72** and **3.73** in yields of 81% and 67%, respectively (Scheme 3.1.11). Likewise, the monomeric Fe(III)porphycenates **3.74a–f** could be prepared from the μ-oxo-diiron(III)porphycenate **3.72** by treating these latter with the appropriate acid (HX).

Scheme 3.1.11

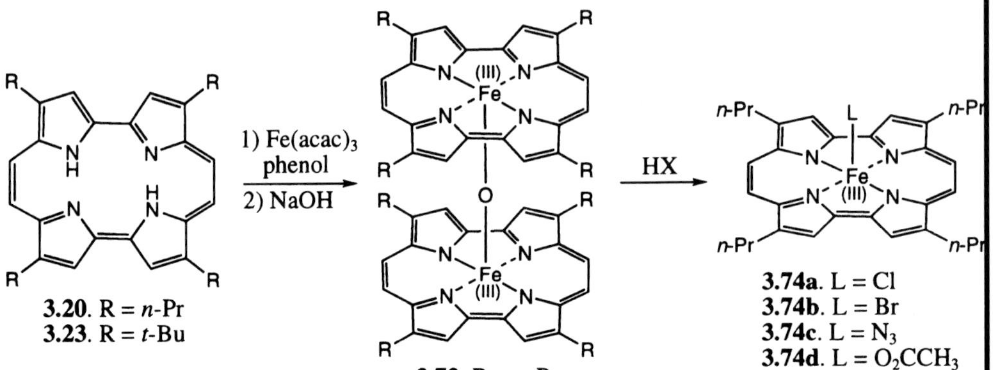

3.20. R = *n*-Pr
3.23. R = *t*-Bu

3.72. R = *n*-Pr
3.73. R = *t*-Bu

3.74a. L = Cl
3.74b. L = Br
3.74c. L = N_3
3.74d. L = O_2CCH_3
3.74e. L = O_2CCF_3
3.74f. L = OC_6H_5

Single crystal X-ray diffraction structural analyses of complexes **3.72** and **3.73** served to confirm that these species are μ-oxo-bridged dimers (Figures 3.1.14 and 3.1.15).[12] Interestingly, the Fe-O-Fe angles in these two structures were found to differ substantially. In **3.72**, the Fe-O-Fe angle is 145.3°, in accord with angles anticipated for such kinds of complexes.[36,37] The Fe-O-Fe angle for the *t*-butyl derivative **3.73**, on the other hand, is nearly linear (178.5°). This linearity was rationalized by the authors in terms of incipient steric interactions between the *t*-butyl groups that serve to preclude close contacts between the two porphycene rings.[12]

Several other examples of porphycene complexes with trivalent metals have also appeared in the literature.[3,4,12,29,29–32] These include the aluminum(III) complexes **3.75a–d** and **3.76**, as well at the manganese(III), cobalt(III) iron(III), indium(III), and gallium(III) complexes **3.77–3.81** (Figure 3.1.16). Interestingly, a μ-oxo-dialuminum(III)porphycenate **3.85** analogous to the μ-oxo-diiron(III)porphycenates **3.72** and **3.73** has also been prepared.[30] This was accomplished by heating the aluminum(III)porphycenato hydroxide complex **3.76** to 350 °C under reduced pressure (Scheme 3.1.12).

Porphycenes have also been shown to form stable complexes with tetravalent metal cations. Examples of tetravalent systems include the tin(IV) and germanium(IV) complexes **3.82–3.84** (Figure 3.1.16).[3,4,34] In each of these complexes, two axial chloride anions were found to be coordinated to the metal atom. While it is presumed that these complexes contain the metal chelated within the porphycene plane, currently the assumption remains unverified as precise X-ray diffraction-derived structural details have yet to appear in the literature.

Figure 3.1.14   Single Crystal X-Ray Diffraction Structure of μ-Oxo-diiron(III) 9,10,19,20-Tetra-*n*-propylporphycene **3.72**.
This figure was generated using information down-loaded from the Cambridge Crystallographic Data Centre and corresponds to a structure originally reported in reference 12. Each of the iron atoms is coordinated in a square-pyramidal geometry. The porphycene ligands are dome shaped (the maximum deviation from the mean macrocyclic plane is 0.40 Å). Each iron atom sits 0.67 Å above the mean N_4 plane and the Fe-O-Fe angle is 145.3°. Atom labeling scheme: iron: ⬤; carbon: ○; nitrogen: ◑. Hydrogen atoms have been omitted for clarity

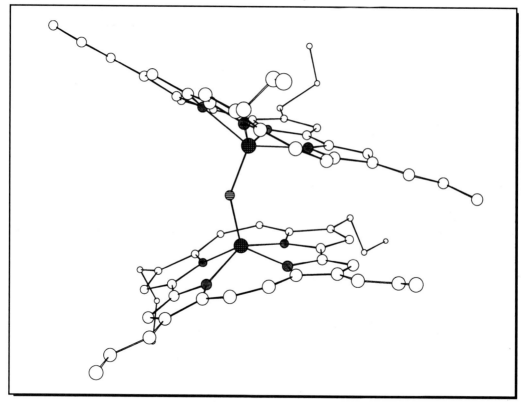

The reaction between porphycene, and second and third row transition-metal carbonyls has also been investigated. Here, it was found, for instance, that the reaction between decacarbonyldirhenium ($Re_2(CO)_{10}$) and tetrapropylporphycene **3.20** in decalin leads to the formation of the bis[tricarbonylrhenium(I)] complex **3.86** (Scheme 3.1.13),[28] wherein the two metal centers are bound in a "sitting-atop" fashion (as judged from a single crystal X-ray diffraction analysis; Figure 3.1.17). On the other hand, when porphycene **3.20** is reacted with $Ru_3(CO)_{12}$ in decalin, a monometalated complex (**3.87**) may be isolated in 71% yield.[33] Similarly, reaction of **3.20** with $Os_3(CO)_{12}$ in diethyleneglycol monomethylether was found to afford a 75% yield of the osmium(II)-porphycenate **3.88** (Scheme 3.1.14).[33] Interestingly, when a dichloromethane solution of this latter osmium(II) complex (**3.88**) was treated with *m*-chloroperoxybenzoic acid (*m*-CPBA), the hexavalent osmium(VI)porphycene(O)$_2$ complex **3.89** was obtained in *c*. 50% yield. A similar reaction could not, however, be effected in the case of the ruthenium complex **3.87**. Nevertheless, the in-plane incorporation of the osmium metal center serves as a cogent demonstration that porphycene is capable of stabilizing cations bound in high oxidation states.

Figure 3.1.15   Single Crystal X-Ray Diffraction Structure of μ-Oxo-diiron(III) 9,10,19,20-Tetra-*t*-butylporphycene **3.73**.
This figure was generated using information down-loaded from the Cambridge Crystallographic Data Centre and corresponds to a structure originally reported in reference 12. Each of the iron atoms is coordinated in a square-pyramidal geometry. The porphycene ligands are slightly dome shaped (the maximum deviation from the mean macrocyclic plane is 0.26 Å). Each iron atom sits 0.60 Å above the mean N$_4$ plane and the Fe-O-Fe angle is 178.5°. Atom labeling scheme: iron: ●; carbon: ○; nitrogen: ◉. Hydrogen atoms have been omitted for clarity

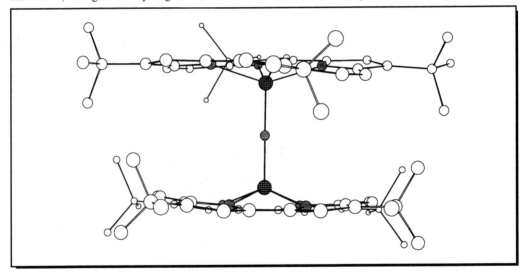

3.1.3   Porphycene Transformations

3.1.3.1   Dihydroporphycene

When subjected to catalytic hydrogenation with 10% Pd/C in ethyl acetate, porphycene **3.2** was found to be reduced to the 2,3-dihydroporphycene **3.90**.[38] Interestingly, this same 2,3-dihydroporphycene **3.90** was also obtained as a side-product (*ca*. 1% yield) when the reductive carbonyl coupling of diformyl bipyrrole

3.7 was effected with TiCl₃/LiAlH₄, rather than TiCl₄/Zn as the reducing agent (Scheme 3.1.15). In the latter instance, it was presumed that when TiCl₃/LiAlH₄ is used, the rate of prototropic rearrangement from **3.8** to **3.90** is faster than the direct dehydrogenation of **3.8** to give **3.2**.

Figure 3.1.16   Structures **3.75–3.84**

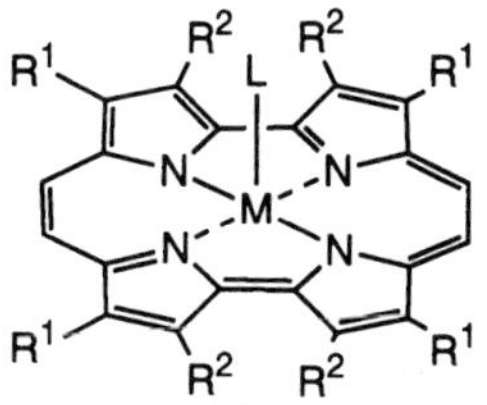

3.75a. R¹ = *n*-Pr, R² = H, M = Al, L = Me
3.75b. R¹ = *n*-Pr, R² = H, M = Al, L = OH
3.75c. R¹ = *n*-Pr, R² = H, M = Al, L = Cl
3.75d. R¹ = *n*-Pr, R² = H, M = Al, L = OAc
3.76. R¹ = Et, R² = Me, M = Al, L = OH
3.77. R¹ = *n*-Pr, R² = H, M = Mn, L = Cl
3.78a. R¹ = *n*-Pr, R² = H, M = Co, L = Cl
3.78b. R¹ = *n*-Pr, R² = H, M = Co, L = Me
3.79a. R¹ = Et, R² = Me, M = Fe, L = Cl
3.79b. R¹ = Et, R² = Me, M = Fe, L = Ph
3.80a. R¹ = Et, R² = Me, M = In, L = Cl
3.80b. R¹ = Et, R² = Me, M = In, L = Ph
3.81. R¹ = *n*-Pr, R² = H, M = Ga, L = Cl

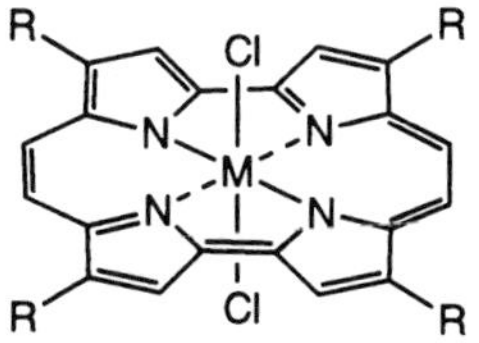

3.82. R = H, M = Sn
3.83. R = *n*-Pr, M = Sn
3.84. R = *n*-Pr, M = Ge

Scheme 3.1.12

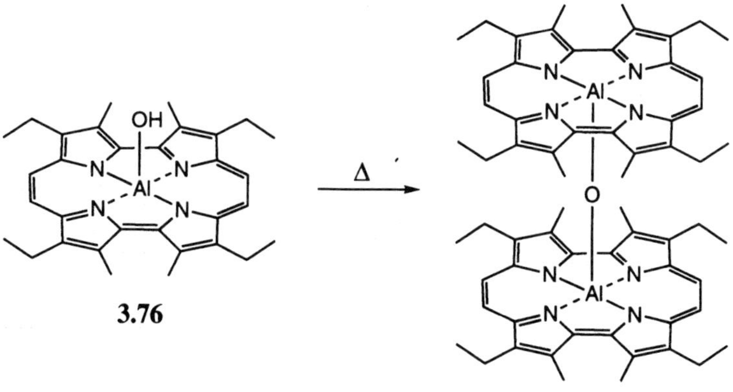

Scheme 3.1.13

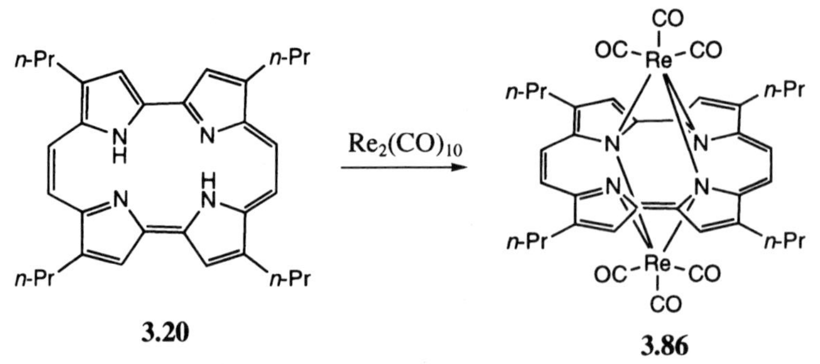

3.20   **3.86**

Scheme 3.1.14

3.20   **3.87**. M = Ru   **3.89**. M = Os
                  3.88. M = Os   (M ≠ Ru)

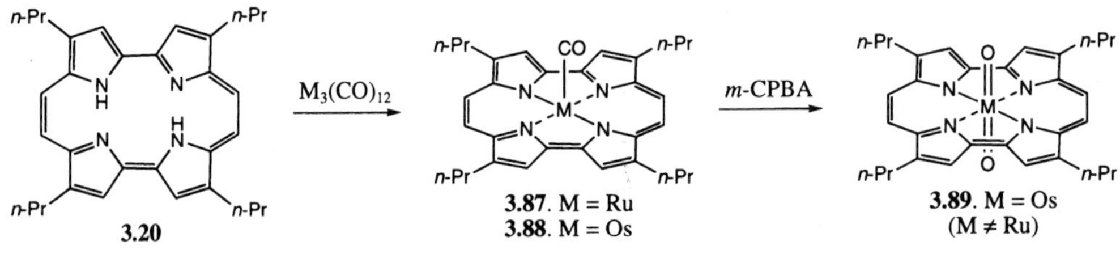

The reduced porphycene **3.90** may be considered as an analog of chlorin **3.91** (Figure 3.1.18), a species which, in turn, may be considered as a dihydro-analog of porphyrin. In the specific case of **3.90**, confirmation of the chlorin-like nature of the system in terms of formal π-electron bookkeeping, was obtained from ^{1}H NMR experiments.[38] These same experiments provided insights into the specific site of reduction. For instance, it was found that the β-pyrrolic CHCH protons resonated at lower field than analogous protons in pyrrole (i.e., at δ = 8.2–9.3 ppm). On the other hand, the CH_2CH_2 protons of the reduced pyrrolic subunit were found at a relatively upfield location (i.e., δ = 4.6–4.9 ppm); this is as one would expect for aliphatic protons of this type.

More definitive support for the assigned structure was obtained from a single crystal X-ray diffraction structure of **3.90** (Figure 3.1.19). This structure, like that of the porphycene **3.2**, revealed a nearly planar tetrapyrrolic skeleton. It also served to establish that the CH_2CH_2 bond length of the erstwhile pyrrole ring was 1.45 Å, while the CHCH bond lengths were on the order of 1.39 Å. These findings, taken together, were considered consistent with the formulation of this species as being the 2,3-dihydroporphycene **3.90**.[38]

Figure 3.1.17   Single Crystal X-Ray Diffraction Structure of
bis[Tricarbonylrhenium(I)] 2,7,12,17-Tetra-*n*-propylporphycene **3.86**.
This figure was generated using information down-loaded from the Cambridge Crystallographic
Data Centre and corresponds to a structure originally reported in reference 28. The N$_4$ core atoms
are in an approximately square arrangement. Each rhenium(I) atom is coordinated to three CO
groups and to three nitrogen atoms of the porphycene ligand. The average displacement of the
rhodium atoms from the mean porphycene plane is 1.50 Å. Atom labeling scheme: rhenium: ◑;
carbon: ○; nitrogen: ◓; oxygen: ⊖. Hydrogen atoms have been omitted for clarity

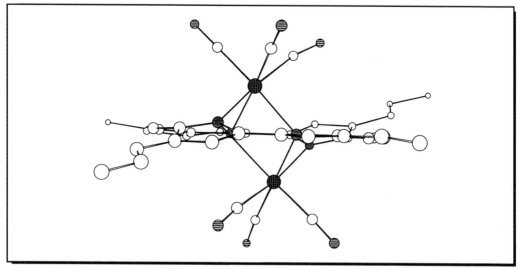

The 9,10,19,20-tetrapropylporphycene **3.39** could also be reduced by treating
with H$_2$ in the presence of Pd-C as the reducing agent.[17] Such a reaction affords the
cyclophane-type *N,N'*-dihydroporphycene **3.36**, a species that proved identical in all
respects to the intermediate cyclophane-type product isolated from the original coupling reaction leading to 9,10,19,20-tetrapropylporphycene **3.39** (see Scheme 3.1.4).
When this *N,N'*-dihydro species **3.36** was treated with a catalytic amount of concentrated sulfuric acid in dichloromethane, tautomerization to the 2,3-dihydroporphy-
cene **3.92** occurred in 52% yield (Scheme 3.1.16). Interestingly, the rearrangement of
3.36 into **3.92** under these reaction conditions was found to be practically irreversible. However, oxidation of the 2,3-dihydroporphycene **3.92** to educt **3.39** proved to
be rather facile and occurred rapidly enough such that 1–2% of **3.39** was always
present as a contaminant.[17]

Reducing porphycene in the presence of alcoholic sodium metal gives rise to a
different reduction product. Under these conditions, it was found, for instance, that
the unsubstituted porphycene **3.2** affords the 9,10-dihydroporphycene **3.93** (Scheme
3.1.17) instead of an *N,N*-dihydro- or 2,3-dihydroporphycenes, such as **3.36** and **3.90**
(afforded by the H$_2$/Pd-C reduction process discussed above).[16] A species such as
3.93 thus bears a structural analogy to the phlorins of the porphyrin series.
Unfortunately, structural information for this class of molecules, including **3.93**, is
currently lacking.

Scheme 3.1.15

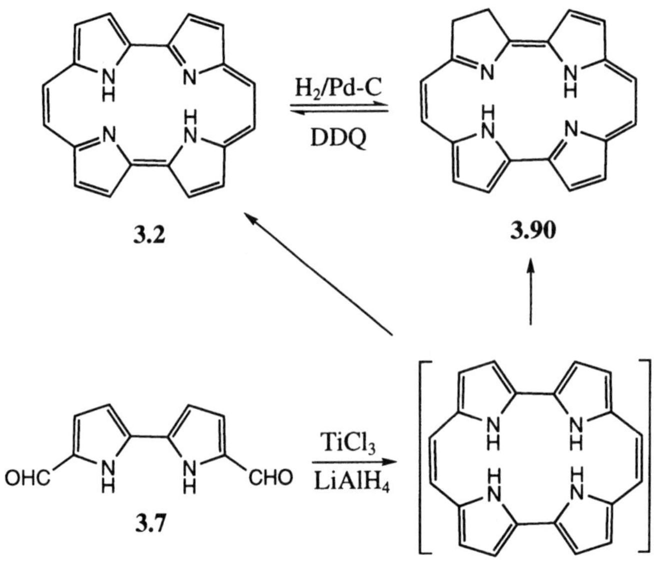

3.2   **3.90**

3.7   **3.8**

Figure 3.1.18 Structure **3.91**

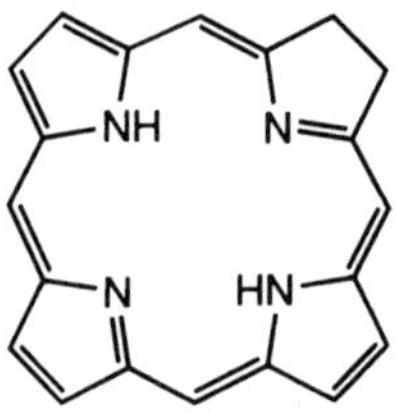

3.91
"chlorin"

3.1.3.2   **Porphycenes for PDT**

Several functionalized porphycenes have been prepared in the context of generating systems that may act as photosensitizers for photodynamic therapy (PDT; discussed in Chapter 10). Specific examples of systems designed for study in this area include the β-methoxyethyl-functionalized porphycenes **3.25** and **3.30** and derivatives derived therefrom (described below).[11] A 9-acetoxy substituted system **3.94** related to tetraether **3.25** has also been made,[39] as has the corresponding tetra-*n*-propyl analog **3.95** (Figure 3.1.20).[11]

Figure 3.1.19   Single Crystal X-Ray Diffraction Structure of 2,3-Dihydroporphycene **3.90**.

This figure was generated using information down-loaded from the Cambridge Crystallographic Data Centre and corresponds to a structure originally reported in reference[38]. The molecule is near-planar in the solid state (the maximum deviation from the mean macrocyclic plane is 0.09 Å). Top: top view; bottom: side view. Atom labeling scheme: carbon: ○; hydrogen: ○; nitrogen: ●. Selected hydrogen atoms have been omitted for clarity

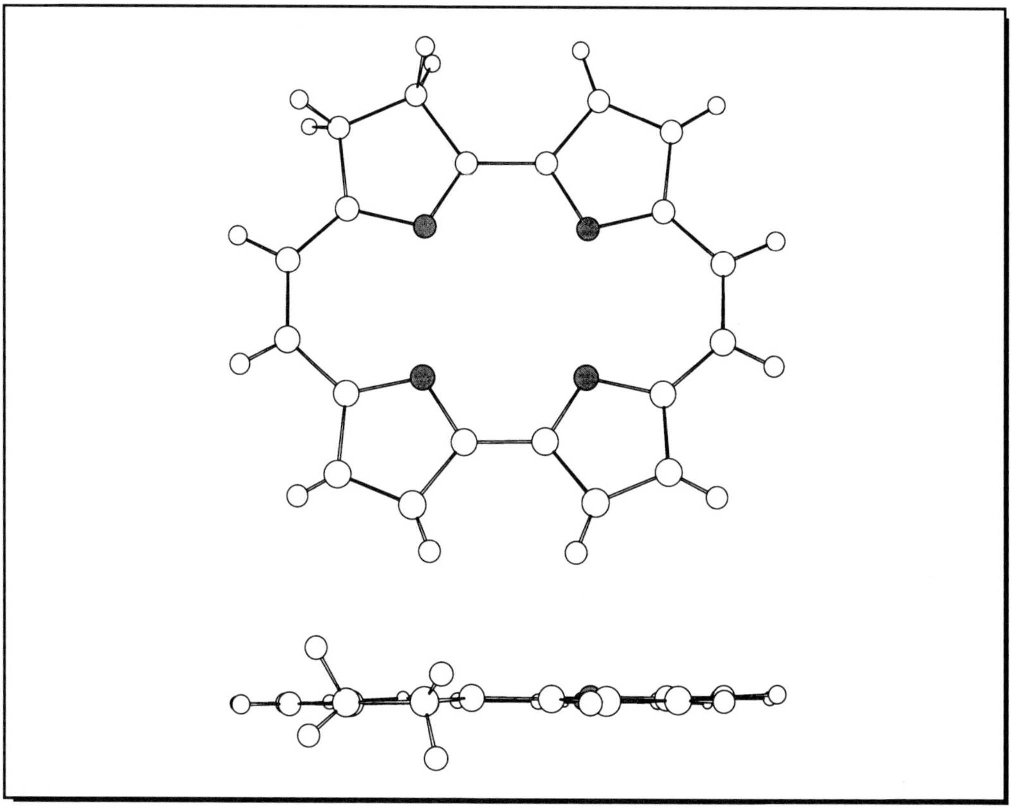

In the case of **3.25** and the difunctionalized derivative **3.30**, Richert, *et al.* have performed various functional group interconversion. This has afforded a number of differentially substituted porphycenes.[11] For instance, it was found that when the porphycene tetraether **3.25** was treated with 0.74 equivalents of BBr_3, the mono-refunctionalized derivative **3.96** could be isolated in 18% yield (Scheme 3.1.18). However, the selectivity in this reaction was low. In fact, it afforded, in addition to **3.96**, a 50–60% return of educt **3.25** along with the three possible isomeric dialcohols **3.97–3.99** (10–12% combined yield) and the trialcohol **3.100** in 4% yield. Fortunately, with the exception of the dialcohol isomers **3.97–3.99**, these products could be readily separated by standard chromatographic means. These types of isomer problems were not encountered in the case of diether **3.30**. Here, treatment with excess BBr$_3$ provided a relatively clean preparation of dialcohol **3.101** in 70% yield (Scheme 3.1.19).[11]

Scheme 3.1.16

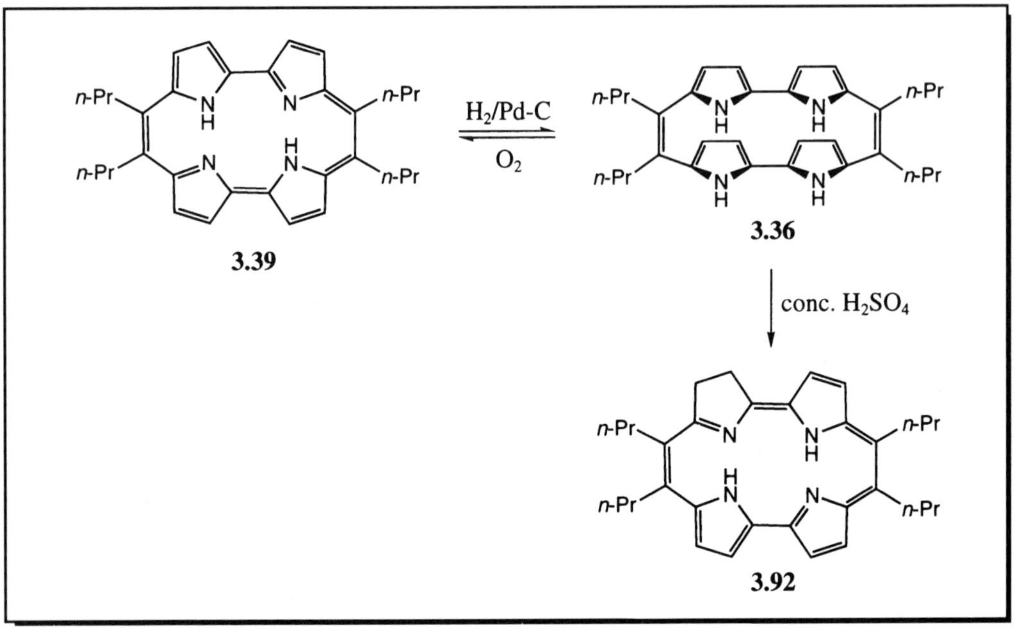

H$_2$/Pd-C
O$_2$

3.39

3.36

conc. H$_2$SO$_4$

3.92

Scheme 3.1.17

Na°
isoamyl alcohol

3.2

3.93

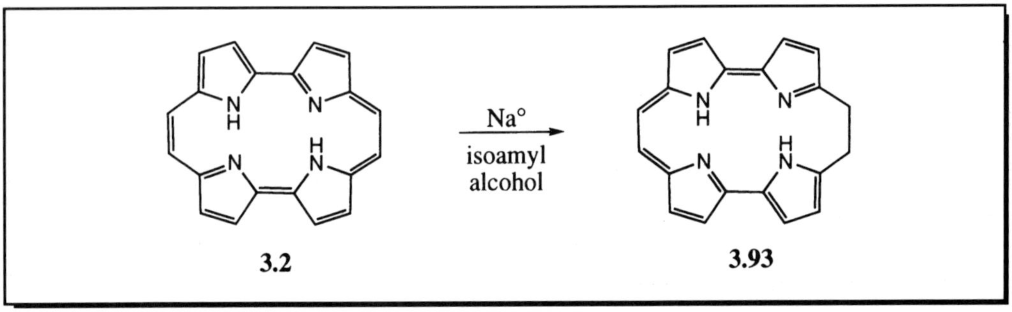

Richert, *et al.* have also synthesized the benzoporphycene derivatives **3.106** and **3.107** starting from tetraether **3.25** (Scheme 3.1.20).[11] Specifically, these workers found that the bromoethyl-functionalized porphycene **3.104** could be prepared in 19% yield by treating the nickel(II) complex of tetraether **3.25** (i.e., **3.102**) with BBr$_3$ and boric acid, followed by demetalation using concentrated sulfuric acid. This allowed access to the somewhat light-sensitive vinyl porphycene **3.105**. It was obtained in 86% yield by treating **3.104** with 1,8-diazabicyclol [S.4.0] Iunadec-7-ene (DBU). Subjecting **3.105** to a Diels–Alder reaction involving dimethylacetylene dicarboxylate then gave the benzoporphycene **3.106** in 60% yield (Scheme 3.1.20). Finally, careful hydrolysis of the benzo diester **3.106** with lithium hydroxide in tetrahydrofuran (THF)/water afforded the monoacid **3.107** in 65% yield.

Figure 3.1.20   Structures **3.25**, **3.30**, **3.94**, and **3.95**

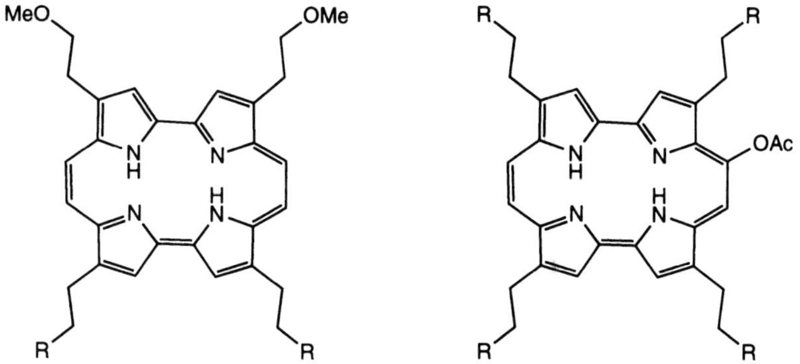

3.25. R = OMe
3.30. R = H

3.94. R = OMe
3.95. R = H

Scheme 3.1.18

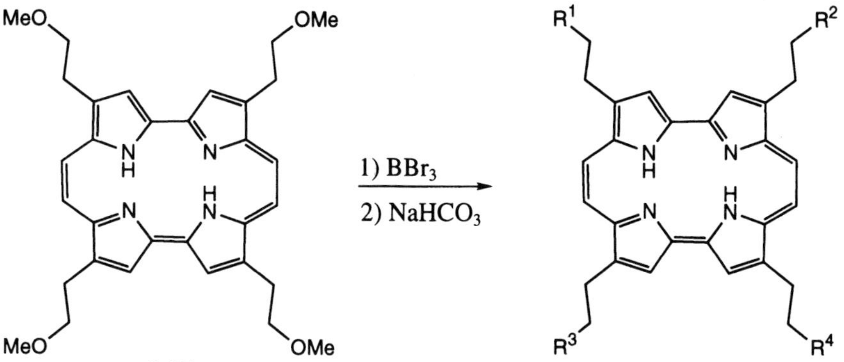

1) BBr₃

2) NaHCO₃

3.25

3.96. R^1 = OH, R^2 = R^3 = R^4 = OMe
3.97. R^1 = R^2 = OH, R^3 = R^4 = OMe
3.98. R^1 = R^3 = OH, R^2 = R^4 = OMe
3.99. R^1 = R^4 = OH, R^2 = R^3 = OMe
3.100. R^1 = R^2 = R^3 = OH, R^4 = OMe

Scheme 3.1.19

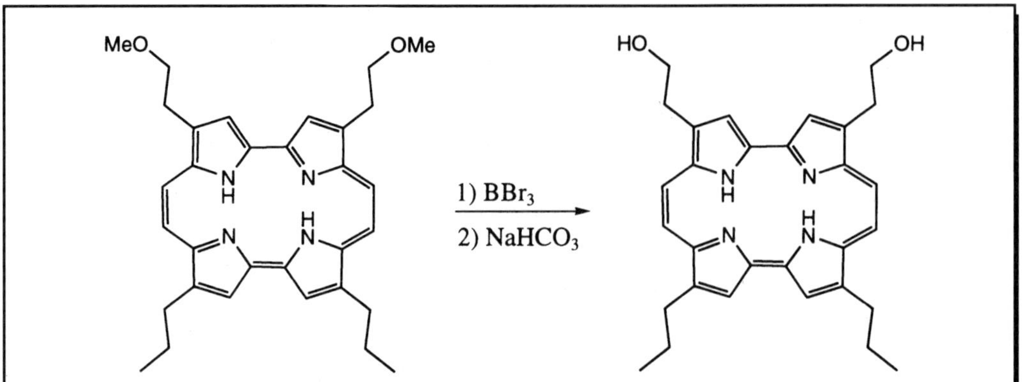

3.30   **3.101**

1) BBr₃
2) NaHCO₃

Scheme 3.1.20

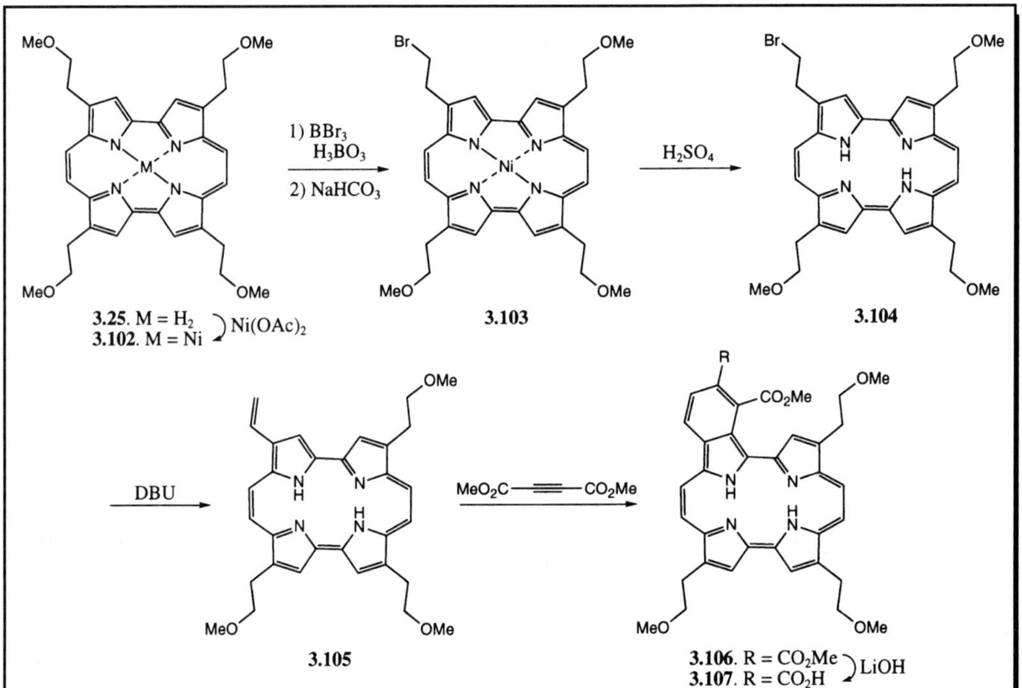

3.25. M = H₂ ⎱ Ni(OAc)₂
3.102. M = Ni ⎰

3.103

3.104

1) BBr₃
   H₃BO₃
2) NaHCO₃

H₂SO₄

DBU

MeO₂C━━━CO₂Me

3.105

3.106. R = CO₂Me ⎱ LiOH
3.107. R = CO₂H ⎰

3.1.3.3   *N*-Substituted Porphycenes

Studies involving the methylation of porphycenes and their analogs have been carried out by the groups of Vogel and Chang. The first example of this came from the group of Vogel and involved methylation of the 20 π-electron dioxa species **3.47**.[20] Specifically, these researchers found that treatment of dihydroporphycene **3.47** with methyliodide using potassium *t*-butoxide as a base afforded the dimethylated species **3.108** in 12% yield. Once isolated, this latter species could be oxidized readily using concentrated nitric acid in 70% aqueous perchloric acid. This gave the aromatic dication **3.109** in good (i.e., 89%) yield (Scheme 3.1.21).

Scheme 3.1.21

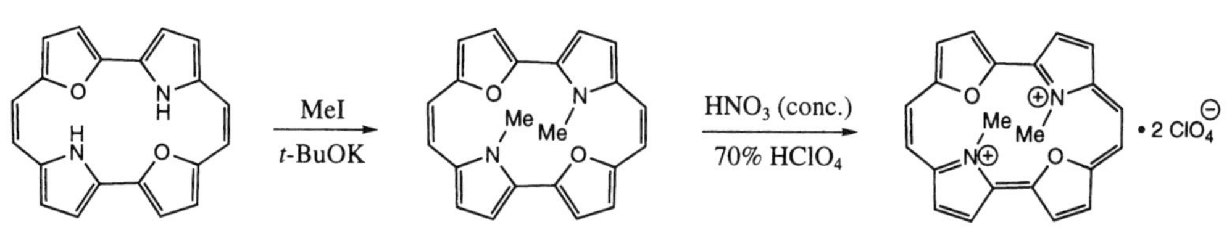

| 3.47 | 3.108 | 3.109 |

In the all-aza series, Vogel and coworkers investigated the methylation of 2,7,12,17-tetrapropylporphycene **3.20**. Here, these researchers found, among other things, that treating this porphycene with dimethylsulfate yields the *N*(1)-methylporphycene derivative **3.110** in 50% yield (Scheme 3.1.22).[40] They also found that subjecting this mono-methylated product to dimethylsulfate in the presence of potassium carbonate yielded the *N*,*N'*-dimethyl porphycene salt **3.111** (Scheme 3.1.23). Interestingly, after acidic workup, it was found that methylation in this case involved addition, as opposed to H-for-CH₃ replacement, and that this addition occurred on the pyrrole ring directly across from the site of initial methylation (i.e., at N(3)). These findings were taken as an indication that N(3) is the most basic nitrogen in the *N*(1)-methylated system **3.110**, a conclusion that was subsequently confirmed *via* direct protonation studies.[40]

In addition to the *N*(1),*N*(3)-dimethyl derivative **3.111**, a small amount of *N*,*N'N''*-trimethylporphycene **3.112** was also isolated from this reaction.[40] This compound was assigned as being the N(1),N(3),N(4)-structural isomer (Figure 3.1.21), and was isolated in the form of its perchlorate salt. Interestingly, under conditions similar to those described above, the corresponding tetra-*N*-methylporphycene derivative could not be obtained. Therefore, an attempt was made to prepare such a species using an alternative route. The specific route chosen involved carrying out a McMurry-type dimerization of the *N*,*N'*-dimethylbipyrrole dialdehyde **3.113**. This reaction was indeed found to produce the desired 20 π-electron species **3.114**, albeit in low (i.e. 2%) yield (Scheme 3.1.24). Unfortunately, however, all attempts to effect oxidation of **3.114** to the corresponding aromatic dication **3.115** failed.[40]

Scheme 3.1.22

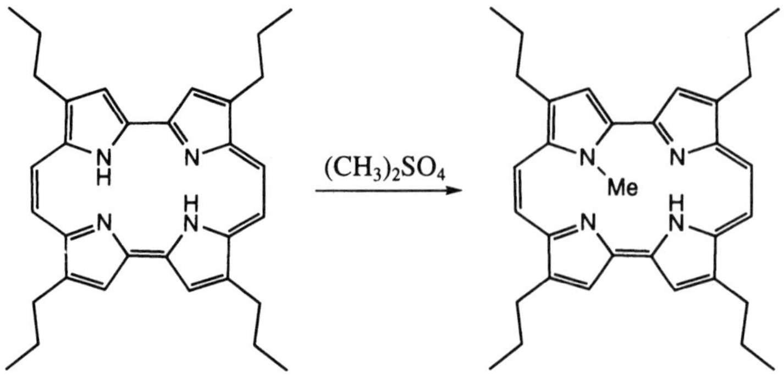

3.20　　　　　　　　　　　　　　　　3.110

Scheme 3.1.23

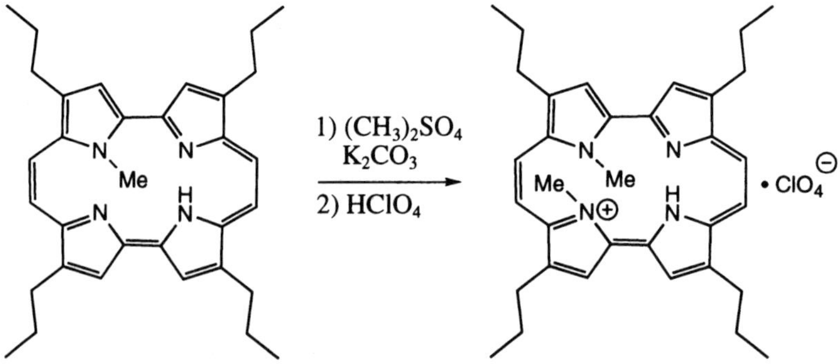

3.110　　　　　　　　　　　　　　　3.111

More recently, the group of Chang has investigated the synthesis of *N,N'*-bridged porphycenes.[41] The first of these, **3.119**, was actually isolated first as an unexpected product during an attempt to carry out a Vilsmeier-type formylation reaction. Specifically, it was found that treatment of the copper(II) complex of tetrapropylporphycene (**3.64**) with Vilsmeier's reagent (POCl₃–DMF complex) did not afford the expected formylated porphycene product. Rather, a compound was isolated that gave spectroscopic data consistent with the author's proposed *N,N*-bridged structure **3.119** (Scheme 3.1.25). It was subsequently determined that this same product could be prepared starting from the metal-free porphycene **3.20** in yields as high as 92%.

Chang and coworkers went on to demonstrate the generality of the above reaction. For instance, when the substituted formamides **3.117** and **3.118** were used in place of *N,N*-dimethylformamide **3.116**, the *N,N'*-bridged porphycenes **3.120** and **3.121** could be prepared in *ca.* 94% yield. These same workers also learned that this type of reaction could be used to generate the doubly bridged product **3.123**.[41] In this instance, the key chemistry involved treating porphycene **3.20** with a 1,4-piper-azinedicarboxaldehyde-POCl$_3$ complex; this afforded **3.123** in 40% yield (Scheme 3.1.26).

Figure 3.1.21   Structure 3.112

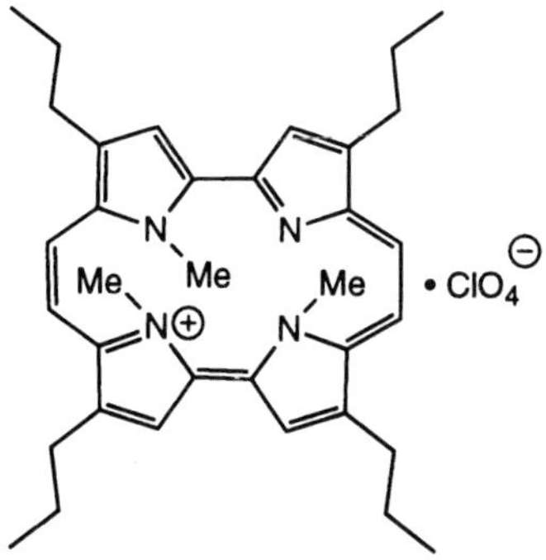

3.112

Scheme 3.1.24

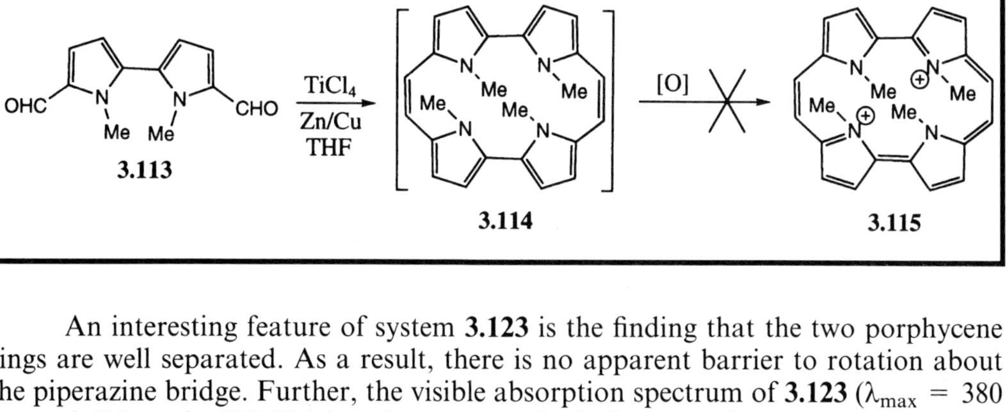

An interesting feature of system **3.123** is the finding that the two porphycene rings are well separated. As a result, there is no apparent barrier to rotation about the piperazine bridge. Further, the visible absorption spectrum of **3.123** (λ_{max} = 380 nm and 618 nm in CH$_2$Cl$_2$) is rather unperturbed relative to the monomer **3.119** and displays no evidence of appreciable exitonic coupling between the rings.[41]

The *N,N'*-bridged porphycenes **3.119–3.121** were found to exhibit little tendency to act as viable metal-chelating agents.[41] By contrast, the mono-*N*-methylpor-phycene **3.110** has been demonstrated to be capable of forming stable metal complexes.[40] For instance, treatment of *N*(1)-methylporphycene **3.110** with a metal chloride salt such as $ZnCl_2$, $CoCl_2$, or $CuCl_2$, was found to result in the formation of the corresponding divalent metal complexes **3.124–3.126** in 60–80% yield (Scheme 3.1.27). Interestingly, however, reaction with divalent nickel or palladium salts resulted in rapid demethylation and metal insertion to afford the metal(II) porphycene complexes **3.65** and **3.66** (Figure 3.1.10). Indeed, it was also found that zinc, cobalt, and copper salts also could effect demethylation, albeit much more slowly than for nickel or palladium.

Scheme 3.1.25

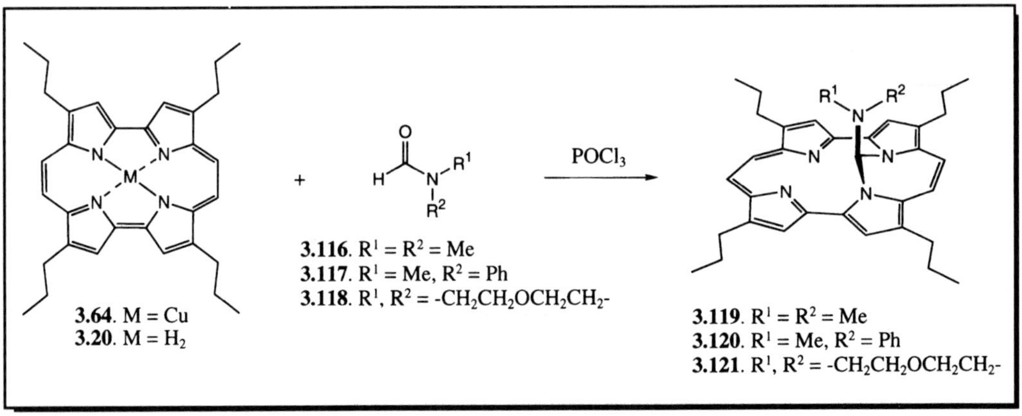
POCl$_3$
3.64. M = Cu
3.20. M = H$_2$
3.116. R^1 = R^2 = Me
3.117. R^1 = Me, R^2 = Ph
3.118. R^1, R^2 = -CH$_2$CH$_2$OCH$_2$CH$_2$-
3.119. R^1 = R^2 = Me
3.120. R^1 = Me, R^2 = Ph
3.121. R^1, R^2 = -CH$_2$CH$_2$OCH$_2$CH$_2$-

Scheme 3.1.26

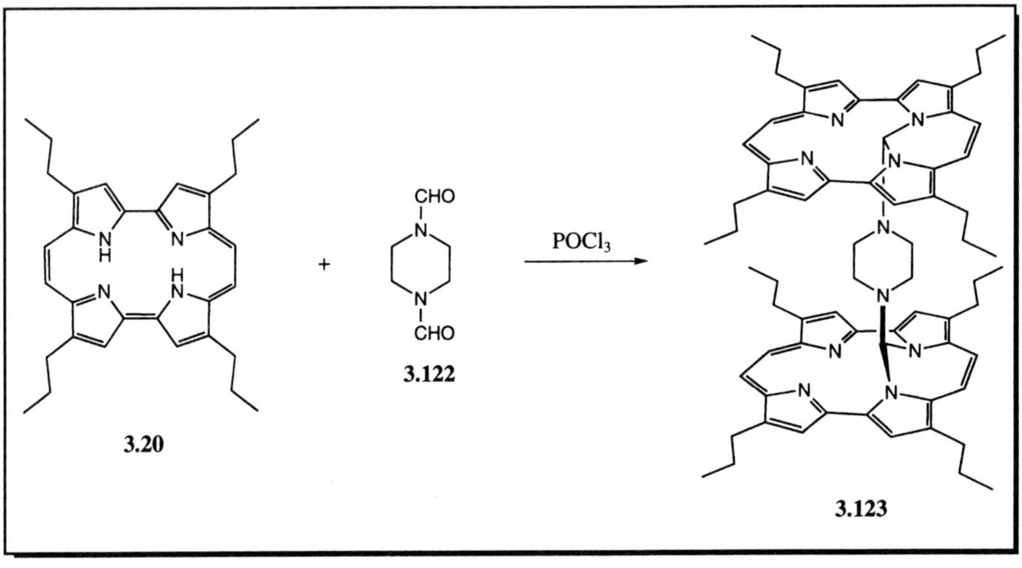
POCl$_3$
CHO
CHO
3.20
3.122
3.123

Scheme 3.1.27

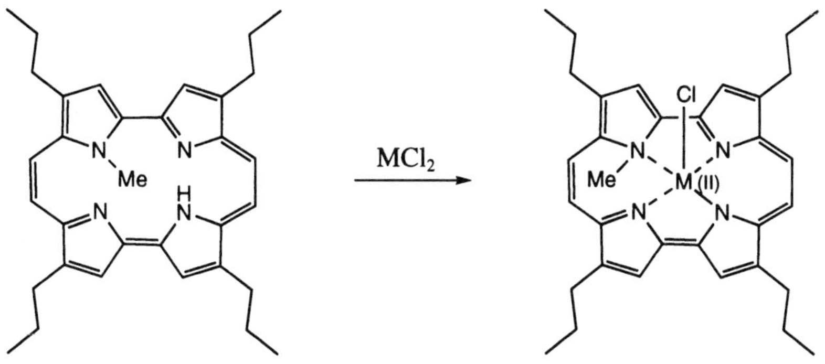

3.110   **3.124**. M = Zn
                                             3.125. M = Co
                                             3.126. M = Cu

Scheme 3.1.28

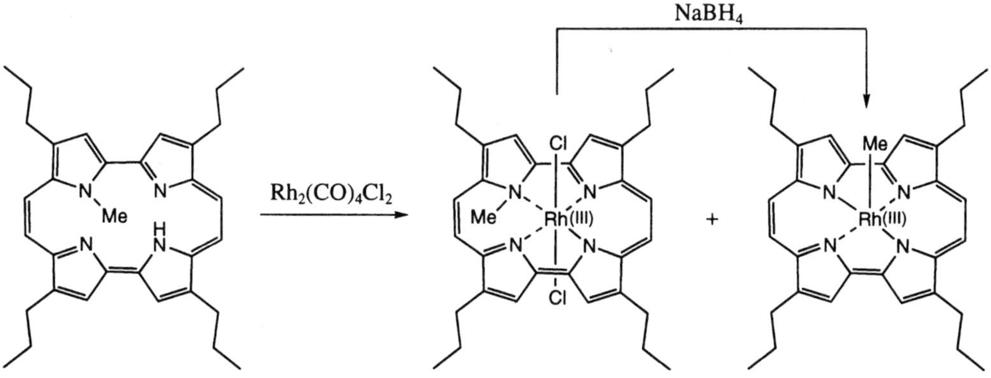

3.110   **3.127**   **3.128**

The reaction of N(1)-methylporphycene **3.110** with one equivalent of $Rh_2(CO)_4Cl_2$ was found to produce a mixture of two products.[40] The first of these was found to be the dichlororhodium(III) complex **3.127**, isolated in *ca.* 70% yield. The other complex, also determined to be a rhodium(III) complex, was isolated in 15–20% yield. Interestingly, in this latter case, it was found that migration of a methyl group (from N(1) to the metal center) had occurred during the course of the reaction. This resulted in the formation of complex **3.128** (Scheme 3.1.28). Interestingly, if two molar equivalents of the metal source were used instead of one, the product ratio between **3.127** and **3.128** is reversed (i.e., a 15–20% yield of

3.127 and a 70% yield of **3.128** is obtained). In either event, the methyl migration that occurs spontaneously during the reaction could also be effected at will by treating the dichlororhodium complex **3.127** with sodium borohydride.

3.2   [18]Porphyrin-(2.1.0.1) ("Corrphycene" or "Porphycerin")

3.2.1   Synthesis

3.2.1.1   The Approach of Vogel, Sessler, and Guilard

Inspired by the rich chemistry of porphycene, Vogel and Sessler took up the challenge of preparing a second completely synthetic porphyrin isomer.[42,43] This culminated in the synthesis of the etio- and octaethyl derivatives of an isomer shown at the beginning of this chapter in its generalized, unsubstituted form **3.3**. This system may be referred to as [18]porphyrin-(2.1.0.1) according to the nomenclature system of Franck. However, because this new isomer contains both a direct pyrrole–pyrrole linkage (such as that seen in corrole) and a dimethine bridge (such as that seen in porphycene), this macrocycle was named "corrphycene" by Vogel and Sessler.[†] Interestingly, nearly concurrent with the original Vogel–Sessler report, Guilard reported a synthesis of this same isomer. It was prepared *via* essentially the same route, and assigned the name "porphycerin".[44] More recently, an improved synthesis of corrphycene (authors' choice of names) was reported by Falk.[45]

For the sake of illustration, the Vogel–Sessler synthesis of the "etio"-derivative of [18]porphyrin-(2.1.0.1) **3.135** is shown in Scheme 3.2.1. It has as its key step the reductive (McMurry) coupling of the formyl groups at the termini of the tetrapyrrolic precursor **3.129**. As is generally true in the case of porphycene syntheses, the tetrahydro intermediate **3.132** was not isolated. Instead, oxidation was immediately effected using either air or $FeCl_3$. This afforded the expected 18 π-electron aromatic porphyrin isomer **3.135** in *ca.* 2% yield. This procedure, which has now been optimized to provide yields in the 10–20% range, has been applied in analogous fashion to the synthesis of the corresponding the octamethyl and octaethyl derivatives **3.136** and **3.137**.[43,46]

3.2.1.2   Falk's Synthesis

The Falk procedure (Scheme 3.2.2), in contrast to the Sessler–Vogel one, relies on an Ullmann-type reaction using a dibromo or diiodo tetrapyrrolic precursor (e.g., **3.138–3.141**) in its key, macrocycle-forming step.[45] In accord with this general strategy, these workers found that heating the dibromo tetrapyrrole **3.138**, or the

---

[†]In the Abstracts of the 207th American Chemical Society National Meeting (Inorg. Abstract no. 295), San Diego, CA, USA, March 13–17, 1994, the name "porphylene" was suggested for this class of compounds. However, the more descriptive name "corrphycene" is now preferred.

diiodo analog **3.139**, in DMF in the presence of metallic copper gave a 50% yield of the corresponding copper(II) corrphycene **3.142**. Similar yields of the methoxy-carbonylethyl-substituted copper(II) corrphycene **3.143** could be obtained starting from either **3.140** or **3.141**. Once the metallocorrphycenes were in hand, demetalation could be effected by treating with 98% H_2SO_4. This afforded the metal-free corrphycenes **3.136** and **3.144** in 80% and 82% yield (based on **3.142** and **3.143**), respectively.

Scheme 3.2.1

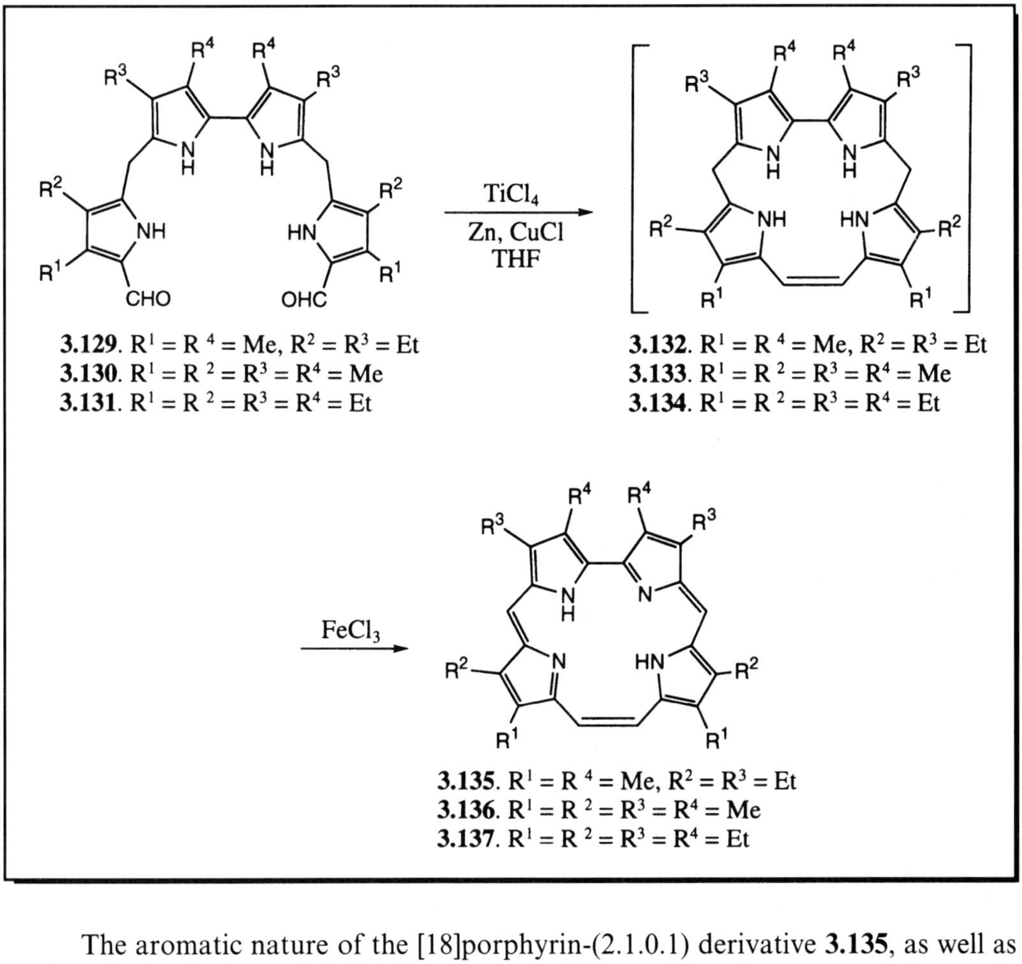

3.129. $R^1 = R^4 = Me, R^2 = R^3 = Et$
3.130. $R^1 = R^2 = R^3 = R^4 = Me$
3.131. $R^1 = R^2 = R^3 = R^4 = Et$

3.132. $R^1 = R^4 = Me, R^2 = R^3 = Et$
3.133. $R^1 = R^2 = R^3 = R^4 = Me$
3.134. $R^1 = R^2 = R^3 = R^4 = Et$

3.135. $R^1 = R^4 = Me, R^2 = R^3 = Et$
3.136. $R^1 = R^2 = R^3 = R^4 = Me$
3.137. $R^1 = R^2 = R^3 = R^4 = Et$

The aromatic nature of the [18]porphyrin-(2.1.0.1) derivative **3.135**, as well as analogs **3.136**, **3.137**, and **3.144** (systems we will hereafter refer to as corrphycenes), was confirmed by the observation of sustained diamagnetic ring current effects in its ^{1}H NMR spectrum.[43] The UV-vis absorption spectrum (in benzene) is also consistent with the proposed aromaticity and with an 18-π-electron formulation. Specifically, a strong Soret-like absorption is observed at 415 nm, along with several weaker Q-type transitions at 511, 540, 577, 584, and 632 nm, respectively.

Scheme 3.2.2

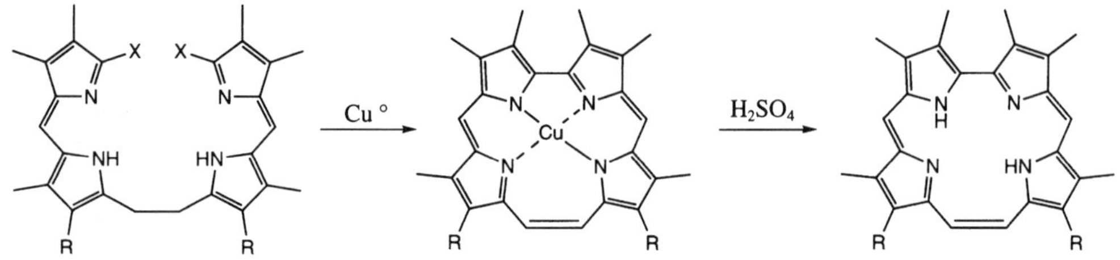

3.138. R = Me, X = Br
3.139. R = Me, X = I
3.140. R = CH$_2$CH$_2$CO$_2$Me, X = Br
3.141. R = CH$_2$CH$_2$CO$_2$Me, X = I

3.142. R = Me
3.143. R = CH$_2$CH$_2$CO$_2$Me

3.136. R = Me
3.144. R = CH$_2$CH$_2$CO$_2$Me

Confirmation of the structural assignment for this class of molecules also came in the form of a single crystal X-ray diffraction structure of the octaethyl corrphycene derivative **3.137**.[43] By inspection of this structure (Figure 3.2.1), it becomes immediately apparent that this porphyrin analog can adopt a nearly planar conformation (the maximum deviation of the corrphycene C and N atoms from the mean plane of the ring is approximately 0.08 Å). From this same structure, it is clear that the four core nitrogen atoms of the macrocycle serve to define the corners of a trapezoid. This is in contrast to the square and rectangular core geometries typically observed for porphyrins and porphycenes, respectively.

3.2.2   *Metal Complexes of Corrphycene*

In spite of the trapezoidal geometry observed for the core, it has been demonstrated that macrocycle **3.137** and other corrphycene derivatives act as remarkably good ligands for a wide range of metal cations. To date, complexes with no less than 19 different metals including Mg(II), Cu(II), Ni(II), Al(III), Fe(III), and Sn(IV) have been isolated using standard porphyrin metalation procedures.[3] Single crystal X-ray structures carried out on several of these complexes revealed in-plane coordination of the metal cation, and retention of the overall planarity observed in the case of the metal-free form of the macrocycle.[46] A case in point is the cobalt complex **3.145** wherein the metal center is found to lie directly within the mean macrocyclic plane (Figures 3.2.2 and 3.2.3). In this structure, the metal is in an octahedral-like ligand environment with two molecules of pyridine serving to complete the coordination sphere. As a result, this structure bears considerable analogy to many Co(II) porphyrinates.[47–50]

Figure 3.2.1   Single Crystal X-Ray Diffraction Structure of 2,3,6,7,11,12,17,18-Octaethylcorrphycene **3.137**.
This figure was generated using information down-loaded from the Cambridge Crystallographic Data Centre and corresponds to a structure originally reported in reference 43. The core nitrogen atoms define a regular trapezoid. The molecule has approximate C_2 molecular symmetry and is nearly planar (the maximum deviation of the corrphycene atoms from the mean macrocyclic plane is 0.08 Å). Top: top view; bottom: side view. Atom labeling scheme: carbon: ○; nitrogen: ●. Hydrogen atoms have been omitted for clarity

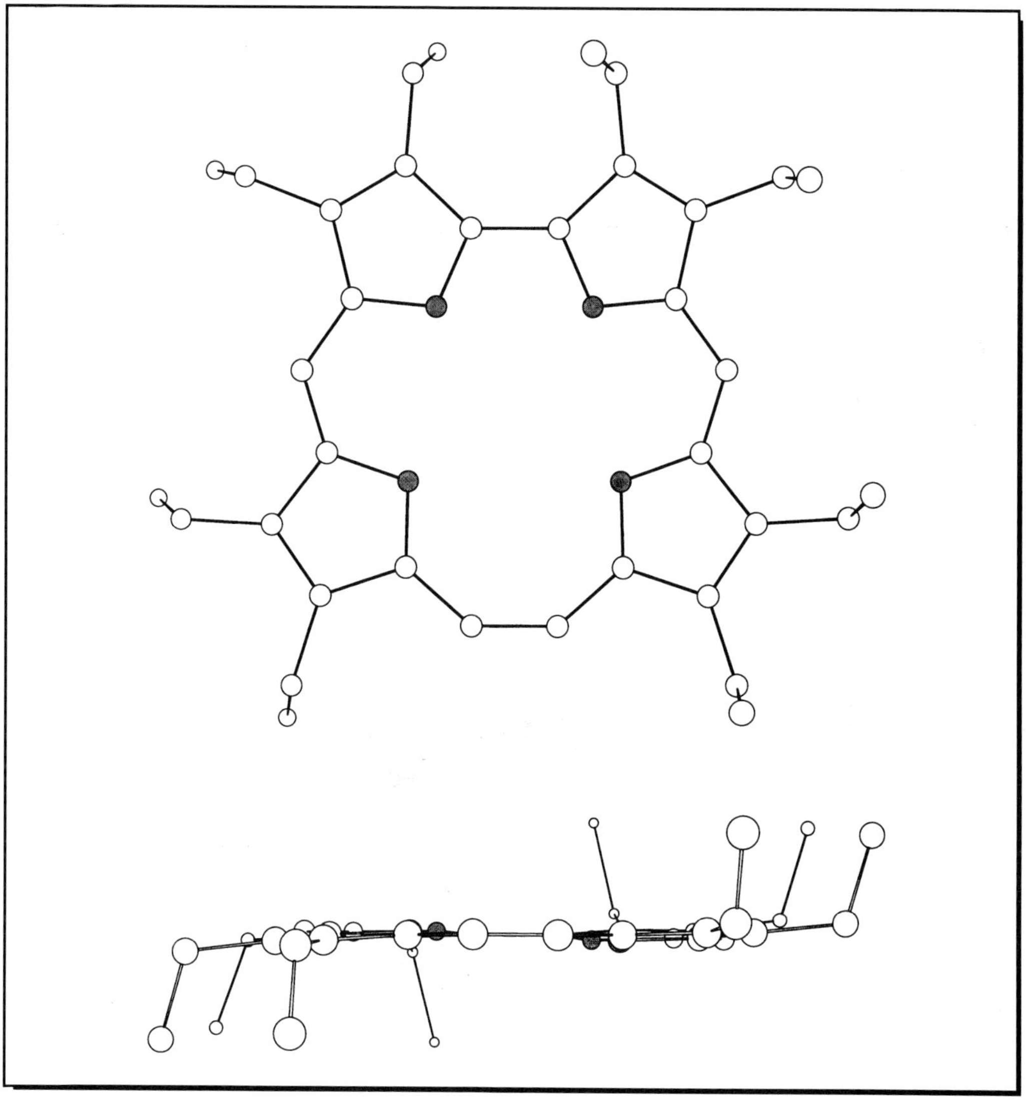

Figure 3.2.2   Structure **3.145**

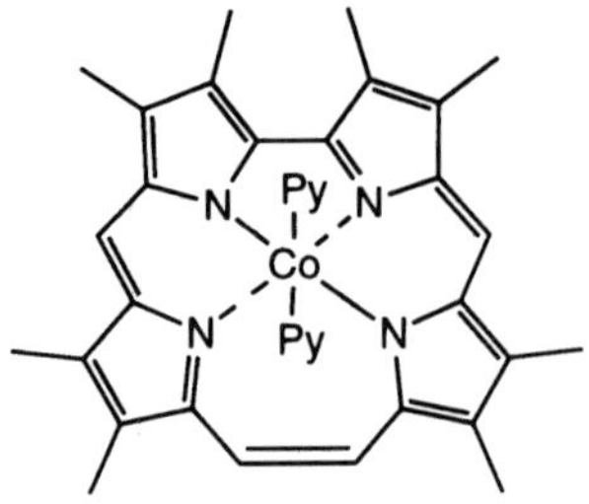

3.145

Figure 3.2.3   Single Crystal X-Ray Diffraction Structure of bis-Pyridine cobalt(II) 2,3,6,7,11,12,17,18-Octamethylcorrphycene **3.145**.

This figure was generated using data provided by Sessler and Vogel, *et al.*[46] The core nitrogen atoms define a regular trapezoid. The molecule is nearly planar (the maximum deviation from the mean macrocyclic plane is 0.14 Å) and the cobalt atom sits directly within the mean macrocyclic plane. The average Co–N(pyridine) distance is 1.95 Å. Atom labeling scheme: cobalt: ◕; carbon: ○; nitrogen: ◓: Hydrogen atoms have been omitted for clarity

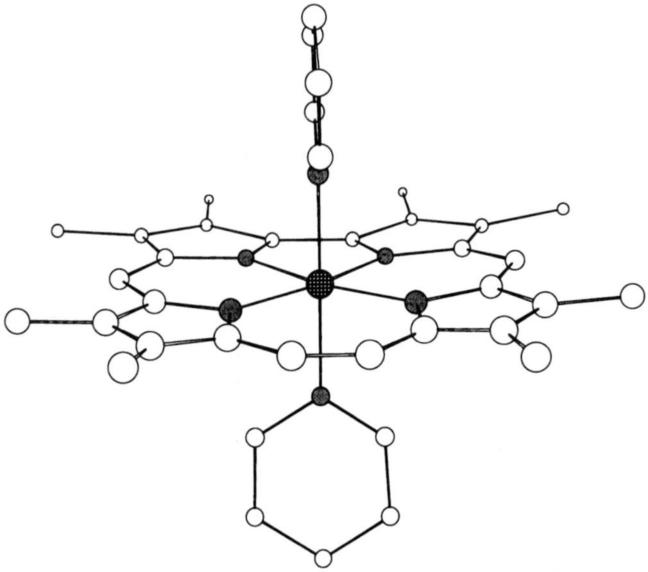

3.3 [18]Porphyrin-(2.1.1.0) ("Hemiporphycene")

Recently, a third kind of porphyrin isomer was reported by Callot and Vogel–Sessler.[43,51,52] This new isomer, [18]porphyrin-(2.1.1.0), given earlier in the form of its substituent-free parent structure (i.e., **3.4**) was first prepared serendipitously by Callot and coworkers in the form of two nickel(II) regioisomeric triphenylcarbethoxyethyl derivatives **3.147a–b** (Scheme 3.3.1).[52] These were isolated as ring-contraction products of a demetalation–metalation sequence involving homoporphyrin **3.146**. A single crystal X-ray structural analysis carried out on isomer **3.147a** revealed the overall connectivity of this species and demonstrated that the nickel(II) ion was bound in an in-plane fashion within the macrocyclic core (Figure 3.3.1).

Scheme 3.3.1

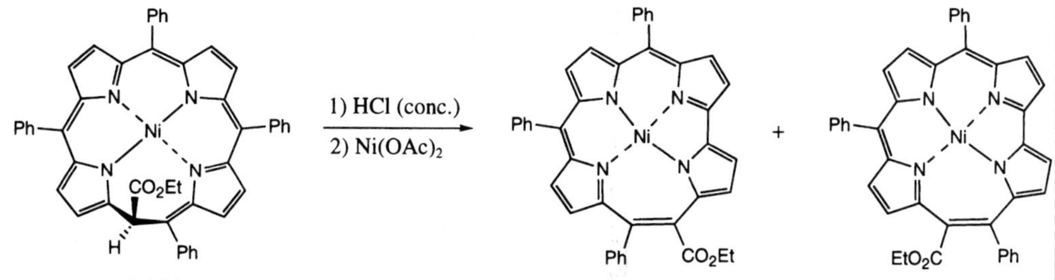

3.146 **3.147a** **3.147b**

The free-base form of isomer **3.147a** was prepared by treating it with concentrated sulfuric acid (to effect demetalation), followed by neutralization with ammonium carbonate (Scheme 3.3.2). Because the structure of **3.148** may be considered as

Scheme 3.3.2

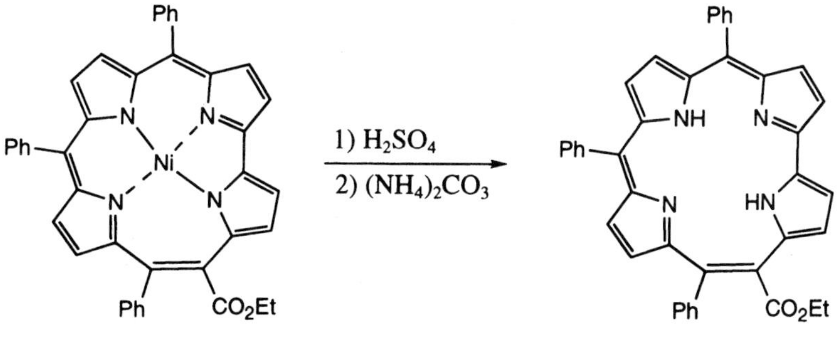

3.147a **3.148**

being "midway" between that of porphyrin and porphycene, the name "hemiporphycene" has been suggested for this class of molecules.[52]

Figure 3.3.1 Single Crystal X-Ray Diffraction Structure of Nickel(II) 9-Ethoxycarbonyl-10,15,20-triphenylhemiporphycene **3.147a**.

This figure was generated using information down-loaded from the Cambridge Crystallographic Data Centre and corresponds to a structure originally reported in reference 52. The coordination geometry about the nickel center is nearly square planar, but the four Ni–N bonds are inequivalent and range from 1.87 Å to 1.95 Å in length. The nickel atom sits 0.04 Å out of the mean plane defined by the four nitrogen atoms. The macrocycle is slightly saddle-shaped, with the maximum deviation from the mean macrocyclic plane being 0.30 Å. Atom labeling scheme: nickel: ◓; carbon: ○; nitrogen: ◉; oxygen: ◒. Hydrogen atoms have been omitted for clarity

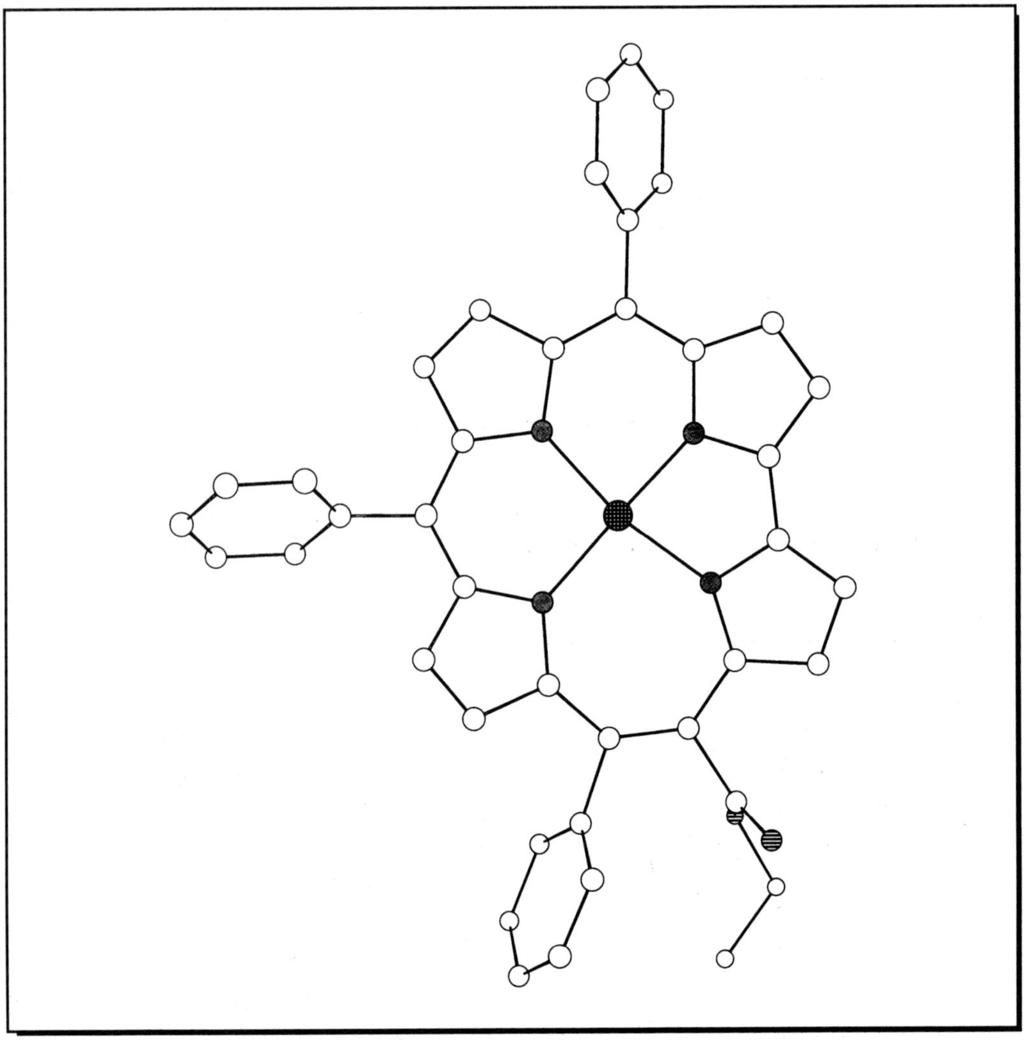

The Vogel–Sessler approach to hemiporphycene synthesis is more rational than the above. It was first reported briefly in footnote 11 of the original Vogel–Sessler corrphycene communication.[43] It has also been mentioned in several account-type reviews.[3,4] In any event, this approach, which has now been fine-tuned,[46] involves firstly the preparation of an α,ω-diformyl linear tetrapyrrolic intermediate such as **3.149**. Reductive carbonyl coupling of the formyl groups contained in this intermediate *via* the use of low-valent titanium then gives what is assumed to be the tetradehydro intermediate **3.151** (Scheme 3.3.3). Finally, oxidation, effected using FeCl₃ affords the hemiporphycene **3.152** in yields as high as 25%. This synthetic pathway thus resembles closely that used in the preparation of corrphycene. However, it contains one "twist" in that the key tetrapyrrolic intermediate **3.149** was found to undergo dimerization readily to give an acyclic octapyrrolic species (i.e., **3.150**) that, interestingly, was found to produce **3.152** when subject to the normal reductive coupling and subsequent oxidation procedures.

Scheme 3.3.3.

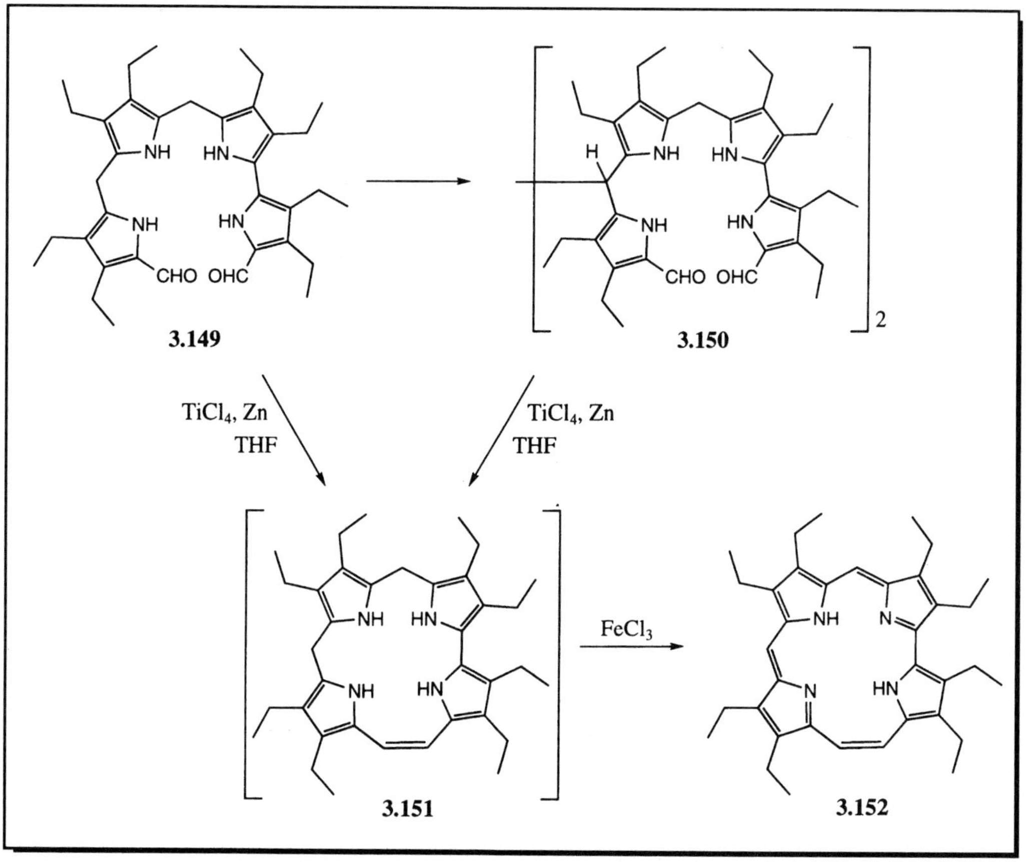

Figure 3.3.2   Single Crystal X-Ray Diffraction Structure of Free-base
2,3,6,7,12,13,17,18-Octaethylhemiporphycene **3.152**.
This figure was generated using data provided by Vogel and Sessler and coworkers.[51] The geometry
of the N_4 core of the macrocycle defines an unsymmetrical, irregular polygon. The macrocycle is,
however, very nearly planar (the maximum deviation of the hemiporphycene atoms from the mean
macrocyclic plane is 0.09 Å). Top: top view; bottom: side view. Atom labeling scheme: carbon: ○;
nitrogen: ◉. Hydrogen atoms have been omitted for clarity

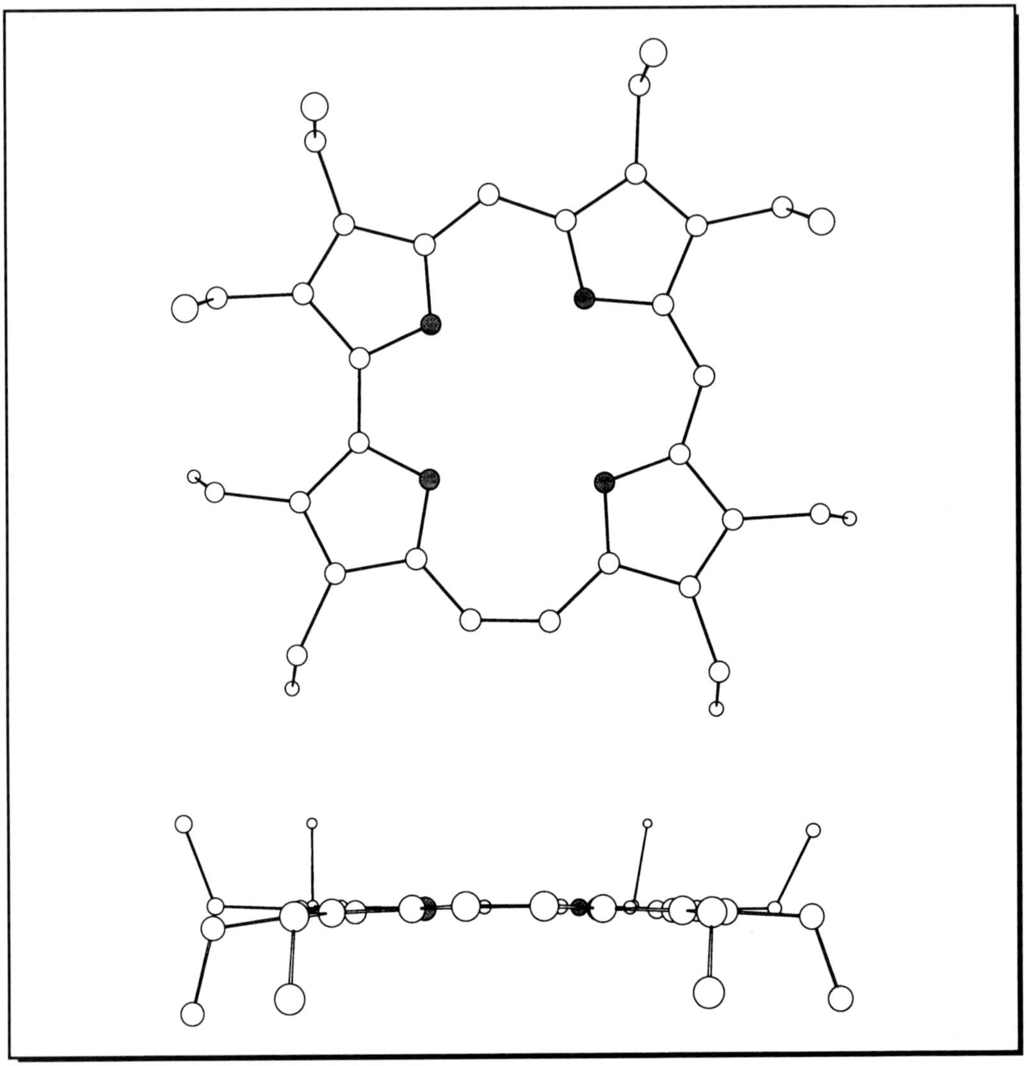

Hemiporphycene, like the porphyrins, may be considered as being elaborated
[18]annulenes. Such a formulation is consistent with the available data. For instance,
the UV-vis absorption spectrum of **3.152**, recorded in CH_2Cl_2, reveals a rather strong
Soret-like absorbance band at 405 nm, as well as four Q-type transitions at 512, 552,

583, and 632 nm, respectively.[51] Likewise, apparent diamagnetic ring current effects are seen in the ^{1}H NMR spectrum of this prototypical species.[51] Finally, a single crystal X-ray diffraction analysis, carried out on the free-base form of hemiporphycene **3.152**, revealed that it, like its isomeric porphyrin "cousins", can adopt a nearly planar conformation in the solid state (Figure 3.3.2).[51]

It is presumed that hemiporphycene will exhibit a rich metal coordination chemistry. Indeed, preliminary reports have indicated that complexes of divalent magnesium, zinc, nickel, and copper, trivalent iron, cobalt and rhodium, and tetravalent tin may readily be prepared.[3] Of particular interest in such metalation studies is the fact that metal complexes of hemiporphycene containing an axial substituent (e.g., **3.153**; Figure 3.3.3) bear metal-centered chirality because of the dissymmetric nature of the ligand. Unfortunately, further details with regard to this point and other aspects of hemiporphycene coordination chemistry are still lacking at this time.

Figure 3.3.3   Structure **3.153**

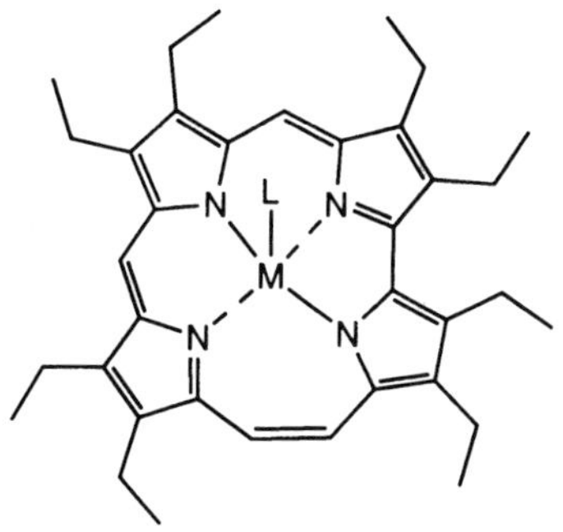

3.153

3.4   [18]Porphyrin-(3.0.1.0) ("Isoporphycene")

Recently, Vogel and coworkers have mentioned the successful isolation of yet another isomer of porphyrin.[3] This system, derived formally from the generalized system **3.5** (Figure 3.0.1), was termed isoporphycene by Vogel. In this work it was found that the key, macrocycle-forming ring closure of tetrapyrrole monoaldehyde **3.154** could be effected in the presence of palladium(II) chloride. Interestingly, three different isoporphycene species were found to form under these conditions. The first two of these were isolated as a mixture of the photochemically equilibrating palladium(II) complexes **3.155a** and **3.155b** (Scheme 3.4.1). The third species isolated was the 15-formyl derivative **3.156**. Unfortunately, at this point not enough is known about the nature of the electronic and coordinative abilities of [18]porphyrin-(3.0.1.0) to allow comparisons to be made between it and its isomers, porphyrin, porphycene, corrphycene and hemiporphycene.

Scheme 3.4.1

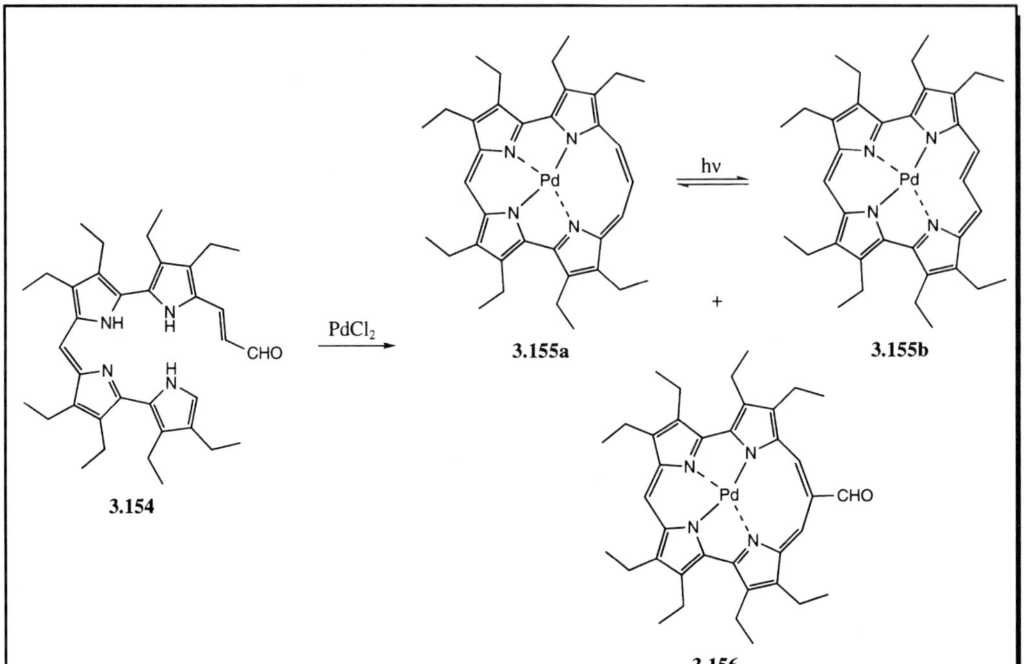

3.154 $\xrightarrow{\text{PdCl}_2}$ **3.155a**   **3.155b**

3.156

3.5   "Mutant" or "N-Confused" Porphyrins

A very different type of porphyrin isomer, containing pyrroles linked in an α–β′, as opposed to α–α′ fashion was reported recently. The first of these, systems **3.160** and **3.161**, appeared almost simultaneously in two separate publications submitted by Japanese and Polish research groups, respectively.[52,53] The Japanese group, led by Furuta, referred to this class of compounds as "N-confused" porphyrins, while the Polish group, led by Latos-Grażyński, termed their system a "mutant" porphyrin. Independent of the choice of name, it is important to appreciate that this class of molecules is characterized by the "inversion" of one of the pyrrolic subunits in the macrocyclic ring, such that the nitrogen atom is located outside of the core of the macrocycle (Scheme 3.5.1). In its place (i.e., within the macrocycle core) is the corresponding β-pyrrolic C-H.

The porphyrin isomers **3.160** and **3.161** were isolated as by-products in the course of carrying out what were ostensibly "normal" Rothemund-type tetraphenyl porphyrin syntheses.[55–57] For instance, the Japanese group isolated **3.160** as a by-product of the HBr-catalyzed reaction between pyrrole and benzaldehyde. The Polish group, on the other hand, used $BF_3 \cdot (OEt)_2$ as the catalyst and p-tolylaldehyde instead of benzaldehyde as the key *meso*-carbon forming precursor. In both cases, the porphyrin isomer is obtained in *ca.* 5% yield.

Scheme 3.5.1

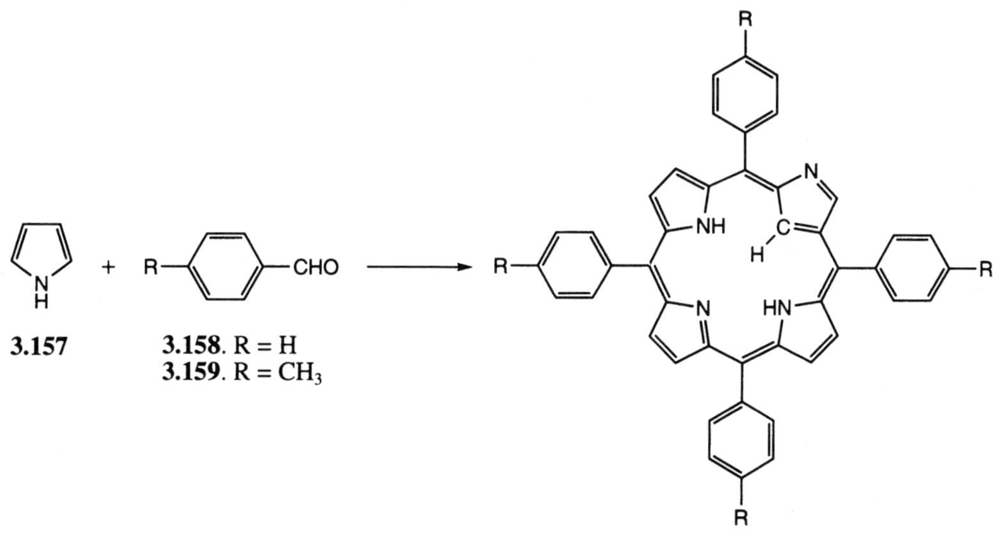

3.157   **3.158**. R = H
                3.159. R = CH$_3$

3.160. R = H
3.161. R = CH$_3$

Based on their findings, Latos-Grażyński and his team have suggested that these "mutant" porphyrins are produced as the result of a final ring-closing condensation in which one of the terminal pyrrole rings has rotated in such a way (see Scheme 3.5.2) that the β-pyrrolic carbon of the pyrrole can condense to form an inverted porphyrinogen.[54] This porphyrinogen then oxidizes to form the "mutant" porphyrin systems in question. This proposal was in accord with the observations of Furuta, *et al.* However, in the latter case, it was found that the course of the reaction to form **3.160** was influenced by the anions present in the reaction. Thus, it was determined that **3.160**, while formed in the presence of Br$^-$ or Cl$^-$ anions, could not be obtained when F$^-$, TFA$^-$, NO$_3$$^-$, and/or H$_2PO_4$$^-$ anions were used instead. These results led the authors to suggest that the formation of their N-confused porphyrin was mediated by an anion template effect of some sort.[53]

Like normal porphyrin, the α–β'-linked porphyrin isomers **3.160** and **3.161** may be considered as being heteroatom-bridged [18]annulenes. They display, for instance, properties consistent with a 4n + 2 aromatic formulation. In particular, ^{1}H NMR spectra of both **3.160** and **3.161** indicate the presence of a sustained diamagnetic ring current.[53,54] This is evident from the upfield shift of the internal NH and CH protons as well as from the corresponding downfield shift of the external CH and NH protons. The UV-vis spectra of these compounds also resemble those of normal porphyrins in that they are characterized by the presence of a strong Soret- and weaker Q-like absorbance transitions.

Scheme 3.5.2

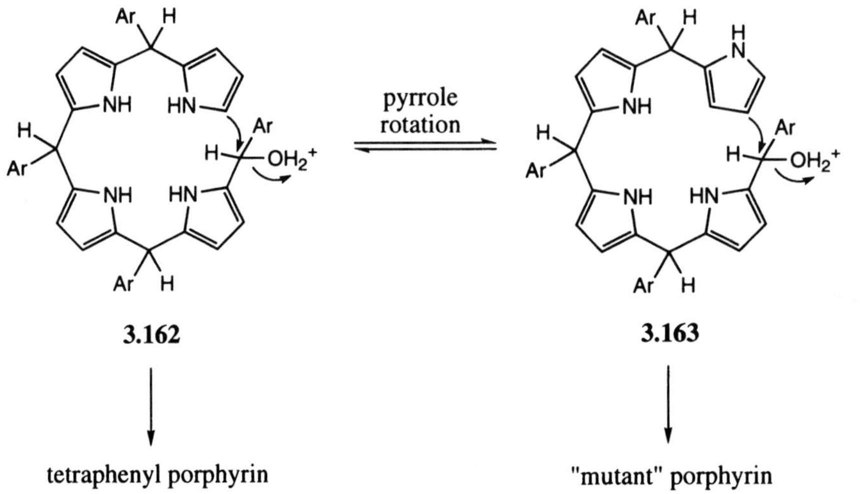

pyrrole rotation

3.162   **3.163**

tetraphenyl porphyrin   "mutant" porphyrin

Single crystal X-ray diffraction structural information for these porphyrin isomers is consistent with the "N-out" formulation of these macrocycles. The structure of **3.160** by Furuta, *et al.* revealed a nearly planar macrocyclic conformation, with only slight deviation from planarity.[53] Latos-Grażyński and coworkers demonstrated that the Ni(II) complex of **3.161** (i.e., **3.164**) was essentially flat, with four nearly equivalent donor atom-to-metal bonds (Figures 3.5.1–3.5.3).[54] This result can be interpreted in terms of an increased acidity (at least in a kinetic sense) for the internal C-H group relative to the external C-H pyrrolic protons. The neutral nature of the Ni(II) complex formed from **3.161** has led Ghosh to suggest that the metal binding in these systems is carbene-like in character.[58] This leads, in turn, leads us to suggest that these and other conceivable α–β′-linked porphyrin isomers could emerge as a new class of interesting ligands.

Since the initial disclosure of the above synthesis of inverted porphyrins **3.160** and **3.161**, two rational synthetic approaches to structures of this type have been devised. The first of these, reported by Lee and coworkers, actually involved the synthesis of the furan-containing system **3.167** (Scheme 3.5.3).[59,60] In this approach, diol **3.165** was condensed with the tripyrrane derivative **3.166** using boron trifluoride etherate as the acid catalyst. After oxidation with DDQ, the inverted monoxaporphyrin **3.167** was isolated in 5.5% yield.

Figure 3.5.1 Structure **3.164**.

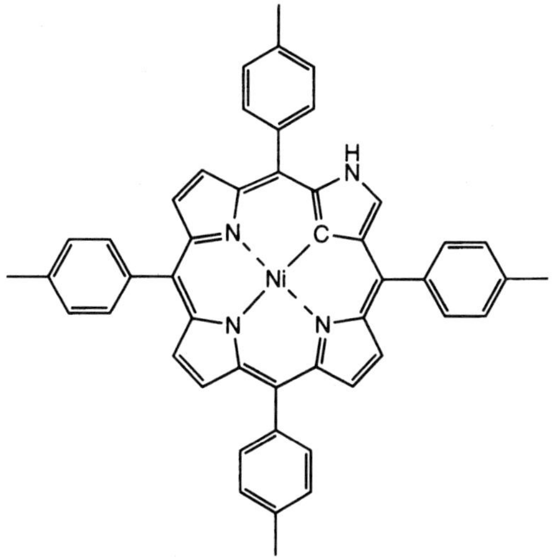

3.164

The other more rational synthesis of inverted porphyrins was reported recently by Dolphin and coworkers.[61] This approach has as its key macrocycle-forming step a McDonald-type "2 + 2" condensation between the α–β-linked dipyrrylmethane **3.168** and the diformyl dipyrrylmethane **3.169**. Following oxidation with air, the heptaalkyl N-inverted porphyrin system **3.170** is isolated in *ca.* 25% yield (Scheme 3.5.4).

Dolphin and coworkers found that the nitrogen on the periphery of macrocycle is the most basic of the four heteroatoms in **3.170**. This they concluded on the basis of a protonation study using trifluoroacetic acid in chloroform. Specifically, it was found in ^{1}H NMR spectroscopic experiments that the addition of an equimolar amount of trifluoroacetate (TFA) resulted in the appearance of an additional broadened signal at 13.5 ppm. Only when several equivalents of TFA were added could protonation of the internal nitrogen atoms be observed to occur.[61]

Perhaps consistent with the conclusion that the external nitrogen atom is the most basic is the fact that methylation of the tetraphenyl-substituted inverted porphyrin **3.160** with methyl iodide was found to occur practically quantitatively at this same peripheral nitrogen atom (affording system **3.171**; Scheme 3.5.5).[62] The high selectivity for methylation at this position was attributed both to the steric protection of the three "normal" nitrogen atoms by the porphyrin pocket, as well as to deformations of the porphyrin ring that would be necessary in order to accommodate an internal methyl group.

The addition of the methyl group at the peripheral nitrogen atom of **3.171** was proposed by Latos-Grażyński and Chmielewski to change the primary tautomeric form of the system from one like that in **3.160**, with three internal protons, to that like **3.171** wherein only two internal protons are found.[62] The formation of such a tautomeric species should, of course, disrupt the ring current of the system. Interestingly, while the *N*-methyl derivative **3.171** remains aromatic, it appears to be less so than the non-methylated parent **3.160**, at least based on an analysis of ^{1}H NMR chemical-shift differences.

Figure 3.5.2 Single Crystal X-Ray Diffraction Structure of Free-base "N-inverted" porphyrin **3.160**.

This figure was generated using information down-loaded from the Cambridge Crystallographic Data Centre and corresponds to a structure originally reported in reference 53. The molecule deviates considerably from planarity (the maximum deviation of the porphyrin-like skeleton from the mean macrocyclic plane is 0.63 Å). The pyrrole ring that is oriented outward is twisted by 26.9° from the mean plane defined by the other three inward-oriented nitrogen atoms. The other three pyrrole rings are rotated 5.8°, 7.8°, and 13.4° out of the plane, respectively. Atom labeling scheme: carbon: ○; nitrogen: ●. Hydrogen atoms have been omitted for clarity

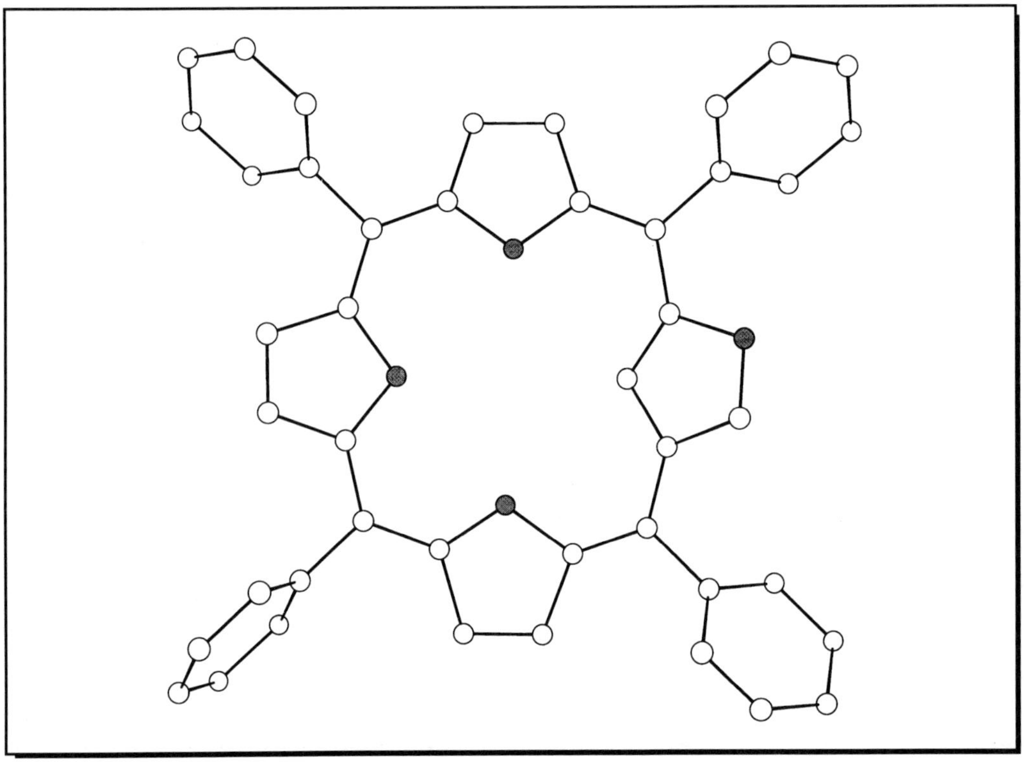

The *N*-methyl inverted porphyrin **3.171** has also been demonstrated as being capable of forming a stable complex with divalent nickel.[62] In this instance, it was found that treatment of **3.171** with nickel(II) acetate tetrahydrate in methanol affords the metal complex **3.172** in 75% yield (Scheme 3.5.6). This complex, like that of the N-unsubstituted inverted porphyrin discussed above, serves as a cogent reminder of just how easy it is to remove the central C-H proton in this kind of system. This, in turn, means that these types of ligands could prove useful in stabilizing a wide range of novel metal complexes.

Figure 3.5.3 Single Crystal X-Ray Diffraction Structure of Nickel(II) "N-inverted" porphyrin **3.164**.

This figure was generated using information down-loaded from the Cambridge Crystallographic Data Centre and corresponds to a structure originally reported in reference 54. The molecule is nearly planar (the maximum deviation of the porphyrin-like skeleton from the mean macrocyclic plane is 0.08 Å). The Ni–N and Ni–C bond lengths are all similar, with the average bond length being 1.96 Å. Atom labeling scheme: nickel: ◓; carbon: ○; nitrogen: ◓. Hydrogen atoms have been omitted for clarity

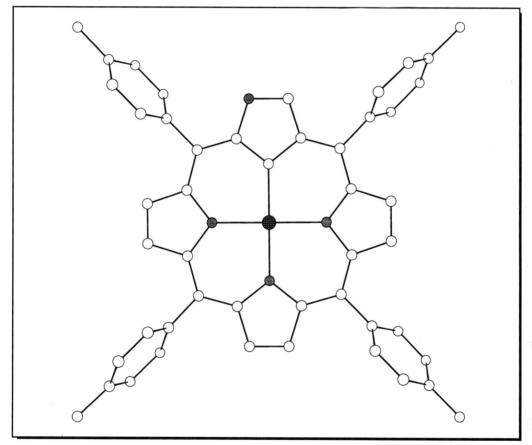

Scheme 3.5.3

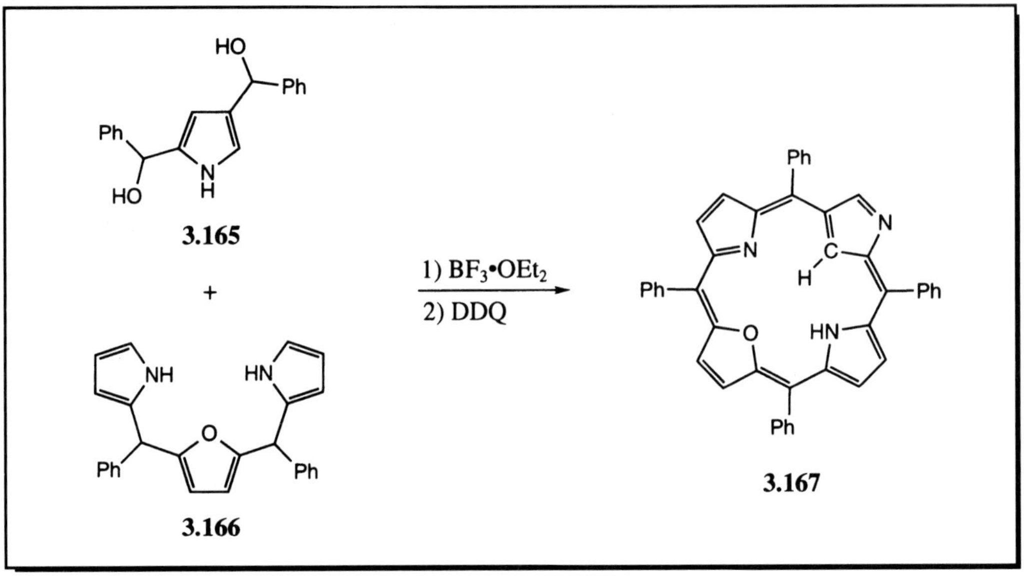

3.165

+

3.166

1) BF₃•OEt₂
2) DDQ

3.167

Scheme 3.5.4

3.168

+

3.169

HCl/HOAc
O₂

3.170

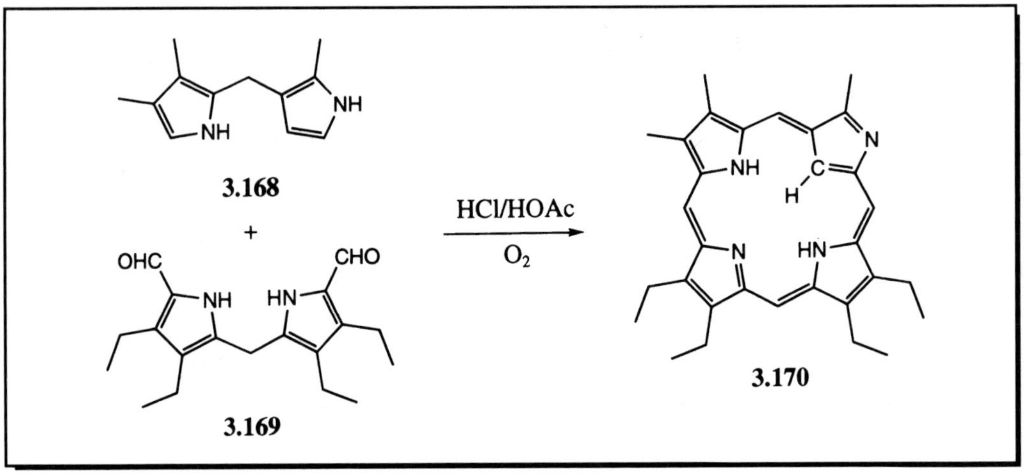

Scheme 3.5.5

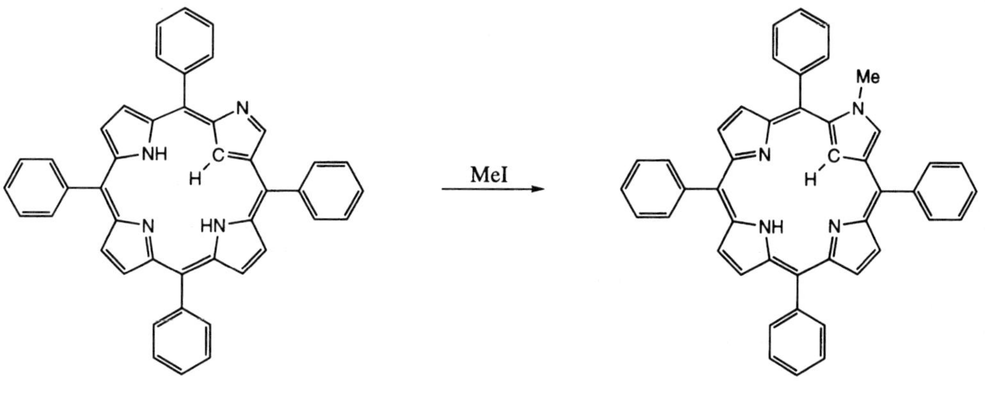

3.160   3.171

Scheme 3.5.6

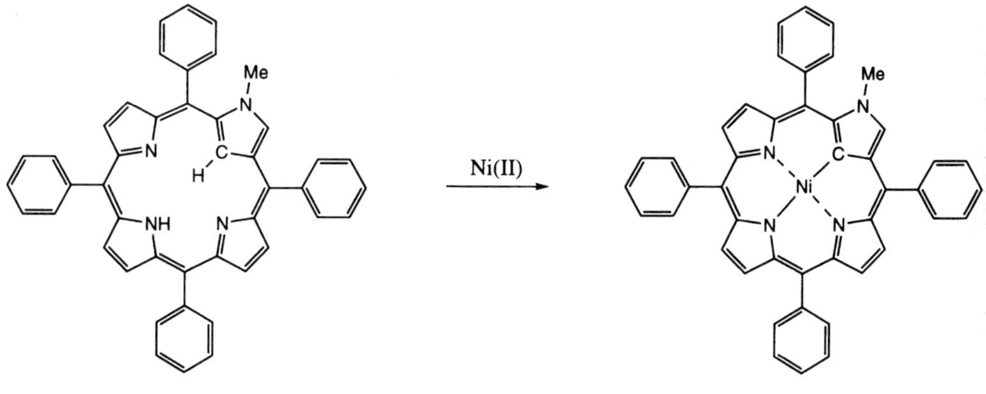

3.171   3.172

3.6   Future Directions in Porphyrin Isomer Chemistry

The intriguing work that has been accomplished to date under the aegis of porphyrin isomer research is surely but a precursor for more interesting things yet to come. Nonetheless, it has served to demonstrate effectively that relatively minor structural differences (i.e., mere connectivity-type modifications in the $C_{20}H_{14}N_4$ porphyrin periphery) can induce dramatic changes in chemical behavior within a series of otherwise very similar molecules. These changes in chemical behavior, in particular those leading to modifications in the basic porphyrin-like coordination or optical characteristics, have led to suggestions that porphyrin isomers could prove useful in a range of biomedical applications, including (in particular) photodynamic therapy. Indeed, as detailed in Chapter 10 of this book, evidence is beginning to accumulate that the prototypical porphycenes show promise in this regard. Such findings, in turn, are providing an important incentive for the preparation and study other isomeric systems.

In terms of possible future work, three pyrrole-in porphyrin isomers (i.e., **3.173–3.175**) look particularly attractive as synthetic targets (Figure 3.6.1). While it is understood that all three of these isomers may prove impossible to isolate, owing to stability and/or synthesis-related constraints,[63] the very difficulty thus imposed makes their synthesis worth attempting. Comparing the metal coordinating properties with the five isomers already known (i.e., **3.1–3.5**) would be particularly fascinating and could serve to reveal how metal chelation effects might act to stabilize otherwise unstable polypyrrolic macrocycles (e.g., **3.155**).[‡]

Figure 3.6.1   Structures **3.173–3.175**

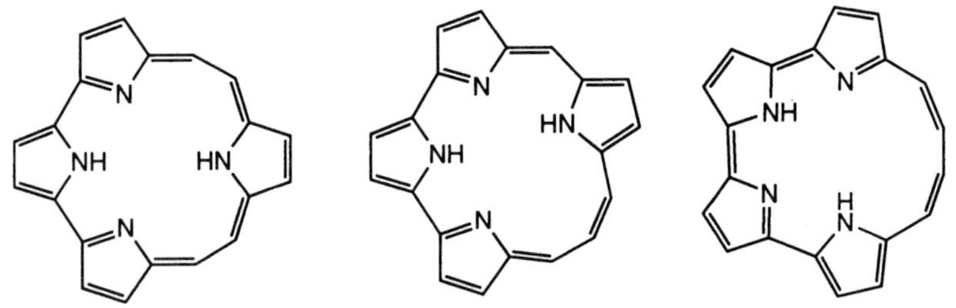

3.173
(2.2.0.0)

3.174
(3.1.0.0)

3.175
(4.0.0.0)

Also remaining as intriguing synthetic targets are further analogs of the mutant or N-confused porphyrins. These include still-unknown porphyrin isomers wherein two, three, or even all four of the pyrroles are α–β' linked (e.g., **3.176, 3.177**), as well as systems such as **3.179–3.182**, wherein the pyrroles are replaced with cyclopentadienyl and/or imidazole subunits (Figure 3.6.2). Given this excitement, it is the opinion of the authors that some or all of these as-yet-unknown porphyrin analogs will undoubtedly be prepared in short order. In each case, systematic investigations into the properties of the compound and comparisons of its chemistry with that of other, earlier-prepared systems are likely to provide a wealth of knowledge. Work in this field should thus prove exciting for years to come.

[‡]A dramatic example of this effect is also provided in the case of the texaphyrins (cf. Chapter 9).

Figure 3.6.2   Structures **3.6** and **3.176–3.182**

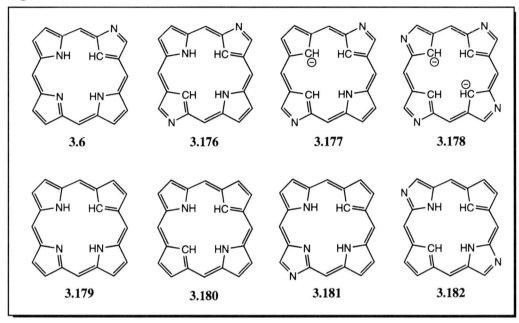

Note added in proof

Subsequent to the submission of this manuscript, a detailed communication describing the synthesis and properties of palladium(II) isoporphycenes appeared in the literature.[64] Also appearing in the literature were two separate reports detailing the synthesis of derivatized forms of so-called carbaporphyrin **3.179**.[65,66]

3.7   References

1.   *Porphyrins and Metalloporphyrins*; Smith, K. M., Ed.; Elsevier: Amsterdam, 1976.
2.   *The Porphyrins*, Vols 1–7; Dolphin, D., Ed.; Academic Press: New York, 1978.
3.   Vogel, E. *J. Heterocyclic Chem.* **1996**, *33*, 1461–1487.
4.   Vogel, E. *Pure. Appl. Chem.* **1996**, *68*, 1355–1360.
5.   Sessler, J. L. *Angew. Chem. Int. Ed. Eng.* **1994**, *33*, 1348–1350.
6.   Vogel, E.; Köcher, M.; Schmickler, H.; Lex, J. *Angew. Chem. Int. Ed. Eng.* **1986**, *25*, 257–259.
7.   Wehrle, B.; Limbach, H.-H.; Köcher, M.; Ermer, O.; Vogel, E. *Angew. Chem. Int. Ed. Eng.* **1987**, *26*, 934–936.
8.   Vogel, E.; Balci, M.; Pramod, K.; Koch, P.; Lex, J.; Ermer, O. *Angew. Chem. Int. Ed. Eng.* **1987**, *26*, 928–931.

9.   Vogel, E.; Koch, P.; Hou, X.-L.; Lex, J.; Lausmann, M.; Kisters, M.; Aukauloo, M. A.; Richard, P.; Guilard, R. *Angew. Chem. Int. Ed. Eng.* **1993**, *32*, 1600–1604.
10.   Barbe, J.-M.; Richard, P.; Aukauloo, M. A.; Lecomte, C.; Petit, P.; Guilard, R. *J. Chem. Soc., Chem. Commun.* **1994**, 2757–2758.
11.   Richert, C.; Wessels, J. M.; Müller, M.; Kisters, M.; Benninghaus, T.; Goetz, A. E. *J. Med. Chem.* **1994**, *37*, 2797–2807.
12.   Lausmann, M.; Zimmer, I.; Lex, J.; Lueken, H.; Wieghardt, K.; Vogel, E. *Angew. Chem. Int. Ed. Eng.* **1994**, *33*, 736–739.
13.   Hennig, W., Ph.D. Dissertation, University of Cologne, Germany, **1992**.
14.   Müller, M., Ph.D. Dissertation, University of Cologne, Germany, **1994**.
15.   Vogel, E.; Köcher, M.; Lex, J.; Ermer, O. *Isr. J. Chem.* **1989**, *29*, 257–266.
16.   Vogel, E. *Pure Appl. Chem.* **1990**, *62*, 557–564.
17.   Vogel, E.; Grigat, I.; Köcher, M.; Lex, J. *Angew. Chem. Int. Ed. Eng.* **1989**, *28*, 1655–1657.
18.   Vogel, E.; Sicken, M.; Röhrig, P.; Schmickler, H.; Lex, J.; Ermer, O. *Angew. Chem. Int. Ed. Eng.* **1988**, *27*, 411–414.
19.   Munno, G. D.; Lucchesini, F.; Neidlein, R. *Tetrahedron* **1993**, *49*, 6863–6872.
20.   Cöln, D., Ph.D. Dissertation, University of Cologne, Germany, **1991**.
21.   Ulman, A.; Manassen, J. *J. Am. Chem. Soc.* **1975**, *97*, 6540–6544.
22.   Hill, R. L.; Gouterman, M.; Ulman, A. *Inorg. Chem.* **1982**, *21*, 1450–1455.
23.   Stein, P.; Ulman, A.; Spiro, T. G. *J. Phys. Chem.* **1984**, *88*, 369–374.
24.   Ellinger, F.; Gieren, A.; Hübner, T.; Lex, J.; Lucchesini, F.; Merz, A.; Neidlein, R.; Salbeck, J. *Monatsh. Chem.* **1993**, *124*, 931–943.
25.   Hu, Z.; Atwood, J. L.; Cava, M. P. *J. Org. Chem.* **1994**, *59*, 8071–8075.
26.   Toporowicz, M.; Ofir, H.; Levanon, H.; Vogel, E.; Köcher, M.; Pramod, K.; Fessenden, R. W. *Photochem. Photobiol.* **1989**, *50*, 37–43.
27.   Renner, M. W.; Forman, A.; Wu, W.; Chang, C. K.; Fajer, J. *J. Am. Chem. Soc.* **1989**, *111*, 8618–8621.
28.   Che, C.-M.; Li, Z.-Y.; Guo, C.-X.; Wong, K.-Y.; Chern, S.-S.; Peng, S.-M. *Inorg. Chem.* **1995**, *34*, 984–987.
29.   Gisselbrecht, J. P.; Gross, M.; Köcher, M.; Lausmann, M.; Vogel, E. *J. Am. Chem. Soc.* **1990**, *112*, 8618–8620.
30.   Kadish, K. M.; Boulas, P.; D'Souza, F.; Aukauloo, A. M.; Guilard, R.; Lausmann, M.; Vogel, E. *Inorg. Chem.* **1994**, *33*, 471–476.
31.   Oertling, W. A.; Wu, W.; López-Garriga, J. J.; Kim, Y.; Chang, C. K. *J. Am. Chem. Soc.* **1991**, *113*, 127–134.
32.   Kadish, K. M.; D'Souza, F.; Van Caemelbecke, E.; Boulas, P.; Vogel, E.; Aukauloo, A. M.; Guilard, R. *Inorg. Chem.* **1994**, *33*, 4474–4479.
33.   Li, Z.-Y.; Huang, J.-S.; Che, C.-M.; Chang, C. K. *Inorg. Chem.* **1992**, *31*, 2670–2672.
34.   Zimmer, I., Ph.D. Dissertation, University of Cologne, Germany, **1993**.
35.   Kurtz, D. M., Jr. *Chem. Rev.* **1990**, *90*, 585–606.
36.   Tatsumi, K.; Hoffmann, R. *J. Am. Chem. Soc.* **1981**, *103*, 3328–3341.

37.   Sanders-Loehr, J.; Wheeler, W. D.; Shiemke, A. K.; Averill, B. A.; Loehr, T. M. *J. Am. Chem. Soc.* **1989**, *111*, 8084–8093.

38.   Vogel, E.; Köcher, M.; Balci, M.; Teichler, I.; Lex, J.; Schmickler, H.; Ermer, O. *Angew. Chem. Int. Ed. Eng.* **1987**, *25*, 931–934.

39.   Szeimies, R.-M.; Karrer, S.; Abels, C.; Steinbach, P.; Fickweiler, S.; Messmann, H.; Baeumler, W.; Landthaler, M. *J. Photochem. Photobiol., B.* **1996**, *34*, 67–72.

40.   Mühlhoff, J., Ph.D. Dissertation, University of Cologne, Germany, **1994**.

41.   Chang, C. K.; Morrison, I.; Wu, W.; Chern, S.-S.; Peng, S.-M. *J. Chem. Soc., Chem. Commun.* **1995**, 1173–1174.

42.   Sessler, J. L.; Brucker, E. A.; Weghorn, S. J.; Vogel, E.; Jux, N.; Neumann, L. Presented at the 207th ACS National Meeting, San Diego, CA (Inorg. Abstr. No. 295), March 13–17, 1994.

43.   Sessler, J. L.; Brucker, E. A.; Weghorn, S. J.; Kisters, M.; Schäfer, M.; Lex, J.; Vogel, E. *Angew. Chem. Int. Ed. Eng.* **1994**, *33*, 2308–2312.

44.   Aukauloo, M. A.; Guilard, R. *New. J. Chem.* **1994**, *18*, 1205–1207.

45.   Falk, H.; Chen, Q.-Q. *Monatsh. Chem.* **1996**, *127*, 69–75.

46.   Sessler, J. L.; Vogel, E. Unpublished results.

47.   Little, R. G.; Ibers, J. A. *J. Am. Chem. Soc.* **1974**, *96*, 4440–4441.

48.   Walker, F. A. *J. Am. Chem. Soc.* **1973**, *95*, 1150–1153.

49.   Walker, F. A. *J. Am. Chem. Soc.* **1973**, *95*, 1154–1159.

50.   Walker, F. A. *J. Am. Chem. Soc.* **1970**, *92*, 4235–4244.

51.   Vogel, E.; Broring, M.; Scholz, P.; Deponte, R.; Lex, J.; Schmickler, H.; Schaffner, K.; Braslavsky, S. E.; Müller, M.; Pörting, S.; Weghorn, S. J.; Fowler, C. J.; Sessler, J. L.; *Angew. Chem. Int. Ed. Engl.*, in press.

52.   Callot, H. J.; Rohrer, A.; Tschamber T. *New. J. Chem.* **1995**, *19*, 155–159.

53.   Furuta, H.; Asano, T.; Ogawa, T. *J. Am. Chem. Soc.* **1994**, *116*, 767–768.

54.   Chmielewski, P. J.; Latos-Grażyński , L.; Rachlewicz, K.; Glowiak, T. *Angew. Chem. Int. Ed. Eng.* **1994**, *33*, 779–781.

55.   Rothemund, P. *J. Am. Chem. Soc.* **1936**, *58*, 625–627.

56.   Rothemund, P.; Menotti, A. R. *J. Am. Chem. Soc.* **1941**, *63*, 267–270.

57.   Kim, J. B.; Adler, A. D.; Longo, F. R. In *The Porphyrins*, Vol. 1; Dolphin, D., Ed.; Academic Press: New York, 1978; pp. 85–100.

58.   Ghosh, A. *Angew. Chem. Int. Ed. Eng.* **1995**, *34*, 943–1042.

59.   Heo, P.-Y.; Shin, K.; Lee, C. H. *Tetrahedron Lett.* **1996**, *37*, 197–200.

60.   Heo, P.-Y.; Shin, K.; Lee, C. H. *Tetrahedron Lett.* **1996**, *37*, 1521.

61.   Liu, B. Y.; Brückner, C.; Dolphin, D. *J. Chem. Soc., Chem. Commun.* **1996**, 2141–2142.

62.   Chmielewski, P. J.; Latos-Grazynski, L. *J. Chem. Soc., Perkin Trans. II* **1995**, 503–509.

63.   Vogel, E.; Bröring, M.; Fink, J.; Rosen, D.; Schmickler, H.; Lex, J.; Chan, K. W. K.; Wu, Y.-D.; Plattner, D. A.; Nendel, M.; Houk, K. N. *Angew. Chem. Int. Ed. Eng.* **1995**, *34*, 2511–2514.

4   Introduction

From a design perspective, perhaps the most obvious way to "expand" a porphyrin is to increase the number of bridging atoms separating the heterocyclic rings. For example, by adding, formally, one atom between the *meso* and α-pyrrolic position of a porphyrin, one obtains a tetrapyrrolic macrocycle such as **4.1** (shown in its generalized, unsubstituted form; Figure 4.0.1)[†] that contains 17 atoms in its smallest macrocyclic ring. A number of such systems, known as "homoporphyrins", have indeed been prepared. These will be discussed in this chapter. Also discussed in this chapter are another kind of "expanded" tetrapyrrolic macrocycles known as vinylogous porphyrins. Finally, several other tetrapyrrolic macrocycles that are not formally either homo- or vinylogous porphyrins will also be reviewed in the context of this chapter.

4.1   Homoporphyrins

As implied above, the homoporphyrins are a class of expanded porphyrins that arise *via* the conceptual addition of one atom into the porphyrin skeleton. In general, homoporphyrins are non-aromatic. This is because they typically contain an sp³-hybridized center that serves to break the π-conjugation pathway. There are some homoporphyrinoid systems, however, that do contain an sp²-hybridized center as their "extra" bridging atom. Interestingly, even in these instances, the systems end up being non-aromatic (or even antiaromatic) because the π-electron pathway contains 20 electrons (Hückel 4n). Nonetheless, in spite of being non-aromatic and non-planar (*vide infra*), the homoporphyrins display a rich metal chelation chemistry. In fact, using various homoporphyrin ligands, stable complexes (e.g., **4.2**) have been made from many of the metals of the first transition metal series.

Much of the initial impetus leading to the preparation of the homoporphyrins derived from a desire to extend comparisons made at the time between the contracted

---

[†]The "wedges" used in structure **4.1** and indeed in all similar homoporphyrin structures in this chapter are used to designate confirmed deviation from macrocyclic planarity at this particular position, and should not be taken as an assignment of absolute stereochemistry. In no case has the separation of homoporphyrin enantiomers been reported. However, racemic epimers that result from substitution at this out-of-"plane" *meso*-like position have in some cases been separated, and solid and "hashed" wedges are used to designate these epimeric forms as appropriate.

porphyrin corrole (Chapter 2) and porphyrin. In such comparisons, the analogy between the corrole–porphyrin relationship relative to the cyclopentadiene–benzene relationship was being investigated.[1–5] In this context, the homoporphyrins were considered as being analogs of cycloheptatriene, the next higher homolog of benzene. It was thus expected that the homoporphyrins, like cycloheptatriene, would be non-aromatic since both classes of molecules contain an "extra" sp^3-hybridized center.

Figure 4.0.1 Structures **4.1** and **4.2**

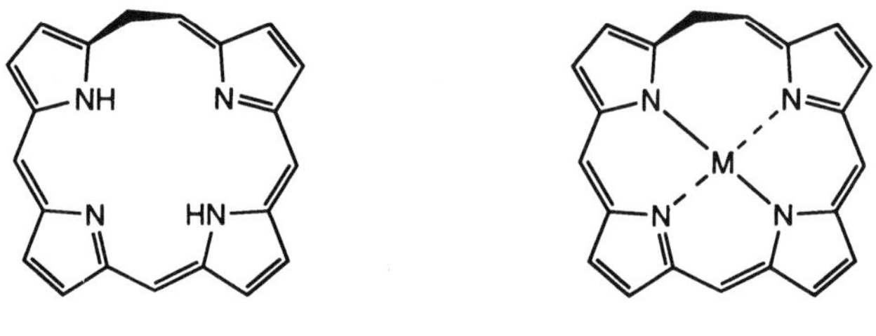

4.1   **4.2**

Although the implication of the above discussion was that the sp^3-hybridized atom being inserted into the porphyrin conjugation pathway would be carbon, the first examples of homoporphyrins were actually *N-meso*-homoporphyrins (homoazaporphyrins) in which a saturated nitrogen center plays this role.[6,7] They were obtained by Grigg in 1967 as the result of treating free-base porphyrins with nitrenes. Specifically, he found that, in the case of etioporphyrin **4.3**, treatment with ethoxycarbonylnitrene results in the ring-expanded homoazaporphyrin **4.5**, a product that was obtained in 29% yield as a 1:1 mixture of regioisomers (**4.5a** and **4.5b**) (Scheme 4.1.1). In the case of octaethylporphyrin **4.4**, the corresponding homoaza-porphyrin **4.6** is obtained in 52% yield. In this instance, because of the molecular symmetry involved, only one product (**4.6**) was obtained.

Proton NMR spectroscopic studies of these products served to confirm their non-aromatic nature (e.g., for the octaethyl derivative **4.6**, the *meso*-like protons resonate between 6.8 and 10.0 ppm). On the other hand, the homoazaporphyrins **4.5** and **4.6** do retain visible spectral features characteristic of the parent porphyrins. For instance, octaethylhomoporphyrin **4.6** exhibits a Soret-like absorbance band in diethylether at 404 nm ($\varepsilon = 69\,000$ M^{-1} cm^{-1}), and two Q-type transitions at 610 and 655 nm ($\varepsilon = 11\,000$ and $12\,800$ M^{-1} cm^{-1}, respectively).

Unfortunately, the homoazaporphyrins proved to be rather unstable. The general instability of the homoazaporphyrins was demonstrated by heating the solid free-base material to 160 °C or by heating a chloroform solution of **4.5** or **4.6** to reflux. Under these conditions, the homoazaporphyrins **4.5** and **4.6** undergo ring contraction to the *meso*-ethoxycarbonylamino porphyrins **4.7** and **4.8** in 53% and 95% yield, respectively (Scheme 4.1.1).

Scheme 4.1.1

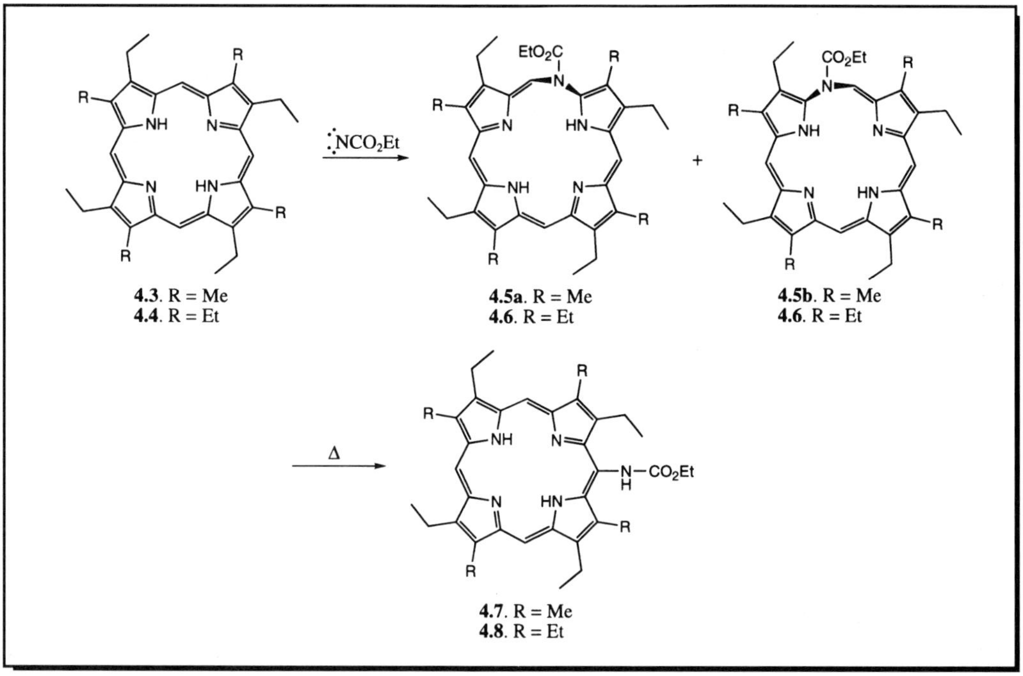

:NCO2Et
Δ
4.3. R = Me
4.4. R = Et
4.5a. R = Me
4.6. R = Et
4.5b. R = Me
4.6. R = Et
4.7. R = Me
4.8. R = Et

The general instability of the homoazaporphyrins has also limited attempts to prepare metalated derivatives. Specifically, efforts to prepare zinc(II) and copper(II) homoazaporphyrins from the metal-free homoazaporphyrin **4.6** resulted only in ring contraction to the corresponding *meso*-ethoxycarbonylamino-substituted metalloporphyrins **4.9** or **4.10**, respectively (Scheme 4.1.2).[7]

Scheme 4.1.2

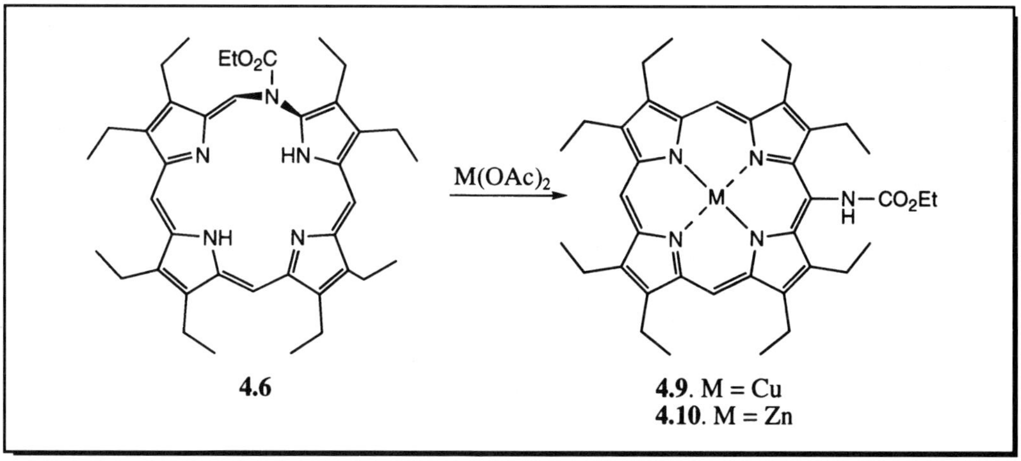

M(OAc)2
4.6
4.9. M = Cu
4.10. M = Zn

Attempts to prepare metalated homoazaporphyrins by effecting the addition of ethoxycarbonyl nitrene into the *meso*-positions of pre-metalated porphyrins (e.g., **4.11** and **4.12**) afforded only the corresponding metallo-derivatives of the *meso*-substituted porphyrins **4.13** and **4.14**, with no traces of homoporphyrins ever being detected (Scheme 4.1.3).[6,7] Further, when nickel(II) or cobalt(III) etio- or octaethyporphyrins were treated with this same carbene, no reaction was observed at all. While these subtle differences in reactivity are interesting, the results in aggregate are perhaps more important. They highlight the fact that the homoazaporphyrins are poor ligands for metal complexation.

Scheme 4.1.3

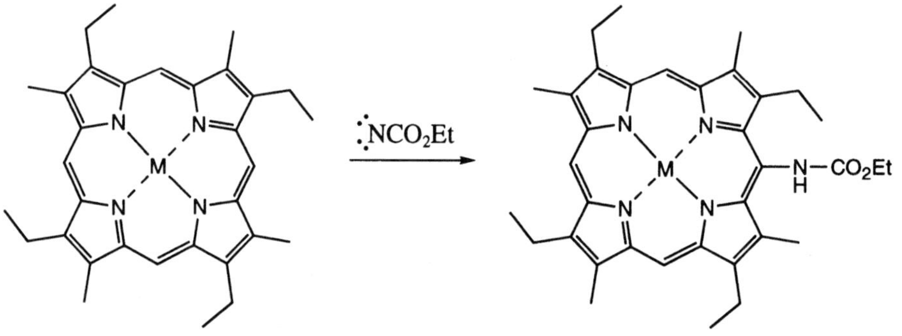

4.11. M = Zn
4.12. M = Cu

4.13. M = Zn
4.14. M = Cu

The limited, perhaps non-existent, metalation chemistry of the homoazaporphyrins coupled with their general instability and propensity to undergo ring contraction prompted efforts to prepare homoporphyrins in which a carbon, rather than a nitrogen, atom is inserted into the porphyrin *meso*-position. This synthetic objective was first achieved by Callot and Tschamber in 1974.[8] They did this by effecting an appropriate rearrangement of an N-substituted tetraarylprophyrin. Specifically, they treated *N*-ethyoxycarbonylmethyltetraphenylporphyrin **4.15** with Ni(II) bis-(acetyl)acetonate (Ni(acac)₂).[8] This afforded a mixture of two epimeric nickel(II) homoporphyrins (**4.16a** and **4.16b**), with the endo epimer predominating (57% endo vs. only 4% exo) (Scheme 4.1.4).[‡] When these same reactants were heated to 110 °C in toluene, it was found that a 60 : 40 equilibrium mixture of the two epimers resulted, with the endo product remaining the dominant one.

Importantly, under neither set of reaction conditions was significant decomposition or ring-contraction chemistry observed. This was taken as an indication that

---

[‡]It was determined that this reaction did not work when nickel(II) salts were replaced by copper(II), zinc(II), or cobalt(II) ones. In these latter instances, only cleavage of the ester group was observed. See reference 8.

these *bona fide* homoporphyrins might prove more stable than their aza analogs. On the other hand, attempts to prepare an octaethylhomoporphyrin derivative starting from the N-substituted porphyrin **4.17** resulted only in the isolation of the *meso*-substituted porphyrin **4.18** (Scheme 4.1.5).[9] Here, what was presumed to be a homoporphyrin intermediate was observed spectroscopically during the course of the reaction. However, all attempts to isolate this putative species failed. This led to the conclusion that the special stability of the homoporphyrins is not universal.

Scheme 4.1.4

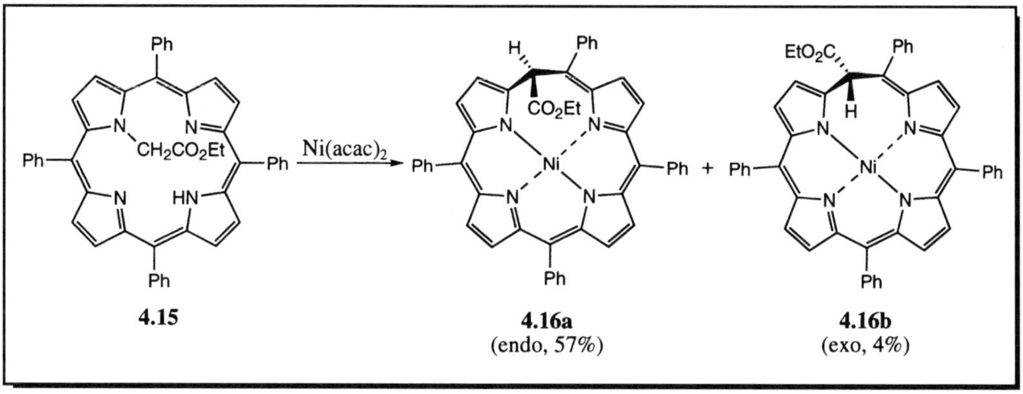

4.15

4.16a
(endo, 57%)

4.16b
(exo, 4%)

Scheme 4.1.5

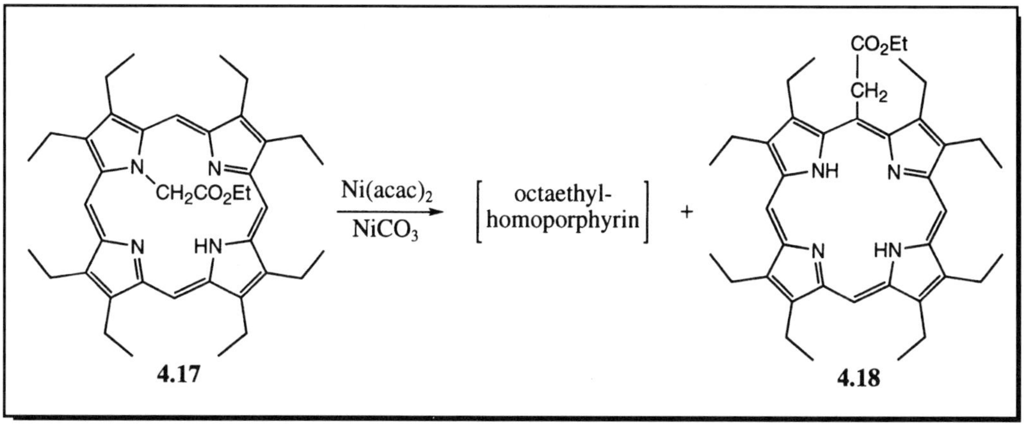

4.17

4.18

The UV-vis absorption spectra of homoporphyrins **4.16a** and **4.16b** resemble those of the parent porphyrin analogs. Both the endo and exo isomers exhibit Soret-like absorbance bands in the 450 nm region ($\varepsilon = 87\,000$ M^{-1} cm^{-1}), and two Q-type transitions at 584 and 688 nm ($\varepsilon = 6350$ and $17\,300$ M^{-1} cm^{-1}, respectively). A single crystal X-ray structure of the endo-epimer **4.16a** confirmed the sp^3-hybridization at the "inserted" *meso*-carbon (Figure 4.1.1). It also revealed a highly distorted

Figure 4.1.1   Single Crystal X-Ray Diffraction Structure of Nickel(II) Homoporphyrin **4.16a**.

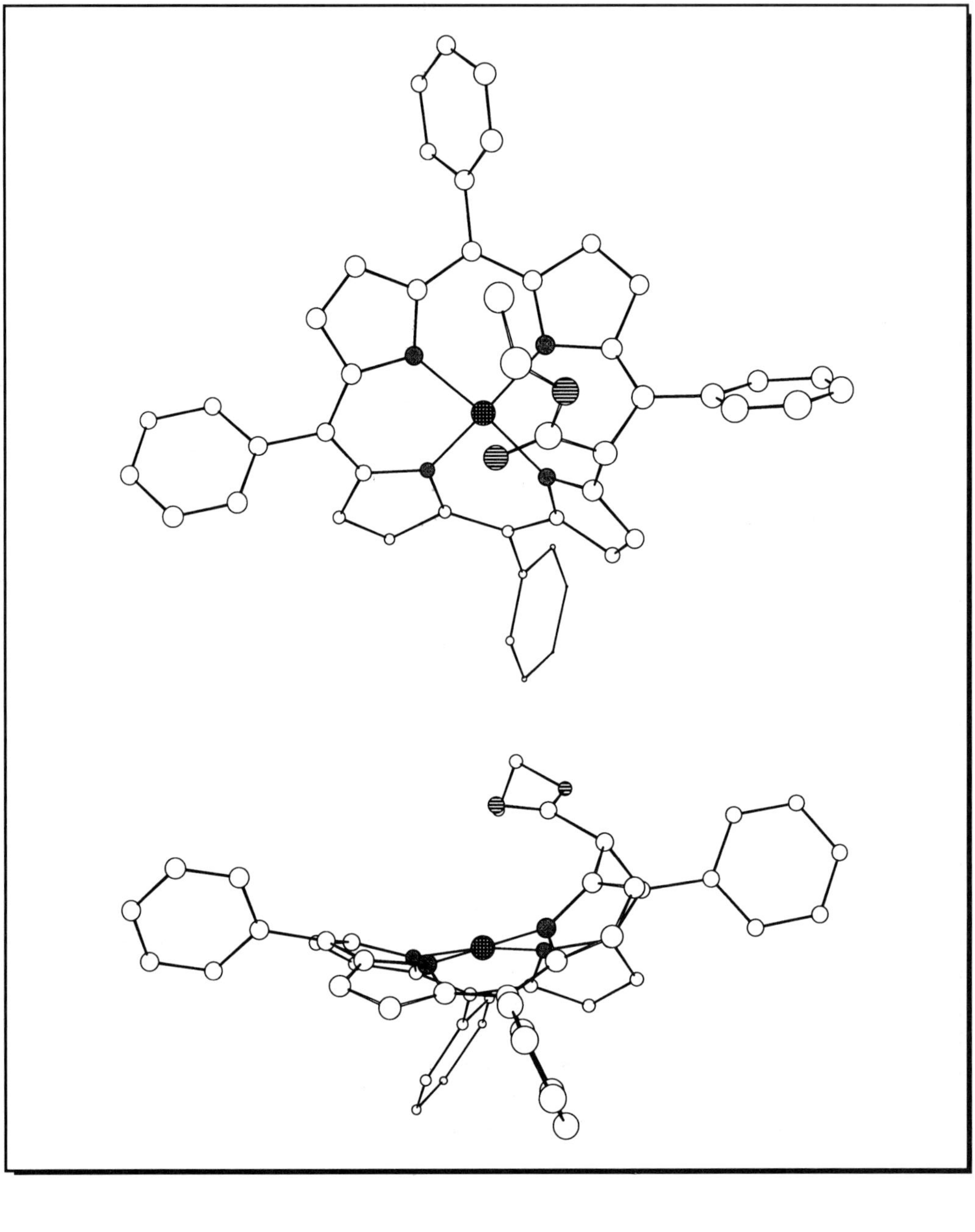

This figure was generated using information down-loaded from the Cambridge Crystallographic Data Centre and corresponds to a structure originally reported in reference 10. The macrocycle is rather non-planar. Most notably, the dihedral angle between the pyrroles adjacent to the *meso*-C_2 macrocyclic bridge is 73.4°. The nickel atom is in an essentially square-planar coordination environment. There is one long Ni–N bond (1.96 Å) and three shorter, equivalent, Ni–N bonds (1.89 Å). Top: top view; bottom: side view. Atom labeling scheme: nickel: ⬤; carbon: ○; nitrogen: ⬤; oxygen: ⊜. Hydrogen atoms have been omitted for clarity

ligand, with the nickel atom sitting in an essentially square planar coordination environment.[10,11] The high degree of distortion observed was presumed to be the result of the insertion of the additional carbon atom into the erstwhile porphyrin ring.

Scheme 4.1.6

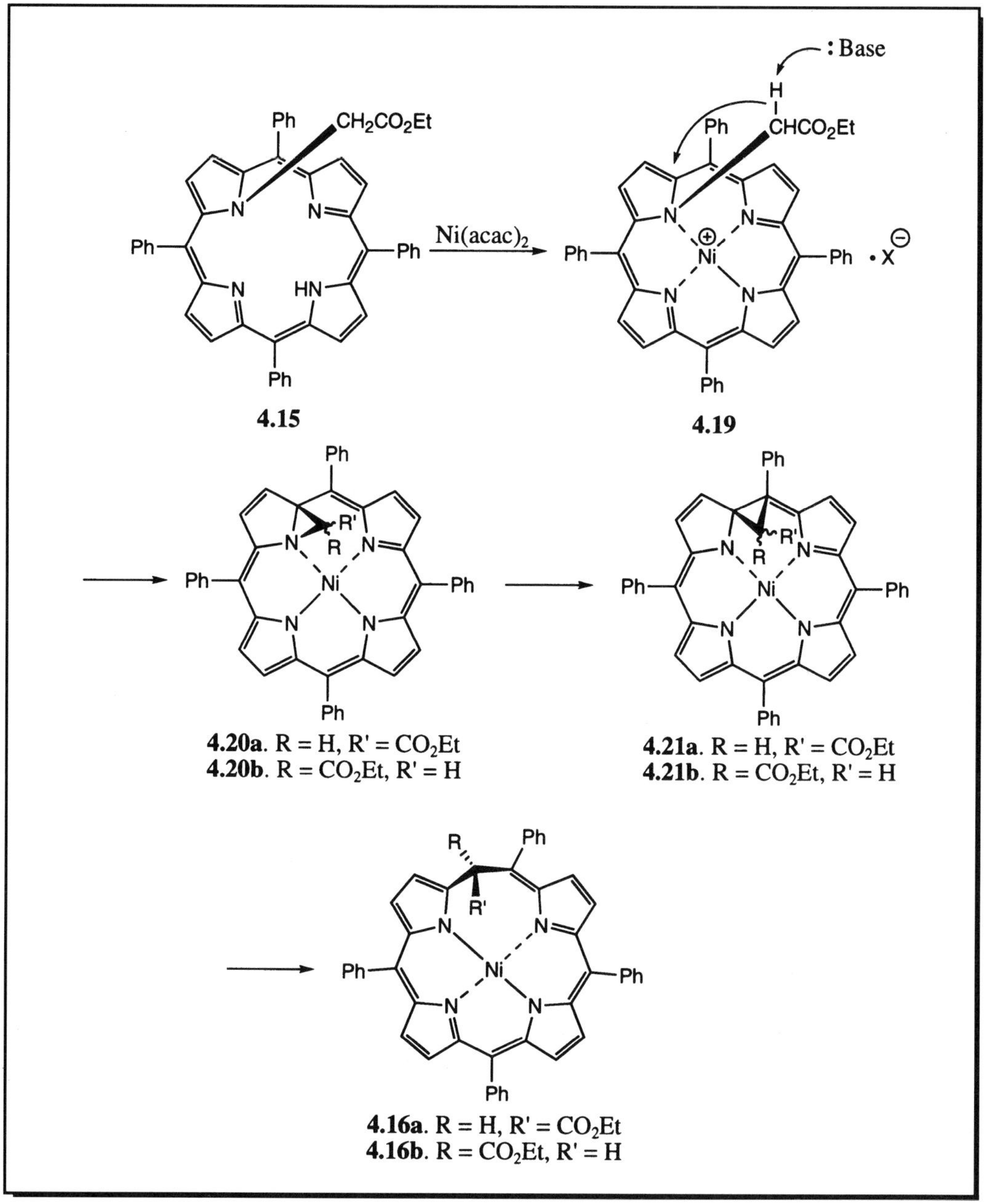
: Base
H
Ph
CH_2CO_2Et
$CHCO_2Et$
N
HN
Ni
Ph
Ni(acac)$_2$
• X$^{\ominus}$
R
R'
4.15
4.19
4.20a. R = H, R' = CO_2Et
4.20b. R = CO_2Et, R' = H
4.21a. R = H, R' = CO_2Et
4.21b. R = CO_2Et, R' = H
4.16a. R = H, R' = CO_2Et
4.16b. R = CO_2Et, R' = H

The proposed mechanism for the ring expansion of tetraphenylporphyrin is shown in Scheme 4.1.6.[8,9] It involves formation of the intermediate Ni(II) cationic complex **4.19**, followed by cyclization to an aziridine fused pyrrole-containing species **4.20**. This is then followed by migration of a C–C bond to the *meso*-position and the electrocyclic ring opening of the resulting cyclopropanic intermediate **4.21**. This sequence of reactions proceeds with a high degree of stereospecificity, as shown by the dominance of the endo epimer formation.

When either the endo- or exo-homoporphyrins **4.16a** or **4.16b** is heated to 160 °C, an equilibrium is obtained between the four products **4.16a, b** and **4.22a, b**.[12] The equilibrium ratio of isomers **4.16a**, **4.16b**, **4.22a**, and **4.22b** is 41%, 29%, 23%, and 7%, respectively (Scheme 4.1.7). When any of these four species is heated to greater than 200 °C, ring contraction occurs to afford a mixture of the four cyclopropyl chlorin derivatives **4.23a**, **4.23b**, **4.24a**, and **4.24b** in 15%, 65%, 5%, and 15% relative abundance, respectively (Scheme 4.1.8).[12]

Scheme 4.1.7

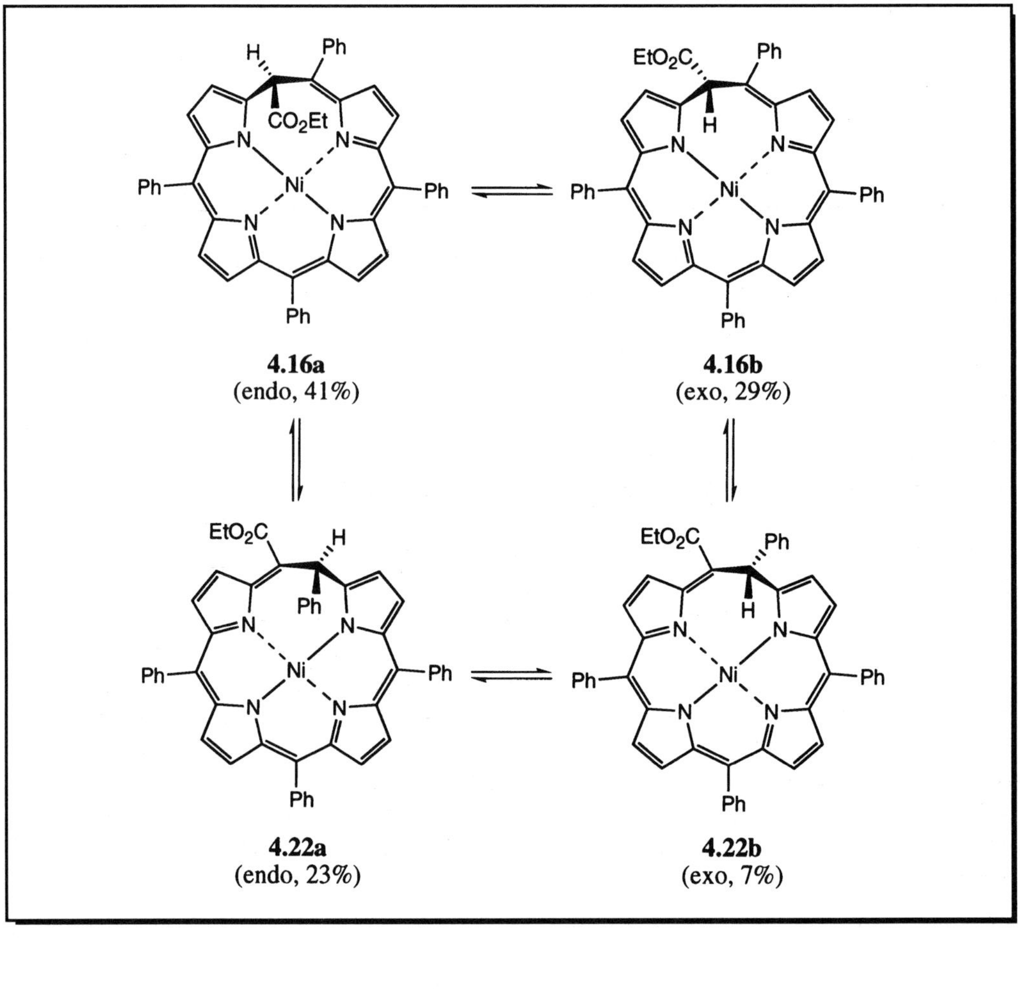

4.16a
(endo, 41%)

4.16b
(exo, 29%)

4.22a
(endo, 23%)

4.22b
(exo, 7%)

Acid-catalyzed equilibration of the four homoporphyrins has also been demonstrated.[13,14] Specifically, when any of the four pure homoporphyrins **4.16a**, **4.16b**, **4.22a**, or **4.22b** is subjected to low TFA concentrations in dichloromethane at room temperature, rapid equilibration occurs to afford a 60 : 25 : 10 : 5 ratio of **4.16a**, **4.16b**, **4.22a**, and **4.22b**, respectively (Scheme 4.1.7). When, however, more strongly acidic conditions are employed (1 M TFA or conc. HCl), demetalation occurs to afford the free-base homoporphyrin **4.25** (Scheme 4.1.9).[14] This free-base macrocycle can be readily remetalated with Ni(II) to afford the starting complex **4.22b**.

Scheme 4.1.8

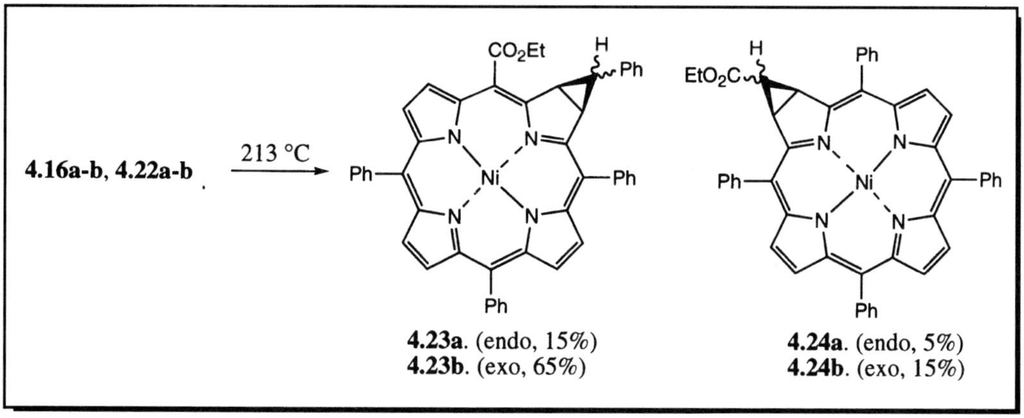

4.16a-b, 4.22a-b $\xrightarrow{213\,^\circ\text{C}}$

4.23a. (endo, 15%)
4.23b. (exo, 65%)

4.24a. (endo, 5%)
4.24b. (exo, 15%)

Interestingly, under the highly acidic demetalation conditions used to generate **4.25**, another metal-free macrocycle, **4.26**, was also isolated.[14] Presumably, this material, formally a 20 π-electron non-aromatic homoporphyrin results from proton migration. Unfortunately, in spite of possessing an sp^2-hybridized conjugation pathway, this product proved unstable. For instance, attempted metalation of **4.26** using Ni(OAc)$_2$ in acetic acid under an inert atmosphere at 60 °C did not give the corresponding metal complex. Instead, it resulted in the regeneration of the "original" nickel(II) homoporphyrins, a mixture of **4.16a, b** and **4.22a, b**, in 30% overall yield (Scheme 4.1.10). Likewise, reaction with Cu(II) salts afforded only various copper(II) homoporphyrins,[15,16] while reaction of **4.26** with cobalt(II) afforded only a ring-contraction mixture of cobalt(II) porphyrins **4.27** and **4.28** (Scheme 4.1.11).[15,16]

By contrast, if attempts are made to insert Ni(II) into **4.26** while this latter is exposed to air in a nucleophilic solvent such as methanol or water, a different reaction occurs. For instance, when the homoporphyrin derivative **4.26** was treated with Ni(OAc)$_2$ in water in the presence of O$_2$, a mixture of two compounds was produced. On the basis of ^{1}H NMR spectra and an X-ray structural analysis (on **4.29b**), these compounds were determined to be the epimeric hydroxyl-substituted compounds **4.29a** and **4.29b** (Scheme 4.1.12).[15,16] Running the reaction in methanol under otherwise identical conditions gave rise to the analogous methoxy-substituted compounds **4.30a** and **4.30b**.

Scheme 4.1.9

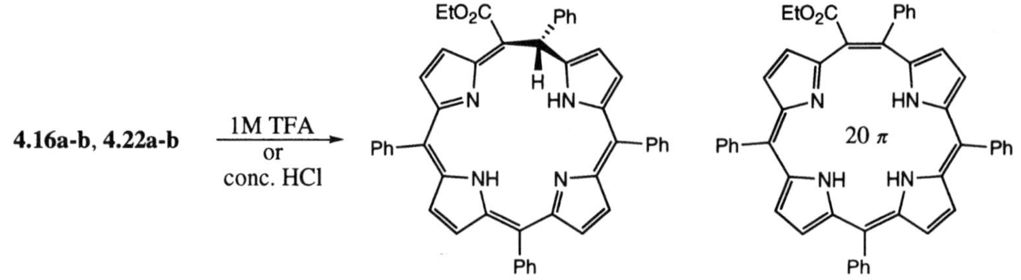

4.16a-b, 4.22a-b

$\xrightarrow[\text{conc. HCl}]{\substack{\text{1M TFA} \\ \text{or}}}$

4.25
(exo)

4.26

Scheme 4.1.10

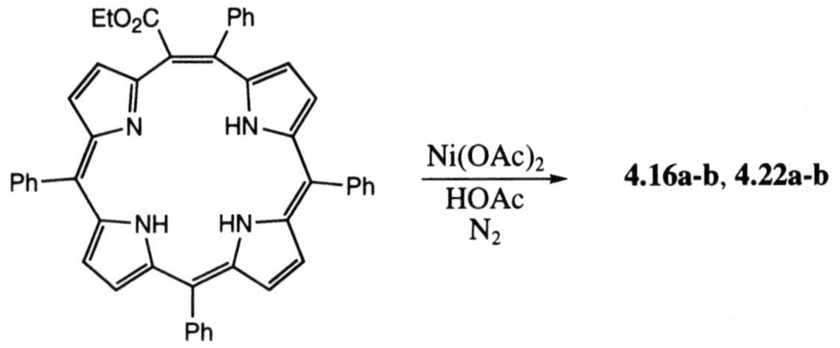

4.26

$\xrightarrow[\substack{\text{HOAc} \\ N_2}]{\text{Ni(OAc)}_2}$

4.16a-b, 4.22a-b

Scheme 4.1.11

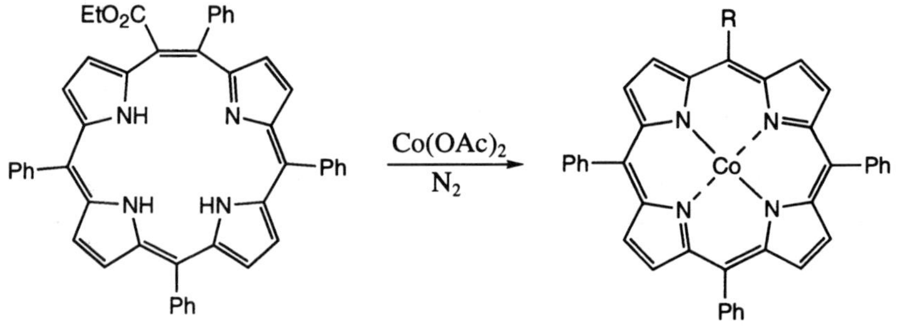

4.26

$\xrightarrow[N_2]{\text{Co(OAc)}_2}$

4.27. R = C$_6$H$_5$
4.28. R = CO$_2$Et

Scheme 4.1.12

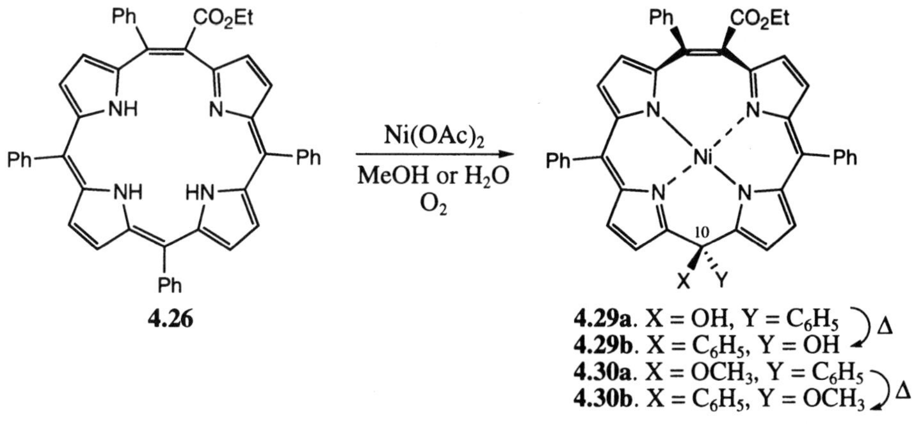

4.26

4.29a. X = OH, Y = C$_6$H$_5$ ⎫
4.29b. X = C$_6$H$_5$, Y = OH ⎭ Δ
4.30a. X = OCH$_3$, Y = C$_6$H$_5$ ⎫
4.30b. X = C$_6$H$_5$, Y = OCH$_3$ ⎭ Δ

A single crystal X-ray diffraction analysis on the hydroxyl-substituted species **4.29b** (Figure 4.1.2) revealed a marked folding of the macrocycle, as represented schematically in Figure 4.1.3.[15–17] This folding results in the formation of distinct "axial" and "equatorial" orientations for the C(10) substituents (cf. Figure 4.1.3). While these orientations are thought to be configurationally distinct, it was also found that heating macrocycles **4.29a** or **4.30a**, in which the phenyl substituents are in equatorial positions, affords the corresponding products in which the phenyl ring resides in the "axial" position (**4.29b** and **4.30b**, respectively).[18] This rearrangement, which proceeds quantitatively, is thought to involve a heterolytic dissociation–reassociation mechanism.

When any of the above C(10)-disubstituted homoporphyrin derivatives is treated with acid (HX), elimination to the conjugated, cationic complexes **4.31a–c** occurs (Scheme 4.1.13).[14,18] Methanolysis of **4.31** results in the exclusive formation of **4.30a** (X = OCH$_3$). Reduction of cation **4.31** with zinc in acetic acid affords derivative **4.32a** as the major product.[19] Heating this isomer (**4.32a**) in *o*-dichlorobenzene at reflux then affords **4.32b**. In the presence of base (e.g., triethylamine), **4.32a** rapidly isomerized to the original endo homoporphyrin **4.16a** (Scheme 4.1.14). By contrast, under identical conditions **4.32b** did not appear to isomerize.[18]

In the fully conjugated homoporphyrin series, an anionic derivative, **4.33a**, has also been produced. It was prepared by treating the Ni(II)-containing endo and exo isomers **4.16a** and **4.16b** with lithium diisopropylamine in THF (Scheme 4.1.15).[14,18] This anionic nickel complex (**4.33a**) was not characterized extensively. However, it was noted that protonation with water or acetic acid affords **4.32a** as the major product.

Chemistry related to the above was also observed when the nickel cation **4.31d** was subject to electrochemical reduction.[19] In this instance, it was determined that one electron reduction afforded the neutral, stable radical species **4.34**. Further reduction afforded the anionic nickel complex **4.33b** (Scheme 4.1.16). While this latter species appears ostensibly identical to **4.33a**, Callot and coworkers claim that protonation of this electrochemically generated anionic species affords **4.32b**. This is in contrast to protonation of the base-produced anion **4.33a**, which affords the opposite stereochemistry on protonation (i.e., it gives **4.32a**). Thus, to the extent these assignments are correct, one is left to conclude that there is a clear incongruence between the chemical and electrochemical results. This is something that might prove interesting to resolve at a later date.

Figure 4.1.2 Single Crystal X-Ray Diffraction Structure of Nickel(II) 10-Hydroxyhomoporphyrin **4.29b**.

This figure was generated using information down-loaded from the Cambridge Crystallographic Data Centre and corresponds to a structure originally reported in reference 17. The macrocycle is far from planar and is highly distorted. The dihedral angles of the four individual pyrrole planes with respect to the mean plane defined by the N_4 core are 45.1°, 43.7°, 44.5°, and 39.0°, respectively (starting at the upper right pyrrole ring and proceeding clockwise around the ring). The nickel atom is in an essentially square-planar coordination environment. All four Ni–N bonds are equal in length (1.88 Å). Atom labeling scheme: nickel: ◓; carbon: ○; nitrogen: ◑; oxygen: ◒. Hydrogen atoms have been omitted for clarity

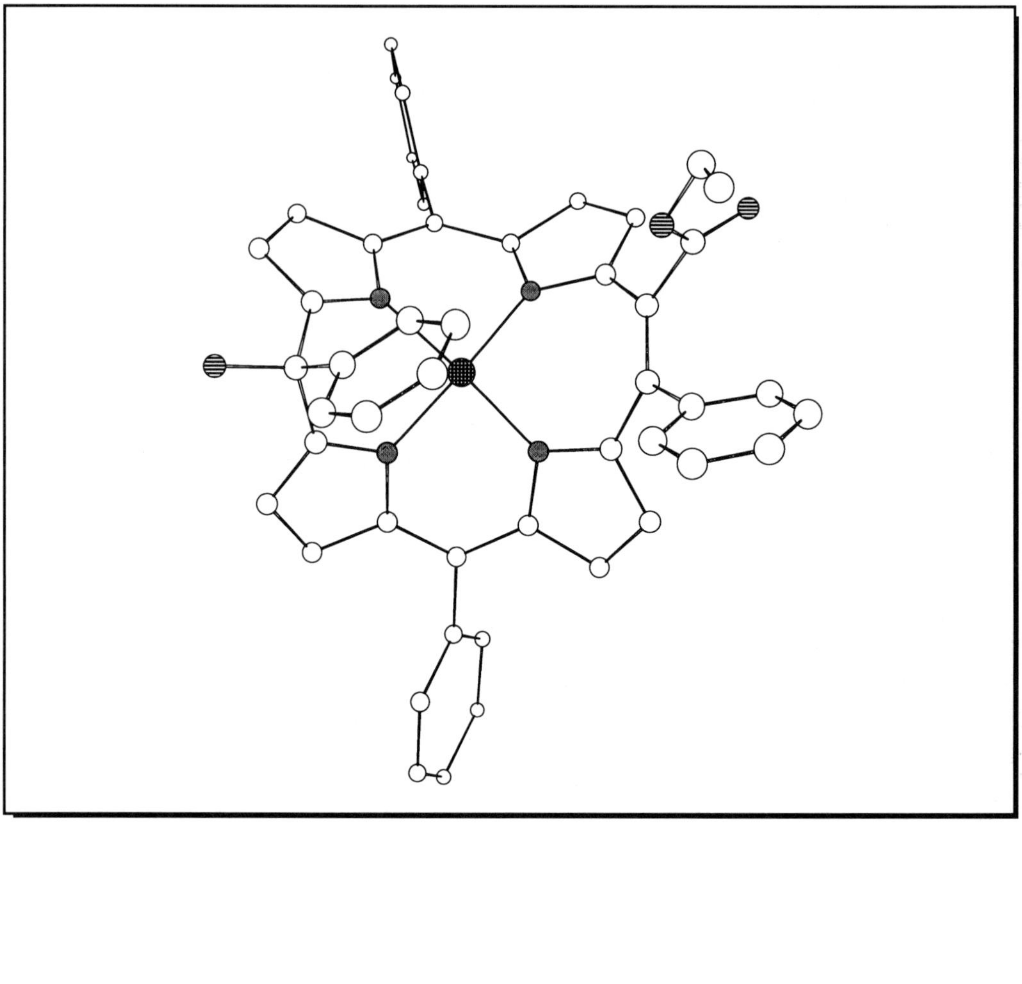

Figure 4.1.3   Schematic Representation Depicting the Conformation of 10-Disubstituted Homoporphyrins **4.29–4.30**

Scheme 4.1.13

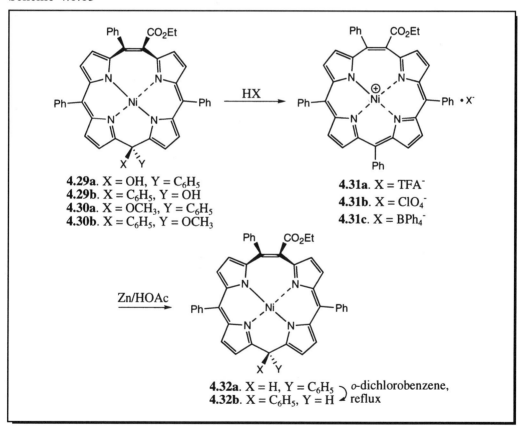

4.29a. X = OH, Y = C$_6$H$_5$
4.29b. X = C$_6$H$_5$, Y = OH
4.30a. X = OCH$_3$, Y = C$_6$H$_5$
4.30b. X = C$_6$H$_5$, Y = OCH$_3$

4.31a. X = TFA$^-$
4.31b. X = ClO$_4^-$
4.31c. X = BPh$_4^-$

4.32a. X = H, Y = C$_6$H$_5$
4.32b. X = C$_6$H$_5$, Y = H

o-dichlorobenzene, reflux

Scheme 4.1.14

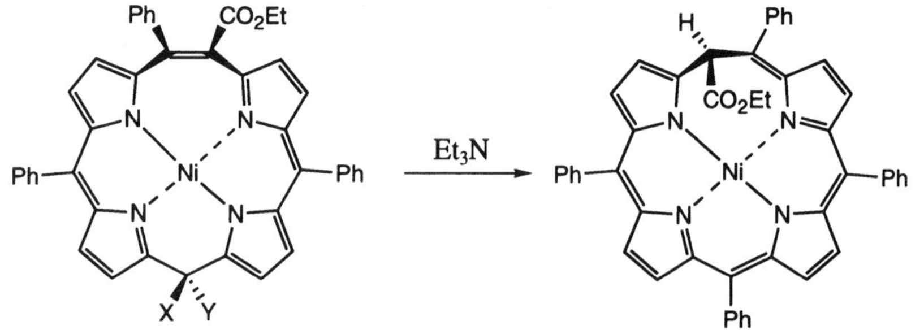

4.32a. X = H, Y = C$_6$H$_5$ **4.16a**

Scheme 4.1.15

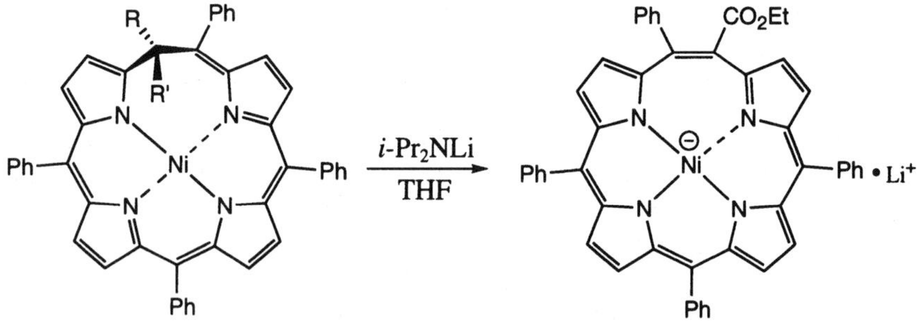

4.16a. R = H, R' = CO$_2$Et
4.16b. R = CO$_2$Et, R' = H

4.33a

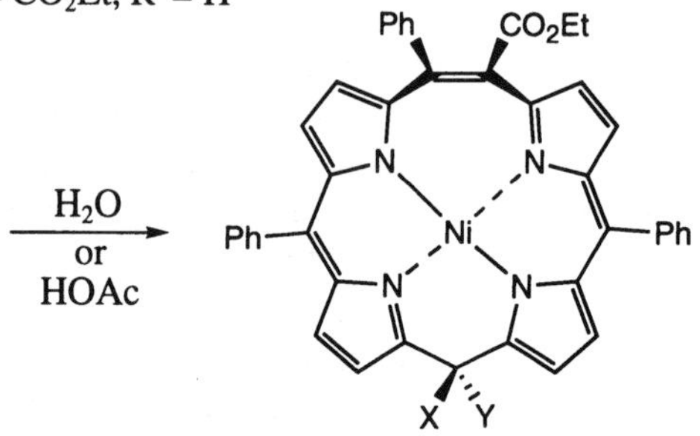

4.32a. X = H, Y = C$_6$H$_5$

Returning back to the main theme of homoporphyrins, it is of interest to note that an alternative approach to their preparation appeared in 1978.[20] In this method, also introduced by Callot, a zinc(II)tetraphenylporphyrin, e.g., **4.35**, is treated with a variety of diazo compounds **4.36–4.40**. What results after workup are the metal-free homoporphyrins **4.41–4.45** (Scheme 4.1.17). In each case, the predominant isomer formed is the "ester-endo" one. It was rationalized that this specificity is ascribable to preorganization resulting from an association between the carbonyl oxygen of the ester and the zinc atom of the porphyrin (see Scheme 4.1.18). After association and loss of a nitrogen molecule, the resulting carbene inserts into the *meso*-position of the porphyrin, affording the observed "ester-endo" orientation.

Scheme 4.1.16

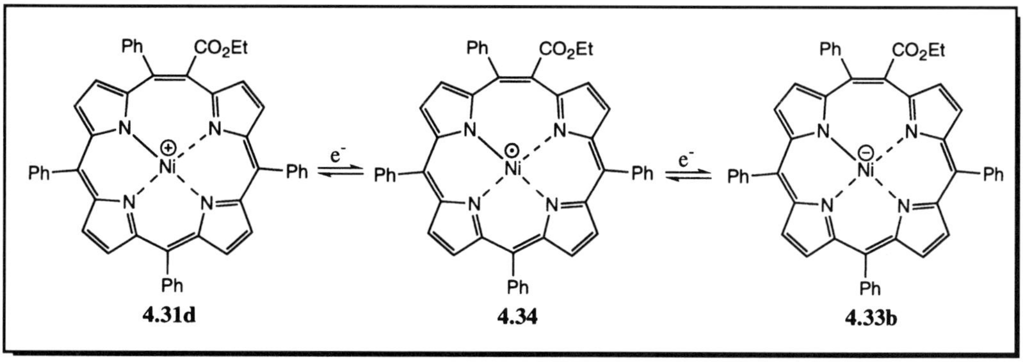

Each of the new homoporphyrin derivatives prepared by this alternative approach was demonstrated to form a stable nickel(II) complex when treated with Ni(OAc)$_2$ (e.g., **4.46–4.50**). Further, the free-base homoporphyrin **4.41** was shown to form the stable complexes **4.51–4.53** when treated with the appropriate divalent metal salt (M = Cu, Zn, Co, X = OAc).[20,21] Additionally, in analogy to the porphyrins, treatment of the free-base homoporphyrin **4.41** with FeCl$_2$ was found to give, after workup, the iron(III) complex **4.54** (Scheme 4.1.19); it presumably results from normal, porphyrin-like autoxidation at the metal.[21]

More recently, homoporphyrins have also been prepared *via* ring-closure reactions involving dideoxybiladienes-ac. Such closures were initially achieved serendipitously during the course of attempts aimed at effecting the demetalation of the copper(II) macrocycle **4.56**, a species that was derived, in turn, from the dideoxybiladiene-ac **4.55**. Specifically, it was found that, using a mixture of TFA and sulfuric acid, to attempt the demetalation of **4.56** afforded the metal-free homoporphyrin **4.57** in 30% yield (Scheme 4.1.20).[22] Here, it was posited that the ring expansion proceeded *via* a mechanism involving first ring opening followed by ring closure. Consistent with this was a single crystal structural analysis which revealed that the product did in fact display the expected "ester-endo" configuration (Figure 4.1.4).

The same homoporphyrin, i.e., **4.57**, could also be prepared directly from dideoxybiladiene-ac dihydrobromide **4.55** *via* anodic oxidation at 800 mV vs. Ag/AgCl in DMF using tetraethylammonium toluenesulfonate as the supporting electrolyte.[23,24] In a similar fashion, dideoxybiladiene-ac dihydrobromide **4.58** could be cyclized *via* anodic oxidation to generate the metal-free homoporphyrin derivative **4.59** (no exo/endo stereochemistry was specified for this compound) (Scheme 4.1.21).

Scheme 4.1.17

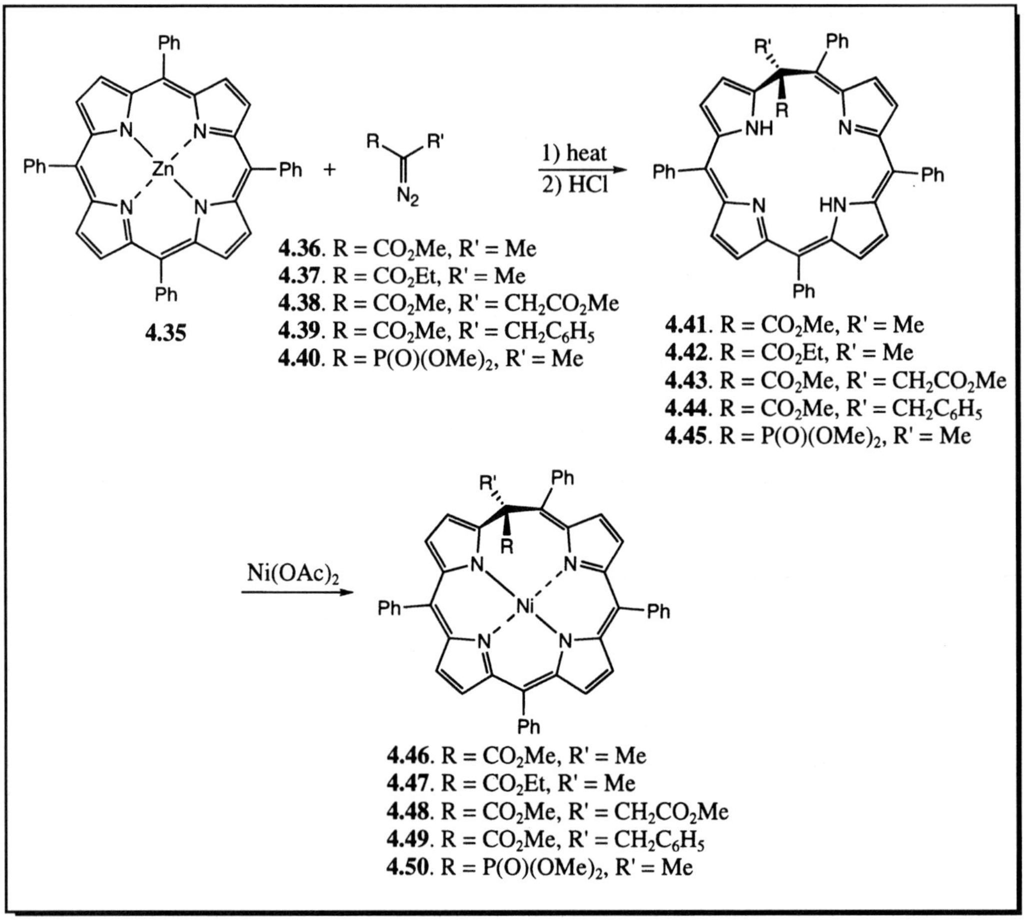
1) heat
2) HCl
4.35
4.36. R = CO_2Me, R' = Me
4.37. R = CO_2Et, R' = Me
4.38. R = CO_2Me, R' = CH_2CO_2Me
4.39. R = CO_2Me, R' = $CH_2C_6H_5$
4.40. R = $P(O)(OMe)_2$, R' = Me
4.41. R = CO_2Me, R' = Me
4.42. R = CO_2Et, R' = Me
4.43. R = CO_2Me, R' = CH_2CO_2Me
4.44. R = CO_2Me, R' = $CH_2C_6H_5$
4.45. R = $P(O)(OMe)_2$, R' = Me
$Ni(OAc)_2$
4.46. R = CO_2Me, R' = Me
4.47. R = CO_2Et, R' = Me
4.48. R = CO_2Me, R' = CH_2CO_2Me
4.49. R = CO_2Me, R' = $CH_2C_6H_5$
4.50. R = $P(O)(OMe)_2$, R' = Me

Scheme 4.1.18

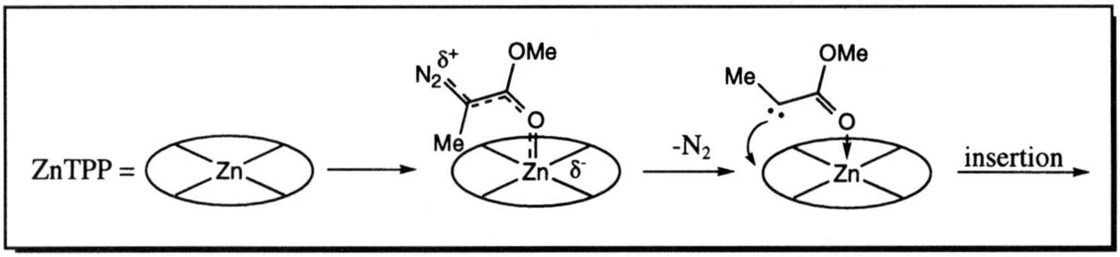
ZnTPP =
$-N_2$
insertion

Heating the dideoxybiladiene-ac dihydrobromide **4.58** with $Cu(OAc)_2$ in DMF was found to afford the metalated homoporphyrin **4.60** directly, along with macrocycle **4.61** (Scheme 4.1.22).[24] This is in contrast to what was observed in the case of the dideoxybiladiene-ac dihydrobromide **4.55** wherein no homoporphyrin was found

Scheme 4.1.19

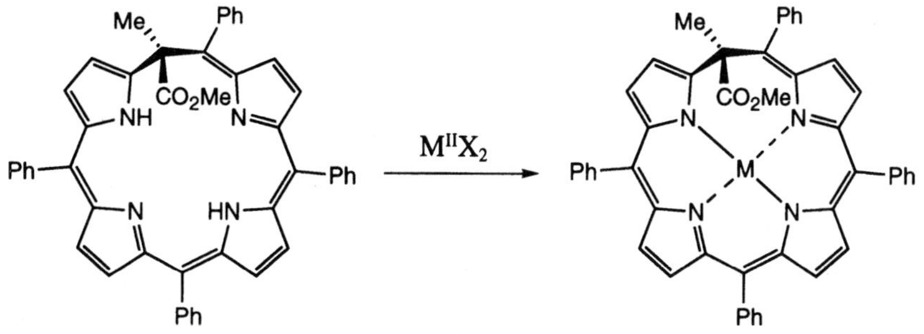

4.41

4.51. M = Cu
4.52. M = Zn
4.53. M = Co
4.54. M = FeCl

Scheme 4.1.20

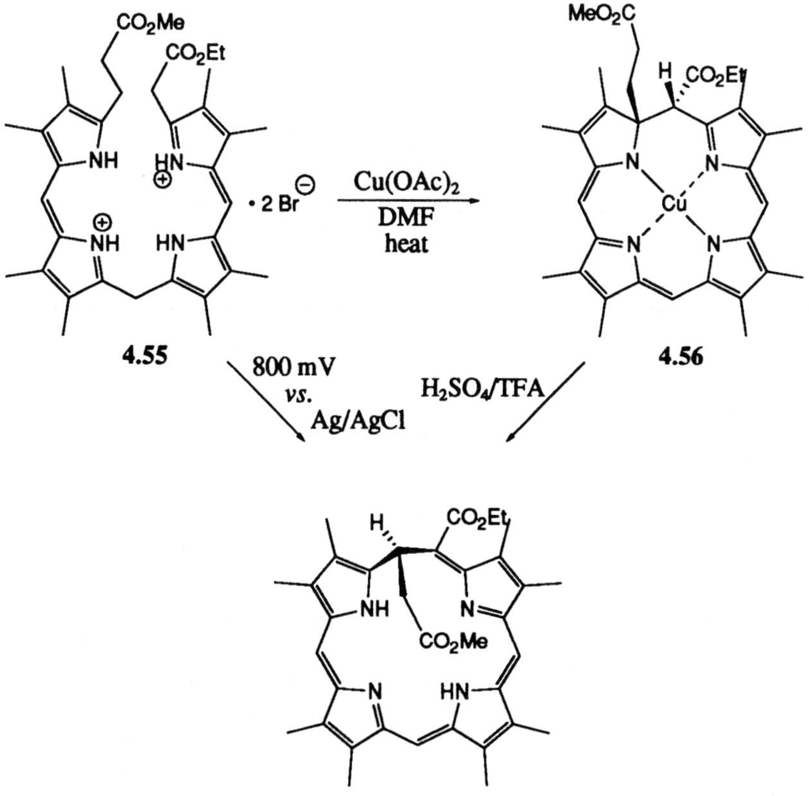

4.55

4.56

4.57

to be formed during the metal-mediated cyclization. Nevertheless, the structural assignment for complex **4.60** as a Cu(II) homoporphyrin derivative is considered secure in that it was confirmed by a single crystal X-ray structural analysis (Figure 4.1.5).

Scheme 4.1.21

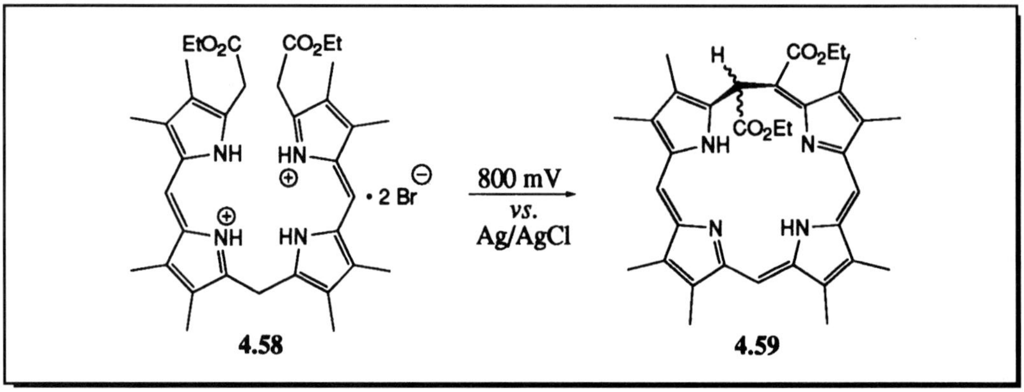

4.58 $\cdot 2$ Br⁻ $\dfrac{800\,\text{mV}}{vs.\ \text{Ag/AgCl}}$ **4.59**

Figure 4.1.4   Single Crystal X-Ray Diffraction Structure of Free-base Homoporphyrin **4.57** Showing the "Ester–Endo" Configuration of the Substituents.

This figure was generated using information down-loaded from the Cambridge Crystallographic Data Centre and corresponds to a structure originally reported in reference 22. The macrocycle is highly non-planar (the average deviation from the mean macrocyclic plane is 0.47 Å). Atom labeling scheme: carbon: ○; nitrogen: ●; oxygen: ⊖. Hydrogen atoms have been omitted for clarity

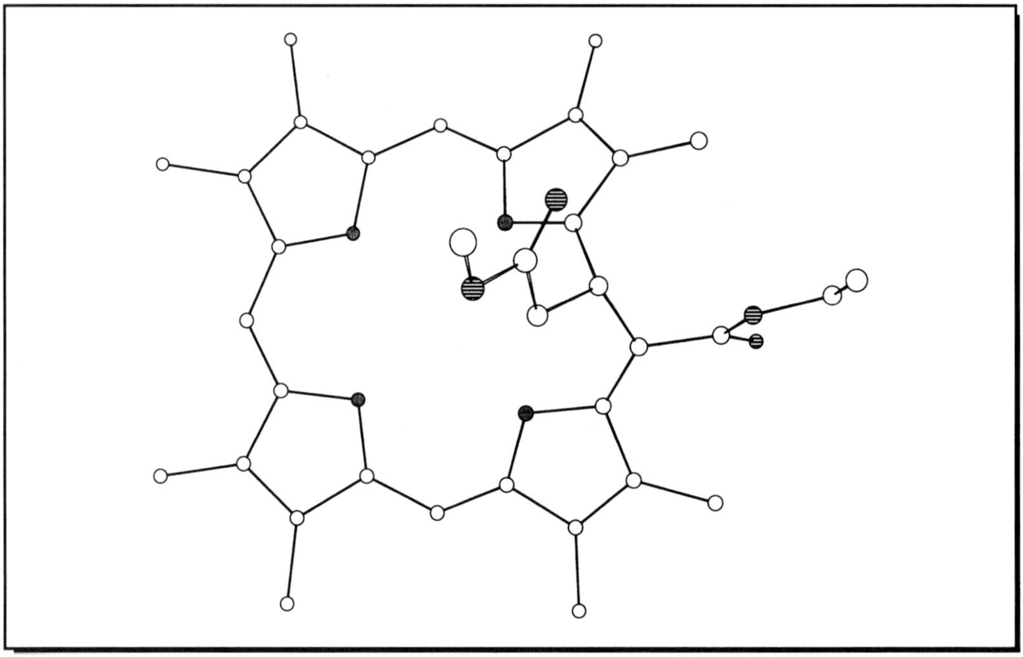

Scheme 4.1.22

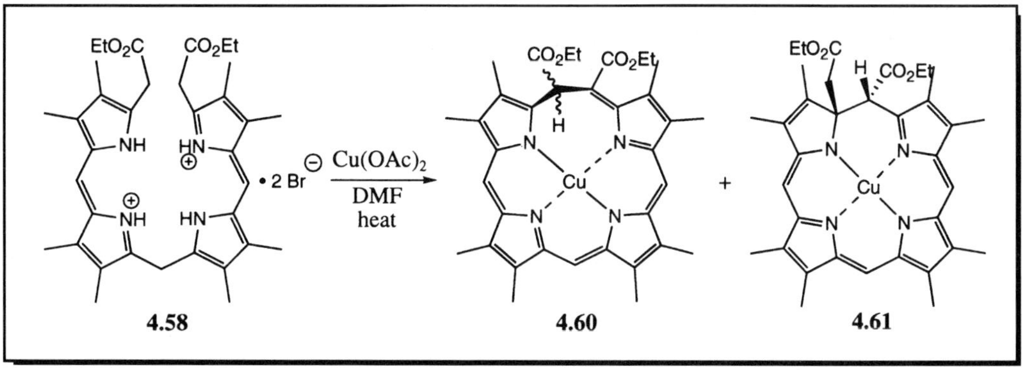

Figure 4.1.5   Single Crystal X-Ray Diffraction Structure of Copper(II)
Homoporphyrin **4.60**.
This figure was generated using information down-loaded from the Cambridge Crystallographic
Data Centre and corresponds to a structure originally reported in reference 24. The copper atom is
held in a distorted square planar environment. The average Cu–N bond distance is 1.99 Å. Atom
labeling scheme: copper: ◉; carbon: ○; nitrogen: ●; oxygen: ◒. Hydrogen atoms have been omitted
for clarity

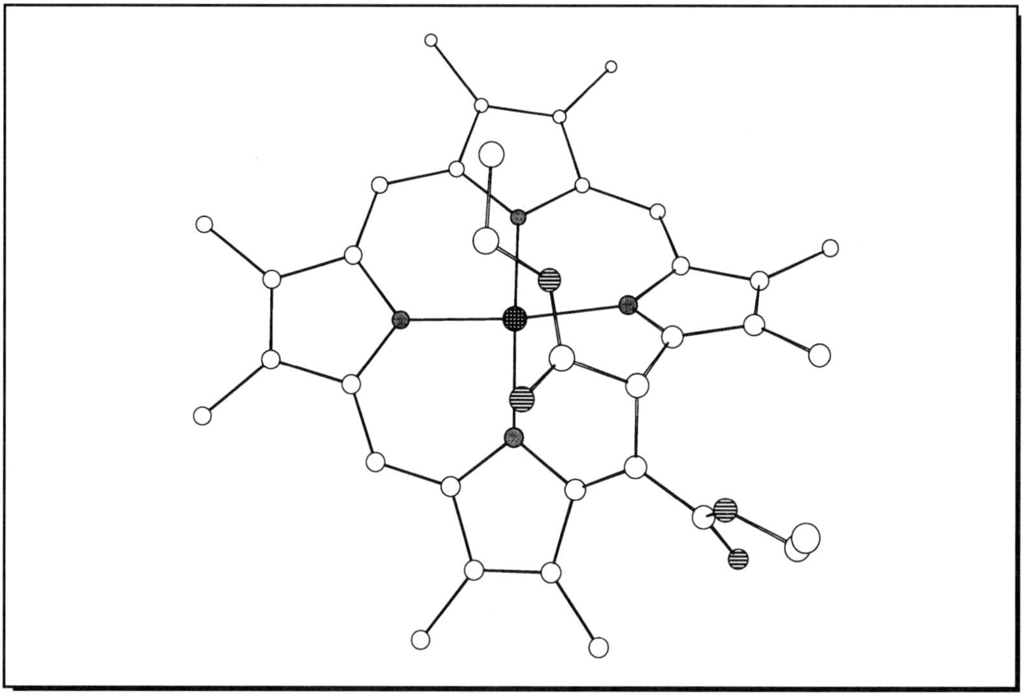

In analogy to the above, dideoxybiladiene-ac **4.62** was cyclized in hot DMF in the presence of $Cu(OAc)_2$. This afforded a copper(II) complex **4.63** of unspecified relative stereochemistry.[24] Demetalation of this macrocycle with TFA/H_2SO_4 afforded a small quantity of a product that was presumed to be the homoporphyrin **4.64** (Scheme 4.1.23). Unfortunately, no conclusive evidence for the formation of this species was ever gathered. Thus, this structural assignment remains insecure at present.

Scheme 4.1.23

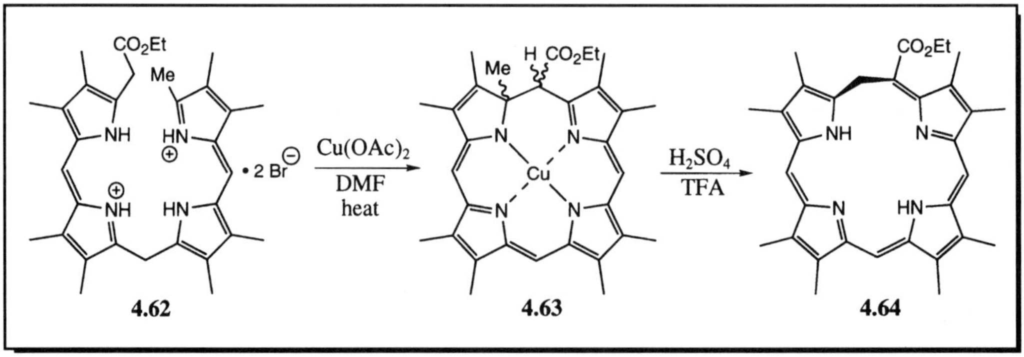

4.2   Vinylogous Porphyrins (Odd–Odd Systems)

The porphyrin skeleton can also be expanded formally so as to produce systems wherein an odd number of carbon atoms serves to link the pyrrole rings. This produces vinylogously enlarged porphyrin-like macrocycles of general structure **4.65** (Figure 4.2.1). More specifically, by inserting two carbon atoms into opposite sides of the porphyrin ($n = m = 1$) skeleton results in a bis-vinylogous structure (e.g., **4.65**, $n = 1$, $m = 3$) which has been elongated at two of the *meso*-sites. Further "expansion" in this way gives rise, at least in principle, to higher order vinylogous systems such as the tetravinylogous (e.g., **4.65**, $n = 1$, $m = 5$) and the octavinylogous (**4.65**, $n = m = 3$) porphyrins. To date a number of such vinylogously enlarged porphyrins have been prepared. Their synthesis and general characteristics will be described in the following section. The interested reader is also referred to a recent account by Franck and Nonn, in which the chemistry of these systems is reviewed.[25]

4.2.1   *Historical Overview*

Much of the original interest in preparing vinylogous porphyrins stemmed from a desire to obtain annulene-like systems with large, extended π-conjugation pathways that could be used both to explore the limits of aromaticity and test for the possible existence of antiaromaticity. As will be discussed later on in this chapter, this promise has largely been realized with these systems. In fact, the diatropic behavior of the vinylogous porphyrins has proved in many instances to be greater not only than that of their tetrapyrrolic [18]porphyrin(1.1.1.1) "parents" as well as most heteroatom-free annulenes. On the other hand, because they typically contain extended conjugation pathways in a rather rigid, planar framework, the vinylogous porphyrins have also stimulated interest that transcends the original issues of aromaticity and antiaromaticity. They have, for instance, been found to exhibit remarkable photochemical and spectroscopic properties. This, as discussed in Chapter 10, has made various vinylogous porphyrins attractive as potential sensitizers for use in various light-based therapeutic applications (e.g., photodynamic therapy, PDT).

Figure 4.2.1   Structure **4.65**

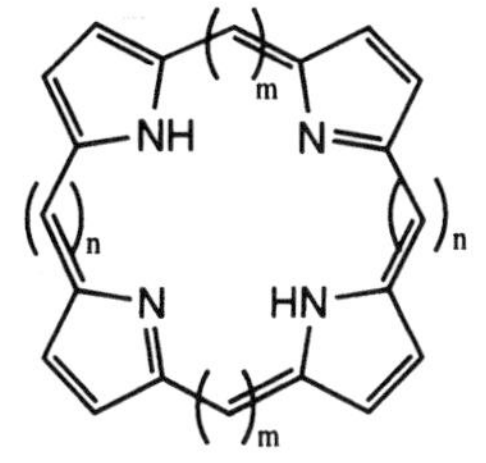

4.65

4.2.2   Bisvinylogous Porphyrins

The first successful synthesis of a vinylogous porphyrin was reported in 1978 by LeGoff and Berger[26] and involved the 22 π-electron aromatic bisvinylogous system **4.68**. Inspired by the Greek "platys" meaning wide or broad, these researchers coined the name "platyrin" for this class of macrocycles. Based on the more general nomenclature used throughout this book, this specific macrocycle may alternatively be referred to as being "[22]porphyrin-(3.1.3.1)".

[22]Porphyrin-(3.1.3.1) **4.68** was prepared by the HBr catalyzed reaction between the "vinylogous" dipyrrylmethane **4.66** and its formyl derivative **4.67**. The presumed intermediate condensation product was not isolated. Instead, it was oxidized with air during the course of the reaction to produce the fully conjugated system directly. After workup and chromatographic purification over neutral alumina, the free-base macrocycle **4.68a** was obtained in 19% yield (Scheme 4.2.1).

The highly conjugated, aromatic nature of this molecule was shown by the intense Soret-like absorbance band at 477 nm (ε = c. 395 000 M^{-1} cm^{-1} in CH_2Cl_2) observable in the UV-vis spectrum of both the free-base form of the macrocycle (**4.68a**) and its TFA salt (**4.68b**). Further, the ^1H NMR spectrum of **4.68b** revealed the apparent presence of a sustained diamagnetic ring-current effect, with chemical shifts resembling those found in octaethylporphyrin and decamethylsapphyrin

(Chapter 5).[26] Specifically, the internal *meso*-like protons in **4.68b** were found to resonate at −8.97 ppm, which constitutes a 16.15 ppm upfield shift compared to the same cyclohexene-like proton in **4.66** (δ_{CH} = 7.18 ppm).

Scheme 4.2.1

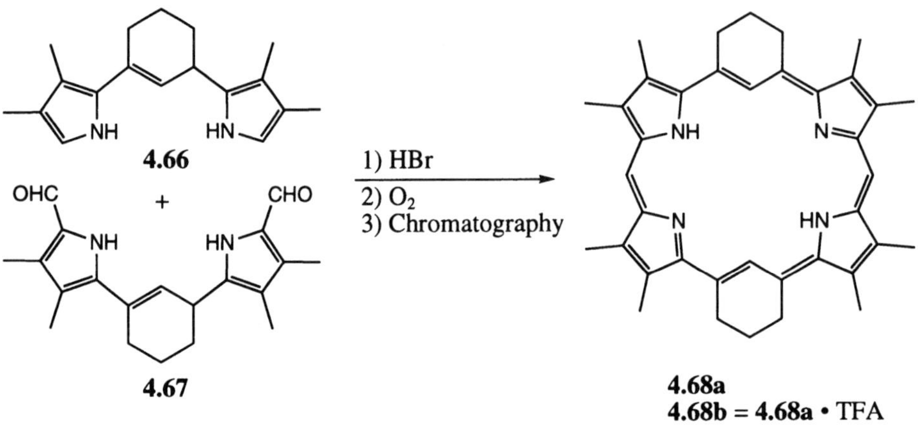

4.66

OHC   +   CHO

4.67

1) HBr
2) O_2
3) Chromatography

4.68a
4.68b = 4.68a • TFA

The ability of [22]porphyrin-(3.1.3.1) **4.68a** to form metal complexes was investigated in the context of the original LeGoff–Berger report. Specifically, these workers found, based on spectral evidence, that this macrocycle did form complexes with Ni(II) and Cu(II).[26] However, these complexes proved to be not only highly insoluble but also rather unstable. Indeed, rapid demetalation occurred on treatment with weak acid. Perhaps as a consequence of these initial findings, no further metalation-related reports have appeared involving this molecule.

Since the initial report of the synthesis of the bisvinylogous porphyrin system **4.68**, two other approaches to structurally similar species were reported by Franck and coworkers. The first of these, appearing in 1990, involved the preparation of the bisvinylogous dideoxybiladiene dihydrobromide **4.69**.[27] This linear tetrapyrrole was then condensed with formaldehyde and oxidized *in situ* with 2,3-dichloro-5,6-dicyano-1,4-benzoquinone (DDQ) to afford the 22 π-electron aromatic macrocycle **4.70a** in 42% yield (Scheme 4.2.2).

The other route to bisvinylogous porphyrins, also reported by Franck in 1990, is shown in Scheme 4.2.3. The key reactions in this sequence involve the preparation of the bis(carbonylvinyl)dipyrrylmethane **4.71** and its subsequent condensation with dipyrrylmethane **4.73**.[28] *In situ* dehydrogenation of the condensation product using DDQ followed by chromatographic workup afforded a 35% yield of the aromatic macrocycle **4.75a**. [22]Porphyrin-(3.1.3.1) **4.76** was also prepared in an analogous fashion.[29]

Proton NMR data for **4.70** were consistent with an aromatic formulation for this molecule. Further, a single crystal X-ray diffraction analysis of the bis-trifluoroacetate salt of this macrocycle (**4.70b**) revealed a nearly planar macrocyclic nucleus (Figure 4.2.2). As predicted for an aromatic molecule, the bond lengths along the 22 π-electron C–C framework were nearly equivalent, with an average length of 1.38 Å. This resembles closely those observed for the dication of octaethylporphyrin[30,31] (see also reference 27).

Scheme 4.2.2.

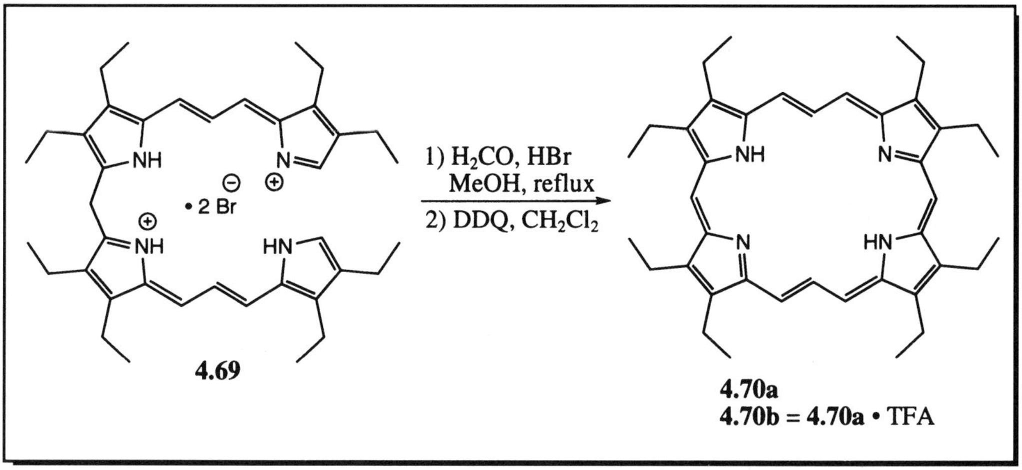

Scheme 4.2.3

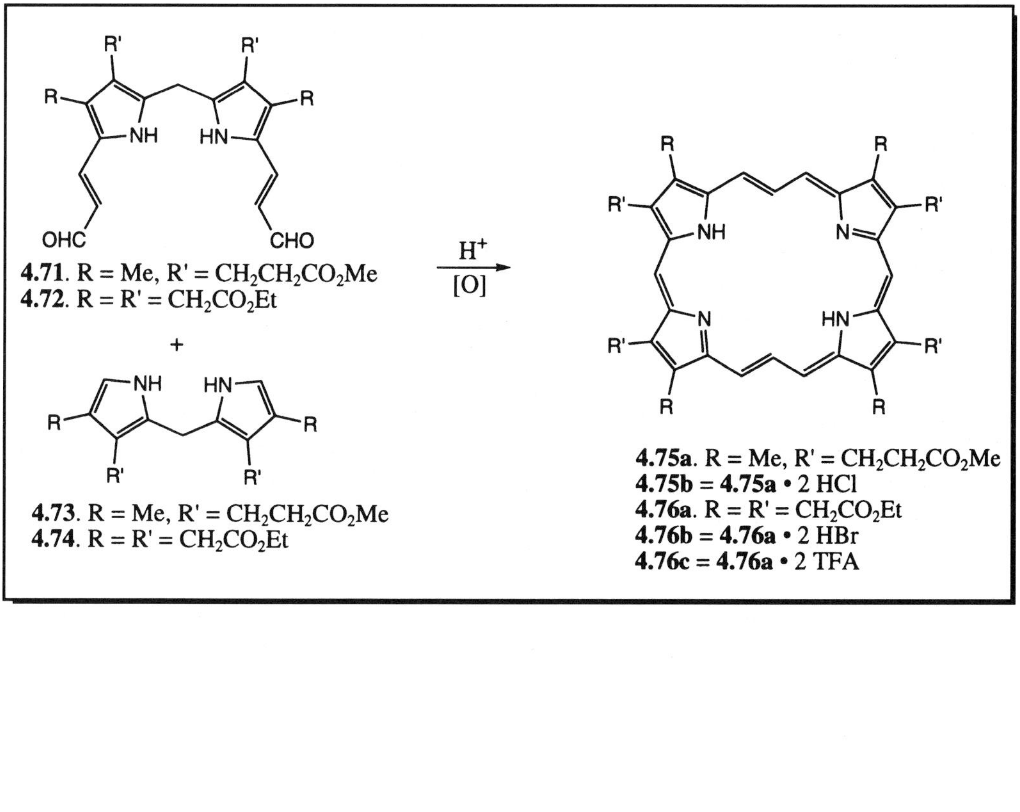

Figure 4.2.2   Single Crystal X-Ray Diffraction Structure of the Bistrifluoroacetate Salt of [22]Porphyrin-(3.1.3.1) **4.70b**.

This figure was generated using information down-loaded from the Cambridge Crystallographic Data Centre and corresponds to a structure originally reported in reference 27. The macrocycle is centrosymmetric in the crystal and is highly planar (the average deviation of the sp^2-hybridized atoms from the mean macrocyclic plane is 0.09 Å). Atom labeling scheme: carbon: ○; nitrogen: ●. The trifluoroacetate counteranions and hydrogen atoms have been omitted for clarity

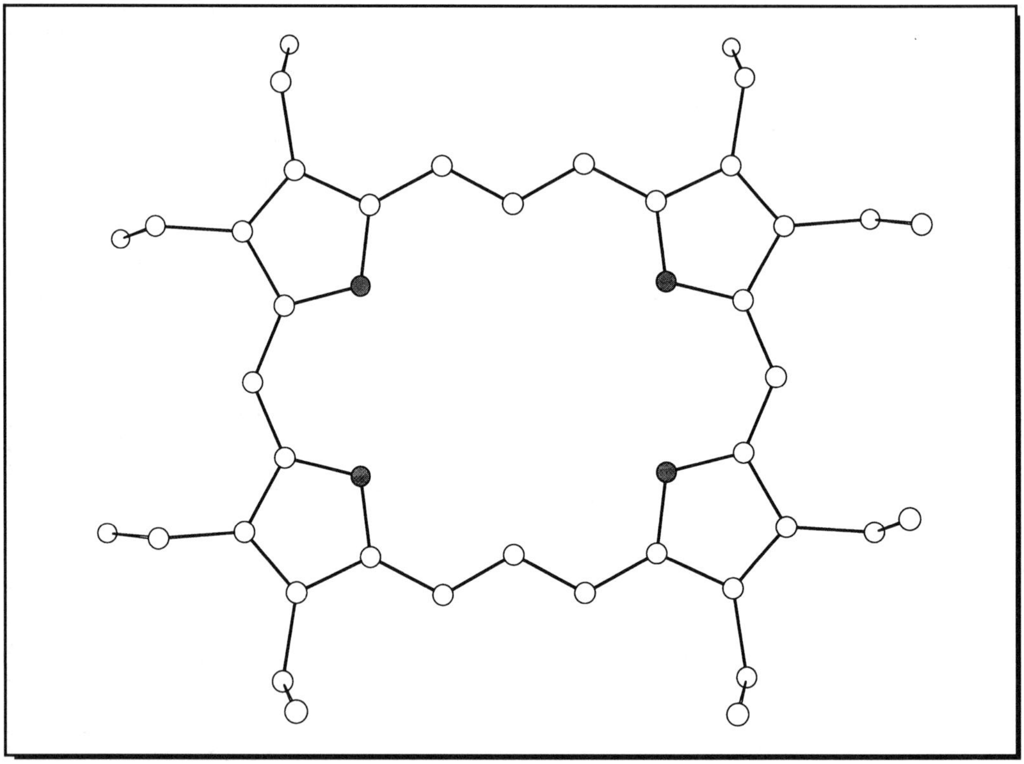

[22]Porphyrins-(3.1.3.1) **4.75** and **4.76** exhibit spectral features characteristic of bisvinylogous porphyrins. Specifically, strong diamagnetic ring-current effects and intense Soret-like absorbance bands are observed in the ^{1}H NMR and UV-vis spectra, respectively. Of particular note is the remarkable light-absorbing ability of the protonated macrocycle **4.76c**. This species displays a Soret band (in 1% TFA in CHCl$_3$) that is both extremely narrow and extremely intense (molar absorptivity is $1\,090\,000$ M^{-1} cm^{-1}). The high molar extinction coefficient and high diatropic ring-current effect observed for this and indeed other vinylogous porphyrins led Franck to coin the term "superarene" for this type of organic pigment.[25]

4.2.3   Tetravinylogous Porphyrins

The successful preparation of the bisvinylogous porphyrins served to inspire the preparation of other, even larger, vinylogous porphyrins. The 26 π-electron aromatic tetravinylogous porphyrin **4.79** was the first example of such a further expanded vinylogous porphyrin to be reported.[32] This macrocycle, termed [26]porphyrin-(5.1.5.1), was reported in 1987 by LeGoff and Weaver. It was prepared in a fashion analogous to that used to prepare its smaller "platyrin" cousin, **4.68** (*vide supra*). Specifically, its synthesis involved the HBr-catalyzed condensation of the "bisvinylogous dipyrrylmethane" **4.77** with its formyl derivative **4.78**. *In situ* dehydrogenation with O₂ followed by chromatographic purification afforded the aromatic system **4.79a** in 2% yield (Scheme 4.2.4). While NMR and UV-vis analyses could be used to confirm that **4.79** displays the spectral features expected of an aromatic macrocycle, the free-base and protonated forms of **4.79** both proved to be rather unstable. Specifically, the half-lives for decomposition were found to range from several hours to several days even when the materials were stored at low temperatures.

Scheme 4.2.4

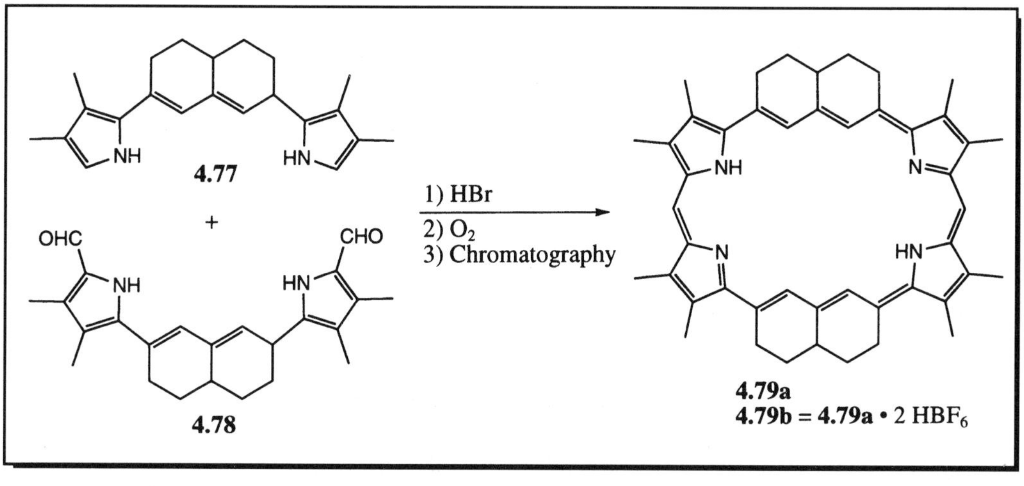

An alternative approach to [26]porphyrins-(5.1.5.1) was devised by Franck and coworkers and was reported in 1993.[33] This approach, outlined in Scheme 4.2.5, involves a more linear synthetic strategy than was employed by LeGoff. Specifically, it involves preparing first the tetravinylogous dideoxybiladiene-ac dihydrobromide precursor **4.82** and then reacting it with formaldehyde. Oxidation *in situ* with DDQ then affords the [26]porphyrin-(5.1.5.1) **4.83a** in 43% yield. The C(14)-substituted [26]porphyrins-(5.1.5.1) **4.84** and **4.85** were prepared in an analogous manner (64% and 5% yield, respectively) by condensing **4.82** with benzaldehyde or propionaldehyde as appropriate.

Scheme 4.2.5

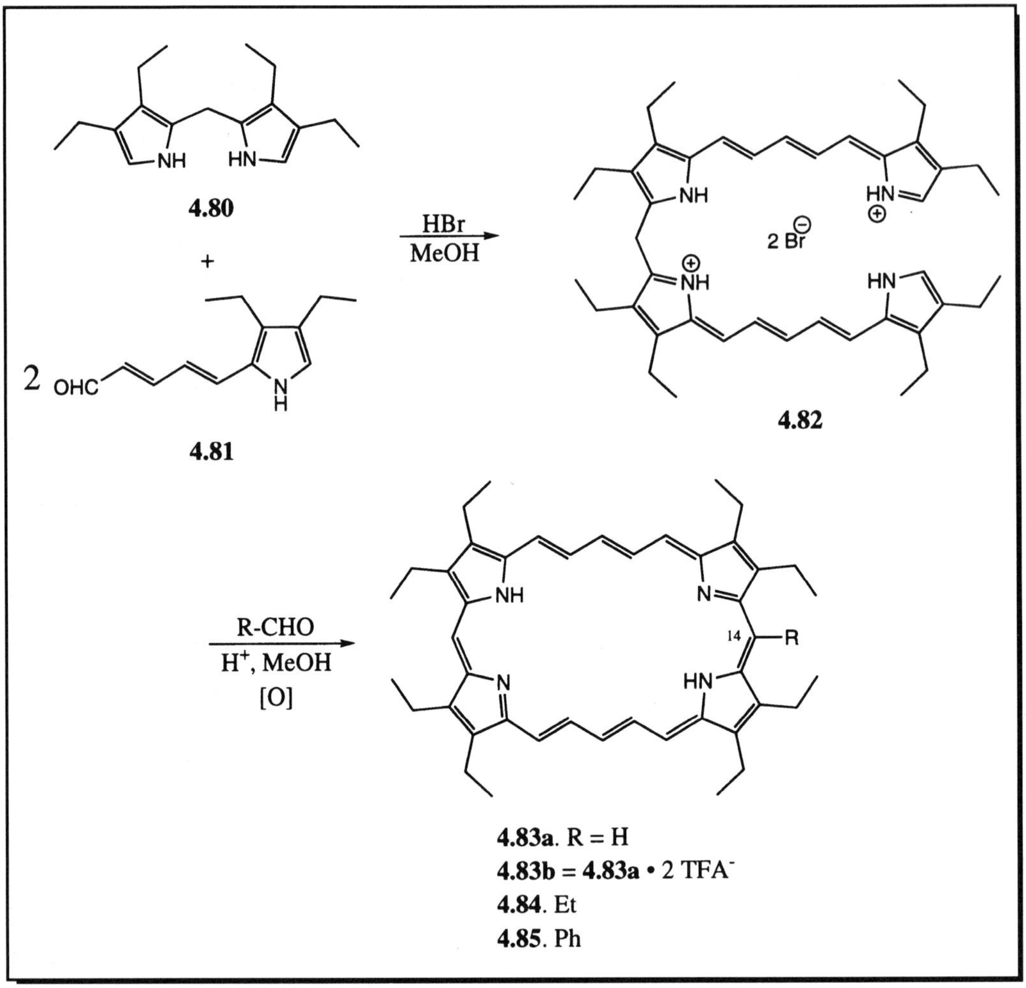

4.80

+

2 OHC

4.81

HBr
MeOH
⟶

2 Br⁻

4.82

R-CHO
H⁺, MeOH
[O]
⟶

14 — R

4.83a. R = H
4.83b = **4.83a** • 2 TFA⁻
4.84. Et
4.85. Ph

Like the related platyrin-type [26]porphyrin(5.1.5.1) **4.79** of LeGoff and Weaver,[32] these latter (5.1.5.1)-type 26 π-electron systems proved to be aromatic (as judged from ¹H NMR spectral analysis). They were also found, in analogy to these earlier reported systems, to exhibit strong Soret-band absorbances in the 510–525 nm region of the electronic spectrum. In marked contrast to **4.79**, however, the [26]porphyrins-(5.1.5.1) **4.83–4.85** proved to be rather stable. This increased stability, coupled with a propensity for crystallinity allowed a single crystal X-ray structural analysis to be performed on the C(14)-unsubstituted macrocycle **4.83b** (Figure 4.2.3). The resulting structure revealed not only a remarkably planar macrocyclic ring, but also C–C bond lengths for the main conjugation pathway that are nearly equal in length. This structure thus supports the claim that system **4.86b**, like its smaller congener **4.70b**, is in fact aromatic in character.

Figure 4.2.3   Single Crystal X-Ray Diffraction Structure of the Bistrifluoroacetate Salt of [26]Porphyrin-(5.1.5.1) **4.83b**.

This figure was generated using information down-loaded from the Cambridge Crystallographic Data Centre and corresponds to a structure originally reported in reference 33. The macrocycle exhibits pseudo-C_2 symmetry in the crystal and is highly planar (the average deviation of the macrocyclic atoms from the mean macrocyclic plane is 0.15 Å). Atom labeling scheme: carbon: ○; nitrogen: ●. The trifluoroacetate counteranions and hydrogen atoms have been omitted for clarity

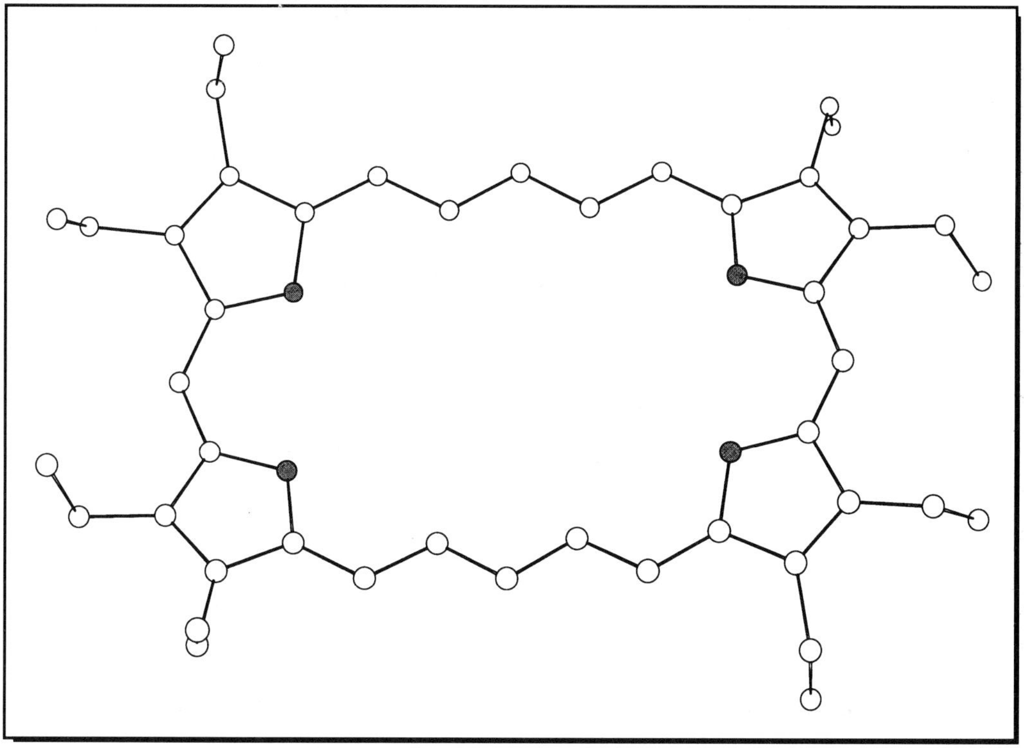

Tetravinylogous porphyrins with an alternative arrangement [(3.3.3.3) vs. (5.1.5.1)] of *meso*-carbons have also been synthesized, as shown in Scheme 4.2.6.[25,34–36] In the initial 1986 communication describing this work,[34] Franck and Gosmann reported that the octaethylporphyrinogen-(3.3.3.3) **4.92** could be obtained in 24% yield from *N*-methylpyrrole-vinyl aldehyde **4.87** by generating the corresponding allyl alcohol *in situ* and then subjecting it to acid-catalyzed self-condensation. Porphyrinogen-(3.3.3.3) **4.92** was then subject to oxidation with six equivalents of Br_2 to afford the dicationic bisquaternary octaethyl [26]porphyrin-(3.3.3.3) **4.96** as its HBr salt in 41% yield. It was later reported[25,35] that the analogous porphyrinogens-(3.3.3.3) **4.91** and **4.93**–**4.95** could be prepared in a similar manner in 0.1%, 35%, 8%, and 52% yields, respectively. These macrocycles, too, with the exception of the β-unsubstituted derivative **4.91**, could be successfully dehydrogenated with Br_2 to afford the corresponding conjugated tetravinylogous porphyrins **4.97–4.99** in 39%, 24%, and 34% yield, respectively.

Scheme 4.2.6

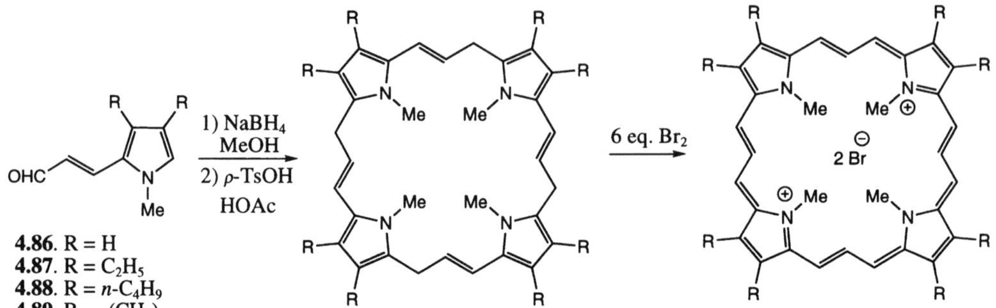

4.86. R = H
4.87. R = C₂H₅
4.88. R = n-C₄H₉
4.89. R = -(CH₂)₄-
4.90. R = CH₂CH(CH₃)₂

4.91. R = H
4.92. R = C₂H₅
4.93. R = n-C₄H₉
4.94. R = -(CH₂)₄-
4.95. R = CH₂CH(CH₃)₂

4.96a. R = C₂H₅
4.97. R = n-C₄H₉
4.98. R = -(CH₂)₄-
4.99. R = CH₂CH(CH₃)₂

The aromatic nature of macrocycles **4.96a–4.99** was inferred from the large diamagnetic ring-current effect observed in their ^{1}H NMR spectra. For instance, in the case of the octaethyl derivative **4.96a**, the shift difference between the internal C–H protons and the external ones was found to be 25.3 ppm. This shift difference is over twice as large as that found for [18]annulene ($\Delta\delta = 12.1$ ppm). The UV-vis absorbance spectrum for **4.96a** exhibits an intense Soret-like absorbance at 547 nm with a molar extinction coefficient of 909 600 $M^{-1} cm^{-1}$. This extremely large extinction coefficient, as well as the large diamagnetic ring current, arise from the presumed planar nature of this macrocycle in solution. Unfortunately, as yet, no solid-state evidence has been obtained to support the presumed planar conformation of this macrocycle.

The tetravinylogous [26]porphyrin-(3.3.3.3) **4.96b** was subject to alkali metal reduction conditions.[36] In this work, it was found that treatment of the bisperchlorate salt **4.96b** (generated from the bisbromide salt by treatment with 70% $HClO_4$) with potassium metal in THF at $-78\,°C$ affords the neutral 28 π-electron macrocycle **4.100** in 10% yield (Scheme 4.2.7). On the other hand, when this same reduction was allowed to continue for 5 days at room temperature, the dianionic 30 π-electron potassium salt **4.101** was observed to form.

The nature of π-conjugation present in the reduced macrocycles **4.100** and **4.101** was probed using proton NMR spectroscopy.[36] While the internal *meso*-like protons in the 26 π-electron dication **4.96b** are found to resonate at -11.98 ppm, the corresponding protons in the 28 π-electron system **4.100** resonate at 16.48 ppm. Upon further reduction to the 30 π-electron dianion **4.101**, the chemical shift of the internal *meso*-protons returns to a remarkably high field location (−12.21 ppm). Taken together, these data provide convincing evidence in support of the validity of the Hückel 4n + 2 rule for aromaticity when it is applied to expanded porphyrins.

Scheme 4.2.7

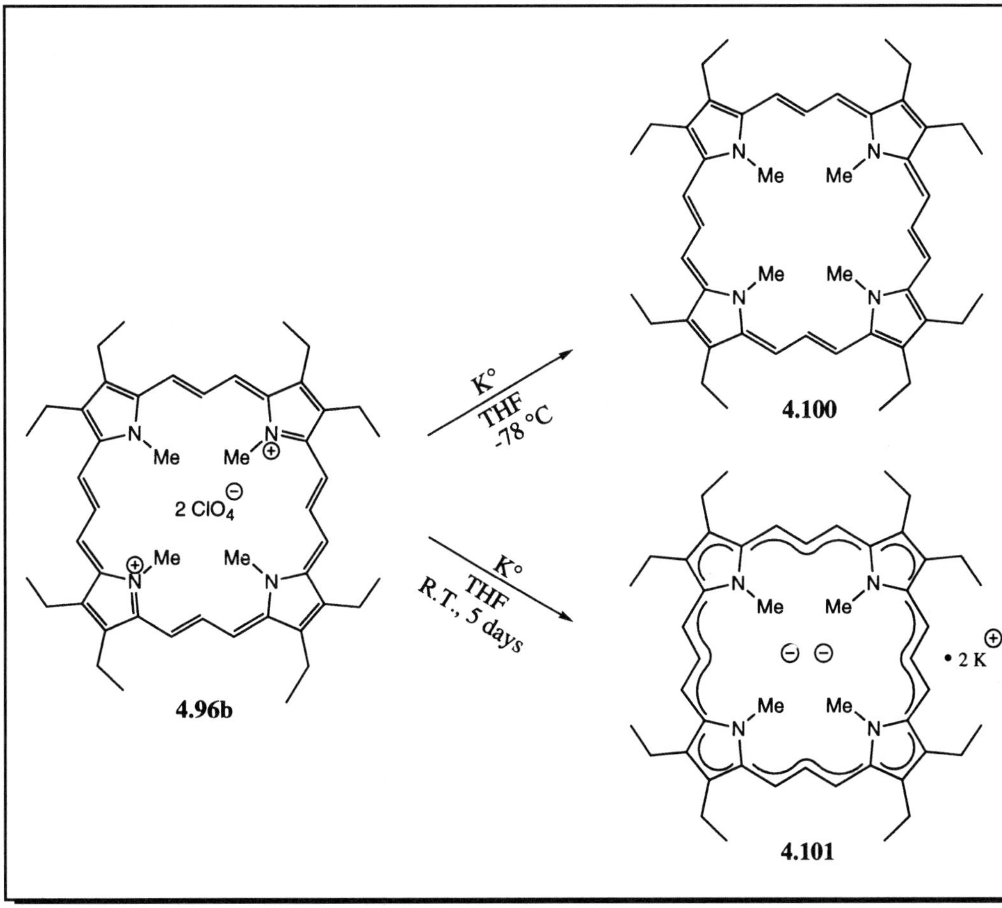

The furan analog of octaethyl-[26]porphyrin-(3.3.3.3) has also been prepared.[37,38] This was accomplished in a fashion analogous to that used for the synthesis of the tetrakis-*N*-methyl derivatives **4.96–4.99**. Specifically, the allyl-alcohol-substituted diethyl furan **4.102** was self-condensed in nitromethane using citric acid as the catalyst. The resulting tetraoxaporphyrinogen **4.103** was obtained from this reaction in 3% yield (Scheme 4.2.8). Oxidation to the fully conjugated 26 π-electron dication **4.104a** was then achieved using bis(trifluoroacetoxy)iodobenzene, followed by treatment with trifluoroacetic acid. The bisperchlorate salt **4.104b** was prepared from the bistrifluoroacetate salt by treatment with 70% HClO$_4$.

Scheme 4.2.8

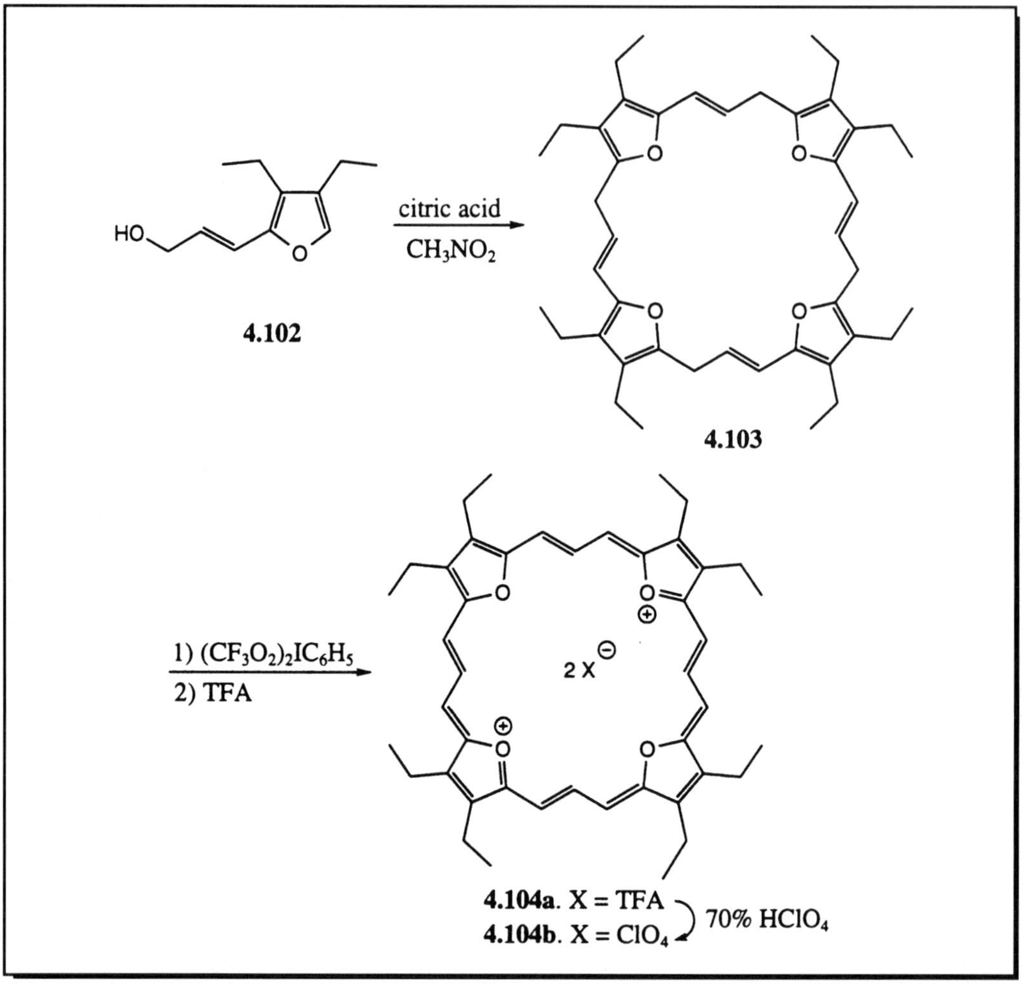

Tetraoxa-[26]porphyrin-(3.3.3.3) **4.104**, like the related *N*-methyl analogs, proved to be aromatic. This was concluded from the large diamagnetic ring-current effect observed in its ^{1}H NMR spectrum ($\Delta\delta$ = 25.53 ppm between the external (14.61 ppm) and internal (−10.92 ppm) *meso*-like proton resonance frequencies). Also like the *N*-methyl derivatives, **4.104** exhibits a remarkably intense absorption in the visible region of the electromagnetic spectrum (λ_{max} for **4.104** = 525 nm). In fact, **4.104b**, as its bisperchlorate salt, displays the largest extinction coefficient ever recorded for an organic pigment. In fact, its ε value of 1 600 000 M^{-1} cm^{-1} is roughly four times as great as that of the corresponding porphyrin, for which maximal molar absorptivities of *ca.* 400 000 M^{-1} cm^{-1} are generally recorded. Interestingly, however, the molar extinction coefficient for the Soret band of the corresponding bistrifluoroacetate salt **4.104a** is roughly half as intense (ε = 800 000 M^{-1} cm^{-1}) as the bisperchlorate salt **4.105b**. This observation leads to the suggestion that the counteranion in this macrocycle has a significant effect on the π-electron system of the macrocycle. However, the exact nature of this effect remains unclear.

The tetraoxa-[26]porphyrin-(3.3.3.3) **4.104b** has also been subject to alkali metal reduction.[38] Like its *N*-methyl counterpart, **4.104b** undergoes reduction to the paratropic 28-π-electron macrocycle **4.105** after a 2-hour treatment with potassium metal at −78 °C (Scheme 4.2.9). The designation as **4.105** paratropic was based on the observed change in the chemical shift of the internal protons from −10.92 ppm in the dication to 15.94 ppm for **4.105**, as well as the change in the shift of the external protons from 14.61 in **4.104b** to 2.58 in **4.105**. Upon extended treatment with potassium (6 weeks instead of 2 hours as above), further reduction of **4.104b** to the dianionic, aromatic 30 π-electron system **4.106** was observed to occur.

Scheme 4.2.9

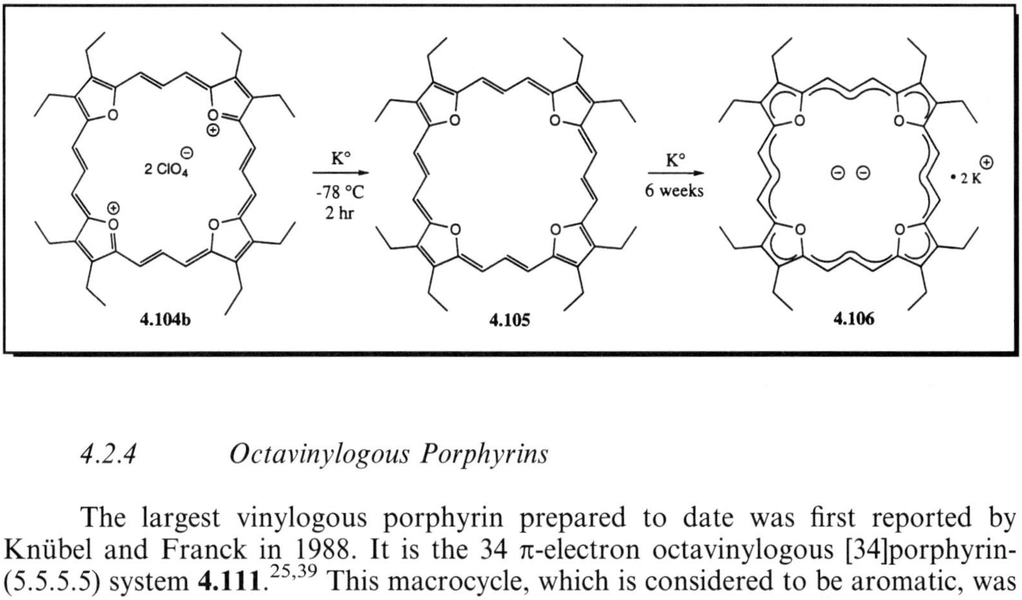

4.104b   4.105   4.106

4.2.4   *Octavinylogous Porphyrins*

The largest vinylogous porphyrin prepared to date was first reported by Knübel and Franck in 1988. It is the 34 π-electron octavinylogous [34]porphyrin-(5.5.5.5) system **4.111**.[25,39] This macrocycle, which is considered to be aromatic, was prepared using an approach similar to that used to prepare the [26]porphyrins-(3.3.3.3) **4.96–4.99** described above. In this case, the key steps in the synthesis involved the reduction of the bisvinyl aldehyde **4.107** with $NaBH_4$, followed by acid-catalyzed self-condensation of the resulting vinylogous allyl alcohol. This afforded the [34]porphyrinogen species **4.109** in yields as high as 20% (Scheme 4.2.10). This macrocycle was then dehydrogenated using 6 equivalents of Br_2 to afford the bisquaternary dibromide **4.111** in 12.5% yield. This same synthetic strategy was later employed to synthesize the isobutyl-substituted porphyrinogen-(5.5.5.5) **4.110** and the corresponding octaisobutyl-[34]porphyrin-(5.5.5.5) derivative **4.112** in 38% and 13% yields, respectively.[25]

The [34]porphyrins-(5.5.5.5) **4.111** and **4.112** each exhibit an extremely large diamagnetic ring current effect in the ¹H NMR spectrum. In particular, the shift difference between the inner and outer proton resonances for **4.111** was found to be 31.5 ppm. This is a chemical shift difference that remains unrivaled for non-organometallic compounds and is thus a result that lends credence to the suggestion that the Hückel (4n + 2) rule remains valid for heteroaromatic systems of this type containing up to at least 34 π-electrons in a conjugation pathway.

Scheme 4.2.10

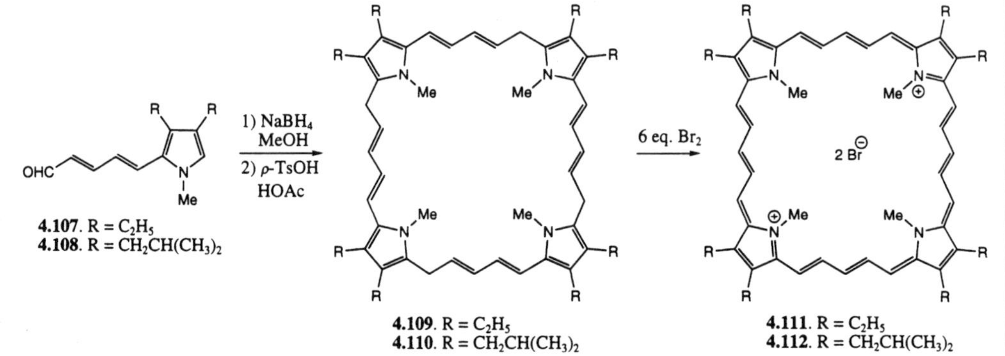

4.107. R = C$_2$H$_5$
4.108. R = CH$_2$CH(CH$_3$)$_2$

4.109. R = C$_2$H$_5$
4.110. R = CH$_2$CH(CH$_3$)$_2$

4.111. R = C$_2$H$_5$
4.112. R = CH$_2$CH(CH$_3$)$_2$

4.3   Stretched Porphycenes (Even–Even Systems)

Vinylogous porphyrins are formally built up of odd-numbered carbon spacers separating four pyrrole rings. There are, however, several expanded porphyrins that contain four pyrrolic or pyrrole-like subunits that contain only even-numbered carbon spacers. These macrocycles will be grouped under the general heading "stretched porphycenes", since they are typically prepared *via* the low-valent titanium mediated coupling of pyrrolic aldehydes just as are the porphycenes (cf. Chapter 3).

The first example of a stretched porphycene was reported in 1990 by Vogel and coworkers.[40] It was the acetylene-cumulene macrocycle **4.115** ([22]tetra-dehydroporphyrin-(2.2.2.2)) prepared *via* the McMurry coupling of the diformyl dipyrryldimethyne derivative **4.113**. As proved true for porphycene, the presumed intermediate species **4.114** could not be isolated. Instead it was allowed to undergo spontaneous oxidation during workup to afford the aromatic macrocycle **4.115** in 18% yield (Scheme 4.3.1).

The structure and aromatic nature of **4.115** was deduced from an analysis of its ^{1}H NMR and ^{13}C NMR spectral properties, and from a single crystal X-ray diffraction study. For instance, in the ^{1}H NMR spectrum both upfield shifts for the internal (NH) protons and downfield shifts for the external *meso*-protons were observed. This led to the conclusion that **4.115** can indeed sustain an induced diamagnetic ring current. The ^{13}C NMR spectrum confirmed the cumulenic nature of the product in that similar chemical shifts were observed for all of the sp-hybridized carbons. The UV-vis absorption spectrum of **4.115** in CH_2Cl_2 showed a striking resemblance to those recorded for the parent porphycenes. However, as expected given the longer π-conjugation pathway that can be drawn for **4.115**, the Soret- and Q-bands were red-shifted significantly compared to those of the porphycenes. Finally, a single crystal X-ray structure of **4.115** revealed, as expected, that the macrocyclic nucleus adopts a nearly perfectly planar conformation (Figure 4.3.1). This, of course, is as expected for an aromatic system.

Scheme 4.3.1

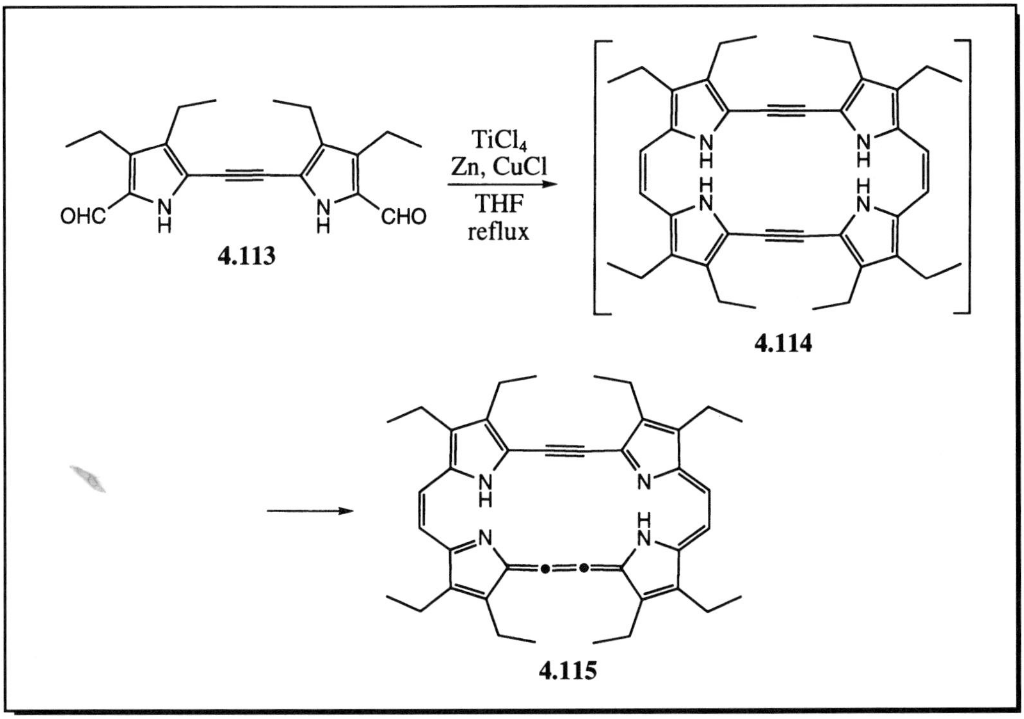

Interestingly, in addition to the desired acetylene-cumulene [22]tetradehydroporphyrin-(2.2.2.2), two other macrocycles were isolated in the course of trying to prepare **4.115**. These macrocycles are the [22]didehydroporphyrin-(2.2.2.2) **4.116**, and the [22]porphyrin-(2.2.2.2) **4.117** (shown in Figure 4.3.2 in their presumed *trans*-configurations). While the events leading to the formation of **4.116** and **4.117** are still not entirely clear, these products were presumed to arise as the result of the strongly reducing conditions associated with the low-valent titanium-mediated coupling reaction.[40]

The [22]porphyrins-(2.2.2.2) **4.116** and **4.117** were also intentionally prepared by Vogel, *et al.* *via* the partial hydrogenation of [22]tetradehydroporphyrin-(2.2.2.2) **4.115** using a lead-poisoned (Lindlar-type) catalyst.[41,42] For instance, the singly-reduced derivative **4.116** was prepared in 5% yield by subjecting **4.115** to Lindlar reduction conditions for two hours. If the reduction is allowed to proceed for five to six hours, a 45% yield of the doubly-reduced [22]porphyrin-(2.2.2.2) **4.117** can be obtained (Scheme 4.3.2). The *cis–trans–cis–trans* orientation of the *meso*-double bonds in product **4.117** was confirmed by a single crystal X-ray diffraction analysis (Figure 4.3.3). This analysis served to show that compound **4.117** has an ability to adopt a planar conformation in the solid state. It thus supported the proposal that this system is aromatic in character.

Figure 4.3.1   Single Crystal X-Ray Diffraction Structure of Free-base [22]Tetradehydroporphyrin-(2.2.2.2) **4.115**.
This figure was generated using information down-loaded from the Cambridge Crystallographic Data Centre and corresponds to a structure originally reported in reference 40. The macrocycle is centrosymmetric in the crystal and is highly planar (the average deviation of the non-substituent atoms from the mean macrocyclic plane is 0.05 Å). Atom labeling scheme: carbon: ○; nitrogen: ●. Hydrogen atoms have been omitted for clarity

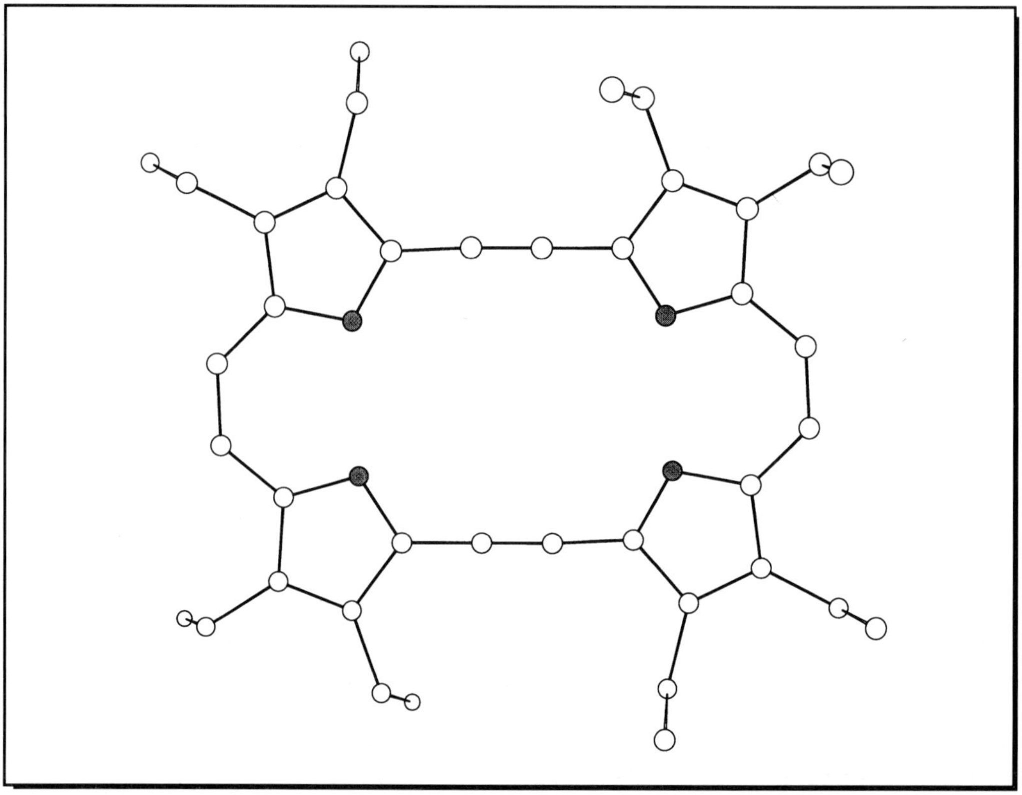

Further evidence for this conclusion came from spectroscopic studies. For instance, the UV-vis spectrum of **4.117** revealed a strong, split-Soret absorbance at 440/464 nm ($\varepsilon = 207\,900/104\,400\,\mathrm{M}^{-1}\,\mathrm{cm}^{-1}$) with Q-bands in the region of 672–790

nm), as might be expected for a highly delocalized, conjugated system. The ^{1}H NMR spectra also revealed a striking similarity in the apparent magnitude of the sustained ring current for **4.117** and Franck's [22]porphyrin-(3.1.3.1) **4.70**. In both cases, the difference in the chemical shifts of the internal and external proton resonances proved to be greater than 20 ppm. Thus, on this basis, **4.117** was assigned as being aromatic.

Figure 4.3.2   Structures **4.116** and **4.117**

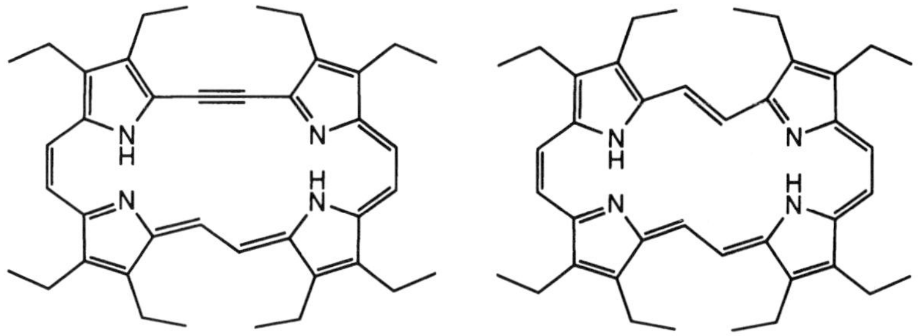

4.116   **4.117**

Scheme 4.3.2

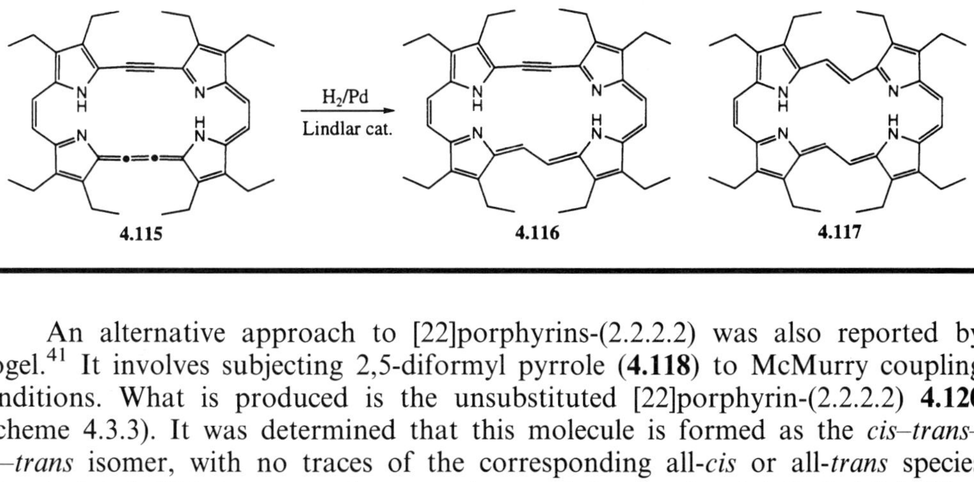

4.115   **4.116**   **4.117**

An alternative approach to [22]porphyrins-(2.2.2.2) was also reported by Vogel.[41] It involves subjecting 2,5-diformyl pyrrole (**4.118**) to McMurry coupling conditions. What is produced is the unsubstituted [22]porphyrin-(2.2.2.2) **4.120** (Scheme 4.3.3). It was determined that this molecule is formed as the *cis–trans–cis–trans* isomer, with no traces of the corresponding all-*cis* or all-*trans* species being isolated. This same reaction was also carried out using 3,4-diethyl-2,5-difor-mylpyrrole (**4.119**). This afforded [22]porphyrin-(2.2.2.2) **4.117**, a product that proved identical to that prepared *via* the Lindlar-like reduction of [22]tetradehydroporphyrin-(2.2.2.2) **4.115**. Unfortunately, however, the yields of these reactions proved very low, being on the order of 0.1–0.2% and 1% for **4.120** and **4.117**, respectively.

Figure 4.3.3 View of the Single Crystal X-Ray Diffraction Structure of Free-base [22]Porphyrin-(2.2.2.2) **4.117** Showing the *cis,trans,cis,trans* Orientation of the *Meso*-like Double Bonds.

This figure was generated using information down-loaded from the Cambridge Crystallographic Data Centre and corresponds to a structure originally reported in reference 40. The four nitrogen atoms define a parallelogram as a result of the *trans*-CH=CH fragments. The macrocycle is centrosymmetric in the crystal and is highly planar (the average deviation of the non-substituent atoms from the mean macrocyclic plane is 0.03 Å. Atom labeling scheme: carbon: ○; nitrogen: ●. Hydrogen atoms have been omitted for clarity

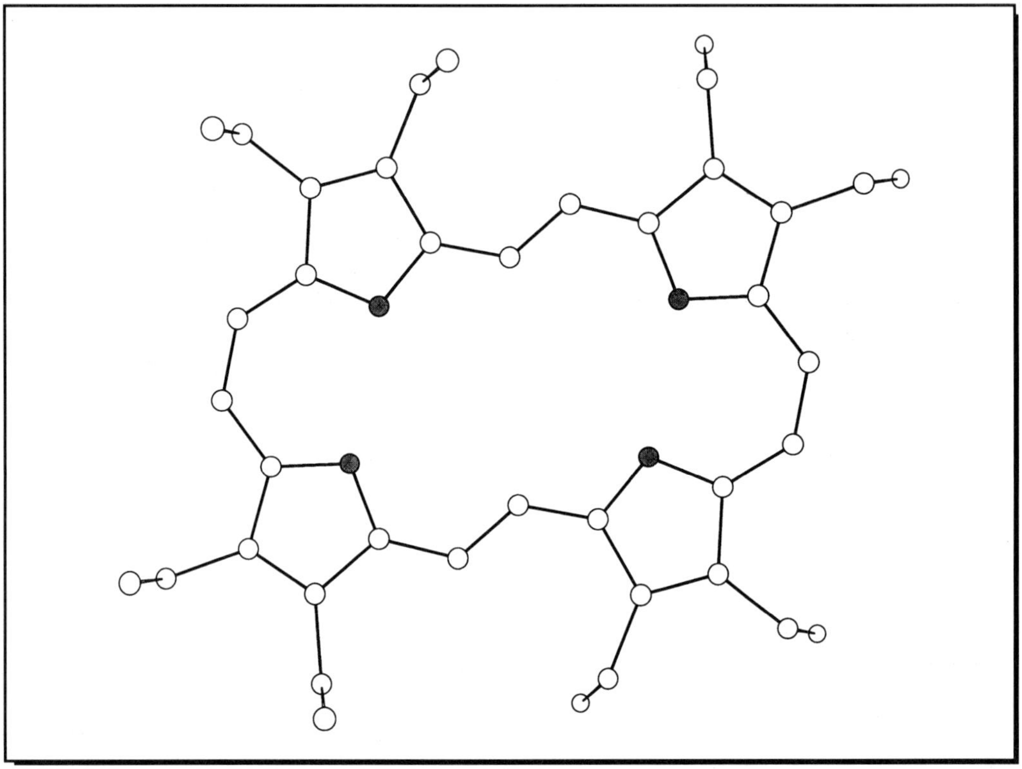

The synthesis of the tetra-*N*-methyl derivative of octaethyl-[22]porphyrin-(2.2.2.2) **4.124** has also been reported.[42] This synthesis is shown in Scheme 4.3.4 and has as its first key step the partial reduction of the diformyldipyrrylethyne **4.121** to the corresponding *cis*-dipyrrylethene **4.122** in 95% yield. This latter species was then subjected to McMurry-type reductive carbonyl coupling conditions. This afforded the somewhat unstable tetra-*N*-methyl-[24]porphyrin-(2.2.2.2) **4.123** in 12% yield. Oxidation of **4.123** to the bisquaternary bisperchlorate [22]porphyrin-(2.2.2.2) **4.124** was achieved using iron(III) chloride followed by treatment with 10% HClO$_4$. Interestingly, salt **4.124** was obtained as a mixture of two isomers, each having a *cis–trans–cis–trans*-orientation of the *meso*-like double bonds. Using spectroscopic means, one of the isomers was determined to possess C_{2h} symmetry and the other to have D_{2h} symmetry (82% and 18% relative abundance, respectively).[§] However, all efforts to separate these isomers have thus far failed. As a consequence, information regarding their absolute structure is not currently available.

Scheme 4.3.3

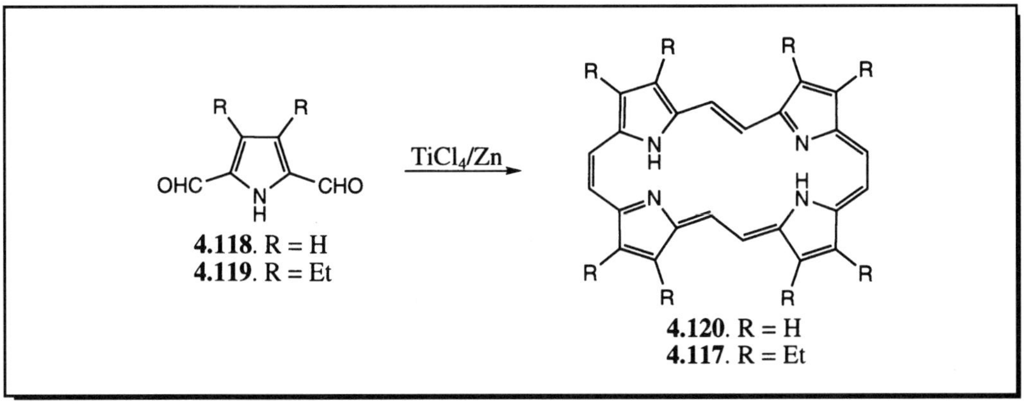

4.118. R = H
4.119. R = Et

4.120. R = H
4.117. R = Et

Scheme 4.3.4

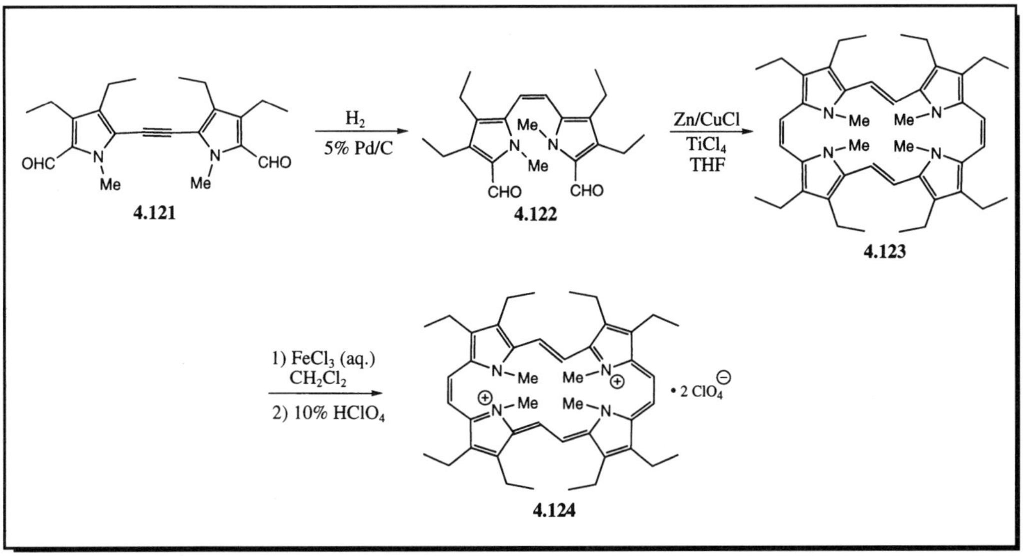

4.121

4.122

4.123

4.124

The stretched porphycenes have shown some promise as ligands for the complexation of transition metal cations. For instance, when [22]tetradehydroporphyrin-(2.2.2.2) **4.115** is treated with di-μ-chlorobis[dicarbonylrhodium(I)] in dichloromethane using potassium carbonate as base, a bis[Rh(CO)₂] complex is formed.[37]

[§]The use of different oxidants and reaction temperatures (e.g., $(CF_3CO_2)_2IC_6H_5$ at $-30°C$) did not alter this isomer ratio.

This complex is actually a mixture of two isomers; one in which the $Rh(CO)_2$ moieties reside on the same face of the macrocycle (*syn* isomer **4.125a**), and one with the metal centers on opposite faces (*anti* isomer **4.125b**) (Scheme 4.3.5). The *syn* isomer **4.125a** was isolated in 20% yield and the *anti* isomer in 12% yield.

Scheme 4.3.5

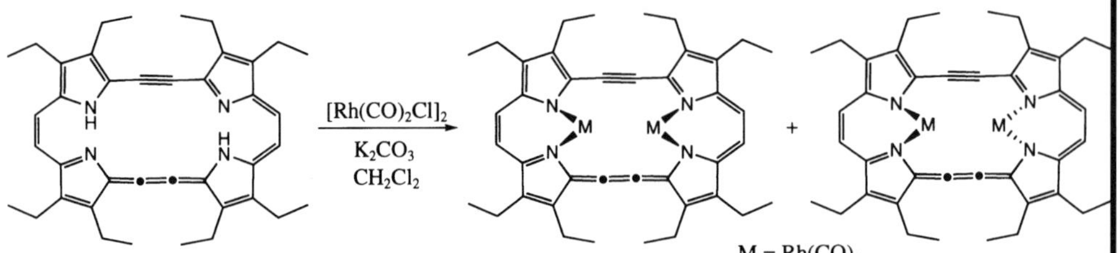

| 4.115 | | [Rh(CO)₂Cl]₂ K₂CO₃ CH₂Cl₂ | | 4.125a | M = Rh(CO)₂ | 4.125b |

The doubly reduced form of **4.115**, the [22]porphyrin-(2.2.2.2) **4.117** has been shown to form stable Rh(I) and Ir(I) metal complexes.[43] In the case of rhodium, treatment of **4.117** with di-μ-chlorobis[dicarbonylrhodium(I)] results in the formation of the mono[Rh(CO)₂] complex **4.126** in 18% yield (Scheme 4.3.6). Under more forcing conditions, the bis[Rh(CO)₂] **4.127** can be obtained in 62% yield. Finally, using similar procedures, the mono[Ir(CO)₂] (**4.128**) and the bis[Ir(CO)₂] (**4.129**) complexes could be generated in 16% and 37% yield, respectively.

Scheme 4.3.6

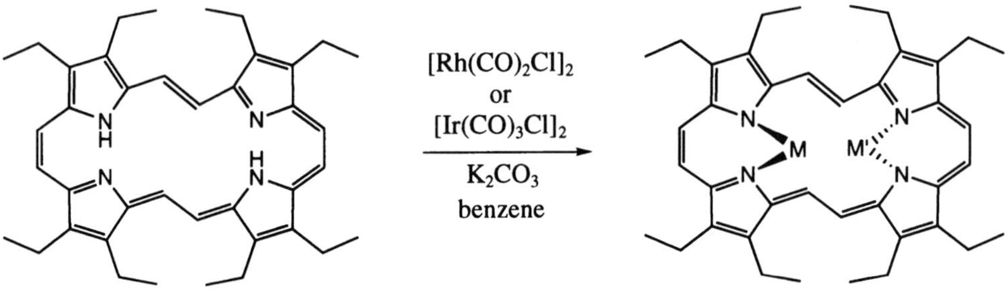

4.117

[Rh(CO)₂Cl]₂
or
[Ir(CO)₃Cl]₂

K₂CO₃
benzene

4.126. M = Rh(CO)₂, M' = H
4.127. M = M' = Rh(CO)₂
4.128. M = Ir(CO)₂, M' = H
4.129. M = M' = Ir(CO)₂

Using the same approach as was used to prepare the [22]tetra-dehydro-porphyrin-(2.2.2.2) species **4.115** above, Vogel and coworkers also succeeded in preparing the next higher aromatic homolog of **4.115**, namely [26]octadehydroporphyrin-(2.4.2.4) **4.132**.[44] This macrocycle was prepared from the diformylbipyrrole-tetramethyne **4.130** *via* the usual low-valent titanium coupling procedure. As normal in the case of these kinds of reactions, the presumed intermediate species **4.131** was not isolated, but was allowed instead to undergo spontaneous oxidation. This gave the 26 π-electron aromatic macrocycle **4.132** in 9% yield after workup (Scheme 4.3.7).

Scheme 4.3.7

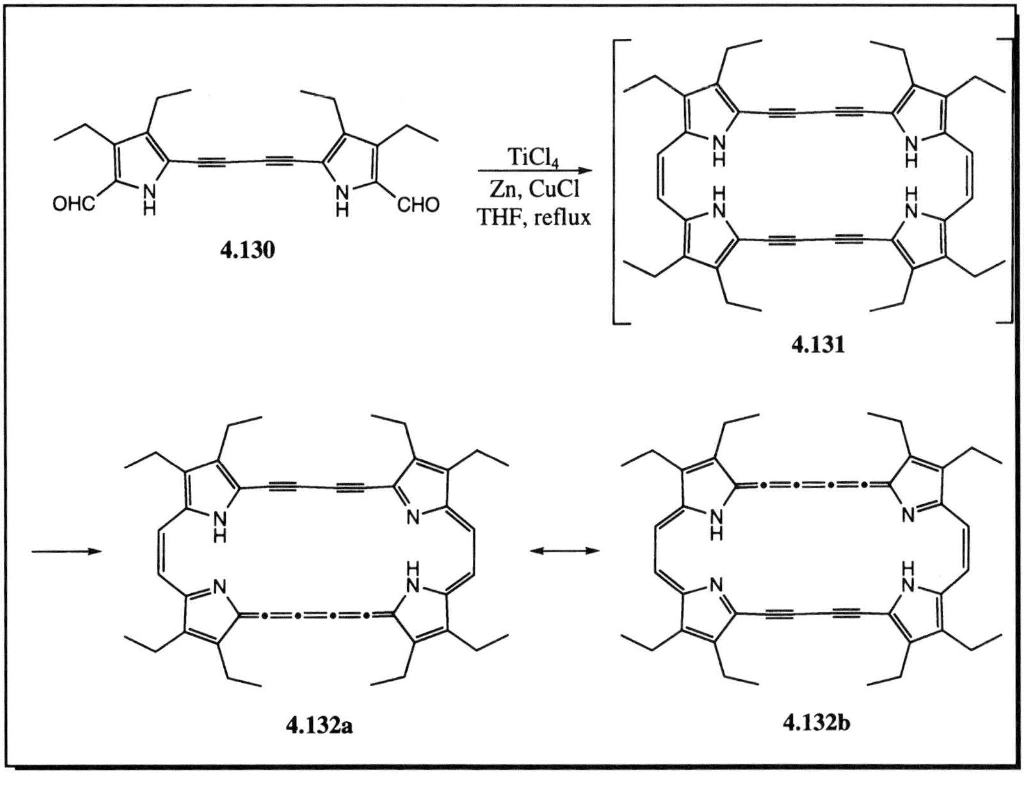

The structure and electronic character of this acetylene-cumulene macrocycle was assigned on the basis of spectroscopic analyses. For instance, the ^{1}H NMR and ^{13}C NMR spectra of **4.132** were found to resemble those of the smaller [22]tetradehydroporphyrin-(2.2.2.2) **4.115**. This was taken as prima-facie evidence that a delocalized 26 π-electron pathway is available to this molecule, and that the system is thus both aromatic and cumulenic in nature. The UV-vis absorbance spectrum of **4.132** was also found to resemble closely that of **4.115.** Compared to this control, however, a significant bathochromic shift in the Soret band (*ca.* 50 nm) was noted. This is as expected given the larger π-electron conjugation pathway available to **4.132**.

Further structural information regarding macrocycle **4.132** was garnered from a single crystal X-ray structural analysis. In this way, it was found that the macrocycle possesses a nearly planar ring framework in the solid state and is centrosymmetric in the crystal lattice (Figure 4.3.4). This latter symmetry observation was considered consistent with the existence of two equivalent $C_{sp^2}(C_{sp})_4C_{sp^2}$ structural units in the molecule and thus supported the conclusion that **4.132** may be considered as being a resonance hybrid of the two structural forms **4.132a** and **4.132b**.

Figure 4.3.4   Single Crystal X-Ray Diffraction Structure of Free-base [26]Octadehydroporphyrin-(4.2.4.2) **4.132**.

This figure was generated using information down-loaded from the Cambridge Crystallographic Data Centre and corresponds to a structure originally reported in reference 44. The macrocycle is centrosymmetric in the crystal and is highly planar (the average deviation from the mean macrocyclic plane is 0.08 Å). Atom labeling scheme: carbon: ○; nitrogen: ●. Hydrogen atoms have been omitted for clarity

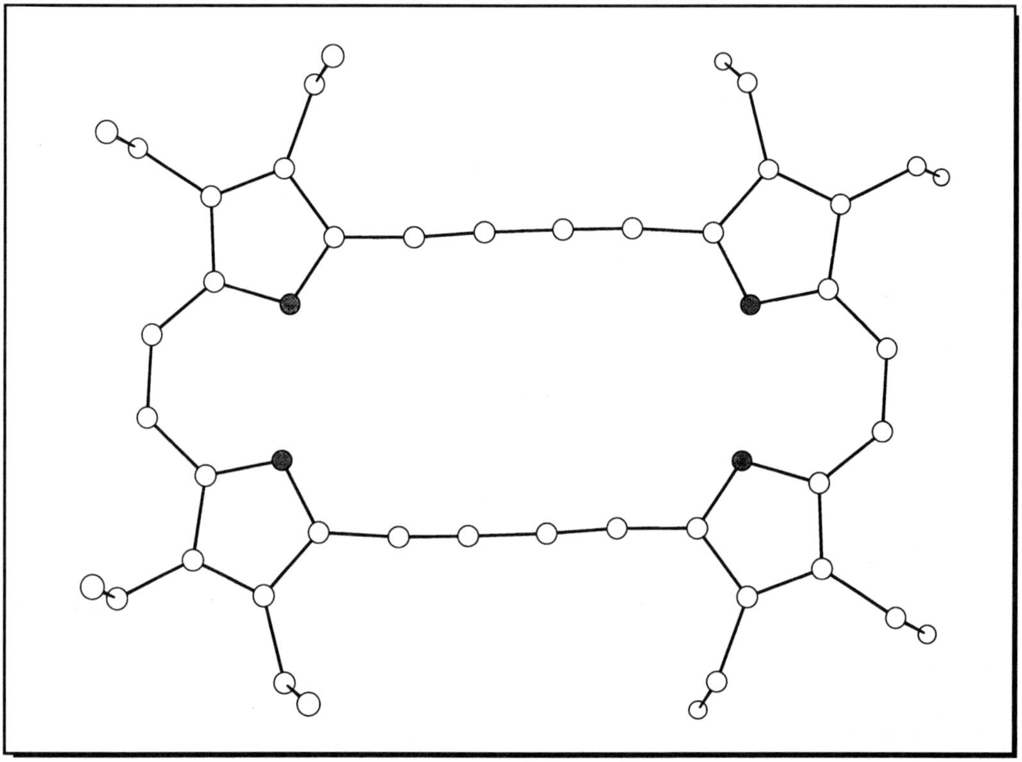

Vogel and coworkers have also prepared other stretched porphycenes derived from 2,2'-bipyrrolic building blocks. The smallest of these are the [22]porphyrins-(4.0.4.0) **4.135** and **4.136**.[36] These macrocycles were prepared from their corresponding formyl-vinylogous-formyl bipyrroles (**4.133** and **4.134**) *via* the usual McMurry-type reductive coupling and subsequent *in situ* air oxidation procedure. The tetra-*n*-propyl derivative **4.135** was isolated in 7% yield with a similar yield also being obtained for the tetra-*t*-butyl derivative **4.136** (Scheme 4.3.8).

Scheme 4.3.8

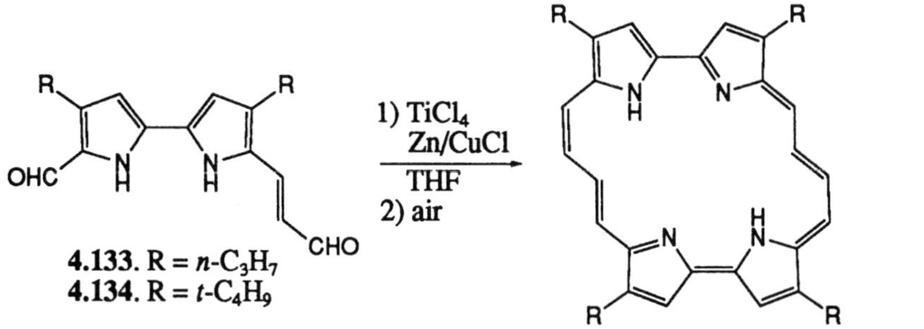

4.133. R = *n*-C₃H₇
4.134. R = *t*-C₄H₉

1) TiCl₄ / Zn/CuCl / THF
2) air

4.135a. R = *n*-C₃H₇
4.135b = **4.135a** · 2 TFA ⟩ TFA

HClO₄ ⟨ **4.136a.** R = *t*-C₄H₉
4.136b = **4.136a** · 2 TFA ⟩ TFA
4.136c = **4.136a** · 2 ClO₄

Both of these macrocycles are aromatic, as shown by the large chemical shift difference between the internal and external *meso*-like protons (e.g., $\Delta\delta$ = 17.89 ppm for **4.135a**). Further, their UV-vis absorption spectral characteristics coincide with what would be expected of macrocycles that possess highly conjugated 22-π-electron frameworks. For instance, the free-base form of the tetrapropyl derivative (**4.135a**) exhibits both an intense Soret-like transition at 445 nm (ε = 179 700 M^{-1} cm^{-1} in CH₂Cl₂) and several less intense, longer-wavelength Q-type transitions.

The tetra-*n*-propyl[22]porphyrin-(4.0.4.0) **4.135a** has also been shown to form a stable rhodium(I) metal complex. Specifically, treatment of **4.135a** with tetracarbonyldi-μ-chlorodirhodium(I) in benzene using potassium carbonate as base results in the formation of the bis[Rh(CO)₂] complex **4.137** (Scheme 4.3.9).[36] In this instance, a 54% yield of the *anti* isomeric complex is isolated, with no *syn* isomer having been observed to form. The *trans–cis–trans–cis* orientation of the *meso*-like double bonds found in the metal-free macrocycle is not changed upon complexation. However, upon metalation, the Soret-like absorbance band experiences a bathochromic shift of *ca.* 40 nm relative to **4.135a**. Also, complex **4.137** appears to sustain a stronger diamagnetic ring current effect than its non-metalated free-base analog, as evidenced by the greater magnitude in the chemical shift difference between the internal and external ring protons (for **4.137** $\Delta\delta$ = 18.47 ppm; for **4.135a**, $\Delta\delta$ = 17.89 ppm). This increase was rationalized, at least in part, in terms of the reduced macrocyclic flexibility that results from metal complexation.

Scheme 4.3.9

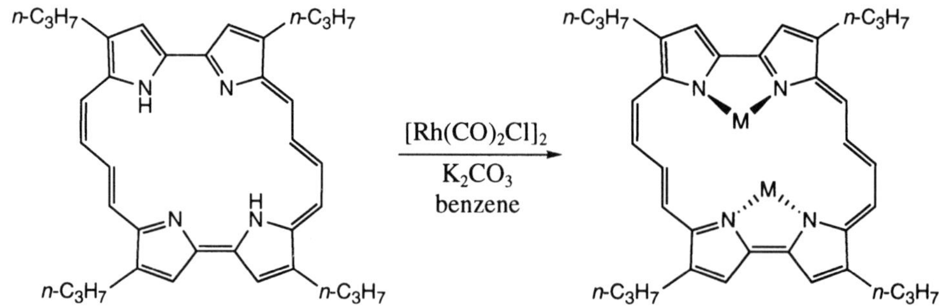

4.135a   **4.137**. M = Rh(CO)$_2$

As a side-product of the reaction leading to the tetra-*n*-propyl[22]porphyrin-(4.0.4.0) **4.135** (Scheme 4.3.8), the tetra-*n*-propyl[22]porphyrin-(6.0.2.0) **4.138**[¶] was also obtained (Scheme 4.3.10).[36] Unfortunately, this compound, formally an isomer of **4.135**, could not be isolated in pure form. In fact, during attempted recrystallization of the macrocycle, the mother liquor was found to contain a small amount of the air- and light-stable electrocyclized product **4.139**. This electrocyclization was also intentionally carried out by heating **4.138** at reflux in benzene for two days. Although conversion to the resulting benzo-porphycene **4.139** was low in this reaction (0.7% yield), it nevertheless provided a novel route to this interesting macrocycle.

Scheme 4.3.10

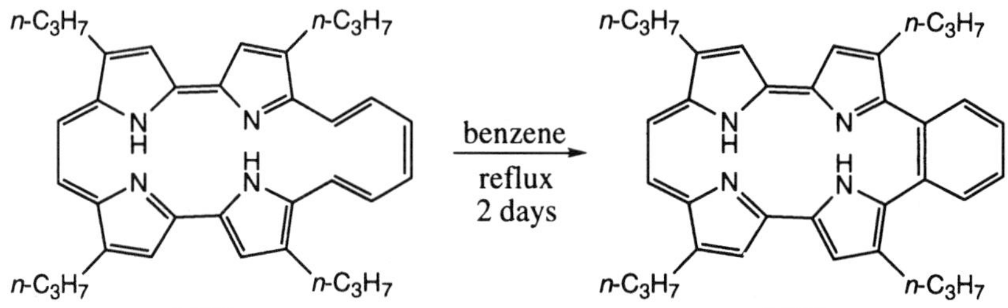

4.138   **4.139**

The stretched porphycene **4.142** ([26]-porphyrin-(6.0.6.0)) was also recently reported by Vogel, *et al.*[36,45] This macrocycle was prepared in 50% yield *via* a McMurry-type dimerization of dialdehyde **4.140**, followed by a spontaneous dehydrogenation of the presumed non-aromatic intermediate (Scheme 4.3.11). The free-base form of [26]-porphyrin-(6.0.6.0) (**4.142a**) proved to be rather fluctional in a variety of solvents over the temperature range of 25 to −50 °C. This hindered

[¶]The formation of the corresponding tetra-*t*butyl derivative was not observed.

efforts to characterize this species by ^{1}H NMR spectroscopy. However, it was found that protonation with trifluoroacetic acid slowed significantly the conformational interconversion processes. This allowed the observation of well-resolved NMR signals and led to the assignment of **4.142** as being an aromatic macrocycle; ring-current effects on the order of those observed for Franck's [26]porphyrins-(3.3.3.3) were readily apparent. In addition, it was discovered that when a small amount of methanol-d_4 was added to a chloroform-d_1 solution of **4.142a**, the ^{1}H NMR signals became well resolved. This is presumably because the methanol solvent interacts with **4.142a** *via* hydrogen bonding, and thus serves to break up the aggregates that apparently form. To the extent this is true, species such as **4.142a** could find potential use in the emerging area of neutral substrate recognition. As yet, however, this possibility does not appear to have been explored.

Scheme 4.3.11

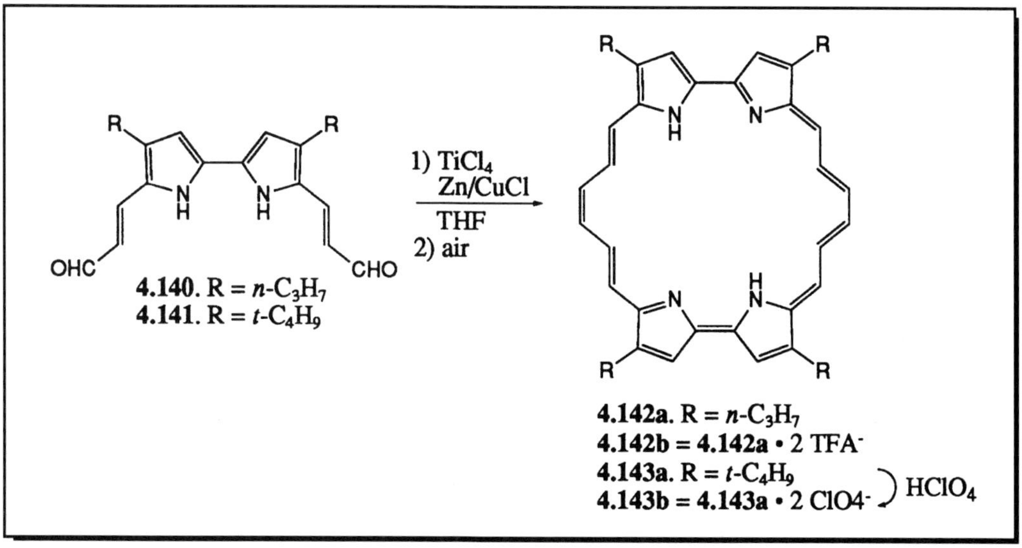

4.140. R = *n*-C$_3$H$_7$
4.141. R = *t*-C$_4$H$_9$

1) TiCl$_4$
   Zn/CuCl
   THF
2) air

4.142a. R = *n*-C$_3$H$_7$
4.142b = **4.142a** • 2 TFA⁻
4.143a. R = *t*-C$_4$H$_9$
4.143b = **4.143a** • 2 ClO4⁻ } HClO$_4$

The tetra-*t*-butyl derivative of [26]porphyrin-(6.0.6.0) (**4.143**) has also been prepared in a fashion analogous to that used to prepare **4.142**, albeit in lower (10%) yield.[42] This macrocycle has been subjected to detailed solution state NMR structural investigations. For instance, it was determined that the macrocycle is highly flexible and dynamic in solution, although a decreased flexibility is observed for the dication **4.143b** relative to the free-base form **4.143a**. It was also determined by ^{1}H NMR studies that different NH-tautomeric forms are present in different solvents. In particular, in chloroform solution at room-temperature tautomerism between structures A and C (Figure 4.3.5) is observed, while in dichloromethane tautomers A and B are present.

Like its smaller 22 π-electron "cousin" **4.135**, [26]porphyrin-(6.0.6.0) has been demonstrated to form stable rhodium(I) complexes.[36] When **4.142a** is treated with di-μ-chlorobis[dicarbonylrhodium(I)], both the *syn* and *anti* forms of the bis[Rh(CO)$_2$] complex **4.144** are found to form in 20% total yield (Scheme 4.3.12). In this case too, metalation is accompanied by a blue-shift in the Soret-like absorbance band in the UV-vis spectrum from 487 nm to 531 nm. The product isomer ratio in this case is 81% to 19%, although it remains unclear as to which isomer actually predominates.

Figure 4.3.5   The three NH-tautomeric forms observed for macrocycle **4.143a**

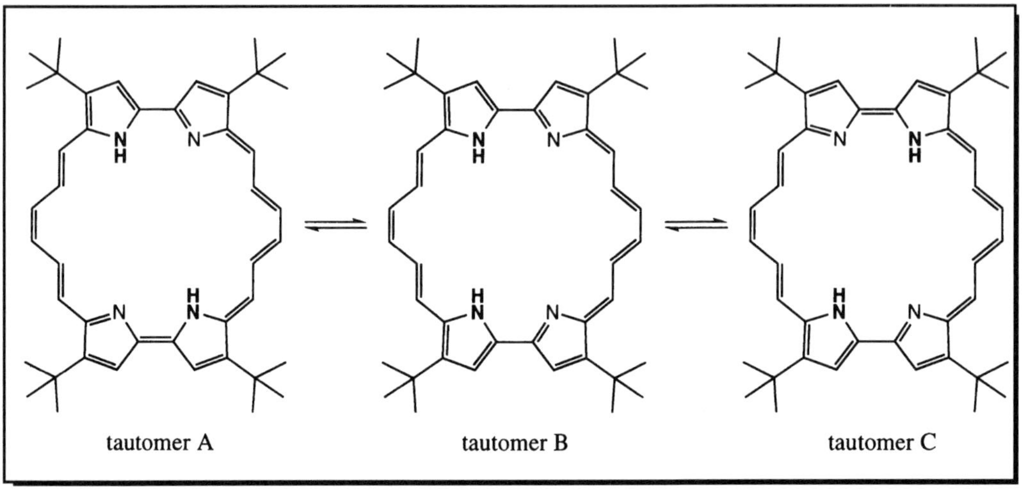

tautomer A   tautomer B   tautomer C

Scheme 4.3.12

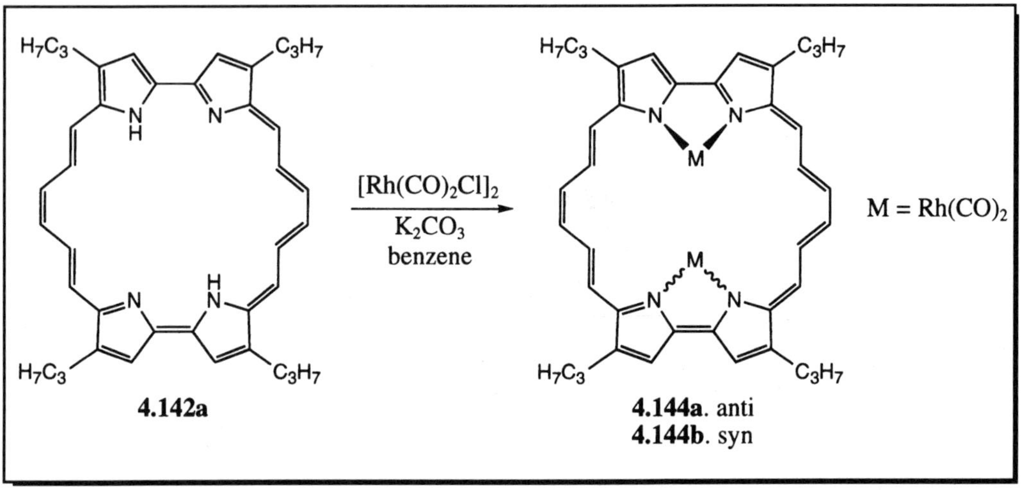

4.142a   **4.144a**. anti
                              4.144b. syn

Tetra-*N*-methyl[26]porphyrin-(6.0.6.0) **4.149** and its tetra-*n*-propyl analog **4.150** have also been prepared.[36] Here, the initial non-aromatic products **4.147** and **4.148** from the McMurry-mediated coupling of the respective bis(vinylaldehydes)

4.145 and **4.146** were readily isolated, each in 20% yield (Scheme 4.3.13). In the case of the unsubstituted macrocycle **4.147**, NMR data showed the existence of three different isomeric products. The main isomer, present in 67% abundance, was found to be highly symmetric, and is presumed to be the *trans–cis–trans–trans–cis–trans* isomer represented by structure **4.147**. The other isomers, present in 21% and 12% abundance, possess much lower degrees of symmetry and remain of unknown absolute structure. Such isomerism is not observed for **4.148**. In this case, where the highly symmetric *trans–cis–trans–trans–cis–trans* isomer predominates.

Scheme 4.3.13

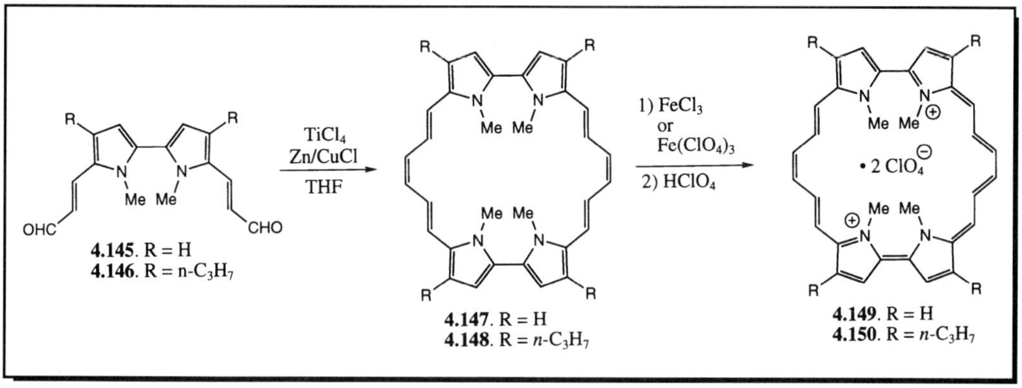

4.145. R = H
4.146. R = n-C$_3$H$_7$

4.147. R = H
4.148. R = *n*-C$_3$H$_7$

4.149. R = H
4.150. R = *n*-C$_3$H$_7$

Conversion of the non-aromatic macrocycles **4.147** and **4.148** into their respective aromatic salts was achieved (in 82% and 90% yield, respectively) *via* iron(III)-mediated oxidation, followed by treatment with perchloric acid.[36] The designation of these compounds as aromatic was based upon ^{1}H NMR experiments whereby, for instance, the maximum difference in the chemical shifts of the internal and external *meso*-like protons for **4.149** was determined to be 20.93 ppm (in DCO$_2$D at room temperature).

The UV-vis absorption spectra (in CH$_2$Cl$_2$) obtained for **4.149** and **4.150** are also suggestive of highly conjugated extended π-electron systems. For instance, for the tetra-*n*-propyl derivative **4.150**, an intense split-Soret-like absorption band is observed at 505/513 nm (ε = 257 500/249 800 M^{-1} cm^{-1}), along with two less intense, longer wavelength Q-type bands at 648 and 776 nm. This is in marked contrast to the spectrum observed for the 28 π-electron non-aromatic species **4.148**, where only relatively weak bands in the 350–480 nm region are observed. The contrasting spectral results obtained for **4.148** and **4.150**, which differ formally only by two conjugated π-electrons, can be taken as evidence for a more rigid and more planar conformation being imposed upon the macrocyclic framework in the case of the aromatic system.

The synthesis of an "ene-diyne"-containing [28]porphyrin-(6.0.6.0) has also been described.[46] This macrocycle, tetra-*N*-methyl[28]octadehydroporphyrin-(6.0.6.0) **4.152** was prepared by the low-valent titanium-mediated dimerization of dialdehyde **4.151** in 2–3% yield (Scheme 4.3.14). Attempts to oxidize this macrocycle to the corresponding aromatic dication have thus far met with failure. Likewise, efforts to prepare the analogous macrocycle free of the protecting methyl groups also proved unsuccessful. Thus, information regarding the nature and properties of this class of macrocycle remain only limited at this time.

Scheme 4.3.14

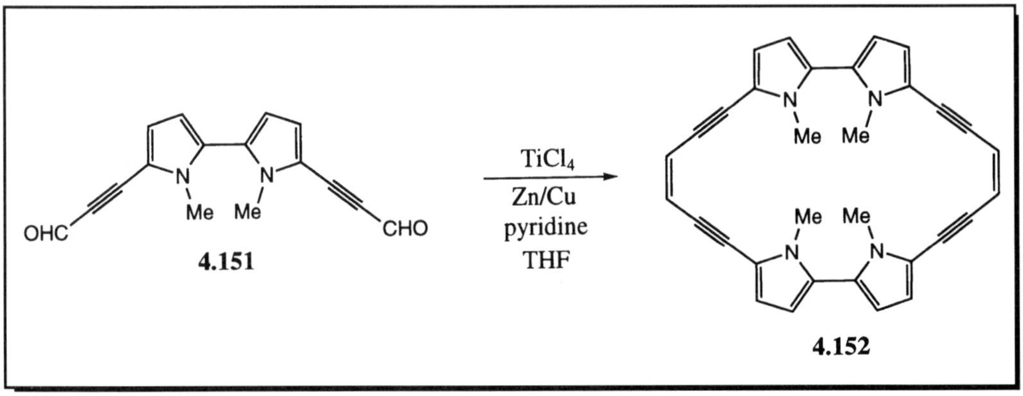

4.4   Heteroatom-containing Stretched Porphycenes

Macrocycles that can be considered as being heteroatom-containing (i.e., containing other than nitrogen) stretched porphycenes are also known. The first of these to be reported are the two isomers of the tetraoxa[24]porphyrin-(2.2.2.2) **4.154** shown in Scheme 4.4.1. They were reported in 1969 by J. A. Elix.[47] This set of isomeric macrocycles was prepared *via* a Wittig-type self-condensation of the ylid obtained from 5-formyl-2-furfuryltriphenylphosphonium chloride **4.153**. The two products obtained, each in 0.7% yield, were assigned as being the *cis–trans–trans–trans* and all-*trans* products **4.154a** and **4.154b**, respectively.

Later, Märkl, *et al.* reported an alternate stepwise approach to the *cis–trans–cis–trans* isomer of **4.154**.[48] This synthesis involved firstly the Wittig condensation of the formylated furan monoacetal **4.155** with the phosphonium salt **4.156**. This afforded, as the major product, the *trans*-isomeric diformyl-difuryldimethene **4.157** (Scheme 4.4.2). McMurry dimerization of **4.157** then afforded the *cis–trans–cis–trans*-tetraoxa[24]porphyrin-(2.2.2.2) **4.154c** in 7% yield. Variable temperature ¹H NMR experiments revealed a high degree of conformational dynamism in the case of this molecule. At low temperature, however, a C_2-symmetric conformation is frozen out; this was assigned the structure represented by formula **4.154c**. A single crystal X-ray structural analysis of **4.154c** further confirmed that, at least in the solid state, this species exists in such a proposed C_2-symmetric conformation (Figure 4.4.1).

The uncharged tetraoxa[24]porphyrin-(2.2.2.2) **4.154c** could also be oxidized to its dicationic, aromatic congener tetraoxa[22]porphyrin-(2.2.2.2) **4.158**.[48] This was done by treating it with Br₂ in CH₂Cl₂ and isolating the product in the form of the bisperchlorate salt **4.158**. Oxidation of **4.154c** to **4.158** was accompanied by a *ca*. 100 nm red shift in the Soret-like absorbance band in the electronic spectrum. Specifically, this band went from being a relatively weak one centered at *ca*. 360 nm (for **4.154c** in CH₂Cl₂) to a split band centered at *ca*. 450 nm (in HClO₄), which is much sharper and more intense in the case of **4.158**. Dication **4.158** also displays several Q-type transitions in the 650–750 nm spectral region that are not observed for **4.154c**.

Scheme 4.4.1

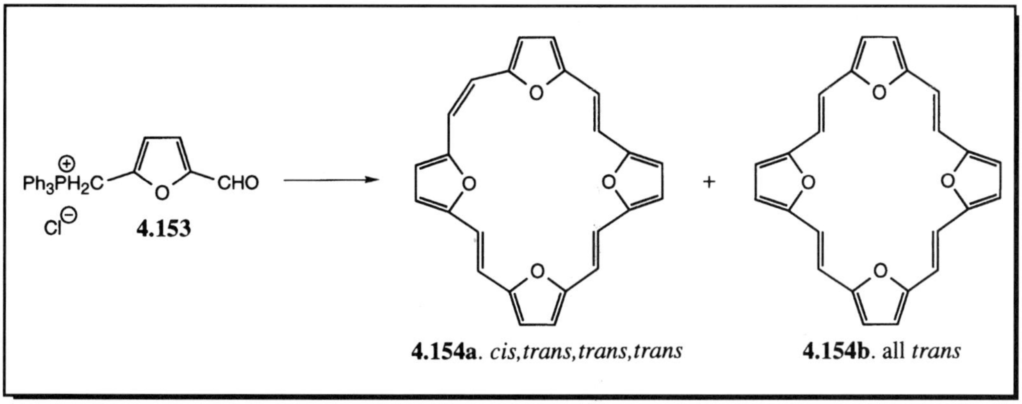

4.154a. *cis,trans,trans,trans*   **4.154b**. all *trans*

Scheme 4.4.2

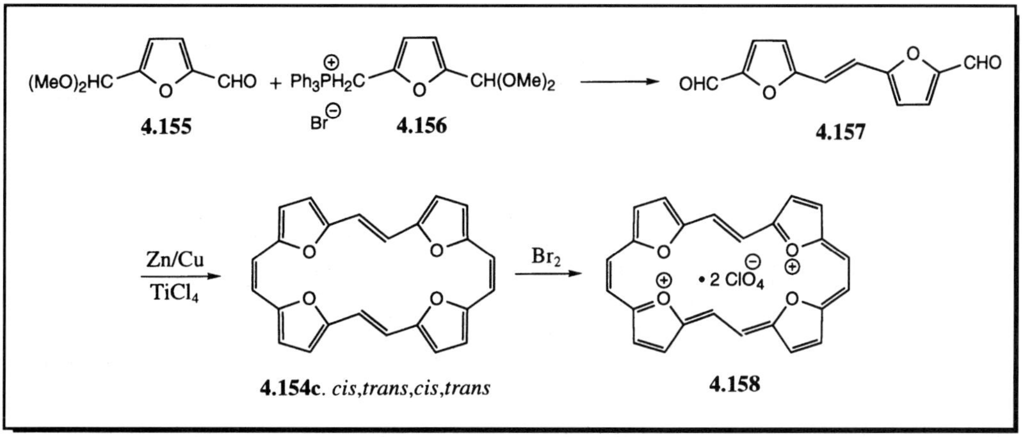

4.155   **4.156**   **4.157**

4.154c. *cis,trans,cis,trans*   **4.158**

Like its non-aromatic precursor, the bisperchlorate salt, **4.158** was found to exist in a *cis–trans–cis–trans* orientation, as judged from ¹H NMR spectroscopic analyses. The relevant spectra also served to confirm the aromatic nature of this compound. Upon aromatization (to give **4.158**), the internal proton resonances of the non-aromatic species **4.154c** shift upfield by 20.46 ppm (i.e., from 12.38 to −8.08 ppm). Similarly, the external protons of **4.154c** shift downfield by 8.36 ppm (i.e., from 5.30 to 13.66 ppm). In **4.158**, the difference in the chemical shifts of the internal and external protons is thus 21.74 ppm. This is a difference which is similar to that observed for Vogel's all-pyrrole-containing [22]porphyrin-(2.2.2.2) **4.117**.[41]

Figure 4.4.1   View of the Single Crystal X-Ray Diffraction Structure of Free-base Tetraoxa[24]porphyrin-(2.2.2.2) **4.154** Showing the *cis,trans,cis,trans* Orientation of the *Meso*-like Double Bonds.

This figure was generated using information down-loaded from the Cambridge Crystallographic Data Centre and corresponds to a structure originally reported in reference 48. The four oxygen atoms define a parallelogram as a result of the *trans*-CH=CH fragments. The macrocycle possesses C_2-symmetry in the crystal and is rather planar (the average deviation from the mean macrocyclic plane is 0.10 Å). Atom labeling scheme: carbon: ○; oxygen: ⊖. Hydrogen atoms have been omitted for clarity

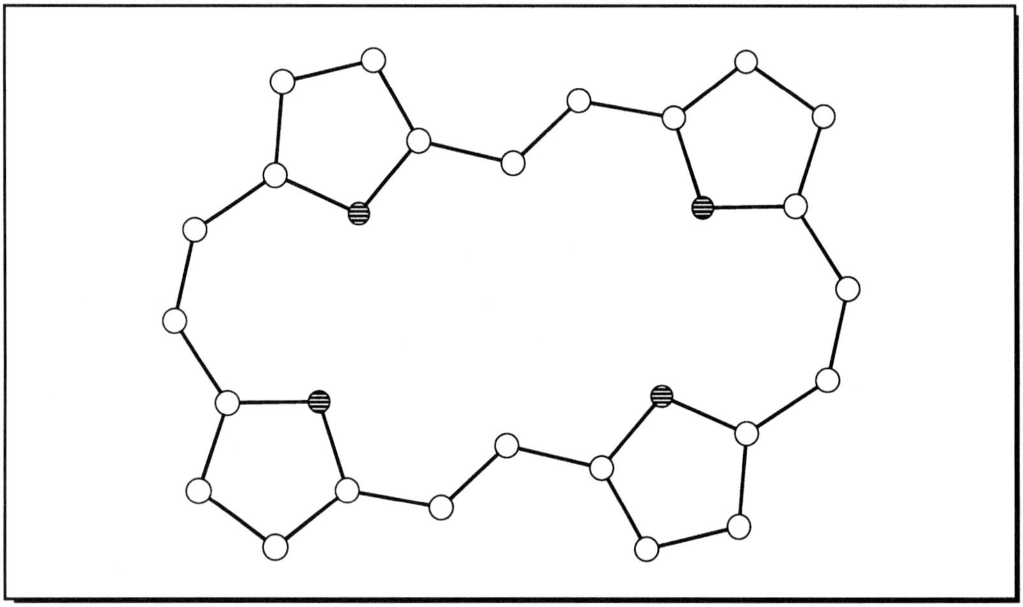

Tetraoxa[22]porphyrin-(2.2.2.2) bisperchlorate **4.158** has also been prepared *via* the McMurry-type condensation of 2,5-diformyl furan **4.159**.[49] Here, the intermediate product of the cyclocondensation was determined to be the dihydro compound **4.160**. It was isolated in 0.8% yield,[||] and was determined to possess a *cis–trans–cis–trans* orientation of the single and double *meso*-like bonds (Scheme 4.4.3). The unsaturated bonds present in **4.160** posed no significant barrier to aromatization, however. Indeed, **4.160** underwent smooth oxidation to the 22 π-electron macrocycle **4.158** upon treatment with DDQ, while maintaining the *cis–trans–cis–trans* conformation found in the dihydro species **4.160**.

[||][18]annulene trioxide, discussed in Chapter 2, was isolated in 4% yield from this reaction.

Scheme 4.4.3

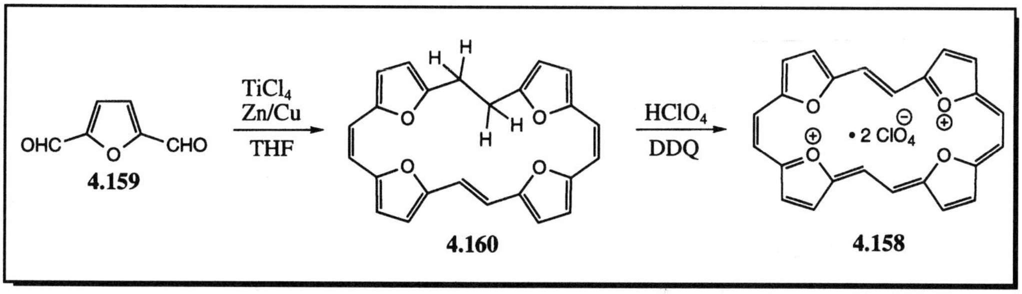

4.159   4.160   4.158

The synthesis of the octaethyl-substituted tetraoxa[22]porphyrin-(2.2.2.2) bis-perchlorate **4.162** has also been realized.[43,50] This was accomplished *via* the McMurry-type reductive coupling of the ethyl-substituted diformyl-difuryldimethene **4.161** (Scheme 4.4.4). In this instance, however, the presumed non-aromatic inter-mediate was not isolated in pure form. Rather, oxidation with bis(trifluoroacet-oxy)iodobenzene followed by treatment with perchloric acid afforded the aromatic bisperchlorate salt **4.162** directly.

Scheme 4.4.4

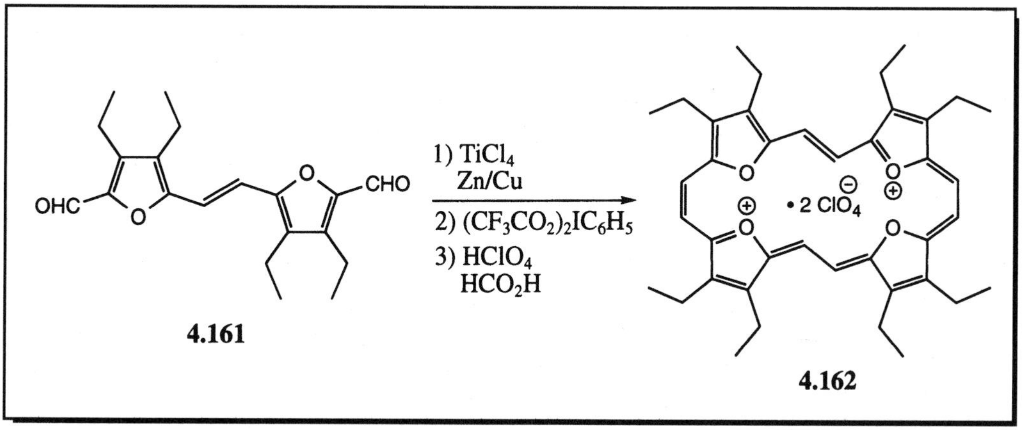

4.161   4.162

The UV-vis spectrum of **4.162** proved, as expected, quite similar to those of the unsubstituted analog **4.158**.[43,48,50] For instance, like **4.158**, the Soret-like visible absorption band observed for **4.162** in formic acid is split into two sharp and intense peaks. However, in the same solvent, these peaks (in **4.162**) are red-shifted by 15–20 nm relative to the unsubstituted analog. Further similarities between **4.158** and **4.162** can be seen in the proton NMR spectra for these compounds. In particular, it was found that the internal *meso*-protons for both compounds resonate at *ca.* −8.25 ppm while the external ones are found at *ca.* 11–13 ppm.

The synthesis of the mixed furan-thiophene-containing macrocycle dioxa-dithia[24]porphyrin-(2.2.2.2) **4.166** was reported by Strand, *et al.* in 1977.[51] This was accomplished using Wittig-type chemistry, and involved the condensation reaction between 2,5-diformylfuran **4.163** and the thiophene-based phosphonium salt **4.165** (Scheme 4.4.5). Proton NMR spectroscopic studies revealed the presence of a highly symmetric macrocycle in solution. These studies further confirmed that **4.166** exists in an all-*cis* conformation. However, because of its inherent instability, more detailed structural information could not be gathered for this molecule.

Scheme 4.4.5

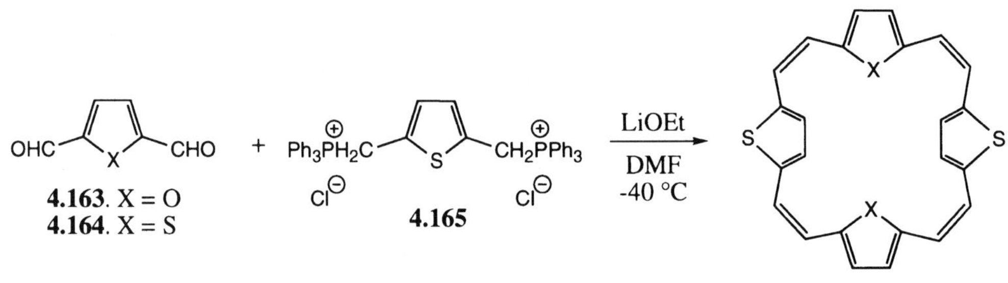

In analogy to the above, the all-*cis* tetrathia[24]porphyrin-(2.2.2.2) **4.167** was prepared from 2,5-diformylthiophene **4.164** and phosphonium salt **4.165** in 4.6% yield.[51] Fortunately, this macrocycle proved to be much more stable than its dioxa congener **4.166**. It could be subject, therefore, to more detailed spectroscopic characterization. For instance, inspection of the ¹H NMR spectrum of **4.167** could be recorded easily. It revealed the presence of a small but significant paramagnetic ring current that is enhanced on cooling. Based upon inspection of molecular models, a conformation in which two of the sulfur atoms point out from the center of the macrocycle and two point inward would be the most sterically favored. Such a conformation, it was reasoned, must be present in order for the macrocycle to achieve sufficient planarity (and the associated π-electron conjugation) necessary for establishing any significant paratropic ring current. Thus, Strand and coworkers suggested that a structure similar to that shown for **4.167** perhaps best represents the dominant solution-state conformation.[51] That this "two-in, two-out" conformation is the preferred one was later confirmed from more detailed variable temperature ¹H NMR studies.[52] In this work, it was also concluded that two degenerate low-energy conformations (for instance, A and B in Figure 4.4.2) are likely available to the molecule and that the two forms readily interconvert on the NMR time scale.

An alternate synthetic approach to tetrathia[24]porphyrin-(2.2.2.2) **4.167** was reported by Cava and coworkers in 1994.[53] It involved the low-valent titanium-mediated cyclization of 2,5-thiophenedicarboxaldehyde **4.164** and afforded **4.167** in

4–5% yield (Scheme 4.4.6).[53] As was true for the Wittig-generated macrocycle above, the tetrathia[24]porphyrin-(2.2.2.2) generated from this reaction was determined to be the all-*cis* isomer. However, in contrast to the conclusions of Strand, *et al.*,[51] these researchers judged **4.164** to be non-aromatic and suggested that it behaves only as an array of isolated thiophene units. Although the reported room temperature NMR data from **4.167** from both groups are nearly identical (two observed signals of equal intensity at 7.34 and 6.19 ppm), these two disparate interpretations serve to highlight the difficulties that can be encountered when trying to assign a molecule as being aromatic, non-aromatic, or antiaromatic.

Figure 4.4.2 Two Low-energy Conformations Presumed Present for Macrocycle **4.167**.

These two forms readily interconvert on the NMR time scale

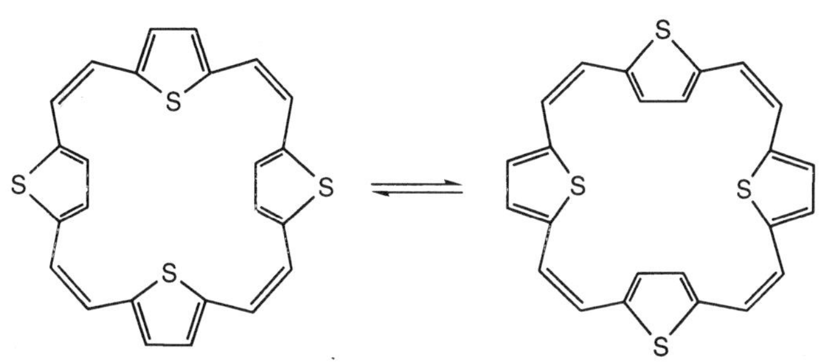

Scheme 4.4.6

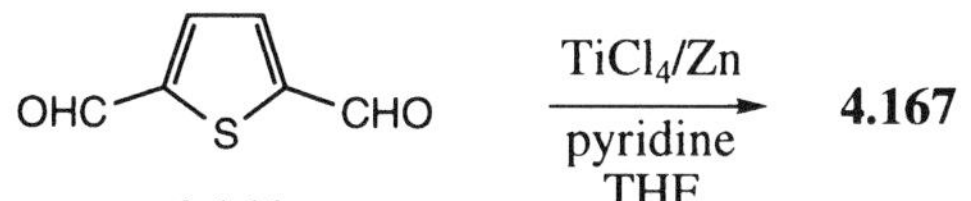

The synthesis of the aromatic analog of [24]tetrathiaporphyrin-(2.2.2.2), namely tetrathia[22]porphyrin-(2.2.2.2) **4.169**, has also recently been recorded.[49,52] It was achieved *via* the DDQ-mediated oxidation of the dihydro tetrathiaporphyr-inogen-(2.2.2.2) **4.168**, a species that was in turn prepared in 0.7% yield from a McMurry-type cyclocondensation of 2,5-thiophenedicarboxaldehyde **4.164** (Scheme 4.4.7).[49] It is of interest to note that the isolation of the dihydro compound **4.168** from this reaction sequence is in direct contrast to the observations of Cava and coworkers discussed above. While it is presumed that **4.168** arises as a result of the highly reductive reaction conditions, it remains unclear as to why these disparate results were obtained. This discrepancy is perhaps attributable to the use of slightly different McMurry-type reductive coupling reaction conditions. Such differences may also account for the fact that Cava and coworkers report a 38% yield of the trimeric [18]annulene trisulfide (cf. Chapter 2) from this reaction, as opposed to the much lower 5% yield in the present instance. However, such an explanation remains only speculative at this point.

Scheme 4.4.7

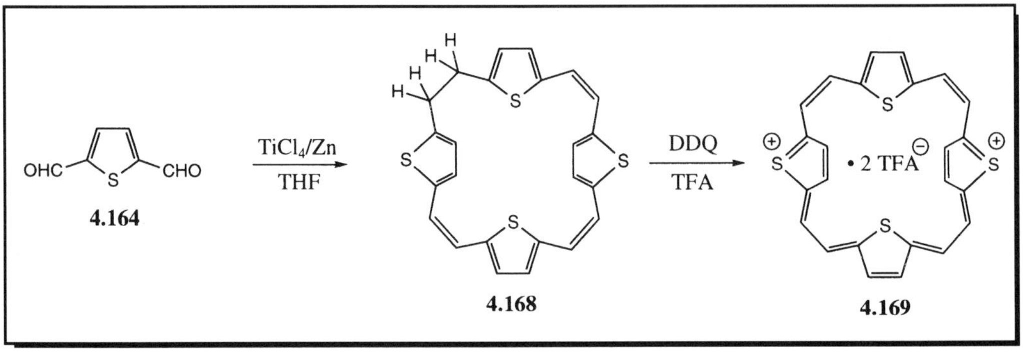

4.164   TiCl$_4$/Zn
              THF   4.168   DDQ
                                          TFA   • 2 TFA$^{\ominus}$ 4.169

Tetrathia[22]porphyrin-(2.2.2.2) **4.169**, like its nitrogen- and oxygen-containing analogs discussed earlier, proved to be aromatic as judged from ^{1}H NMR spectroscopic analyses.[49] These same studies also revealed an all-*cis* configuration of the *meso*-like double bonds and that two of the four thiophene rings are oriented such that their sulfur atoms point out from the macrocyclic core. This latter assessment was based upon the finding that the β-thiophenic protons on two of the thiophene rings resonate at −2.13 ppm, whereas the analogous protons in the corresponding pyrrole- and furan-based macrocycles experience downfield shifts to 9.5–12 ppm. As was argued for tetrathia[24]porphyrin-(2.2.2.2) **4.167**, it is presumed that this "two-in, two-out" conformation, not seen in the case of the nitrogen and oxygen analogs **4.117** and **4.158**, is the favored one because of the greater steric demands imposed by the larger sulfur atoms.

The octaethyl-derivative of [24]tetrathiaporphyrin-(2.2.2.2) has also been prepared *via* two differing synthetic routes.[49] In the first of these, 3,4-diethyl-2,5-thiophenedicarboxaldehyde **4.170** was cyclo-condensed under reductive McMurry-type conditions to afford **4.171** in 0.3% yield (Scheme 4.4.8).** The other approach involved the reductive carbonyl dimerization of the diformyl dithiophenylethene **4.172** to afford macrocycle **4.171** in 0.6% yield.††

Octaethyl[24]tetrathiaporphyrin-(2.2.2.2) proved to be a conjugated but non-aromatic macrocycle. However, unlike the unsubstituted macrocycle **4.167**, ^{1}H NMR experiments revealed that conformations in which two of the thiophene rings are rotated by 180° are not observed for this macrocycle. This is presumably due to the unfavorable ethyl group-derived steric interactions that would arise as a consequence

---

In addition to the cyclotetramer **4.171, a significant yield (16%) of the cyclotrimeric [18]annulene trisulfide (discussed in Chapter 2) was obtained. Also generated, albeit in lower yields (on the order of 1–1.5%), were two isomeric cyclopentamers (Chapter 6) and two isomeric hexamers (Chapter 7).
††This reaction affords a 1.3% yield of a cyclic hexamer. This product is discussed in Chapter 7.

of this sort of alternating-thiophene orientation. In any event, it is tempting to speculate that the inability of **4.171** to adopt such a conformation readily accounts for the fact that it has not yet proved possible to generate an aromatic (and hence nearly planar) form of this macrocycle.

Scheme 4.4.8

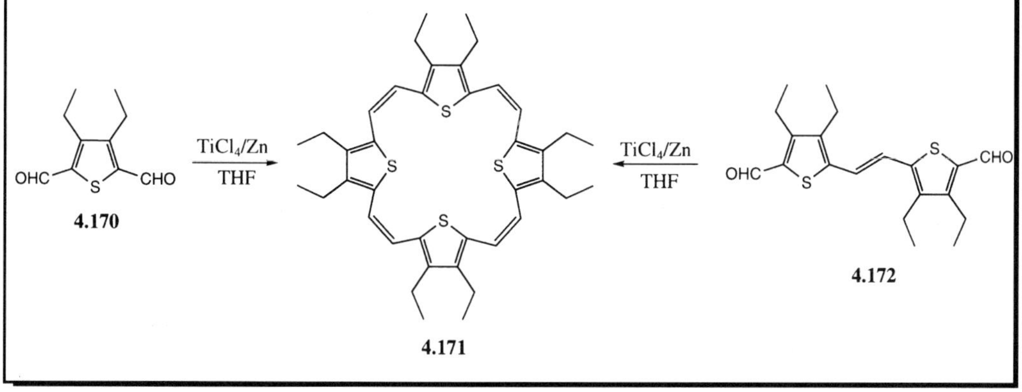

4.170

4.171

4.172

4.5   Odd–Even Systems

A number of tetrapyrrole-type macrocyclic systems have been prepared that contain both an even and an odd number of bridging atoms between the heterocyclic subunits. The first of these to be reported was the thiacycloheneicosin **4.175**.[54] This macrocycle was prepared in 1.3% yield *via* a Wittig reaction between 5,5'-thiodi-2-furaldehyde **4.173** and the phosphonium salt **4.174** (Scheme 4.5.1). While little in the way of information is available for this macrocycle, ^{1}H NMR spectroscopic analyses led to the conclusion that, while formally a 22 π-electron system, compound **4.175** will not apparently sustain a diamagnetic ring current. Why this is so is not clear at present.

The synthesis of the novel tetrafuran-containing macrocycle **4.178** was reported in 1992.[55] This macrocycle was prepared by the aldol-type condensation between diketone **4.176** and dialdehyde **4.177** in 20% yield (Scheme 4.5.2). Subjecting this macrocycle to reductive deoxygenation using lithium aluminum hydride and aluminum trichloride afforded the tetraoxaporphyrinogens-(3.0.3.0) **4.179** and **4.180** in combined 35% yield (Scheme 4.5.3). These two isomeric products, formed in 55% and 45% relative abundance, respectively, could not be readily separated. Instead, they were treated together (i.e., as a mixture) with ρ-chloranil to afford the conjugated tetraoxa[22]porphyrin-(3.0.3.0) species **4.181** in 15% yield (Scheme 4.5.4).

Scheme 4.5.1

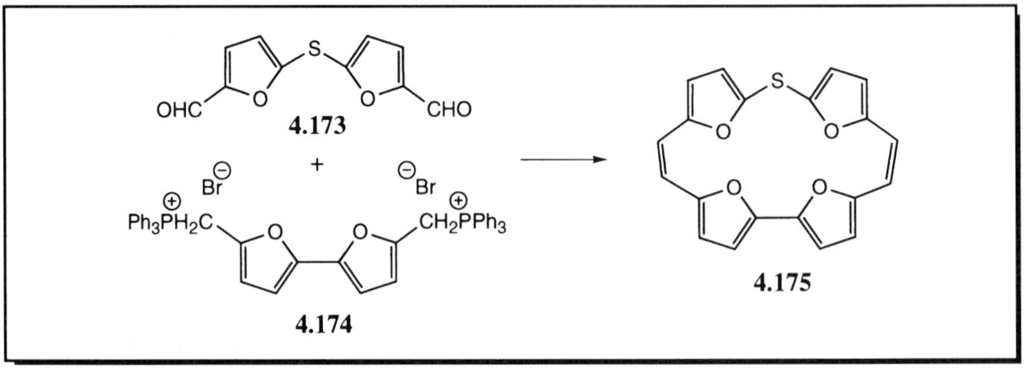

4.173

+

4.174

4.175

Scheme 4.5.2

4.176

+

4.177

$\xrightarrow[\text{r.t. 24hr}]{\begin{array}{c}\text{NaOH}\\\text{MeOH}\end{array}}$

4.178

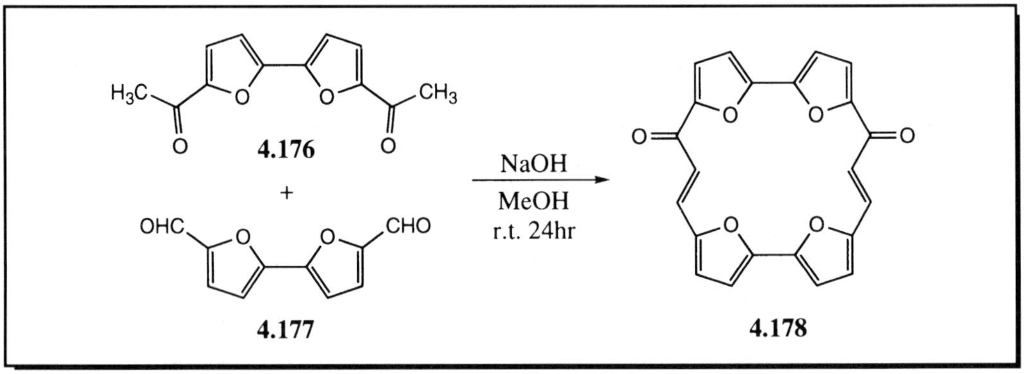

Scheme 4.5.3

4.178

$\xrightarrow[\text{Et}_2\text{O}]{\begin{array}{c}\text{LiAlH}_4\\\text{AlCl}_3\end{array}}$

4.179
55%

+

4.180
45%

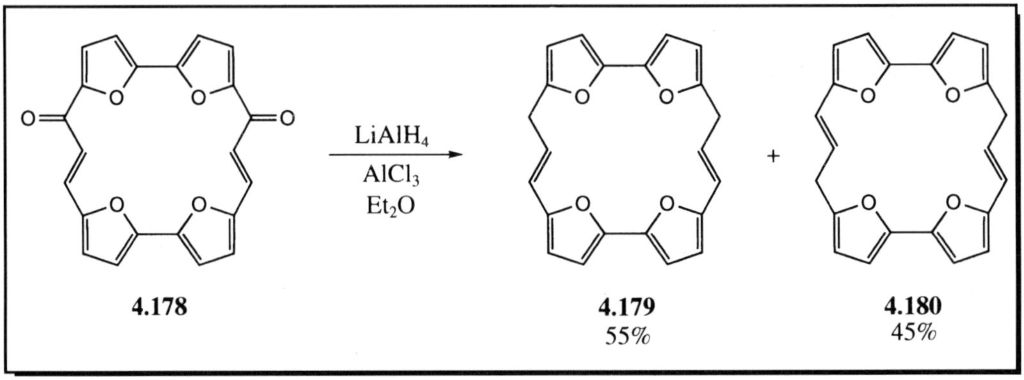

Scheme 4.5.4

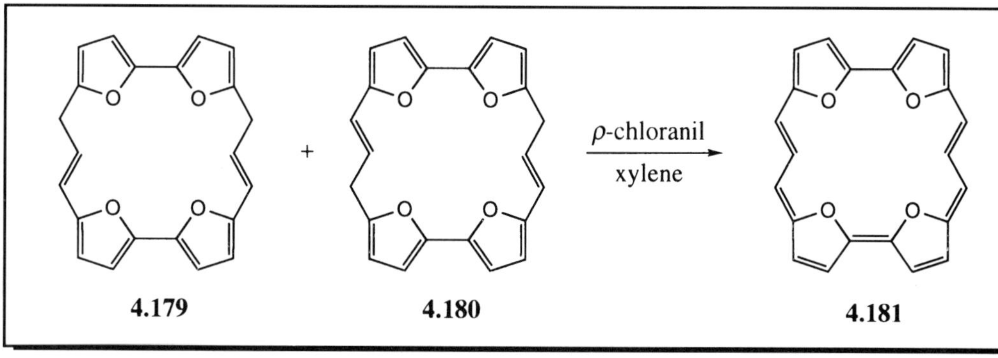

4.179   **4.180**   **4.181**

Macrocycle **4.181** exhibits a UV-vis absorption spectrum characteristic for highly delocalized extended π-electron conjugation systems. Its Soret-like absorbance band is relatively sharp and is split, exhibiting maxima at 395 and 402 nm (ε = 145 300 and 268 100 $M^{-1} cm^{-1}$ in CH_2Cl_2). System **4.181** also displays several Q-type transitions in the 560–800 nm region. That **4.181** is aromatic is evident from the apparent sustained diamagnetic ring-current effects visible in its ^1H NMR spectrum. Here, the internal *meso*-like protons were found to resonate at -5.92 ppm while the external ones were found to resonate at 10.40 ppm. However, in spite of this indication of apparent aromatic character, tetraoxa[22]porphyrin-(3.0.3.0) was found to be somewhat unstable and to decompose rapidly under even weakly acidic conditions.[55]

In 1994 Cava and coworkers reported the synthesis of the tetrathiophene-containing systems **4.183** and **4.184**.[53,56] The first of these was the non-conjugated tetrathiaporphyrinogen-(2.1.2.1) **4.183**. This species was isolated in remarkably high yield (78%) from the reductive McMurry coupling of dialdehyde **4.182** (Scheme 4.5.5). Dehydrogenation of this macrocycle was effected by treatment with DDQ, followed by hydrazine. This afforded the neutral aromatic tetrathia[22]porphyrin-(2.1.2.1) systems **4.184** in 82% yield.

Scheme 4.5.5

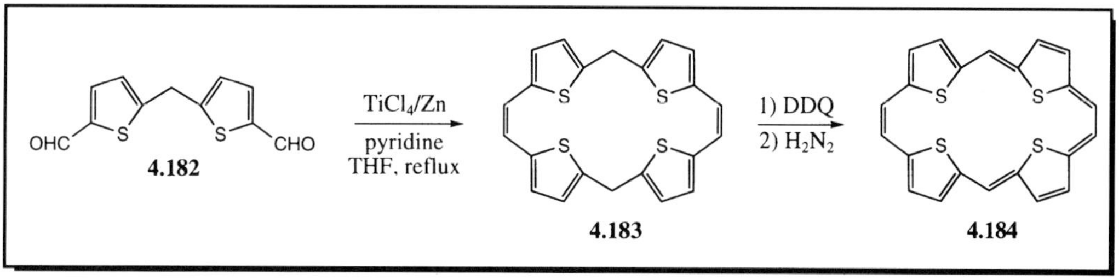

4.182   TiCl₄/Zn   **4.183**   1) DDQ   **4.184**
                   pyridine   2) H₂N₂
                   THF, reflux

The aromaticity of **4.184** was confirmed by ^{1}H NMR spectral analyses. In particular, the chemical shift of the bridging *meso*-like ethene protons were found to undergo a downfield shift of *ca.* 5 ppm on aromatization. The high degree of π-electron delocalization assumed for **4.184** is also evident in its UV-vis spectrum. Specifically, in CH_2Cl_2, [22]tetrathiaporphyrin-(2.1.2.1) **4.184** exhibits a strong, sharp Soret-like absorbance band at 417 nm along with weaker Q-type transitions at 503, 540, 579, and 771 nm. These porphyrin-like spectral features are in contrast to that which is observed in the spectrum of the non-aromatic precursor tetrathiaporphyrinogen-(2.1.2.1) **4.183**. This latter species displays only three absorbance bands at much shorter wavelengths (i.e., < 350 nm).

4.6   Miscellaneous Systems of Interest

A new class of non-aromatic macrocycles that is structurally related to the vinylogous porphyrins was reported in 1993 by Corriu and Moreau and coworkers.[57] These systems, which contain aryl spacer units, are represented by structures **4.187** and **4.188** in Scheme 4.6.1. The first of these, the fused-benzene-containing bisvinylogous porphyrin **4.187**, was prepared in 92% yield from the condensation of the 1,5-bis (pyrrolyl)benzene derivative **4.185** with benzaldehyde followed by oxidation with DDQ. As might be anticipated, during the oxidation step, the aromaticity of the benzene rings is not disturbed. Thus, it is only the non-aromatic macrocycle **4.187** that is isolated. The pyridine-containing macrocycle **4.188**, also non-aromatic, was likewise prepared from 2,6-bis(pyrrolyl)pyridine **4.186** and benzaldehyde in 10% yield.

Scheme 4.6.1

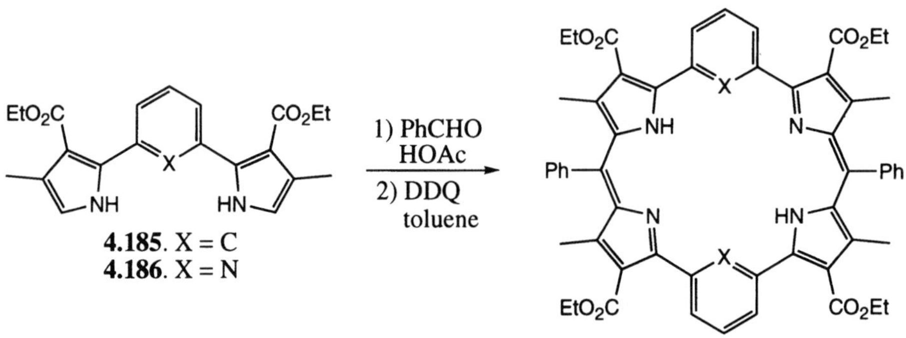

In spite of the fact that it is not formally aromatic, the bisvinylogous porphyrin **4.187** has shown promise as a ligand for the coordination of two metal centers. For instance, treatment of **4.187** with either Pd(acac)$_2$ or Ni(acac)$_2$ affords the dimetalated macrocycles **4.189** and **4.190** in 91% and 61% yield, respectively (Scheme 4.6.2). Additionally, treatment of **4.187** with Rh$_2$(CO)$_4$Cl$_2$ in benzene affords the bis-rhodium complex **4.191** (Scheme 4.6.3), a species that has been characterized structurally *via* single crystal X-ray diffraction analysis (Figure 4.6.1). Surprisingly, and in contrast to what is seen for dirhodium complexes of octaethylporphyrin,[58,59] in this structure it was shown that both of the rhodium atoms were on the same, convex side of what proved to be a bowl-shaped macrocycle. In spite of this, the two rhodium centers were found to be separated by *ca.* 4.4 Å, thus ruling out any strong interaction between the metal atoms.

Scheme 4.6.2

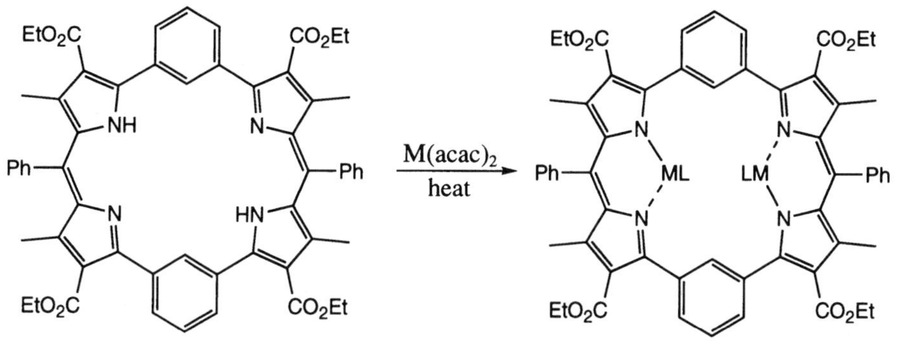

4.187

4.189. M = Pd, L = acac
4.190. M = Ni, L = acac

Scheme 4.6.3

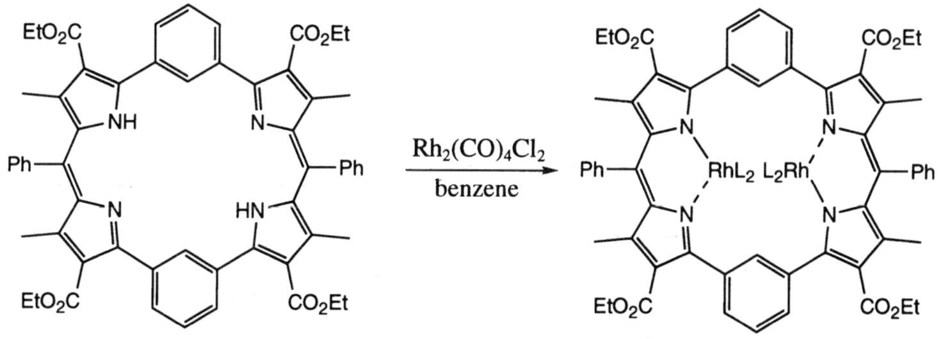

4.187

4.191. L = CO

Figure 4.6.1   Single Crystal X-Ray Diffraction Structure of the Bis[dicarbonylrhodium(I)] complex **4.191**.
This figure was generated using information down-loaded from the Cambridge Crystallographic Data Centre and corresponds to a structure originally reported in reference 57. The macrocycle adopts a saddle shape in the crystal with the rhodium centers being bound on the same side of the macrocycle. The Rh–Rh interatomic distance is 3.31 Å, and the mean Rh–N bond distance is 2.07 Å. Atom labeling scheme: rhodium: ⬤; carbon: ○; nitrogen: ⬤; oxygen: ⊖. Hydrogen atoms have been omitted for clarity

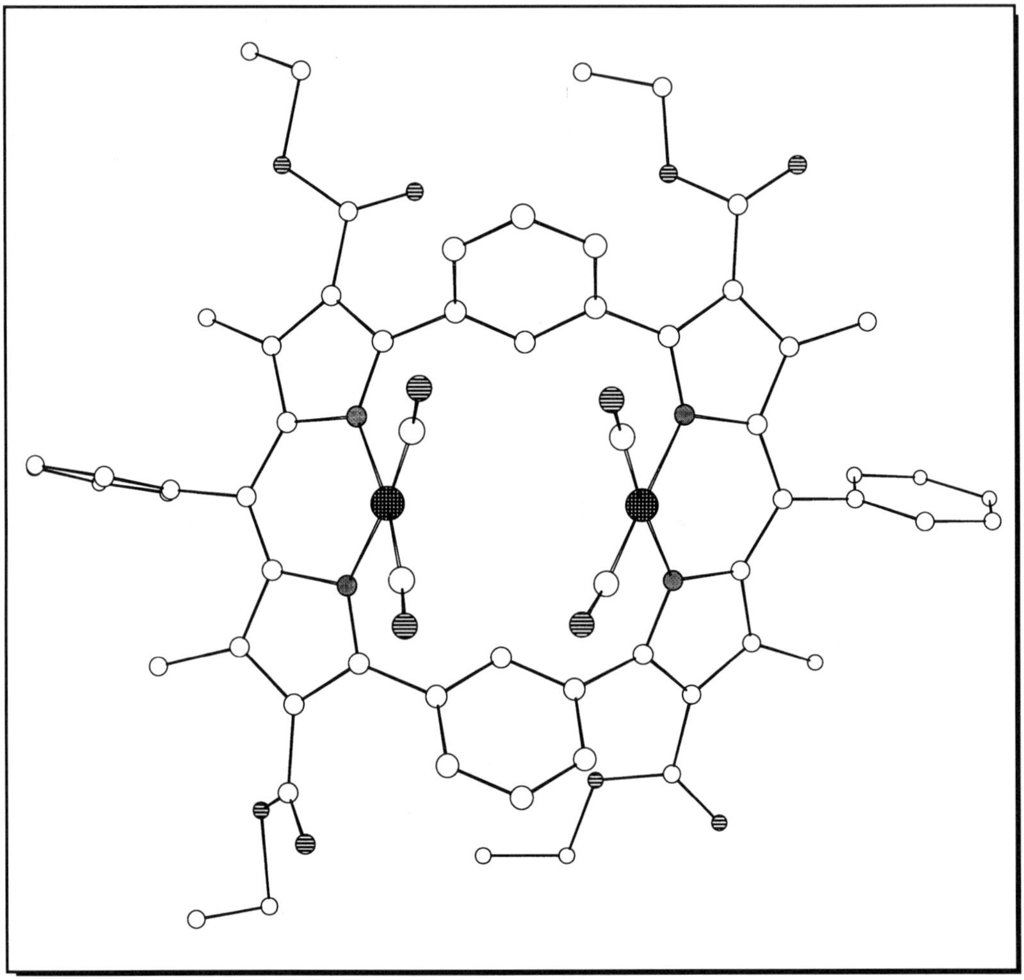

The synthesis of the pyrrole–pyridine-containing expanded porphyrins **4.196** and **4.197** was reported by Bell, *et al.* in 1993.[60] These macrocycles, termed "torands" because of their rigid torroidal structure,[61] were prepared *via* an elaborate linear synthesis, part of which is shown in Scheme 4.6.4. The key macrocyclization step in the synthesis involves the pyrolysis of the bis(semicarbazone) **4.195** to afford torand **4.196** in up to 69% yield. This torand may then be dehydrogenated with

Scheme 4.6.4

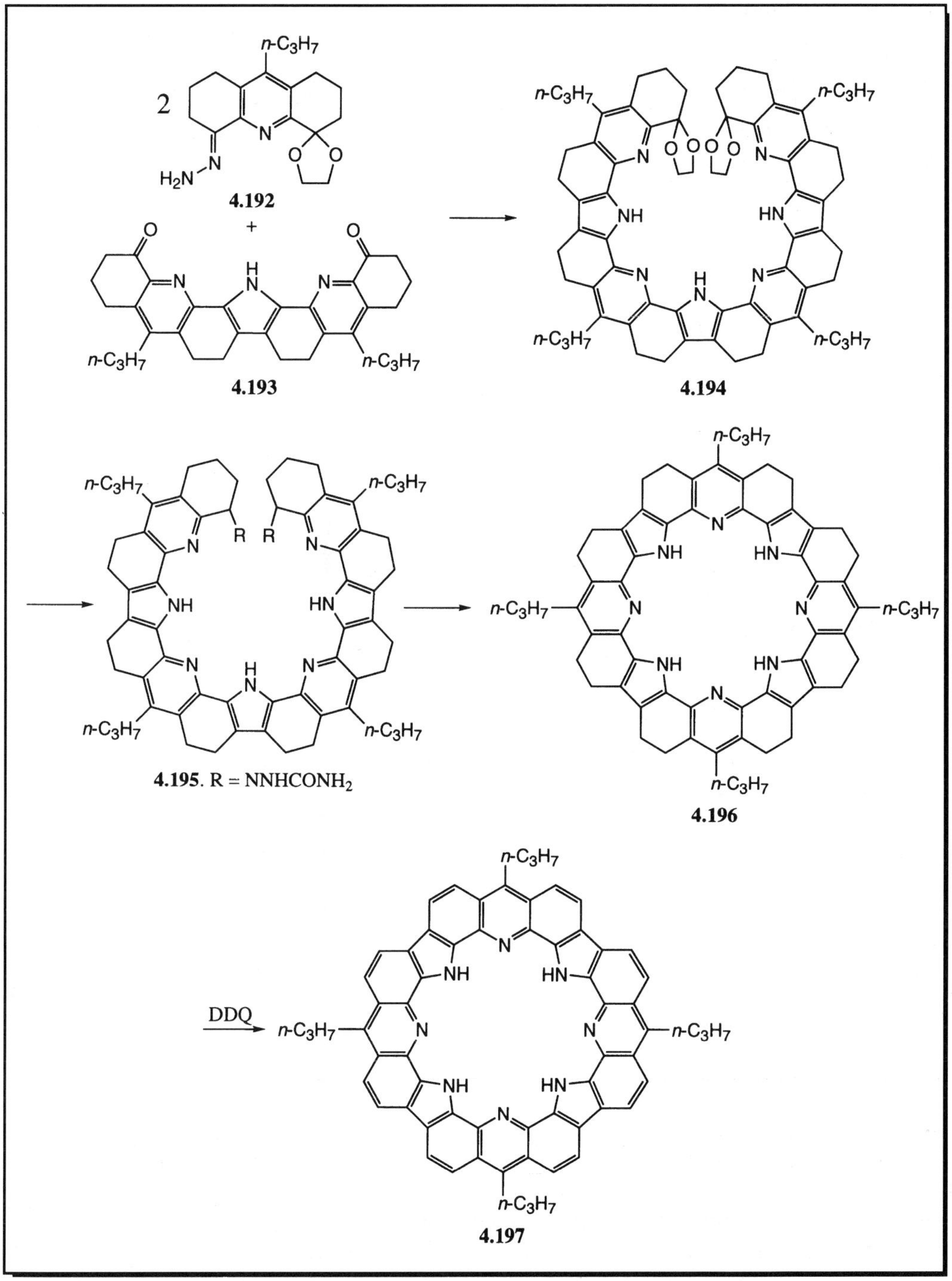

4.192

+

4.193

4.194

4.195. R = NNHCONH$_2$

4.196

DDQ

4.197

DDQ to afford the fully aromatic macrocycle **4.197**. Both of these macrocycles, namely **4.196** and **4.197**, are anticipated to display rich metalation and anion binding chemistry. However, to date, little insight into these potential applications or, indeed other aspects of these systems has been forthcoming.

A very different kind of tetrapyrrolic expanded porphyrin is the [22]annulenoquinone **4.201**.[62] This tetrafuran-containing macrocycle was prepared in 10% yield *via* a Wittig condensation between the ketodialdehyde **4.198** and the phosphonate salt **4.200** (Scheme 4.6.5). A related difuran-dithiophene-containing [22]annulenoquinone **4.202** was likewise prepared *via* reaction of the corresponding ketodialdehyde **4.199** with **4.200**.

Scheme 4.6.5

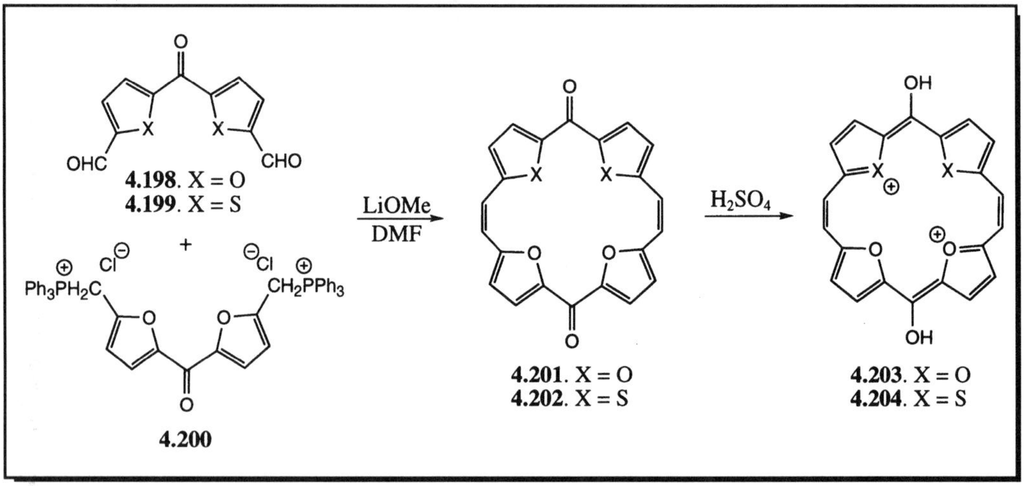
O
X
X
OHC
CHO
4.198. X = O
4.199. X = S
+
Cl
Cl
Ph3PH2C
CH2PPh3
O
O
O
4.200
LiOMe
DMF
O
X
X
O
O
O
4.201. X = O
4.202. X = S
H2SO4
OH
X
X
O
O
OH
4.203. X = O
4.204. X = S

The ^{1}H NMR spectra of **4.201** and **4.202** are consistent with dynamic structures in solution, which, on time average, are planar and highly symmetric. By contrast, a single crystal X-ray structural analysis of **4.202** showed that, in the solid state, the molecule is not planar and that one of the furan rings is twisted significantly out of the mean molecular plane (Figure 4.6.2).

The UV-vis absorption spectra of **4.201** and **4.202** recorded in chloroform are characterized by the presence of broad, relatively weak absorbance bands in the 300–450 nm spectral region (only). The spectrum of **4.201** in sulfuric acid, however, is considerably different. Under these conditions, the absorbance bands become both bathochromically shifted and are more intense. These findings were rationalized by assuming that treatment of **4.201** and **4.202** with sulfuric acid generates the dicationic enol-tautomeric species **4.203** and **4.204**.

An interesting feature of the assumed products **4.203** and **4.204** is that they contain what are formally cyclic-conjugated, 20 π-electron pathways. Nonetheless, the recorded ^{1}H NMR spectra served to reveal resonances that were shifted to a higher field in comparison to the corresponding signals for **4.201** and **4.202**. Such high field resonances are generally considered as being more characteristic of aromatic, as opposed to non- or antiaromatic materials. In the present instance, however, such explanations do not hold. It was thus assumed that the unusual shifts more likely reflect the effects of the double-positive charge present in the macrocycles, rather than some unusual electronic phenomenon. Still, the very fact that high field shifts were observed in this instance underscores an important caveat and warning. Just because upfield shifts are observed does not necessarily mean that one is dealing with an expanded porphyrin that is aromatic.

Figure 4.6.2   Single Crystal X-Ray Diffraction Structure of the [22]Annulenoquinone **4.202**.
This figure was generated using information down-loaded from the Cambridge Crystallographic Data Centre and corresponds to a structure originally reported in reference 62. The macrocycle is not planar and one furan ring is twisted significantly out of the molecular plane. Atom labeling scheme: carbon: ○; sulfur: ◐; oxygen: ◓. Hydrogen atoms have been omitted for clarity

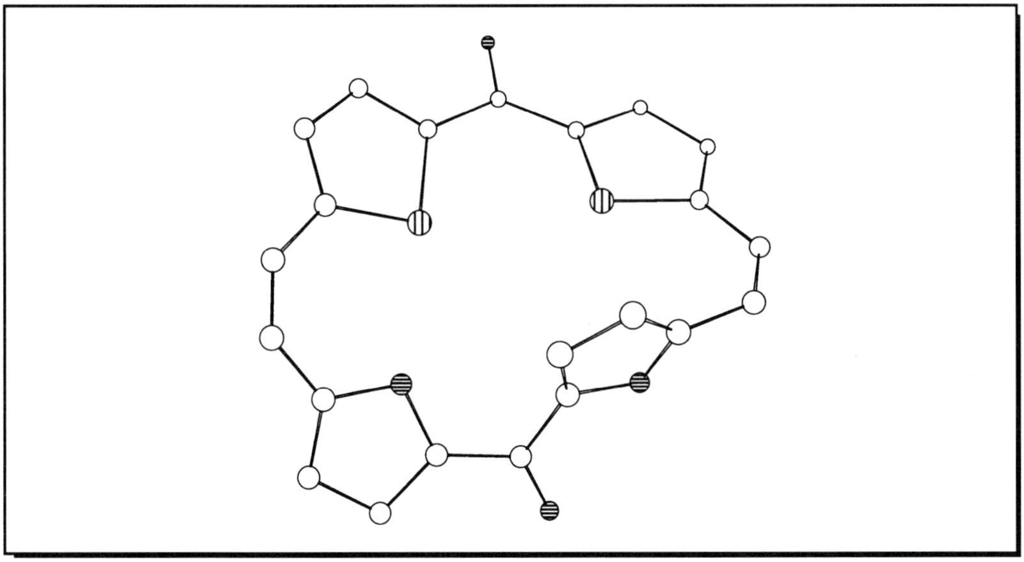

A class of "cavitand" macrocycles with a formal structural core related to that of tetraoxa[24]porphyrin-(2.2.2.2) **4.154** (*vide supra*) was reported by Cram and co-workers in 1983 (Scheme 4.6.6).[63] The first of these was prepared from the dibromo-bis(benzofuran) species **4.205** by treatment with first *s*-BuLi followed by Fe(acac)$_3$. Evidence supporting the formation of macrocycle **4.208** was gleaned from a preliminary mass spectrometric analysis. However, **4.208** exhibited extremely low solubility and could, therefore, not be purified or characterized further. To overcome this solubility problem, Cram prepared the ethyl- and trimethylsilyl-substituted congeners **4.209** and **4.210**.[63,64] These macrocycles were prepared *via* the condensation of the corresponding dibenzofuran dimers **4.206** and **4.207** in 11% and 7% yield, respectively (Scheme 4.6.6).‡‡

Scheme 4.6.6

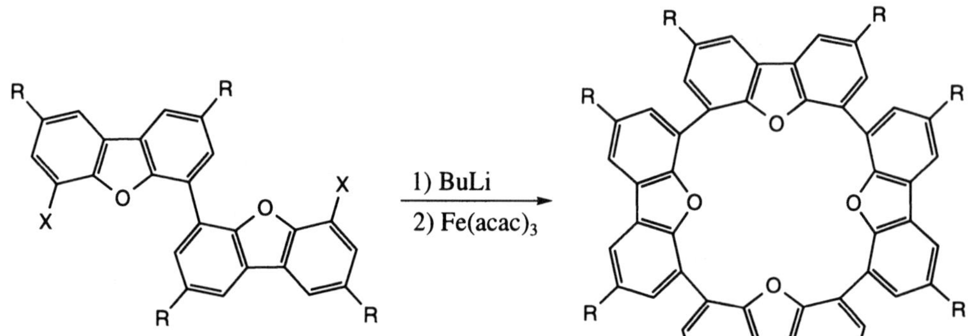

4.205. R = H, X = Br
4.206. R = C₂H₅, X = H
4.207. R = Si(CH₃)₃, X = H

4.208. R = H
4.209. R = C₂H₅
4.210. R = Si(CH₃)₃

In the case of **4.209** and **4.210** structural assignments could be made based on NMR and single crystal X-ray structural analyses. For instance, as judged from ¹H and ¹³C NMR spectral analyses, **4.209** and **4.210** were considered to possess a high degree of symmetry. This observation is in fact consistent with molecular modeling (CPK) studies, which predict the presence of a highly symmetric (D_{2h}) framework. These same model studies also revealed the presence of two cleft-shaped cavities. This was confirmed *via* a single crystal X-ray structural analysis of the trimethylsilyl-substituted macrocycle **4.210** (Figure 4.6.3).

In addition to confirming the overall structure of the macrocycle, the above X-ray structure showed that **4.210** crystallized with one benzene molecule located inside each of the two clefts formed by the macrocycle. This host–guest binding was considered to be of the π–π attraction variety. It was, therefore, considered that this cavitand might act as an efficient host for the complexation of neutral aromatic substrates. However, experiments along these lines revealed that only π-acids such as 1,2-(NC)₂C₆H₄ and 1,3-(O₂N)₂C₆H₄ were bound with affinities sufficient to allow for a binding constant determination.[64] In these instances, the average association constants were determined to be 1.45 ± 0.88 M⁻¹ and 3.72 ± 0.45 M⁻¹ in chloroform, respectively.

---

‡‡Dibenzofuran-derived cyclohexamers were also detected in these syntheses. See Chapter 7 for a discussion of these compounds.

Figure 4.6.3   Single Crystal X-Ray Diffraction Structure of the Cavitand **4.210**.
This figure was generated using information down-loaded from the Cambridge Crystallographic
Data Centre and corresponds to a structure originally reported in reference 64. The macrocycle
crystallizes with two moles of benzene per host molecule (not shown). The four oxygen atoms are
essentially coplanar (the average deviation from the mean O_4-plane is 0.35 Å). Atom labeling
scheme: carbon: ○; oxygen: ⊜; silicon: ⊘. The bound benzene molecules and hydrogen atoms have
been omitted for clarity

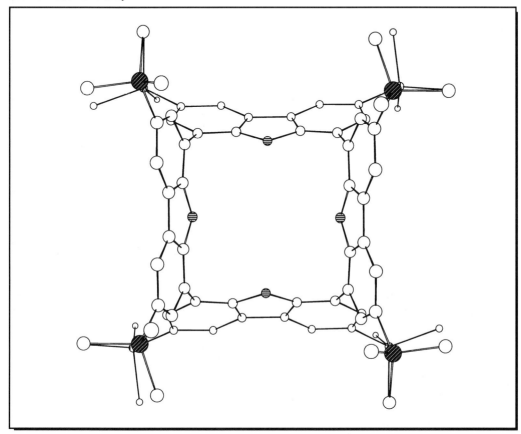

The crystal structure of **4.210** also revealed that the four furan oxygen atoms lie in an
essentially coplanar array (±0.35 Å) and that they serve to define a cavity that is
approximately 5.08 Å in diameter. This observation led to the prediction that cavi-
tands **4.209** and **4.210** might act as potential ligands for large, geometrically complementary metal cations. So far, this promise has not been realized, however.
Indeed, in initial two-phase extraction experiments, macrocycle **4.210** failed to
extract either Cs^+ or Rb^+ cations from a 50 mM aqueous solution into a 50 mM
solution of the macrocycle in $CDCl_3$. It was postulated that the apparent inability of
4.210 to bind strongly to these metal cations was the result of the fact that of the
oxygen lone-pair electrons, being delocalized into the aromatic system, are poor
electron donors.[64]

Scheme 4.6.7

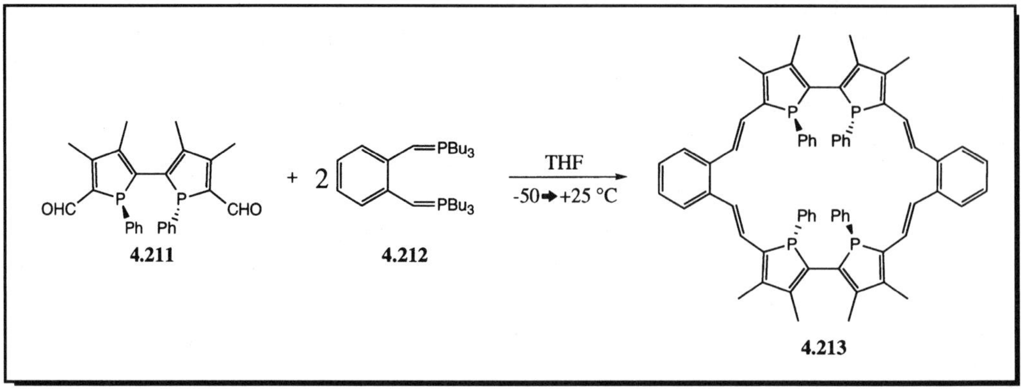

Figure 4.6.4   Single Crystal X-Ray Diffraction Structure of the Tetraphosphole Macrocycle **4.213**.

This figure was generated using information down-loaded from the Cambridge Crystallographic Data Centre and corresponds to a structure originally reported in reference 65. The four phosphorus–phenyl substituents are in an all-*trans* disposition. The diagonal P–P distance is *ca.* 6.1 Å. Atom labeling scheme: carbon: ○; phosphorus: ①. Hydrogen atoms have been omitted for clarity

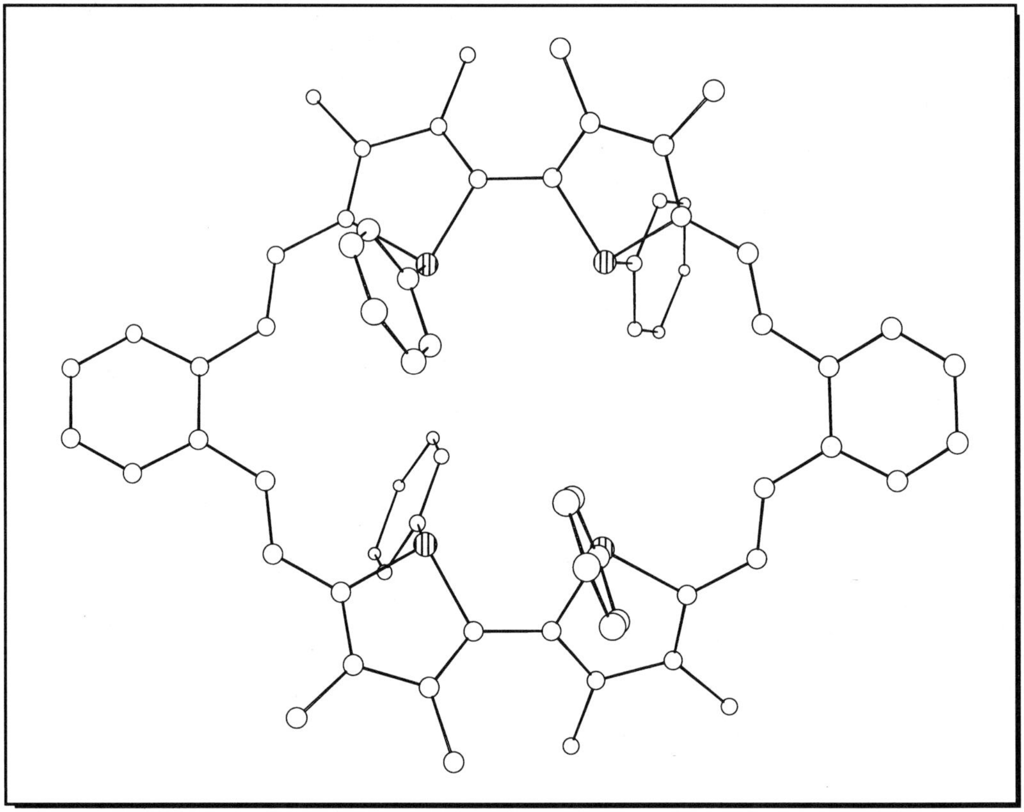

One final example of a "miscellaneous" expanded porphyrin is the 24-membered tetraphosphole macrocycle **4.213** reported by Mathey and coworkers in 1995.[65] It was synthesized in 60% yield from the Wittig-type reaction between the phenyl-substituted 5,5′-bis(formyl)-2,2′-biphosphole **4.211** and the *o*-xylylidene bis-ylid **4.212** (Scheme 4.6.7). A single crystal X-ray structural analysis carried out on **4.213** revealed an all-*trans* orientation of the four phosphorus lone pairs, and a macrocyclic diameter of *ca.* 6 Å (Figure 4.6.4). Because of the steric influence of the phenyl substituents, macrocycle **4.213** is decidedly non-planar. Nevertheless, this macrocycle is expected to exhibit a rich coordination chemistry, particularly towards soft, heavy transition elements. Unfortunately, no such putative metal complexes have as yet been described in the literature.

4.7   References

1.   Johnson, A. W.; Kay, I. T. *J. Chem. Soc.* **1965**, 1620–1629.
2.   Broadhurst, M. J.; Grigg, R.; Shelton, G.; Johnson, A. W. *J. Chem. Soc. Chem. Commun.* **1970**, 231–233.
3.   Broadhurst, M. J.; Grigg, R.; Shelton, G.; Johnson, A. W. *J. Chem. Soc., Perkin Trans. I* **1972**, 143–151.
4.   Grigg, R.; Hamilton, R. J.; Jozefowicz, M. L.; Rochester, C. H.; Terrell, R. J.; Wickwar, H. *J. Chem. Soc., Perkin Trans. II* **1973**, 407–413.
5.   Licoccia, S.; Paci, M.; Paolesse, R.; Boschi, T. *J. Chem. Soc. Dalton Trans.* **1991**, 461–466.
6.   Grigg, R. *J. Chem. Soc. Chem. Commun.* **1967**, 1238–1239.
7.   Grigg, R. *J. Chem. Soc. (C)* **1971**, 3664–3668.
8.   Callot, H. J.; Tschamber, T. *Tetrahedron Lett.* **1974**, *36*, 3155–3158.
9.   Callot, H. J.; Tschamber, T. *J. Am. Chem. Soc.* **1975**, *97*, 6175–6178.
10.   Chevrier, B.; Weiss, R. *J. Chem. Soc., Chem. Commun.* **1974**, 884–885.
11.   Chevrier, B.; Weiss, R. *J. Am. Chem. Soc.* **1975**, *97*, 1416–1421.
12.   Callot, H. J.; Tschamber, T. *Tetrahedron Lett.* **1974**, *36*, 3159–3162.
13.   Callot, H. J.; Tschamber, T.; Schaeffer, E. *J. Am. Chem. Soc.* **1975**, *97*, 6178–6180.
14.   Callot, H. J.; Tschamber, T.; Schaeffer, E. *Tetrahedron Lett.* **1975**, *33*, 2919–2922.
15.   Callot, H. J.; Schaeffer, E. *J. Chem. Res. (M)* **1978**, 0690–0697.
16.   Callot, H. J.; Schaeffer, E. *J. Chem. Res. (S)* **1978**, 51.
17.   Chevrier, B.; Weiss, R. *Inorg. Chem.* **1976**, *4*, 770–774.
18.   Callot, H. J.; Tschamber, T.; Schaeffer, E. *J. Org. Chem.* **1977**, *42*, 1567–1570.
19.   Louati, A.; Schaeffer, E.; Callot, H. J.; Gross, M. *Nouv. J. Chem. 3*, **1979**, 191–194.
20.   Callot, H. J.; Schaeffer, E. *Tetrahedron* **1978**, *34*, 2295–2300.
21.   Louati, A.; Schaeffer, E.; Callot, H. J.; Gross, M. *Nouv. J. Chem. 2* **1978**, 163–168.

22.   Liddell, P. A.; Olmstead, M. M.; Smith, K. M. *J. Am. Chem. Soc.* **1990**, *112*, 2038–2040.

23.   Swanson, K. L.; Snow, K. M.; Jeyakumar, D.; Smith, K. M. *Tetrahedron* **1991**, *47*, 685–696.

24.   Liddell, P. A.; Gerzevske, K. R.; Lin, J. J.; Olmstead, M. M.; Smith, K. M. *J. Org. Chem.* **1993**, *58*, 6681–6691.

25.   Franck, B.; Nonn, A. *Angew. Chem. Int. Ed. Eng.* **1995**, *34*, 1795–1811.

26.   Berger, R. A.; LeGoff, E. *Tetrahedron Lett.* **1978**, *44*, 4225–4228.

27.   König, H.; Eickemeier, C.; Möller, M.; Rodewald, U.; Franck, B. *Angew. Chem. Int. Ed. Engl.* **1990**, *29*, 1393–1395.

28.   Beckmann, S.; Wessel, T.; Franck, B.; Hönle, W.; Borrmann, H.; von Schnering, H.-G. *Angew. Chem. Int. Ed. Eng.* **1990**, *29*, 1395–1397.

29.   Gosmann, M.; Vogt, A.; Franck, B. *Liebigs Ann. Chem.* **1990**, 163–168.

30.   Cetinkaya, E.; Johnson, A. W.; Lappert, M. F.; MacLaughlin, G. M.; Muir, K. W. *J. Chem. Soc. Dalton Trans. 1* **1974**, 1236–1243.

31.   Senge, M. O.; Forsyth, T. P.; Nguyen, L. T.; Smith, K. M. *Angew. Chem. Int. Ed. Eng.* **1994**, *33*, 2485–2487.

32.   LeGoff, E.; Weaver, O. G. *J. Org. Chem.* **1987**, *52*, 710–711.

33.   Wessel, T.; Franck, B.; Möller, M.; Rodewald, U.; Läge, M. *Angew. Chem. Int. Ed. Eng.* **1993**, *32*, 1148–1151.

34.   Gosmann, M.; Franck, B. *Angew. Chem. Int. Ed. Eng.* **1986**, *25*, 1100–1101.

35.   Franck, B.; Nonn, A.; Fuchs, K.; Gosmann, M. *Liebigs Ann. Chem.* **1994**, 503–510.

36.   Dietrich, H.-J., Ph.D. Dissertation, University of Cologne, Germany, **1994**.

37.   Jux, N., Ph.D. Dissertation, University of Cologne, Germany, **1994**.

38.   Dörr, J., Ph.D. Dissertation, University of Cologne, Germany, **1994**.

39.   Knübel, G.; Franck, B. *Angew. Chem. Int. Ed. Eng.* **1988**, *27*, 1170–1172.

40.   Jux, N.; Koch, P.; Schmickler, H.; Lex, J.; Vogel, E. *Angew. Chem., Int. Ed. Eng.* **1990**, *29*, 1385–1387.

41.   Vogel, E.; Jux, N.; Rodriguez-Val, E.; Lex, J.; Schmickler, H. *Angew. Chem., Int. Ed. Eng.* **1990**, *29*, 1387–1390.

42.   Rodriguez-Val, R., Ph.D. Dissertation, University of Cologne, Germany, **1994**.

43.   Böhn, H.-S., Ph.D. Dissertation, University of Cologne, Germany, **1993**.

44.   Mártire, D. O.; Jux, N.; Aramendía, P. F.; Negri, R. M.; Lex, J.; Braslavsky, S. E.; Schaffner, K.; Vogel, E. *J. Am. Chem. Soc.* **1992**, *114*, 9969–9978.

45.   Vogel, E. *Pure Appl. Chem.* **1993**, *65*, 143–152.

46.   Sauer, F., Ph.D. Dissertation, University of Cologne, Germany, **1993**.

47.   Elix, J. A. *Aust. J. Chem.* **1969**, *22*, 1951–1962.

48.   Märkl, G.; Sauer, H.; Kreitmeier, P.; Burgemeister, T.; Kastner, F.; Adolin, G.; Nöth, H.; Polborn, K. *Angew. Chem. Int. Ed. Eng.* **1994**, *33*, 1151–1153.

49.   Schall, R., Ph.D. Dissertation, University of Cologne, Germany, **1993**.

50.   Behrens, F., Ph.D. Dissertation, University of Cologne, Germany, **1994**.

51.   Strand, A.; Thulin, B.; Wennerström, O. *Acta Chem. Scand.* **1977**, *B31*, 521–523.

52. Jörrens, F., Ph.D. Dissertation, University of Cologne, Germany, **1994**.

53. Hu, Z.; Atwood, J. L.; Cava, M. P. *J. Org. Chem.* **1994**, *59*, 8071–8075.

54. Cresp, T. M.; Sargent, M. V. *J. Chem. Soc., Perkin Trans. I* **1973**, 1786–1790.

55. Scharfe, O., Ph.D. Dissertation, University of Cologne, Germany, **1992**.

56. Hu, Z.; Cava, M. P. *Tetrahedron Lett.* **1994**, *35*, 3493–3496.

57. Carré, F. H.; Corriu, R. J. P.; Bolin, G.; Moreau, J. J. E.; Vernhet, C. *Organometallics* **1993**, *12*, 2478–2486.

58. Takenaka, A.; Sasada, Y. *J. Chem. Soc. Chem. Commun.* **1973**, 792–793.

59. Takenaka, A.; Sasada, Y.; Ogoshi, H.; Omura, T.; Yoshida, Z.-I. *Acta Cryst.* **1975**, *B31*, 1–6.

60. Bell, T. W.; Cragg, P. J.; Drew, M. G. B.; Firestone, A.; Kwok, A. D.-I.; Liu, J.; Ludwig, R. T.; Papoulis, A. T. *Pure Appl. Chem.* **1993**, *65*, 361–366.

61. Bell, T. W.; Firestone, A.; Guzzo, F.; Hu, L.-Y. *J. Inclu. Phenom.* **1987**, *5*, 149–152.

62. Märkl, G.; Striebl, U. *Angew. Chem. Int. Ed. Eng.* **1993**, *32*, 1333–1335.

63. Helgeson, R. C.; Lauer, M.; Cram, D. J. *J. Chem. Soc. Chem. Commun.* **1983**, 101–103.

64. Schwartz, E. B.; Knobler, C. B.; Cram, D. J. *J. Am. Chem. Soc.* **1992**, *114*, 10775–10784.

65. Deschamps, E.; Ricard, L.; Mathey, F. *J. Chem. Soc. Chem. Commun.* **1995**, 1561–1562.

5   Introduction

As discussed in Chapter 4, one way to "expand" a porphyrin is to add "extra" *meso*-like bridging carbon atoms into the porphyrin skeleton. Another way to expand a porphyrin is to insert, formally, additional heterocyclic rings into the erstwhile tetrapyrrolic framework. This approach to expanded porphyrins generally provides macrocycles with extended π-electron conjugation pathways and ones that are often aromatic. It also generates systems that contain a greater number of heteroatoms than do the porphyrins. This feature makes such molecules of interest as potential metal-chelating ligands and, in many instances, as anion and neutral substrate receptors. In this chapter, the synthesis-related chemistry of two pentapyrrolic systems, sapphyrin and smaragdyrin, is discussed. These macrocycles contain five pyrroles, but only four and three bridging *meso* carbons, respectively. Other expanded porphyrins containing five pyrroles are discussed in the following chapter.

5.1   Sapphyrins

Sapphyrins are a class of "heterocycle-inserted" expanded porphyrins that contain a (1.1.1.1.0) arrangement of four *meso*-carbon atoms. They were discovered serendipitously by R. B. Woodward, *et al.*[1-3] during the course of investigations directed toward the synthesis of vitamin B_{12}, and actually represent the first expanded porphyrin (of any ilk) to have been identified and characterized. The basic, substituent-free form of the sapphyrin molecule is shown below in Figure 5.1.1. It is formally derived from the parent porphyrins by virtue of replacing one pyrrole ring with a bipyrrole unit. Sapphyrin thus contains five pyrrole rings and four bridging carbon atoms within its macrocyclic periphery. Because sapphyrin is deep blue in the solid state and intense green in organic solution, R. B. Woodward assigned the trivial name "sapphyrin" to this class of macrocycle.

As implied above, the first synthesis of a sapphyrin was purely accidental and came about as the result of the reaction of the linear tetrapyrrolic precursor **5.1** with formic acid and HBr.[2] Following oxidation with iodine, instead of obtaining the desired tetrapyrrolic corrole macrocycle, a different macrocycle was obtained (Scheme 5.1.1). This unexpected by-product was best formulated as being the decaalkyl-substituted aromatic pentapyrrolic macrocycle **5.2**. The unexpected formation of **5.2** was rationalized in terms of two molecules of tetrapyrrole **5.1** undergoing condensation with formic acid to form first a linear octapyrrolic intermediate. This later species was then assumed to undergo acid-promoted cleavage to form a linear pentapyrrolic intermediate that subsequently cyclized to give a pentapyrrolic macrocycle. Oxidation with iodine then produced a product that was thought to be the 22 π-electron macrocycle **5.2** (isolated as its bisperchlorate salt **5.2b**). In order to confirm that a sapphyrin was in fact being formed under the reaction conditions, tetrapyrrole **5.1** was reacted with 2,5-diformyl-3,4-dimethylpyrrole **5.3** under acidic conditions.[2] As shown in Scheme 5.1.2, this resulted in the formation of the more highly symmetrical sapphyrin **5.4**.

Figure 5.1.1   Generalized Structure of Sapphyrin Showing the Atom Numbering Scheme

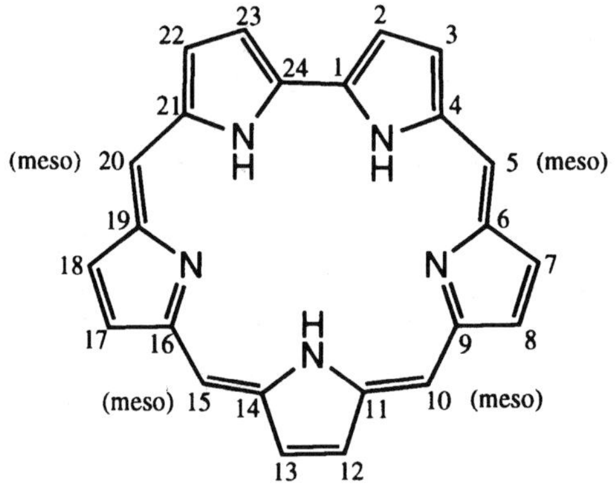

Scheme 5.1.1

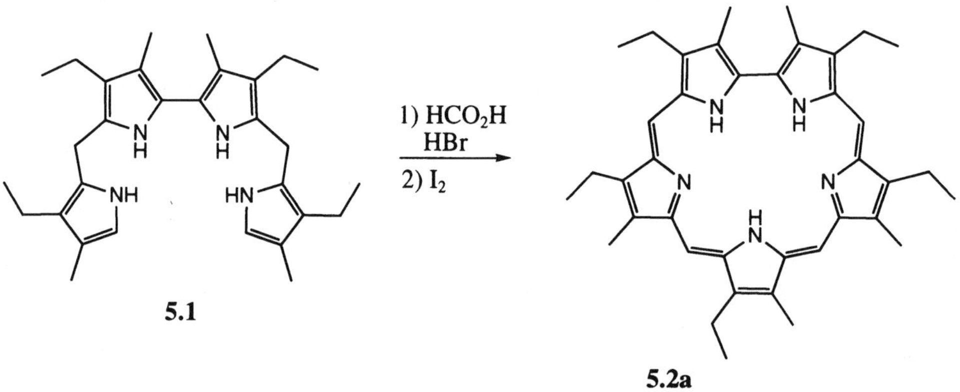

Scheme 5.1.2

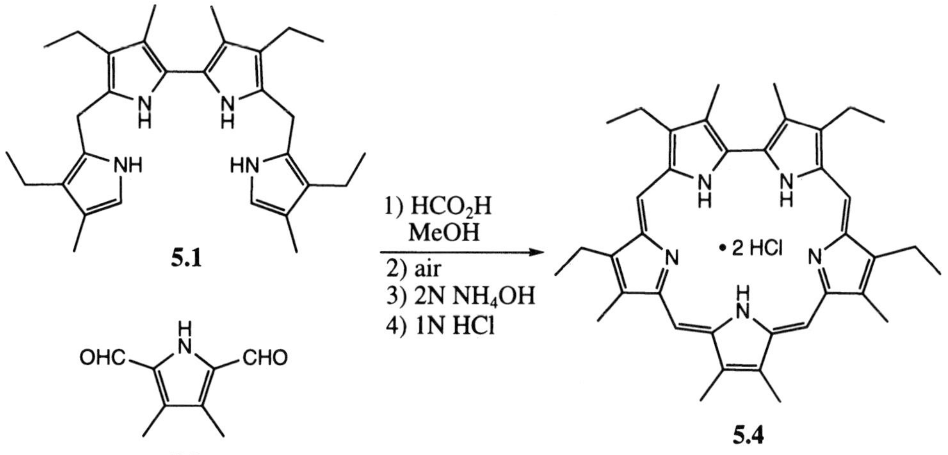

In the full paper describing Woodward's sapphyrin-directed studies, it was noted that small amounts of sapphyrins could also be prepared *via* the HBr-catalyzed reaction between a bipyrrole dicarboxaldehyde and a dipyrrylmethane dicarboxylic acid.[2] Additionally, the HBr-catalyzed reaction of 3,4-dimethylpyrrole **5.5** with 2,5-diformyl-3,4-dimethylpyrrole **5.3** was reported to afford small yields of decamethyl-sapphyrin **5.6** (Scheme 5.1.3).[2] The fact that any sapphyrin could be isolated from these reactions, or even from the procedures of Schemes 5.1.1 and 5.1.2, attests to the stability of this pentapyrrolic macrocycle.

Scheme 5.1.3

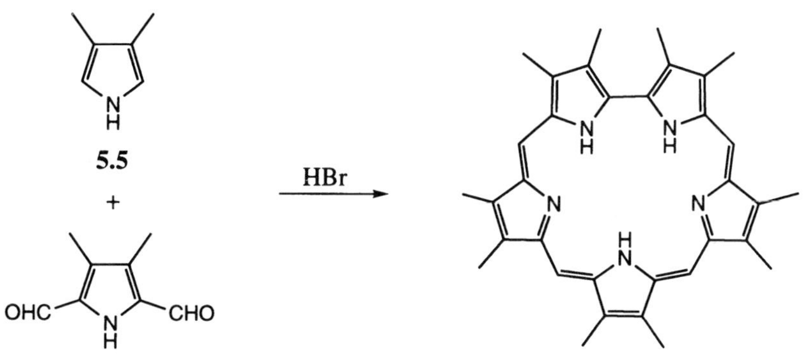

The excitement aroused by the unexpected formation of sapphyrin inspired a search for a more rational syntheses. Here, the impetus came from a recognition that while the "4 + 1" approach of Scheme 5.1.2 was workable, the key linear tetrapyrrolic precursors (e.g., **5.1**) were hard to make. As a result, considerable synthetic effort has focused on making workable a convergent MacDonald-type "3 + 2" approach. This approach was first implemented by Woodward[2] and, independently, Grigg and Johnson[4-6] and others.[7] Later it was also used by Sessler.[8-10] Briefly, this "3 + 2" strategy involves carrying out an acid-catalyzed condensation between a diformyl bipyrrole, such as **5.7–5.11**, and a diacid tripyrrane, such as **5.12–5.18**. After oxidation, the corresponding sapphyrins (e.g., **5.6** and **5.19–5.26**) are afforded (Scheme 5.1.4).

Scheme 5.1.4

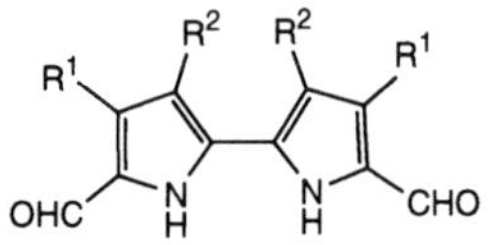

5.7. $R^1 = R^2 = H$
5.8. $R^1 = R^2 = Me$
5.9. $R^1 = Me, R^2 = Et$
5.10. $R^1 = Et, R^2 = Me$
5.11. $R^1 = n\text{-Pr}, R^2 = H$

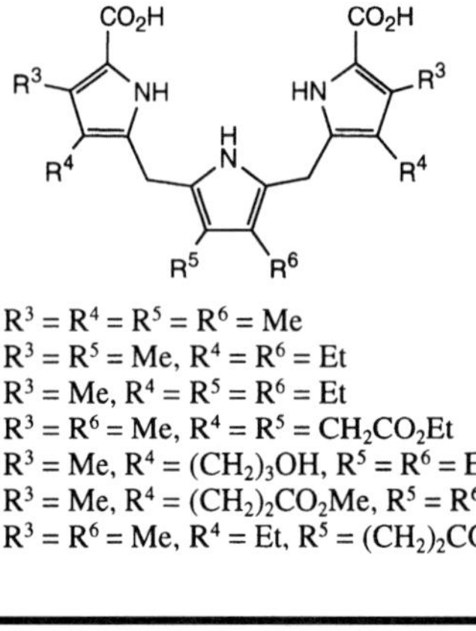

5.12. $R^3 = R^4 = R^5 = R^6 = Me$
5.13. $R^3 = R^5 = Me, R^4 = R^6 = Et$
5.14. $R^3 = Me, R^4 = R^5 = R^6 = Et$
5.15. $R^3 = R^6 = Me, R^4 = R^5 = CH_2CO_2Et$
5.16. $R^3 = Me, R^4 = (CH_2)_3OH, R^5 = R^6 = Et$
5.17. $R^3 = Me, R^4 = (CH_2)_2CO_2Me, R^5 = R^6 = Et$
5.18. $R^3 = R^6 = Me, R^4 = Et, R^5 = (CH_2)_2CO_2Me$

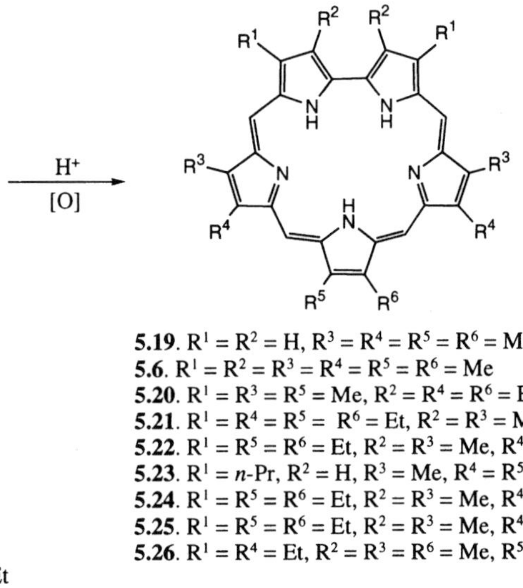

5.19. $R^1 = R^2 = H, R^3 = R^4 = R^5 = R^6 = Me$
5.6. $R^1 = R^2 = R^3 = R^4 = R^5 = R^6 = Me$
5.20. $R^1 = R^3 = R^5 = Me, R^2 = R^4 = R^6 = Et$
5.21. $R^1 = R^4 = R^5 = R^6 = Et, R^2 = R^3 = Me$
5.22. $R^1 = R^5 = R^6 = Et, R^2 = R^3 = Me, R^4 = CH_2CO_2Et$
5.23. $R^1 = n\text{-Pr}, R^2 = H, R^3 = Me, R^4 = R^5 = R^6 = Et$
5.24. $R^1 = R^5 = R^6 = Et, R^2 = R^3 = Me, R^4 = (CH_2)_3OH$
5.25. $R^1 = R^5 = R^6 = Et, R^2 = R^3 = Me, R^4 = (CH_2)_2CO_2Me$
5.26. $R^1 = R^4 = Et, R^2 = R^3 = R^6 = Me, R^5 = (CH_2)_2CO_2Me$

Although more practical than the "4 + 1" approach, the "3 + 2" route was also found early on to be limited by difficulties associated with obtaining the requisite tripyrrolic ("3") and bipyrrolic ("2") precursors. Thus, even though the initial reports of Woodward and Johnson served to pique the curiosity of the porphyrin community, the chemistry of sapphyrins remained undeveloped until 1991, when Sessler and coworkers introduced an improved synthesis of sapphyrins.[9] This new synthesis is based on simple high-yielding routes to both dicarboxyl-substituted tripyrranes and β-substituted bipyrroles.[8–13] Interestingly, these precursor syntheses of Sessler, *et al.* are made efficient in part by the α-free pyrrole synthesis of Barton and Zard.[14] In any event, these improvements have made practical the preparation of C_2-symmetric sapphyrins.

It is worth mentioning that an alternative "3 + 2" synthetic approach to sapphyrin has been reported. Specifically, it was found that the alkyl-substituted sapphyrin **5.19** could be prepared *via* the acid-catalyzed condensation of pyrrolyl-bipyrrole **5.27** with diformyl dipyrrylmethane **5.28** (Scheme 5.1.5).[3] Following oxidation with air and chromatographic purification, this method affords a 35% yield of the desired sapphyrin. Unfortunately, this method has yet to find any kind of general applicability. This is most likely a reflection of the fact that the needed pyrrolyl-bipyrrolic precursors are hard to make.

Scheme 5.1.5

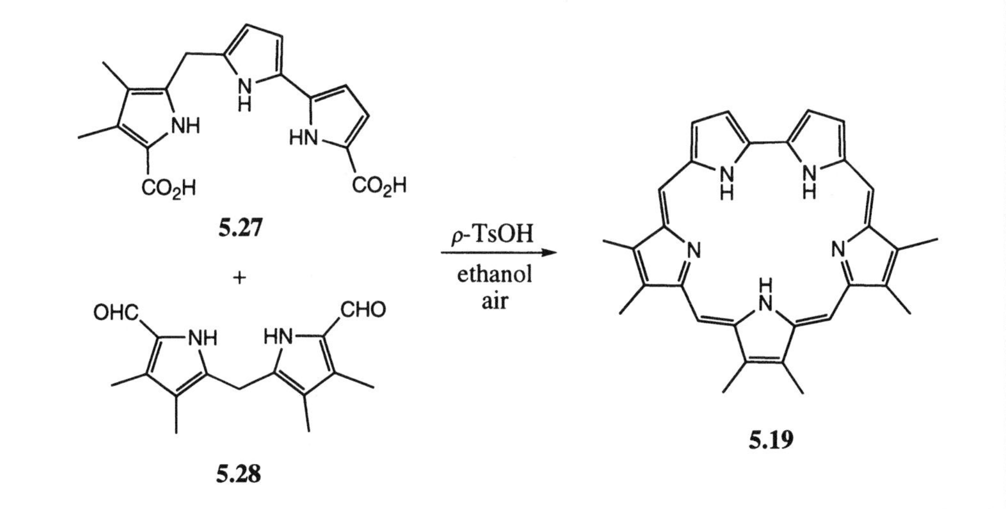

The increased availability of sapphyrins, as the result of Sessler's 1991 report,[8] had the immediate effect of allowing these species to be characterized crystallographically. The first X-ray structure to be reported was the mixed HF-PF_6 salt of 3,8,12,13,17,22-hexaethyl-2,7,18,23-tetramethylsapphyrin (**5.21**).[8] In this structure, the sapphyrin nucleus was found to be nearly flat with a fluoride anion being held by hydrogen bonds inside the central cavity (Figure 5.1.2). Subsequent X-ray diffraction studies showed that several other anions, including Cl^-, N_3^- and a wide variety of phosphate derivatives could also be bound in the solid state (Figure 5.1.3).[9,15–18] These crystallographic findings, in turn, sparked intense interest into the question of whether sapphyrins could be used to bind and transport biologically relevant anions under solution phase conditions. This aspect of sapphyrin chemistry, which is being generalized slowly to include other expanded porphyrin species, is discussed further in Chapter 10.

Figure 5.1.2   Single Crystal X-Ray Diffraction Structure of the Mixed HF/HPF_6 Salt of Sapphyrin **5.21**.

This figure was generated using information down-loaded from the Cambridge Crystallographic Data Centre and corresponds to a structure originally reported in reference 8. The mean deviation from planarity of the five nitrogen atoms and the central fluorine atom is 0.03 Å. Top: top view; bottom, side view. Atom labeling scheme: carbon: ◯; nitrogen: ⬤; fluorine: ⊗. The PF_6-anion and the hydrogen atoms have been omitted for clarity

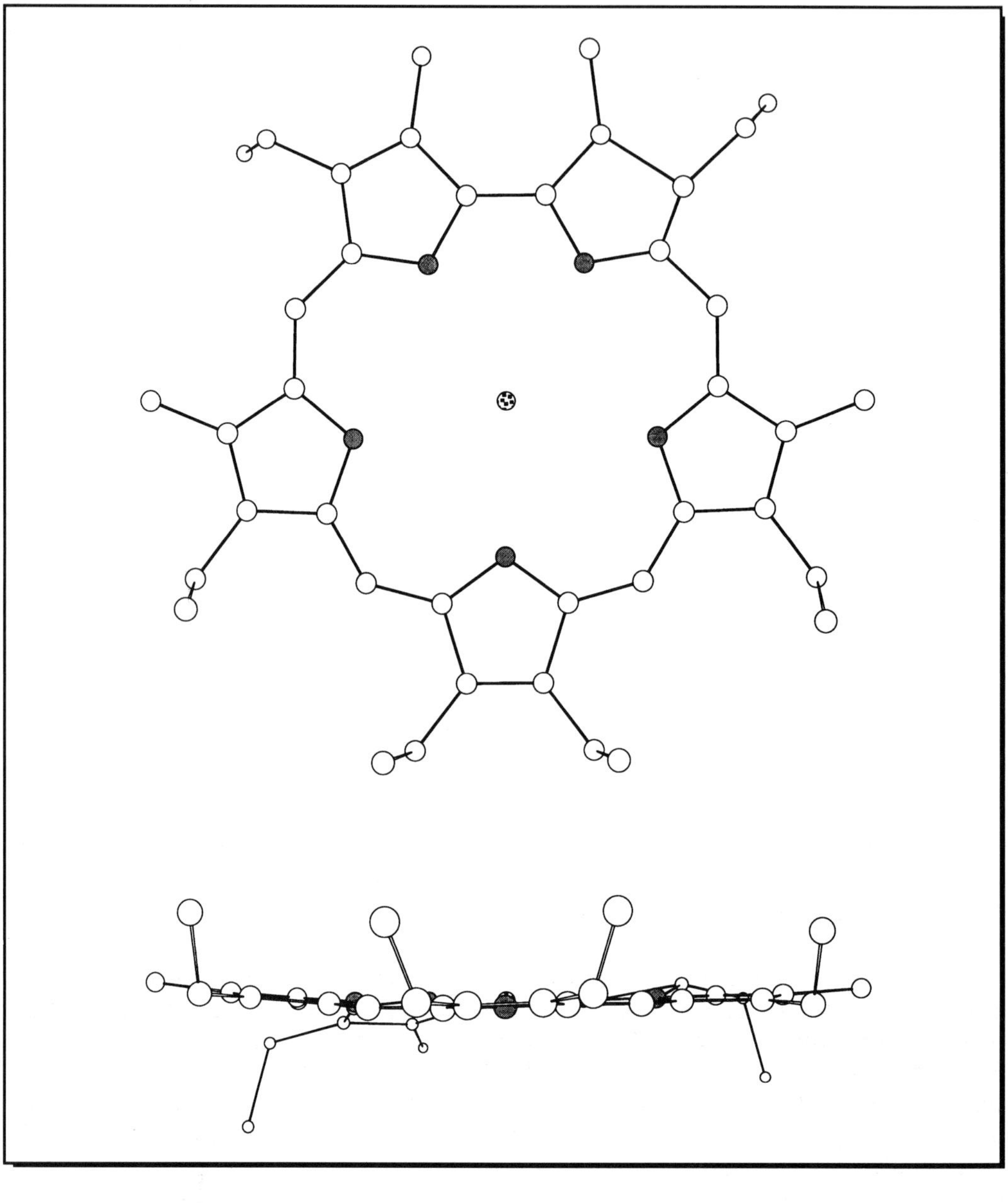

Figure 5.1.3   Single Crystal X-Ray Diffraction Structure of the Bis-HCl Salt of Sapphyrin **5.21**.

This figure was generated using information down-loaded from the Cambridge Crystallographic Data Centre and corresponds to a structure originally reported in reference 15. One chloride anion is located above the mean macrocyclic plane and the other below. The chloride anions lie *ca.* 1.8 Å out of the mean N_5 plane. Atom labeling scheme: carbon: ○; nitrogen: ◉; chlorine: ⊗. Hydrogen atoms have been omitted for clarity

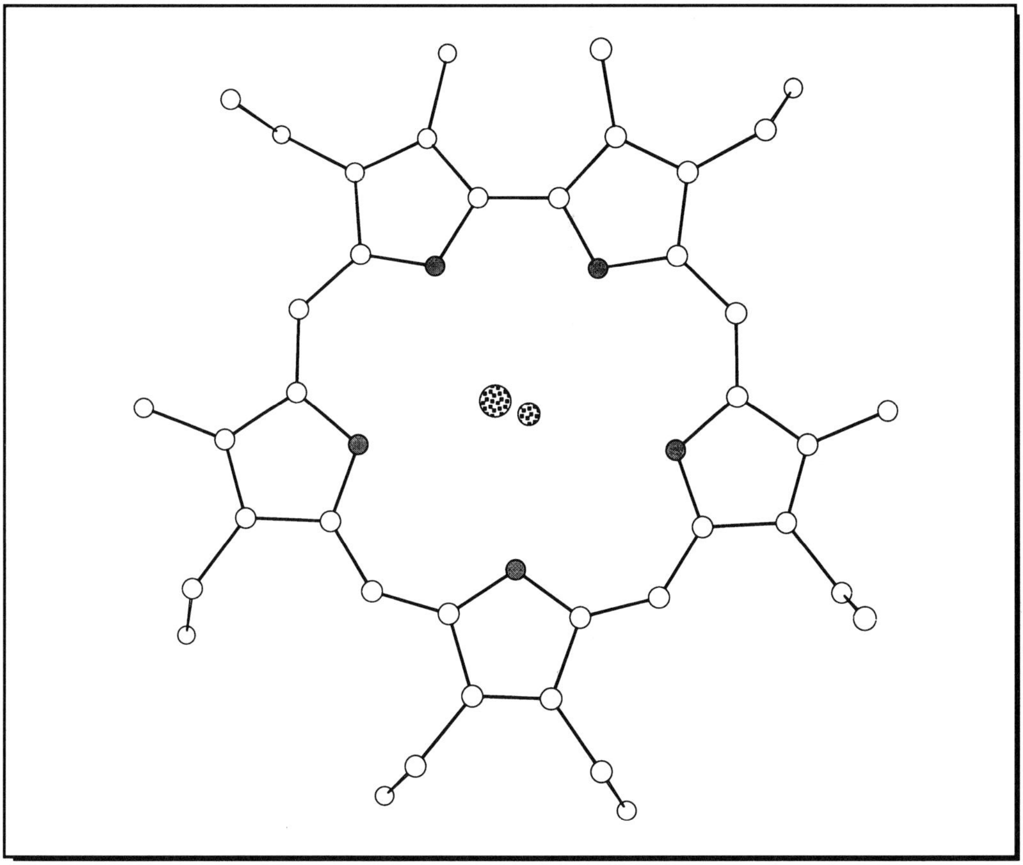

5.2   *meso*-Aryl-substituted Sapphyrins

A potentially practical synthesis of *meso*-diarylsapphyrins has recently been introduced by Sessler and Kodadek.[19] It involves the one-pot condensation reaction between diformyl bipyrrole **5.10**, pyrrole **5.29**, and benzaldehyde **5.30** using $BF_3 \cdot MeOH$ as the Lewis acid catalyst. In this "2 + 1 + 1 + 1" approach, careful control of the ratio of starting materials is required. When run properly, however, this one-step procedure allows for the preparation of 10,15-diphenylsapphyrin **5.33** in *ca.* 10% yield (Scheme 5.2.1). In a similar fashion, the ρ-CH_3- and ρ-CN-derivatives **5.34** and **5.35** were prepared using the appropriately substituted benzaldehyde derivatives **5.31** and **5.32**. Also, using this approach, the first example of a chiral sapphyrin derivative (**5.37**) was obtained. Here, instead of using the simple benzaldehyde derivatives **5.30–5.32**, the non-racemic aldehyde **5.36** was employed (Scheme 5.2.2). This aldehyde, which is derived from 1R-(+)-nopinone, has also been used by Kodadek, *et al.* to prepare non-racemic porphyrins.[20]

Scheme 5.2.1

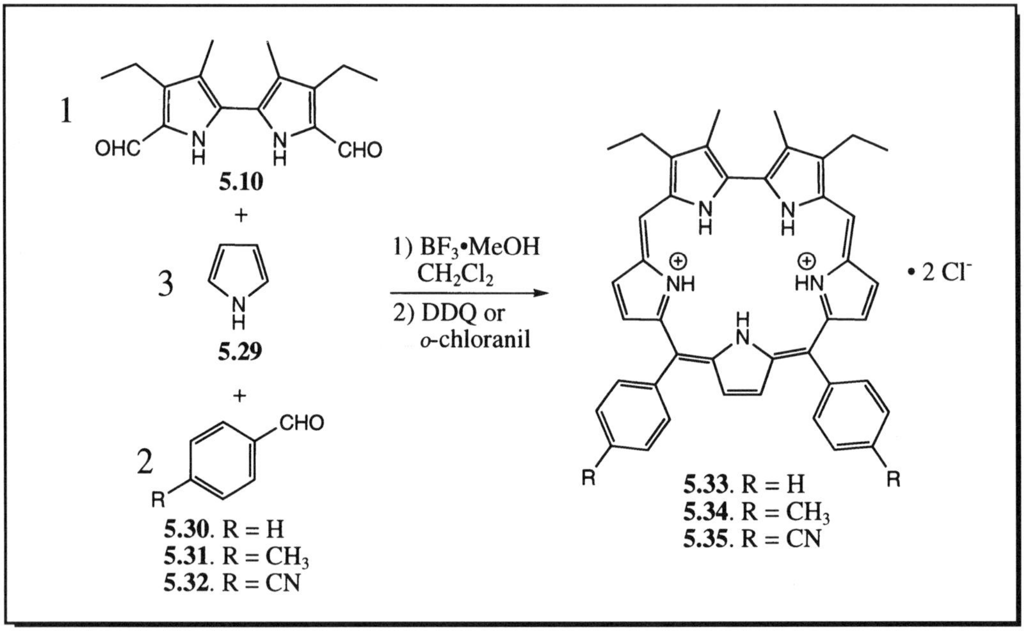

A single crystal X-ray structure of the bis-HCl salt of 10,15-diphenylsapphyrin **5.33** revealed a ring system that deviates only minimally from planarity (Figure 5.2.1). The two chloride counteranions are held above and below the mean macrocyclic plane by hydrogen bonding interactions. Thus, the solid-state structure of sapphyrin **5.33**·2HCl closely resembles the bis-HCl salt of 3,8,12,13,17,22-hexaethyl-2,7,18,23-tetramethylsapphyrin (**5.21**). Because of this, it is currently being considered that system **5.37** could be used as a stereo-differentiating receptor for racemic or non-racemic anions.

Prior to developing the above synthesis, Sessler and coworkers discovered a different route to *meso*-aryl sapphyrins.[21] This synthetic strategy, like so many in the area of sapphyrin-related research, is based upon a reaction that was not initially intended to produce sapphyrins. Specifically, it involves the reaction between bis(pyrrolyl)bipyrrole **5.38**[22] and diformyl dipyrrylmethane **5.39**, and was found to produce, quite unexpectedly, sapphyrin **5.41** in *ca.* 5% yield, with none of the intended hexapyrrolic macrocycle **5.40** being detected (Scheme 5.2.3). It was thus considered that such a "4 + 2 = 5" approach might prove useful for the synthesis of *meso*-phenyl sapphyrins. This indeed proved to be the case. In fact, this strategy has been used to synthesize the 10-phenyl sapphyrins **5.45–5.47** starting from tetrapyrrole **5.38** and the corresponding phenyl-substituted dipyrrylmethane precursors **5.43** and **5.44** (Scheme 5.2.4). In all cases, the reaction yields were on the order of 5%.

Scheme 5.2.2

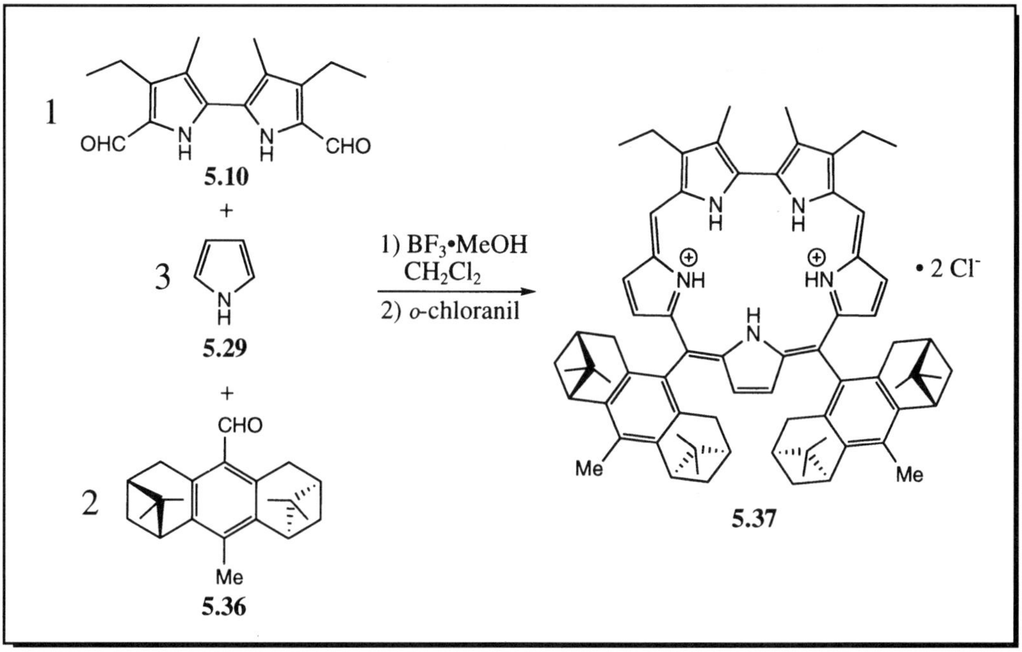

Sessler and coworkers have also developed a "3 + 2" approach to *meso*-diarylsapphyrins.[21] This strategy, which is based on those used by Ogoshi[23,24] and Smith[25,26] to generate *meso*-arylporphyrins, is outlined in Scheme 5.2.5. It involves the acid-catalyzed condensation of the diol **5.48** with the diacid tripyrrane **5.14**. After oxidation with *o*-chloranil and chromatographic workup, diphenylsapphyrin **5.49** was obtained in *ca.* 3% yield. While less direct than the "one-pot" approach to 10,15-diarylsapphyrins outlined earlier (Schemes 5.1.6 and 5.1.7), this strategy allows synthetic access to isomeric 5,20-substituted diarylsapphyrins (e.g., **5.49**).

In 1995, Latos-Grażyński and coworkers reported an efficient, albeit somewhat low-yielding, synthesis of tetraphenylsapphyrin **5.50**.[27] In this work, it was found that the acid-catalyzed, oxidative reaction of benzaldehyde (**5.30**) with pyrrole (**5.29**) in a 1 : 3 molar ratio gave, as expected, tetraphenylporphyrin and an isomeric N-inverted porphyrin as the main macrocyclic products (cf. discussion in Chapter 3). However, careful examination of the minor fractions revealed that one side-product of this reaction was in fact the pentapyrrolic tetraphenylsapphyrin **5.50a** (Scheme 5.2.6). Unfortunately, this species could only be isolated in 1.1% yield. Nonetheless, this synthesis represented the first synthetic entry into the area of tetrakis-*meso*-substituted sapphyrins.

Figure 5.2.1   Single Crystal X-Ray Diffraction Structure of the Bis-HCl Salt of 10,15-Diphenylsapphyrin **5.33**.
This figure was generated using information down-loaded from the Cambridge Crystallographic Data Centre and corresponds to a structure originally reported in reference 19. The sapphyrin skeleton deviates somewhat from planarity (the average deviation from the mean macrocyclic plane is 0.41 Å). One chloride anion is located 1.85 Å above the mean N_5 macrocyclic plane and the other 1.82 Å below. Atom labeling scheme: carbon: ○; nitrogen: ●; chlorine: ⊘. Hydrogen atoms have been omitted for clarity

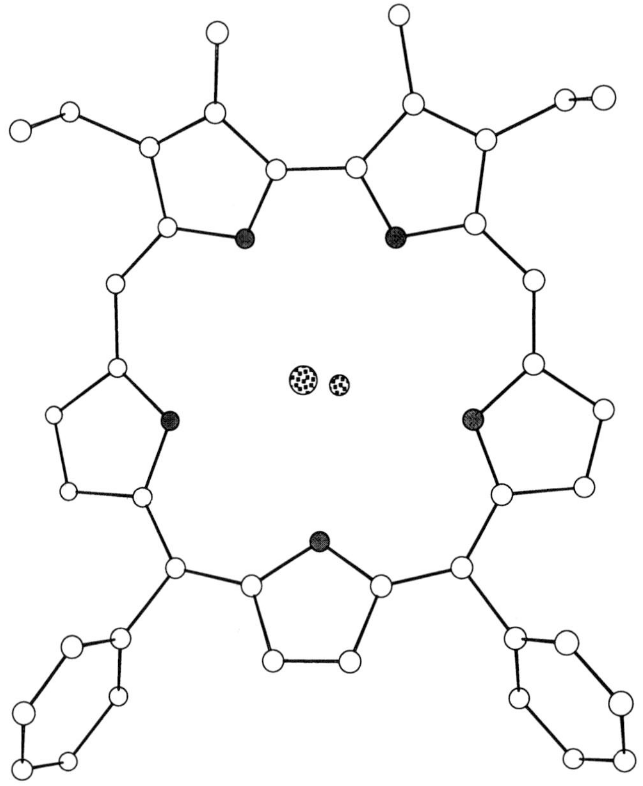

Tetraphenylsapphyrin **5.50a** proved to be aromatic, as evidenced by ^{1}H NMR spectroscopic analysis. In spite of this, Latos-Grażyński and coworkers determined that the free-base form of tetraphenylsapphyrin **5.50a** exists in a conformation wherein the pyrrole ring opposite the bipyrrole moiety is rotationally inverted (i.e., the nitrogen faces "out").[27] This conclusion was based on the observation that the hydrogen atom associated with the inverted pyrrole ring nitrogen resonates at 11.75 ppm while the β-pyrrolic protons of the same pyrrole ring resonate at −1.21 ppm. The location of these chemical shifts is consistent with an aromatic macrocycle in which these β-pyrrolic protons are located on the interior of an aromatic ring current and the NH proton is on the periphery. Such an orientation is in contrast to that observed in the case of *meso*-unsubstituted decaalkylsapphyrins wherein all five of the pyrrole nitrogen atoms point into the macrocyclic core.

Scheme 5.2.3

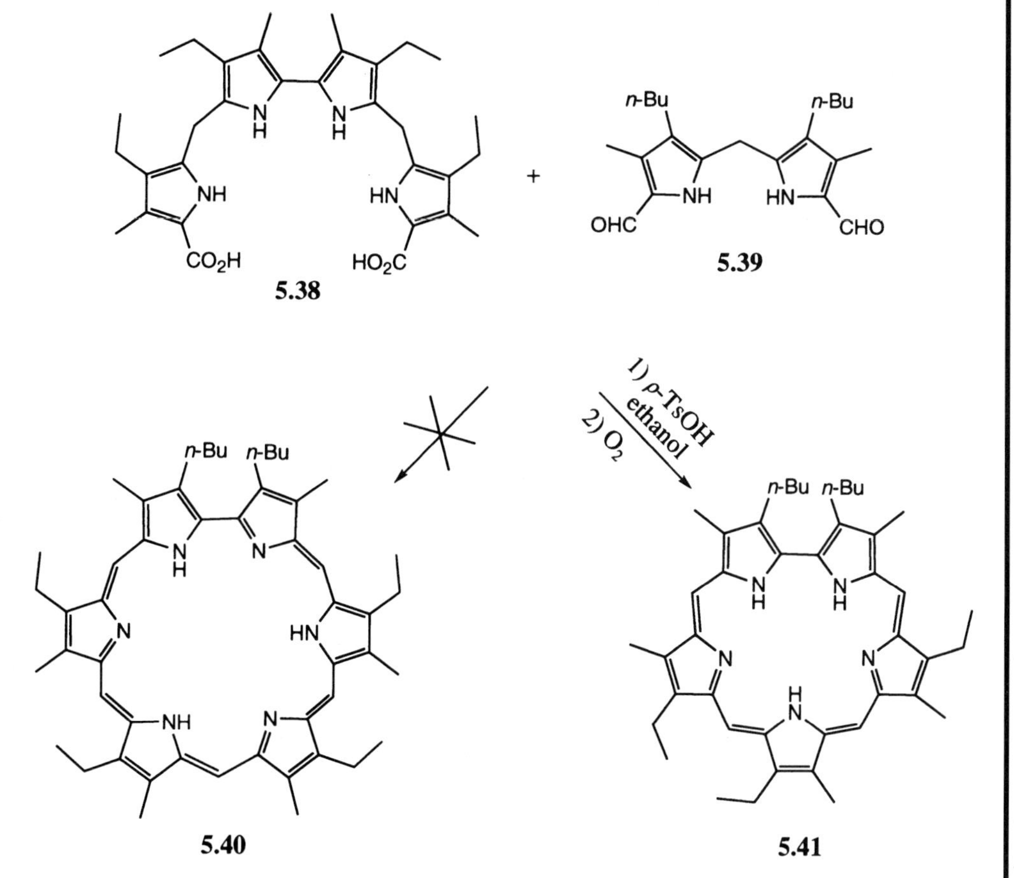

The flexibility of tetraphenylsapphyrin **5.50** was inferred from an analysis of the ^{1}H NMR spectrum of the fully protonated form (e.g., bis-HCl salt) of the macrocycle. Here, for instance, it was found that the erstwhile downfield NH proton resonance could now be seen at $\delta = -2.74$ ppm. This represents an upfield chemical shift "jump" for this proton of 14.49 ppm. A similar, but inverse upfield–downfield displacement was noted for the β-pyrrolic protons; these resonances move from -1.21 ppm to 8.87 ppm ($\Delta\delta = 10.08$ ppm). These spectral changes, it was concluded, result from pyrrole-ring rotations that lead to the stabilization of a more "traditional" conformation in which all five of the pyrrolic nitrogen atoms are pointed into the macrocycle core (e.g., **5.50b**, Scheme 5.2.7).

Scheme 5.2.4

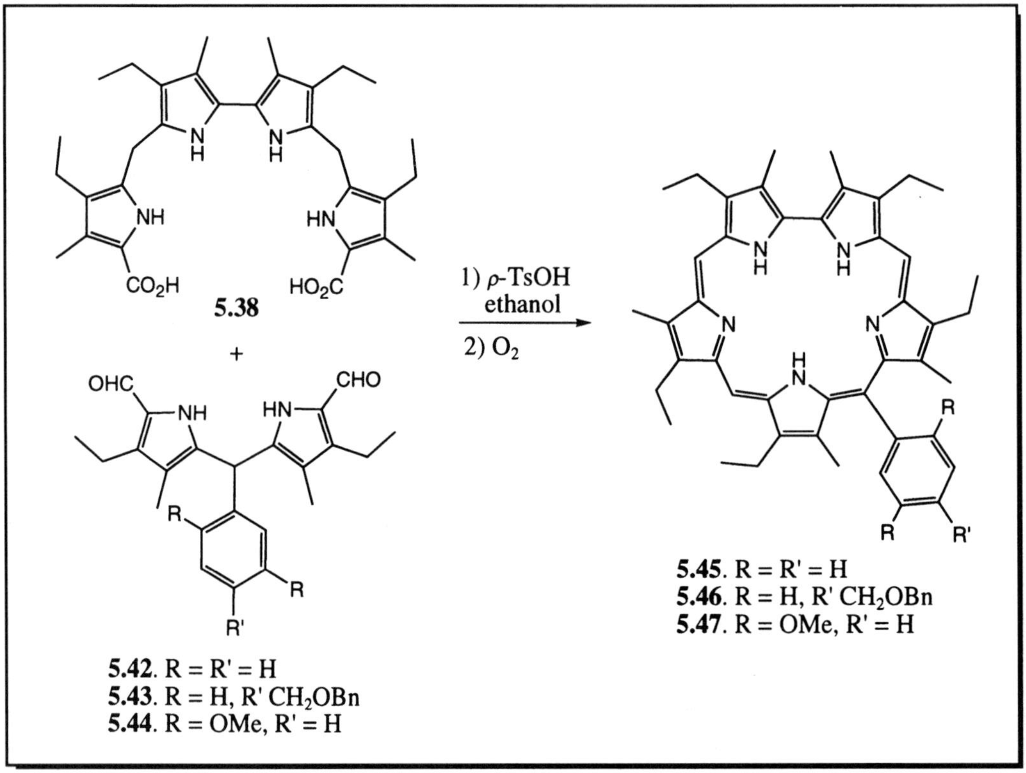

5.38

1) *p*-TsOH
   ethanol

2) O$_2$

5.42. R = R' = H
5.43. R = H, R' CH$_2$OBn
5.44. R = OMe, R' = H

5.45. R = R' = H
5.46. R = H, R' CH$_2$OBn
5.47. R = OMe, R' = H

Scheme 5.2.5

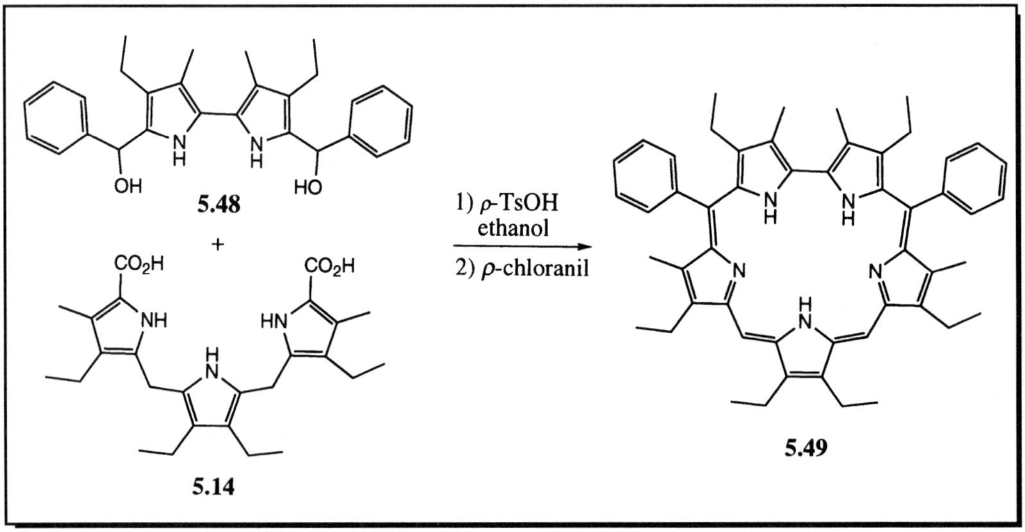

5.48

5.14

1) *p*-TsOH
   ethanol

2) *p*-chloranil

5.49

Scheme 5.2.6

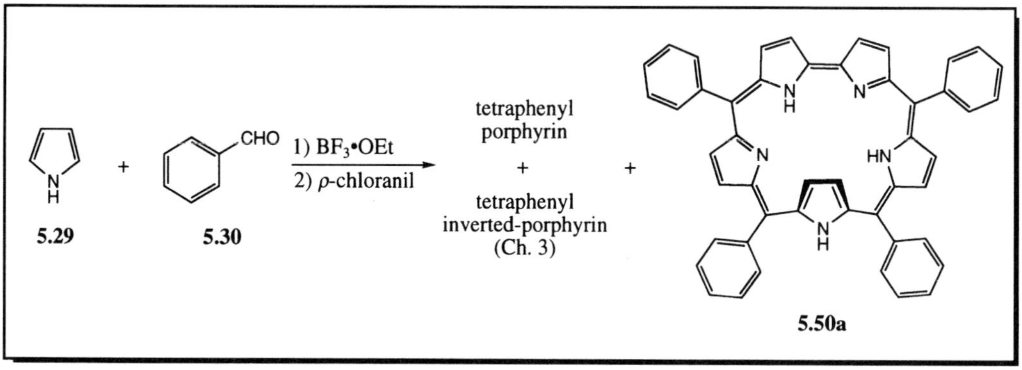

Scheme 5.2.7

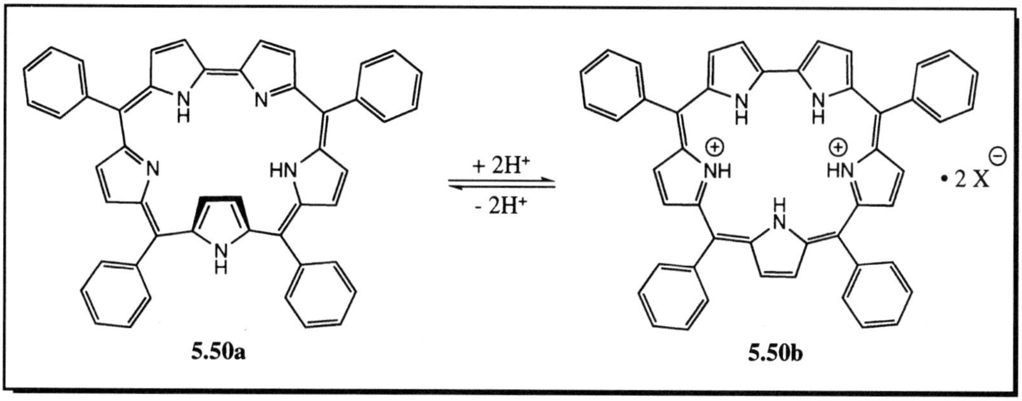

The methods described above for the synthesis of *meso*-arylsapphyrins are relatively new. This newness has so far limited the availability, and therefore the study, of aryl-substituted sapphyrin systems. This, however, is something that is likely to change in the near future, especially when one considers the important role that the "parent" *meso*-phenyl-substituted porphyrins have played in the development of numerous biological and materials science model systems.[28–53]

5.3   Heterosapphyrins

For the purposes of this review, heterosapphyrins are defined as being sapphyrin-like macrocycles in which a furan, thiophene, or selenophene subunit serves to replace one or more of the pyrrole rings. Interestingly, like the pentaazasapphyrins, this class of macrocycles has its origins in a serendipitous discovery. It occurred during efforts directed toward the synthesis of the heteroatom analogs of corrole. Specifically, as first reported by Johnson and coworkers in 1969,[54] reaction of diformylbifuran **5.51** with either of the dicarboxyl-substituted dipyrrylmethanes **5.52** or **5.53** leads not only to the "expected" dioxacorroles **5.56** or **5.57**, but also to small quantities (5–9%) of the corresponding dioxasapphyrins **5.54** and **5.55** (Scheme 5.3.1). The formation of sapphyrin derivative during these reactions, which was not at all anticipated, was rationalized after the fact as coming from a cleavage-recombination process analogous to that known to take place in simple dipyrrylmethanes.[55]

Scheme 5.3.1

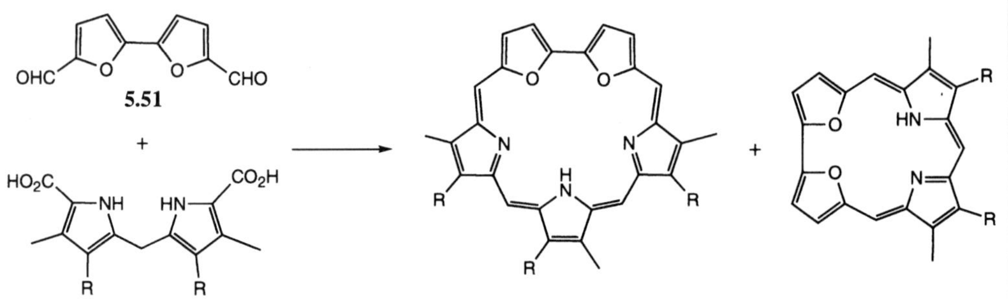

5.52. R = Me	**5.54**. R = Me	**5.56**. R = Me
5.53. R = Et	**5.55**. R = Et	**5.57**. R = Et

     In order to confirm the structure of the heterosapphyrins that were presumed to have formed during the course of the above reactions, Johnson and coworkers carried out a "3 + 2" reaction analogous to that now commonly used to prepare pentaazasapphyrins (*vide supra*). Here, it was found that reacting diformylbifuran **5.51** with the dicarboxyl-substituted tripyrrane **5.58** did indeed produce the hexaalkyl sapphyrin derivative **5.60** (Scheme 5.3.2).[54–56] Johnson, *et al.* also used this approach to prepare the thiasapphyrin analog **5.61**.[4] This sapphyrin was obtained in 19.5% yield from diformyl bipyrrole **5.9** and the dipyrrolylthiophene **5.59**.

     Recently, the above "3 + 2" approach has been used to prepare a wide range of heterosapphyrins.[9,10,57–59] In fact, in direct analogy to the early work of Johnson and Grigg, sapphyrins **5.68–5.73**[†] containing furan-, thiophene-, and selenophene rings were prepared from the corresponding diformyl precursors **5.51** and **5.10** and tripyrrane intermediates **5.14**, **5.63–5.66** (Scheme 5.3.3).[9,10,57,58] Unfortunately, as was true for the original heterosapphyrins of Johnson and coworkers these new products in all cases were prepared using precursors that gave rise to final heterosapphyrin products lacking β-substituents on the heterorings. While this made the syntheses more streamlined, it precluded direct comparisons to the extensively studied all-alkyl pentaazasapphyrin parents. To address this issue, considerable effort has been invested of late in an effort to prepare decaalkyl heterosapphyrins containing one, two, and three furan rings (e.g., **5.74–5.76**).[59] This was accomplished by

---

[†]The structures of precursor **5.63** and sapphyrins **5.71** and **5.72** were shown incorrectly in reference 10; they are shown correctly in Scheme 5.3.3.

using the β-alkyl-substituted furan precursors **5.62** and **5.67**, either together to afford **5.76**, or in conjunction with the appropriate pyrrole-containing educt (i.e., **5.10** and **5.67**) to afford **5.74** and **5.75**, respectively. As expected, complete β-substitution provides a significant increase in the stability and solubility of these latter hetero-sapphyrins compared to their β-unsubstituted congeners.

Scheme 5.3.2

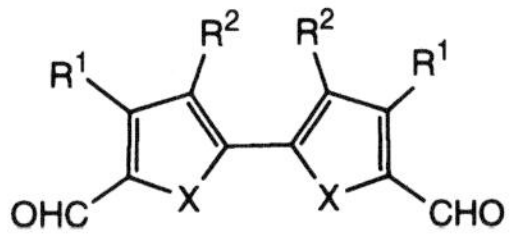

5.51. X = O, R^1 = R^2 = H
5.9. X = NH, R^1 = Me, R^2 = Et

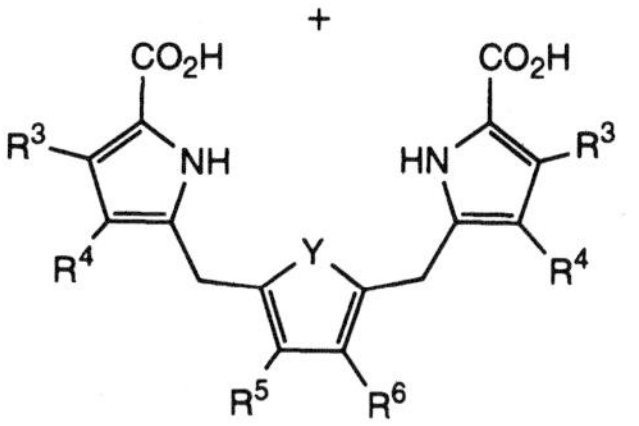

5.58. Y = NH, R^3 = R^6 = Me, R^4 = R^5 = Et
5.59. Y = S, R^3 = Me, R^4 = Et, R^5 = R^6 = H

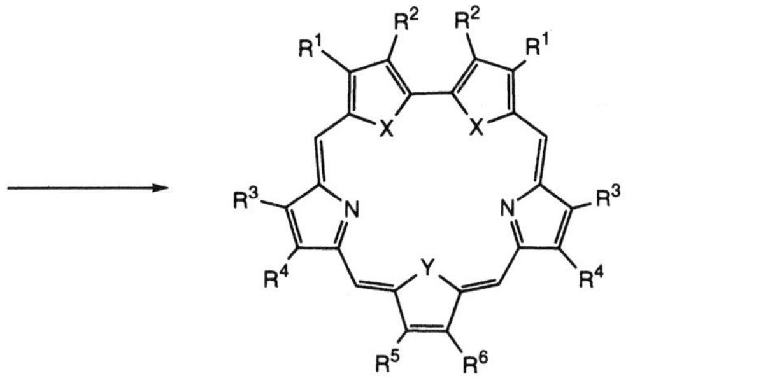

5.60. X = O, Y = NH, R^1 = R^2 = H, R^3 = R^6 = Me, R^4 = R^5 = Et
5.61. X = NH, Y = S, R^1 = R^3 = Me, R^2 = R^4 = Et, R^5 = R^6 = H

Scheme 5.3.3

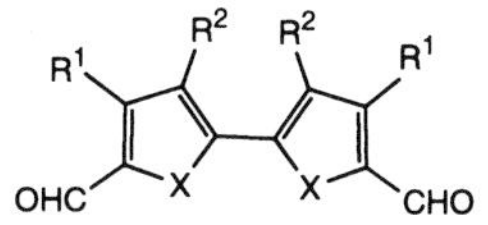

5.51. X = O, R^1 = R^2 = H
5.10. X = NH, R^1 = Et, R^2 = Me
5.62. X = O, R^1 = Et, R^2 = Me

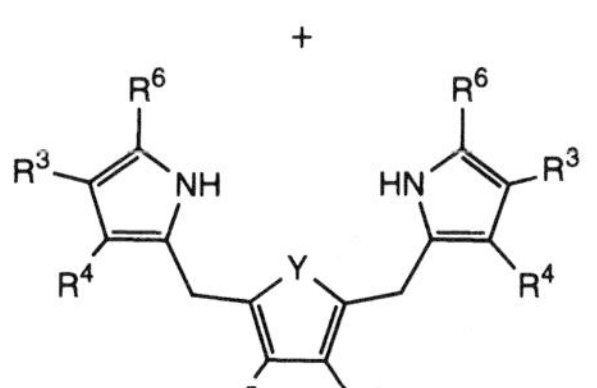

5.14. Y = NH, R^3 = Me, R^4 = R^5 = R^6 = Et, R^6 = CO$_2$H
5.63. Y = S, R^3 = Et, R^4 = Me, R^5 = R^6 = H
5.64. Y = S, R^3 = Et, R^4 = Me, R^5 = H, R^6 = CO$_2$H
5.65. Y = O, R^3 = Et, R^4 = Me, R^5 = H, R^6 = CO$_2$H
5.66. Y = Se, R^3 = Et, R^4 = Me, R^5 = R^6 = H
5.67. Y = O, R^3 = Et, R^4 = R^5 = Me, R^6 = CO$_2$H

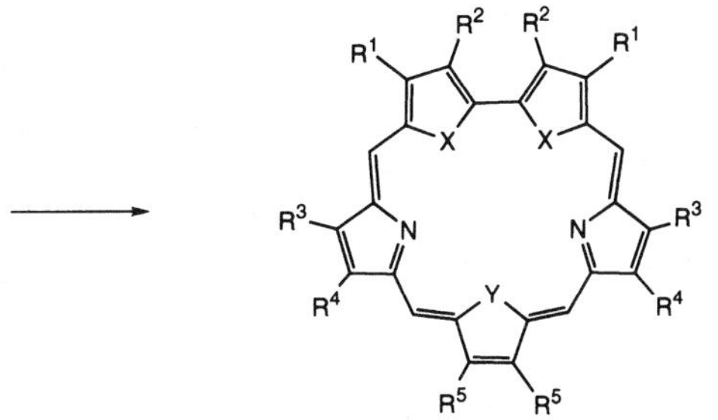

5.68. X = O, Y = NH, R^1 = R^2 = H, R^3 = Me, R^4 = R^5 = Et
5.69. X = O, Y = S, R^1 = R^2 = R^5 = H, R^3 = Et, R^4 = Me
5.70. X = Y = O, R^1 = R^2 = R^5 = H, R^3 = Et, R^4 = Me
5.71. X = NH, Y = S, R^1 = R^3 = Et, R^2 = R^4 = Me, R^5 = H
5.72. X = NH, Y = O, R^1 = R^3 = Et, R^2 = R^4 = Me, R^5 = H
5.73. X = NH, Y = Se, R^1 = R^3 = Et, R^2 = R^4 = Me, R^5 = H
5.74. X = NH, Y = O, R^1 = R^3 = Et, R^2 = R^4 = R^5 = Me
5.75. X = O, Y = NH, R^1 = R^4 = R^5 = Et, R^2 = R^3 = Me
5.76. X = Y = O, R^1 = R^3 = Et, R^2 = R^4 = R^5 = Me

Scheme 5.3.4

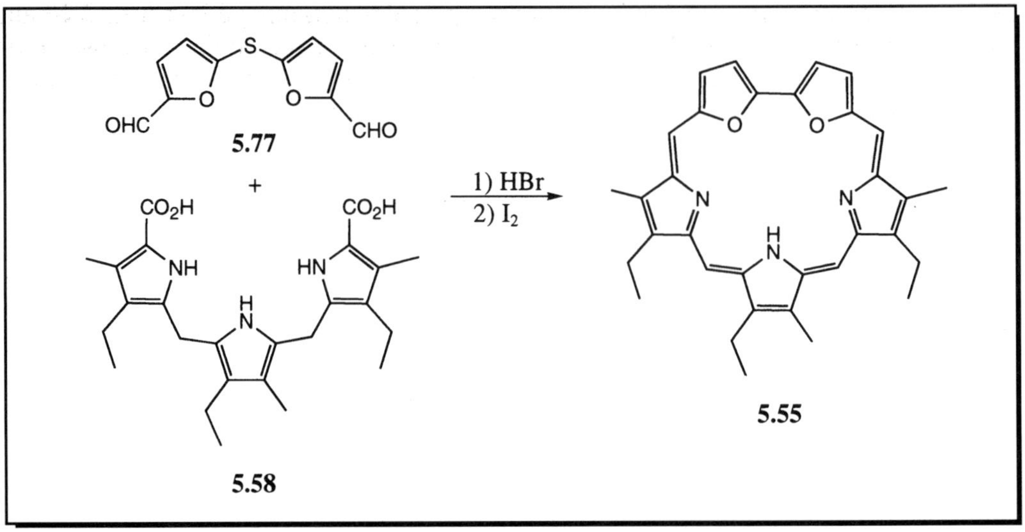

5.77

+

5.58

1) HBr
2) I$_2$

5.55

Scheme 5.3.5

5.78

5.79

[O]

5.80

5.81

5.55

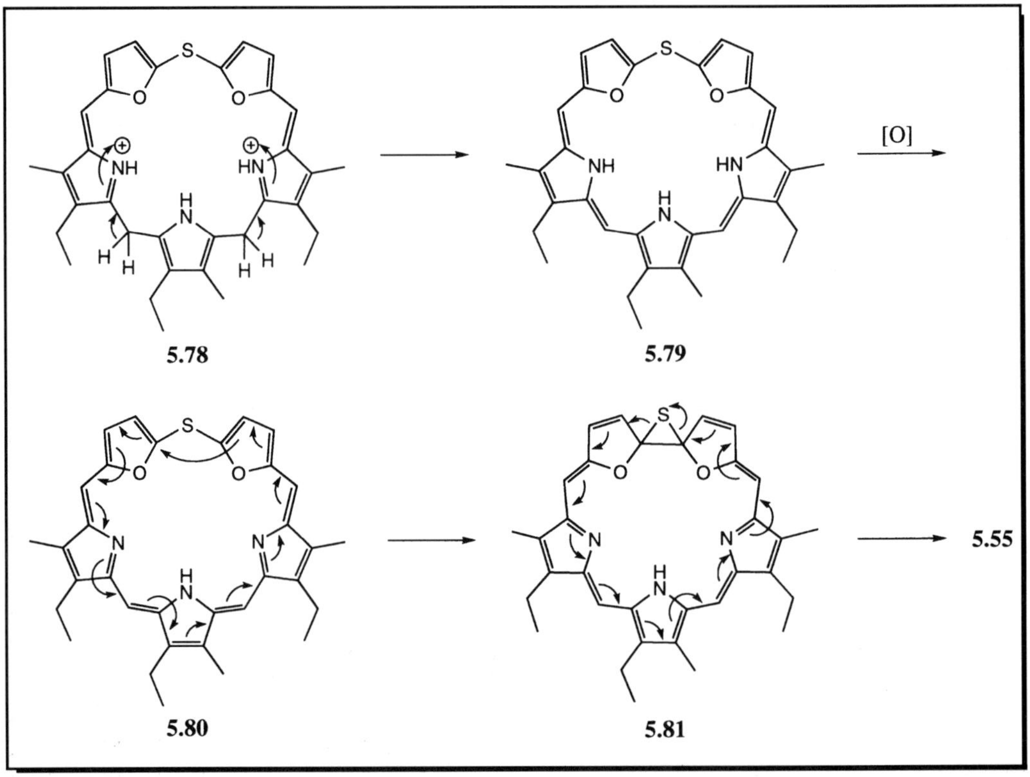

While the "3 + 2" route is now firmly entrenched as being the approach of choice, one other approach to heterosapphyrins recorded in the literature is worth noting. It involves the acid-catalyzed condensation between the bis(formylfuryl) sulfide **5.77** and the dicarboxyl tripyrrane **5.58**.[4,60] Following oxidation of the reaction mixture with I_2, the dioxasapphyrin **5.55** is obtained in 22% yield (Scheme 5.3.4). The presumed sulfur-bridged macrocycle **5.78** is not isolated in the course of this reaction sequence. Rather, extrusion of the sulfur atom was proposed as proceeding spontaneously upon oxidation of the reaction mixture. The mechanism of the proposed extrusion (Scheme 5.3.5) is thus the same as that evoked to account for the formation of corrole from *meso*-thiaphlorin under similar reaction conditions.[4]

5.4   Optical Properties of Sapphyrins and Heterosapphyrins

5.4.1   *Sapphyrins and* meso-*Aryl-substituted Sapphyrins*

In its free-base form (e.g., **5.21**), pentaazasapphyrin has three pyrrole-like nitrogen atoms that are protonated and two pyrrolidine (i.e., pyridine-like) nitrogen atoms that are unprotonated. These latter two nitrogen atoms can be sequentially protonated by weak acids to form the mono- and diprotonated salts **5.21a** and **5.21b**, respectively (Scheme 5.4.1). Precise pK_{a1} and pK_{a2} values have been determined for each of these two pyridine-like nitrogen atoms in the case of the water-soluble sapphyrin derivative **5.82** (Figure 5.4.1).[17] These values were found to be 4.8 and 8.8, respectively, using a procedure that involved titrating an aqueous 2.5% Tween 20 solution of the bis-HCl salt of sapphyrin **5.82** with NaOH. Prior to this recent study, approximate pK_a were determined by monitoring the absorbance change of the Soret transition in the visible spectrum (at $\lambda = 451.5$ nm) upon mixing a CH_2Cl_2 solution of the bis-HCl salt of **5.21** with a pH-adjusted NaCl aqueous solution. Using this less precise procedure, pK_{a1} and pK_{a2} values of $\cong 3.5$ and $\cong 9.5$, respectively, were recorded.[61]

Scheme 5.4.1

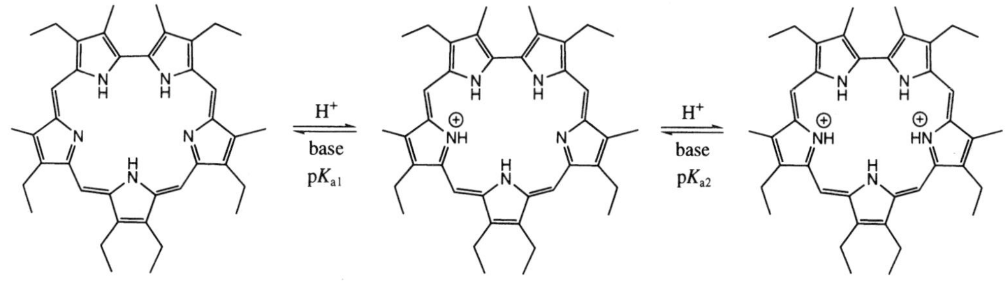

Figure 5.4.1   Structure **5.82**

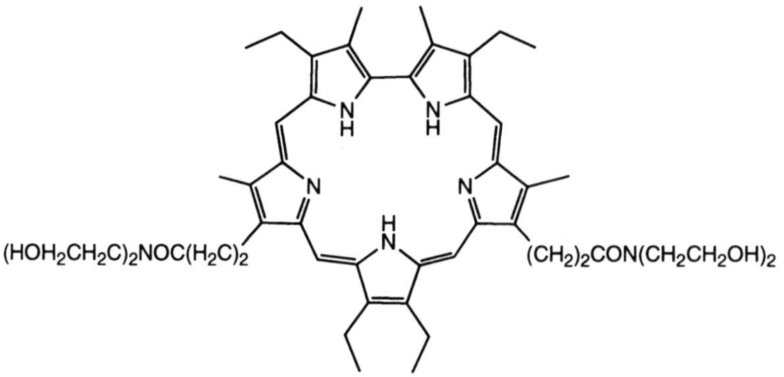

5.82

Like porphyrins, the sapphyrins are highly colored. This is true both for the free-base and for the various protonated salts. Unlike the porphyrins, however, the sapphyrins are not reddish purple. Rather, they are bright green when made up as organic solutions. From a more quantitative perspective, the free-base form of sapphyrins generally display an intense Soret-like absorbance band at *ca.* 450 nm.[10] This band is approximately 50 nm red-shifted relative to that of the porphyrins. This, of course, is consistent with the aromatic 22 π-electron formulation of sapphyrin (vs. 18 π-electrons in the case of porphyrin). Interestingly, the Soret-like band of sapphyrin shifts under conditions where dimerization or aggregation of the macrocycle is extensive. As discussed extensively in a recent publication, a Soret band at *ca.* 420 nm is considered indicative of a dimer, whereas one at *ca.* 445 nm reflects aggregation, a state that generally dominates at concentrations above *ca.* 1 mM.[17] On protonation, the Soret band of sapphyrin shifts slightly either to the red or blue, depending upon which acid is used, and its intensity nearly doubles relative to the free-base form (e.g., the molar extinction coefficient of the bis-HCl salt of 3,8,12,13,17,22-hexaethyl-2,7,18,23-tetramethylsapphyrin   is   537 000 $M^{-1} cm^{-1}$, whereas that of the free base is 281 800 $M^{-1} cm^{-1}$). However, the aggregation observed to occur for the free-base form of sapphyrin is less pronounced in the case of the dication.

The free-base form of sapphyrin also displays three relatively weak Q-type transitions in the 620–710 nm region of the visible spectrum. On protonation, these bands increase in intensity, and are generally blue-shifted to between 615 and 690 nm. Depending upon which acid is used, the number of Q-bands observed for protonated sapphyrins varies from two to four in number. These bands too are red-shifted relative to the Q-bands of porphyrins. In fact, because of this bathochromic shifting, the sapphyrins actually absorb light at the edge of the physiological "window of transparency"[62] that occurs between *ca.* 700 and 900 nm. This makes sapphyrin and its derivatives attractive candidates for use as PDT photosensitizers.[62-67] This potential application of expanded porphyrins is discussed in Chapter 10.

The *meso*-phenyl-substituted sapphyrins **5.33–5.35**, **5.37**, **5.45–5.47**, and **5.49–5.50** (as HCl salts) display UV-vis absorption bands that resemble qualitatively those of the better studied all β-alkyl-substituted sapphyrins.[19] For instance, 10,15-diphenylsapphyrin **5.33**, when dissolved in CH_2Cl_2, exhibits a Soret band that is bathochromically shifted by *ca.* 5–10 nm relative to that of the *meso*-unsubstituted alkylsapphyrins to 462 nm. A further bathochromic shift of *ca.* 30 nm (out to λ_{max} = 484 nm) is observed for the Soret band of the bis-HCl salt of *meso*-tetraphenylsapphyrin **5.50**.[27] Each of the phenyl-substituted sapphyrin salts also exhibit two to three (as opposed to four for alkyl-sapphyrin salts) weak Q-type bands in their absorption spectra. These bands fall in the 625–780 nm region and are thus red-shifted relative to the corresponding bands of alkyl-substituted sapphyrins.

5.4.2   *Heterosapphyrins*

Generally speaking, the spectral properties of the heterosapphyrins resemble those of the pentaaza "parents". That is to say, each of these species displays intense Soret- and Q-bands that are red-shifted relative to those of the porphyrins. For instance, when one, two, or even three of the pyrrolic subunits of sapphyrin is formally replaced by a 3,4-dialkyl furan (as in **5.74–5.76**), the absorption spectrum is relatively unaffected (λ_{max} for the free-base form is *ca.* 450 nm as compared to the λ_{max} value of 445 nm for typical decaalkyl sapphyrins such as **5.21**). Treatment of these oxasapphyrins with an acid such as HCl or HBr has only a minimal effect on the location of the Soret band; it is also found at *ca.* 450 nm in the case of these protonated systems.[59]

Removal (formally) of the β-furan substituents has some effect on the spectral characteristics of the oxasapphyrins. In the case of the monoxasapphyrin **5.72**, the position of the Soret band of the free-base appears relatively unchanged when compared to **5.21** or **5.74**. However, protonation with HCl results in a slight red-shift of the Soret bands (to 445/453 nm), while protonation with HF results in a more pronounced Soret band splitting (λ_{max} = 436 and 460 nm).[10] If two of the pyrroles of pentaazasapphyrin are formally replaced with β-unsubstituted furans (e.g., **5.68**), the Soret bands of both the free-base and protonated forms of the macrocycle shift to the blue by *ca.* 15 nm, relative to the β-substituted dioxa system **5.75**. In this latter instance, however, the Soret band does not split upon protonation.

Formal substitution of one of the pyrrole rings of pentaazasapphyrin with a β-unsubstituted thiophene (e.g., **5.71**) or selenophene (e.g., **5.73**) subunit results in a red-shifted (*ca.* 10 nm) Soret band (λ_{max} = 463 and 464 nm for the free-base forms of **5.71** and **5.73**, respectively). By contrast, the mixed heterosapphyrin **5.69** in which the bipyrrolic subunit of monothiasapphyrin **5.71** is formally replaced by a β-unsubstituted bifuran displays a Soret band at *ca.* 440 nm.[10]

Generally speaking, the UV-vis absorption spectrum of each of the sapphyrins and heterosapphyrins is similar, yet unique. While the precise reasons for the subtle difference in absorption spectral properties are not yet fully clear, it is likely that the effects of heteroatom substitution, aggregation, counter ion, and substitution all come into play. However, the relative extent to which each of these factors contributes in each particular instance remains to be investigated systematically.

5.5   Metal Complexes of Sapphyrins

Because of their basic resemblance to porphyrins, it was initially expected that the sapphyrins would mimic, at least on some level, the rich coordination chemistry displayed by the porphyrins. However, the larger core size (*ca.* 5.5 Å inner N–N diameter vs. *ca.* 4.0 Å for porphyrins), the greater number of potentially chelating heteroatom centers, and the fact that pentaazasapphyrins when fully deprotonated are potentially trianionic ligands made sapphyrin a likely candidate for large metal chelation, particularly as a potential ligand for the trivalent lanthanides and actinides. Unfortunately, in spite of extensive effort, this hope remains largely unrealized. Nonetheless, some metal complexes of sapphyrins and heterosapphyrins have been successfully prepared and characterized. Their preparation and properties are reviewed in this section.

5.5.1   *Complexes with First-row Transition Elements*

Quite early on, investigations into the metal-chelating chemistry of the sapphyrins were carried out by Woodward and his group.[2] However, only a few poorly characterized metal complexes resulted from these efforts, and these primarily involved first-row transition elements. For instance, these workers found that, when decamethyl sapphyrin was treated with the acetate salts of Ni^{2+}, Fe^{2+}, Cd^{2+}, Mn^{2+}, Co^{2+}, and Zn^{2+} in the presence of sodium acetate, changes occurred in the visible spectra that could be interpreted in terms of complexes being formed from each of these metals. However, only the Co^{2+} and Zn^{2+}-containing species could be isolated. These species displayed Soret bands that were 10–15 nm bathochromically shifted as compared to the free-base sapphyrin (to $\lambda_{max} = 462$ and 468 for Co^{2+} and Zn^{2+}, respectively).

Unfortunately, crystals of X-ray diffraction quality for the above cobalt and zinc derivatives could not be obtained. Thus, tentative assignments of structure were made on the basis of mass spectrometric analysis. Such studies were consistent with the apparent complexation of only one metal atom by the sapphyrin core. They also revealed a parent ion for each of these complexes that was considered most in accord with the symmetrical structures represented by **5.83** and **5.84** (Figure 5.5.1). In other words, based on these analyses, Woodward and coworkers proposed cobalt(II) and zinc(II) sapphyrin complexes wherein only four of the five possible nitrogen centers interacted with the metal center, and only two of the three possible NH protons became lost upon complexation.[2]

Later, a more detailed study of the way in which zinc(II) is ligated by sapphyrin was carried out by Sessler, *et al.*[9] This was done *via* ^{1}H NMR spectroscopic
analysis. While the data obtained supported the general conclusion that zinc(II)
sapphyrins are in fact tetraligated monometalated species, they were also found to
be more precisely consistent with the presence of two isomeric tetraligated complexes
in equilibrium with one another. Thus, Sessler and coworkers suggested that the
static structural representation of Woodward (**5.83** above) might be better depicted
in terms of two isomeric and interconverting forms, namely **5.85a** and **5.85b** (Figure
5.5.2).

Figure 5.5.1   Structures **5.83** and **5.84**

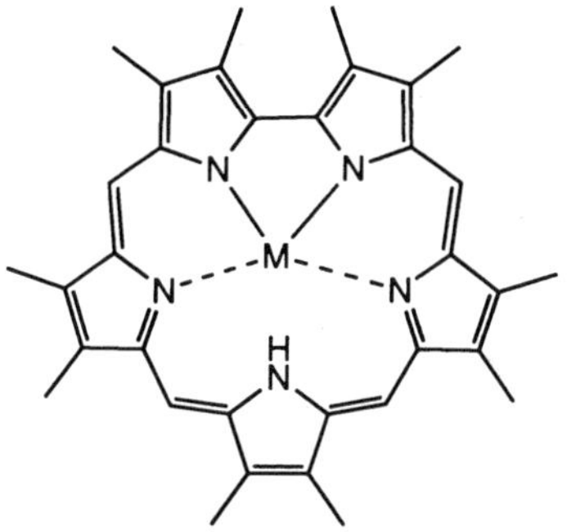

5.83. M = Zn
5.84. M = Co

Figure 5.5.2   Structures **5.85a** and **5.85b** showing the two observed isomers of the
tetraligated zinc(II) complex of sapphyrin

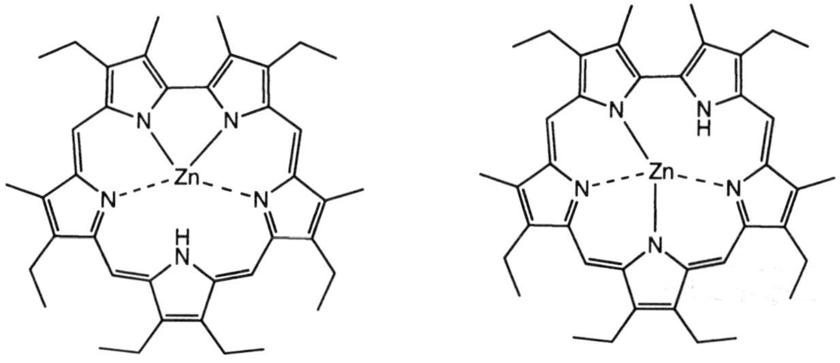

5.85a   **5.85b**

The tetraligation observed to occur for the zinc(II) and cobalt(II) sapphyrins **5.83–5.85** above was presumed to be the result of the size mismatch between the large core of sapphyrin and the relatively small size of these first-row transition metal cations. This led Sessler, *et al.* to direct efforts toward the complexation of second- and third-row transition metals. Initial reactions using metal halides, such as $HgCl_2$, $CdCl_2$, $PdCl_2$, $RhCl_3$, $IrCl_3$, and $RuCl_3$, resulted only in decomposition of the sapphyrin.[9] When $PdCl_2(CH_3CN)_2$ was used, however, a Pd(II)sapphyrin complex appeared to form. Unfortunately, this complex was found to undergo slow decomposition. Therefore, it could never be adequately characterized, nor could a precise structural assignment ever be made.

Successful metalation of sapphyrin using second- and third-row transition metals was finally achieved by Sessler, *et al.* by using the carbonyl salts of rhodium(I) and iridium(I) as the metalating agents.[68] Although no in-plane complexes were observed to form, a wide variety of new complexes did result. For instance, it was found that the mono-metalated sapphyrin adducts **5.86** and **5.87** are formed by treating the free-base form of 3,8,12,13,17,22-hexaethyl-2,7,18,23-tetramethylsapphyrin (**5.21**) with 0.5 equivalents of $Rh_2(CO)_4Cl_2$ or 1 equivalent of $Ir(CO)_2(Py)Cl$ in dichloromethane with triethyl amine as an added base (Scheme 5.5.1). Likewise, the $Rh(CO)_2$ complex **5.88** could be obtained by reacting 7,18-dimethyl-3,22-dipropyl-8,12,13,17-tetraethylsapphyrin (**5.23**) with $Rh_2(CO)_4Cl_2$. In general, it was the protonated, cationic forms of these complexes **5.89–5.91** that were isolated during chromatographic purification over silica gel. However, treatment of these adducts with triethyl amine served in all cases to regenerate the neutral complexes **5.86–5.88**.

Proton NMR spectral analysis of the protonated complexes **5.89–5.91** revealed highly unsymmetrical structures, each of which appeared (on the basis of integration studies) to contain three internal NH protons. For these reasons, a "sitting-a-top" structure was proposed for these complexes. Subsequently, the out-of-plane metal ligation that these structures required was confirmed *via* a single crystal X-ray diffraction study of complex **5.90**. The resulting structure is shown in Figure 5.5.3.[68]

Treating the mono-metalated adducts **5.86–5.88** with a further molar equivalent of the respective metal carbonyl salts affords the bis-metalated species **5.92–5.94**.[68] Alternatively, these same bimetallic complexes could be prepared directly from the free-base sapphyrins **5.21** and **5.23** by treatment with an excess of the metal carbonyl precursor. The structures of these bimetallic complexes were elucidated by spectroscopic means, and were confirmed *via* single crystal X-ray diffraction analysis of the $[Rh(CO)_2]_2$ sapphyrin complex **5.92** (Figure 5.5.4). Interestingly, the heterobimetallic complex **5.95** can also be prepared by treating either $Rh(CO)_2$ sapphyrin **5.86** with $Ir(CO)_2(Py)Cl$ or by treating $Ir(CO)_2$ sapphyrin **5.87** with $Rh_2(CO)_4Cl_2$. In this instance, the structural assignment was made on the basis of spectroscopic, mass spectrometric, and microanalytic means.

With the exception of the mono-rhodium complexes of the *n*-propylsapphyrins (i.e., **5.88** and **5.91**), each of the above mono- and bimetallic complexes proved to be

Scheme 5.5.1

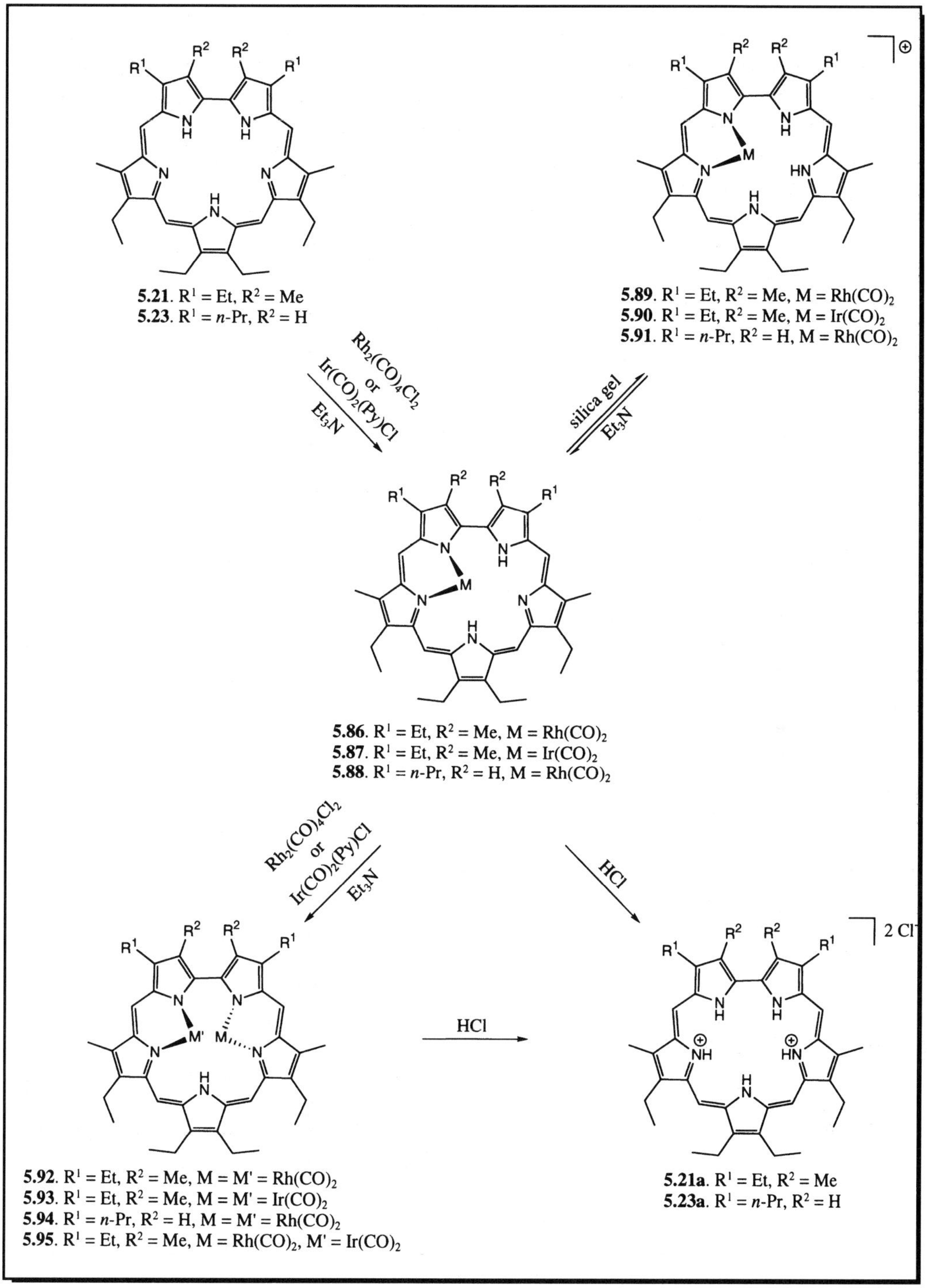
5.21. R^1 = Et, R^2 = Me
5.23. R^1 = *n*-Pr, R^2 = H
5.89. R^1 = Et, R^2 = Me, M = $Rh(CO)_2$
5.90. R^1 = Et, R^2 = Me, M = $Ir(CO)_2$
5.91. R^1 = *n*-Pr, R^2 = H, M = $Rh(CO)_2$
$Rh_2(CO)_4Cl_2$ or $Ir(CO)_2(Py)Cl$
Et_3N
silica gel
Et_3N
5.86. R^1 = Et, R^2 = Me, M = $Rh(CO)_2$
5.87. R^1 = Et, R^2 = Me, M = $Ir(CO)_2$
5.88. R^1 = *n*-Pr, R^2 = H, M = $Rh(CO)_2$
$Rh_2(CO)_4Cl_2$ or $Ir(CO)_2(Py)Cl$
Et_3N
HCl
HCl
2 Cl⁻
5.92. R^1 = Et, R^2 = Me, M = M' = $Rh(CO)_2$
5.93. R^1 = Et, R^2 = Me, M = M' = $Ir(CO)_2$
5.94. R^1 = *n*-Pr, R^2 = H, M = M' = $Rh(CO)_2$
5.95. R^1 = Et, R^2 = Me, M = $Rh(CO)_2$, M' = $Ir(CO)_2$
5.21a. R^1 = Et, R^2 = Me
5.23a. R^1 = *n*-Pr, R^2 = H

remarkably stable.[68] For instance, no thermal or photolytic loss of CO could be effected from complexes **5.89, 5.91, 5.92**, and **5.94**. Also, complexes **5.92–5.94** were found to react only slowly with I_2. Interestingly, after reaction with I_2, the products were found to have metal carbonyl moieties removed from the starting complex. Complete demetalation, if desired, could be achieved by treating complexes **5.86–5.95** with highly acidic reagents such as HCl. Such a procedure results in "regeneration" of the diprotonated forms of the starting sapphyrins (i.e., **5.21a** and **5.23a**).

Figure 5.5.3   Single Crystal X-Ray Diffraction Structure of the Sapphyrin-Dicarbonyliridium(I) Complex **5.90**.

This figure was generated using information down-loaded from the Cambridge Crystallographic Data Centre and corresponds to a structure originally reported in reference 68. The two Ir–N bonds are roughly equivalent in this complex and are 2.08 Å in length. Atom labeling scheme: iridium: ◍; carbon: ○; nitrogen: ●; oxygen: ⊖. Hydrogen atoms have been omitted for clarity

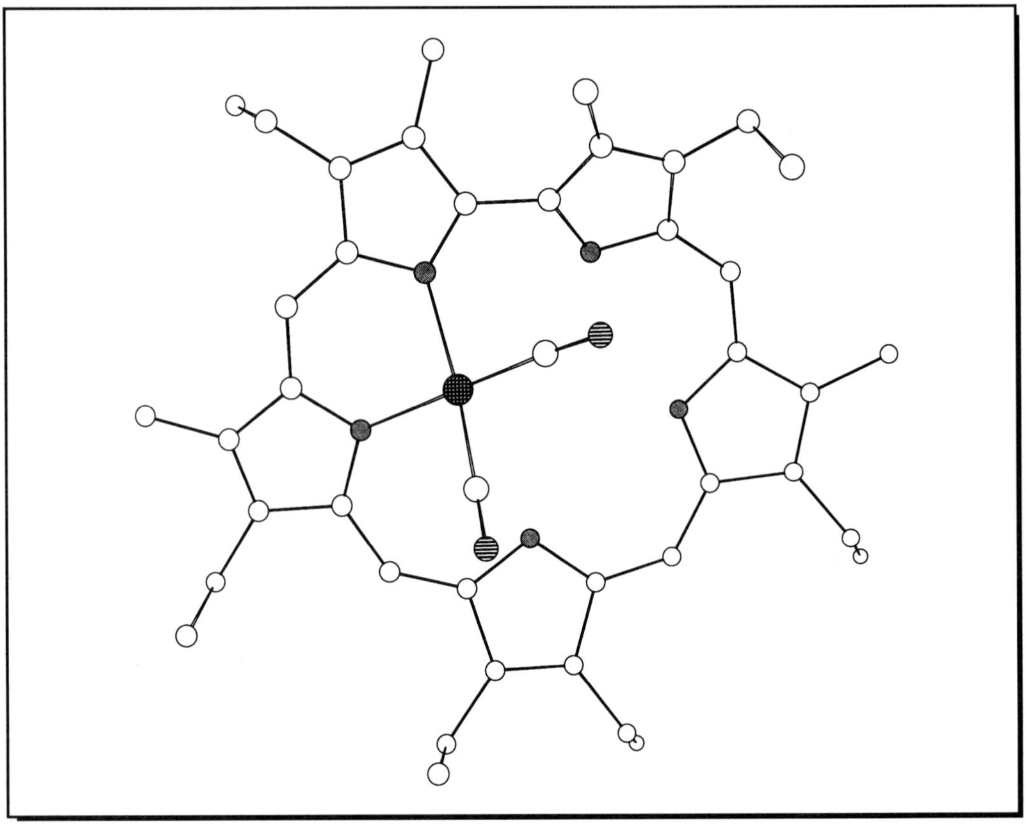

Figure 5.5.4   Single Crystal X-Ray Diffraction Structure of the Sapphyrin–Bis[dicarbonylrhodium(I)] Complex **5.92**.
This figure was generated using information down-loaded from the Cambridge Crystallographic Data Centre and corresponds to a structure originally reported in reference 68. The $Rh(CO)_2$ moieties are found on opposite faces of the macrocyclic plane. Each rhodium center is coordinated to two pyrrole nitrogen atoms and the average Rh–N distance is 2.08 Å. Atom labeling scheme: rhodium: ◕; carbon: ○; nitrogen: ●; oxygen: ◒. Hydrogen atoms have been omitted for clarity

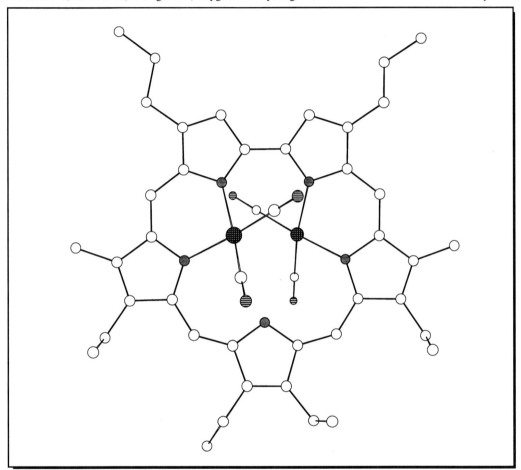

5.5.3   *Uranyl Cation Coordination by Modified Sapphyrins*

Consistent with the original supposition that sapphyrins should be good ligands for large cations, Woodward predicted that sapphyrins might form particularly stable chelate complexes with uranyl cation (UO_2^{2+}), a species that is known to favor pentagonal planar coordination environments.[69-71] Unfortunately, this promise could never be realized. Specifically, Woodward and coworkers failed in their efforts to prepare a *bona fide* sapphyrin·UO_2 complex. (They also proved unable to insert other large divalent metals, such as Pb^{2+}, into the sapphyrin core.[2])

Several years after the initial report of Woodward, *et al.*, Sessler and co-workers found that sapphyrin **5.21** would in fact form a stable complex with uranyl cation.[72] It was determined, however, that UO_2^{2+} complexation was accompanied by reduction of the sapphyrin macrocycle. Specifically, it was found that metal insertion occurred concurrent with addition of methanol to one of the *meso*-like carbon centers. The net result was an overall neutral complex of a modified, non-aromatic sapphyrin-like system (Scheme 5.5.2).

Scheme 5.5.2

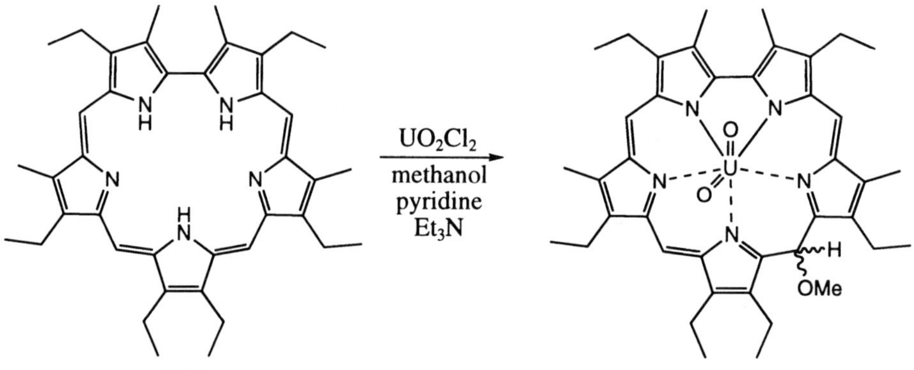

5.21　　　　　　　　　　　**5.96**

Although unexpected, the above reduction and methanol addition process does appear to be well established. For instance, the relevant product, **5.96**, has been subject to X-ray diffraction analysis (Figure 5.5.5). The structure so obtained revealed that complex **5.96** is "bowl" shaped, and does, indeed, possess a *meso*-position that has been converted into a methoxy-bearing sp^3-hybridized center. Consistent with this structure and the addition of methanol it implies, is the observed loss of a Soret-like band in the visible spectrum of **5.96**. Also consistent with it is the complex nature of the ^1H NMR spectrum. Complex **5.96** actually exists as four stereoisomers (an enantiomeric set of diastereomers). These stereoisomers were found to interconvert readily in solution. This interconversion is slow on the NMR time scale. Thus, all species are seen in the ^1H NMR spectrum and this, as well as the loss in symmetry, contributes to the complexity of the ^1H NMR spectrum.

Through a series of experiments, Sessler and coworkers were able to determine a likely mechanism for the formation of complex **5.96**. This mechanism,[72] shown in Scheme 5.5.3, involves the initial coordination of uranyl cation to sapphyrin giving a complex such as **5.97** or **5.98**. Addition of methanol (or methoxide anion) to one of the *meso*-positions then results in the formal reduction of the U(VI) to a presumed U(IV) complex such as **5.99**. Finally, oxidation by air gives the U(VI) complex **5.96**.

Figure 5.5.5 Single Crystal X-Ray Diffraction Structure of Uranyl Complex **5.96** Confirming the Formal Attack by Methoxide at the Macrocyclic C(10) Position. This figure was generated using information down-loaded from the Cambridge Crystallographic Data Centre and corresponds to a structure originally reported in reference 72. The macrocycle is severely distorted from planarity in the crystal and is bowl-shaped. Two diastereomeric forms of this complex are found in the crystal lattice in a 7:3 ratio. The average U–N bond length is 2.50 Å. Atom labeling scheme: uranium: ◉; carbon: ○; nitrogen: ◉; oxygen: ◒. Hydrogen atoms have been omitted for clarity

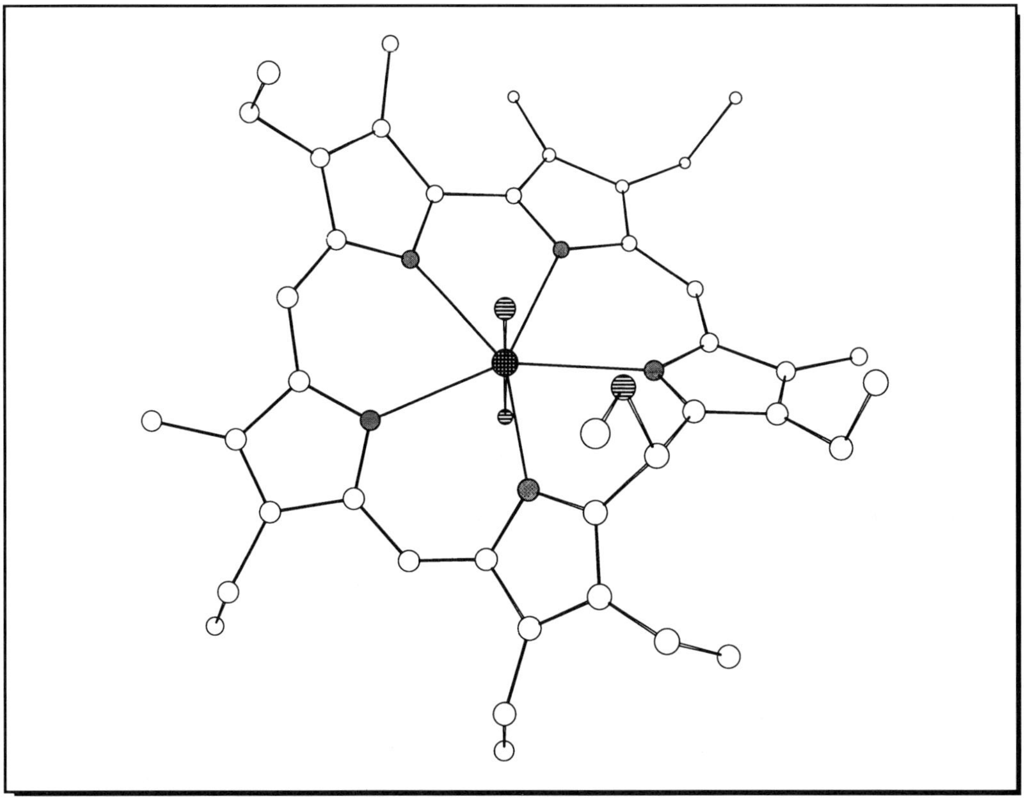

Sapphyrins bearing peripheral substituents other than those present in **5.21** also react with uranyl cation in methanol to form complexes akin to **5.96**.[72] Analogous reactions, wherein the methoxide nucleophile is replaced by cyanide were also described by Sessler and coworkers. Unfortunately, when cyanide anion is used as the nucleophile, the reaction yield was very poor. It thus proved impossible to isolate and characterize fully the resulting uranyl complex.

Scheme 5.5.3

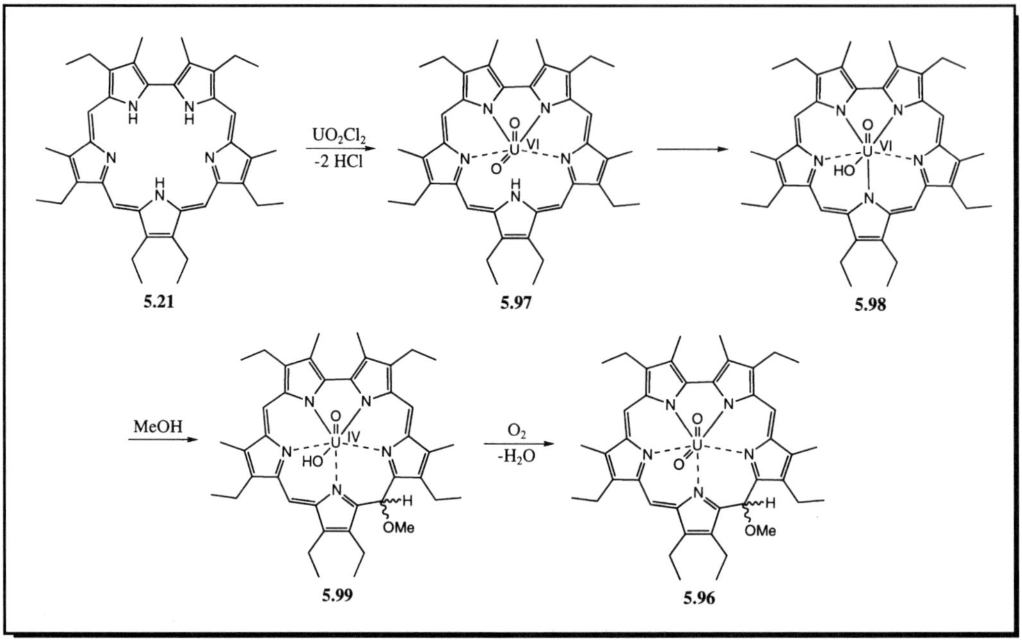

5.5.4   *Metal Complexes of Heterosapphyrins*

Because the heterosapphyrins contain five potentially ligating heteroatoms, they have also been studied as possible pentaligating ligands. Unfortunately, these efforts, like those with the pentaazasapphyrin "parents", have proved disappointing. For instance, in what was the first report of an attempt to prepare a metallo-hetero-sapphyrin, Johnson and coworkers found in 1972 that attempted metalation of dioxasapphyrin **5.55** (Scheme 5.3.1) with nickel(II) or zinc(II) salts did not give rise to the formation of a stable complex.[4] It was presumed that the larger ring size had a destabilizing effect on metal complexation.

Perhaps as a consequence of Johnson's failure, 20 years elapsed before Sessler and coworkers reported that certain η^2-type complexes could in fact be formed with heterosapphyrins.[10,57] These latter workers were clearly inspired by their earlier successful syntheses of η^2-type rhodium(I) and iridium(I) carbonyl complexes of pentaazasapphyrins **5.21** and **5.23** (*vide supra*).[68] Thus, in a first experiment, they treated the monothiasapphyrin **5.71** with $Rh_2(CO)_4Cl_2$ (Scheme 5.5.4). This afforded the structurally characterized $[Rh(CO)_2]_2$ monothiasapphyrin complex **5.100** (Figure 5.5.6).[58] Subsequently, they prepared the bis-iridium complex **5.101** of the mono-selenasapphyrin **5.73**.[57] This complex was also characterized by X-ray diffraction analysis. The resulting structure then served to confirm the expected "sitting-a-top" binding mode (Figure 5.5.7).

Scheme 5.5.4

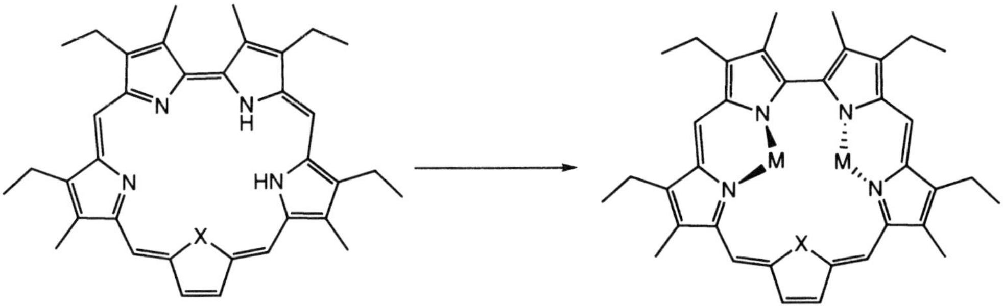

5.71. X = S
5.73. X = Se

5.100. X = S, M = Rh(CO)$_2$
5.101. X = Se, M = Ir(CO)$_2$

Figure 5.5.6   Single Crystal X-Ray Diffraction Structure of the
Monothiasapphyrin-Bis[dicarbonylrhodium(I)] Complex **5.100**.
This figure was generated using information down-loaded from the Cambridge Crystallographic
Data Centre and corresponds to a structure originally reported in reference 58. The Rh(CO)$_2$
moieties are found on opposite faces of the macrocyclic plane. Each rhodium center is coordinated
to two pyrrole nitrogen atoms and the average Rh–N distance is 2.08 Å. The Rh–S contact distance
is 3.16 Å. Atom labeling scheme: rhodium: ◕; carbon: ○; nitrogen: ◉; oxygen: ⊖; sulfur: ⓪.
Hydrogen atoms have been omitted for clarity

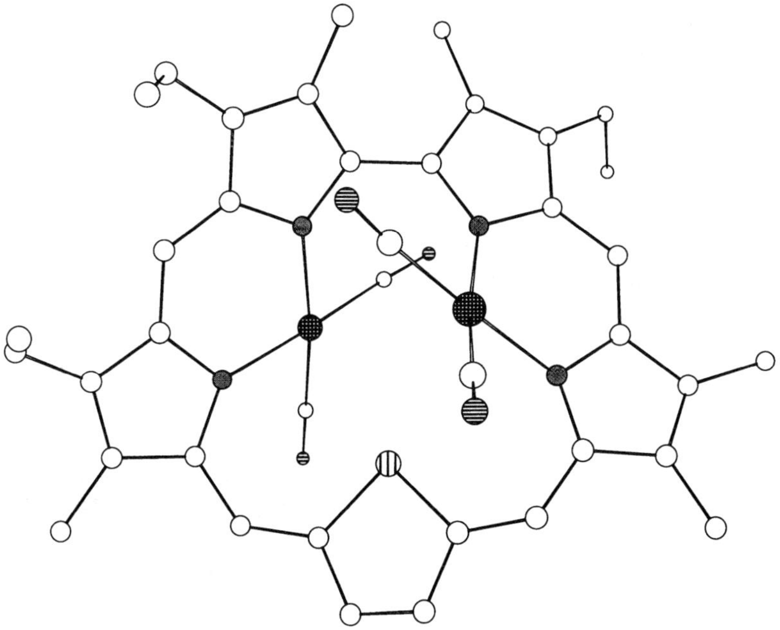

Figure 5.5.7   Single Crystal X-Ray Diffraction Structure of the
Monoselenasapphyrin-Bis[dicarbonyliridium(I)] Complex **5.101**.
This figure was generated using information down-loaded from the Cambridge Crystallographic
Data Centre and corresponds to a structure originally reported in reference 57. The Ir(CO)₂ moieties
are found on opposite faces of the macrocyclic plane. Each iridium center is coordinated to two
pyrrole nitrogen atoms and the average Ir–N distance is 2.09 Å. The Rh–S contact distance is 3.10
Å. Atom labeling scheme: iridium: ◕; carbon: ○; nitrogen: ●; oxygen: ◒; selenium: ⊘. Hydrogen
atoms have been omitted for clarity

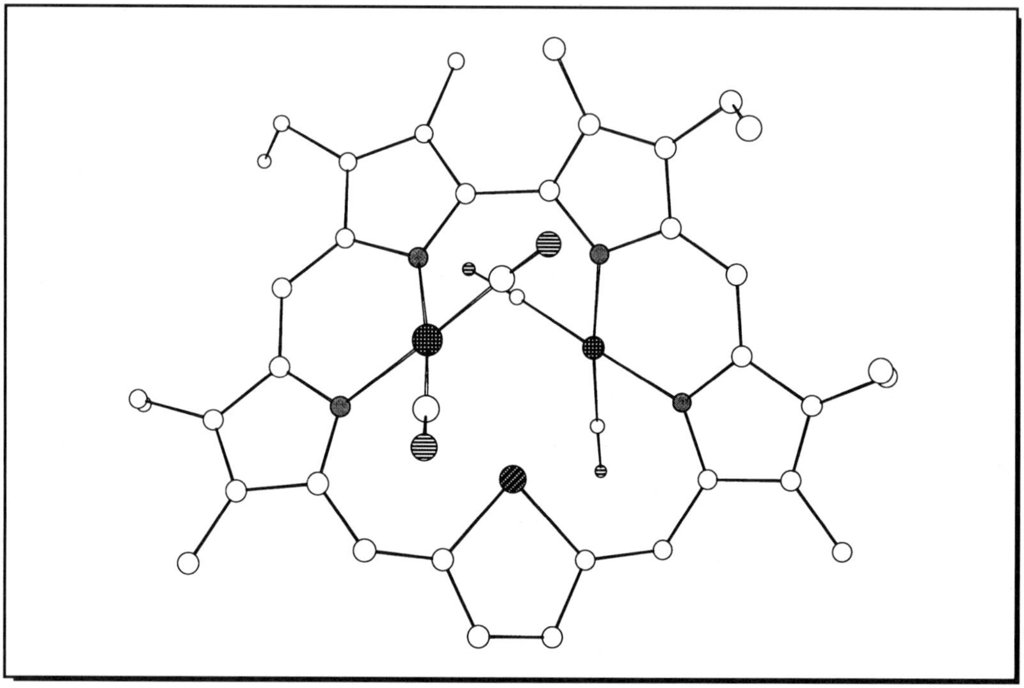

Finally, in work that deviates at least in terms of structural precedent from that
which was found in the case of pentaazasapphyrins, Sessler, *et al.* successfully prepared and characterized three cobalt(II) heterosapphyrin complexes.[58] The first of
these (complex **5.102**) was prepared by treating the monothiasapphyrin **5.71** with
cobalt(II) chloride (Scheme 5.5.5). In exact analogy, the cobalt(II) monoxasapphyrin
complexes **5.103** and **5.104** could be prepared from monoxasapphyrin **5.72** by reaction
with the appropriate cobalt(II) salt. In each of these cases, single crystal X-ray diffraction analysis established that the metal atom was indeed held within the mean
plane of what was essentially a planar macrocyclic frame (Figures 5.5.8 and 5.5.9).
These same analyses also served to reveal, however, that only two nitrogen atoms, of
the five potentially donating heteroatoms, were actually coordinated to the metal
center. Of further interest was the finding that the sapphyrin ligand in these complexes
is formally neutral. Thus, no deprotonation of the nitrogen atoms occurs upon metal
chelation. This, in turn, means that the heterosapphyrin ligand is only serving to fill a
tetrahedral coordination environment about the cobalt atom. In the case of the

Scheme 5.5.5

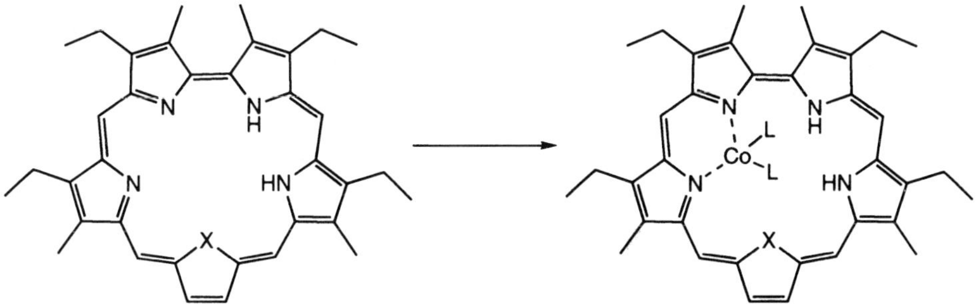

5.71. X = S
5.72. X = O

5.102. X = S, L = Cl⁻
5.103. X = O, L = Cl⁻
5.104. X = O, L = OAc⁻

Figure 5.5.8   Single Crystal X-Ray Diffraction Structure of the
Monothiasapphyrin-Cobalt(II) Dichloride Complex **5.102**.
This figure was generated using information down-loaded from the Cambridge Crystallographic
Data Centre and corresponds to a structure originally reported in reference 58. The cobalt atom sits
0.01 Å above the mean macrocyclic plane. The average Co–N distance: is 2.01 Å and the Co–S
distance is 2.23 Å. Atom labeling scheme: cobalt: ●; carbon: ○; nitrogen: ●; chlorine: ⊙; sulfur: ⊕.
Hydrogen atoms have been omitted for clarity

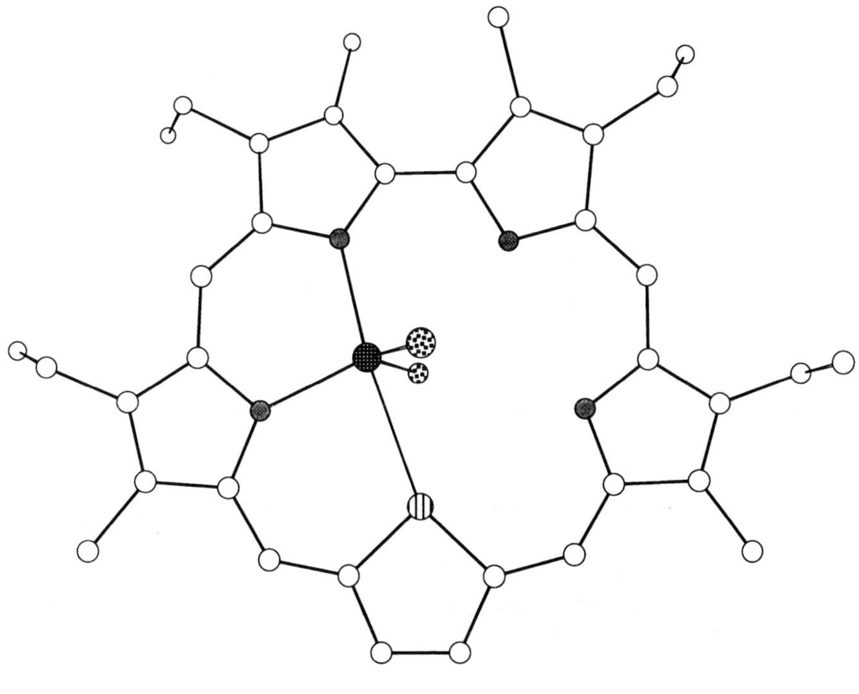

cobalt(II) heterosapphyrins **5.102–5.104**, it appears that the respective oxygen or sulfur atom participates in the coordination to the cobalt, as judged by contact distances in the crystal structure (cobalt–sulfur distance in **5.102** is 2.725 Å, and the corresponding cobalt–oxygen distance in **5.104** is 2.950 Å). Whether this apparent coordination is "real", however, is subject to debate; it could simply reflect constraints of the macrocyclic geometry that force these atoms into contact with the cobalt(II) cation.

Figure 5.5.9   Single Crystal X-Ray Diffraction Structure of the Monoxasapphyrin-Cobalt(II) Diacetate Complex **5.104**.

This figure was generated using information down-loaded from the Cambridge Crystallographic Data Centre and corresponds to a structure originally reported in reference 58. The cobalt atom sits 0.02 Å above the mean macrocyclic plane. The two relevant Co–N bonds are 1.96 Å and 2.02 Å, and the Co–O distance is 2.95 Å. The acetate anions are stabilized by hydrogen bonds (not shown) between the coordinated acetate groups and the NH-groups of the pyrroles not participating in metal ion coordination. Atom labeling scheme: cobalt: ●; carbon: ○; nitrogen: ●; chlorine: ⊗; sulfur: ⦻. Hydrogen atoms have been omitted for clarity

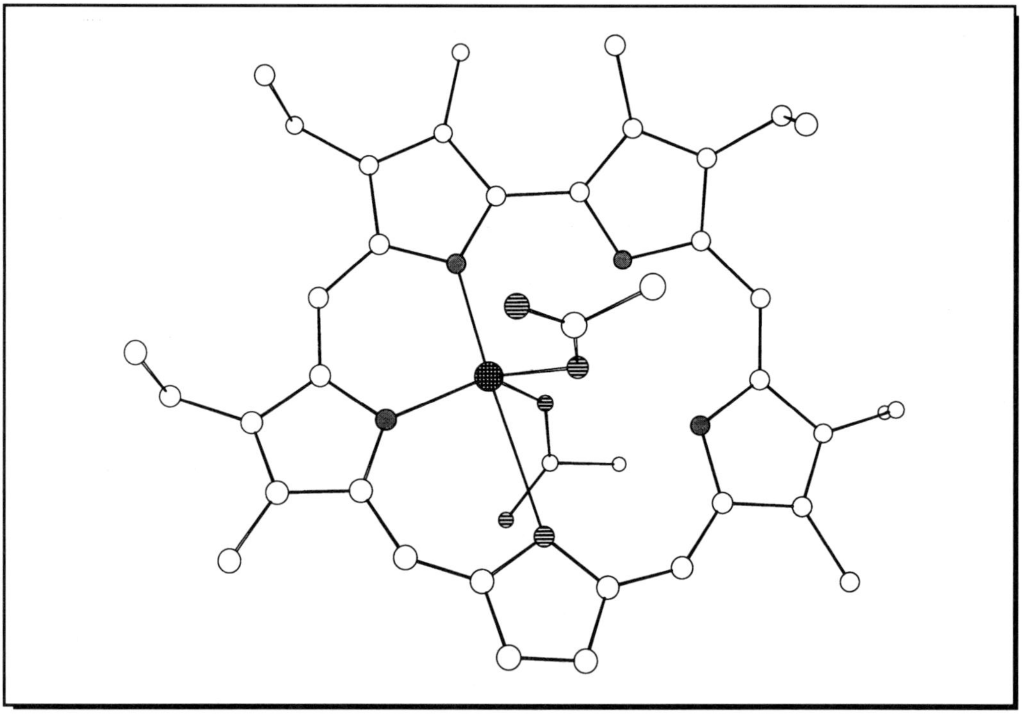

5.6   Three-dimensional Sapphyrins and Sapphyrin Conjugates

Enormous amounts of research over the past two decades have been devoted to the preparation of "three-dimensional" porphyrin-based systems such as the so-called "capped" porphyrins (e.g., **5.105**), "picket-fence" porphyrins (e.g., **5.106**), and porphyrin–porphyrin dimers and oligomers (e.g., **5.107** and **5.108**) (Figure 5.6.1).[28-53,73] Because of this interest, the chemistry of these systems is exceedingly well developed. Such systems have been studied as mimics of heme-based proteins, as models for the study of electron processes, reaction catalysts, and as molecular recognition units, among other things. By contrast, the chemistry of the "three-dimensional" and oligomeric expanded porphyrins is only in its infancy. In fact at present, only sapphyrin-based "capped" expanded porphyrins or expanded porphyrin oligomers have received significant synthetic and investigative attention. In spite of this, the very fact that such systems exist serves as an indication that "three-dimensional" expanded porphyrins can be made and could prove extremely interesting. In this section, the synthesis of the two known "capped" sapphyrins will be reviewed. First, however, the chemistry of the conceptually related "sapphyrin conjugates" will be described. These latter systems bear one or more functionalized "tails" and are thus intermediate between "capped" and unfunctionalized sapphyrins in terms of their categorization.

Figure 5.6.1   Structures **5.105–5.108**

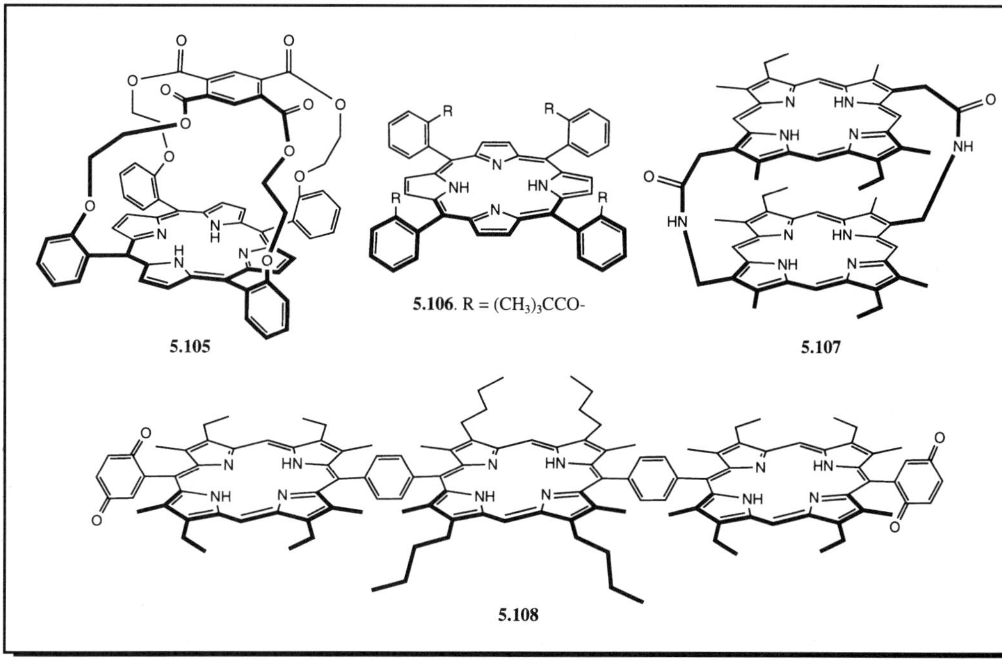

5.106. R = (CH₃)₃CCO-

5.105

5.107

5.108

5.6.1   *Sapphyrin Conjugates*

5.6.1.1   Sapphyrin-Nucleobase Conjugates

So far, several different kinds of sapphyrin conjugates have appeared in the literature.[74-78] The first of these to be reported were the sapphyrin-cytosine conjugates **5.112** and **5.118**.[74-76] These were prepared by reacting 1-(2-aminoethyl)-4-[triphenylmethyl)amino]pyrimidin-2-one **5.115** with an activated form of the appropriate sapphyrin mono- or dicarboxylic acid (**5.109** and **5.110**; Figure 5.6.2), followed by deprotection of the trityl groups with trifluoroacetic acid (Scheme 5.6.1). Activation of the carboxylic acid functional groups was achieved through the use of acid chlorides or *via* the use of carbodiimide or carbonyldiimidazole. The sapphyrin-guanosine conjugates **5.114** and **5.120** were subsequently prepared using sapphyrins **5.109** and **5.110** and benzoyl-protected aminoethylguanosine **5.116**. Here, procedures analogous to those used to prepare **5.112** and **5.118** were employed. The only major difference was that the protecting groups were removed from the guanosine subunits using ammonia.

Figure 5.6.2   Structures **5.109** and **5.110**

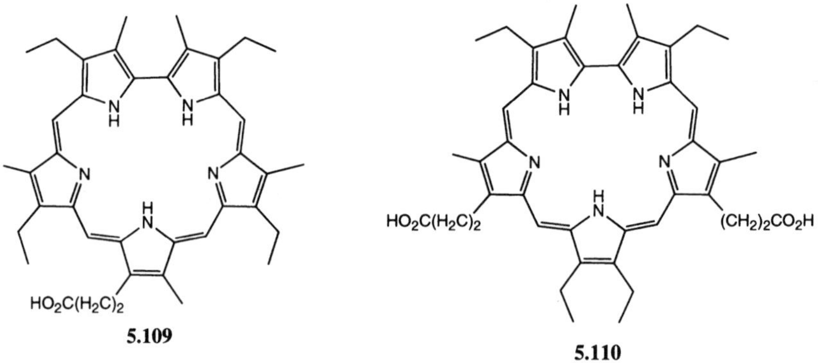

The above cytosine and the guanosine sapphyrin conjugates were designed to act as receptors and carriers for monophosphorylated nucleotides and nucleotide analogs.[74-76] The ditopic species **5.112** and **5.114** were intended to exploit both the general phosphate binding of the sapphyrin center and the specific base-pairing interactions made possible by the conjugated nucleoside moiety so as to bind strongly and specifically the targeted nucleotide substrate. The tritopic receptors **5.118** and **5.120** were designed to allow for the formation of triple-helix-like C-G-C and G-C-G complexation motifs. As it transpired, these presumptions proved to be borne out by experiment. Appropriate nucleobase-substituted sapphyrin derivatives did in fact act as efficient systems for the specific through-CH_2Cl_2 transport of complementary nucleotide monophosphates at neutral pH. The interested reader is directed to a recent review covering anion binding for a more thorough description of these results.[77]

Scheme 5.6.1

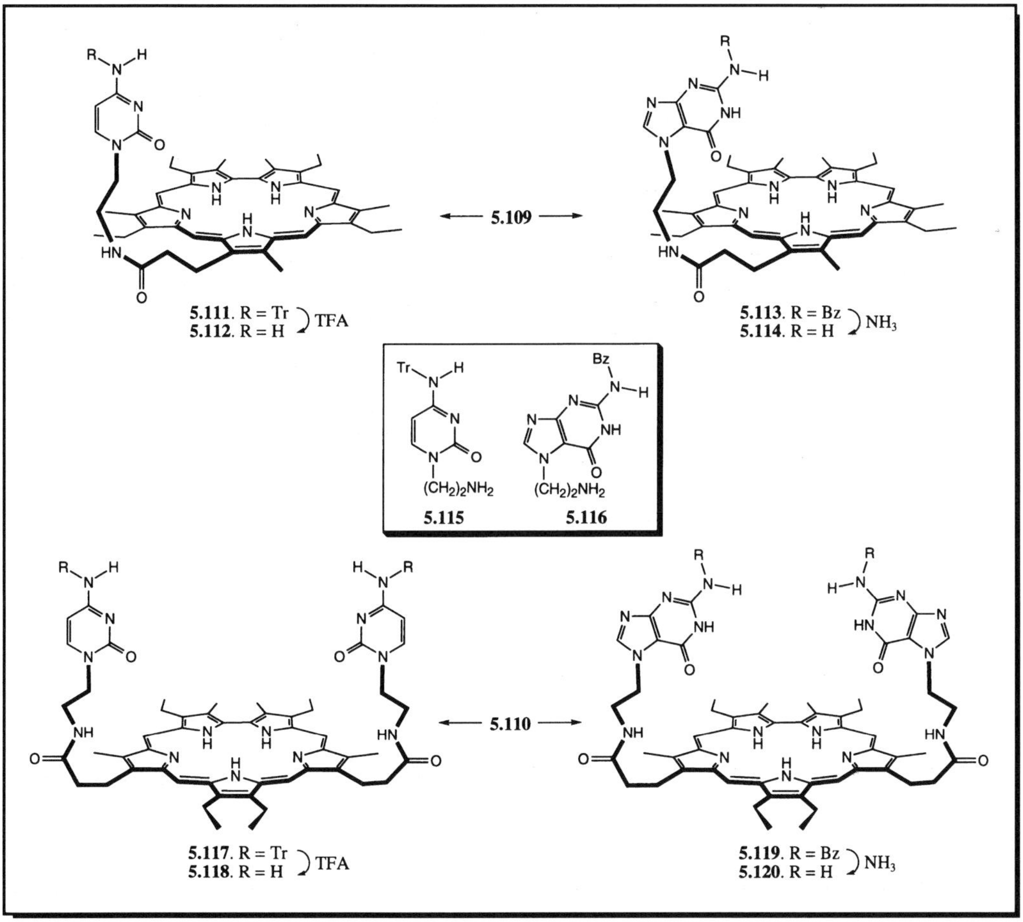

5.111. R = Tr ⎞ TFA
5.112. R = H ⎠

5.113. R = Bz ⎞ NH₃
5.114. R = H ⎠

5.115   5.116

5.117. R = Tr ⎞ TFA
5.118. R = H ⎠

5.119. R = Bz ⎞ NH₃
5.120. R = H ⎠

5.6.1.2   Sapphyrin–Oligonucleotide Conjugates

Sessler and coworkers have also prepared a first, prototypic sapphyrin–oligo-nucleotide conjugate.[78] It is the dodecadeoxythymidine-containing sapphyrin **5.123**. This material was prepared from the sapphyrin derivative **5.121** by first activating with pivaloyl chloride and then coupling to the 5′-end of a $(dT)_{12}$ oligonucleotide supported on control pore glass (CPG) beads (Scheme 5.6.2). This gave intermediate **5.122** that was then deprotected at the 5′-end by treatment with dichloroacetic acid. The resulting H-phosphonate-linked conjugate was then oxidized with basic aqueous iodine. Next, the phosphodiester linkages on the $(dT)_{12}$ portion of the molecule were deprotected and, finally, the product (**5.123**) was cleaved from the CPG with concentrated aqueous NH_4OH; it was produced in an overall yield of 56%.

Scheme 5.6.2

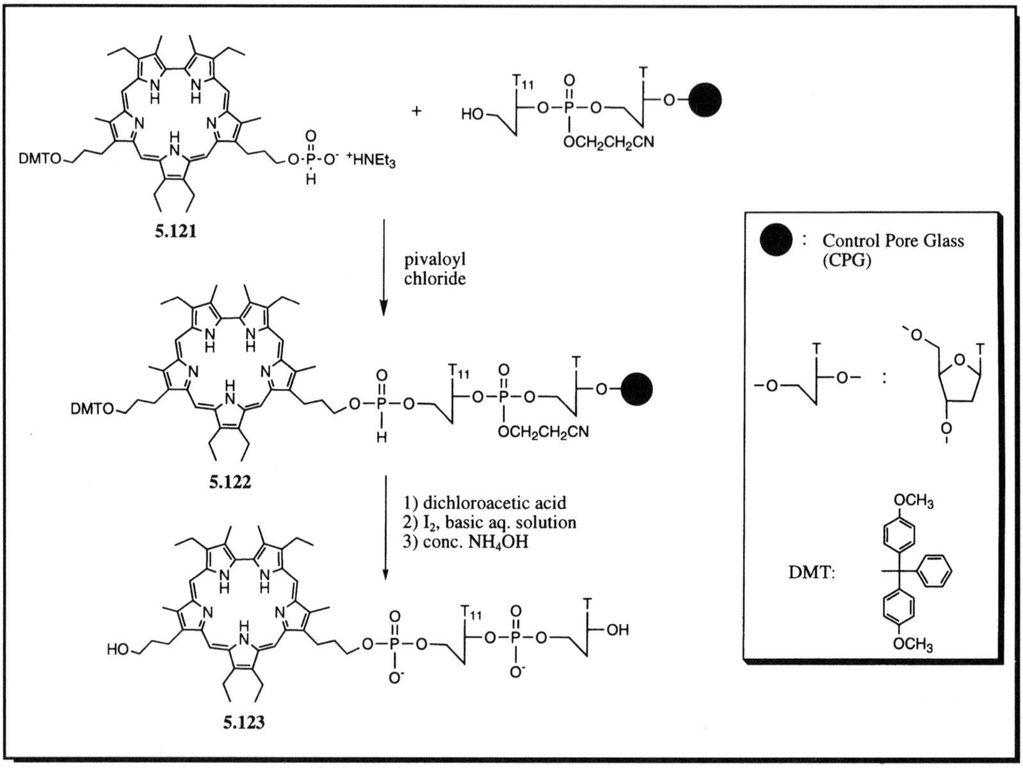

Sapphyrin–oligonucleotide conjugate **5.123** was designed to act as a site-selective DNA photocleavage agent and, in fact was found capable of effecting sequence-specific photomodification of complementary DNA targets upon irradiation with wavelengths above 620 nm.[78] The results obtained with **5.123** are thus considered extremely encouraging. Nonetheless, much remains to be done in terms of designing and synthesizing additional sapphyrin–oligonucleotide conjugates.

5.6.1.3   Sapphyrin–Lasalocid Conjugate

The sapphyrin–lasalocid conjugate **5.125** was reported recently.[79] It was synthesized under amide-forming conditions similar to those used to prepare the sapphyrin nucleobase conjugates **5.111**, **5.113**, **5.117**, and **5.119**. Specifically, the sapphyrin acid **5.109** was activated using 1-(3-dimethylaminopropy)-3-ethylcarbodiimide hydrochloride (EDC) and coupled with the lasalocid derivative **5.124**. The resulting chiral conjugate **5.125** was intended to act as a zwitterionic, ditopic receptor at neutral pH that would serve to bind selectively certain amino acids and mediate their transport across a model dichloromethane membrane (Scheme 5.6.3).

Scheme 5.6.3

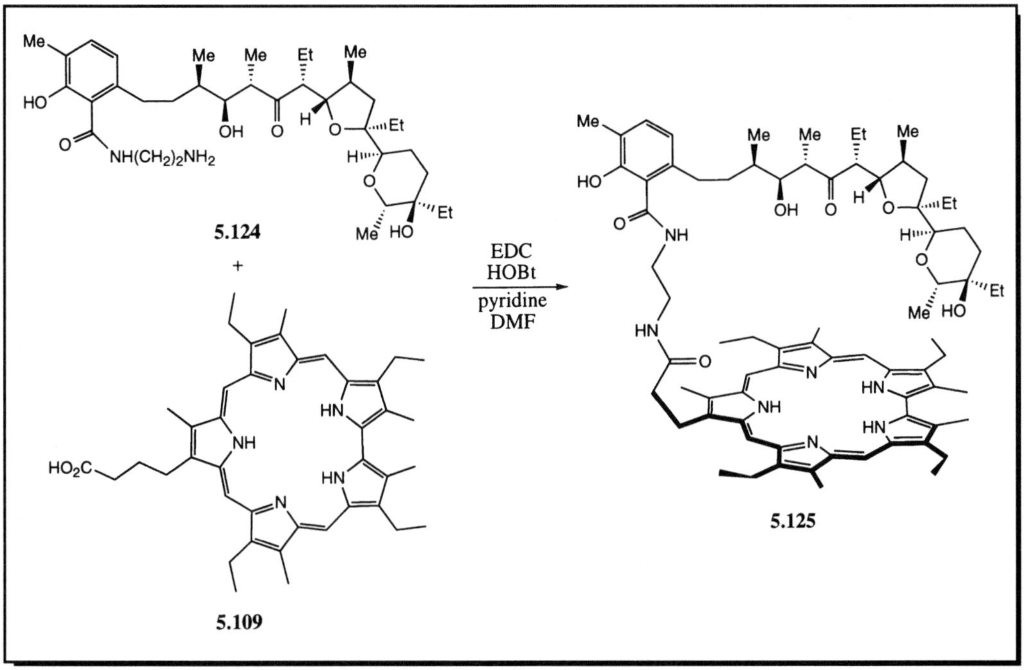

The above hope was in fact realized. It was found, for instance, that conjugate **5.125** forms 1:1 complexes with both phenylalanine and tryptophan. It was also found that this same species mediates the transport of L-phenylalanine some four times faster than L-tryptophan, and 1000 times faster than L-tyrosine. While taken together these results are considered to be extremely promising, it was also found that the L-enantiomers of these amino acids are bound with the same affinity (within error) as the D-antipodes. While the reasons for this remain obscure, it is nevertheless encouraging that sapphyrin conjugates, such as **5.125**, can indeed be designed that are capable of binding and transporting amino acids in their normal, hard-to-solubilize zwitterionic forms with which it is hard to work.

5.6.2   *"Capped" Sapphyrins*

The first example of a "capped" sapphyrin to be reported was the face-to-face porphyrin–sapphyrin pseudo dimer **5.127**. It was reported by Sessler, *et al.* in 1994[80]

and was prepared in 11% yield *via* the condensation of α,α-bis(aminophenyl) porphyrin **5.126** with the bis(acid chloride) form of sapphyrin **5.110** (Scheme 5.6.4). The
UV-vis absorption spectrum **5.127** is characterized by the presence of absorbance
bands corresponding to both the porphyrin and the sapphyrin chromophores (i.e.,
Soret bands were observed both at 409 and 457 nm). Interestingly, however, the
molar extinction coefficients of these bands were found to be significantly lower in
magnitude than were found for the monomeric subunits. On the basis of fluorescence
emission spectroscopic analyses, it was concluded that this is the result of efficient
intramolecular singlet energy transfer between the two chromophoric units. Thus,
5.127 was put forward by the authors as being a good model for the final steps in
natural light-harvesting arrays.[80]

Scheme 5.6.4

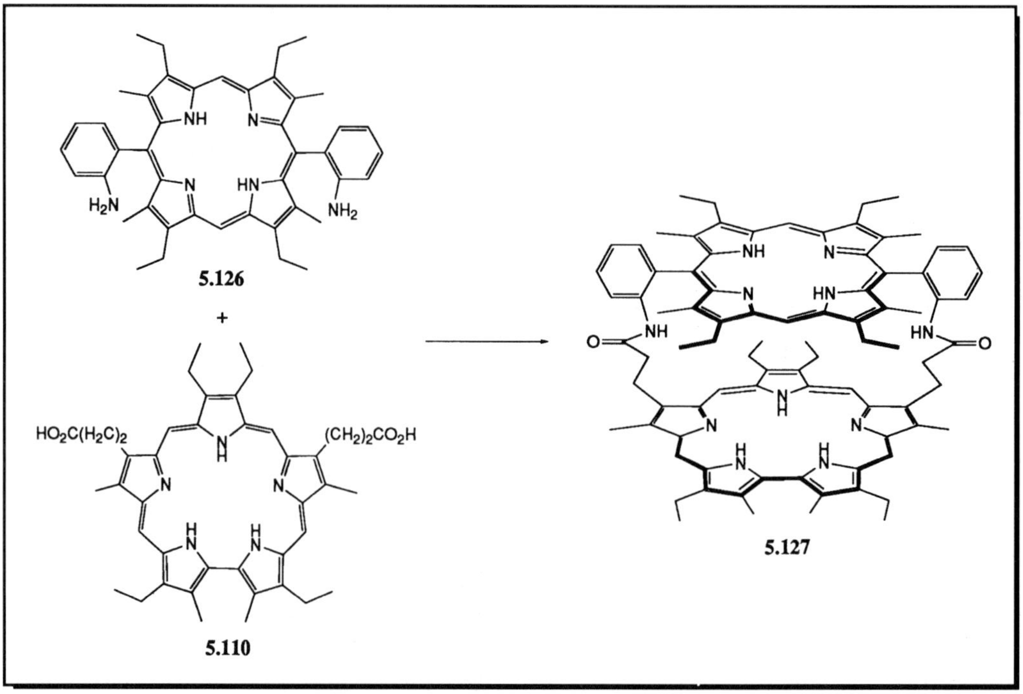

A second example of a capped sapphyrin is the azacrown sapphyrin **5.129**.[81]
This system was prepared in 23% yield *via* the condensation reaction between the
bis(acid chloride) form of sapphyrin **5.110** and bis(aminopropyl)diaza-18-crown-6
(**5.128**) (Scheme 5.6.5). It was hoped that this crown sapphyrin system might function
as a ditopic ligand, capable of coordinating simultaneously both cationic and anionic
substrates. To date, some preliminary evidence in favor of this proposal has been
obtained. For instance, in work involving NH_4F as the substrate, it was found that
NH_4^+ and F^- are simultaneously bound by the crown sapphyrin receptor. However,
at present the structural nature of this proposed interaction remains indeterminate.

Scheme 5.6.5

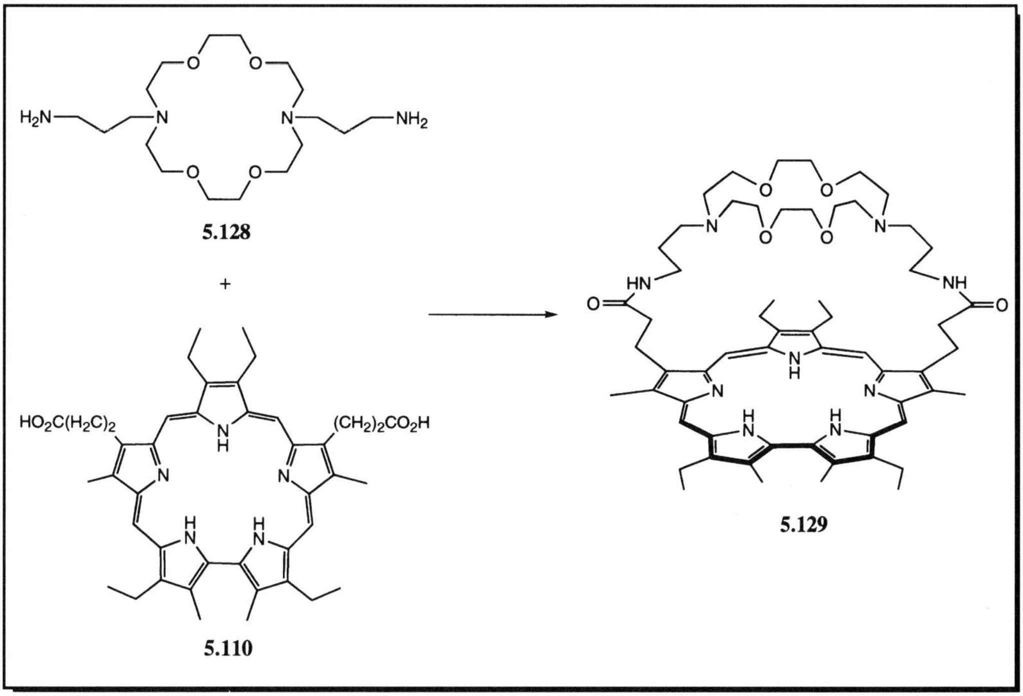

5.128

+

5.110

5.129

5.6.3   *Sapphyrin Oligomers*

In addition to the capped sapphyrins discussed above, a variety of oligomeric sapphyrin systems are now known. The first of these to be presented, system **5.130**, was prepared by treating two equivalents of the activated form of sapphyrin carboxylic acid **5.109** with one equivalent of 1,3-diaminopropane (Scheme 5.6.6).[82] In a similar manner, the trimer **5.131** and tetramer **5.132** were prepared by reacting a 30% excess of sapphyrin **5.109** (after activation) with the corresponding polyamine.[83] In each case, reaction yields were reported to be on the order of 50–80%.

The synthesis of the linear trimeric oligosapphyrin **5.134** was also reported by Sessler, *et al.*[83] Here, sapphyrin monoacid **5.109** was first activated (i.e., converted into any of the following forms: acid chloride, acylimidazole, mixed anhydride, or activated ester) and then reacted with the bis-amino-functionalized sapphyrin **5.133**. Here, too, reaction yields on the order of 50–80% were routinely achieved (Scheme 5.6.7).

Dimeric sapphyrins **5.138** and **5.139**, containing chiral bridging units have also been synthesized.[77] These were obtained by treating the activated form of the sapphyrin mono-acid **5.109** with a chiral diamine (i.e., **5.135** or **5.136**; Figure 5.6.3). This afforded the corresponding sapphyrin dimers **5.138** and **5.139** in yields as high as 60% (Schemes 5.6.8 and 5.6.9). The cyclic sapphyrin dimer **5.140** was also prepared from the chiral diamine **5.137** and bis-acid sapphyrin **5.109**. However, in this latter instance a stepwise synthetic approach was required. It involved reacting two equiva-

Scheme 5.6.6

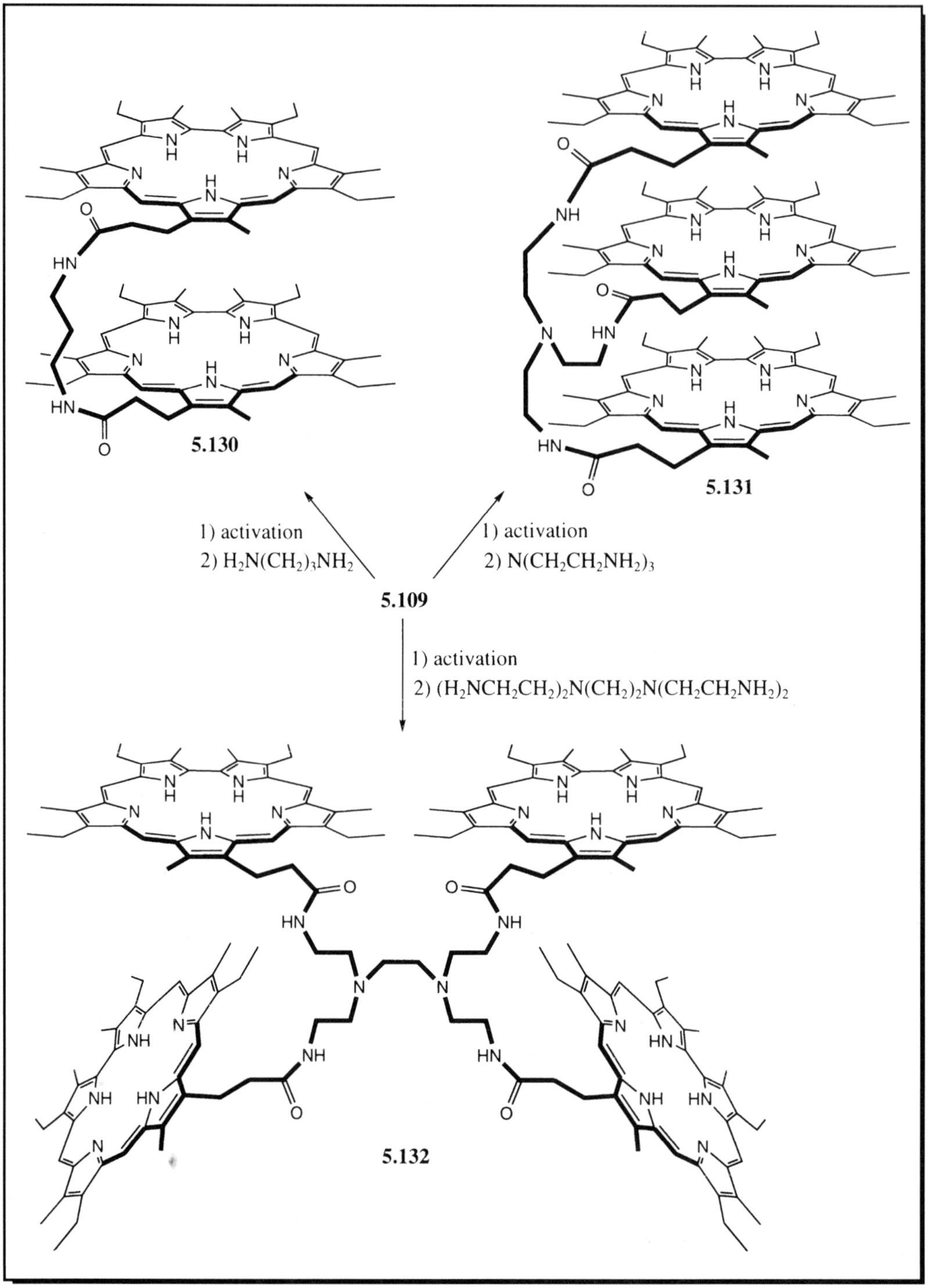

5.130

5.131

1) activation
2) H₂N(CH₂)₃NH₂

1) activation
2) N(CH₂CH₂NH₂)₃

5.109

1) activation
2) (H₂NCH₂CH₂)₂N(CH₂)₂N(CH₂CH₂NH₂)₂

5.132

lents of the mono-butoxycarbonyl (BOC)-protected diamine **5.137** with the activated diacid sapphyrin **5.109**. Deprotection of the BOC-group was effected using TFA to afford sapphyrin diamine **5.140** (Scheme 5.6.10). Finally, sapphyrin diamine **5.140** was reacted with a second equivalent of an activated form of sapphyrin **5.109** to afford the cyclic dimer **5.141** in yields as high as 60% yield (based on **5.109**).

Scheme 5.6.7

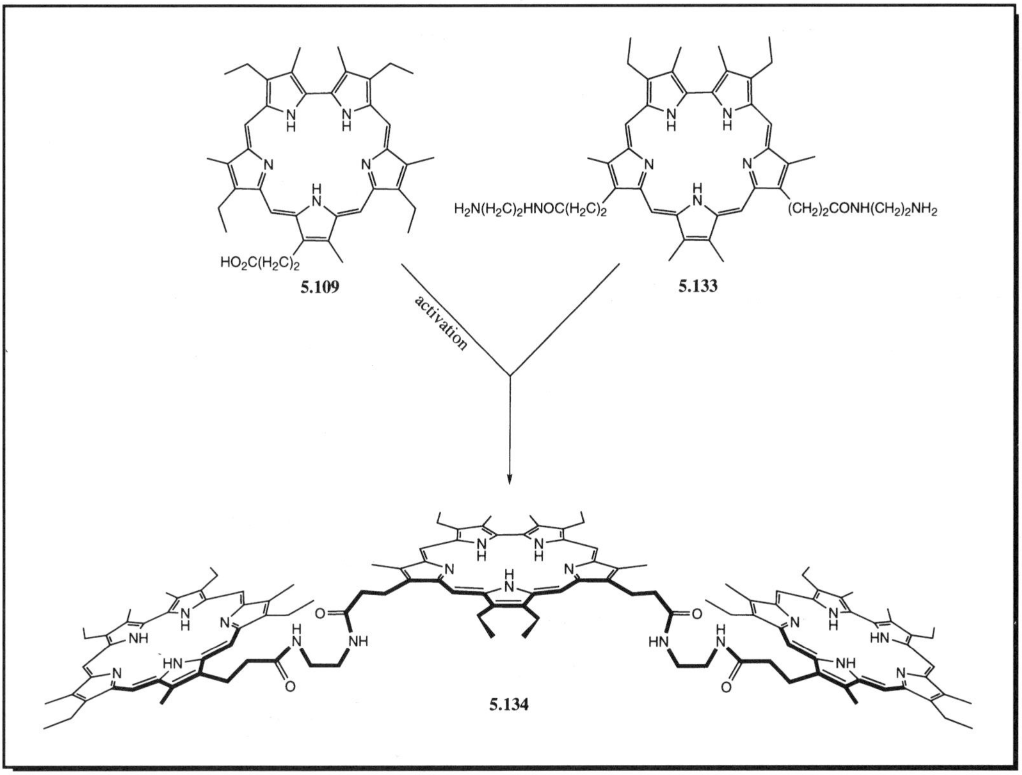

Figure 5.6.3   Structures **5.135–5.137**

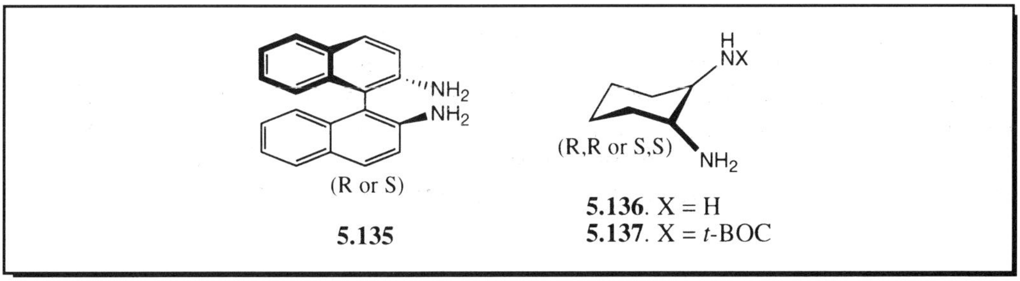

Scheme 5.6.8

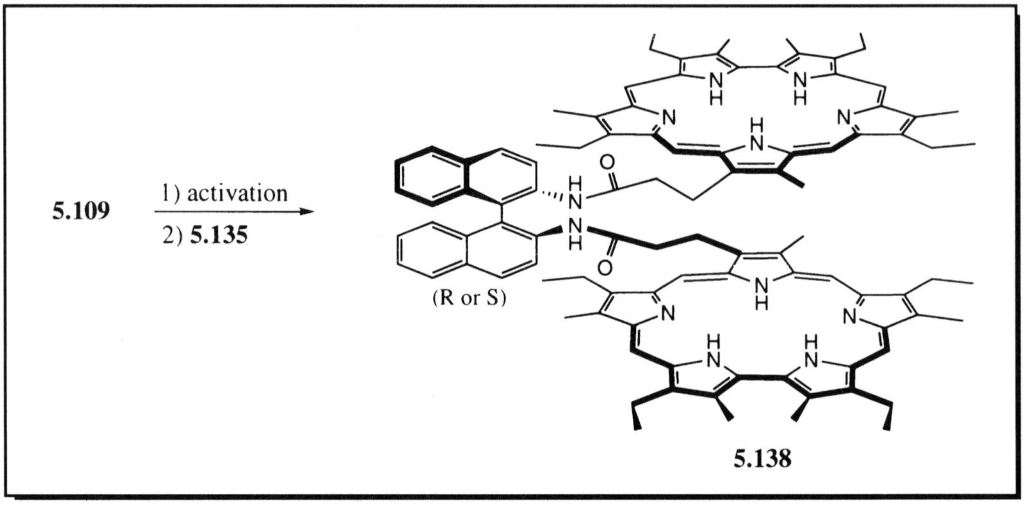

Scheme 5.6.9

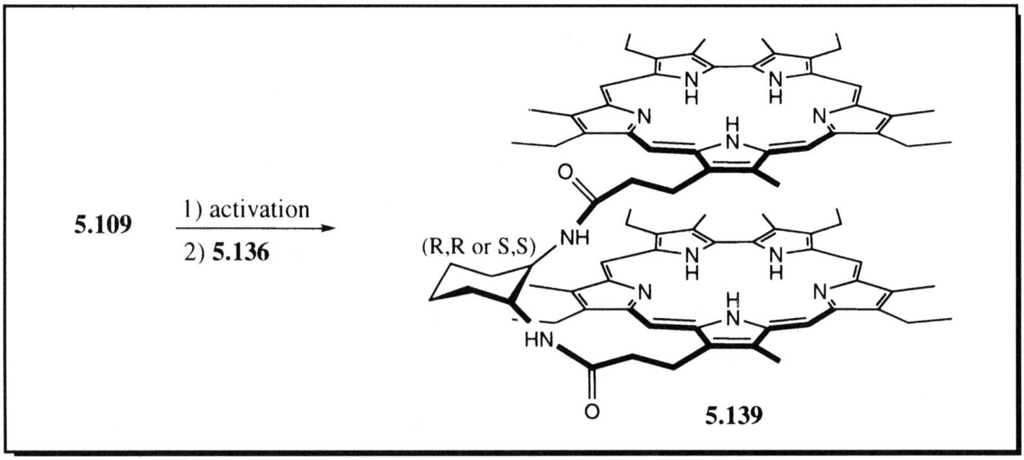

The primary impetus for preparing each of the oligosapphyrins described above, and indeed all of the three-dimensional sapphyrins, was to obtain systems capable of binding multiply charged anionic species with high specificity and affinity. This was obtained, with remarkable success, in the case of the sapphyrin oligomers. These systems were found to function as remarkably good receptors of dicarboxylate anions, nucleotide di- and triphosphates, and amino acids.[79,82,83]

Scheme 5.6.10

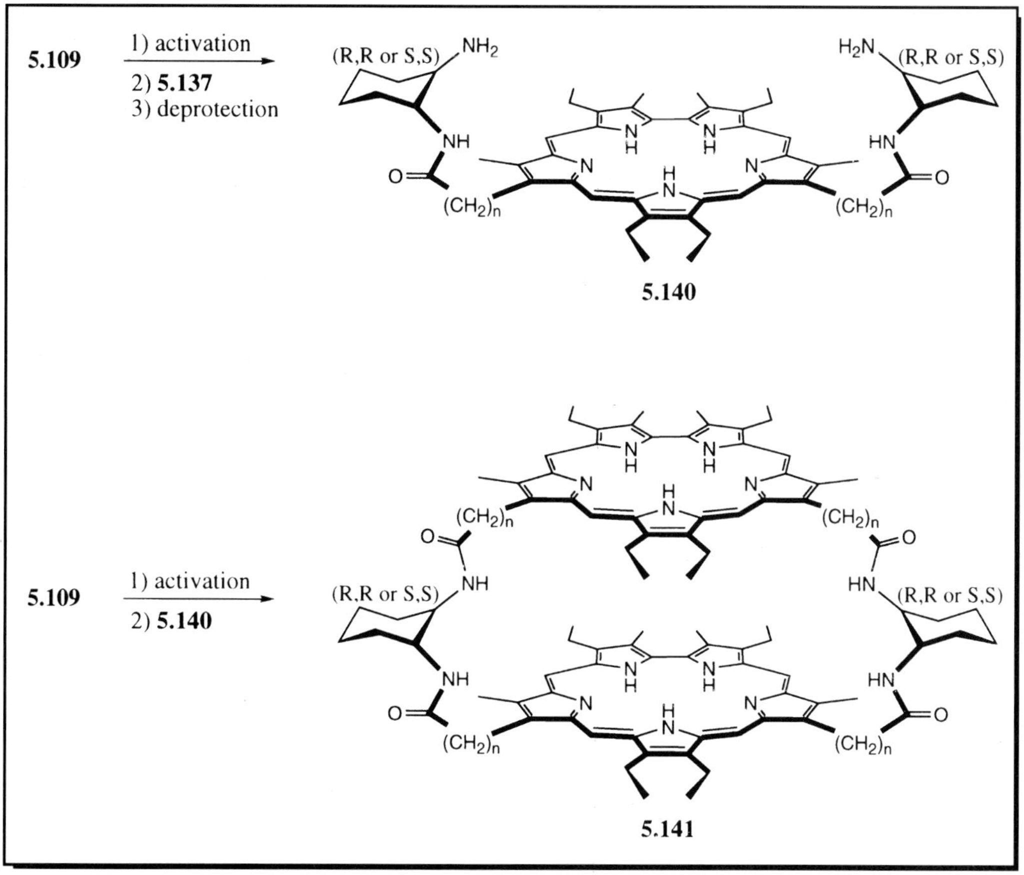

5.109 $\xrightarrow{\text{1) activation} \atop \text{2) 5.137} \atop \text{3) deprotection}}$

(R,R or S,S) NH$_2$ H$_2$N (R,R or S,S)

NH   HN

O=   =O

(CH$_2$)$_n$ (CH$_2$)$_n$

5.140

5.109 $\xrightarrow{\text{1) activation} \atop \text{2) 5.140}}$

(CH$_2$)$_n$ (CH$_2$)$_n$

(R,R or S,S) NH   HN (R,R or S,S)

NH   HN

O=   =O

(CH$_2$)$_n$ (CH$_2$)$_n$

5.141

5.7   Smaragdyrins and Heterosmaragdyrins (Norsapphyrins)

5.7.1   Contributions from Grigg and Johnson

The interest and excitement aroused by the successful synthesis of sapphyrin led Grigg and Johnson to suggest that a pentapyrrolic macrocycle containing two directly linked pyrroles (e.g., **5.142**, shown in its unsubstituted, free-base form; Figure 5.7.1) could exist in a planar, strain-free conformation.[4] They referred to this class of molecules as "norsapphyrins" since they are formally derived from sapphyrin by the removal of one *meso*-carbon from its macrocyclic nucleus. Thus, the norsapphyrins (later called "smaragdyrins" by Woodward; see below) would be expected to bear the same relation to sapphyrin as corrole does to porphyrin.

In their initial report, Grigg and Johnson described the synthesis of the "hetero-norsapphyrins" **5.146** and **5.147**. These were prepared in 15–20% yield using a "3 + 2" MacDonald-type HBr-catalyzed condensation between diformyl bifuran **5.51** and the pyrrolyl-bipyrrole derivatives **5.144** and **5.145** (Scheme 5.7.1).[56] Like sapphyrin, these macrocycles were found to display reversible protonation properties. However, in this instance a monocationic species is afforded on protonation, as shown explicitly for **5.146** in Scheme 5.7.2.

Figure 5.7.1   Structure **5.142**

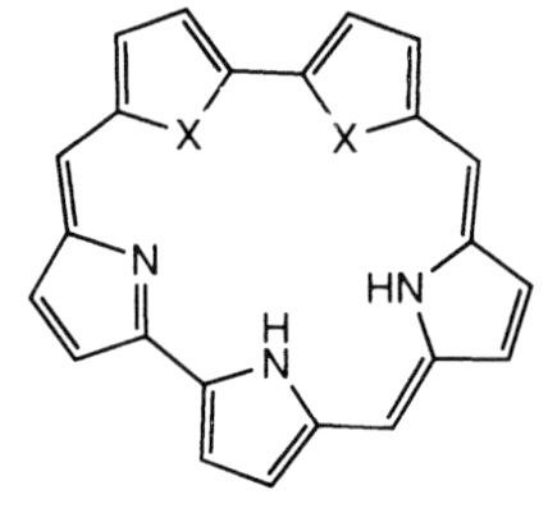

5.142

Scheme 5.7.1

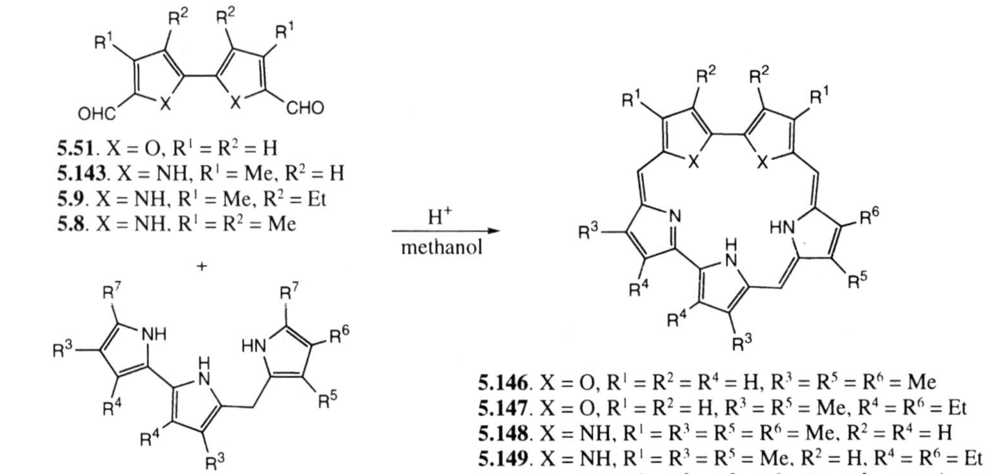

5.51. X = O, R^1 = R^2 = H
5.143. X = NH, R^1 = Me, R^2 = H
5.9. X = NH, R^1 = Me, R^2 = Et
5.8. X = NH. R^1 = R^2 = Me

$$\xrightarrow[\text{methanol}]{H^+}$$

5.144. R^3 = R^5 = R^6 = Me, R^4 = R^7 = H
5.145. R^3 = R^5 = Me, R^4 = R^6 = Et, R^7 = H
5.27. R^3 = R^4 = H, R^5 = R^6 = Me, R^7 = CO$_2$H

5.146. X = O, R^1 = R^2 = R^4 = H, R^3 = R^5 = R^6 = Me
5.147. X = O, R^1 = R^2 = H, R^3 = R^5 = Me, R^4 = R^6 = Et
5.148. X = NH, R^1 = R^3 = R^5 = R^6 = Me, R^2 = R^4 = H
5.149. X = NH, R^1 = R^3 = R^5 = Me. R^2 = H. R^4 = R^6 = Et
5.150. X = NH, R^1 = R^3 = R^5 = R^6 = Me, R^2 = Et, R^4 = H
5.151. X = NH, R^1 = R^3 = R^5 = Me, R^2 = R^4 = R^6 = Et
5.152 X = NH. R^1 = R^2 = R^5 = R^6 = Me, R^3 = R^4 = H

The UV-vis absorption spectra of these dioxanorsapphyrins was considered consistent with the proposed aromatic formulation. For instance, in pyridine, compound **5.146** exhibits an intense Soret-like absorbance band at 442 nm (ε = 201 200 M^{-1} cm^{-1}) and several Q-like absorbance bands at 518, 547, 591, 655, 666, and 730 nm, respectively. Interestingly, the monocation of the dioxanorsapphyrins **5.146** and **5.147** display a sharp, but split Soret-like absorbance band centered at *ca.* 454 nm. While the origins of this splitting remain obscure, the position and intensity of the bands is certainly in accord with what would be expected for an aromatic system.

Scheme 5.7.2

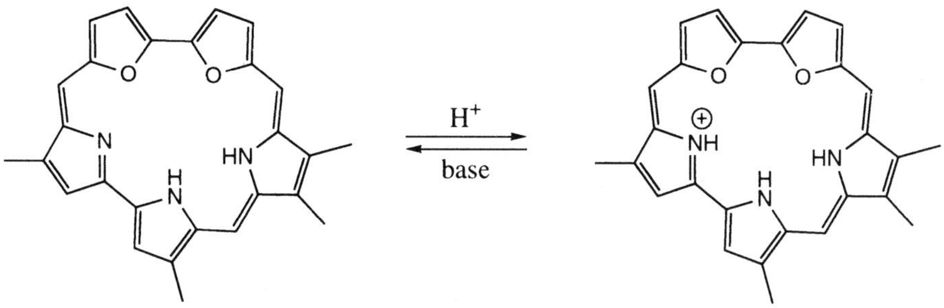

Further support for the aromatic nature of the dioxanorsapphyrins came from a proton NMR spectroscopic analysis of the monocation of **5.147**. In the spectrum of this species, the internal NH signals appear strongly shielded while those of the *meso-* and β-protons are strongly deshielded. This is as would be expected for an aromatic macrocycle of this type.

In later work, Grigg and Johnson reported on their efforts to synthesize pentaazanorsapphyrins **5.148–5.151**. This they attempted *via* the acid-catalyzed condensation of the diformyl bipyrrole derivatives **5.143** and **5.9** with the pyrrolylbipyrroles **5.144** and **5.145**.[4] While UV-vis spectroscopic analyses of the crude reaction mixtures revealed sharp absorption maxima at *ca.* 450 nm (and could thus be interpreted in terms of the desired pentaaza systems having been formed), these putative norsapphyrins proved too unstable to isolate. Thus, they could never be fully characterized nor could the success of this reaction sequence ever be confirmed.

5.7.2   *Contributions from Woodward and coworkers*

Concurrent with Grigg and Johnson, R. B. Woodward's group was also working on the synthesis of similar norsapphyrin-type macrocycles.[2,3] After an attempted "4 + 1" approach (involving the condensation of a 2,5-diformylpyrrole with a suitable dipyrrolyldipyrrylmethane) to these macrocycles failed, a "3 + 2" MacDonald-type strategy similar to that of Grigg and Johnson was settled on. Using this latter approach, macrocycle **5.152** was prepared from pyrrolylbipyrrole **5.27** and diformyl bipyrrole **5.8** using camphorsulfonic acid as the catalyst. Because of the brilliant emerald green color of this macrocycle, Woodward coined the term "smaragdyrin" (from the Greek *smaragdos*, meaning emerald) for this class of macrocycles. Currently, it is this name, rather than norsapphyrin, that is most widely accepted in the literature.

As proved true for Grigg and Johnson, Woodward and coworkers found that the pentaazasmaragdyrin system **5.152** was rather unstable. It underwent rapid decomposition both in the presence of light and during attempted chromatographic purification on neutral alumina. Nonetheless, Woodward and coworkers could achieve reproducible 27.5% yields of smaragdyrin **5.152** by (a) carrying out the reaction and subsequent work-up operations in the absence of light, and (b) using florisil instead of alumina as the chromatographic support. Unfortunately, both the free-base and protonated forms of smaragdyrin **5.152** were found to be unstable not only in solution but also in the solid state. This finding obviously limited the extent to which this prototypic smaragdyrin could be studied.

Nonetheless, Woodward and his group did attempt to prepare metal derivatives of **5.152**.[3] These attempts were made in the hope of stabilizing the basic smaragdyrin skeleton. Sadly, however, the various metalation reactions they attempted either failed to effect metal insertion (giving back unchanged smaragdyrin) or resulted in decomposition of the macrocycle.

In spite of its basic instability, the Woodward team was able to record the ^{1}H NMR spectrum of **5.152**.[3] They were also able to study its optical properties. They found that smaragdyrin **5.152**, like sapphyrin, exhibits an intense Soret-like absorbance band at 453 nm that is shifted roughly 50 nm to the red of the Soret-type bands exhibited by either porphyrin or corrole. Interestingly, however, the general shape of the Soret band of hexamethylsmaragdyrin **5.152** resembles more closely that of octamethylcorrole than that of decamethylsapphyrin. As opposed to the single sharp peak observed for sapphyrin, the Soret band of smaragdyrin **5.152** is characterized by an inflection on the short-wavelength side, and a shoulder on the long-wavelength side. Smaragdyrin **5.152** also displays several long-wavelength Q-type absorption bands in the 600–700 nm region, which, like the Soret band, are consistent with the proposed 22 π-electron aromatic formulation.

5.7.3   *Contributions from Sessler and coworkers*

Recently, Sessler and coworkers have taken interest in the heterosmaragdyrins.[59] The reason for this is that they postulated that fully decaalkyl-substituted dioxasmaragdyrins (also available from a "3 + 2" approach) might prove more stable and have more favorable solubility and crystallinity properties than the congeneric tetra- or hexaalkylsmaragdyrins **5.146** and **5.147** of Grigg. To test this hypothesis, diformyl tetraalkylbifuran **5.153** was reacted with the pyrrolylbipyrrole derivative **5.154**. As expected, the decaalkyl-dioxasmaragdyrin **5.155** was formed in reasonable (26%) yield from this reaction (Scheme 5.7.3). Fortunately, this product did indeed prove to be quite stable in solution and in the solid state (**5.155** is soluble in most common organic solvents). Like its lesser-substituted congeners, decaalkyl-dioxasmaragdyrin **5.155** is aromatic (as inferred from the apparent diamagnetic ring-current effects observed in the ^{1}H NMR spectrum). It also displays the expected single, sharp Soret-like absorbance band at 464 nm as well as two Q-bands at 658 and 725 nm (in CH_2Cl_2). Thus compound **5.155** does appear to be a *bona fide* smaragdyrin. Unfortunately, however, the higher degree of alkyl substitution it bears has as yet to be translated into single crystals suitable for X-ray diffraction analysis. Thus, the goal of obtaining a solid-state structure for this, or indeed any, smaragdyrin remains unattained.

Scheme 5.7.3

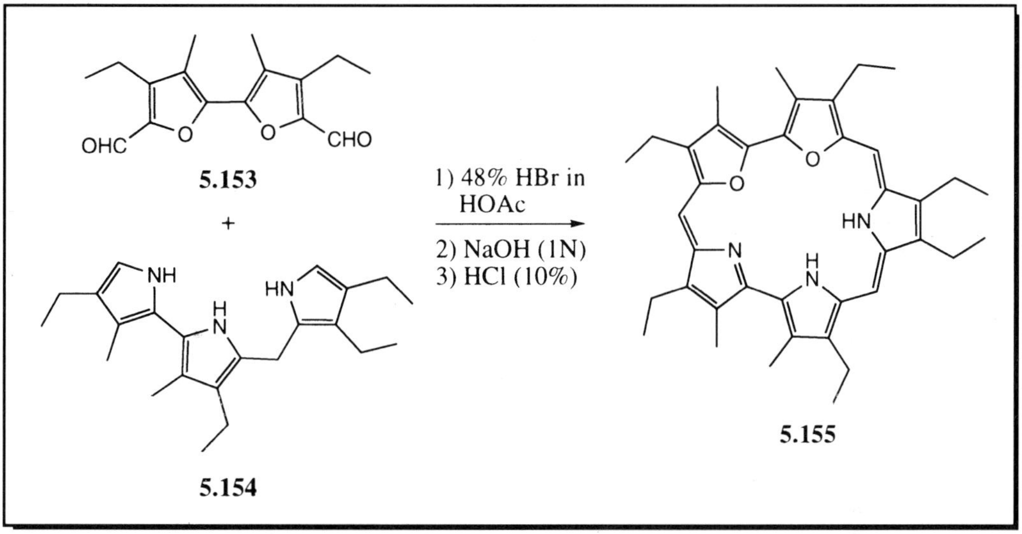

Note in proof

Subsequent to submitting this manuscript, a single crystal x-ray diffraction structure of the neutral uranyl complex of monoxasapphyrin **5.74** was obtained. The uranyl cation is pentaligated (with a long furan oxygen-uranium bond) and resides within the macrocyclic plane.[84]

5.8   References

1.   First reported at the aromaticity conference, Sheffield UK, 1966.
2.   Bauer, V. J.; Clive, D. L. J.; Dolphin, D.; Paine III, J. B.; Harris, F. L.; King, M. M.; Loder, J.; Wang, S.-W. C.; Woodward, R. B. *J. Am. Chem. Soc.* **1983**, *105*, 6429–6436.
3.   King, M. M., Ph.D. Dissertation, Harvard University, Cambridge, MA, USA, **1970**.
4.   Broadhurst, M. J.; Grigg, R.; Johnson, A. W. *J. Chem. Soc., Perkin Trans. I* **1972**, 2111–2116.
5.   Johnson, A. W. In *Porphyrins and Metalloporphyrins*; Smith, K. M., Ed.; Elsevier: Amsterdam, 1976; p. 750.

6.   Grigg, R. In *The Porphyrins*, Vol. 2; Dolphin, D., Ed.; Academic Press: New York, 1978; pp. 327–391.

7.   Shiau, F.-Y.; Liddell, P. A.; Vicente, G. H.; Ramana, N. V.; Ramachandran, K.; Lee, S.-J.; Pandey, R. K.; Dougherty, T. J.; Smith, K. M. *SPIE Future Directions and Applications in Photodynamic Therapy, IS* **1989**, *6*, 71.

8.   Sessler, J. L.; Cyr, M. J.; Lynch, V.; McGhee, E.; Ibers, J. A. *J. Am. Chem. Soc.* **1990**, *112*, 2810–2813.

9.   Sessler, J. L.; Cyr, M. J.; Burrell, A. K. *Synlett* **1991**, 127–134.

10.   Sessler, J. L.; Cyr, M.; Burrell, A. K. *Tetrahedron* **1992**, *48*, 9661–9672.

11.   Sessler, J. L.; Murai, T.; Lynch, V.; Cyr, M. *J. Am. Chem. Soc.* **1988**, *110*, 5586–5588.

12.   Sessler, J. L.; Murai, T.; Lynch, V. *Inorg. Chem.* **1989**, *28*, 1333–1341.

13.   Sessler, J. L.; Murai, T.; Hemmi, G. *Inorg. Chem.* **1989**, *28*, 3390–3393.

14.   Barton, D. H. R.; Zard, S. Z. *J. Chem. Soc., Chem. Commun.* **1985**, 1098–1100.

15.   Shionoya, M.; Furuta, H.; Lynch, V.; Harriman, A.; Sessler, J. L. *J. Am. Chem. Soc.* **1992**, *114*, 5714–5722.

16.   Iverson, B. L.; Shreder, K.; Král, V.; Sessler, J. L. *J. Am. Chem. Soc.* **1993**, *115*, 11022–11023.

17.   Král, V.; Furuta, H.; Shreder, K.; Lynch, V.; Sessler, J. L. *J. Am. Chem. Soc.* **1996**, *118*, 1595–1607.

18.   Iverson, B. L.; Shreder, K.; Král, V.; Sansom, P.; Lynch, V.; Sessler, J. L. *J. Am. Chem. Soc.* **1996**, *118*, 1608–1616.

19.   Sessler, J. L.; Lisowski, J.; Boudreaux, K. A.; Lynch, V.; Barry, J.; Kodadek, T. J. *J. Org. Chem.* **1995**, *60*, 5975–5978.

20.   Barry, J.; Kodadek, T. *Tetrahedron Lett.* **1994**, *35*, 2465–2468.

21.   Boudreaux, K. B., M.S. Thesis, University of Texas at Austin, Austin, TX, USA, **1995**.

22.   Sessler, J. L.; Morishima, T.; Lynch, V. *Angew. Chem. Int. Ed. Eng.* **1991**, *30*, 977–980.

23.   Kuroda, Y.; Murase, H.; Suzuki, Y.; Ogoshi, H. *Tetrahedron Lett.* **1989**, *30*, 2411–2412.

24.   Ema, T.; Kuroda, Y.; Ogoshi, H. *Tetrahedron Lett.* **1991**, *35*, 4529–4532.

25.   Wallace, D. M.; Smith, K. M. *Tetrahedron Lett.* **1990**, *31*, 7265–7268.

26.   Wallace, D. M.; Leung, S. H.; Senge, M. O.; Smith, K. M. *J. Org. Chem.* **1993**, *58*, 7245–7257.

27.   Chmielewski, P. J.; Latos-Grażyński, L.; Rachlewicz, K. *Chem. Eur. J.* **1995**, *1*, 68–73.

28.   Momenteau, M.; Mispelter, J.; Loock, B.; Bisagni, E. *J. Chem. Soc. Perkin Trans. I*, **1983**, 189–196.

29.   Collman, J. P.; Brauman, J. I.; Iverson, B. L.; Sessler, J. L.; Morris, R. M.; Gibson, Q. H. *J. Am. Chem. Soc.* **1983**, *105*, 3052–3064.

30.   Collman, J. P.; Brauman, J. I.; Doxsee, K. M.; Sessler, J. L.; Morris, R. M.; Gibson, Q. H. *Inorg. Chem.* **1983**, *22*, 1427–1432.

31.   Sessler, J. L.; Capuano, V. L. *Angew. Chem. Int. Ed. Eng.* **1990**, *29*, 1134–1137.

32.   Sessler, J. L.; Capuano, V. L. *Tetrahedron Lett.* **1993**, *34*, 2287–2290.

33. Sessler, J. L.; Capuano, V. L.; Harriman, A. *J. Am. Chem. Soc.* **1993**, *115*, 4618–4628.

34. Giraudeau, A.; Ruhlmann, L.; El Kahef, L.; Gross, M. *J. Am. Chem. Soc.* **1996**, *118*, 2969–2979.

35. Momenteau, M.; Reed, C. A. *Chem. Rev.* **1994**, *94*, 659–698.

36   Rose, E.; Kossanyi, A.; Quelquejeu, M.; Soleilhavoup, M.; Duwavran, F.; Bernard, N.; Lecas, A. *J. Am. Chem. Soc.* **1996**, *118*, 1567–1568.

37. Wagner, R. W.; Lindsey, J. S. *J. Am. Chem. Soc.* **1994**, *116*, 9759–9760.

38. Seth, J.; Palaniappan, V.; Johnson, T. E.; Prathapan, S.; Lindsey, J. S.; Bocian, D. F. *J. Am. Chem. Soc.* **1994**, *116*, 10578–10592.

39. Wagner, R. W.; Johnson, T. E.; Li, F.; Lindsey, J. S. *J. Org. Chem.* **1995**, *60*, 5266–5273.

40. Schwarz, F. P.; Gouterman, M.; Muljiani, Z.; Dolphin, D. H. *Bioinorg. Chem.* **1972**, *2*, 1–32.

41. Kagan, N. E.; Mauzerall, D.; Merrifield, R. B. *J. Am. Chem. Soc.* **1977**, *99*, 5484.

42. Ogoshi, H.; Sugimoto, S.; Yoshida, Z. *Tetrahedron Lett.* **1977**, 169–172.

43. Netzel, T. L.; Droger, P.; Chang, C. K.; Fujita, I.; Fajer, J. *Chem. Phys. Lett.* **1979**, *67*, 223–228.

44. Dalton, J.; Milgrom, L. R. *J. Chem. Soc., Chem. Commun.* **1979**, 609–610.

45. Heiler, D.; McLendon, G. L.; Rogalskyj, P. *J. Am. Chem. Soc.* **1987**, *109*, 604–606.

46. Cowan, J. A.; Sanders, J. K. M.; Beddard, G. S.; Harrison, R. J. *J. Chem. Soc., Chem. Commun.* **1987**, 55–58.

47. Chardon-Noblat, S.; Sauvage, J.-P.; Mathis, P. *Angew. Chem. Int. Ed. Eng.* **1989**, *28*, 593–595.

48. Helms, A.; Heiler, D.; McLendon, G. L. *J. Am. Chem. Soc.* **1991**, *113*, 4325–4327.

49. Helms, A.; Heiler, D.; McLendon, G. L. *J. Am. Chem. Soc.* **1992**, *114*, 6227–6638.

50. Senge, M. O.; Gerzevske, K. R.; Vicente, M. G. H.; Forsyth, T. P.; Smith, K. M. *Angew. Chem., Int. Ed. Eng.* **1993**, *32*, 750–753.

51. Almog, J.; Baldwin, J. E.; Dyer, R. L.; Peters, M. *J. Am. Chem. Soc.* **1975**, *97*, 226–227.

52. Hamilton, A. D.; Lehn, J.-M.; Sessler, J. L. *J. Chem. Soc., Chem. Commun.* **1984**, 311–312.

53. Gubelmann, M.; Harriman, A.; Lehn, J.-M.; Sessler, J. L. *J. Chem. Soc., Chem. Commun.* **1988**, 77–79.

54. Broadhurst, M. J.; Grigg, R.; Johnson, A. W. *J. Chem. Soc., Chem. Commun.* **1969**, 23–24.

55. Broadhurst, M. J.; Grigg, R.; Johnson, A. W. *J. Chem. Soc., Perkin Trans. I* **1972**, 1124–1135.

56. Broadhurst, M. J.; Grigg, R.; Johnson, A. W. *J. Chem. Soc., Chem. Commun.* **1969**, 1480–1482.

57. Lisowski, J.; Sessler, J. L.; Lynch, V. *Inorg. Chem.* **1995**, *34*, 3567–3571.

58.   Sessler, J. L.; Burrell, A. K.; Lisowski, J.; Gebauer, A.; Cyr, M. J.; Lynch, V. *Bull. Soc. Chem. France* **1996**, *133*, 725–734.

59.   Hoehner, M. C., Ph.D. Dissertation, University of Texas at Austin, Austin, TX, USA, **1996**.

60.   Broadhurst, M. J.; Grigg, R.; Johnson, A. W. *J. Chem. Soc., Chem. Commun.* **1970**, 807–809.

61.   Furuta, H.; Cyr, M. J.; Sessler, J. L. *J. Am. Chem. Soc.* **1991**, *113*, 6677–6678.

62.   For a discussion of the desirability of long-wavelength photosensitizers and their relevance to the so-called physiological 'window of transparency', see: Kreimer-Birnbaum, M. *Semin. Hematol.* **1989**, *26*, 157–173.

63.   Gomer, C. J. *Photochem. Photobiol.* **1987**, *46*, 561–562.

64.   Dahlman, A.; Wile, A. G.; Burns, R. G.; Mason, G. R.; Johnson, F. M.; Berns, M. W. *Cancer Res.* **1983**, *43*, 430–434.

65.   Dougherty, T. J. In *Methods in Porphyrin Photosensitization*; Kessel, D., Ed.; Plenum Press: New York, 1985; pp. 313–328.

66.   Dougherty, T. J. *Photochem. Photobiol.* **1987**, *45*, 879–889.

67.   Gomer, C. J. *Semin. Hematol.* **1989**, *26*, 27–34.

68.   Burrell, A. K.; Sessler, J. L.; Cyr, M. J.; McGhee, E.; Ibers, J. A. *Angew. Chem. Int. Ed. Eng.* **1991**, *30*, 91–93.

69.   Muetterties, E. L.; Wright, C. M. *Quart. Rev.* **1967**, *21*, 109.

70.   Day, V. W.; Marks, T. J.; Wachter, W. A. *J. Am. Chem. Soc.* **1975**, *97*, 4519–4527.

71.   Burrell, A. K.; Hemmi, G.; Lynch, V.; Sessler, J. L. *J. Am. Chem. Soc.* **1991**, *113*, 4690–4692.

72.   Burrell, A. K.; Cyr, M. J.; Lynch, V.; Sessler, J. L. *J. Chem. Soc. Chem. Commun.* **1991**, 1710–1713.

73.   Vögtle, F. *Cyclophanes*; J. Wiley: Chichester, 1993; pp. 375–404.

74.   Král, V.; Sessler, J. L.; Furuta, H. *J. Am. Chem. Soc.* **1992**, *114*, 8704–8705.

75.   Sessler, J. L.; Furuta, H.; Král, V. *Supramol. Chem.* **1993**, *1*, 209–220.

76.   Král, V.; Sessler, J. L. *Tetrahedron* **1995**, *51*, 539–554.

77.   Sessler, J. L.; Sansom, P. I.; Andrievsky, A.; Král, V. In *The Supramolecular Chemistry of Anions*; Bianchi, A.; Bowman-James, K.; Garcia-España, E., Eds; VCH: Weinheim, in press.

78.   Sessler, J. L.; Sansom, P. I.; Král, V.; O'Connor, D.; Iverson, B. L. *J. Am. Chem. Soc.* **1996**, *118*, 12322–12330.

79.   Sessler, J. L.; Andrievsky, A. *J. Chem. Soc., Chem. Commun.* **1996**, 1119–1120.

80.   Sessler, J. L.; Brucker, E. A.; Král, V.; Harriman, A. *Supramol. Chem.* **1994**, *4*, 35–41.

81.   Sessler, J. L.; Brucker, E. A. *Tetrahedron Lett.* **1995**, *36*, 1175–1176.

82.   Král, V.; Andrievsky, A.; Sessler, J. L. *J. Am. Chem. Soc.* **1995**, *117*, 2953–2954.

83.   Král, V.; Andrievsky, A.; Sessler, J. L. *J. Chem. Soc. Chem. Commun.* **1995**, 2349–2351.

84.   Sessler, J. L.; Gebauer, A.; Hoehner, M.; Lynch, V., unpublished results.

6   Introduction

In the preceding chapter the principal player in the area of pentapyrrolic expanded porphyrins, namely sapphyrin, was introduced. In this chapter, a number of other pentapyrrole-derived macrocycles are discussed that are less well studied than this prototypic system. However, most of these other, less well-studied penta-pyrrolic macrocycles were only discovered recently and, as a result, have not been examined in detail. In fact, nearly half of the new pentapyrrolic expanded porphyrins that have been reported in the literature have appeared in the past 2 years. This dramatic increase in number reflects, to a certain extent, the fact that the field of expanded porphyrins is one that is still emerging.

To provide an intellectual framework for the chemistry of pentapyrrolic expanded porphyrins, it is convenient to sketch some generalized structures of poss-ible five-pyrrole-containing macrocycles that can formed by "using" between zero and five *meso*-like bridging carbons. This sketch, which constitutes Figure 6.0.1 (**6.1**–**6.6**), includes idealized forms of smaragdyrin (**6.4**) and sapphyrin (**6.5**).[†] These two systems were discussed in Chapter 5, and are thus omitted from explicit discussion here. Each macrocycle is shown with a 4n + 2 π electron conjugation pathway, although it is to be understood that other oxidation states could be possible for any of these molecules. Simple CPK modeling studies suggest that directly linked pentapyrrolic macrocycles, such as **6.1** or macrocycles with only one *meso*-like carbon (e.g., **6.2**) would likely be too strained to form stable, isolable systems. Perhaps consistent with this conclusion is the observation that no such systems have as yet been reported in the literature. On the other hand, numerous pentapyrrolic systems containing two or more *meso*-carbons have been reported. These macrocycles are the subject of this chapter. With the exception of smaragdyrin and sapphyrin noted above, they will be pre-sented and discussed in order of increasing *meso*-like carbon number.

6.1   Orangarin: Two Bridging Carbons

The smallest pentapyrrolic expanded porphyrin to be reported to date is the so-called "orangarin" macrocycle (alternatively referred to as "[20]pentaphyrin-(2.1.0.0.1)"). It was prepared in the form of the decaalkyl-substituted system **6.9** *via* an HCl-catalyzed condensation between the α-free terpyrrole **6.7** and the diformyl

---

[†]The structures shown in Figure 6.0.1 do not depict any of the numerous isomeric (i.e., "rearranged") forms of systems **6.3**–**6.6** that are conceivable.

bipyrrole derivative **6.8** (Scheme 6.1.1).[1] Under the conditions of this condensation, orangarin **6.9** is obtained in reasonable yield (59%) as a macrocycle that formulates as a fully conjugated, but non-aromatic 20 π-electron species. Thus, the specific orangarin system obtained has the same basic connectivity as the generalized structure **6.3** (Figure 6.0.1). It differs from this still-unknown parent, however, by virtue of being a two-electron oxidized analog.

The non-aromatic nature of orangarin **6.9** is evident upon inspection of the

Figure 6.0.1   Structures **6.1–6.6**

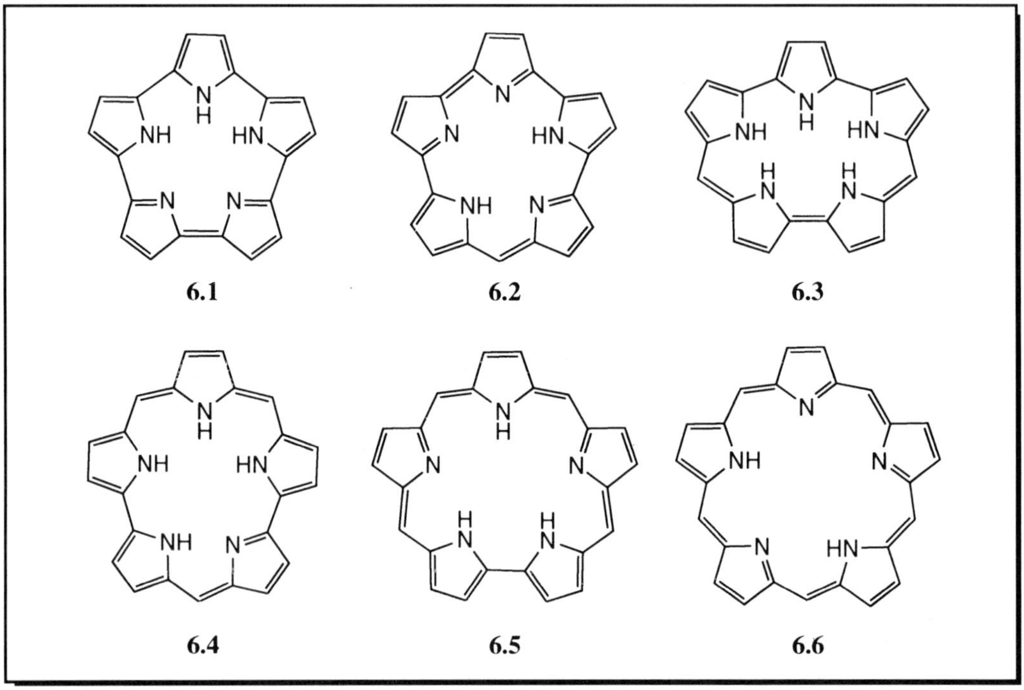

proton NMR spectra of the free-base (**6.9a**) and protonated forms (**6.9b**). In the case of **6.9b**, the *meso*-like protons were found to resonate at 2.37 ppm, while the internal NH protons were found to resonate at remarkably low field (*ca.* 30 ppm, in CDCl$_3$).[‡] The maximum difference in chemical shift between the internal NH protons and external *meso*-like proton is thus 28.08 ppm. This difference, in conjunction with the unusual location of these proton resonance frequencies, might be argued as an indication that macrocycle **6.9b** is antiaromatic. However, the possibility that the unusual chemical shifts are the result of some type of local anisotropic effects cannot as yet be ruled out.[1]

[‡]In the original orangarin report (reference 1), the NH proton chemical shifts for **6.9b** were reported, incorrectly, to be in the *ca.* −0.2 to +0.2 ppm region. Because the actual chemical shift of these protons is at *ca.* 30 ppm, an insufficiently large NMR instrumental sweep width resulted in these signals "wrapping around" to the apparent, erroneous high-field location.

Scheme 6.1.1

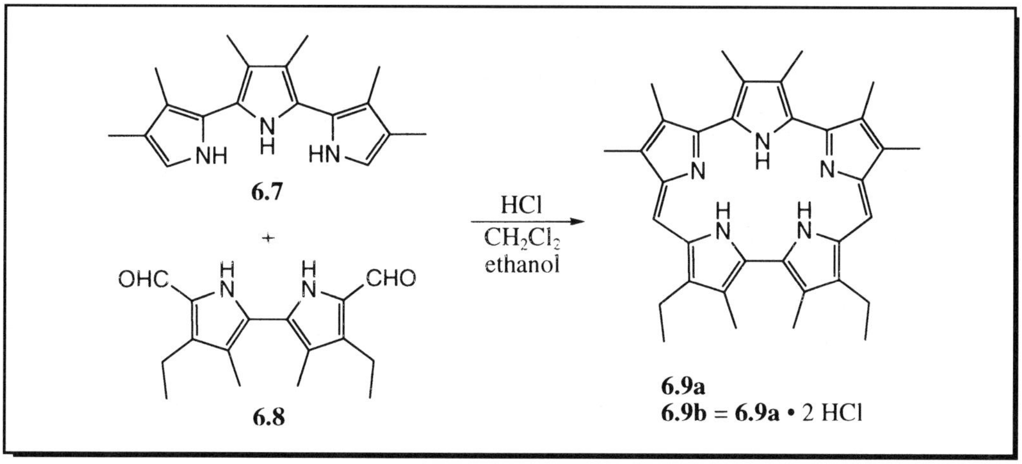

Figure 6.1.1   Single Crystal X-Ray Diffraction Structure of Free-base Orangarin **6.9a**.

This figure was generated using information down-loaded from the Cambridge Crystallographic Data Centre and corresponds to a structure originally reported in reference 1. The mean deviation of the orangarin atoms from macrocyclic planarity is 0.19 Å. The bipyrrole moiety is severely twisted with a dihedral angle of 30.4° existing between the pyrrole planes. Atom labeling scheme: carbon: ○; nitrogen: ●. Hydrogen atoms have been omitted for clarity

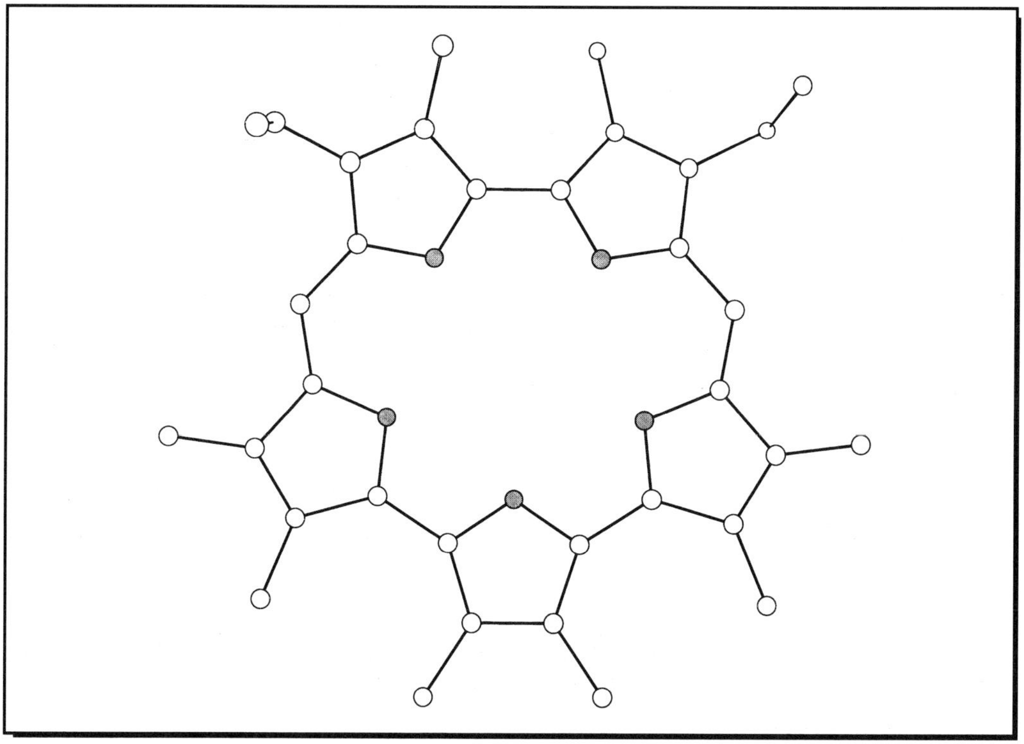

The visible absorption spectrum of **6.9b** is consistent with the lack of aromaticity. Specifically, it is characterized by an extremely broad Soret-like absorbance band at 463 nm ($\varepsilon = 33\,000\,M^{-1}\,cm^{-1}$ in CH_2Cl_2) and one sharper Q-like transition at 552 nm ($\varepsilon = 18\,100\,M^{-1}\,cm^{-1}$). A possible explanation for the broad, relatively weak absorbance bands observed in the electronic spectrum of this macrocycle was inferred from a single crystal X-ray diffraction structure of the free-base form of the macrocycle (**6.9a**) (Figure 6.1.1). In this structure it was observed that the bipyrrole moiety is highly twisted, with a dihedral angle of 30.4° being defined by the two pyrroles of this subunit. This twisting, which was rationalized in terms of unfavorable methyl–methyl steric interactions at the β-positions of the bipyrrole subunit, likely serves to disrupt appreciably the significant π-electron delocalization presumed necessary to produce a "normal" sharp Soret-like absorption band.

6.2   Isosmaragdyrin: Three Bridging Carbons

The smaragdyrins, discussed in Chapter 5, are macrocycles that contain five pyrrole or pyrrole-like subunits and three *meso*-like methine linkages arranged in a (1.1.0.1.0) fashion. While these systems are among the earliest expanded porphyrins to be prepared, it is only recently that the first isomer of smaragdyrin has been prepared.[2] This system, macrocycle (**6.11**), was prepared *via* a "3 + 2" TFA-catalyzed condensation between the α-free terpyrrole **6.7** and diformyl dipyrrylmethane **6.10** (Scheme 6.2.1). As an inspection of its structure shows, system **6.11** may be considered as being an isomer of smaragdyrin because it consists of five pyrrolic subunits and three *meso*-like carbon bridges. However, unlike smaragdyrin itself, the methine groups of this "isosmaragdyrin" are linked together in a (1.1.1.0.0) fashion.

Scheme 6.2.1

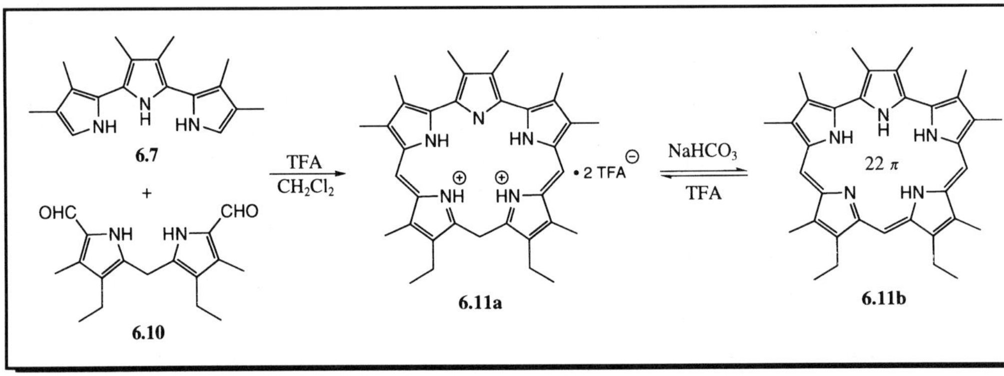

An alternative synthesis of this type of molecule was also attempted. This approach, shown in Scheme 6.2.2, involved "reversing" the nucleophilic and electrophilic building blocks relative to those shown in Scheme 6.2.1. Thus, the diformyl terpyrrole **6.12** was condensed with the dicarboxyl-substituted dipyrrylmethane **6.13**.

Unfortunately, this approach gave only small quantities of the desired macrocycle **6.14**. Thus, while formally viable, this approach was abandoned in favor of that shown in Scheme 6.2.1.

Scheme 6.2.2

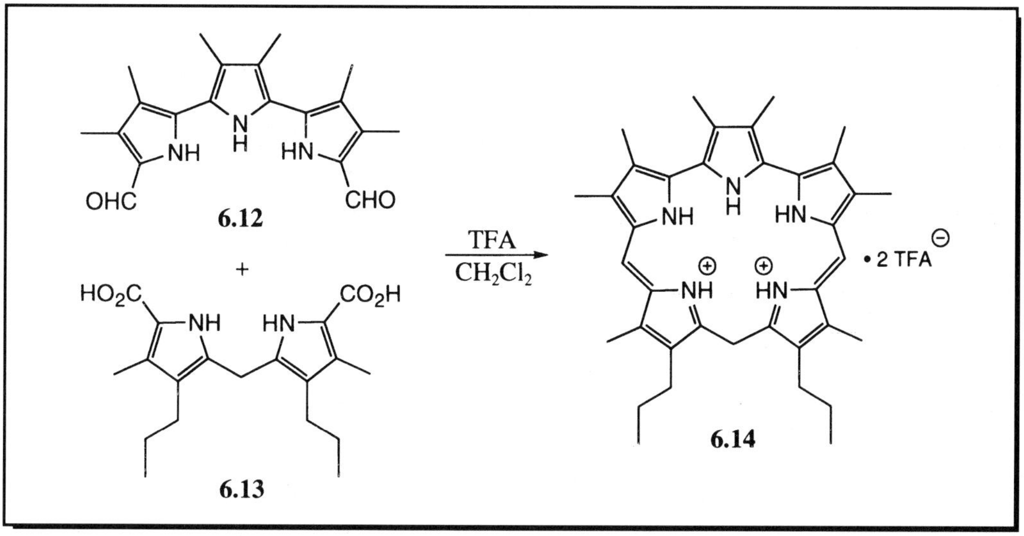

The isosmaragdyrins can be reversibly deprotonated. For instance, treating the protonated macrocycle **6.11a** with saturated aqueous NaHCO$_3$ affords the free-base macrocycle **6.11b**. The protonated form may be regenerated by adding TFA. When protonated, macrocycle **6.11** is reddish in color in organic solution, and displays a single, sharp absorbance band at 478 nm in its visible spectrum. The free-base form (e.g., **6.11b**), on the other hand, is deep green, and it exhibits an intense Soret-like absorbance band at 450 nm, along with two rather strong Q-type transitions at 663 nm and 695 nm in the visible spectrum.

Consistent with its 22 π-electron formulation, the free-base form of isosmaragdyrin (**6.11b**) appears to be aromatic, as judged from a preliminary ^{1}H NMR spectral analysis. On the other hand, the protonated form (**6.11a**) appears to be non-aromatic, as evidenced by the lack of any diamagnetic ring-current effects in its ^{1}H NMR spectrum. Based upon integration of the NMR signals, the lack of aromaticity for **6.11a** appears to be the result of protonation at one of the *meso*-positions of the macrocycle. This protonation effect thus resembles that observed for the 18 π-electron corroles discussed in Chapter 2.

Unfortunately, isosmaragdyrin **6.11**, like the "parent" pentaazasmaragdyrins appears to be inherently unstable, with partial decomposition occurring over a period of days. At the time this text was being produced, attempts were being made to stabilize this macrocycle through the formation of metal complexes and *via* the synthesis of heteroanalogs. In analogy to what was observed in the smaragdyrin series, it is hoped that these species will prove more stable than the all-aza system **6.11**.

6.3   Sapphyrin "Isomers": Four Bridging Carbons

6.3.1   *Ozaphyrins*

The furan-containing macrocycle **6.20**, recently reported by Johnson and Ibers and coworkers,[3] is formally a heterosapphyrin isomer. It is prepared in 5% yield by coupling bis(pyrrolyl)furan species **6.15** with a diformyl bipyrrole (e.g., **6.17**) under McMurry-type low-valent titanium coupling conditions (Scheme 6.3.1). In this reaction, the presumed 24 π-electron intermediate **6.18** is not isolated. Rather, it is presumably oxidized *in situ* to give the aromatic 22 π-electron species **6.20** directly. Using an analogous procedure, the thiophene-containing macrocycle **6.21** could also be prepared in *ca.* 2.6% yield.[4] Based on the emerald-green color displayed by **6.20** in CHCl$_3$ solution, this class of macrocycles has been assigned the name "ozaphyrin" (from the Emerald City of Oz in L. F. Baum's *The Wonderful Wizard of Oz*).

Scheme 6.3.1

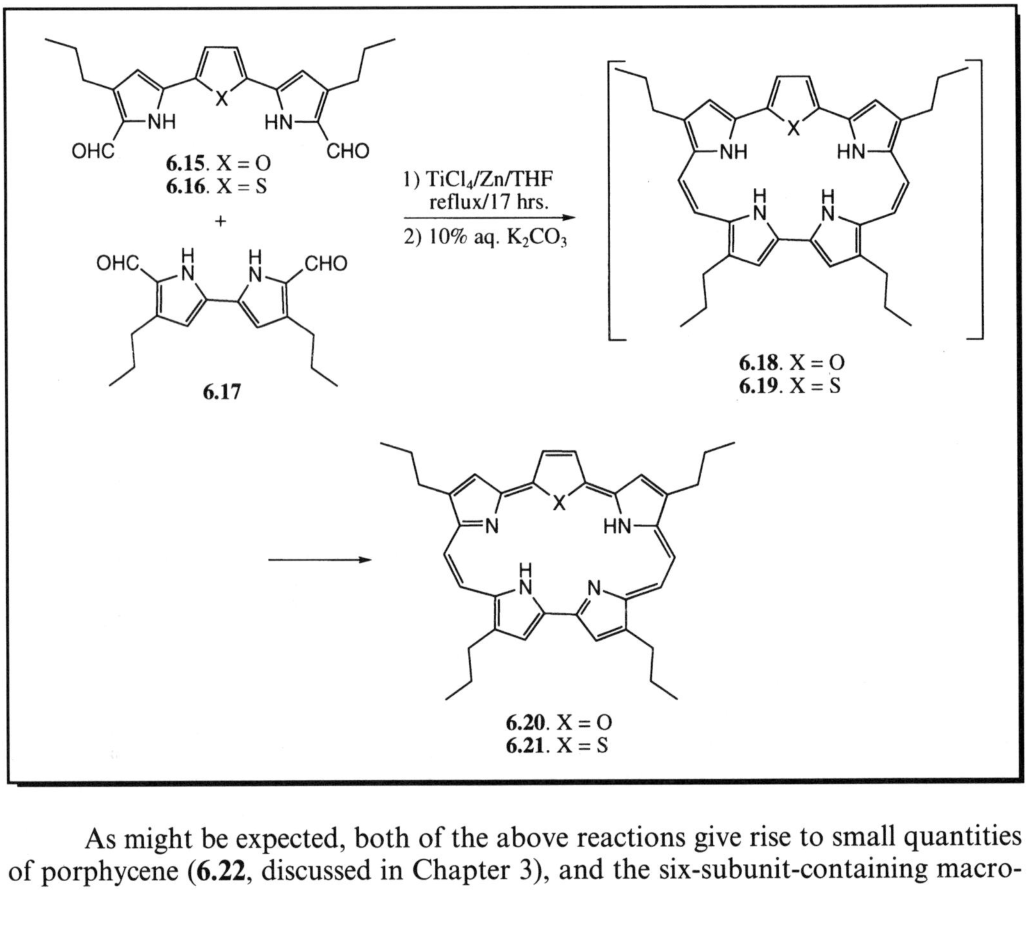

As might be expected, both of the above reactions give rise to small quantities of porphycene (**6.22**, discussed in Chapter 3), and the six-subunit-containing macrocycles **6.23** and **6.24** (termed "bronzaphyrins"; cf. Chapter 7) (Figure 6.3.1) as the result of the self-dimerization of the appropriate precursors (i.e., **6.17**, **6.15**, and **6.16**, respectively). Unfortunately, the bisoxabronzaphyrin **6.23** proved too unstable to isolate.

Figure 6.3.1 Structures **6.22–6.24**

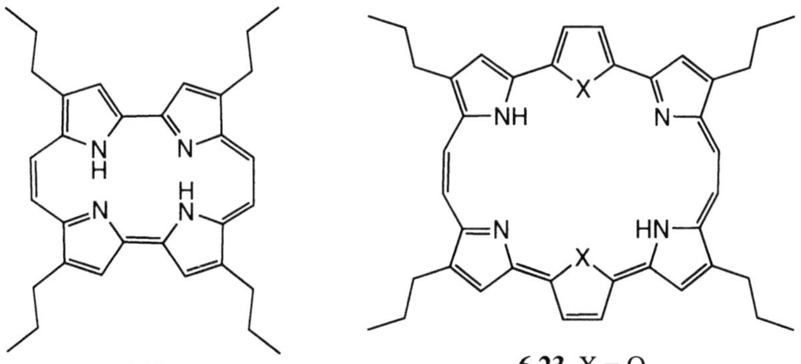

6.22

6.23. X = O
6.24. X = S

As implied above, the ozaphyrins may be considered as being isomers of heterosapphyrins. This is because they are penta-heterocyclic macrocycles that contain four *meso*-like bridging carbon atoms. However, in ozaphyrin, the methine-bridges are arranged in a (2.0.2.0.0) fashion, whereas in sapphyrin the corresponding methine bridges are arranged in a (1.1.1.1.0) fashion. Ozaphyrin may thus be referred to as being [22]pentaphyrin-(2.0.2.0.0) in accord with the Franck-type notation used throughout this book.

Like sapphyrins, the ozaphyrins contain a 22 π-electron conjugation pathway. The aromatic nature of ozaphyrins **6.20** and **6.21** is reflected in their respective ^{1}H NMR spectra. For instance, the internal NH protons of ozaphyrin **6.20** resonate at relatively high field ($\delta = -2.16$ ppm) whereas the external *meso*-like protons resonate at low field ($\delta = 10.31$ ppm); this is as would be expected for an aromatic macrocycle. Also, as expected for highly conjugated systems, the ozaphyrins display intense absorptions in the visible region of their electronic spectra. For instance, the mono-xaozaphyrin **6.20** displays a split Soret band at 414 and 430 nm ($\varepsilon = 120\,000$, and $99\,000\,M^{-1}\,cm^{-1}$, respectively), as well as three less intense Q-bands at 640, 677, and 735 nm ($\varepsilon = 33\,800$, 21 400, and $57\,500\,M^{-1}\,cm^{-1}$, respectively). The visible spectrum of monothiaozaphyrin **6.21** is similar. However, the principal absorption bands are slightly red-shifted relative to those of the furan-containing analog (**6.20**).

A single crystal X-ray structural analysis of the furan-containing macrocycle **6.20** revealed that it adopts a nearly planar conformation in the solid state (Figure 6.3.2). Indeed, the average deviation from the mean least-squares plane of all non-hydrogen atoms was found to be only 0.04 Å.[3] The planarity of **6.20** results in the core heteroatoms defining a pentagonal geometry. This finding, along with an appreciation that the core size of ozaphyrin is larger than that of the porphyrins ($r = 5.0$ Å for **6.20** vs. ≈ 4.0 Å for porphyrins), has led to the expectation that ozaphyrin might well be suited to act as highly effective mono- or even dinuclear complexing agents for first-row transition metals. However, at present no results have been published that would serve to confirm or refute this supposition.

Figure 6.3.2 Single Crystal X-Ray Diffraction Structure of Free-base Ozaphyrin **6.20**.

This figure was generated using information down-loaded from the Cambridge Crystallographic Data Centre and corresponds to a structure originally reported in reference 3. The macrocycle is very nearly planar (the mean deviation from planarity is 0.04 Å). The heteroatoms of the molecule define a distorted pentagon. Atom labeling scheme: carbon: O; nitrogen: ●; oxygen ⊖. Hydrogen atoms have been omitted for clarity

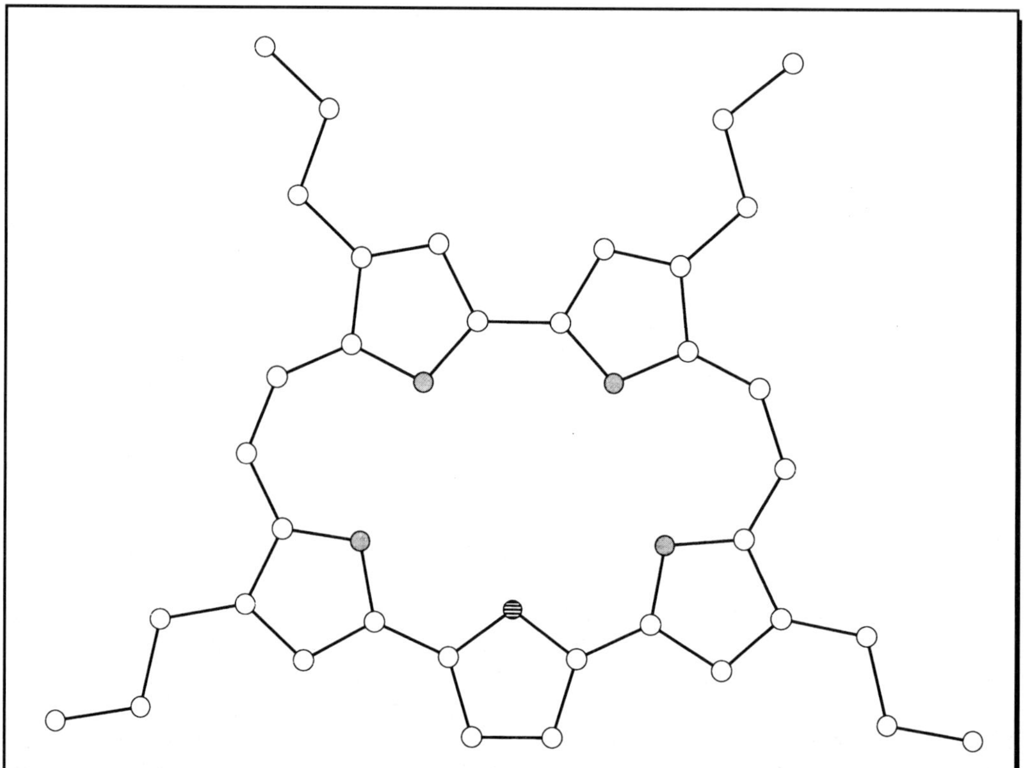

6.3.2   *[22]Dehydropentaphyrin-(2.1.0.0.1) and*
             [22]Pentaphyrin-(2.1.0.0.1)

Two recently reported macrocycles prepared by Sesslcr and coworkers may also formally be considered as being analogs of sapphyrin.[5] The first of these is the alkyne-containing expanded porphyrin **6.26** shown in Scheme 6.3.2. Macrocycle **6.26** was synthesized in 22% yield by condensing the α-free terpyrrole **6.7** with the diformyl "alkyne-stretched" bipyrrole **6.25** under HCl-catalyzed conditions. Macrocycle **6.26b** forms directly upon condensation as a 22 π-electron species. It contains five pyrroles with four *meso*-like bridging carbon atoms arranged in a (2.1.0.0.1) fashion. However, two of the four *meso* carbons are sp-hybridized. Thus, macrocycle **6.26** may be regarded as being the dehydro-form of a (2.1.0.0.1)-sapphyrin isomer. It may, therefore, be referred to as [22]dehydrosapphyrin-(2.1.0.0.1) or, under the nomenclature scheme used in this book, [22]dehydro*pentaphyrin*-(2.1.0.0.1).

Scheme 6.3.2

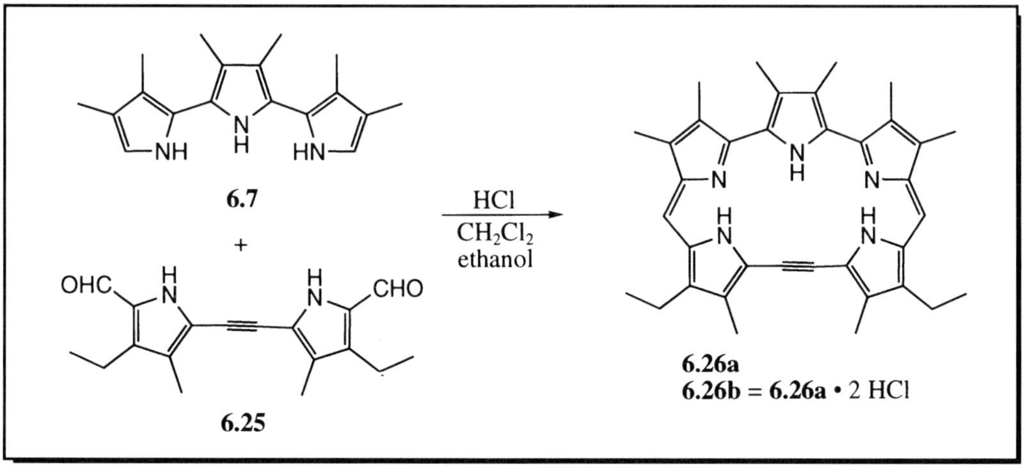

The second sapphyrin analog reported by Sessler and coworkers is derived directly from macrocycle **6.26**.[5] Here, subjecting the alkyne-containing macrocycle **6.26a** to Lindlar reduction conditions was found to afford a near quantitative conversion to the partially reduced macrocycle **6.27a** (Scheme 6.3.3). As inferred from ^{1}H NMR spectroscopic analysis, the specific structure of this reduction product is the *trans*-alkene **6.27**. This compound may formally be regarded as being a true isomer of pentaazasapphyrin, and is thus referred to as [22]sapphyrin-(2.1.0.0.1) or [22]*pentaphyrin*-(2.1.0.0.1).

In spite of the obvious structural differences, macrocycles **6.26** and **6.27** exhibit physical and spectral properties that are similar to those of the parent sapphyrins. For example, organic solutions of these macrocycles are green, while as solids they are metallic blue. The bishydrochloride salt of the alkyne-containing macrocycle (**6.26b**) in CH_2Cl_2 displays a rather intense Soret-like absorbance band at 476 nm

Scheme 6.3.3

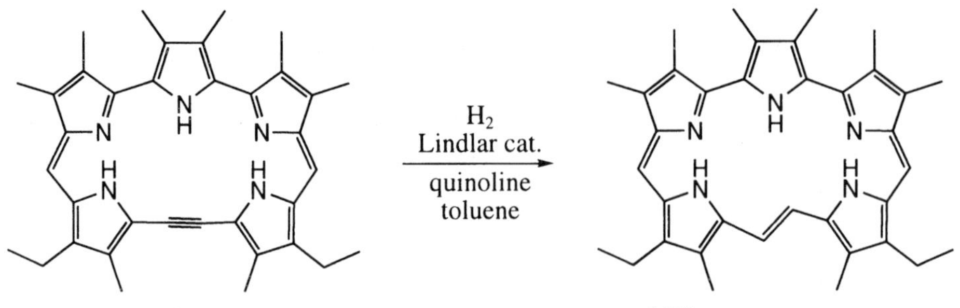

6.26a

H₂
Lindlar cat.

quinoline
toluene

6.27a
6.27b = **6.27a** • 2 HCl
6.27c = **6.27a** • 2 HClO₄

Figure 6.3.3 Single Crystal X-Ray Diffraction Structure of Free-base
[22]Dehydropentaphyrin-(2.1.0.0.1) **6.26a**.
This figure was generated using information down-loaded from the Cambridge Crystallographic
Data Centre and corresponds to a structure originally reported in reference 5. The macrocycle
deviates somewhat from planarity (the mean deviation from planarity is 0.34 Å). The alkyne subunit
is 1.214 Å in length. Atom labeling scheme: carbon: ○; nitrogen: ●. Hydrogen atoms have been
omitted for clarity

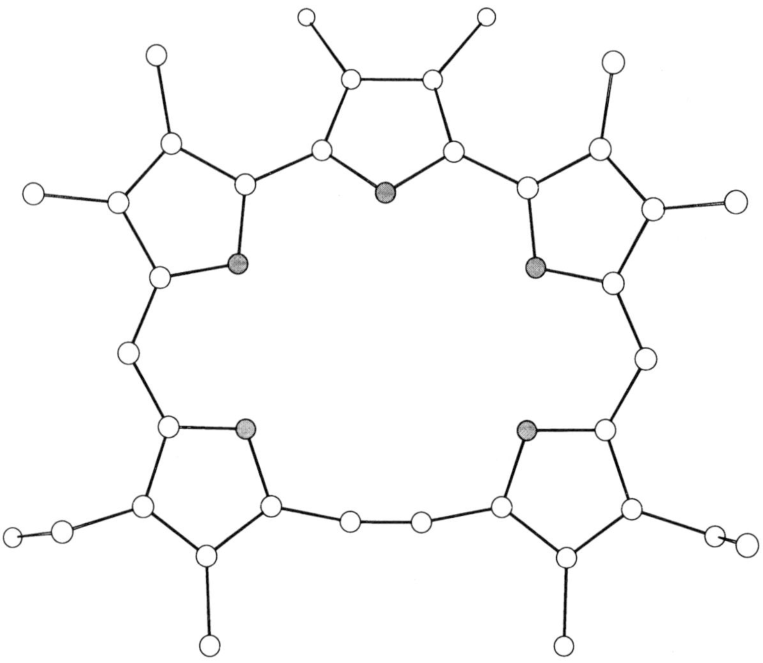

(ε = 406 200 M^{-1}cm^{-1}) in its visible spectrum along with two less-intense Q-type bands at 617 and 665 nm (ε = 15 300 and 11 300 M^{-1}cm^{-1}, respectively). Similarly, the protonated form of [22]sapphyrin-(2.1.0.0.1) (**6.27c**) displays a sharp Soret band at 481 nm (ε = 349 200 M^{-1}cm^{-1}) and two Q-type bands at 642 and 700 nm (ε = 11 700 and 15 900 M^{-1}cm^{-1}, respectively).

    A single crystal X-ray structural analysis of the alkyne-containing product **6.26a** revealed that the macrocycle adopts a near-planar conformation in the solid state (Figure 6.3.3). Deviation from planarity results, presumably, from the methyl–methyl steric interactions within the terpyrrolic subunit. The crystallographic data also served to support the contention that the macrocycle does indeed contain a true an alkyne-like (C-C≡C-C) fragment. Such a bond isolation, although expected in the present instance, stands in contrast to what is seen in the case of Vogel's "stretched porphycene" precursor.[6] In this latter instance cumulene-type (C=C=C=C) bond delocalization is observed.

Figure 6.3.4   Single Crystal X-Ray Diffraction Structure of the Bis-HCl Salt of [22]Pentaphyrin-(2.1.0.0.1) **6.27b** Confirming the *Trans* Orientation of the CH=CH *Meso*-like Bridge.

This figure was generated using information down-loaded from the Cambridge Crystallographic Data Centre and corresponds to a structure originally reported in reference 5. The macrocycle deviates somewhat from planarity (the mean deviation from planarity is 0.29 Å). One chloride anion sits above the mean macrocyclic plane and the other sits below. Atom labeling scheme: carbon: ○; nitrogen: ◉; chlorine: ⊛. Hydrogen atoms have been omitted for clarity

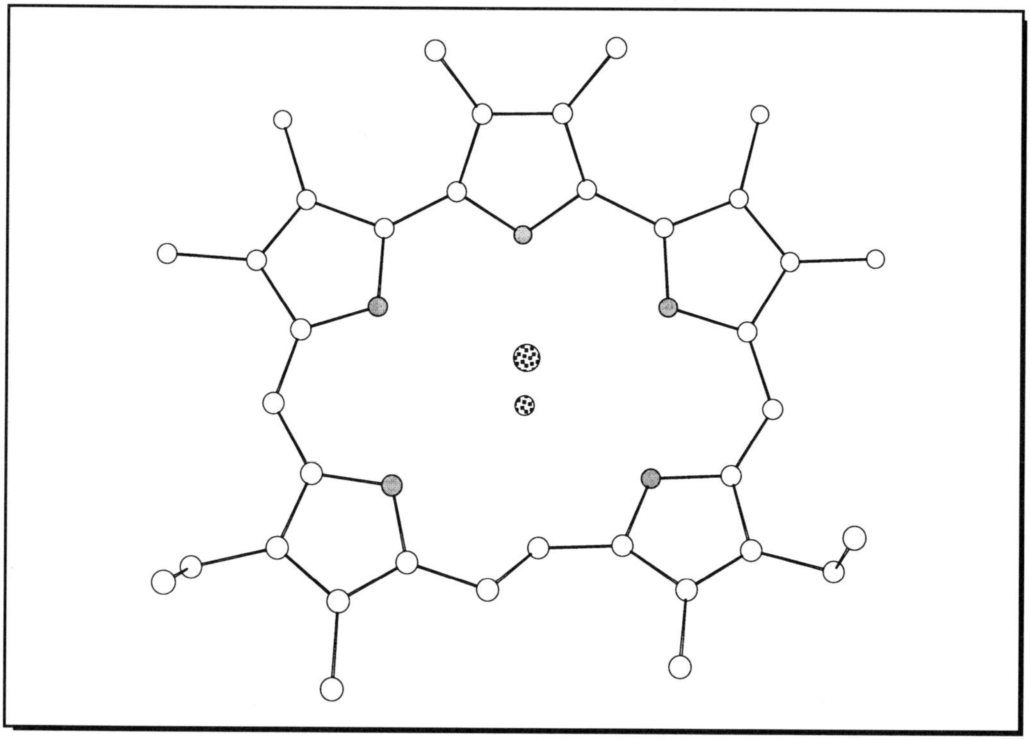

A separate single crystal X-ray structural analysis revealed that the partially reduced macrocycle **6.27b** also adopts a near-planar conformation in the solid state (Figure 6.3.4). It also confirmed the anticipated trans-stereochemistry about the *meso*-C=C bond in that one of the protons was found to be directed into the core of the macrocycle, while the other points outward. Taken together these solid-state structural findings are thus consistent with the proposal that **6.27** is a true isomer of sapphyrin.

6.4   Pentaphyrins

The sapphyrin class of expanded porphyrins discussed in Chapter 5 contains an additional pyrrole-like heterocycle as compared to the porphyrins but contains only four, as opposed to five, *meso*-like bridging carbons. From a conceptual perspective, therefore, the next logical step in expanded porphyrin development (after sapphyrin) would be to expand the porphyrin skeleton by adding an additional *meso*-bridging carbon atom into the basic sapphyrin skeleton. This would result in an expanded porphyrin containing five pyrrolic subunits linked *via* five alternating *meso*-like bridges. Such systems (e.g. **6.6**), termed pentaphyrins (or more precisely [22]pentaphyrins-(1.1.1.1.1)), have indeed been prepared and a summary of their synthesis now follows. The chemistry of a structurally related system, superphthalocyanine **6.28** (Figure 6.4.1), is reviewed in Chapter 9.

Figure 6.4.1   Structure **6.28**

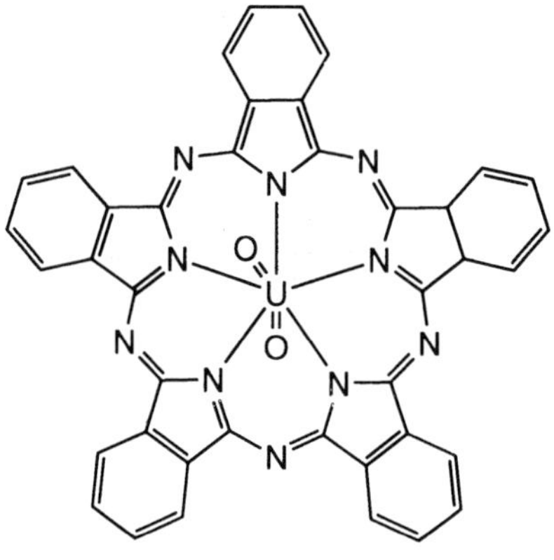

Uranyl Superphthalocyanine
6.28

6.4.1   Pentaazapentaphyrins

6.4.1.1   Synthesis

The first synthesis of pentaphyrin was reported by Gossauer in 1983.[7–10] He used an HBr-catalyzed "2 + 3" MacDonald-type condensation between the diformyl tripyrrane **6.29** and the α-free dipyrrylmethane derivative **6.32** to establish the basic macrocyclic framework. Oxidation with chloranil then gave pentaphyrin **6.35** in 31% overall yield (Scheme 6.4.1). Subsequent to this original disclosure,[7] syntheses of pentaphyrins with various other peripheral substituents appeared in the literature (e.g., **6.38** and **6.39**). They were all made using this same general procedure.[11,12] Interestingly, it was found that pentaphyrins could not be prepared when the nucleophilic and electrophilic nature of the reactant dipyrrylmethane and tripyrrane (the "2" and "3" components, respectively) were reversed. This became evident when the diacid tripyrrane **6.38** and the diformyl dipyrrylmethane **6.39** were reacted under conditions identical to those used to prepare pentaphyrin **6.35**. Here, only porphyrins, and not pentaphyrins, were obtained (Scheme 6.4.2).[7,13,14]

Scheme 6.4.1

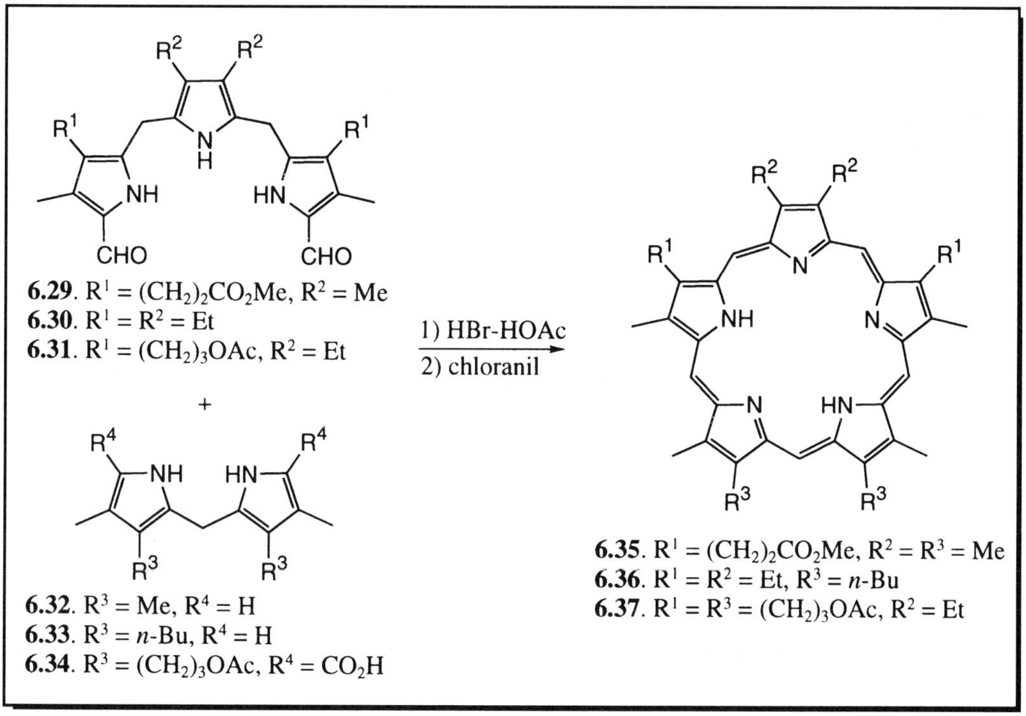

6.29. $R^1 = (CH_2)_2CO_2Me$, $R^2 = Me$
6.30. $R^1 = R^2 = Et$
6.31. $R^1 = (CH_2)_3OAc$, $R^2 = Et$

1) HBr-HOAc
2) chloranil

6.32. $R^3 = Me$, $R^4 = H$
6.33. $R^3 = n\text{-}Bu$, $R^4 = H$
6.34. $R^3 = (CH_2)_3OAc$, $R^4 = CO_2H$

6.35. $R^1 = (CH_2)_2CO_2Me$, $R^2 = R^3 = Me$
6.36. $R^1 = R^2 = Et$, $R^3 = n\text{-}Bu$
6.37. $R^1 = R^3 = (CH_2)_3OAc$, $R^2 = Et$

Pentaphyrins have also been isolated as side-products of "3 + 3" reactions used to prepare a class of hexapyrrolic macrocycles called hexaphyrins (discussed in

Chapter 7).[15] For instance, when α-free tripyrrane **6.40** is condensed with the diformyl tripyrrane **6.29** under HBr-catalyzed conditions, in addition to the desired hexaphyrin **6.41**, an 8% yield of pentaphyrin **6.42** results (Scheme 6.4.3). This pentaphyrin-yielding ring-contraction process thus formally resembles the "4 + 2" rearrangement procedure used to obtain mono *meso*-phenyl-substituted sapphyrins (cf. Chapter 5).

Scheme 6.4.2

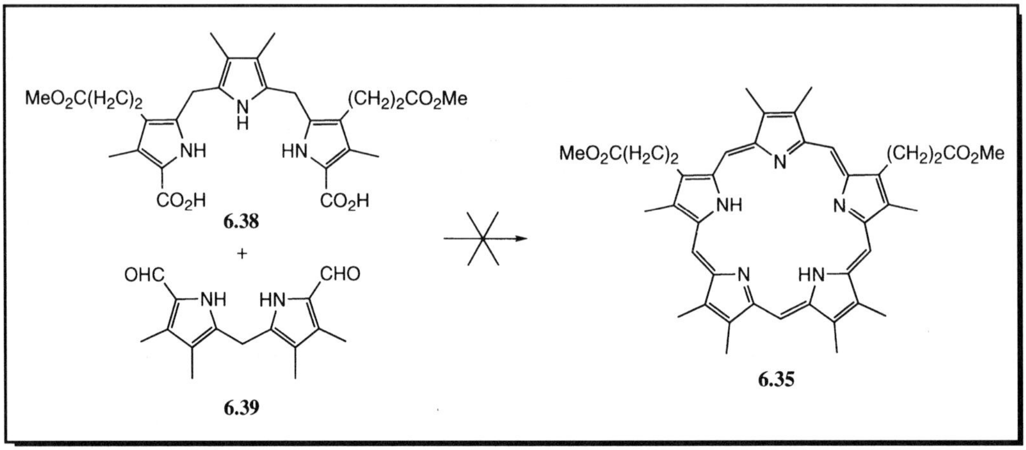

6.4.1.2 General Physical Characteristics

As is true for sapphyrin and smaragdyrin, it is possible to identify a delocalized 22 π-electron periphery in pentaphyrin. As such, members of this class of macrocycles are formally regarded as being aromatic. In general, the physical and spectroscopic properties of pentaphyrin are consistent with this assignment. For instance, free-base pentaphyrins are green in methanolic solution and are yellow–green in CH_2Cl_2. In the solid state, the free-base forms of pentaphyrin are orange, not deep blue in color as is true for the sapphyrins and smaragdyrins. Since three of the five nitrogen atoms of the pentaphyrin core are effectively pyridine-like, they are readily protonated upon addition of acid (Figure 6.4.2). Protonated pentaphyrins are deep green in solution.

As would be expected for an aromatic macrocycle, pentaphyrins display exceptionally strong absorbance bands in the visible portion of the electronic spectrum. For instance, pentaphyrin **6.35** displays a sharp Soret-like absorbance band at 458 nm (CH_2Cl_2, $\varepsilon = 238\,000\ \mathrm{M^{-1}\,cm^{-1}}$), and two Q-like absorbances at 642 and 695 nm ($\varepsilon = 5990$ and $3540\ \mathrm{M^{-1}\,cm^{-1}}$, respectively).[7] They thus resemble spectroscopically both the sapphyrins and smaragdyrins, as well as the tetrapyrrolic porphyrins from which they are conceptually derived. Like these various aromatic "cousins", the pentaphyrins display features in their ^{1}H NMR spectra that are characteristic of strong diamagnetic ring-current effects. The internal NH protons, for instance, reso-

Scheme 6.4.3

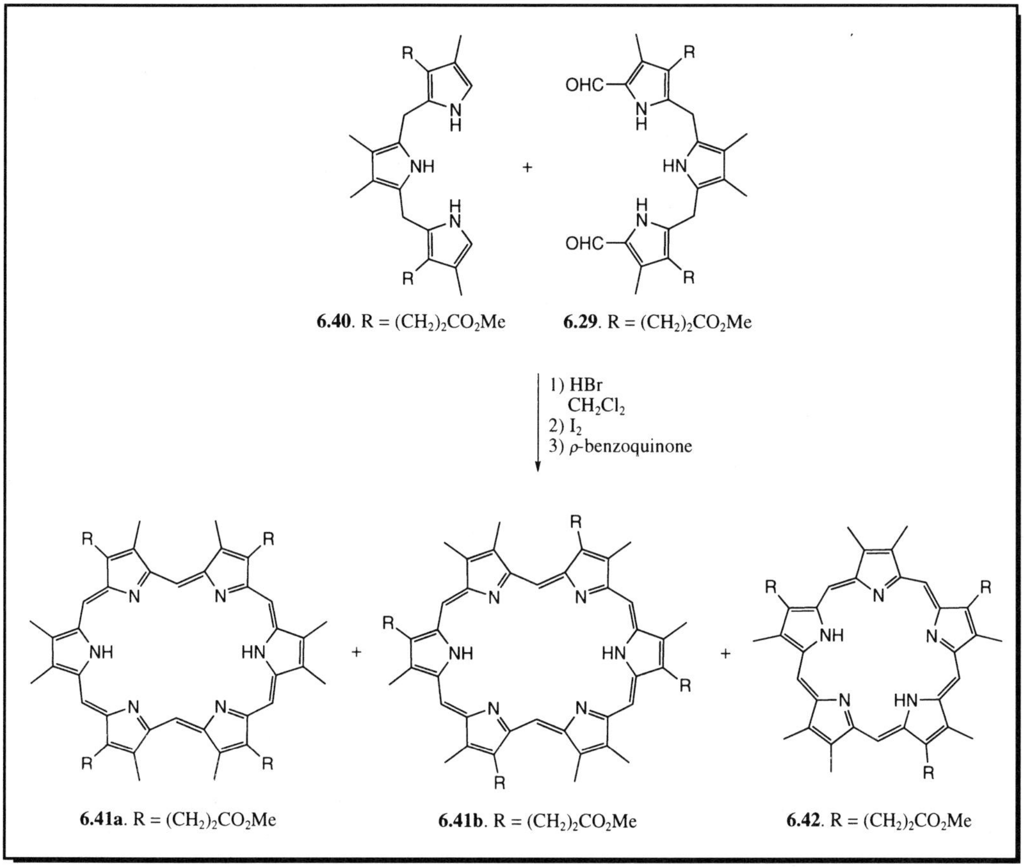

6.40. R = (CH$_2$)$_2$CO$_2$Me   **6.29**. R = (CH$_2$)$_2$CO$_2$Me

1) HBr
   CH$_2$Cl$_2$
2) I$_2$
3) p-benzoquinone

6.41a. R = (CH$_2$)$_2$CO$_2$Me   **6.41b**. R = (CH$_2$)$_2$CO$_2$Me   **6.42**. R = (CH$_2$)$_2$CO$_2$Me

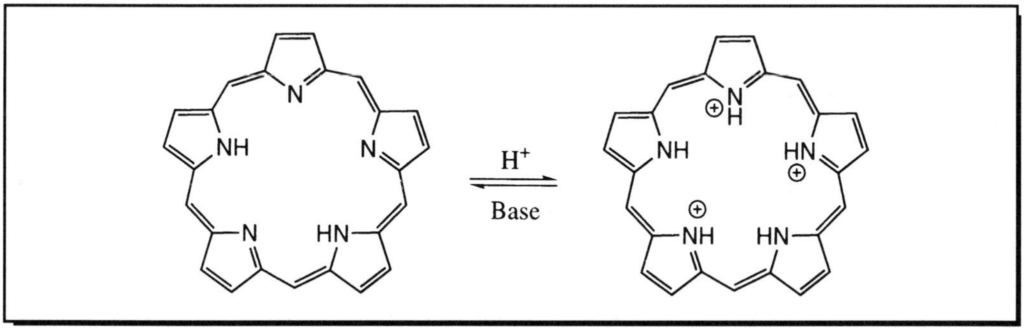

Figure 6.4.2   Schematic Representation of the Relevant Protonation–Deprotonation Equilibria for Pentaphyrins

nate at *ca.* -5 ppm, while the external methine protons resonate at *ca.* 12.5 ppm (in CDCl$_3$ containing 1% trifluoroacetic acid). Such findings are, of course, fully in accord with the presumed aromatic nature of pentaphyrin.

6.4.1.3   Water-soluble Pentaphyrin

In 1995, Sessler and coworkers reported the preparation of a water-soluble pentaphyrin, namely the tetrahydroxypentaphyrin **6.43**.[12] It was synthesized from pentaphyrin **6.37** by deprotection of the acetoxy groups with methanolic HCl (Scheme 6.4.4). Using this procedure, tetrahydroxypentaphyrin **6.43** could be isolated in yields as high as 92% (based on **6.37**; overall yield based on precursors **6.31** and **6.34** $=$ 56%).

Scheme 6.4.4

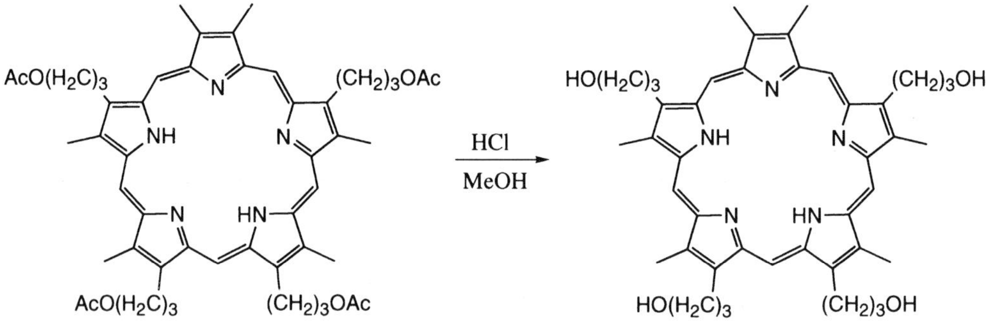

Water-soluble pentaphyrin **6.43** was designed to be studied as a possible cytotoxic agent. Towards this end, the cytotoxicity of **6.43** was evaluated using a cultured tumor cell line.[12] In this study, it was found that this water-soluble pentaphyrin does indeed exert a significant cytotoxic effect, even at low concentrations. The observed ID$_{50}$, the dose at which a 50% reduction in cancerous cells after 48 hours was achieved, was found to be 1.37 µM. This value compares favorably with those of other compounds of the same general macrocyclic class (e.g., pyridinium-substituted metallo-porphyrins). However, unlike other systems, pentaphyrin **6.43** shows an approximate degradation half-life of only 2–3 days under physiological conditions, making it and its congeners attractive targets for future study.

6.4.1.4   Metalation Chemistry

A major motivation leading to the development of pentaphyrin chemistry was the hope that these species would act as pentadentate ligands and thus allow the coordination of a range of both small and large metal cations. While initial reports[16] indicated that pentaphyrin **6.35** was capable of forming complexes with Zn(II), Co(III), and Hg(II) (e.g., **6.44–6.46**), it was speculated that pentaphyrin was acting as a dianionic ligand, and that the bound metal center was coordinated *via* but two of the five available inward-pointing nitrogen atoms (Scheme 6.4.5). Unfortunately, no X-ray diffraction quality crystals were ever obtained for any of these species. Thus, their exact nature remains indeterminate at present.

Scheme 6.4.5

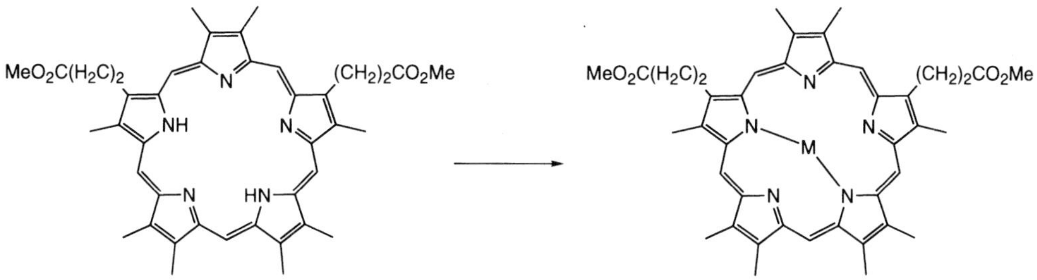

6.35

6.44. M = Zn(II)
6.45. M = Co(III)-CN
6.46. M = Hg(II)

Gossauer also demonstrated that pentaphyrin formed a stable complex with uranyl (UO_2^{2+}) cation (i.e., **6.47**) (Scheme 6.4.6).[10] While the exact nature of this complex was not fully delineated, it was found that the uranyl center could be readily displaced upon treatment with acid without destruction of the macrocycle. Later,

Scheme 6.4.6

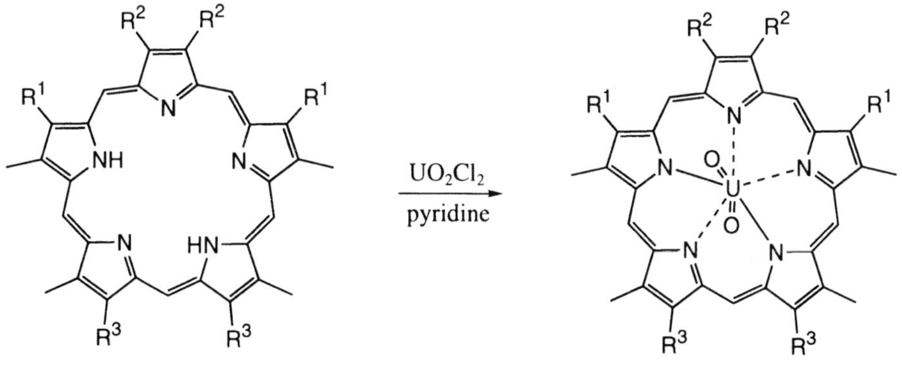

UO_2Cl_2
pyridine

6.35. $R^1 = CH_2CH_2CO_2Me$, $R^2 = R^3 = Me$
6.36. $R^1 = Me$, $R^2 = Et$, $R^3 = n\text{-}Bu$
6.43. $R^1 = R^3 = CH_2CH_2CH_2OH$, $R^2 = Et$

6.47. $R^1 = CH_2CH_2CO_2Me$, $R^2 = R^3 = Me$
6.48. $R^1 = Me$, $R^2 = Et$, $R^3 = n\text{-}Bu$
6.49. $R^1 = R^3 = CH_2CH_2CH_2OH$, $R^2 = Et$

Sessler and coworkers prepared two similar uranyl complexes **6.48**[11] and **6.49**.[12] A single crystal X-ray diffraction study of the former revealed a saddle-shaped penta-phyrin macrocycle with the uranyl cation centrally coordinated in a nearly pentago-nal bipyramidal fashion (Figure 6.4.3). Distortion from planarity arises, presumably,

Figure 6.4.3 Single Crystal X-Ray Diffraction Structure of the Uranylpentaphyrin **6.48**.

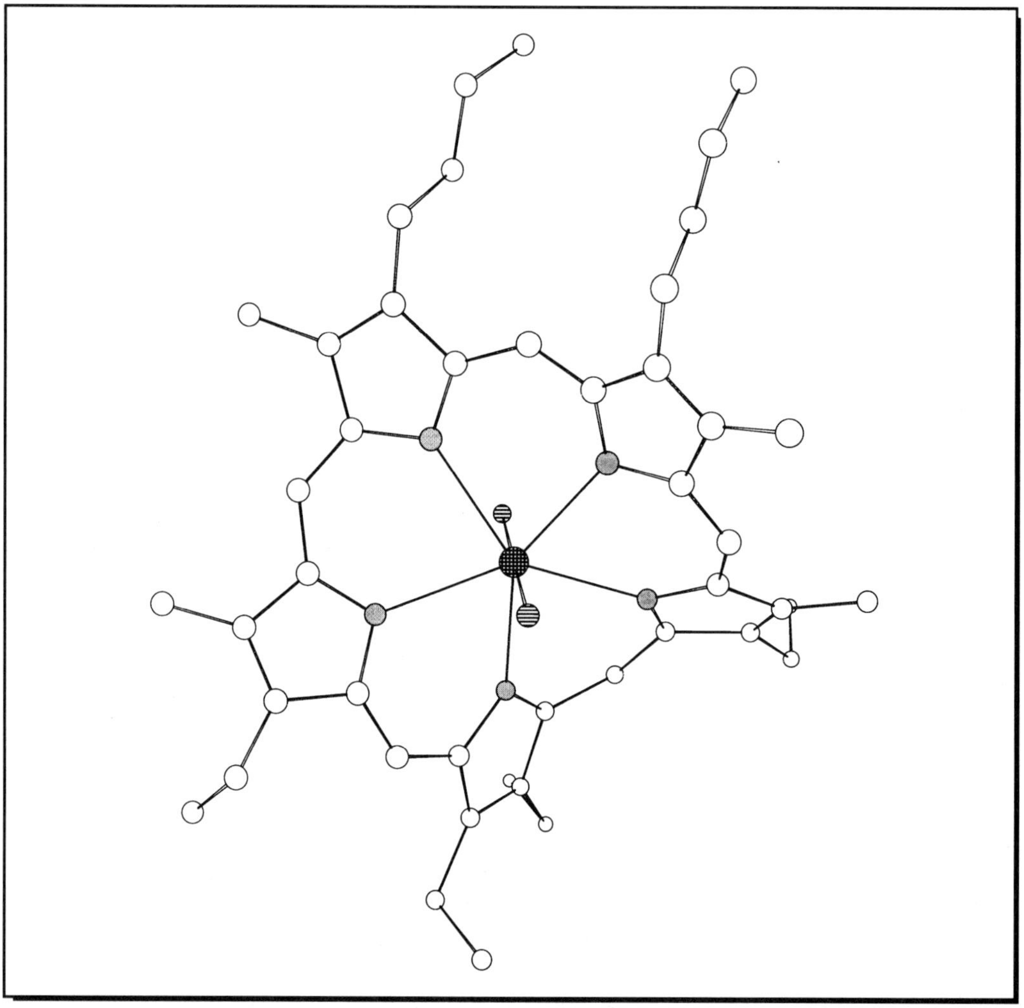

This figure was generated using information down-loaded from the Cambridge Crystallographic Data Centre and corresponds to a structure originally reported in reference 11. The macrocycle is distorted considerably from planarity and is saddle-shaped in the crystal. The macrocyclic core nitrogen atoms and the axial oxygen atoms define an essentially symmetrical pentagonal bipyramidal coordination environment for the uranium(VI) atom. The average U–N bond length is 2.54 Å and the average U–N–U angle for adjacent nitrogen atoms is 73.1°. Atom labeling scheme: uranium: ◉; carbon: ○; nitrogen: ●; oxygen: ⊖. Hydrogen atoms have been omitted for clarity

from the bonding demands of the uranyl cation within a coordinating cavity that is slightly too large. This macrocyclic distortion results in five nearly equivalent N–U bonds with an average bond length of 2.541 Å. Thus, in spite of the apparent size mismatch, this complex is interesting in that it stands as a near unique example of a pyrrole-derived expanded porphyrin wherein well-defined actinide cation coordination is achieved.

6.4.2   Heteropentaphyrins

Vogel and coworkers have recently reported the synthesis of a pentathiophene analog of pentaphyrin, namely decaethyl[22]pentathiapentaphyrin-(1.1.1.1.1) **6.51**.[17–19] During efforts directed toward the preparation of octaethyltetrathiaporphyrin, Vogel found that the acid-catalyzed condensation of 2-hydroxymethyl-3,4-diethylthiophene (**6.50**) afforded not only the intended tetrathiapphyrinogen in 30% yield, but also a 19% yield of the cyclic pentamer **6.51** and a 0.5–2% yield of a cyclic hexamer (described in Chapter 7) (Scheme 6.4.7).

Scheme 6.4.7

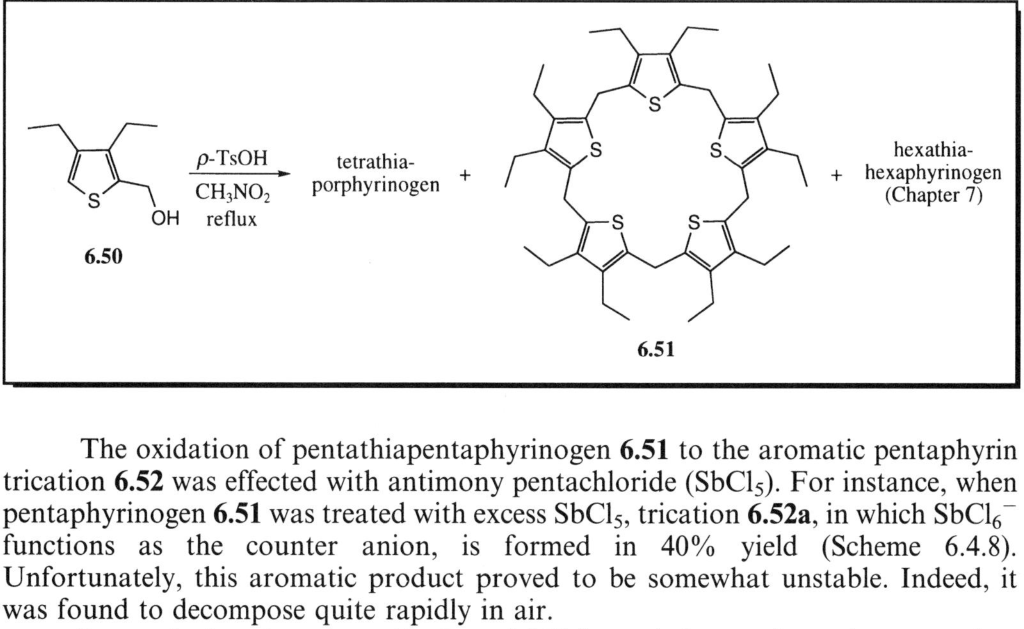

The oxidation of pentathiapentaphyrinogen **6.51** to the aromatic pentaphyrin trication **6.52** was effected with antimony pentachloride ($SbCl_5$). For instance, when pentaphyrinogen **6.51** was treated with excess $SbCl_5$, trication **6.52a**, in which $SbCl_6^-$ functions as the counter anion, is formed in 40% yield (Scheme 6.4.8). Unfortunately, this aromatic product proved to be somewhat unstable. Indeed, it was found to decompose quite rapidly in air.

Interestingly, when the oxidation of **6.51** is carried out using only one equivalent of $SbCl_5$, trication **6.52b**, in which the $Sb_6Cl_{21}^{3-}$ trianion serves as the counterion, is obtained in 7.5% yield. Although the yield in this latter oxidation reaction is significantly lower, the product (**6.52b**) proved to be much more stable in air than **6.52a**.[17] It could thus be subject to X-ray diffraction analysis. The resulting structure revealed that the ring skeleton is near-planar and of approximate D_{5h} symmetry. Proton NMR spectroscopic studies (in CD_3NO_2) were consistent with this planar, highly symmetric arrangement **6.25b** being retained in solution.[17] Further, diamagnetic ring-current effects were also revealed by these studies. Such ring-current effects were most evident in the downfield shifting of the *meso*-like protons (δ = 14.74 ppm). They were also seen in the chemical shift positions of the ethyl substituents; the methylene protons resonate at 5.73 ppm whereas the methyl protons resonate at 2.60 ppm. The ^{1}H NMR spectrum of trication **6.52b** is nearly identical to that of **6.52a**.

Scheme 6.4.8

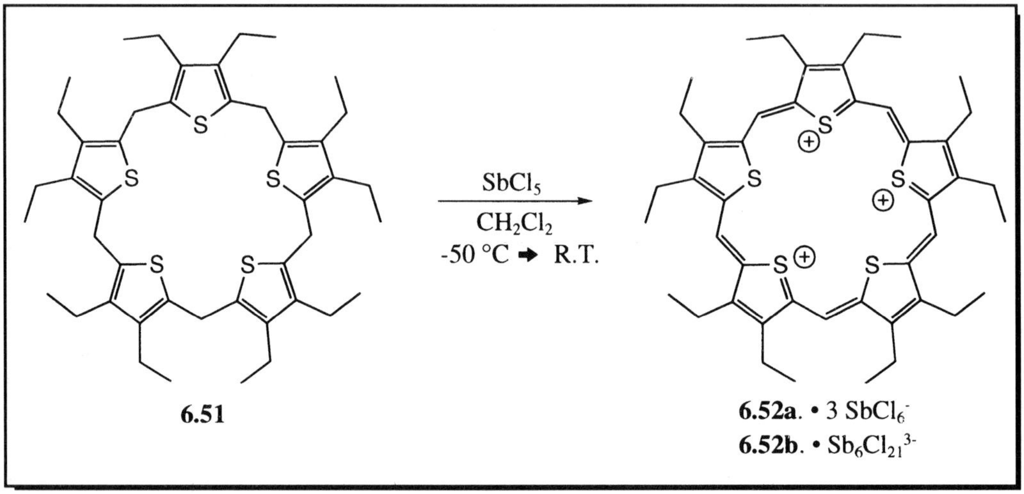

6.51

6.52a. • 3 SbCl$_6^-$
6.52b. • Sb$_6$Cl$_{21}^{3-}$

Like its pentaaza congeners, pentathiapentaphyrin **6.52** exhibits a decidedly porphyrin-like UV-vis spectrum.[17] However, as would be expected for a (22 π-electron) conjugation pathway, the absorbance bands are bathochromically shifted relative to analogous 18 π-electron systems.[19,20] Interestingly, the peaks of **6.52** were found to be red-shifted (by *ca.* 60–100 nm) relative to those of the triply protonated form of pentaazapentaphyrins. For instance, trication **6.52b** displays an intense Soret-like absorbance at λ = 525 nm (ε = 317 000 $M^{-1}cm^{-1}$) and two main Q-type absorbances at 731 and 801 nm (ε = 33 400 and 30 400 $M^{-1}cm^{-1}$, respectively) in its electronic spectrum. By comparison, the typical pentaaza pentaphyrin trication gives rise to a Soret band at *ca.* 460 nm and Q-bands in the *ca.* 640–695 nm spectral region.[7]

Pentathiapentaphyrinogen **6.51** has also been subject to oxidation using iron trichloride (FeCl$_3$) as the oxidizing agent.[17] Interestingly, in this instance, it is not an aromatic trication (e.g. **6.52**) that is formed. Rather, the fully conjugated 24 π-electron monocation **6.53** is obtained in 82% yield (Scheme 6.4.9). On the other hand, when this same starting pentathiapentaphyrinogen **6.51** is first oxidized with FeCl$_3$ or DDQ, and then subsequently treated with the reducing agent sodium borohydride, the partially conjugated macrocycle **6.54** is produced in 50% yield (Scheme 6.4.10).[17]

Treating this latter macrocycle with potassium hydride in THF at $-78\,^\circ$C generates the monoanion **6.55** in the form of its potassium salt.

Scheme 6.4.9

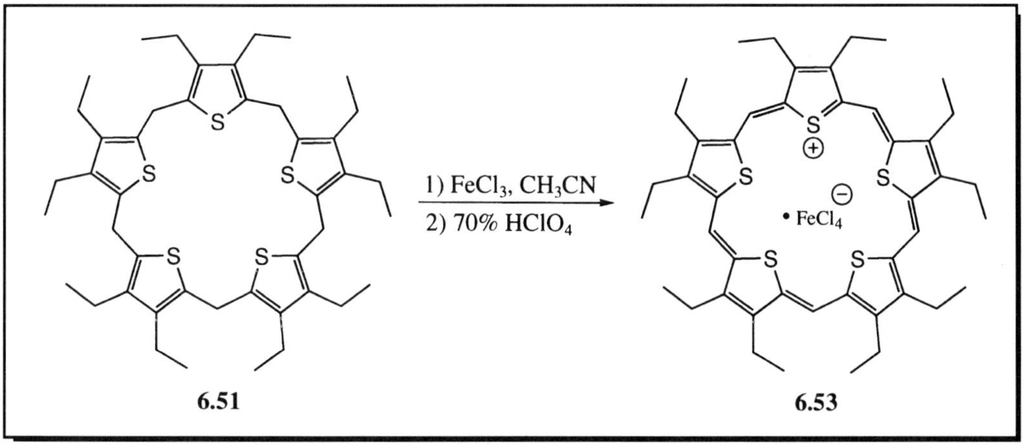

6.51   6.53

Scheme 6.4.10

6.54   6.55

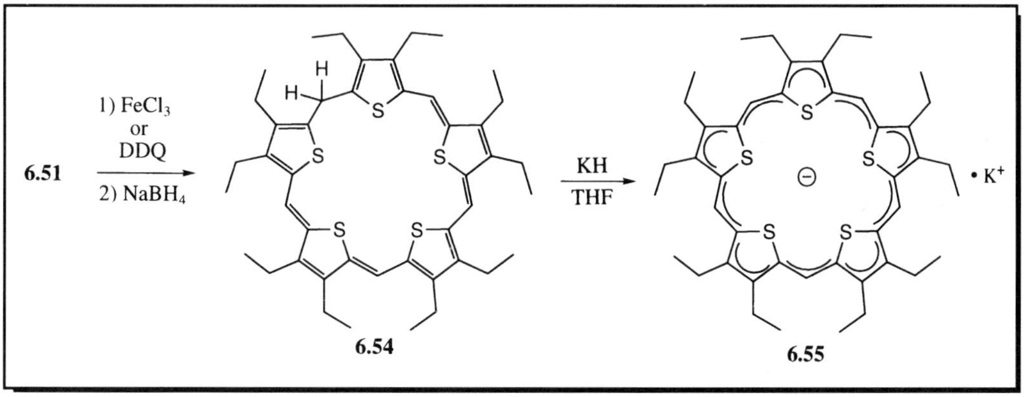

Macrocycle **6.55**, with its 26 π-electron conjugation pathway, was considered to be aromatic. This was based on an analysis of its ^{1}H NMR spectrum. The latter revealed signals, ascribable to the *meso*-like protons, at 13.27 ppm (corresponding to a 6.62 ppm downfield shift as compared to **6.54**). As expected, the UV-vis absorption spectrum of **6.55** was also found to be characteristic of an aromatic species. For instance, a very intense absorption band was observed at λ = 536 nm (ε = 503 000 $M^{-1}\,cm^{-1}$). This corresponds to a Soret band that is red-shifted by 11 nm relative to trication **6.52b**.

6.5   Other Systems

6.5.1   Pentaoxa[30]pentaphyrin-(2.2.2.2.2)

In 1969, J. A. Elix showed that the Wittig-type self-condensation of the furyl-phosphonium salt **6.56** gave a wide mixture of products, most of which (*ca.* 85%) were polymeric in nature.[21] However, small quantities of lower-molecular-weight substances could be isolated that were determined to be cyclic tetramers, pentamers, and hexamers. (The cyclic tetrameric and hexameric products are discussed in Chapters 4 and 7, respectively. Here, a brief overview of the pentameric product, pentaoxa[30]pentaphyrin-(2.2.2.2.2) **6.57** (referred to as "[30]annulene pentoxide" by Elix[21]) is presented. This pentamer was actually isolated from the reaction mixture as two easily separated geometrical isomers (**6.57a** and **6.57b**). They were obtained in a combined yield of *ca.* 0.9% (Scheme 6.5.1). Although represented here as the all-*cis* isomer, the exact configurations of these isomers was not determined. However, IR spectra of both compounds revealed the presence of *trans*-double bonds in both isomers, with one of the isomers (assigned as **6.57b**) containing more *trans*-double bonds than the other (**6.57a**). The pentamer containing more *trans*-double bonds (**6.57b**) was found to be the thermodynamically more stable isomer. This was determined by noting that solutions of **6.57a** stored at 0° C slowly isomerized so as to produce **6.57b**.[21]

Scheme 6.5.1

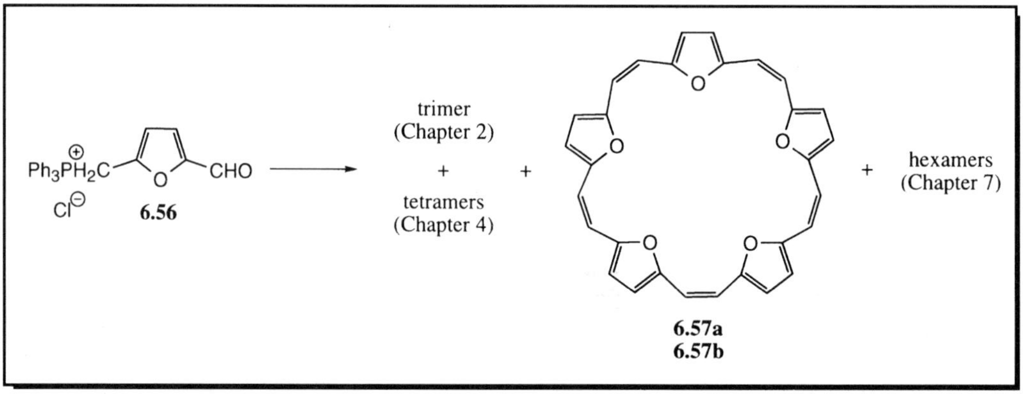

The UV-vis spectra of both of the [30]annulene pentoxide isomers **6.57a** and **6.57b** were found to resemble closely those of the related annulenes. Specifically, they revealed several absorbance bands in the 230–500 nm spectral region. It was inferred from the appearance of these high-intensity, long-wavelength bands that the pentaoxa[30]pentaphyrin-(2.2.2.2.2) isomers did not deviate significantly from planarity in solution. In spite of this assumed planarity, however, ¹H NMR spectral analyses failed to reveal any substantial macrocyclic ring-current effects in either of these isomers.

6.5.2   Pentathia[30]pentaphyrin-(2.2.2.2.2)

Vogel and coworkers have succeeded in preparing two geometrically isomeric [30]annulene pentasulfide macrocycles, namely **6.59a** and **6.59b**.[22,23] These two macrocycles were prepared *via* the McMurry-type reductive carbonyl coupling of 3,4-diethyl-2,5-diformylthiophene (Scheme 6.5.2).[§] They were isolated in a combined yield of 1.5% from a mixture of products that included a cyclic trimer, tetramer, and two isomeric cyclic hexamers (discussed in Chapters 2, 4, and 7, respectively). Separation of isomer **6.59a** from **6.59b** was achieved by fractional crystallization. However, this separation was complicated by the fact that the two isomers obtained appeared to interconvert readily under normal laboratory conditions.[23]

Scheme 6.5.2

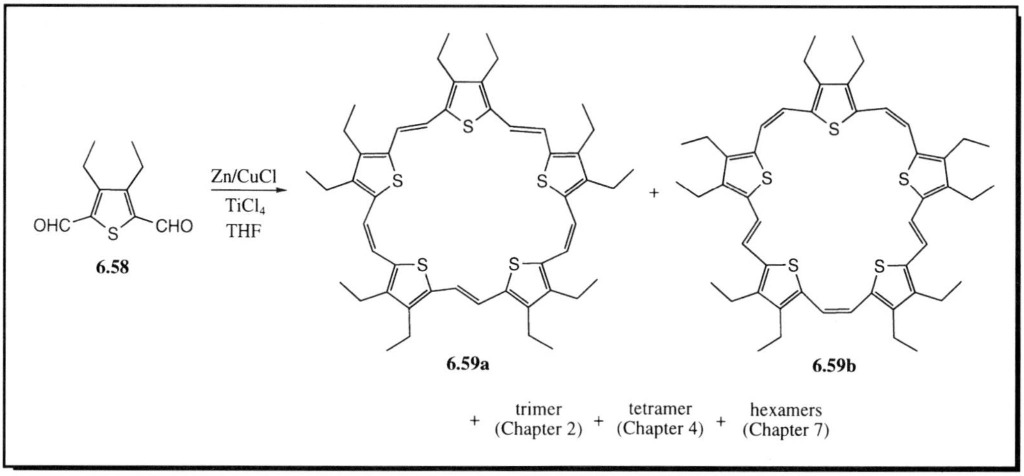

A single crystal X-ray diffraction analysis carried out on **6.59a** showed that the macrocycle rests in a conformation in which two of the thiophene units occupy an approximate plane that is nearly perpendicular to a plane defined by the other three thiophene subunits.[23] This conformation, it was postulated, precludes extended π-conjugation and, therefore, also any diamagnetic (aromatic) ring-current effects. Certainly, this is consistent with what was seen in the course of normal ^{1}H NMR spectroscopic analysis. Specifically, no clear diamagnetic effects were apparent in the spectra of either **6.59a** or **6.59b**. In fact, for both of these species, ^{1}H NMR analyses revealed the existence of very complex solution-state morphologies that likely disrupt π-delocalization. As a consequence of this disrupted conjugation, the UV-vis absorbance spectra recorded for **6.59a** and **6.59b** in CH_2Cl_2 failed to show the classic intense peaks expected for an aromatic expanded porphyrin. Rather, they revealed only broad and relatively weak absorbance bands in the 300–500 nm spectral region.[23]

[§]The analogous reaction with diformyl pyrroles or furans does not lead to cyclic pentamers or other higher order systems. It is assumed that this reflects the steric effects of the larger atomic radius of sulfur relative to either nitrogen or pyrrole.

6.5.3   *Decavinylogous Pentaphyrinogen*

In 1988, Franck, *et al.* reported the synthesis of an octavinylogous porphyrin that was obtained from the cyclotetramerization of the pyrrylpentadienol **6.60** (discussed earlier in Chapter 4).[24] Interestingly, the cyclic pentameric macrocycle **6.61**, derived from *five* pyrrylpentadienol subunits, was isolated as a side-product of this reaction (Scheme 6.5.3). This macrocycle may be regarded as being a decavinylogously enlarged pentaphyrin macrocycle with its *meso*-like carbon atoms arranged in a (5.5.5.5.5) fashion. Unfortunately, however, mass spectral data was the only evidence that could be gathered in support of the formation of this macrocycle. Thus, the exact nature of **6.61** remains recondite at present. Further, it is not yet known whether this species can be oxidized to an aromatic system.

Scheme 6.5.3

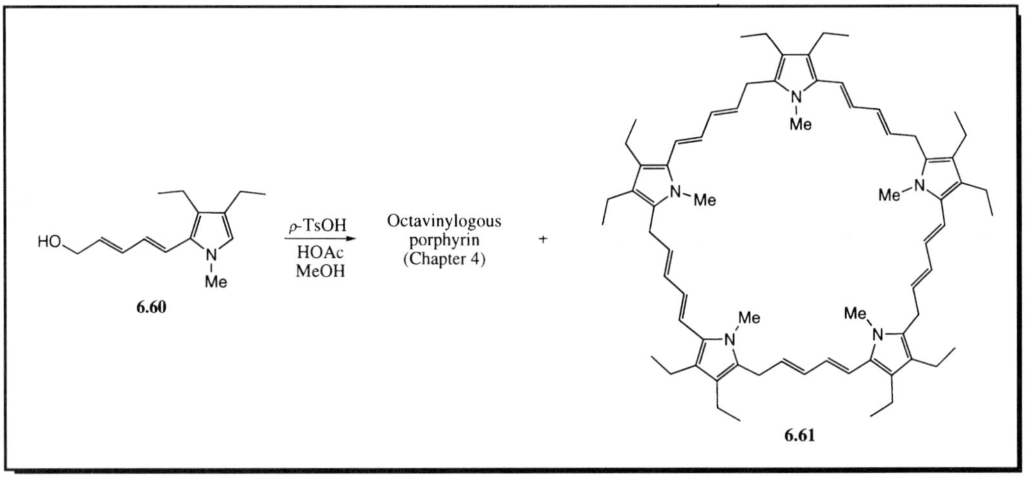

6.5.4   *"Inverted" (Non-conjugated) Pentaphyrin*

The last pentapyrrolic expanded porphyrin to be discussed in this chapter is represented by structure **6.63**.[25] Reported in 1989 by Schumacher and Franck, this macrocycle is prepared in 71% yield *via* the cyclopentamerization of the N-benzyl-protected pyrrole monomer **6.62** (Scheme 6.5.4). The five pyrrolic subunits and five bridging carbon atoms of macrocycle **6.63** are arranged in a (1.1.1.1.1) fashion. However, each pyrrolic subunit is incorporated into the macrocycle *via* one α- and one β-position. This type of connectivity is similar to that discussed in Chapter 3 for one of the pyrroles of the "N-confused" or "mutant" porphyrin isomers.[26–30] Interestingly, the inversion of the pyrrole-rings enforced by allowing reactivity at disparate positions on the pyrrole precursor results in the highly selective formation of the pentapyrrolic macrocycle **6.63**, instead of the corresponding tetrapyrrolic one.

Macrocycle **6.63** is not fully conjugated, and thus may better be regarded as being an inverted "pentaphyrinogen". Unfortunately, attempts to prepare the fully conjugated and potentially aromatic inverted pentaphyrin **6.64** met with failure (Scheme 6.5.5). They led instead to decomposition of the starting macrocycle. After the fact, this failure was rationalized in terms of the destabilizing accumulation of charge that would result upon the formation of a 22 π-electron pathway within the macrocycle.[25]

Scheme 6.5.4

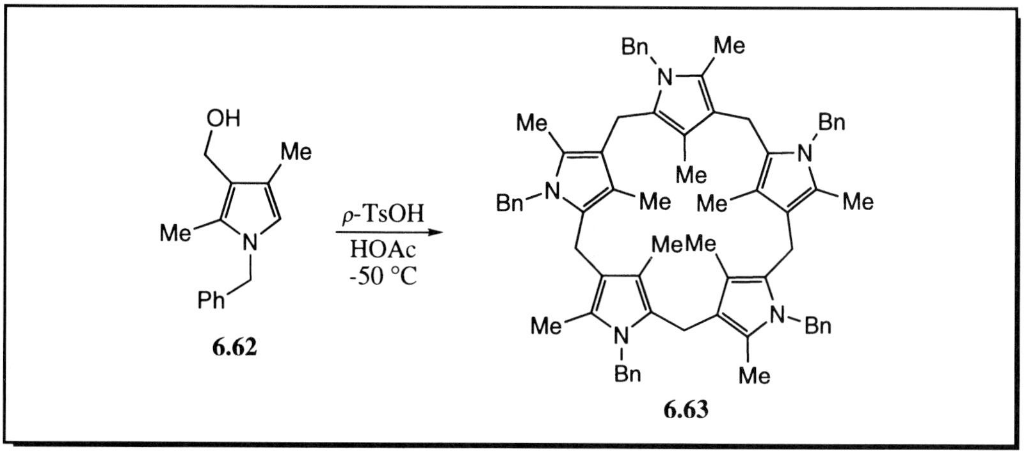

6.62   **6.63**

Scheme 6.5.5.

6.63   **6.64**

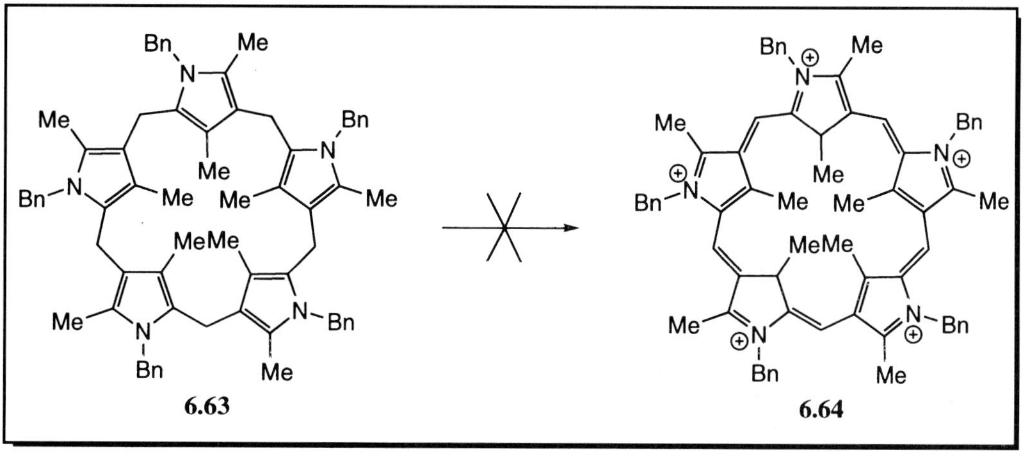

Note in proof

Subsequent to the submission of this manuscript, single crystal x-ray diffraction structures of isosmaragdyrin **6.14** and a monoxa analog were obtained.[2]

6.6 References

1. Sessler, J. L.; Weghorn, S. J.; Hiseada, Y.; Lynch, V. *Chem. Eur. J.* **1995**, *1*, 56–67.
2. Sessler, J. L.; Weghorn, S. J.; Davis, J.; Lynch, V. Unpublished results.
3. Miller, D. C.; Johnson, M. R.; Becker, J. J.; Ibers, J. A. *J. Heterocyclic Chem.* **1993**, *30*, 1485–1490.
4. Miller, D. C.; Johnson, M. R.; Ibers, J. A. *J. Org. Chem.* **1994**, *59*, 2877–2879.
5. Weghorn, S. J.; Lynch, V.; Sessler, J. L. *Tetrahedron Lett.* **1995**, *36*, 4713–4716.
6. Jux, N.; Koch, P., Schmickler, H.; Lex, J.; Vogel, E. *Angew. Chem. Int. Ed. Eng.* **1990**, *29*, 1385–1387.
7. Rexhausen, H.; Gossauer, A. *J. Chem. Soc., Chem. Commun.* **1983**, 275.
8. Gossauer, A. *Bull. Soc. Chim. Belg.* **1983**, *92*, 793–795.
9. Gossauer, A. *Chimia*, **1983**, *37*, 341–342.
10. Gossauer, A. *Chimia*, **1984**, *38*, 45–46.
11. Burrell, A. K.; Hemmi, G.; Lynch, V.; Sessler, J. L. *J. Am. Chem. Soc.* **1991**, *113*, 4690–4692.
12. Král, V.; Brucker, E. A.; Hemmi, G.; Sessler, J. L.; Králová, J.; Bose, H. Jr. *Bioorganic Med. Chem.* **1995**, *3*, 573–578.
13. Franck, B. *Angew. Chem. Int. Ed. Eng.* **1982**, *21*, 343–353.
14. Bringmann, G.; Franck, B. *Liebigs Ann. Chem.* **1982**, 1272–1279.
15. Charrière, R.; Jenny, T. A.; Rexhausen, H.; Gossauer, A. *Heterocycles*, **1993**, *36*, 1561–1575.
16. Rexhausen, H., Ph.D. Dissertation, University of Berlin, Germany, **1984**.
17. Wiß, T., Ph.D. Dissertation, University of Cologne, Germany, **1995**.
18. Vogel, E. *J. Heterocyclic Chem.* **1996**, *33*, 1461–1487.
19. Vogel, E.; Pohl, M.; Hermann, A.; Wiß, T.; König, C.; Lex, J.; Gross, M.; Gisselbrecht, J. P. *Angew. Chem. Int. Ed. Eng.* **1996**, *35*, 1520–1524.
20. Vogel, E.; Röhrig, P.; Sicken, M.; Knipp, B.; Herrmann, A.; Pohl, M.; Schmickler, H.; Lex, J. *Angew. Chem. Int. Ed. Eng.* **1989**, *28*, 1651–1655.
21. Elix, J. A. *Aust. J. Chem.* **1969**, *22*, 1951–1962.
22. Schall, R., Ph.D. Dissertation, University of Cologne, Germany, **1993**.
23. Jörrens, F., Ph.D. Dissertation, University of Cologne, Germany, **1994**.
24. Knübel, G.; Franck, B. *Angew. Chem. Int. Ed. Eng.* **1988**, *27*, 1170–1172.
25. Schumacher, K.-H.; Franck, B. *Angew. Chem. Int. Ed. Eng.* **1989**, *28*, 1243–1245.
26. Furuta, H.; Asano, T.; Ogawa, T. *J. Am. Chem. Soc.* **1994**, *116*, 767–768.
27. Chmielewski, P. J.; Latos-Grażyński, L.; Rachlewicz, K.; Glowiak, T. *Angew. Chem. Int. Ed. Eng.* **1994**, *33*, 779–781.
28. Sessler, J. L. *Angew. Chem. Int. Ed. Eng.* **1994**, *33*, 1348–1350.
29. Ghosh, A. *Angew. Chem. Int. Ed. Eng.* **1995**, *34*, 1028–1030.
30. Chmielewski, P. J.; Latos-Grażyński, L. *J. Chem. Soc., Perkin Trans. 2* **1995**, 503–509.

7   Introduction

The unique physical and chemical properties displayed by the sapphyrins,
pentaphyrins, and other pentapyrrolic expanded porphyrins described in the previous chapters has, in recent years, inspired a new body of work devoted to the
synthesis and study of even larger, hexapyrrolic macrocycles. Here, as has generally
been true in the case of other expanded porphyrins, much of the initial motivation
for this work derived from a desire to test further the limits of aromaticity. Impetus
for this work has also come, especially recently, from an interest in obtaining both
ligands with ever-larger metal chelating cores and structured anion binding receptors
with more readily protonated nitrogen-rich interiors.

7.1   Hexaphyrins

7.1.1   Synthesis

Inspired by their successful "3 + 2" synthesis of pentaphyrin (Chapter 6),
Gossauer and coworkers set out to prepare a somewhat homologous hexapyrrolic
macrocycle *via* an analogous "3 + 3" approach.[1-4] The successful realization of this
strategy first came in 1983.[12] In this instance, Gossauer, *et al.* carried out an HBr-
catalyzed condensation between the diformyl tripyrrane **7.1** and the α-free tripyrrane
7.5. They found that after oxidation with ρ-benzoquinone, a violet-colored product
was obtained that formulated as being the 26 π-electron aromatic species **7.9** (Scheme
7.1.1). Gossauer and coworkers also found that macrocycles **7.10–7.12** could be
prepared in similar fashion from the corresponding tripyrrane derivatives **7.2–7.4**
and **7.6–7.8**. Typically, the yields in these reactions were found to range from 13%
to 29%.

Analysis of the products indicated that these new macrocycles consisted of six
pyrrolic subunits linked in a (1.1.1.1.1.1) fashion by six methine subunits. As such,
they may be referred to as being [26]hexaphyrins-(1.1.1.1.1.1) in the Franck-type
nomenclature used in this book. Here, the key macrocycle name, hexaphyrin,
came from Gossauer. Indeed, because this investigator viewed products **7.9–7.12**
as being the next higher porphyrin homolog in the porphyrin-pentaphyrin series
coined the term "hexaphyrin" for this class of macrocycles.

Scheme 7.1.1

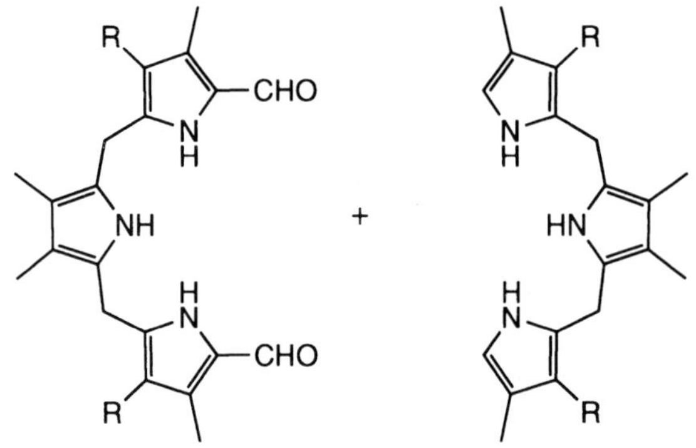

7.1. R = Me
7.2. R = CH$_2$CO$_2$Me
7.3. R = CH$_2$CH$_2$CO$_2$Me
7.4. R = C$_8$H$_{17}$

7.5. R = Me
7.6. R = CH$_2$CO$_2$Me
7.7. R = CH$_2$CH$_2$CO$_2$Me
7.8. R = C$_8$H$_{17}$

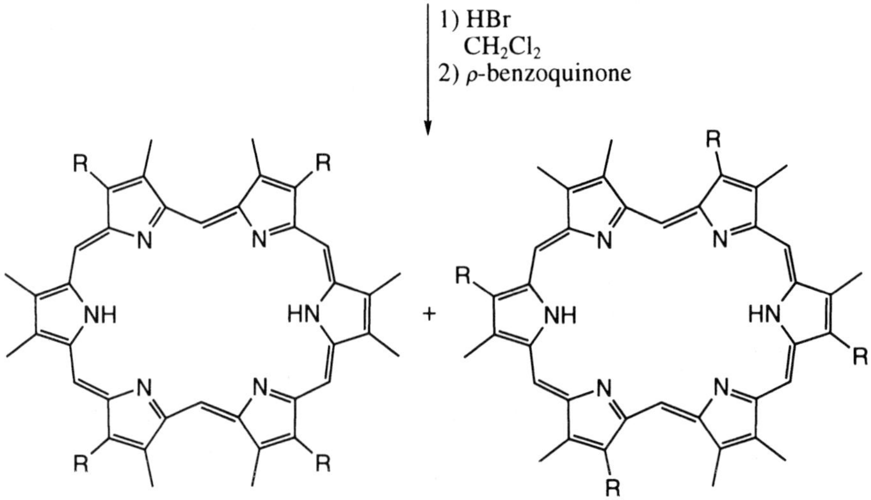

1) HBr
   CH$_2$Cl$_2$
2) p-benzoquinone

7.9a. R = Me
7.10a. R = CH$_2$CO$_2$Me
7.11a. R = CH$_2$CH$_2$CO$_2$Me
7.12a. R = C$_8$H$_{17}$

7.9b. R = Me
7.10b. R = CH$_2$CO$_2$Me
7.11b. R = CH$_2$CH$_2$CO$_2$Me
7.12b. R = C$_8$H$_{17}$

Early NMR spectroscopic studies provided support for the conclusion that hexaphyrin is a fluctional molecule that is able to adopt a nearly planar conformation in solution.[1] However, according to molecular models, planarity can only be achieved when two opposing sets of exopyrrolic double bonds of the macrocycle adopt an *E*-configuration. Forcing the molecule into this configuration has the effect of orienting two of the *meso*-like hydrogen atoms into the macrocyclic core. For this reason, it was expected that these two protons would experience an upfield-inducing ring-current effect, as indeed was seen by experiment ($\delta_{CH\text{-inner}}$ for **7.9** = -7.3 ppm in the ^{1}H NMR spectrum). Interestingly, the adoption of this configuration also results in the formation of two isomeric structures. In the case of **7.9**, these isomers, **7.9a** and **7.9b**, are identical, and cannot be distinguished by ^{1}H NMR spectroscopy. However, in the case of hexaphyrin **7.10**, the two isomers **7.10a** and **7.10b** are not the same, and indeed they were observed to exist in a nearly equal ratio in solution as determined by ^{1}H NMR spectroscopic analysis. The same proved true for hexaphy-rins **7.11** and **7.12**.

The fact that they can sustain an induced diamagnetic ring current as judged by ^{1}H NMR spectroscopy led Gossauer and coworkers to suggest that, like their smaller aromatic congeners the pentaphyrins, the 26 π-electron hexaphyrins are best considered as being aromatic macrocycles. Consistent with this supposition, the hexaphyrins were also found to absorb strongly in the visible region of the electronic spectrum; this is as would be expected for aromatic porphyrin-like materials. Specifically, the free-base decamethylhexaphyrin **7.9** was found to exhibit three strong absorption bands (at *ca.* 569, 588, and 790 nm; ε = 57 500, 34 700, and 4000 M^{-1}cm^{-1}, respectively) when its electronic spectrum was recorded in CH$_2$Cl$_2$. When fully protonated, hexaphyrin **7.9** displays only two absorption bands (at 553 and 785 nm; ε = 74 100, and 4000 M^{-1}cm^{-1}, respectively).[3]

7.1.2   Metalation Chemistry

The metal coordination chemistry of the hexaphyrins, while not fully devel-oped, was nonetheless explored considerably by Gossauer and coworkers.[4] They found, for instance, that treating hexaphyrin **7.11** with nickel(II) chloride gives rise to a single product to which the structure **7.13** was assigned (Scheme 7.1.2). When either hexaphyrin **7.11** or **7.12** is treated with zinc(II) chloride, two isomeric bis-zinc(II) adducts are formed, as determined by ^{1}H NMR spectral analysis of the crude reaction mixtures. In each case, however, as a consequence of the purification process (repeated recrystallizations), only one of the two isomers, namely **7.14** and **7.15** is isolated (Scheme 7.1.3). On the other hand, when hexaphyrin **7.10** is treated with ZnCl$_2$ under analogous conditions, a bis-Zn adduct is isolated, which is con-formationally "opposite" to those obtained with hexaphyrins **7.11** and **7.12** (Scheme 7.1.4). Specifically, the zinc adducts isolated from **7.11** and **7.12** (**7.14** and **7.15**) were found to belong to the symmetry group *C*$_{2v}$, whereas **7.16** belongs to the less-symmetric *C*$_{2h}$ symmetry group. Interestingly, in the case of **7.11** and **7.12**, both of the possible isomers are formed during the reaction, but only one, namely **7.16**, is actually isolated.[4]

When either hexaphyrin **7.10** or **7.11** is treated with Pd(NH$_4$)$_2$Cl$_2$, the bis-metalated species **7.17a** and **7.18** are obtained (Scheme 7.1.5). These complexes are of interest in that two of the pyrrole rings rotate outward by 180° prior to coordination to the two extra-macrocyclic metal centers.[4] In the resulting complexes an ammonia ligand serves to complete the coordination sphere at each of the palladium centers. This NH_3 moiety can be readily exchanged for pyridine by treating with excess pyridine. Such an exchange has been used to afford the pyridine-containing complex **7.17b** in quantitative yield.

Scheme 7.1.2

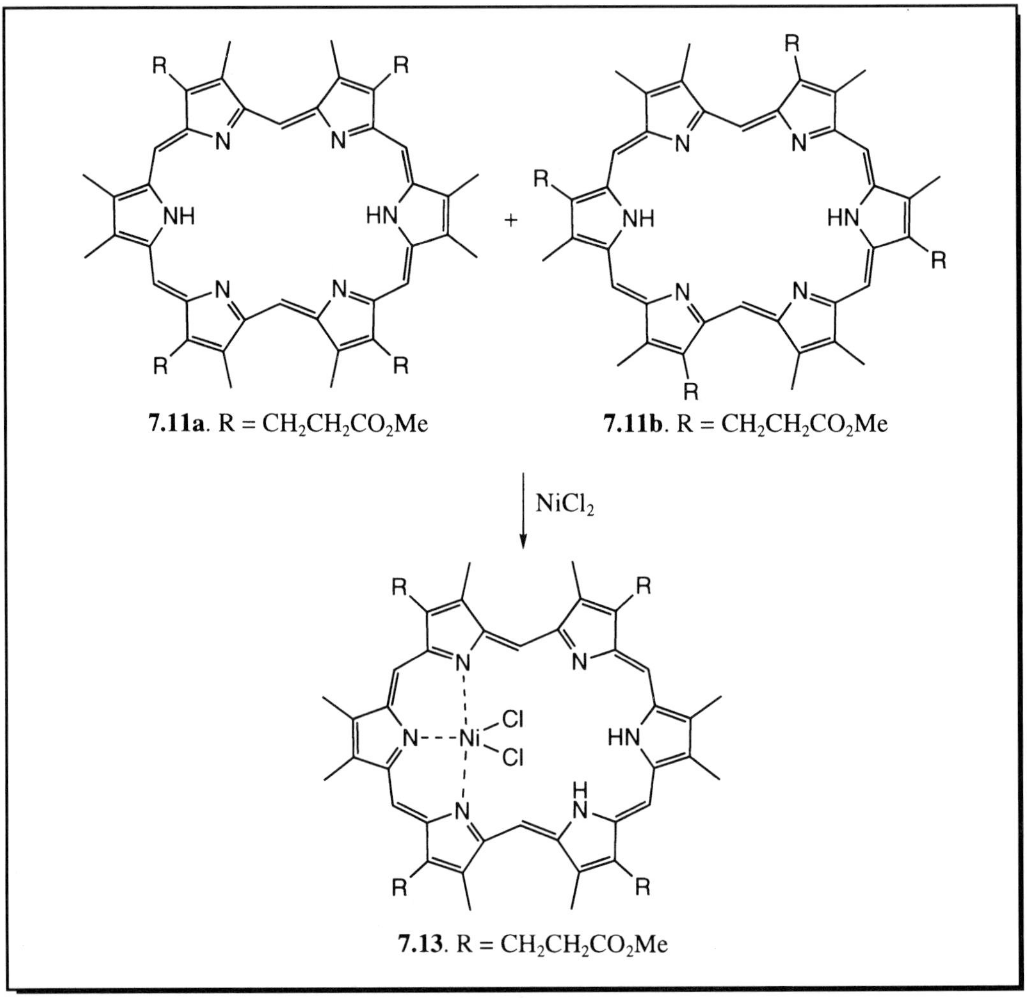

7.11a. R = $CH_2CH_2CO_2Me$ **7.11b**. R = $CH_2CH_2CO_2Me$

NiCl$_2$

7.13. R = $CH_2CH_2CO_2Me$

The conformation of the above palladium(II) hexaphyrin complexes was assigned on the basis of n.O.e. measurements. Once their analysis was complete, it was suggested that the observed outward rotation of two of the pyrrole rings and concomitant Z–E isomerization of two formal C=C bonds occurs, so as to avoid forcing the Pd(II) center into a square planar coordination geometry. While this explanation may or may not be chemically correct, Gossauer's results do serve to highlight the remarkable flexibility of this kind of hexaphyrin macrocycle. They also serve as a siren call suggesting that large polypyrrolic macrocycles may have interesting and non-classic metal coordination properties that are worth exploring. In this context, it is worth noting that Gossauer's hexaphyrins appear to be the first expanded porphyrin-like systems that may be used for the coordination of two metal cations. Unfortunately, however, no solid-state structural data were ever reported. In the absence of such data the suggested solution-state structural assignments remain uncorroborated. This means that the metal coordination chemistry of hexaphyrins has yet to be worked out and defined in a fully rigorous way.

Scheme 7.1.3

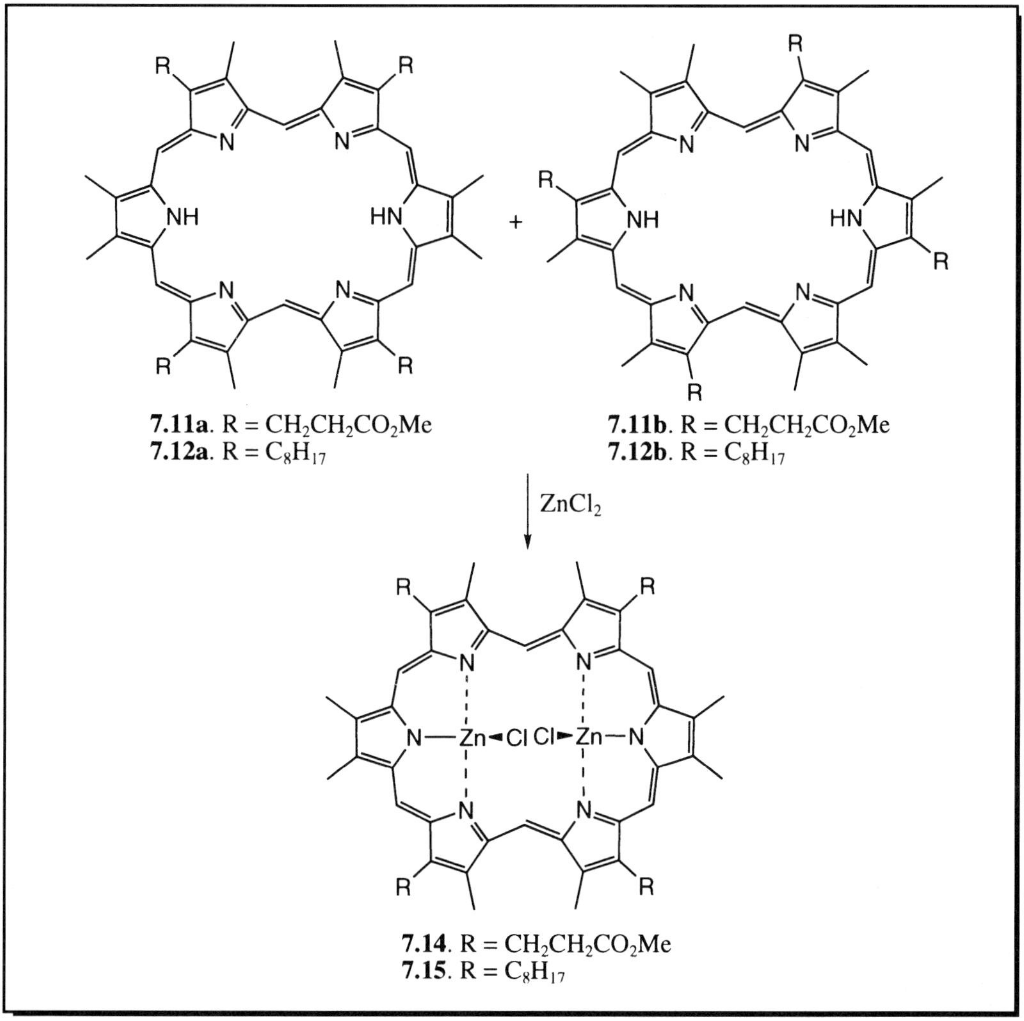

7.11a. R = CH$_2$CH$_2$CO$_2$Me
7.12a. R = C$_8$H$_{17}$

7.11b. R = CH$_2$CH$_2$CO$_2$Me
7.12b. R = C$_8$H$_{17}$

ZnCl$_2$

7.14. R = CH$_2$CH$_2$CO$_2$Me
7.15. R = C$_8$H$_{17}$

Scheme 7.1.4

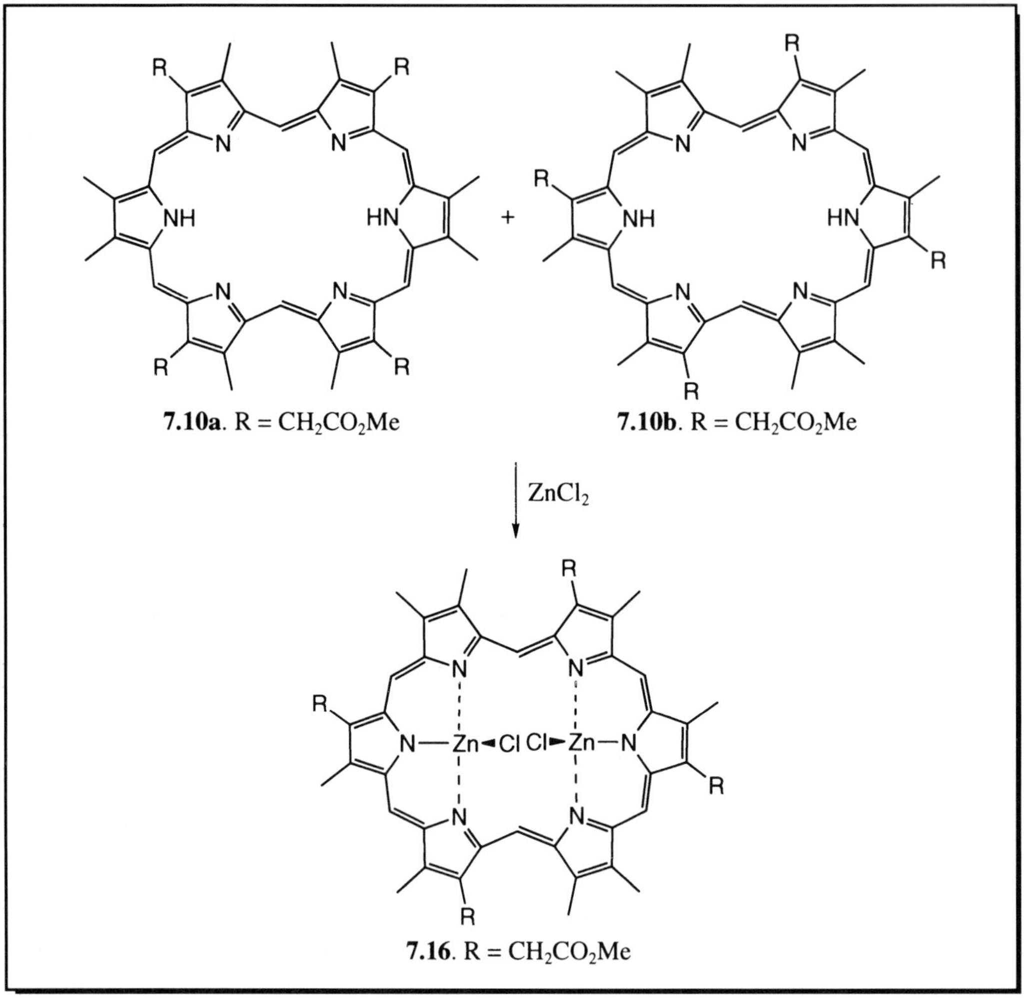

7.10a. R = CH$_2$CO$_2$Me   +   **7.10b.** R = CH$_2$CO$_2$Me

ZnCl$_2$

7.16. R = CH$_2$CO$_2$Me

7.1.3   *Hexathiahexaphyrinogen*

There exists currently only one mention in the literature of the synthesis of a heterohexaphyrin analog.[5] This is the hexathiahexaphyrinogen species **7.20**. This macrocycle was synthesized from 2-hydroxymethyl-3,4-diethylthiophene (**7.19**) by heating a nitromethane solution of this monomer to reflux in the presence of the acid catalyst ρ-toluenesulfonic acid. In addition to a cyclic tetramer and pentamer (cf. Chapter 6), the cyclohexamer **7.20** could be isolated in yields ranging from 0.5% to 2% (Scheme 7.1.6). Unfortunately, it was not reported whether **7.20** could be oxidized to give the corresponding aromatic derivative.

Scheme 7.1.5

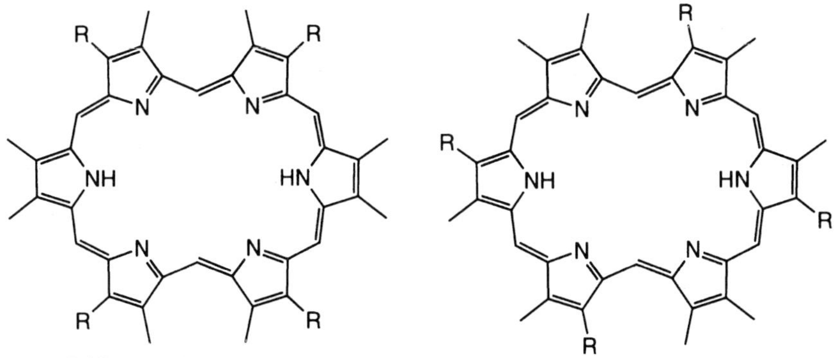

7.10a. R = CH$_2$CO$_2$Me
7.11a. R = CH$_2$CH$_2$CO$_2$Me

7.10b. R = CH$_2$CO$_2$Me
7.11b. R = CH$_2$CH$_2$CO$_2$Me

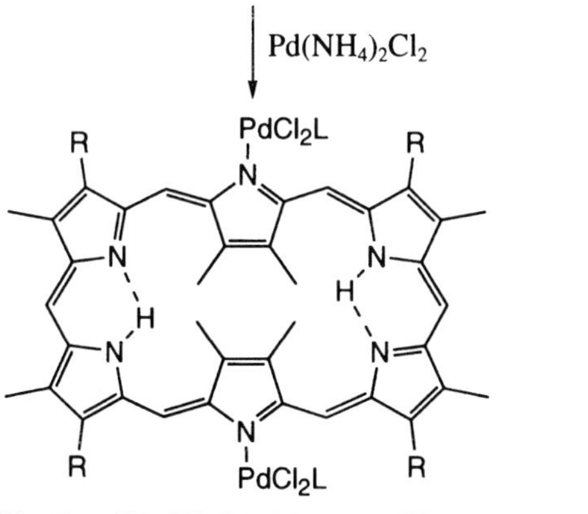

Pd(NH$_4$)$_2$Cl$_2$

7.17a. R = CH$_2$CH$_2$CO$_2$Me; L = NH$_3$ $\rangle$ pyridine
7.17b. R = CH$_2$CH$_2$CO$_2$Me; L = Py
7.18.   R = CH$_2$CO$_2$Me; L = NH$_3$

Scheme 7.1.6

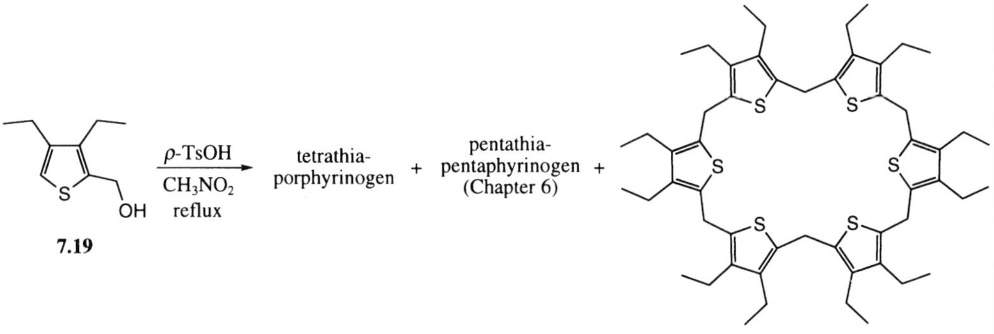

7.19

p-TsOH / CH$_3$NO$_2$ reflux → tetrathia-porphyrinogen + pentathia-pentaphyrinogen (Chapter 6) +

7.20

7.2   Rubyrins and Heterorubyrins

7.2.1   Rubyrins

A second class of aza-containing hexapyrrolic expanded porphyrins was reported in 1991 by Sessler, *et al.*[6] They were assigned the trivial name rubyrin because of the orange–red color displayed by the prototype **7.25** in organic solution. This color reflects the fact, presumably, that the rubyrins are 26 π-electron macrocycles. As such, like the hexaphyrins, they are aromatic, as evidenced by the diamagnetic ring-current effects in the ^{1}H NMR spectra. Unlike the hexaphyrins, however, the methine carbon spacers in this rubyrin ([26]hexaphyrin-(1.1.0.1.1.0)) are arranged in a (1.1.0.1.1.0) fashion. They can thus be thought of as being hexaphyrins in which two of the *meso*-like carbon atoms have been formally removed.

The synthesis of rubyrin **7.25** involves an acid-catalyzed "4 + 2" MacDonald-type condensation between the diformyl bipyrrole **7.21** and the diacid tetrapyrrane **7.23** (Scheme 7.2.1). After oxidation and purification, a roughly 20% yield of a blue crystalline solid is obtained, which, on the basis of solution-phase spectroscopic and solid-state structural studies (*vide infra*), was assigned the hexapyrrolic structure **7.25**. Using a similar procedure, rubyrins **7.26** and **7.27** were also prepared. While the first of these (like **7.25**) proved stable, rubyrin **7.27** was found to decompose during chromatographic purification. It was thus characterized only by UV-vis spectroscopy and mass spectrometry.[7]

Scheme 7.2.1

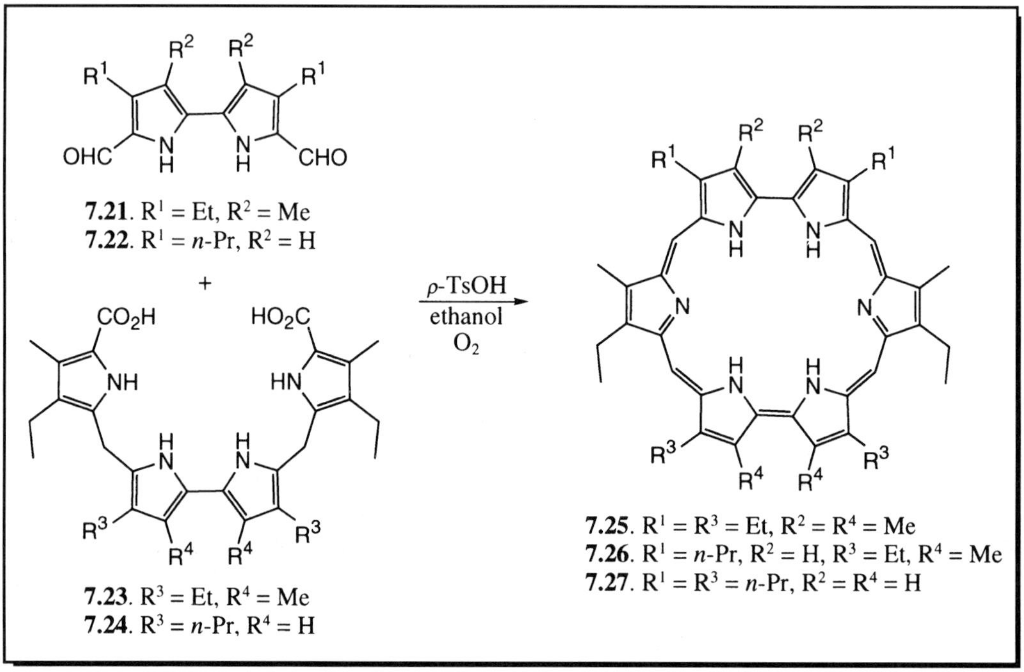

7.21. R^1 = Et, R^2 = Me
7.22. R^1 = *n*-Pr, R^2 = H

7.23. R^3 = Et, R^4 = Me
7.24. R^3 = *n*-Pr, R^4 = H

p-TsOH
ethanol
O_2

7.25. R^1 = R^3 = Et, R^2 = R^4 = Me
7.26. R^1 = *n*-Pr, R^2 = H, R^3 = Et, R^4 = Me
7.27. R^1 = R^3 = *n*-Pr, R^2 = R^4 = H

As expected for a delocalized 26 π-electron aromatic system, the rubyrins, when protonated, display an intense Soret-like absorbance band in the visible spectrum that is considerably red-shifted (by *ca.* 100 nm) relative to that of the porphyrins (Figure 7.1.1). For instance, the Soret band of the bis-HCl salt of dodecaalkylrubyrin **7.25** appears at 505 nm (ε = 302 000 M^{-1} cm^{-1} in CH$_2$Cl$_2$). The rubyrins also exhibit three long-wavelength Q-like transitions in their dicationic forms. These latter bands also show a marked red shift when compared to those of the porphyrins. For instance, the Q-bands of **7.25** appear at 711, 791, and 850 nm (ε = 11 000, 15 500, and 38 000 M^{-1} cm^{-1}, respectively).[6]

Figure 7.1.1   The UV-vis Absorption Spectra of the Bis-HCl Salt of Rubyrin **7.25** (—), the Bis-HCl Salt of 3,8,12,13,17,22-Hexaethyl-2,7,18,23-Tetramethylsapphyrin (—; see Chapter 5), and the Bis-trifluoroacetate Salt of 2,3,7,8,12,13,17,18-Octaethylporphyrin (. . .), Each Recorded in Dilute CH$_2$Cl$_2$ Solution.
This figure was redrawn with permission from reference 6

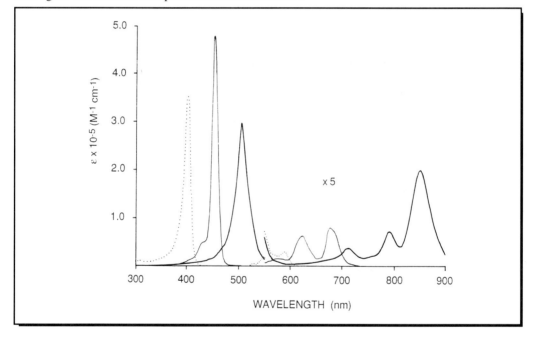

The rather high extinction coefficients recorded for the Soret bands of the diprotonated forms of rubyrins **7.25** and **7.26** are consistent with the idea that these systems are planar or near planar in solution. In the case of the bis-HCl salt of rubyrin **7.25**, this proposed planarity was confirmed in the solid state *via* single crystal X-ray structural analysis. Specifically, this analysis revealed a planar macrocyclic framework (Figure 7.1.2) wherein the only deviations from planarity arise, presumably, as the result of unfavorable methyl–methyl steric interactions at the β-positions of the bipyrrolic subunits.[6]

Figure 7.1.2 Single Crystal X-Ray Diffraction Structure of the Bis-HCl Salt of Rubyrin **7.25**.

This figure was generated using information down-loaded from the Cambridge Crystallographic Data Centre and corresponds to a structure originally reported in reference 6. The macrocycle deviates slightly from planarity (the mean deviation from macrocyclic planarity is 0.20 Å). One chloride anion sits above the mean macrocyclic plane and the other sits below. Atom labeling scheme: carbon: ○; nitrogen: ●; chlorine: ⊘. Hydrogen atoms have been omitted for clarity

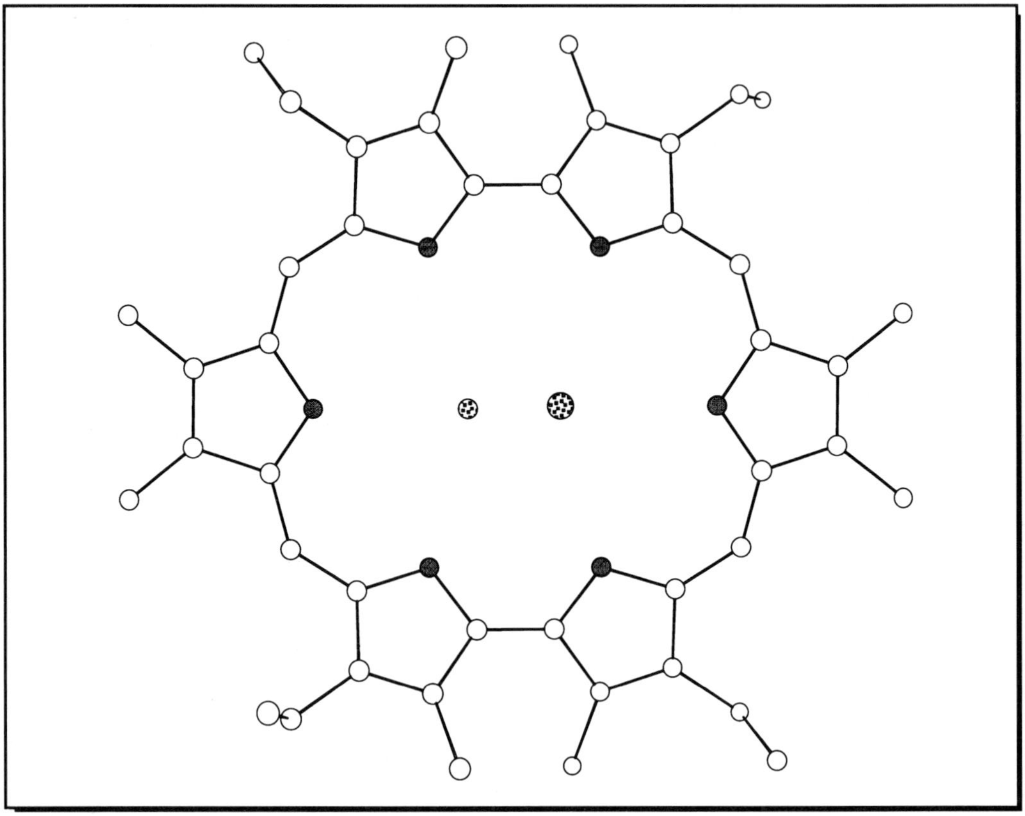

The two chloride counteranions in the crystal structure of the bis-HCl salt of **7.25** are held within hydrogen bonding distance to the core nitrogen atoms. This fact, coupled with earlier findings that sapphyrin may be used as an anion binding agent (see Chapters 5 and 10), led to a consideration that rubyrin analogs could also be used to chelate anions. In preliminary work,[7] it was found that rubyrin **7.25** does indeed bind both fluoride and phosphate anions in a strong and non-labile manner. Later, it was established that rubyrin can also be used to effect the efficient through-model-membrane transport of charged species such as guanosine 5′-monophosphate (GMP).[8] Although this latter transport was found to be most efficient when a cocarrier such as tri-isopropylsilyl-substituted cytosine (C-TIPS) was added, it was nonetheless noted that when rubyrin, but not sapphyrin was used, GMP transport could be effected at near-neutral (physiological) pH. This is important since most perceived applications of expanded porphyrin-mediated phosphate transport (e.g., antiviral drug delivery) would require an ability to function at or near neutral pH.

While rubyrin appears to show promise as an anion binding agent, these macrocycles appear rather ineffectual when it comes to the coordination of metal cations. To date, for instance, no structurally characterized metal complexes of rubyrin have been reported. In fact, only a seemingly unstable bis-zinc complex has been obtained so far. Thus, there remains a need for further metalation-related experimentation. Only then will it prove possible to judge whether or not the rubyrins have a role to play as metal-coordinating ligands.

7.2.2   Heterorubyrins

One of the attractive aspects of rubyrin is that it is built up from bipyrrolic precursors. Thus, in principle, it should be possible to use bifuran and/or bithiophene subunits as building blocks to generate dioxa, dithia, dithia-dioxa, tetraoxa, etc. rubyrins. This, however, appears not yet to have been done. Recently, however, Vogel and Gebauer succeeded in preparing a hexathia analog of rubyrin.[9] This macrocycle, **7.30**, is produced *via* the acid-catalyzed reaction between the bis(hydroxymethyl)bithiophene **7.28** and 3,4-diethylthiophene **7.29** (Scheme 7.2.2). After oxidation with 2,3-dichloro-5,6-dicyano-1,4-benzoquinone (DDQ), the hexathiarubyrin is isolated, as the bisperchlorate salt **7.30**.

Scheme 7.2.2

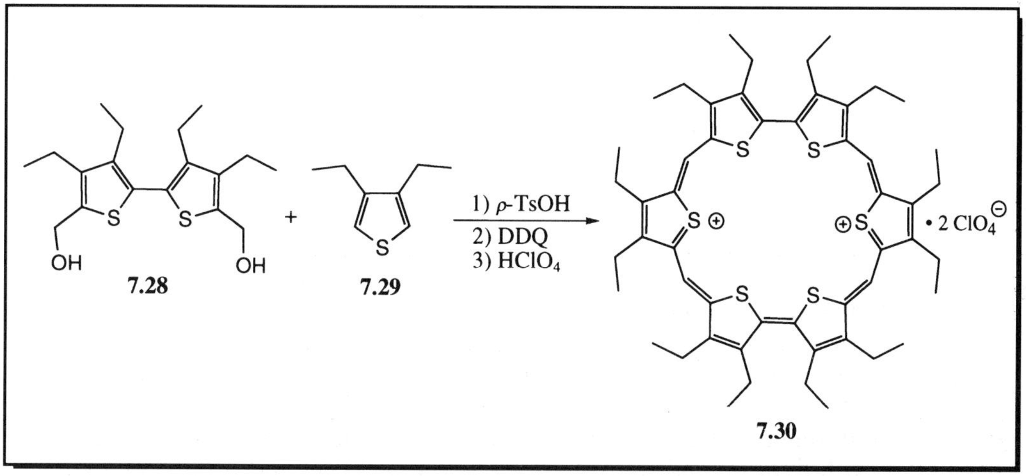

Like its hexaaza-analogs, hexathiarubyrin **7.30** is aromatic. Not surprisingly, therefore, **7.30** gives rise to a UV-vis spectrum that resembles that of the diprotonated form of prototypic hexaazarubyrin **7.25**. Specifically, it was found that compound **7.30** displays a Soret band at 514 nm ($\varepsilon = 120\,800\,\mathrm{M}^{-1}\,\mathrm{cm}^{-1}$) and Q-bands at 697, 788, 889, and 976 nm, which are only slightly red-shifted relative to those of **7.25**

(λ_{max} = 505 nm, and 711, 791, and 850 nm). Further, like its hexaaza "parent", the hexathiarubyrin **7.30** was found to be remarkably planar in the solid state, as judged from X-ray diffraction analysis. This same X-ray diffraction analysis also revealed, surprisingly, that two molecules of hydroquinone (derived from DDQ) cocrystallize with each hexathiarubyrin macrocycle. One of these hydroquinone molecules is found to reside above the macrocyclic plane, and the other below.[9]

7.3   Rosarins

7.3.1   Synthesis

A third class of a hexapyrrolic expanded porphyrins (e.g., **7.36**) was reported in 1992 by Sessler, *et al.*[10] These macrocycles contain three fewer *meso*-like bridging carbons than do the hexaphyrins, and may thus be formally considered as being triply contracted hexaphyrins. However, they are more commonly referred to as being "rosarins". This is because the protonated derivatives appear pink to red in organic solution.

In terms of synthesis, the rosarins are most easily prepared *via* the condensation of three α-free bipyrroles with three equivalents of an aromatic aldehyde. For instance, rosarin **7.36** was prepared in > 70% yield *via* the condensation of bipyrrole **7.31** with benzaldehyde, followed by oxidation with DDQ (Scheme 7.3.1). In a similar way, the rosarins **7.37–7.39** could also be prepared in good yield. The preparation of rosarins thus bears formal analogy to the Rothemund–Adler-type methodology used to prepare *meso*-substituted porphyrins.[11–13] In marked contrast to what is observed under these latter porphyrin-forming conditions, the rosarins are obtained as 24 π-electron, non-aromatic 4n π-electron species. According to the Franck notation, therefore, the rosarins are formally [24]hexaphyrins-(1.0.1.0.1.0).

While the non-aromatic nature of the rosarins was originally unexpected, it is now well established. For instance, detailed solution state ^{1}H and ^{13}C NMR spectral studies of rosarins **7.36–7.39** led to the appreciation that these macrocycles contain a C_3 symmetry element. Further, when protonated, the rosarins (in CH_2Cl_2) were found to display a rather intense Soret-like absorbance band at *ca.* 550 nm in the visible portion of the electronic spectrum (along with several weak Q-type transitions in the 645–845 nm region). These bands were deemed "too red-shifted" for a simple 22 π-electron system and provided a first "hint" the rosarins are not aromatic. A subsequent single crystal X-ray structure of the trihydrochloride salt of tris(phenyl)-rosarin **7.36c** (Figure 7.3.1) then revealed that the molecule is both: (1) capable of being triply protonated (as would be expected for an expanded porphyrin of a [24]-hexaphyrin-(1.0.1.0.1.0) formulation); and (2) found in a highly distorted, non-planar conformation in the solid state. On this basis it was concluded that even though rosarins are highly fluctional in solution on the NMR time scale (giving rise to a time-averaged structure containing a C_3 symmetry element), they are definitely not aromatic in character.

Scheme 7.3.1

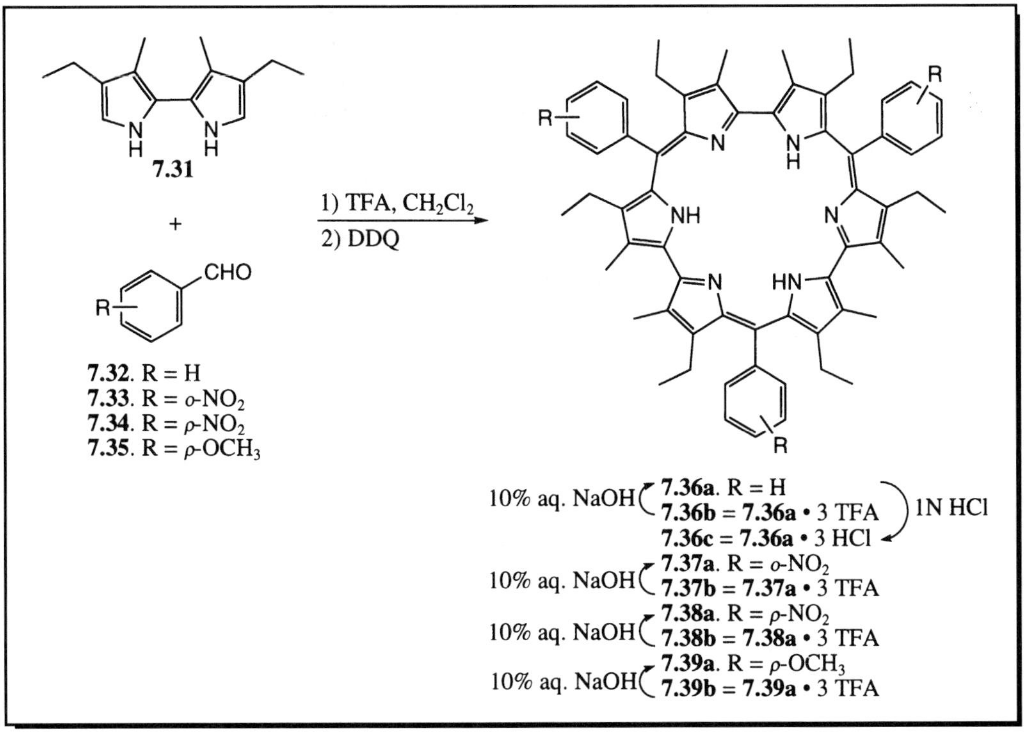

7.31

+

7.32. R = H
7.33. R = o-NO$_2$
7.34. R = p-NO$_2$
7.35. R = p-OCH$_3$

1) TFA, CH$_2$Cl$_2$
2) DDQ

10% aq. NaOH (7.36a. R = H
7.36b = 7.36a • 3 TFA) 1N HCl
7.36c = 7.36a • 3 HCl

10% aq. NaOH (7.37a. R = o-NO$_2$
7.37b = 7.37a • 3 TFA

10% aq. NaOH (7.38a. R = p-NO$_2$
7.38b = 7.38a • 3 TFA

10% aq. NaOH (7.39a. R = p-OCH$_3$
7.39b = 7.39a • 3 TFA

7.3.2   Metalation Chemistry

The rosarins, in spite of being formally 24 π-electron and non-aromatic macro-cycles, are fully conjugated. They are also remarkably stable both in their free-base and protonated forms. Thus, it is realistic to consider exploring these species in terms of their potential anion binding or metal-chelation chemistry. While the rosarins have yet to be explored in terms of the former possibility, they have shown some promise as metal-chelating agents. For instance, it was found early on that the rosarins, like the sapphyrins (see Chapter 5), can form rather stable rhodium and iridium carbonyl complexes.[14] For instance, treating rosarin **7.36c** with Rh$_2$(CO)$_4$Cl$_2$ in dichloro-methane and triethyl amine results in the formation of the stable [Rh(CO)$_2$]$_3$ rosarin complex **7.40** (Scheme 7.3.2). Treatment with Ir(CO)$_2$Cl(p-NH$_2$-C$_6$H$_4$-CH$_3$), on the other hand, gives rise to the bis-iridium complex [Ir(CO)$_2$]$_2$ rosarin **7.41** (Scheme 7.3.3). Unfortunately, in both cases poor crystallinity precluded the obtainment of X-ray diffraction-based structural data. Thus, the structures proposed for these com-plexes (i.e., **7.40** and **7.41**) are still subject to debate.

Under more strenuous deprotonation conditions, rosarin **7.36** has been shown to form samarium(III) and gold(III) complexes of tentative structure **7.42** and **7.43** (Scheme 7.3.4).[14] Unfortunately, these complexes have proved to be rather unstable.

Thus like the Rh(I) and Ir(I) rosarin complexes discussed above, the actual structure of these products remains unknown at present.

Figure 7.3.1   Single Crystal X-Ray Diffraction Structure of the Tri-HCl Salt of Rosarin (**7.36c**).

This figure was generated using information down-loaded from the Cambridge Crystallographic Data Centre and corresponds to a structure originally reported in reference 10. The macrocycle is highly non-planar. One bipyrrole unit (the bottom one) is twisted (dihedral angle = 123.3°) so that the two NH groups are oriented in nearly opposite directions. Atom labeling scheme: carbon: ○; nitrogen: ●; chlorine: ◎. One chloride anion not proximate to the macrocycle, and two water molecules involved in hydrogen bonding within the macrocyclic core are not shown. Hydrogen atoms have been omitted for clarity

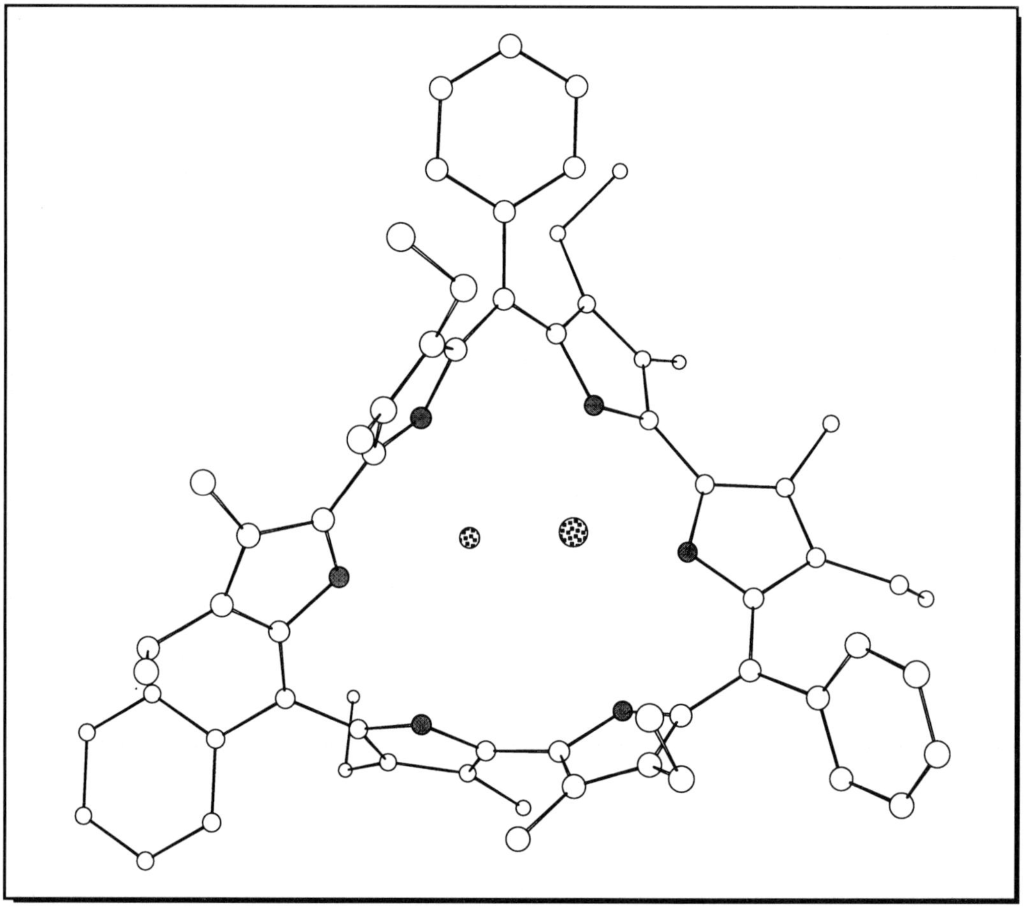

Scheme 7.3.2

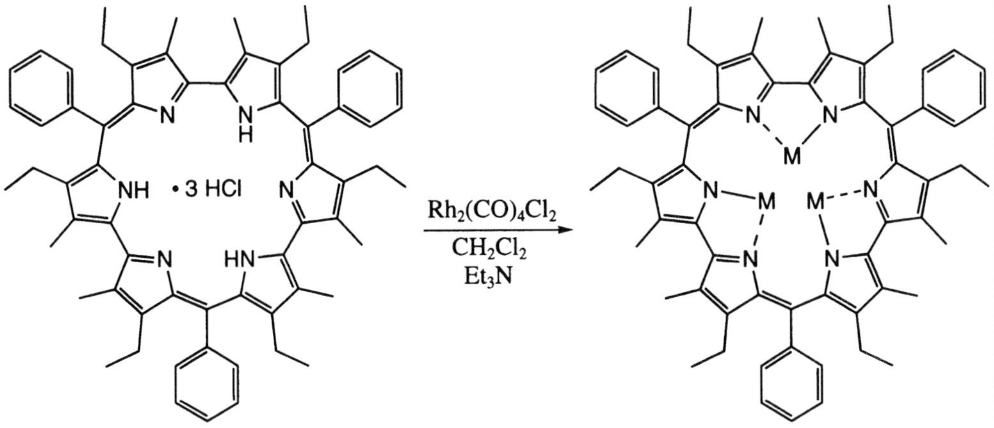

7.36c

Rh$_2$(CO)$_4$Cl$_2$
CH$_2$Cl$_2$
Et$_3$N

7.40. M = Rh(CO)$_2$

Scheme 7.3.3

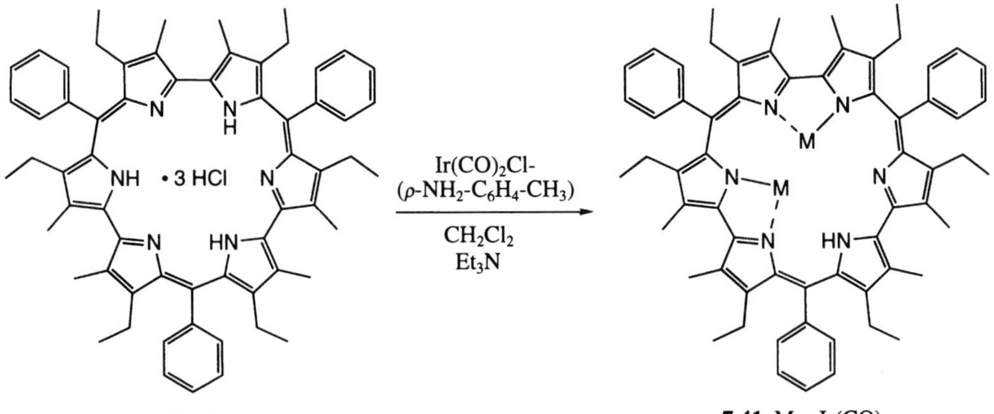

7.36c

Ir(CO)$_2$Cl-
(*p*-NH$_2$-C$_6$H$_4$-CH$_3$)

CH$_2$Cl$_2$
Et$_3$N

7.41. M = Ir(CO)$_2$

Scheme 7.3.4

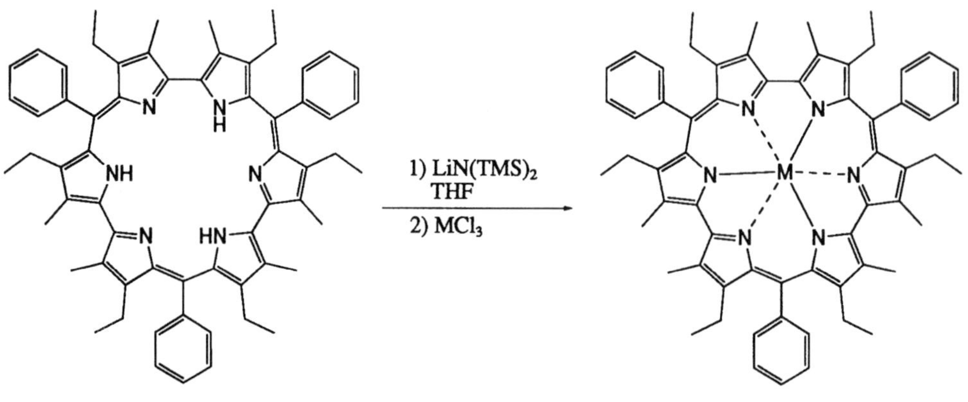

7.36a

1) LiN(TMS)₂
   THF

2) MCl₃

7.42. M = Sm
7.43. M = Au

7.4   Rosarinogens and Heterorosarinogens

The tris(phenyl)rosarins described in the previous section could only be isolated as 24 π-electron, non-aromatic species. While the reasons for this remain recondite, one possibility is that some type of steric barrier involving interactions between the "*meso*" and β-pyrrolic substitutes prevents adoption of a planar geometry. To the extent this is true, it was considered likely that less sterically encumbered systems would be more prone to undergo the planarization process associated with oxidation from a net 24 π- to a net 22 π-electron periphery. Towards this end the mono-aryl-substituted rosarinogen macrocycles **7.45** and **7.46** were prepared. They were generated in high yield *via* the ring-closure cyclization of the linear 5,15,25-tris-nor-hexapyrrin **7.44** with the appropriate aldehyde (Scheme 7.4.1).[14–17] In an analogous fashion, the *meso*-unsubstituted- and mono-phenyl-substituted rosarinogen macrocycles **7.48** and **7.49** were prepared from the corresponding dodecaalkyl-substituted 5,15,25-tris-nor-hexapyrrin **7.47** (Scheme 7.4.2).[14]

An alternative synthesis of these mono-phenyl-substituted rosarinogens was also developed by Sessler and his group.[18] In this instance, the phenyl-substituted tetrapyrrane derivative **7.50** was reacted with the diformyl bipyrrole **7.21**. While this approach afforded the rosarinogen macrocycle **7.49** in only moderate yield, an advantage of this method is that it is potentially amenable to heteroatom substitution (Scheme 7.4.3). Indeed, it has already been found that using the diformyl bifuran **7.51** in place of the diformyl bipyrrole **7.21** affords the dioxarosarinogen **7.52**.[18]

To date, several attempts to oxidize the above-described rosarinogens to the corresponding aromatic derivatives have been made. Unfortunately, none proved successful.[14,18] Thus, while these mono-functionalized rosarinogens could emerge as being of possible interest as potential anion-binding agents, they have so far proved of little use for the purpose for which they were originally designed.

Scheme 7.4.1

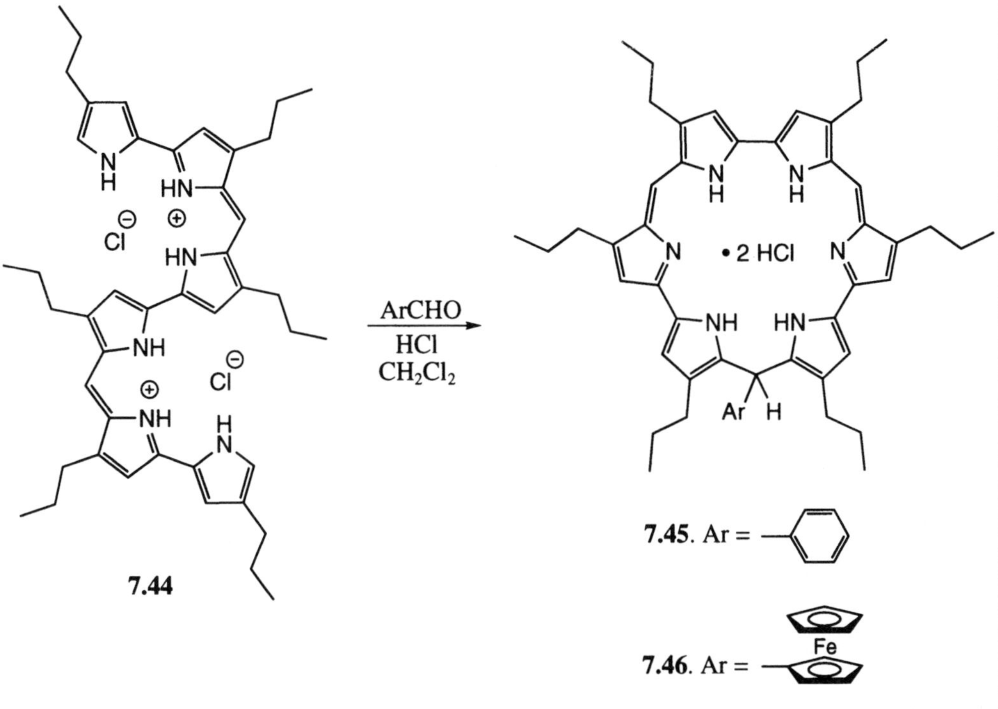

7.44

ArCHO
HCl
CH$_2$Cl$_2$

• 2 HCl

Ar H

7.45. Ar = —⟨phenyl⟩

7.46. Ar = —⟨ferrocenyl, Fe⟩

Scheme 7.4.2

7.47

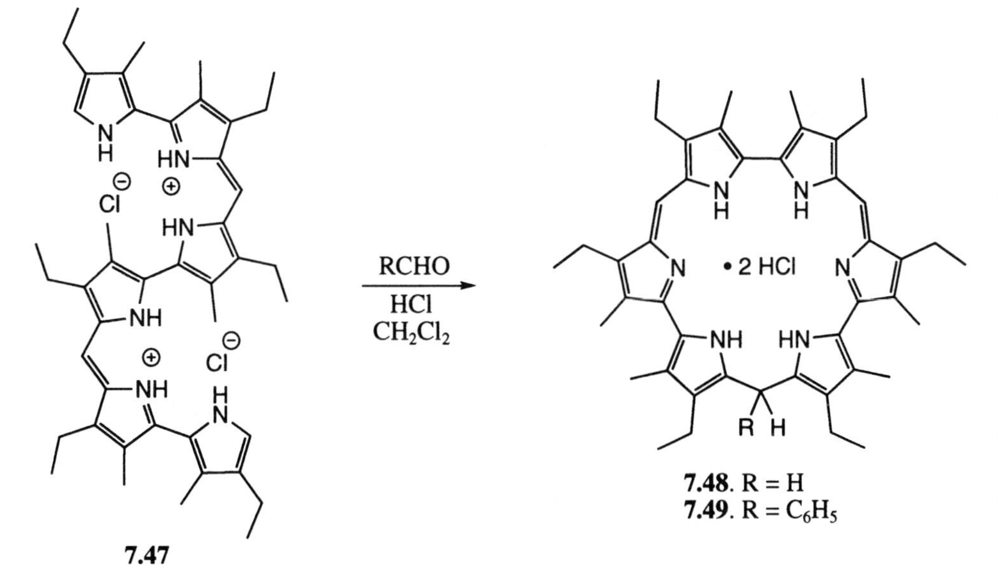

RCHO
HCl
CH$_2$Cl$_2$

• 2 HCl

R H

7.48. R = H
7.49. R = C$_6$H$_5$

Scheme 7.4.3

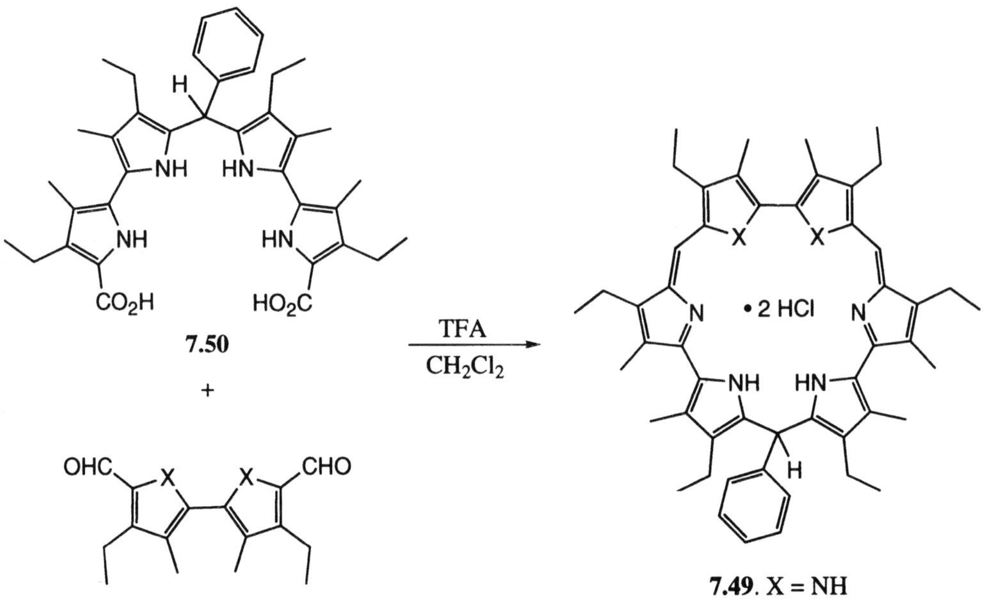

7.5   Bronzaphyrins

A reductive McMurry-type approach similar to that used for the synthesis of porphycenes (Chapter 3) and ozaphyrins (Chapter 6) has also been applied to the synthesis of hexapyrrolic macrocycles. This approach has been explored independently by Johnson and Ibers[19,20] Cava[21–23] and Merz and Neidlein[24] and has led to a range of hexaphyrin-(2.0.0.2.0.0)-type expanded porphyrins (e.g. **7.55**) that are generically known as homoporphycenes or bronzaphyrins. As is often the case, the trivial name bronzaphyrin, originally coined by Johnson and Ibers[19,20] reflects the color of the macrocycle observed in organic media. Indeed, in solution, these macrocycles typically exhibit a distinctive bronze color. For the purpose of continuity in this chapter, the name bronzaphyrin will be universally employed for all macrocycles of the hexaphyrin-(2.0.0.2.0.0) type.

The first example of this class of macrocycles was reported in 1992 by Johnson and Ibers.[19] In this particular instance, the authors effected the reductive self-dimerization of the bis(pyrrolyl)thiophene derivative **7.53**. Following work-up under aerobic conditions, they obtained the [26]bronzaphyrin **7.55** in 28% yield. The presumed 28 π-electron intermediate species **7.54** was not isolated, but rather was, presumably, oxidized directly to the aromatic 26 π-electron bronzaphyrin **7.55** (Scheme 7.5.1). Johnson and Ibers later synthesized bronzaphyrin **7.58** *via* the heterodimerization of the bis(pyrrolyl)thiophene derivative **7.53** and the bis(pyrrolyl)furan intermediate **7.56** (Scheme 7.5.2).[20] Here too, the presumed 28 π-electron intermediate species **7.57** was not isolated. Rather, it was oxidized *in situ* to afford the final product **7.58**.

Scheme 7.5.1

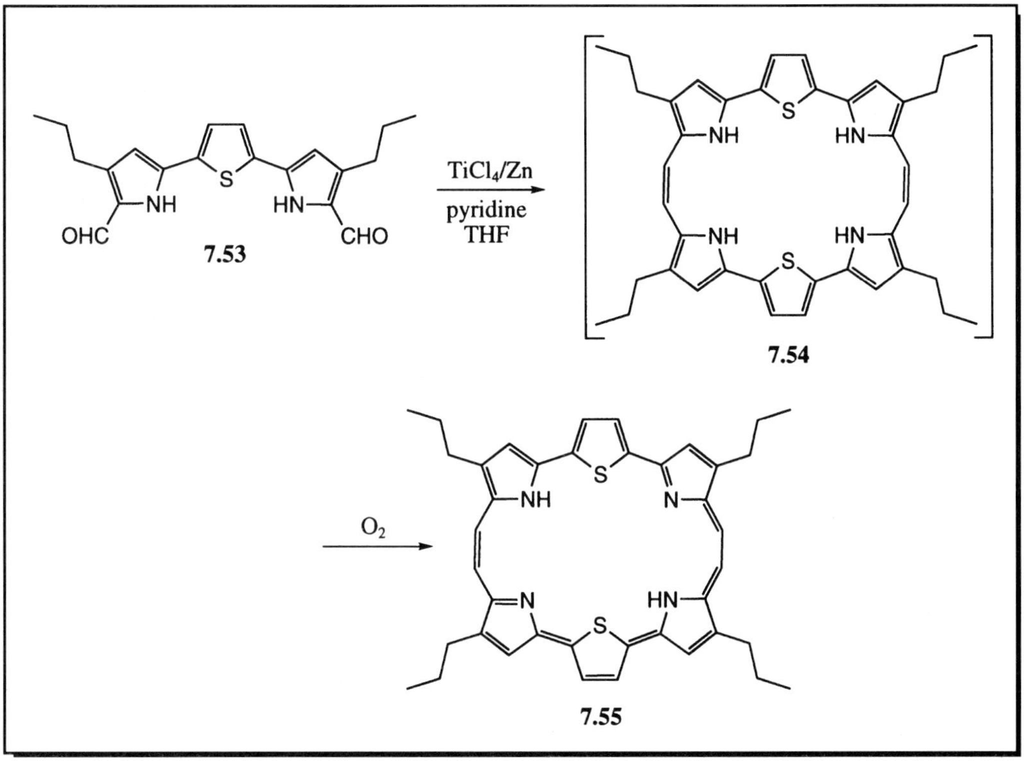

Shortly after the original work of Johnson and Ibers,[19,20] Cava and coworkers reported the low-valent titanium mediated synthesis of the 28 π-electron hexathia-[28]bronzaphyrin **7.62**.[21,22] Concurrently, Cava described the synthesis of the corresponding 28 π-electron *N*-alkyl derivatives **7.63** and **7.64** (Scheme 7.5.3).[21] These *N*-alkyl derivatives, like hexathia[28]bronzaphyrin **7.62**, are formed in remarkably high yield (> 69%). Subsequent to this original work, Cava and coworkers reported the high-yielding oxidative conversion of the hexathia-bronzaphyrinogen **7.62** to the corresponding 26 π-electron, aromatic bronzaphyrin **7.65**. Here, air-saturated concentrated sulfuric acid was used as the oxidant (Scheme 7.5.4).[22] While these conditions proved effective for **7.62**, the *N*-alkyl-analogs **7.63** and **7.64** proved resistant to oxidation under these or other conditions sampled. The inability of these 28 π-electron *N*-alkyl derivatives to undergo chemically mediated oxidation to the corresponding 26 π-electron aromatic macrocycles is presumably the result of the increased steric constraints imposed by *N*-alkyl groups within the macrocycle core relative to the simple thiophene system **7.65**, or the thiophene/pyrrole and thiophene/furan/pyrrole systems **7.55** and **7.58**. On the other hand, electrochemical studies, by virtue of exhibiting two quasi-reversible oxidative waves, provide a "hint" that it may prove possible to oxidize systems such as **7.63** and **7.64** under appropriate conditions. So far, however, this lead has not yet been followed up in terms of a *bona fide* chemical conversion.

Scheme 7.5.2

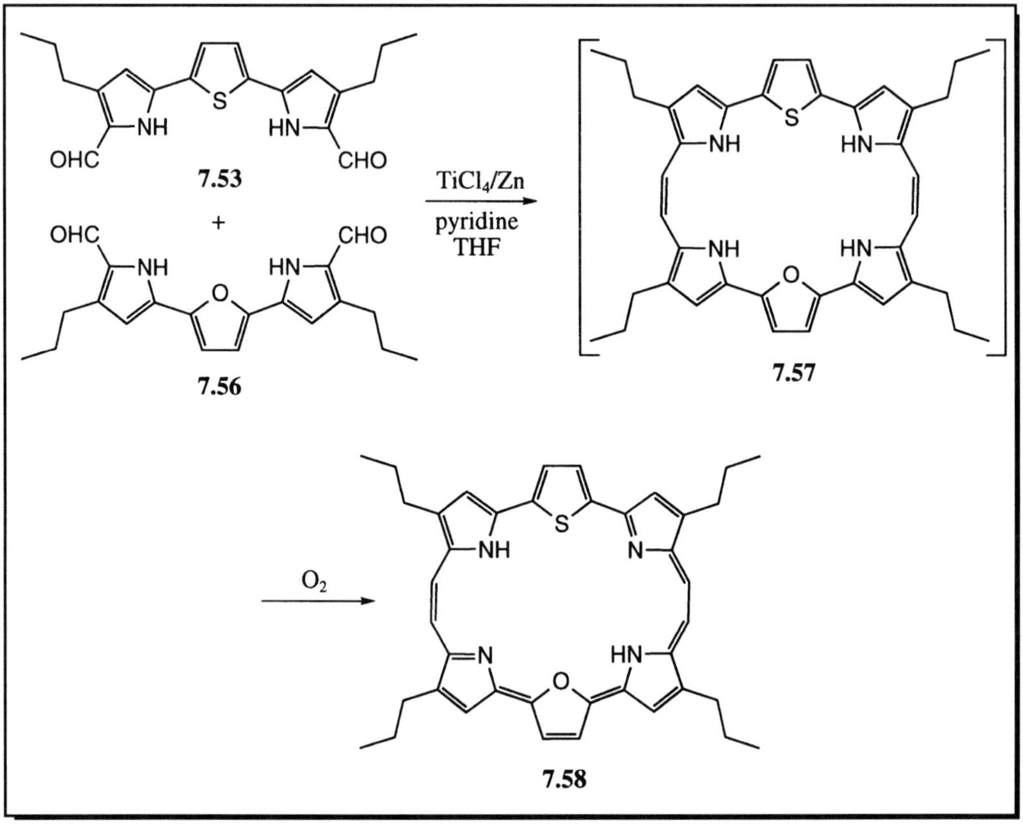

Nearly concurrent with Cava and coworkers, Merz and Neidlein also attempted the synthesis of the hexathia-[28]bronzaphyrin **7.62**.[24] Surprisingly, however, despite using reaction conditions analogous to Cava (i.e., low-valent titanium generated from TiCl4/Zn), the team of Merz and Neidlein did not obtain the expected macrocycle **7.62**. Rather, they obtained the hydrogenated product **7.66**, albeit in minuscule (1%) yield (Scheme 7.5.5). Nonetheless, Merz and Neidlein did succeed in preparing the 28 π-electron hexathia-[28]bronzaphyrin **7.62** when they generated the critical low-valent titanium species needed for coupling by treating TiCl3 with LiAlH4.[24] Even in this case, however, the desired macrocycle could only be isolated in 1% yield (Scheme 7.5.6).

Scheme 7.5.3

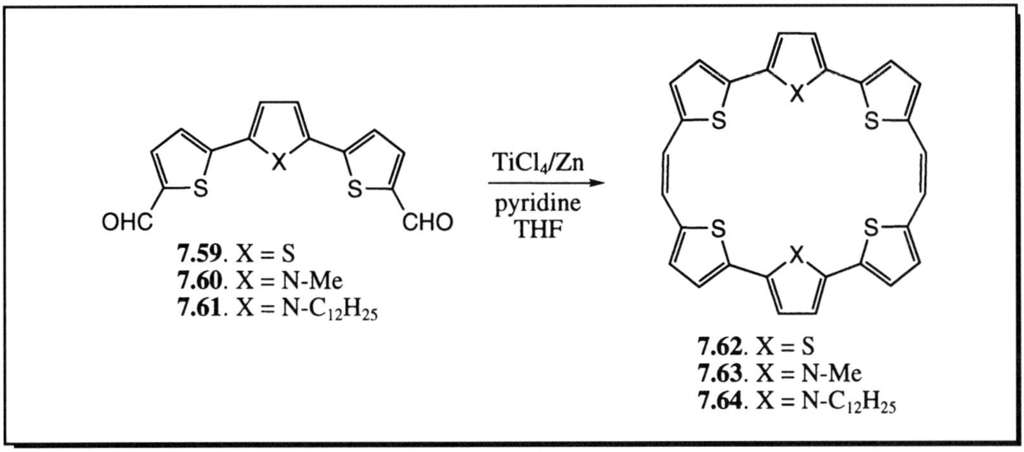

7.59. X = S
7.60. X = N-Me
7.61. X = N-C$_{12}$H$_{25}$

TiCl$_4$/Zn
pyridine
THF

7.62. X = S
7.63. X = N-Me
7.64. X = N-C$_{12}$H$_{25}$

Scheme 7.5.4

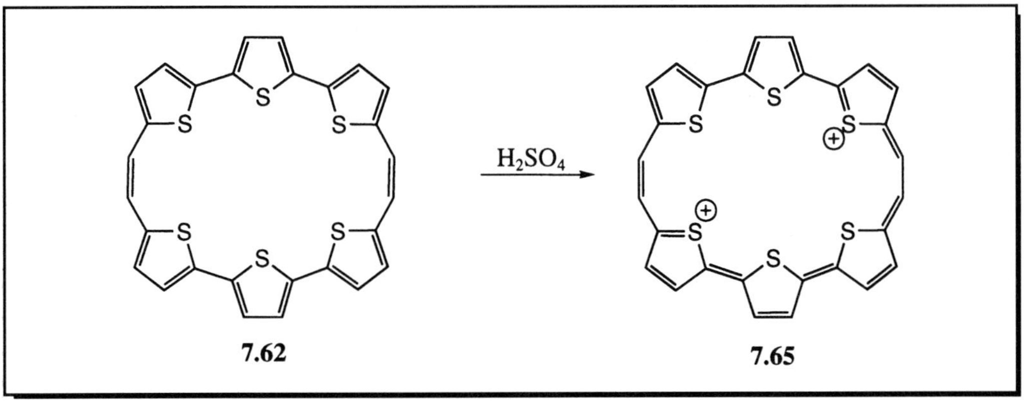

7.62

H$_2$SO$_4$

7.65

Scheme 7.5.5

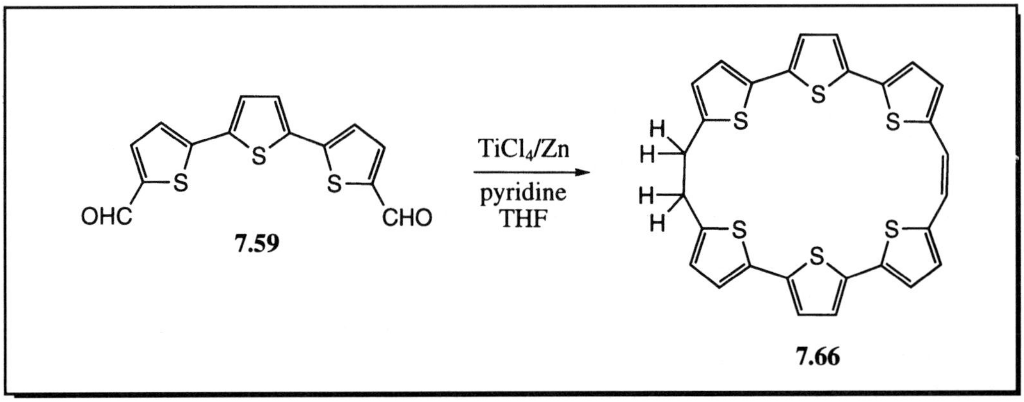

OHC   CHO
7.59

TiCl$_4$/Zn
pyridine
THF

7.66

Scheme 7.5.6

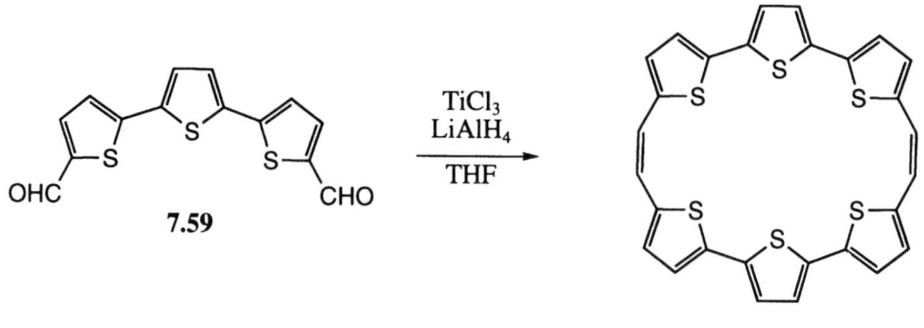

[28]Bronzaphyrins containing various bridging groups spanning the macro-cyclic have also been prepared.[23] This work, reported in 1996, involved, as a first example, the intramolecular McMurry-like ring closure of the N,N'-bridged tetra-aldehyde 7.67 to afford [28]bronzaphyrin 7.70 containing a C_4 alkyl bridge (Scheme 7.5.7). The C_6-bridged and the polyether-bridged [28]bronzaphyrins 7.71 and 7.72 were then also prepared in an analogous way. The yields of final product obtained in these reactions ranged from as low as 24% for the C_6-bridged macrocycle 7.71 to as high as 51% for the ether-linked system 7.72.

Scheme 7.5.7

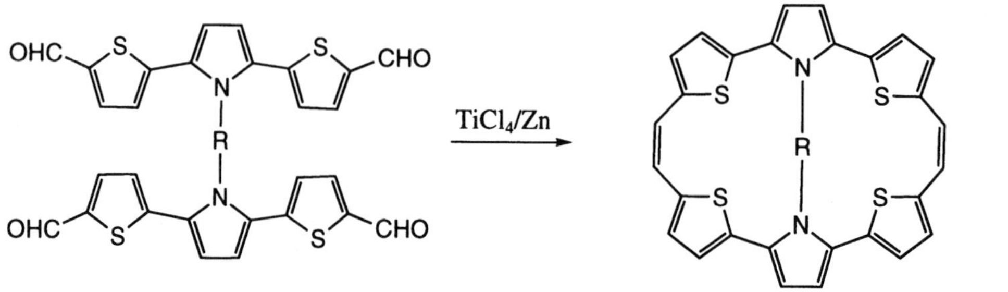

7.67. R = -(CH$_2$)$_4$-
7.68. R = -(CH$_2$)$_6$-
7.69. R = -(CH$_2$)$_3$O(CH$_2$)$_2$O(CH$_2$)$_2$O(CH$_2$)$_3$-

7.70. R = -(CH$_2$)$_4$-
7.71. R = -(CH$_2$)$_6$-
7.72. R = -(CH$_2$)$_3$[O(CH$_2$)$_2$]$_2$O(CH$_2$)$_3$-

The various [28]bronzaphyrins generated by different groups have all been demonstrated using ^{1}H NMR spectroscopy to lack any kind of significant ring-current effect.[21–24] Conversely, the [26]bronzaphyrins were found to display diamag-netic ring-current effects. These species were, therefore, judged to be aromatic. Further, as would be expected of such highly delocalized aromatic systems, the

[26]bronzaphyrins were all found to absorb strongly in the visible portion of the electronic spectrum. For instance, bronzaphyrin **7.51** shows an intense, widely split Soret-like band (460 and 501 nm; ε = 190 500 and 89 100 $M^{-1} cm^{-1}$) in THF and four accompanying Q-like transitions at 745, 780, 790, and 859 nm (ε = 69 200, 60 300, 63 100, and 83 200 $M^{-1} cm^{-1}$, respectively).[19] The [28]bronzaphyrins, on the other hand, exhibit less intense and generally blue-shifted UV-vis absorbance bands relative to their aromatic congeners.

7.6 Amethyrins

7.6.1 Synthesis

Sessler and coworkers have recently reported the synthesis of a new class of (1.0.0.1.0.0)-type hexaphyrin macrocycles that contain but two *meso*-like methine bridges (e.g., **7.74**).[25] These macrocycles, called "amethyrins" because of their amethyst-like color in dilute organic solution, are synthesized *via* the TFA-catalyzed condensation reaction of an α-free hexaalkylterpyrrole, such as **7.73**, and an aromatic or aliphatic aldehyde, followed by ρ-chloranil or DDQ-mediated oxidation. In this way the *meso*-unsubstituted amethyrin **7.74** was prepared in 64% yield starting from terpyrrole **7.73** and formaldehyde, and using ρ-chloranil as the oxidant (Scheme 7.6.1). In a similar fashion, the ρ-NO_2-phenyl-substituted amethyrin **7.75** can be prepared in 29% yield from ρ-nitrobenzaldehyde and terpyrrole **7.73**, although in this case, the stronger oxidant DDQ is required to effect complete oxidation to the desired fully conjugated product.

Scheme 7.6.1

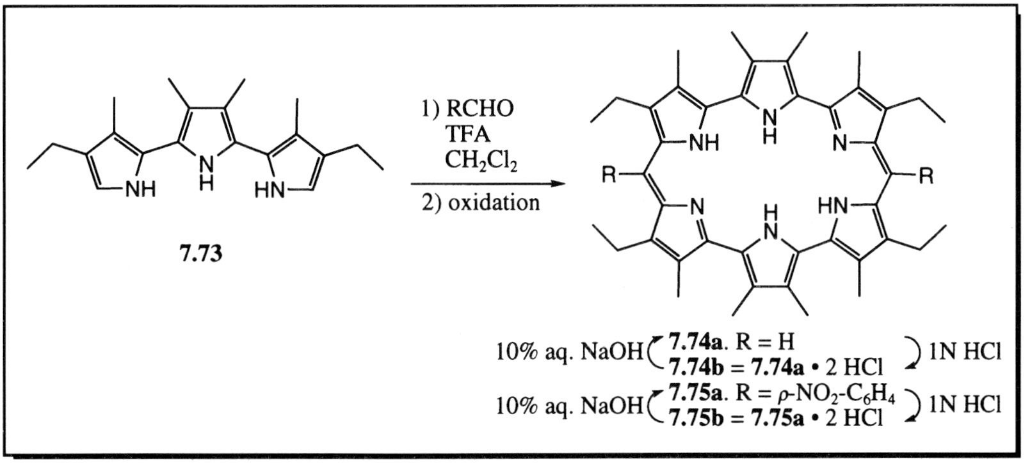
NH
H
HN
7.73
1) RCHO
TFA
CH2Cl2
2) oxidation
R
R
10% aq. NaOH
7.74a. R = H
7.74b = 7.74a • 2 HCl
1N HCl
10% aq. NaOH
7.75a. R = p-NO2-C6H4
7.75b = 7.75a • 2 HCl
1N HCl

Because these reactions involve the direct acid-mediated condensation between an aldehyde and a pyrrolic precursor, they bear direct analogy to the Rothemund–Adler-type procedures used to prepare tetraarylporphyrins,[11–13] as well as to the methods used to prepare the hexapyrrolic rosarins (*vide supra*).[10] In the case of the amethyrins, however, only two terpyrrolic subunits couple together, whereas in the case of porphyrin and rosarin, four and three mono- and bipyrrolic precursors are joined, respectively.

As proved true for the rosarins, the amethyrins prepared to date have so far only been isolated in their 24 π-electron non-aromatic forms. In spite of this, the amethyrins exhibit spectral features typical of highly conjugated macrocycles. For instance, amethyrin **7.74b** exhibits an intense Soret-like absorbance in CH_2Cl_2 at 493 nm ($\varepsilon = 110\,000\,M^{-1}\,cm^{-1}$), with a weaker but still rather intense Q-type transition at 597 nm ($\varepsilon = 56\,500\,M^{-1}\,cm^{-1}$). Further, the protonated form of this prototypic amethyrin and its analog **7.75b** display internal NH protons in their ^{1}H NMR spectra that are found to resonate at a remarkably low field (i.e., at *ca.* 23–25 ppm). They also display *meso*-like protons that are upfield-shifted by *ca.* 3.4 ppm relative to those of normal alkenes. While a complete explanation for these peak positions has yet to be put forward, Sessler, *et al.* have suggested that these effects are the result, at least in part, of steric crowding within the macrocyclic core.[25]

The oxidation state and structure of amethyrins **7.74b** and **7.75b** was confirmed by single crystal X-ray diffraction analyses of the corresponding bis-HCl salts (Figures 7.6.1 and 7.6.2). In both cases, the diprotonated macrocycle was shown to exist in a nearly planar conformation in the solid state, with the two chloride counteranions being held within hydrogen-bonding distance of the ligating internal pyrrole NH groups. Deviation from planarity appears, as expected, to be the result of unfavorable methyl–methyl interactions between substituents of the central pyrrole ring and those on the two terminal pyrroles of each terpyrrolic subunit. These buttressing-type interactions apparently serve to push the NH groups of the central pyrroles in opposite directions above and below the mean plane of the macrocycle. These proposed steric effects thus account nicely for the observed solid-state geometries.

7.6.2   *Metalation Chemistry*

In contrast to a number of other expanded porphyrins, the amethyrins appear to be metal coordinating ligands *par excellence*. This point was firmly established in a first round of studies that served to show that amethyrin can act not only as a mono-metallic receptor, but also as a bona fide binucleating ligand.[25] Treatment of amethyrin **7.74b** with $CoCl_2$ for instance, was found to afford the mono cobalt(II) adduct **7.76** (Scheme 7.6.2, Figure 7.6.3). By contrast, reaction with $ZnCl_2$ was found to give the bis-Zn adduct **7.77** (Scheme 7.6.3, Figure 7.6.4). In both cases, single crystal X-ray diffraction methods were used to confirm the proposed stoichiometries and to demonstrate that the metal centers in both instances were in fact bound in an in-plane fashion.

In the two cases referred to above, inspection of the X-ray structures revealed that only four of the six possible inward-pointing nitrogen atoms were involved in metal binding. The question thus arose whether an amethyrin complex could be prepared wherein all six nitrogen atoms become involved in binding. Recently, it has been found that this can, in fact, be done. Specifically, Weghorn and Sessler found that when amethyrin **7.74b** is treated with CuCl, a copper(I) complex of as yet indeterminate structure is formed.[26] When this species, dissolved in methanol, is exposed to air, oxidation takes place to afford the bis-copper(II) complex **7.78** (Scheme 7.6.4). Here, crystallographic analysis revealed not only that two metal cations are bound, but also that all six of the pyrrolic nitrogen atoms actually participate in metal ligation (three per copper atom) (Figure 7.6.5). Interestingly, this same X-ray analysis reveals that metal complexation is accompanied by significant distortion in the macrocyclic framework. Thus, although complex **7.78** stands as a unique example of a "porphyrin-like" binuclear metal complex, it also serves to show, because of the distortion, that it may be possible to prepare yet-improved expanded porphyrin systems wherein the ligation geometry and metal coordination characteristics are perhaps matched.

Figure 7.6.1   Single Crystal X-Ray Diffraction Structure of the Amethyrin Bishydrochloride **7.74b**.
This figure was generated using information down-loaded from the Cambridge Crystallographic Data Centre and corresponds to a structure originally reported in reference 25. The macrocycle is somewhat non-planar (the average deviation of the amethyrin atoms from the mean macrocyclic plane is 0.28 Å). One chloride anion sits above the mean macrocyclic plane and the other sits below. Atom labeling scheme: carbon: ○; nitrogen: ●; chlorine: ⊙. Hydrogen atoms have been omitted for clarity

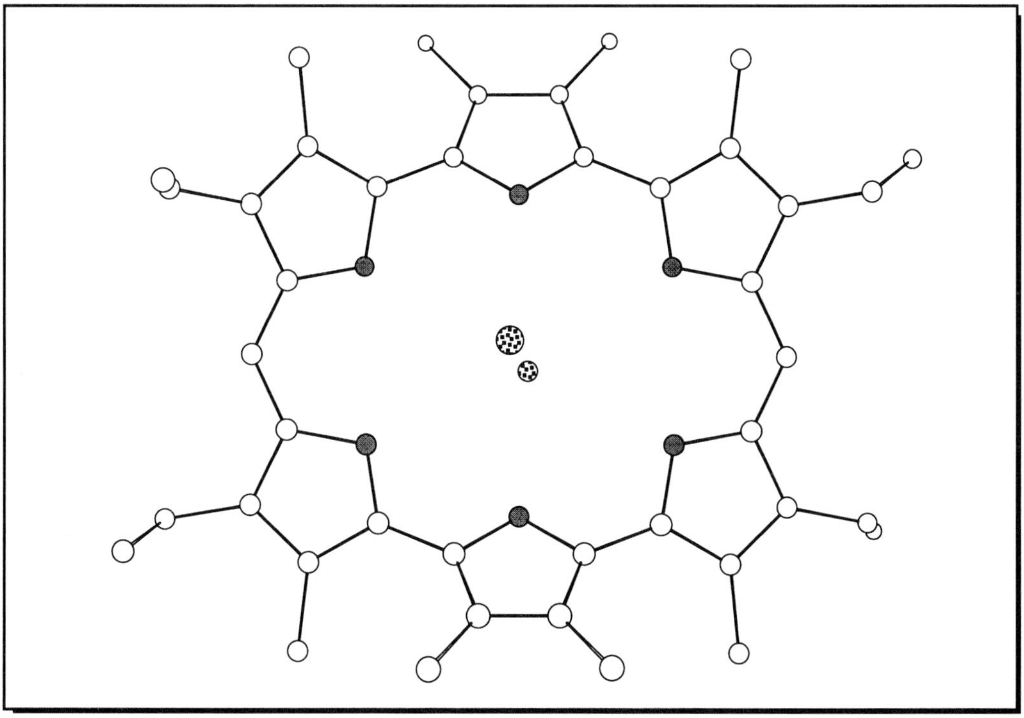

Figure 7.6.2 Single Crystal X-Ray Diffraction Structure of the Bis(ρ-NO$_2$-phenyl) Amethyrin Bishydrochloride 7.75b.

This figure was generated using information down-loaded from the Cambridge Crystallographic Data Centre and corresponds to a structure originally reported in reference 25. The macrocycle is somewhat non-planar (the average deviation of the amethyrin atoms from the mean macrocyclic plane is 0.42 Å). One chloride anion sits above the mean macrocyclic plane and the other sits below. Atom labeling scheme: carbon: $\bigcirc$; nitrogen: $\bullet$; oxygen: $\ominus$; chlorine: $\odot$. Hydrogen atoms have been omitted for clarity

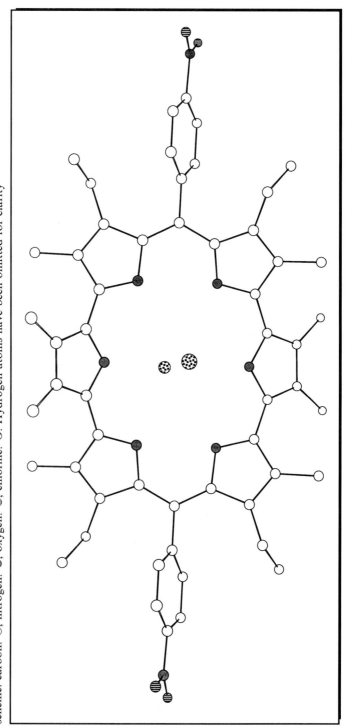

Scheme 7.6.2

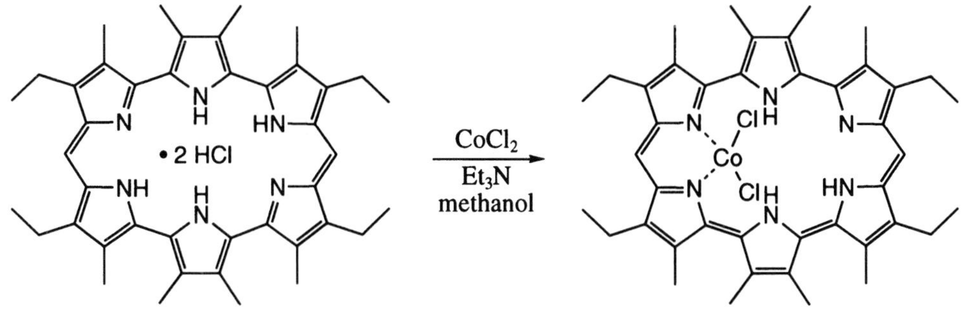

7.74b   **7.76**

Figure 7.6.3   Single Crystal X-Ray Diffraction Structure of the Dichlorocobalt(II) Amethyrin **7.76**.

This figure was generated using information down-loaded from the Cambridge Crystallographic Data Centre and corresponds to a structure originally reported in reference 25. The cobalt ion is disordered about two equivalent positions in the crystal, but sits directly within the mean macrocyclic plane. One of the central pyrrole NH groups is oriented toward the coordinated chloride anion above the macrocyclic plane and the other central pyrrole is oriented toward the chloride anion below the plane. Atom labeling scheme: cobalt: ●; carbon: ○; nitrogen: ●; chlorine: ⊙. Hydrogen atoms have been omitted for clarity

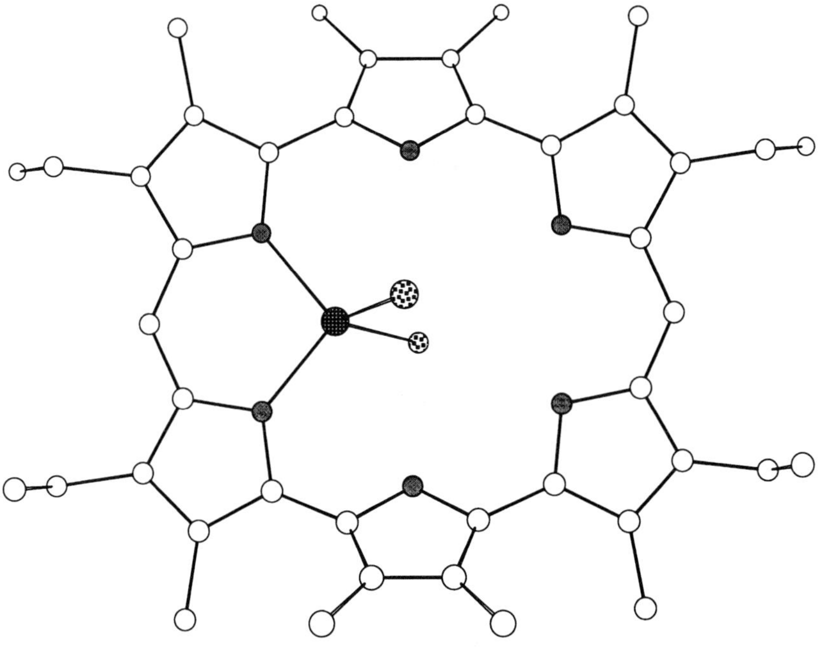

Scheme 7.6.3

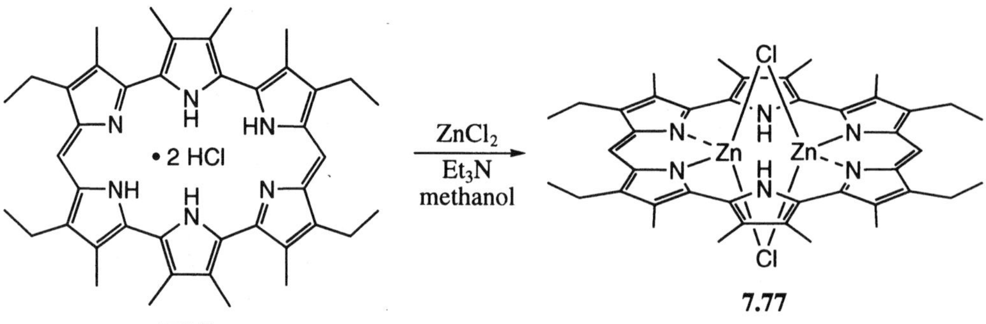

Figure 7.6.4 Single Crystal X-Ray Diffraction Structure of the Bis[chlorozinc(II)]
Amethyrin **7.77**.

This figure was generated using information down-loaded from the Cambridge Crystallographic
Data Centre and corresponds to a structure originally reported in reference 25. The zinc ions sit
directly within the mean macrocyclic plane. The bridging site is partially occupied by Cl⁻ and OH⁻
ions. One of the central pyrrole NH groups is oriented toward the bridging anion above the
macrocyclic plane and the other central pyrrole is oriented toward the anion below the plane. The
Zn–Zn distance is 2.82 Å. Atom labeling scheme: zinc: ◉; carbon: ○; nitrogen: ◉; chlorine: ⊙.
Hydrogen atoms have been omitted for clarity

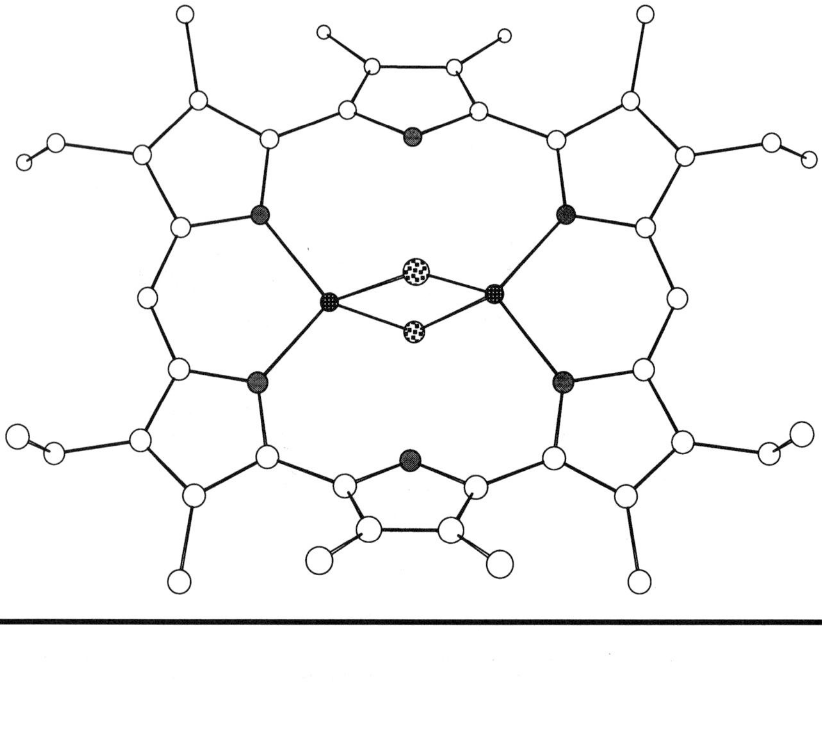

Scheme 7.6.4

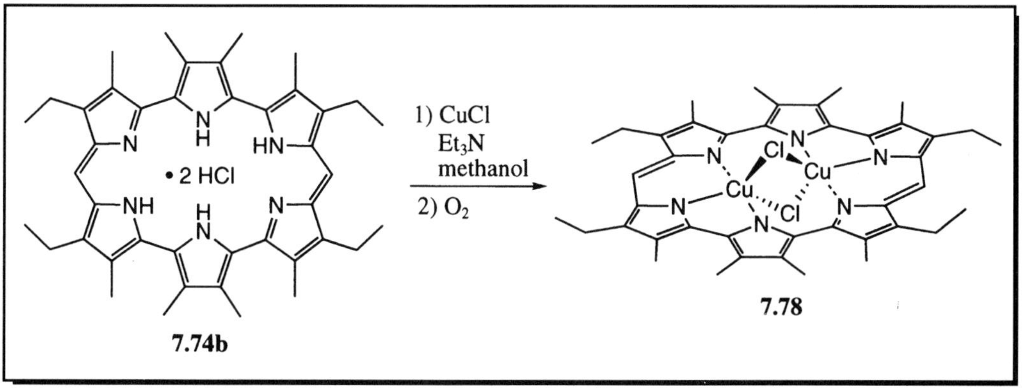

7.74b

1) CuCl
   Et₃N
   methanol

2) O₂

7.78

Figure 7.6.5   Single Crystal X-Ray Diffraction Structure of the
Bis[chlorocopper(II)]Amethyrin **7.78.**
This figure was generated using information down-loaded from the Cambridge Crystallographic
Data Centre and corresponds to a structure originally reported in reference 26. The complex exists in
the solid state in a slightly bowl-shaped conformation. The four Cu–Cl bonds are all nonequivalent
(the Cu–Cl distances for one copper atom are 2.48 Å and 2.37 Å, and those of the other copper
atom are 2.46 Å and 2.39 Å). The Cu–Cu interatomic separation is 2.76 Å. Atom labeling scheme:
copper: ◉; carbon: ○; nitrogen: ●; chlorine: ⊗. Hydrogen atoms have been omitted for clarity

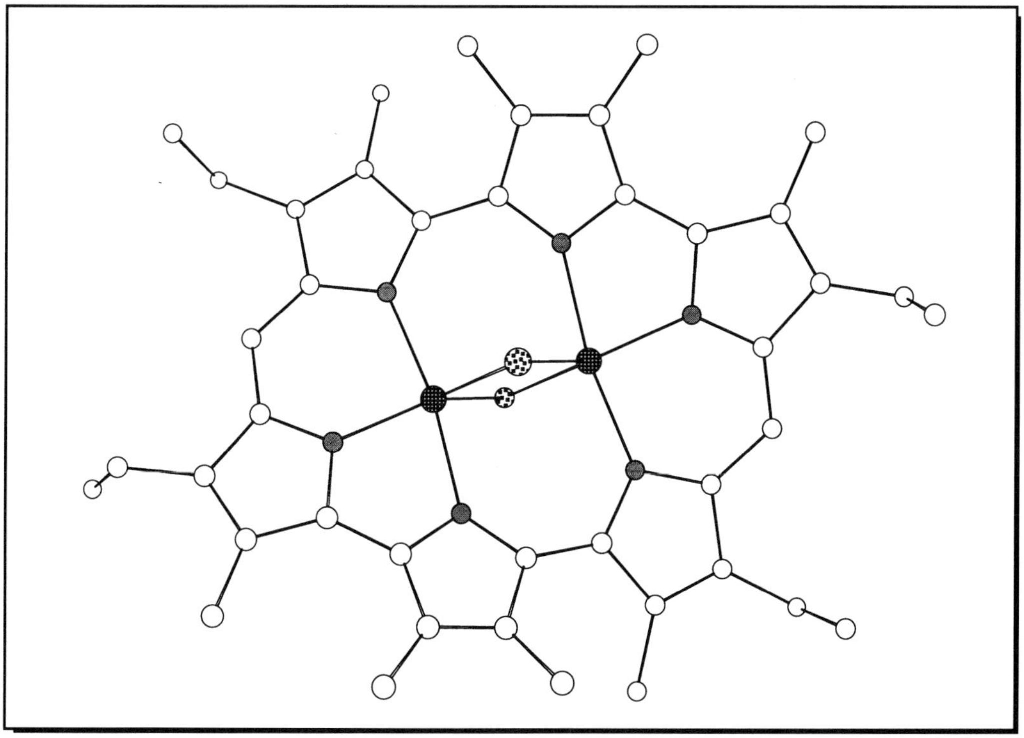

Complex **7.78** was also subject to electron paramagnetic resonance (epr) spectroscopic analysis.[26] This was done in an effort to probe the nature and degree of the metal–metal interaction (if any) and establish more precisely the oxidation state(s) of the metal centers. From epr spectra recorded at 77 K, it was confirmed that it is indeed a Cu(II)–Cu(II) dimer. However, the epr spectrum did not distinguish between ferromagnetic or weak antiferromagnetic exchange between the two metal ions. To distinguish between these two possible modes of interaction, complex **7.78** was subject to magnetic susceptibility analysis. Here, the room temperature magnetic moment per copper ion was found to be $\mu_{eff} = 1.57\ \mu_B$, a value that is slightly below that of an uncoupled copper(II) ion. At 30 K, however, the μ_{eff} per Cu(II) decreases to 1.33 μ_B. Based on these findings, it was concluded that the two copper(II) ions in **7.78** are weakly antiferromagnetically coupled.

7.7   Other Systems of Interest

7.7.1   *Hexaoxa- and Hexathia[36]hexaphyrin-(2.2.2.2.2.2)*

In the late 1960s, as part of an effort devoted to the preparation of high-order annulene polyoxides, J. A. Elix attempted a Wittig-type self-condensation of 5-formyl-2-furfurylphosphonium chloride **7.79**.[27] While the products that resulted were predominantly polymeric, small quantities of macrocyclic products were isolated. These latter consisted of trimers, tetramers, and pentamers (discussed in Chapters 2, 4, and 6, respectively), as well as a hexamer. Unfortunately, this hexameric [36]annulene hexoxide product represented by structure **7.80**, was only obtained in 0.05% yield (Scheme 7.7.1). Thus, Elix was unable to elucidate the exact configurational arrangement of the double bonds in this product. Still, as discussed below, this issue could be addressed at least in part, by spectroscopic means.

Scheme 7.7.1

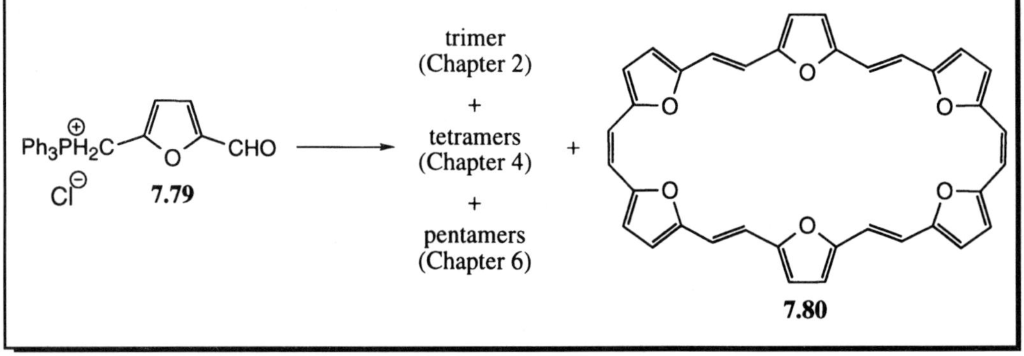

The UV-vis spectrum of **7.80** closely resembles those of the smaller pentaoxa[30]pentaphyrin-(2.2.2.2.2) isomers that were also isolated from the original Wittig-type self-condensation (see Chapter 6). It was found, however, that the hexaoxa[36]hexaphyrin-(2.2.2.2.2.2) **7.80** displays absorbances in its UV-vis spectrum that are substantially red-shifted compared to those of pentaoxa[30]pentaphyrin-(2.2.2.2.2), or its smaller tetrameric and trimeric congeners (see Chapters 4 and 2, respectively). This was rationalized in terms of the greater π-conjugation present in **7.80**.

The ^{1}H NMR spectrum of **7.80** was also recorded by Elix.[27] While it was of insufficient quality to allow an extract assignment of structure (*vide supra*), it did allow Elix to conclude that, as expected given its 4n π-electron formulation, compound **7.80** is not aromatic as judged by the absence of any substantial macrocyclic ring-current effects. This conclusion reached, it should be noted, that the formally 4n + 2 pentaoxa[30]pentaphyrin-(2.2.2.2.2) isomers also failed to display any kind of substantial ring-current effects. This led Elix to conclude that there is no fundamental difference in electronic character between large 4n and 4n + 2 π-electron oxido-bridged annulenes.[27]

Syntheses of a thiophene analog of the above hexafuran-containing macrocycle (i.e., **7.82**) have also been reported.[28,29] In one approach it was found that the reaction of 3,4-diethyl-2,5-diformylthiophene under McMurry-like conditions afforded, among other things,† the cyclohexameric macrocycle **7.82** as a mixture of two geometrical isomers (Scheme 7.7.2). The major isomer (**7.82a**), isolated in 0.5% yield, was determined to possess a *cis,trans,trans,cis,trans,trans* orientation of its bridging *meso*-like C=C bonds. The other isomer (**7.82b**), isolated in 0.3% yield, was found to have a *cis,trans,cic,trans,cis,trans* configuration.

An alternative route to these same two isomeric hexathia[36]hexaphyrin-(2.2.2.2.2.2) macrocycles (**7.82a** and **7.82b**) has also been reported.[29] Using Wittig-type chemistry, 3,4-diethyl-2,5-diformylthiophene **7.81** was reacted with the diethyl-substituted bisphosphonium salt **7.83** to afford macrocycles **7.82a** and **7.82b** in 0.4% and 0.1% yield, respectively (Scheme 7.7.3). It is of interest to note that in this reaction the formation of cyclotetramers was not observed. This is in direct contrast to the analogous reaction using 3,4-unsubstituted thiophene starting materials wherein only cyclic tetramers were found to form (see Chapter 4). This discrepancy is most likely directly attributable to the steric effects caused by the presence of the thiophene β-ethyl substituents.

The viability of yet another approach to hexathia[36]hexaphyrin-(2.2.2.2.2.2) macrocycles has also been demonstrated.[29] In this instance, the diformyl-*trans*-bithiophenylethene **7.84** is used as the starting material with a standard McMurry-type coupling providing the hexathia[36]hexaphyrin-(2.2.2.2.2.2) **7.82b** in 1.3% yield (Scheme 7.7.4).[29] This reaction was found to produce only one hexameric product. On the basis of spectroscopic studies, this single product was found to be identical to the *cis,trans,cis,trans,cis,trans* isomer **7.82b** prepared and isolated as discussed above.

†A cyclic trimer, tetramer, and two isomeric pentamers, as well as two linear oligothiophenes were also isolated from this reaction.

Interestingly, the cyclotetramer isolated from this latter reaction (discussed in Chapter 4) was found to possess an all-*cis* configuration about the *meso*-like double bonds. Unfortunately, this tetrameric product was isolated in the comparatively low yield of 0.6%.

Scheme 7.7.2

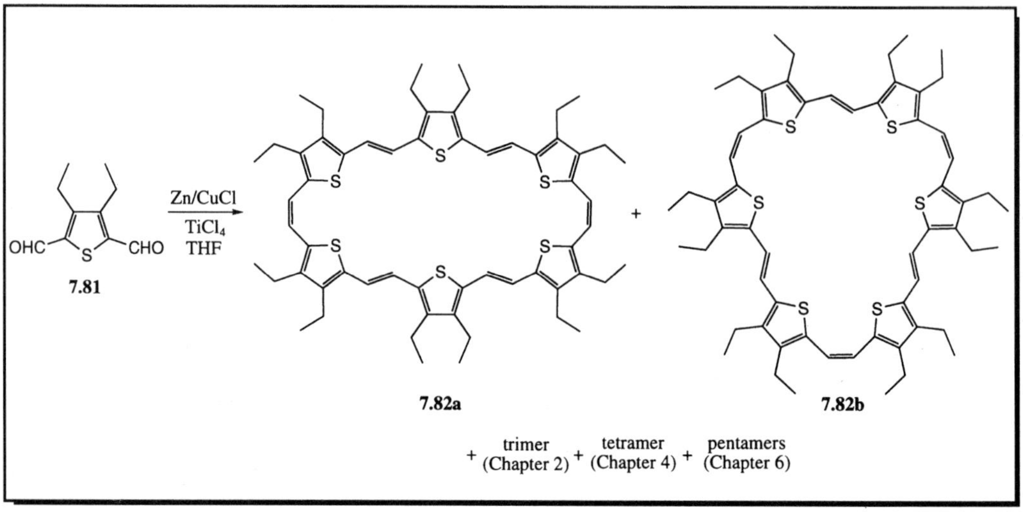

Scheme 7.7.3

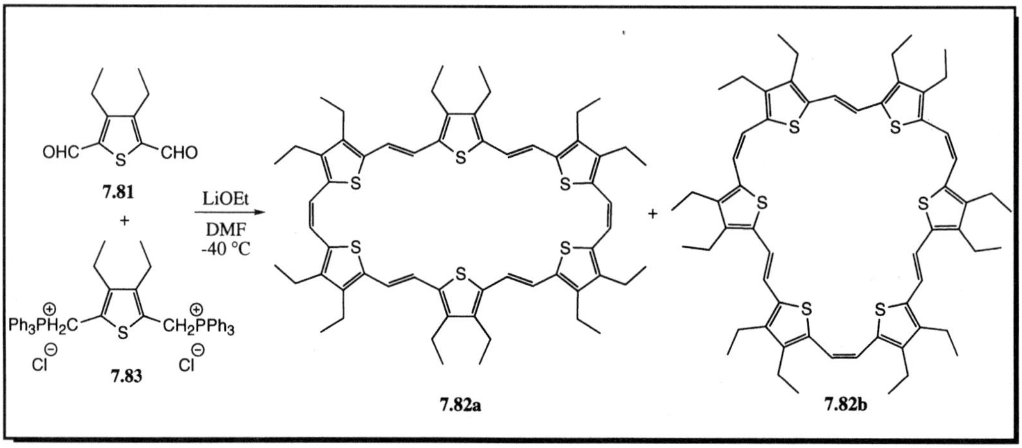

Attempts at oxidizing the hexathia[36]hexaphyrin-(2.2.2.2.2.2) isomers **7.82a** and **7.82b** have been made.[28,29] In the case of isomer **7.82a**, treatment with bis(trifluoroacetoxy)iodobenzene in benzene and TFA was found to effect oxidation to the [34]hexaphyrin-(2.2.2.2.2.2) species **7.85** in 50% yield (Scheme 7.7.5). This macrocycle, isolated as its bistrifluoroacetate salt, was found to be aromatic, as judged from

^{1}H NMR spectroscopic analysis (e.g., $\delta_{CH\text{-inner}}$ = -16.0 ppm; $\delta_{CH\text{-outer}}$ = 13.8 and 14.7 ppm, in CF$_3$CO$_2$D). Macrocycle **7.85** also exhibits UV-vis absorption bands characteristic of highly conjugated porphyrin-related annulenes. For instance, in 5% TFA in CH$_2$Cl$_2$, system **7.85** exhibits one fairly intense Soret-like band at 667 nm (ε = 193 000 M^{-1} cm^{-1}) and two near-IR bands at 1151 and 1190 nm (ε = 86 600 and 74 000 M^{-1} cm^{-1}, respectively). This represents a significant spectral change relative to the 36 π-electron system **7.82a**, whose UV-vis absorption spectrum shows broad, relatively weak maxima at *ca.* 450 and 500 nm.

Scheme 7.7.4

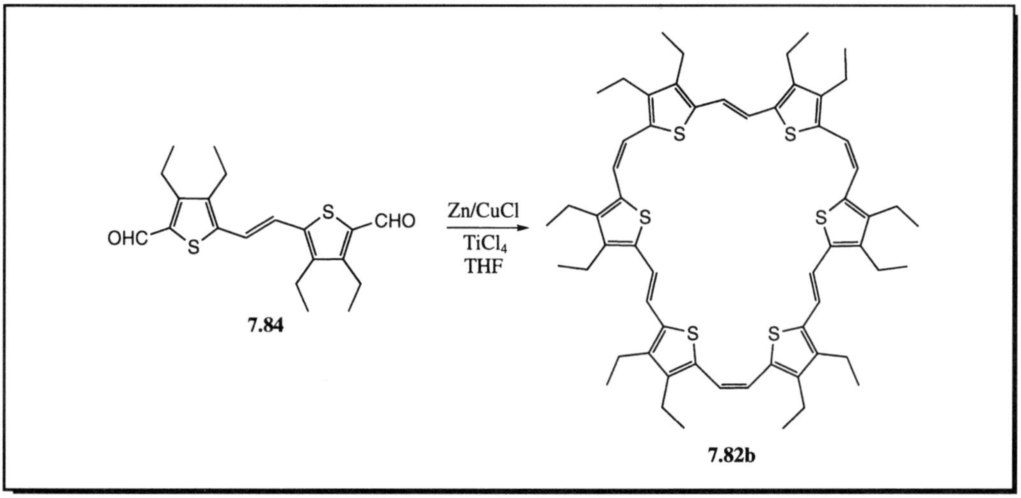

Scheme 7.75

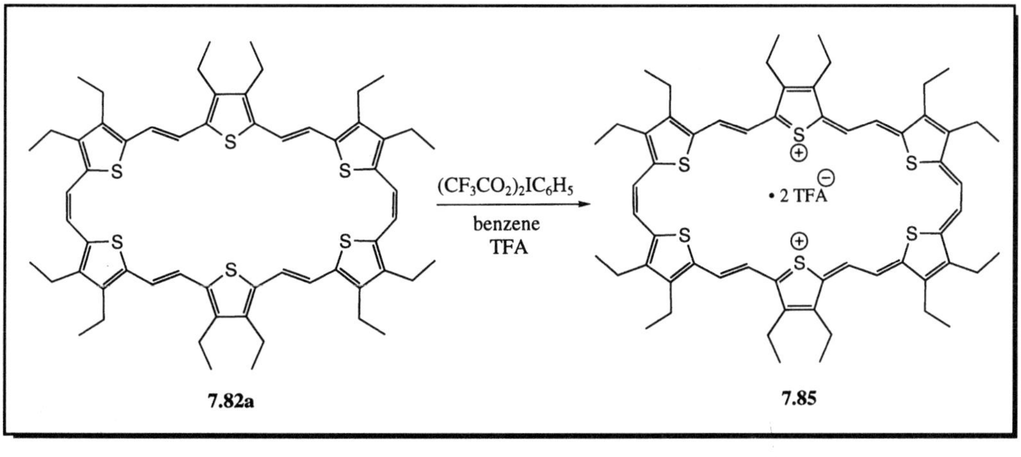

In contrast to the oxidation behavior observed for **7.82a**, isomer **7.82b** was found to decompose in the presence of oxidants.[28] In spite of this, oxidation to **7.86**

could be effected simply by treating **7.82b** with TFA (Scheme 7.7.6). Although the resulting product proved to be less stable than its more symmetrical counterpart **7.85**, sufficient data were gathered to support the claim that it had indeed been formed. For instance, ^{1}H NMR spectroscopic analyses were carried out. These revealed a solution-state structure very similar to that of **7.85**. These same studies provided evidence of an aromatic ring current. In a different vein, the UV-vis absorption spectrum of **7.86** was recorded. It showed some resemblance to that of **7.85**. However, in the case **7.86**, the main absorbance band is both broadened and blue-shifted (λ_{max} = 606 nm for **7.86** vs. 667 nm for the more symmetric isomer **7.85**).

Scheme 7.7.6

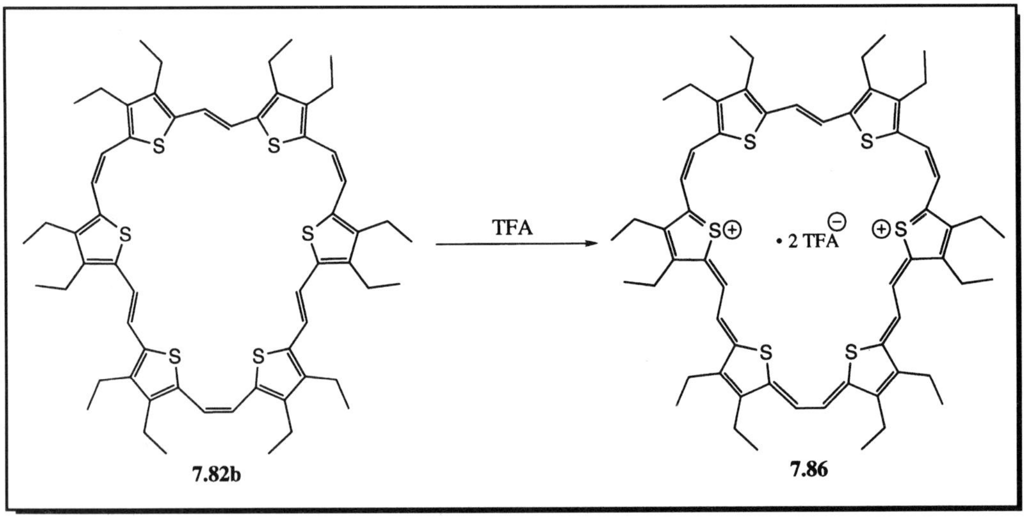

7.82b   **7.86**

7.7.2   Cram's Cavitand

A macrocycle containing a formal structural core related to that of the hex-aoxa- and hexathia[36]porphyrin-(2.2.2.2.2.2) macrocycles **7.80** and **7.82** has been reported by Cram and coworkers.[30,31] This product, the so-called "cavitand" macro-cycle **7.88**, was prepared by treating the dimeric dibenzofuran species **7.87** with *s*-BuLi followed by Fe(acac)$_3$ (Scheme 7.7.7). After fractional crystallization and chro-matography, **7.88** was isolated in 1.6% yield.

Based on molecular modeling experiments, cavitand **7.88** was considered to possess D_{3d} symmetry.[30] This proposed structural assignment was supported by ^{1}H NMR spectroscopic studies. In the conformation giving rise to the D_{3d} symmetry, **7.88** would contain a cavity of approximate 11 × 7 × 7 Å dimensions, with all six oxygen atoms lying in one plane. It would also result in the formation of smaller clefts within the macrocycle. This led to the prediction that **7.88** might serve as a receptor capable of forming host–guest-type complexes with aromatic hydrocarbons through π- π stacking interactions. In particular, modeling studies revealed that **7.88** should be capable of accommodating up to seven benzene molecules. Consistent with this hypothesis is the fact that cavitand **7.88** proved to be more soluble in benzene, toluene, or the xylenes than in chloroform or dichloromethane. Further, when crystallized from benzene, **7.88** was found to form heavily solvated crystals. Unfortunately, however, these crystals released the trapped benzene too rapidly to permit an X-ray crystallographic analysis to be performed. Also, to date, no solution-state binding studies have been reported. Thus, the exact nature of the proposed hydrocarbon recognition process and receptor-to-substrate geometry remains rather undefined.

Scheme 7.7.7

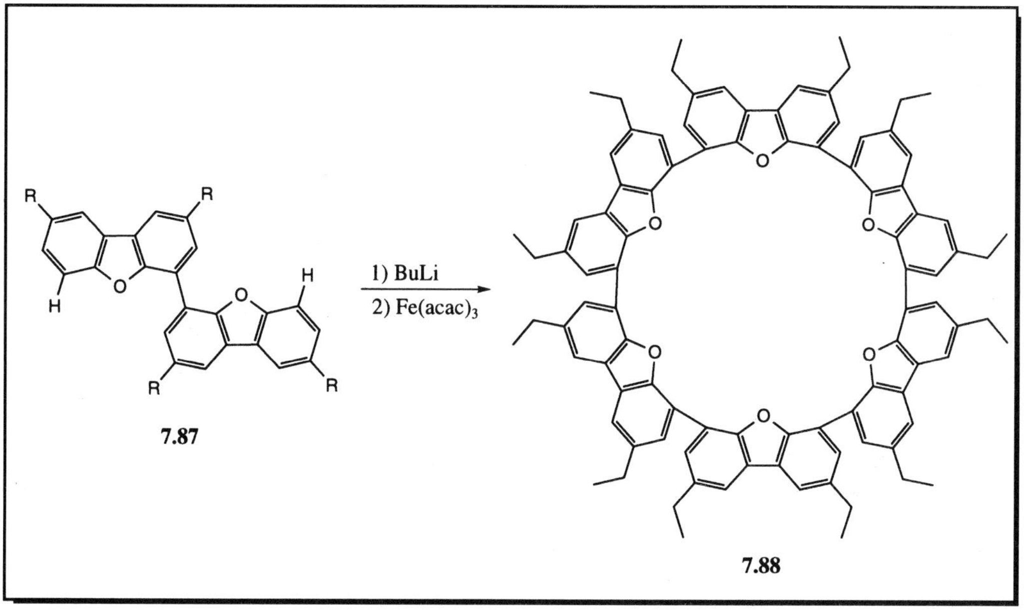

7.87

1) BuLi
2) Fe(acac)₃

7.88

7.7.3 Hexathia[30]hexaphyrin-(2.0.2.0.2.0)

Another example of a hexaphyrin-like macrocycle is the hexathiophene-system **7.90**. This macrocycle was first reported by Merz and Neidlein in 1993.[24] It was isolated in 2.9% yield as a trimeric by-product from the reductive McMurry-type self-coupling of 2,5-diformyl bithiophene **7.89** (Scheme 7.7.8), a reaction that was originally intended to produce the tetrathiaporphycenogen **7.91** (see Chapter 3).[‡]

[‡]Merz and Neidlein also reported the formation (0.7%) of a dihydro-derivative of **7.90** during this same reaction. However, this latter compound could not be isolated in a pure form, and structural information is therefore lacking.

Later, as a result of independent work, Cava reported a 13.6% yield of this same product (i.e., **7.90**) as the result of this same reaction sequence.[22]

Scheme 7.7.8

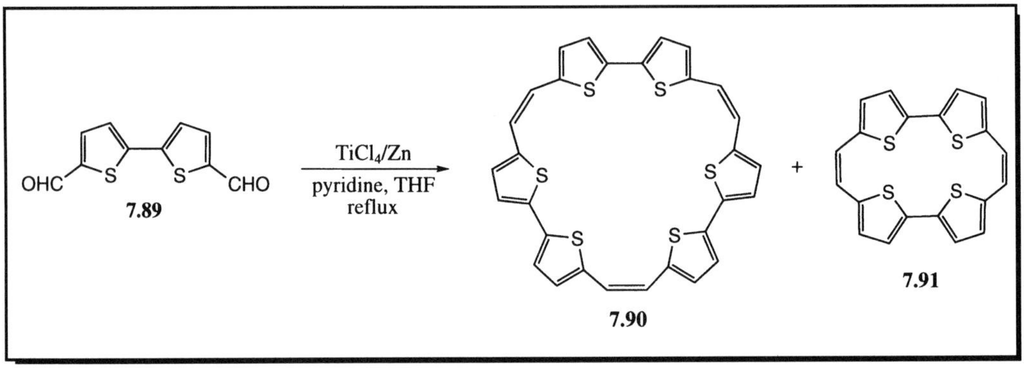

UV-vis spectroscopic studies of **7.90**, carried out by Merz and Neidlein[24] as well as Cava[22] were considered consistent with conjugation between the thiophene subunits being minimal. This is because only weak transitions (between 210 and 340 nm) were observed in the visible spectra region. Further, although formally a 30 π-electron 4n + 2 macrocycle, an analysis of the ^{1}H NMR spectrum of **7.90** failed to reveal any kind of substantial macrocyclic diamagnetic ring-current effect. This fact, considered in conjunction with the UV-vis data, led to the conclusion that system **7.90** must possess a very non-planar structure. This tentative conclusion was confirmed by a single crystal X-ray structure of **7.90**. This structure, which was reported by the group of Merz and Neidlein revealed an overall non-planar geometry for **7.90** in the solid state.[24] While one of the three bithiophene units and its neighboring C=C bridges are nearly planar, the other two bithiophenes suffer a significant twisting about their thiophene–thiophene bond. System **7.90** with its larger sulfur atoms, thus serves to illustrate the fine balance that can exist between steric and electronic effects, and the play-off between factors militating for and against overall macrocyclic planarity. As such, it helps to highlight issues that are even more critical in the case of the still-larger systems discussed in the next chapter.

7.7.4   *Pyrazole-containing Hexaphyrin-like Systems*

A final example of a macrocyclic system containing six five-membered hetero-cycles comes from Lind and LeGoff.[32] These workers devoted considerable effort to the problem of making an analog of hexaphyrin wherein two of the pyrrole rings are replaced formally by pyrazole subunits. While they did not succeed in attaining this goal, they made noteworthy progress. Some of this work is summarized in Scheme 7.7.9. For instance, they found that an HBr-catalyzed reaction could be carried out between the diformyl pyrazole-containing tripyrrane analog **7.92** and the corresponding bis-α-free species **7.93**. Unfortunately, this reaction produced a mixture of products. While it was thought that these products consisted of only two of the four theoretically possible isomers (represented by structures **7.94a–d** in Scheme 7.7.9), the fact of the matter was that these putative isomers could not be separated from one another. Therefore, their exact structure was never determined. Also, attempts to

Scheme 7.7.9

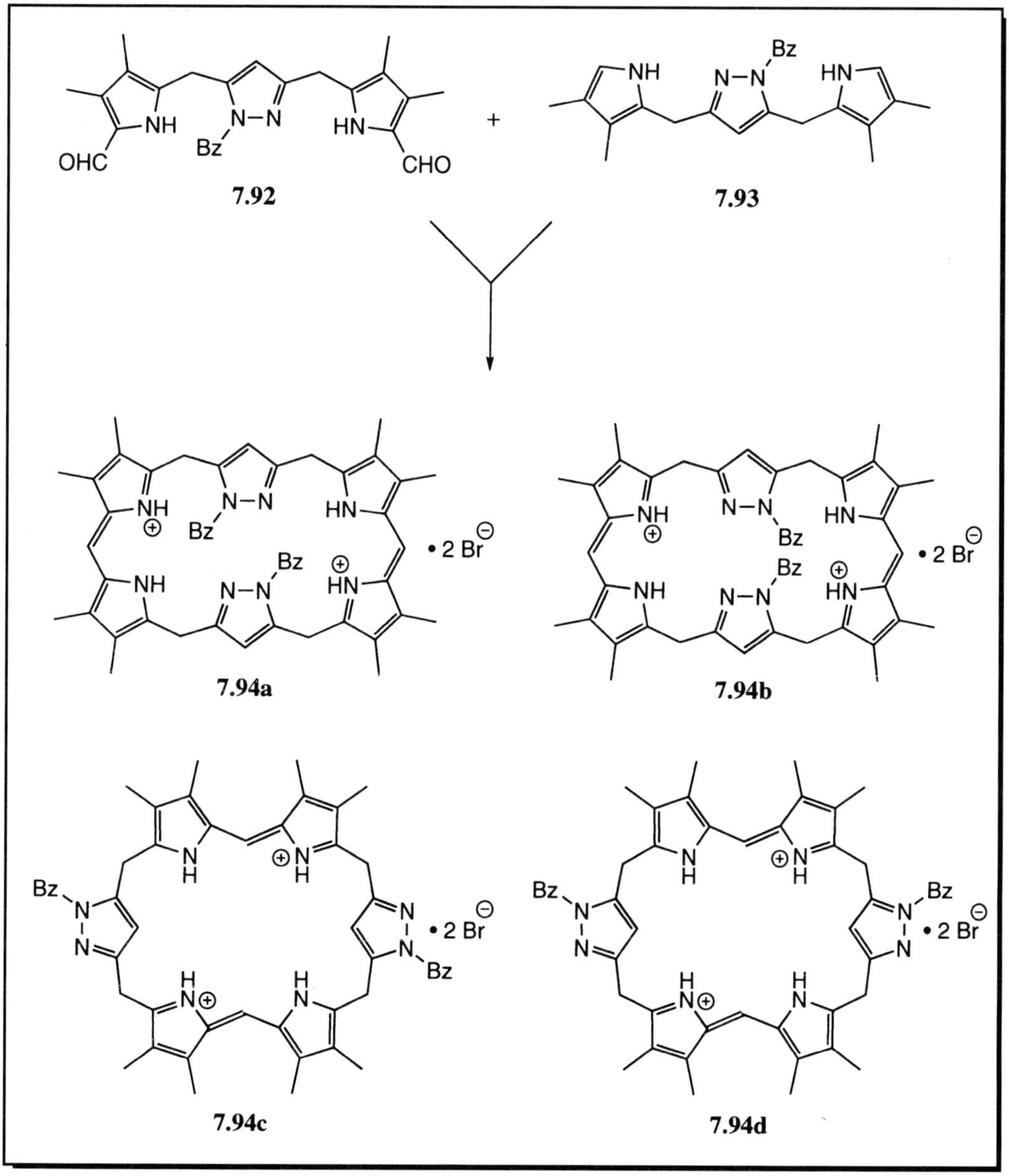

7.92   +   7.93

7.94a   7.94b

7.94c   7.94d

remove the benzyl protecting group and effect oxidation to the corresponding aromatic analogs proved unsuccessful. Thus, it proved impossible to characterize isomers **7.94a–d** *via* this indirect approach. At present, therefore, these materials stand as tantalizing way stations on the road to interesting, but as yet unknown hexaphyrin-like products.

Note in proof

Subsequent to the completion of this manuscript, a communication describing the synthesis of a pentaaza-monothiabronzaphyrin appeared.[33]

7.8   References

1. Gossauer, A. *Bull. Soc. Chim. Belg.* **1983**, *92*, 793–795.
2. Gossauer, A.; *Chimia*, **1983**, *37*, 341–342.
3. Gossauer, A.; *Chimia*, **1984**, *38*, 45–46.
4. Charrière, R.; Jenny, T. A.; Rexhausen, H.; Gossauer, A. *Heterocycles*, **1993**, *36*, 1561–1575.
5. Wiß, T., Ph.D. Dissertation, University of Cologne, Germany, **1995**.
6. Sessler, J. L.; Morishima, T.; Lynch, V. *Angew. Chem. Int. Ed. Eng.* **1991**, *30*, 977–980.
7. Sessler, J. L.; Morishima, T. Unpublished results.
8. Sessler, J. L.; Furuta, H.; Král, V. *Supramol. Chem.* **1993**, *1*, 209–220.
9. Gebauer, A., Diplomarbeit, University of Cologne, Germany, **1993**.
10. Sessler, J. L.; Weghorn, S. J.; Morishima, T.; Rosingana, M.; Lynch, V.; Lee, V. *J. Am. Chem. Soc.* **1992**, *114*, 8306–8307.
11. Rothemund, P. *J. Am. Chem. Soc.* **1936**, *58*, 625–627.
12. Rothemund, P.; Menotti, A. R. *J. Am. Chem. Soc.* **1941**, *63*, 267–270.
13. Kim, J. B.; Adler, A. D.; Longo, F. R. In *The Porphyrins*, Vol. 1; Dolphin, D., Ed.; Academic Press: New York, 1978; pp. 85–100.
14. Weghorn, S. J., Ph.D. Dissertation, University of Texas at Austin, Austin, TX, USA, **1994**.
15. Sessler, J. L.; Weghorn, S. J.; Lynch, V.; Fransson, K. *J. Chem. Soc. Chem. Commun.* **1994**, 1289–1290.
16. Sessler, J. L.; Weghorn, S. J.; Lynch, V.; Fransson, K. *J. Chem. Soc. Chem. Commun.* **1994**, 2737.
17. Sessler, J. L.; Weghorn, S. J.; Lynch, V.; Fransson, K. *J. Chem. Soc. Chem. Commun.* **1995**, 801.
18. Hoehner, M. C., Ph.D. Dissertation, University of Texas at Austin, Austin, TX, USA, **1996**.
19. Johnson, M. R.; Miller, D. C.; Bush, K.; Becker, J. J.; Ibers, J. A. *J. Org. Chem.* **1992**, *57*, 4414–4417.
20. Miller, D. C.; Johnson, M. R.; Ibers, J. A. *J. Org. Chem.* **1994**, *59*, 2877–2879.

21. Hu, Z.; Scordilis-Kelley, C.; Cava, M. P. *Tetrahedron Lett.* **1993**, *34*, 1879–1882.
22. Hu, Z.; Atwood, J. L.; Cava, M. P. *J. Org. Chem.* **1994**, *59*, 8071–8075.
23. Kozaki, M.; Parakka, J. P.; Cava, M. P. *J. Org. Chem.* **1996**, *61*, 3657–3661.
24. Ellinger, F.; Gieren, A.; Hübner, Th.; Lex, J.; Merz, A.; Neidlein, R.; Salbeck, J. *Monatsh. Chem.* **1993**, *124*, 931–943.
25. Sessler, J. L.; Weghorn, S. J.; Hiseada, Y.; Lynch, V. *Chem. Eur. J.* **1995**, *1*, 56–67.
26. Weghorn, S. J.; Sessler, J. L.; Lynch, V.; Baumann, T. F.; Sibert, J. W. *Inorg. Chem.* **1996**, *35*, 1089–1090.
27. Elix, J. A. *Aust. J. Chem.* **1969**, *22*, 1951–1962.
28. Schall, R., Ph.D. Dissertation, University of Cologne, Germany, **1993**.
29. Jörrens, F., Ph.D. Dissertation, University of Cologne, Germany, **1994**.
30. Helgeson, R. C.; Lauer, M.; Cram, D. J. *J. Chem. Soc. Chem. Commun.* **1983**, 101–103.
31. Schwartz, E. B.; Knobler, C. B.; Cram, D. J. *J. Am. Chem. Soc.* **1992**, *114*, 10775–10784.
32. Lind, E. M.Sc. Thesis, Michigan State University, East Lansing, MI, USA, **1987**.
33. Johnson, M. R. *J. Org. Chem.* **1997**, *62*, 1168–1172.

8   Introduction

In recent years, the desire to generate new expanded porphyrins has led to the preparation of ever larger macrocycles. Most of these have consisted of tetra-, penta-, and hexapyrrole-type macrocycles. However, a few "higher order" systems, expanded porphyrins containing more than six pyrrole-type subunits, are now known. It is a review of these systems, still limited to macrocycles containing eight and ten pyrrole-type subunits, that is the subject of the present chapter.

8.1   Turcasarin: Decaphyrin-(1.0.1.0.0.1.0.1.0.0)

Interestingly, and perhaps surprisingly, it was a system with ten, not seven, eight, or nine, pyrroles that was the first "higher order" expanded porphyrin to be reported in the literature.[1] This system was first synthesized in serendipitous fashion as the result of a "[2 + 2]" condensation between the bis-α-free terpyrrole **8.1** and the diformyl bipyrrole **8.3**. While this condensation reaction was originally expected to produce the "[1 + 1]" "orangarin", a pentapyrrolic material that can indeed be isolated under certain conditions (see Chapter 6), it was found by Sessler, *et al.* to produce only the decapyrrolic macrocyclic product **8.5** when run as indicated in Scheme 8.1.1. The fact that it was this "[2 + 2]" product that was obtained represented an important milestone in expanded porphyrin chemistry. It showed clearly that higher order expanded porphyrins could be made without having to construct overly large polypyrrolic (e.g., linear) building blocks.

Macrocycle **8.5b** is formally a non-aromatic system that contains 40 π-electrons in its primary conjugation pathway. In spite of this, the molecule is brightly colored and displays a strong absorbance band in the visible spectrum at 642 nm ($\varepsilon = 312\,500$ $M^{-1}cm^{-1}$ in CH_2Cl_2) (Figure 8.1.1). In fact, the beautiful turquoise color displayed by organic solutions of **8.5** led the original authors to name this class of expanded porphyrins "turcasarins".

Solid-state X-ray structural studies of the tetrakis HCl salt of **8.5** (i.e., **8.5b**) revealed that the molecule does not exist in a planar, circular conformation as represented in Scheme 8.1.1. Rather, as depicted in Figures 8.1.2 and 8.1.3, the molecule adopts a helical twist and overall "figure-eight" conformation.[1] This "twisting" makes **8.5b** chiral by virtue of conformation and serves to define a pair of enantiomeric atropisomers, at least in the solid state. In solution, detailed [1]H

NMR analyses served to reveal that the basic conformational features of **8.5b** are retained. However, here, the two limiting enantiomeric forms of **8.5b** interconvert readily, albeit slowly on the NMR time scale at room temperature.

Scheme 8.1.1

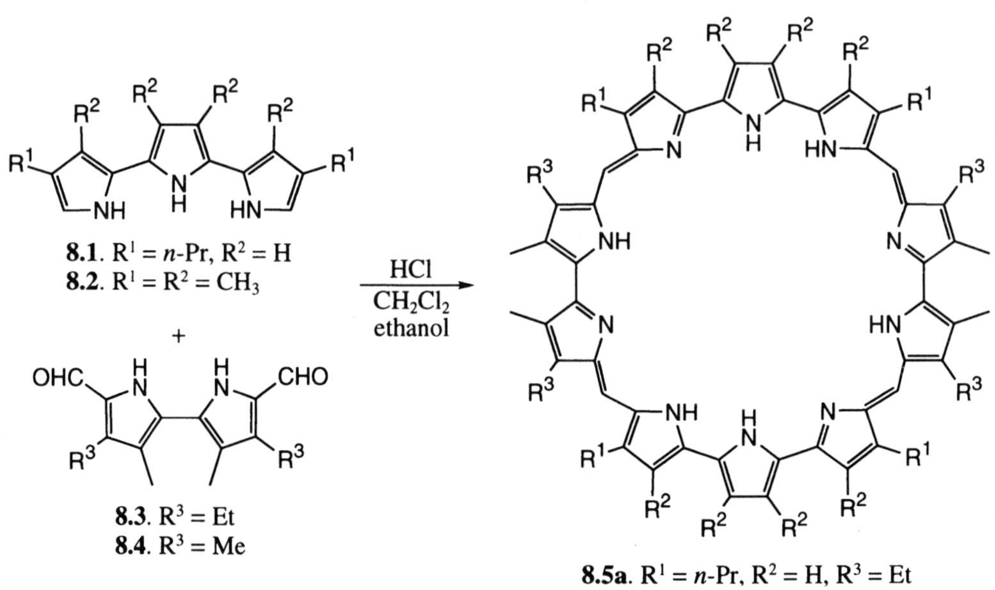

8.1. $R^1 = n\text{-Pr}, R^2 = H$
8.2. $R^1 = R^2 = CH_3$

+

8.3. $R^3 = Et$
8.4. $R^3 = Me$

$\xrightarrow[\substack{CH_2Cl_2 \\ ethanol}]{HCl}$

8.5a. $R^1 = n\text{-Pr}, R^2 = H, R^3 = Et$
8.5b = **8.5a** • 4 HCl
8.6a. $R^1 = R^2 = R^3 = CH_3$
8.6b = **8.6a** • 4 HCl

The figure-eight conformation adopted by turcasarin results in the formation of two "hemipentaphyrin" cavities within one molecule. That is to say, each "half" of the molecule has five inward-pointing nitrogen atoms that could serve, at least conceptually, to coordinate a metal cation. It was thus hoped that turcasarin could function as a "two-times pentadentate", bimetallic receptor. In preliminary investigations involving coordination of uranyl cation (UO_2^{2+}), some support for this hope has in fact been obtained.[1] Specifically, it proved possible to prepare the bis-uranyl chelate **8.7** (Scheme 8.1.2). Although an X-ray crystal structure is still lacking for this complex, mass spectrometric and NMR spectroscopic evidence were gathered that together confirmed the formation of this bimetallic product. Interestingly, this binuclear product (i.e., **8.7**) appears to retain the twisted figure-eight conformation observed for the starting tetrakis-HCl salt **8.5b**.

Presently, the hydrochloride salt **8.5b** remains the best studied of the turcasarins. This class of macrocycles is not limited, however, to this single protonation state or even to this one system. Indeed, Sessler and coworkers have recently prepared the all-methyl system **8.6** by condensing hexamethylterpyrrole **8.2** with diformyl tetra-methylbipyrrole **8.4**.[2] Preliminary studies of this per-methylated macrocycle suggest that it also exists in a figure-eight conformation in solution. However, these same studies indicate that **8.6b** undergoes conformational "flipping" even more slowly in solution than does **8.5**.

Figure 8.1.1   UV-vis Absorption Spectrum of Turcasarin·4HCl **8.5b** in Dichloromethane

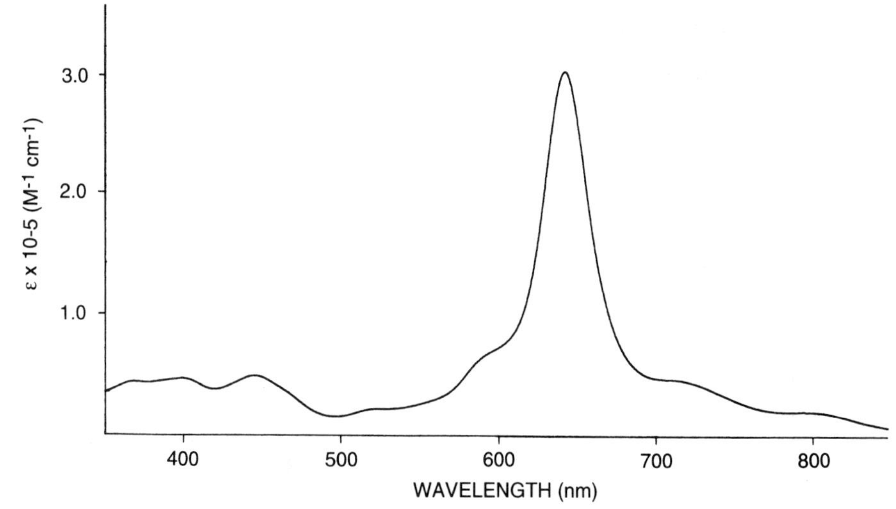

8.2   Cyclooctapyrroles

Subsequent to the discovery of turcasarin by Sessler, *et al.*, Vogel and co-workers reported the syntheses of four different octapyrrolic macrocycles.[3,4] Like turcasarin, these octapyrrolic systems had their genesis in the serendipitous discovery that certain condensations involving bipyrrolic precursors will favor "[2 + 2]" products over "[1 + 1]" ones. In further analogy to the turcasarins, it was also found that these systems generally adopt a chiral figure-eight conformation in solution and in the solid state. This provides an important link between these two classes of "higher order" expanded porphyrins. Nonetheless, there are some subtle but important differences between Sessler's turcasarins and each of Vogel's octapyrroles. These are detailed below.

> *8.2.1   Tetrahydrooctaphyrin-(2.1.0.1.2.1.0.1) and*
> *Octaphyrin-(2.1.0.1.2.1.0.1)*

The first example of a cyclooctapyrrolic macrocycle to be prepared by Vogel and coworkers was the tetrahydrooctaphyrin-(2.1.0.1.2.1.0.1) system **8.12**.[3] This macrocycle was isolated during the course of efforts directed toward the synthesis of the *trans*-isomer of corrphycene (the *cis*-isomer of corrphycene, compound **8.8** (Figure 8.2.1), is discussed in Chapter 3). Specifically, these workers found that when the dipyrrylethane derivative **8.9** was condensed with the diformyl bipyrrole **8.10**, the cyclic octamer **8.12** was isolated (in 30% yield) as the exclusive product of the reaction (i.e., none of the "originally desired" tetrapyrrolic species **8.11** was obtained) (Scheme 8.2.1). By heating this octapyrrolic material in the presence of 10% palladium on carbon, dehydrogenation to the formally conjugated 36 π-electron macrocycle **8.13** could be achieved in 80% yield. In this way, the same basic sequence used to obtain **8.12** was made to yield a second octapyrrolic macrocyclic product.

Figure 8.1.2   Schematic Representation of Structure **8.5a** in a Figure-eight Conformation

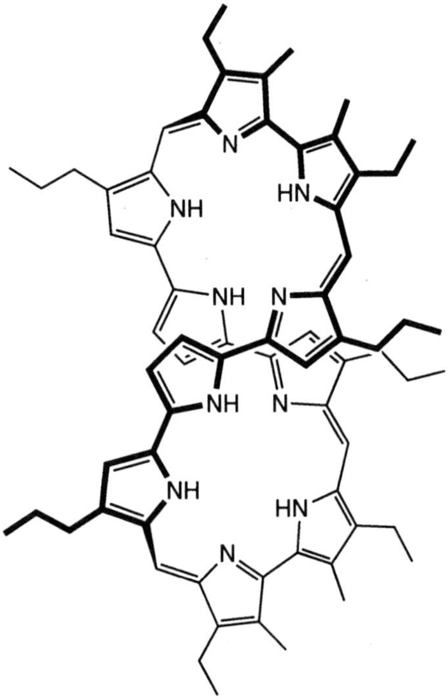

8.5a

Proton NMR spectroscopic studies of the non-conjugated macrocycle **8.12** revealed the existence of a C_2-symmetry element within the macrocycle. However, on the basis of the diastereotopicity observed for the ethyl side-chains on the periphery of the macrocycle, it was concluded that this molecule, like turcasarin, preferentially adopts a chiral figure-eight conformation. The same conclusion was reached for the oxidized derivative **8.13**. In both cases, variable temperature ¹H NMR spectral studies revealed only a minimal temperature dependence. These data were interpreted as indicating that these molecules are relatively rigid in solution, with the conformational inversion barriers between the two enantiomeric atropisomers being higher than 20 kcal mol⁻¹, presumably.[3]

Figure 8.1.3   Single Crystal X-Ray Diffraction Structure of Turcasarin·4HCl **8.5b**.
This figure was generated using information down-loaded from the Cambridge Crystallographic Data Centre and corresponds to a structure originally reported in reference 1. The macrocycle adopts a nearly C_2-symmetric, twisted figure-eight conformation. Each of four pockets defined by the twisted macrocycle contains one hydrogen-bound chloride anion. At the point of crossover, generated by the macrocyclic twist, the interplane separation is 3.27 Å. Atom labeling scheme: carbon: ○; nitrogen: ●; chlorine: ⊘. Hydrogen atoms have been omitted for clarity

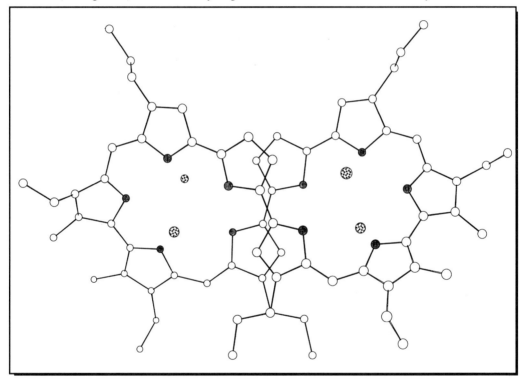

The figure-eight conformation of the conjugated macrocycle **8.13** was confirmed in the solid state *via* single crystal X-ray diffraction analyses of the bishydrochloride and tetrakishydroperchlorate salts. The two structures obtained are reproduced in Figure 8.2.2. The first, involving the bis-HCl salt **8.13b** (Figure 8.2.2, left) confirms the overall helical nature of the system and reveals a macrocycle with near D_2 symmetry overall. The second structure, that of the tetrakis-HClO₄ salt **8.13c** (Figure 8.2.2, right), was found to be similar. In this case, however, the macrocyclic frame is seen to experience an accordion-like stretching as compared to the corresponding HCl salt. This stretching was ascribed to the presence of two additional counteranions within the rather complex overall structure.

Scheme 8.1.2

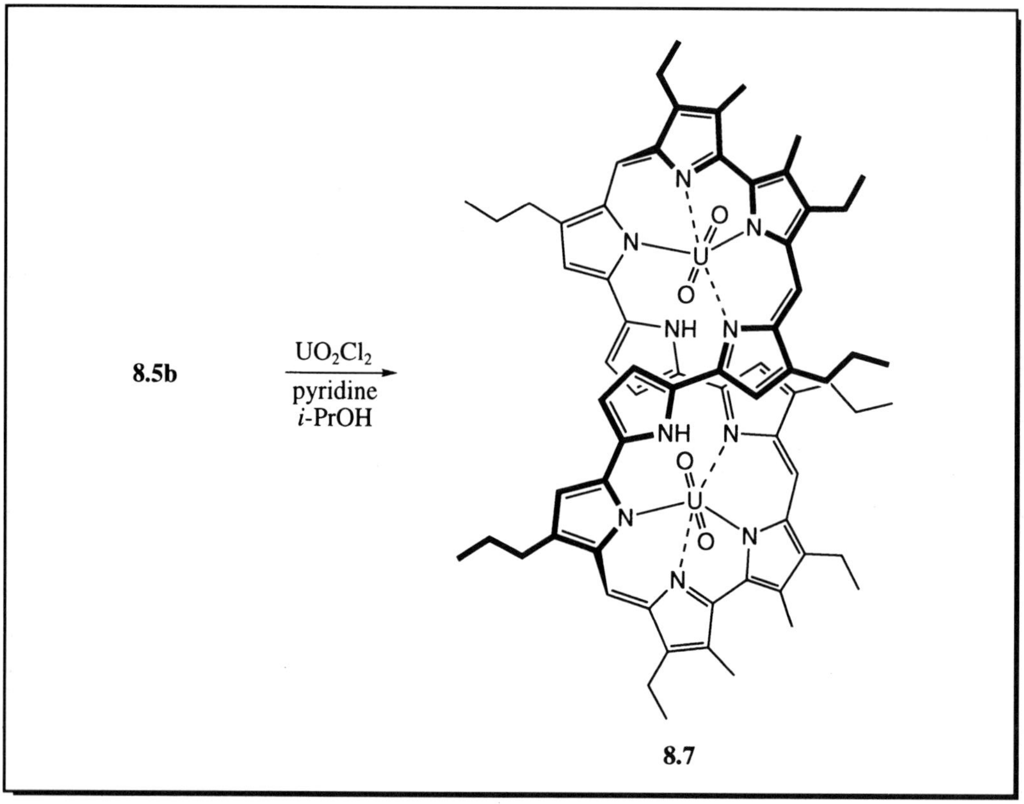

8.5b $\dfrac{UO_2Cl_2}{\text{pyridine} \atop \text{i-PrOH}}$

8.7

Figure 8.2.1   Structure **8.8**

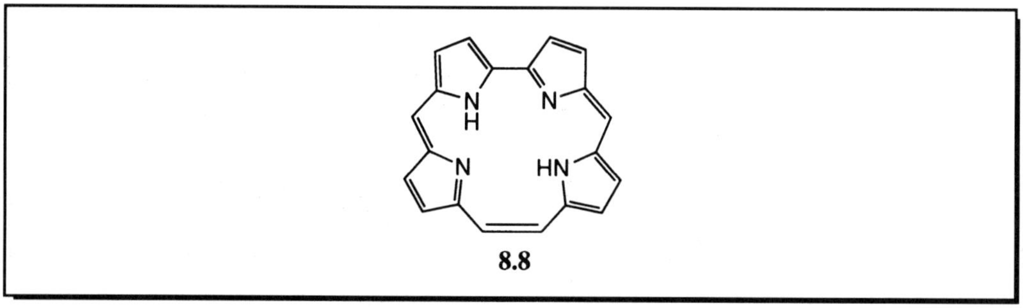

8.8

The free-base form of **8.13a** was also characterized *via* UV-vis spectroscopy. These studies revealed a notable lack of porphyrin-like character. Specifically, no pronounced Soret- or Q-like transitions were seen in the electronic spectrum. While initially unexpected, such findings were considered consistent with the only limited conjugation that would be expected within this non-planar macrocyclic frame.

Scheme 8.2.1

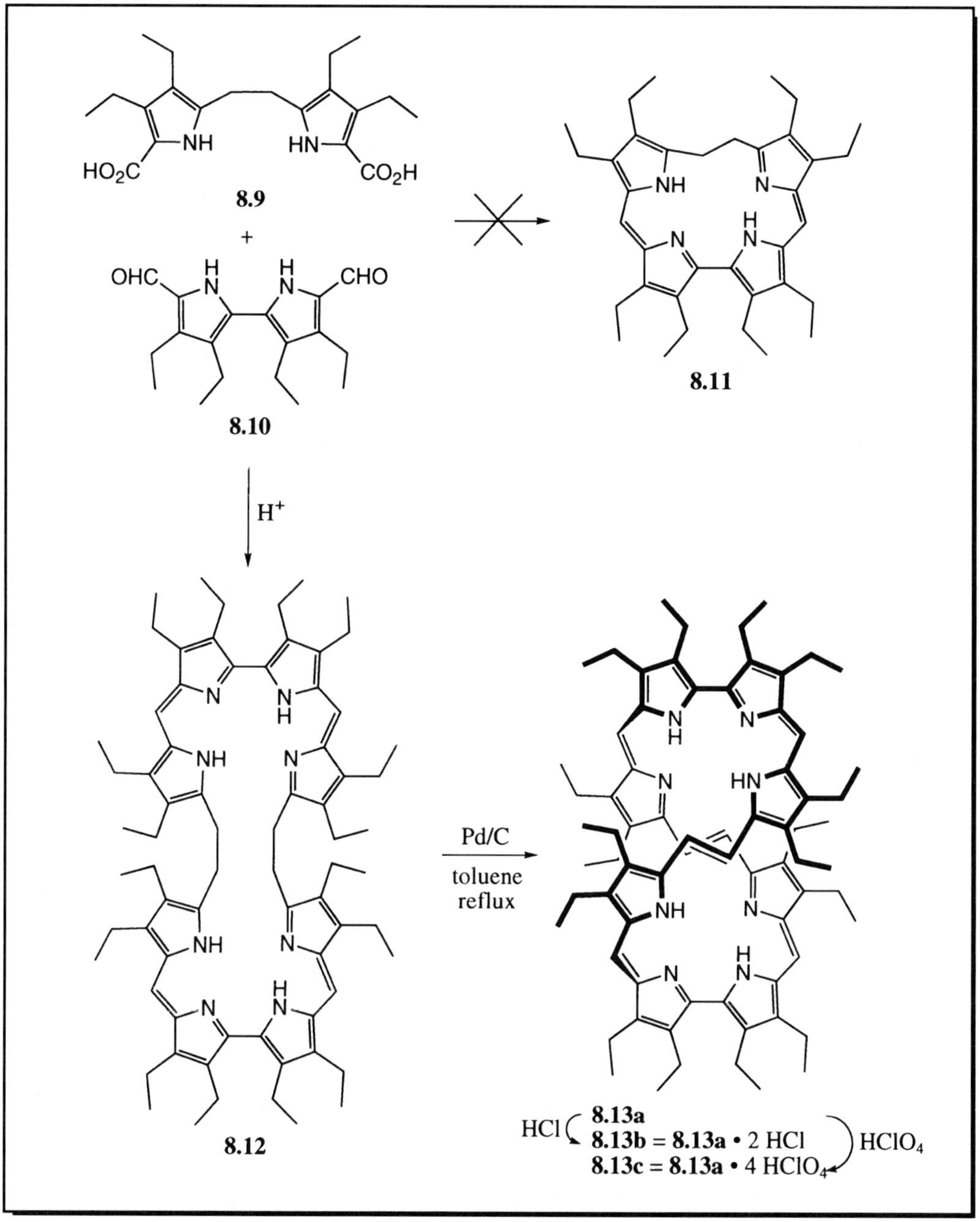

Figure 8.2.2   Single Crystal X-Ray Diffraction Structures (Side Views) of the Bis-HCl (**8.13b**, left) and Tetrakis-HClO4 (**8.13c**, right) Salts of Octaphyrin-(2.1.0.1.2.1.0.1).

This figure was generated using information down-loaded from the Cambridge Crystallographic Data Centre and corresponds to a structure originally reported in reference 3. In both cases the macrocycle adopts a twisted figure-eight conformation with approximate D_2 symmetry. The incorporation of four counter anions in **8.13c** results in an accordion-like stretching relative to the dicationic salt **8.13b**. At the point of crossover, generated by the macrocyclic twist, the interplane separation for **8.13b** and **8.13c** is 3.56 Å and 5.99 Å, respectively. Atom labeling scheme: carbon: $\bigcirc$; nitrogen: $\bullet$; oxygen: $\ominus$; chlorine: $\otimes$. The ethyl substituents and hydrogen atoms have been omitted for clarity

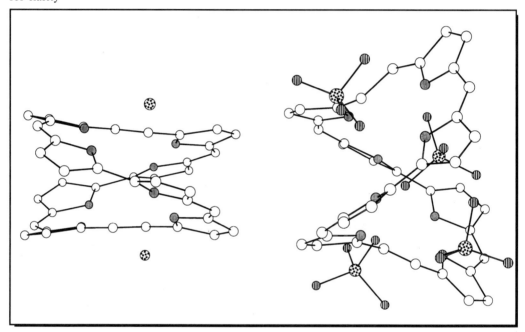

8.2.2   Octaphyrin-(1.1.1.0.1.1.1.0)

Inspired by the unexpected isolation of octaphyrin-(2.1.0.1.2.1.0.1) **8.13**, Vogel and coworkers began investigating whether "[2 + 2]" condensations of bipyrroles might not represent a general synthetic paradigm. In a first effort designed to explore this issue, a MacDonald-type condensation between the diacid dipyrrylmethane **8.14** and the diformyl bipyrrole **8.10** was carried out (Scheme 8.2.2). Although this was a reaction that had been carried out earlier and had been reported to afford only linear condensation products,[5] Vogel found that a "[2 + 2]" macrocyclic condensation product could in fact be isolated in 10% yield.[3] Interestingly, in this case (by contrast to what proved true for octaphyrin-(2.1.0.1.2.1.0.1), **8.13**), the presumed intermediate species **8.16** could not be isolated. Rather, spontaneous dehydrogenation occurred. As a result, it was the conjugated species **8.17** (octaphyrin-(1.1.1.0.1.1.1.0)) that was obtained.

Scheme 8.2.2

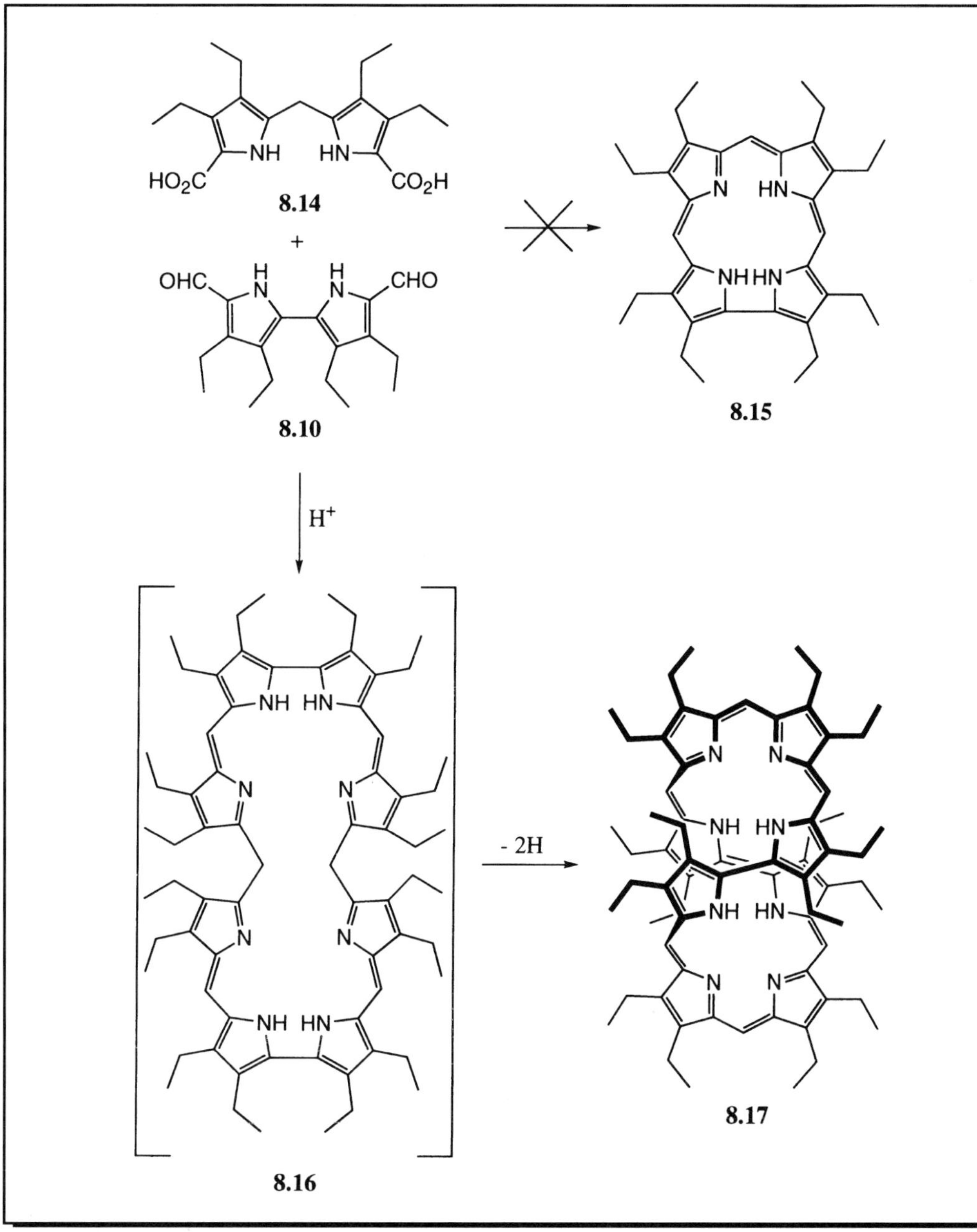

Compound **8.17** contains 34 π-electrons in its principal conjugation pathway. As such, it was perhaps to be expected that this macrocycle might adopt a near planar conformation and be generally "aromatic" in character. However, as was true for octaphyrin **8.13**, NMR spectroscopic studies combined with data from a preliminary single crystal X-ray structural analysis (Figure 8.2.3) quickly served to confirm that this new octaphyrin (**8.17**) adopts a non-planar figure-eight conformation both in solution and in the solid state.[3] In reaching this conclusion, the ^{1}H NMR spectrum of **8.17** was considered to be particularly diagnostic. For instance, in spite of the formal 4n + 2 Hückel-type π-electron formulation, no characteristic diamagnetic ring-current effects were observed in the ^{1}H NMR spectrum of **8.17**. While this absence certainly may be interpreted in terms of a lack of overall aromaticity, it remains an open question whether the observed effect is the result of macrocyclic non-planarity or other more subtle factors such as ring size or self-canceling magnetic moments within each "half" of the macrocycle. This issue aside, it is important to note that variable temperature ^{1}H NMR spectral studies of **8.17** were carried out

Figure 8.2.3   Single Crystal X-Ray Diffraction Structure of Free-base Octaphyrin-(1.1.1.0.1.1.1.0) **8.17**.
This figure was generated using information down-loaded from the Cambridge Crystallographic Data Centre and corresponds to a structure originally reported in reference 3. The macrocycle adopts a twisted figure-eight conformation with approximate D_2 symmetry. At the point of crossover, generated by the macrocyclic twist, the interplane separation is 3.28 Å. Atom labeling scheme: carbon: ○; nitrogen: ●. Hydrogen atoms have been omitted for clarity

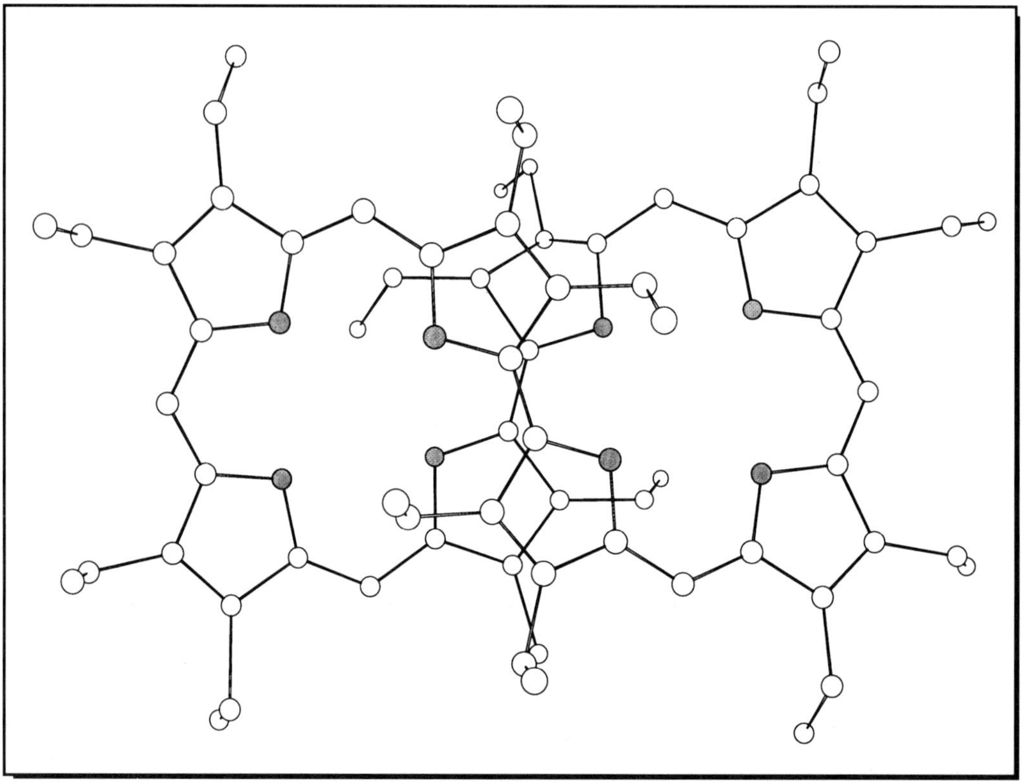

and that these confirmed a lack of conformational movement on the NMR time scale. Thus, like **8.13**, octaphyrin **8.17** appears to be "locked" in shape as a result of its intrinsic structure. In marked contrast to **8.13**, however, octaphyrin **8.17**, while non-aromatic, appears to be highly conjugated. This is evident from the UV-vis absorption spectrum of free-base **8.17**, which shows a strong absorbance band at 656 nm (ε = 175 900 $M^{-1}cm^{-1}$ in CH_2Cl_2) that in some respects appears to be almost Soret-like in character.

8.2.3   Octaphyrin-(1.0.1.0.1.0.1.0)

The final cyclooctapyrrole reported by Vogel and coworkers is the 32 π-electron (therefore, formally non-aromatic) octaphyrin-(1.0.1.0.1.0.1.0) derivative **8.20**.[4] This macrocycle was first synthesized in 7% yield as the result of a "[1 + 1]" acid-catalyzed condensation between the linear tetrapyrroles **8.18** and **8.19**. Subsequently, this same macrocycle was synthesized from bipyrrole **8.21** and diformyl bipyrrole **8.10** in higher yield (11%) than originally obtained *via* the "[1 + 1]" approach (Scheme 8.2.3). This finding attests to the generality of the "[2 + 2]" strategy; certainly, it appears to be a good synthetic route to cyclic octapyrroles.

Scheme 8.2.3

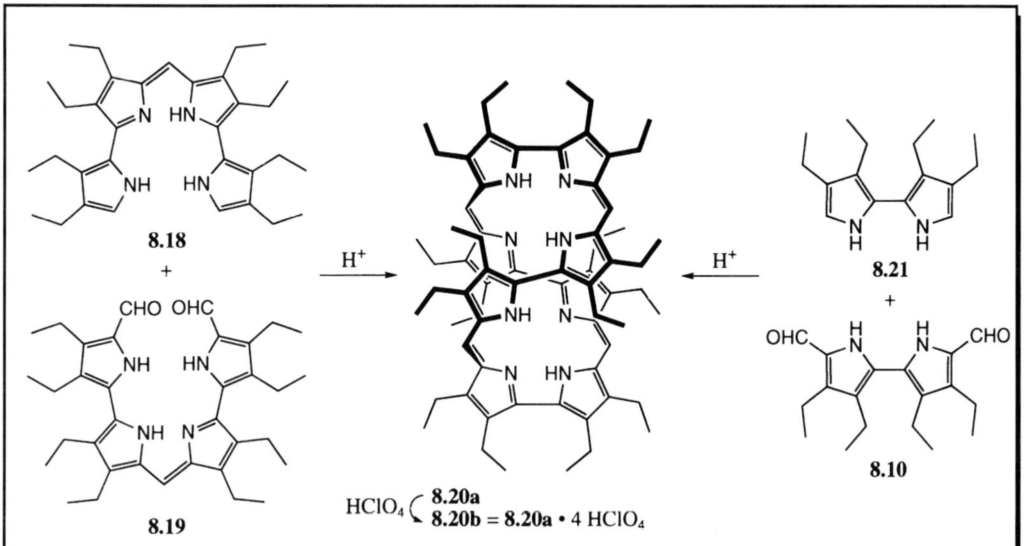

A single crystal X-ray structure (Figure 8.2.4) of **8.20a** confirmed that the proposed figure-eight conformation is manifest in the solid state. In this case, however, ^{1}H NMR spectroscopic studies revealed that this molecule undergoes rapid conformational inversion in solution (i.e., the enantiomeric forms interconvert on

Figure 8.2.4   Single Crystal X-Ray Diffraction Structure of Free-base Octaphyrin-
(1.0.1.0.1.0.1.0) **8.20a**.
This figure was generated using information down-loaded from the Cambridge Crystallographic
Data Centre and corresponds to a structure originally reported in reference 4. The macrocycle
adopts a twisted figure-eight conformation with approximate D_2 symmetry. The bipyrrole units
suffer significant rotation about their linking bond (the tortional angles are in the range of 43° to
52°). At the point of crossover, generated by the macrocyclic twist, the interplane separation is 4.57
Å. Atom labeling scheme: carbon: ○; nitrogen: ●. Hydrogen atoms have been omitted for clarity

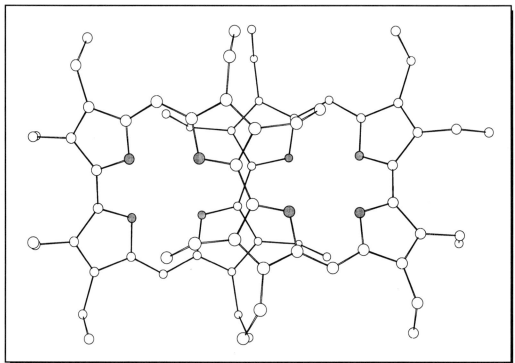

the NMR time scale presumably *via* an intermediate bowl-like conformation). Such
findings stand in interesting contradistinction with what was observed for **8.12**, **8.13**,
and **8.17** in the octapyrrolic series as well as turcasarins **8.5** and **8.6** in the decapyrrolic
one. Nonetheless, like these previously described systems, compound **8.20** is highly
colored and exhibits a rather strong absorption band at 548 nm ($\varepsilon = 104\ 000\ \text{M}^{-1}\text{cm}^{-1}$
in CH_2Cl_2) in its visible spectrum.

**8.3   Pyrrole–Thiophene Decamers: Hexathiatetraaza[44]decaphyrins-
(2.0.0.0.0.2.0.0.0.0)**

Using a synthetic strategy analogous to that used for the preparation of the
hexaheterocyclic bronzaphyrins (e.g., **8.22**, discussed in Chapter 7; Figure 8.2.5),
Cava and coworkers recently succeeded in synthesizing the decaheterocyclic

Figure 8.2.5   Structure **8.22**

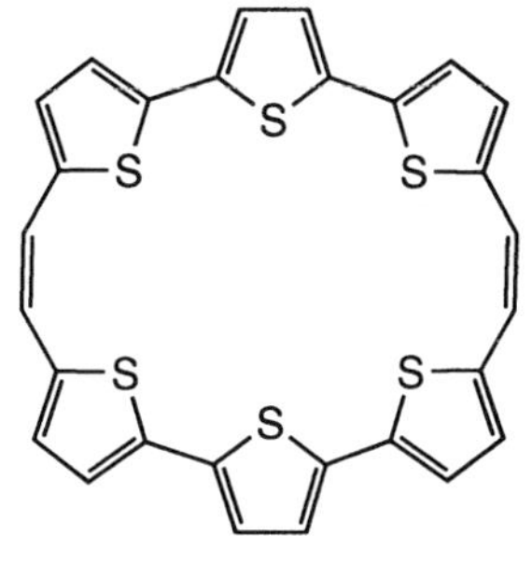

8.22

Scheme 8.2.4

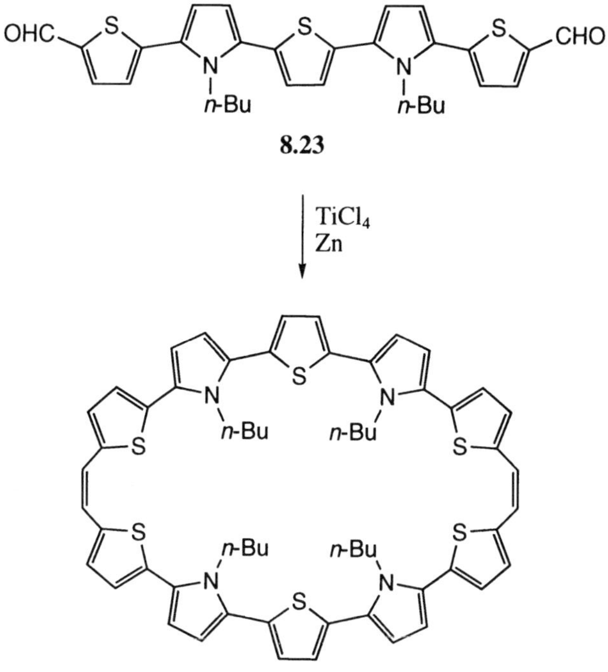

8.23

TiCl$_4$
Zn

8.24

macrocycle **8.24**.[6] This new "higher order" target was prepared by carrying out a McMurry-type dimerization of the diformyl-*N,N'*-dibutyl linear pentamer **8.23**. This afforded system **8.24** in 34% yield (Scheme 8.2.4). Using an analogous procedure, the

N,N'-linked macrocycles **8.27** and **8.28** were prepared in 14% and 37% yield from the corresponding dialdehyde, namely **8.25** and **8.26**, respectively (Scheme 8.2.5).

Scheme 8.2.5

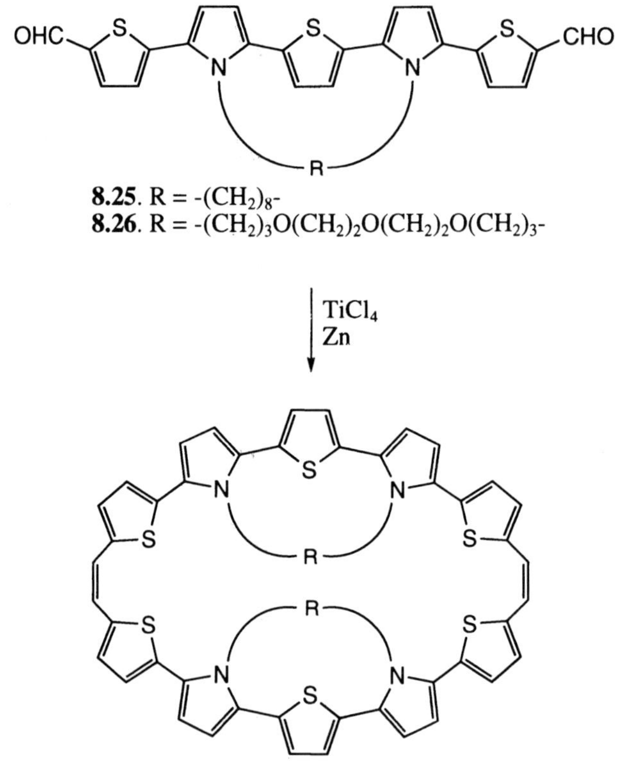

8.25. R = -(CH$_2$)$_8$-
8.26. R = -(CH$_2$)$_3$O(CH$_2$)$_2$O(CH$_2$)$_2$O(CH$_2$)$_3$-

TiCl$_4$
Zn

8.27. R = -(CH$_2$)$_8$-
8.28. R = -(CH$_2$)$_3$O(CH$_2$)$_2$O(CH$_2$)$_2$O(CH$_2$)$_3$-

The decaheterocyclic macrocycles **8.24** and **8.28–8.28**, like their smaller congeners in the hexameric series, failed to display any type of overall macrocyclic ring-current effects as judged from ^{1}H NMR spectroscopic analyses. They may thus be considered as being neither aromatic nor antiaromatic. UV-vis absorption spectroscopic analyses revealed the presence of only relatively weak, short-wavelength absorbance bands at λ_{max} = *ca.* 360 nm and 240 nm for each of these macrocycles. This was taken as being indicative of the presence of little or no extended π-electron conjugation in these materials. This, in turn, can be rationalized by assuming that the overall geometry for these macrocycles is decidedly non-planar. Unfortunately, at present no structural information is available for these still-new and obviously interesting macrocycles.

8.4   References

1. Sessler, J. L.; Weghorn, S. J.; Lynch, V.; Johnson, M. R. *Angew. Chem. Int. Ed. Eng.* **1994**, *33*, 1509–1512.
2. Sessler, J. L.; Guba, A.; Gebauer, A. Unpublished results.
3. Vogel, E.; Bröring, M.; Fink, J.; Rosen, D.; Schmickler, H.; Lex, J.; Chan, K. W. K.; Wu, Y.-D.; Plattner, D. A.; Nendel, M.; Houk, K. N. *Angew. Chem. Int. Ed. Eng.* **1995**, *34*, 2511–2514.
4. Bröring, M.; Jendrny, J.; Zander, L.; Schmickler, H.; Lex, J.; Wu, Y.-D.; Nendel, M.; Chen, J.; Plattner, D. A.; Houk, K. N.; Vogel, E. *Angew. Chem. Int. Ed. Eng.* **1995**, *34*, 2515–2517.
5. Conlon, M.; Johnson, A. W.; Overend, W. R.; Rajapaksa, D.; Elson, C. M. *J. Chem. Soc. Perkin Trans. 1* **1973**, 2281–2288.
6. Kozaki, M.; Parakka, J. P.; Cava, M. P. *J. Org. Chem.* **1996**, *61*, 3657–3661.

Chapter Nine Nitrogen-bridged Expanded Porphyrins

9 Introduction

With the exception of the subphthalocyanines and homoazaporphyrins discussed in Chapters 2 and 4, the macrocycles discussed to this point have all consisted of systems bridged by carbon atoms. In this and the next chapter, the synthesis of expanded porphyrins containing at least one bridging nitrogen atom in place of the "normal" *meso*-carbon atom(s) will be highlighted. It should be noted, however, that the scope of the discussion here will be more limited than that which appeared in the original review of expanded porphyrins by Burrell and Sessler.[1] This limitation results in part from the more restricted definition of expanded porphyrins employed presently. Thus, systems such as **9.1–9.5**, which were previously covered under the general category of Schiff base expanded porphyrins, will not be discussed here (Figure 9.0.1).[2–15] Nonetheless, an occasional macrocycle will be mentioned in this chapter that does not strictly fall under the current working definition of an expanded porphyrin. This will be done when the systems in question are considered to be of particular historic, synthetic, or structural significance.

Figure 9.0.1 Structures **9.1–9.5**

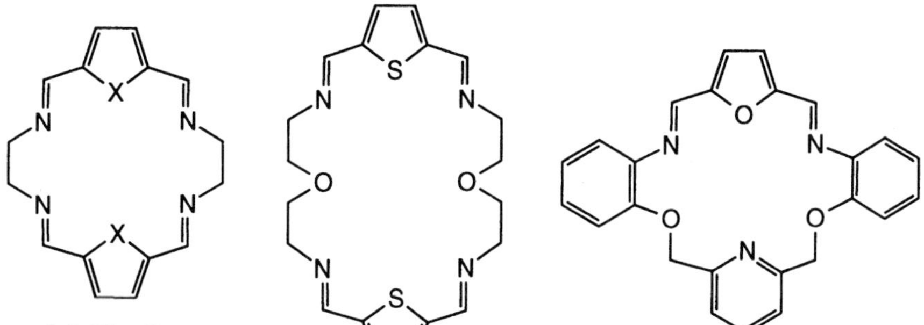

9.1. X = O
9.2. X = S
9.3. X = NH

9.4

9.5

9.1   Schiff base-derived Expanded Porphyrins

9.1.1   The "[2 + 2]" Approach

9.1.1.1   Accordion Macrocycles

The use of Schiff base chemistry for the preparation of expanded porphyrins had its antecedents in the condensation reactions between 2,5-diformylfuran, 2,5-diformylthiophene, and 2,5-diformylpyrrole, and a variety of aliphatic diamines to afford macrocycles such as **9.1–9.5** (Figure 9.0.1).[2–15] In more recent times this basic approach has been applied to the preparation of a wide range of expanded porphyrins. The first of these to be reported was the so-called "accordion" system **9.8**.[16,17] This macrocycle, which was initially described in 1984 by Mertes and Acholla, can be prepared in the form of its free base *via* the condensation of two equivalents of diformyl dipyrrylmethane **9.6** with two equivalents of diamino propane (**9.7**), provided Ba^{2+}, Sr^{2+}, Ca^{2+}, or Mg^{2+} salts are used as templating cations (Scheme 9.1.1).[17] Alternatively, the metalated macrocycles **9.9** and **9.10** can by synthesized by carrying out the above condensation using $Pb(SCN)_2$ or $ZnCl_2$ as the templating ions, respectively (Scheme 9.1.2). In addition to these products, the bis-copper complexes **9.11–9.13** could be prepared, either from the free ligand **9.8** or from the two metalated species **9.9** or **9.10** *via* metathesis (Figure 9.1.1). In both cases, CuX_2 salts ($X = Cl^-$, ClO_4^-, or BF_4^-) were used. Copper complexes with azide and isocyanate counter anions, namely **9.14** and **9.15**, could also be prepared by ligand exchange starting from complex **9.13**.

Scheme 9.1.1

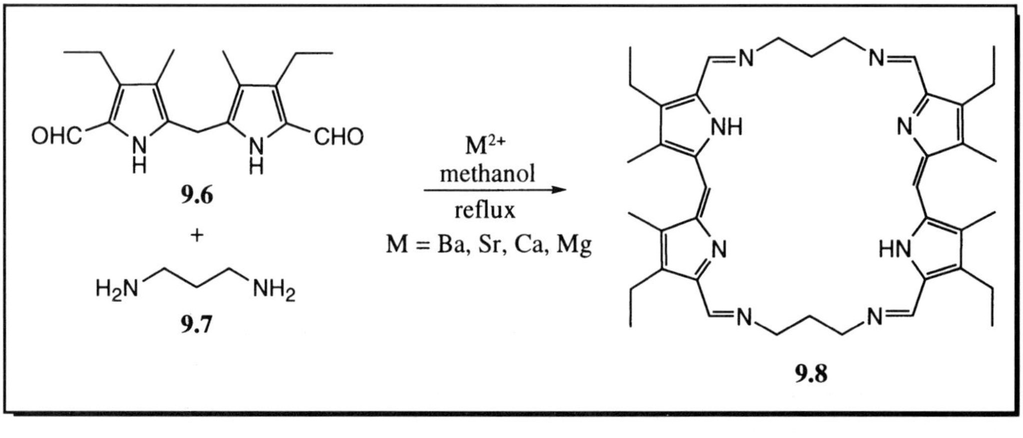

A smaller "accordion" macrocycle, prepared in the form of its metal complexes **9.17** and **9.18**, was also prepared by Mertes and coworkers.[17] This macrocycle was prepared as its bis-Zn chelate **9.17** by treating a solution of ethylenediamine (**9.16**) and dipyrrylmethane **9.6** with $ZnCl_2$ (Scheme 9.1.3). The corresponding copper complex **9.18** could then be prepared by treating this product (**9.17**) with Cu(ClO$_4$)$_2$·6H$_2$O.

Scheme 9.1.2

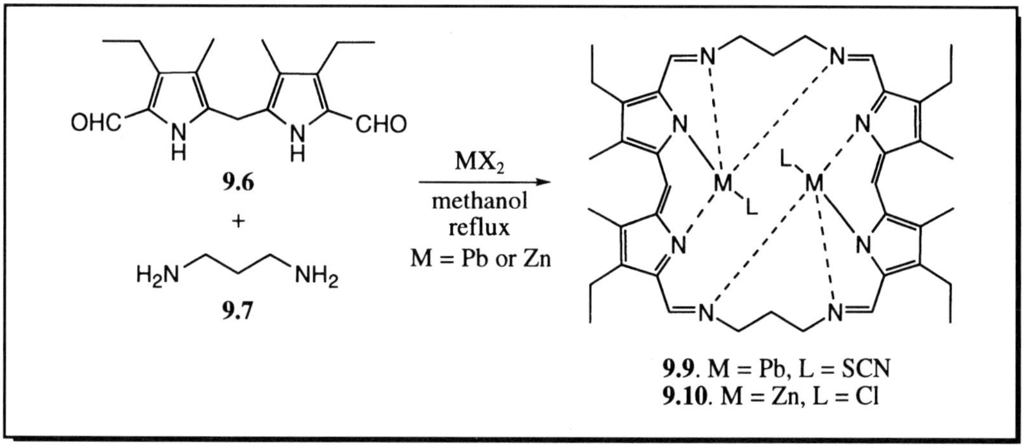

9.9. M = Pb, L = SCN
9.10. M = Zn, L = Cl

Figure 9.1.1   Structures **9.11–9.15**

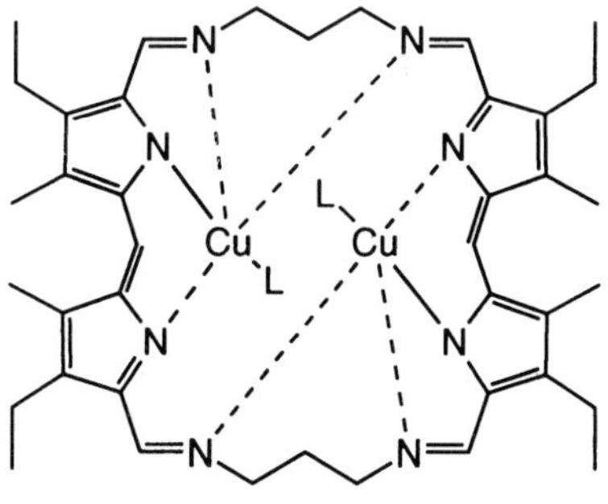

9.11. L = BF$_4^-$
9.12. L = Cl$^-$
9.13. L = ClO$_4^-$
9.14. L = N$_3^-$
9.15. L = SCN$^-$

Infrared spectroscopic analyses of the above metal complexes indicated that the "accordion" skeleton in these complexes can exist in more than one conformation.[17] This general conclusion was supported by a single crystal X-ray diffraction study of the bis-copper adduct **9.14**. Based on this solid state study, the copper ions in complex **9.14** were found to be separated by *ca.* 5.4 Å (Figure 9.1.2).

Scheme 9.1.3

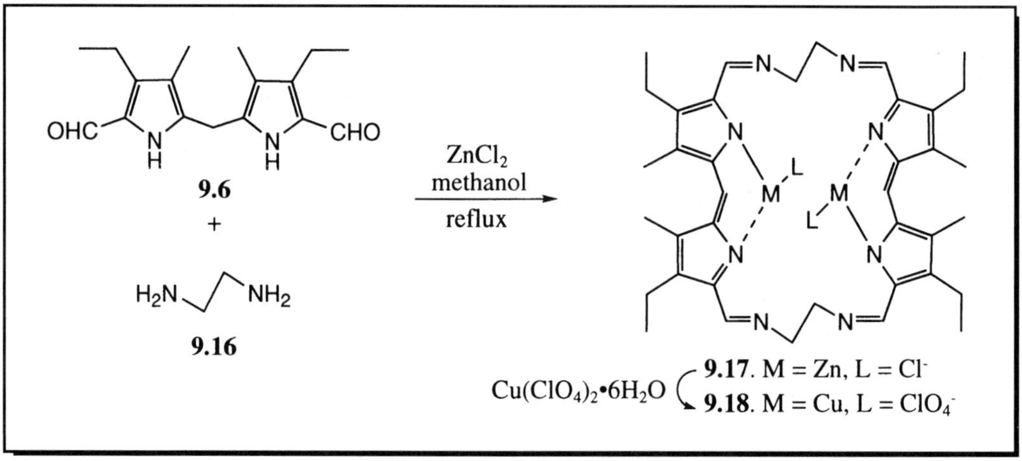

Figure 9.1.2 Single Crystal X-Ray Diffraction Structure of the Bis[azidocopper(II)] Complex **9.14**.

This figure was generated using information down-loaded from the Cambridge Crystallographic Data Centre and corresponds to a structure originally reported in reference 17. The coordination sphere of each copper ion is distorted trigonal bipyramidal and consists of four ligand nitrogen atoms and a nitrogen atom provided by the azide counter anion. The copper–copper interatomic separation is 5.39 Å. Atom labeling scheme: copper: ◉; carbon: ○; nitrogen: ●. Hydrogen atoms have been omitted for clarity

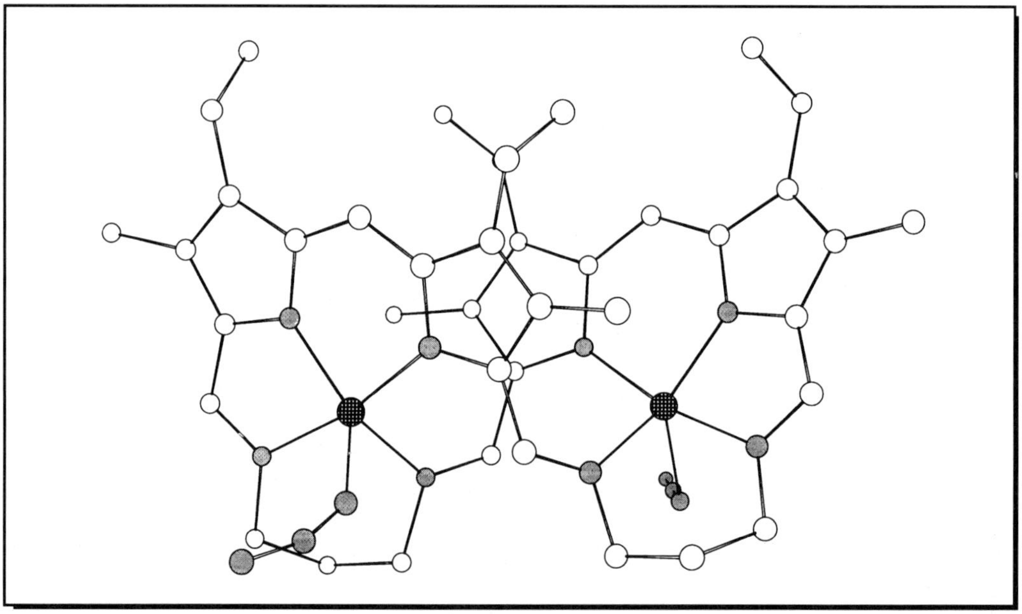

9.1.1.2   Diformylbipyrrole-derived Systems

Subsequent to the work of Mertes, a "[2 + 2]" Schiff base approach was employed by Sessler and coworkers to prepare a series of bipyrrole-containing macro-cycles **9.24–9.27** (Scheme 9.1.4).[18,19] The first of these to be reported, macrocycle **9.24**, was prepared *via* the acid-catalyzed reaction of *o*-phenylenediamine (**9.19**) with 4,4'-diethyl-5,5'-diformyl-3,3'-dimethyl-2,2'-bipyrrole **9.22**.[18] Macrocycles **9.25–9.27** were prepared in a similar fashion by treating the *o*-phenylenediamine derivatives **9.20** and **9.21** with either bipyrrole **9.22** or its *n*-propyl analog **9.23**. Inspiration for the synth-esis of these macrocycles came from the finding that the smaller, mono-pyrrole-derived analog **9.28** was effective as a uranyl cation chelating agent (Figure 9.1.3).[15,20]

Scheme 9.1.4

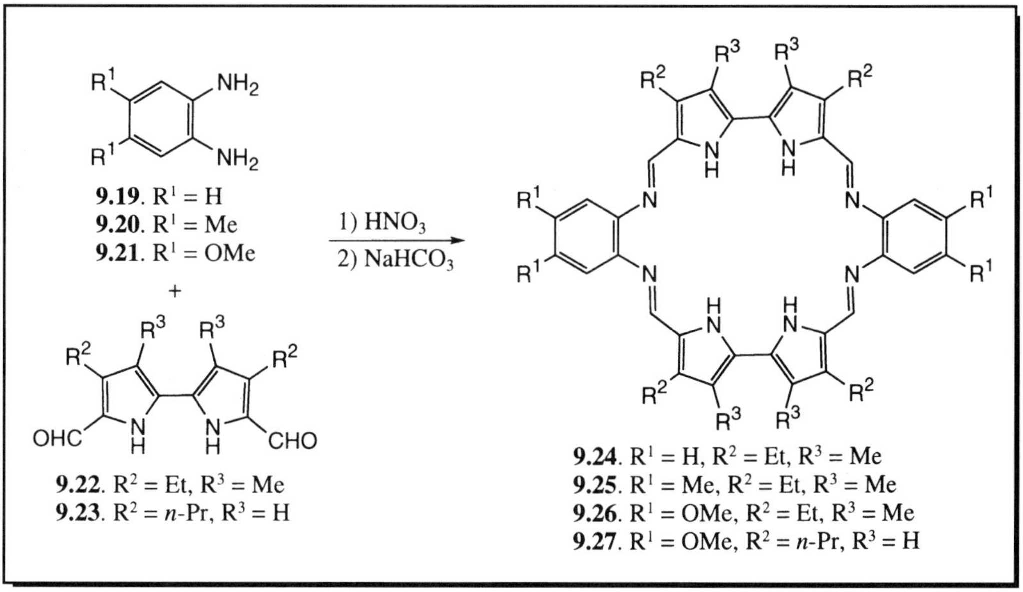

9.19. R¹ = H
9.20. R¹ = Me
9.21. R¹ = OMe

1) HNO₃
2) NaHCO₃

9.22. R² = Et, R³ = Me
9.23. R² = n-Pr, R³ = H

9.24. R¹ = H, R² = Et, R³ = Me
9.25. R¹ = Me, R² = Et, R³ = Me
9.26. R¹ = OMe, R² = Et, R³ = Me
9.27. R¹ = OMe, R² = n-Pr, R³ = H

While the synthesis of macrocycles **9.24–9.27** can be considered as being con-ceptually straightforward in terms of experiment, one fact proved unexpected and interesting. This was the finding that nitric acid proved to be the most efficient catalyst for macrocycle formation. Indeed, attempts to prepare system **9.26**, for instance, using other acid catalysts (e.g., HCl, HBr, H₂SO₄) led to considerably lower yields. This led Sessler and coworkers to suggest that the formation of this macrocycle was mediated *via* an anion template effect. This was a rather new con-ception at the time it was put forward.[18]

Macrocycles **9.24–9.27** contain fully conjugated π-electron frameworks, but are formally non-aromatic. Nevertheless, they each exhibit a rather strong absorption band in the visible spectrum (e.g., λ_{max} for **9.26** = 384.5 nm, ε = 87 100 M⁻¹cm⁻¹). Moreover, these molecules adopt a highly planar conformation in the solid state, as inferred from a single crystal structure of the bis-methanol adduct of the free-base form of **9.26** (Figure 9.1.4).[18] Taken together, these findings led to the conclusion that these systems are highly conjugated and truly "expanded porphyrin-like" in character, even if not truly aromatic.[18,19]

Figure 9.1.3 Structure **9.28**

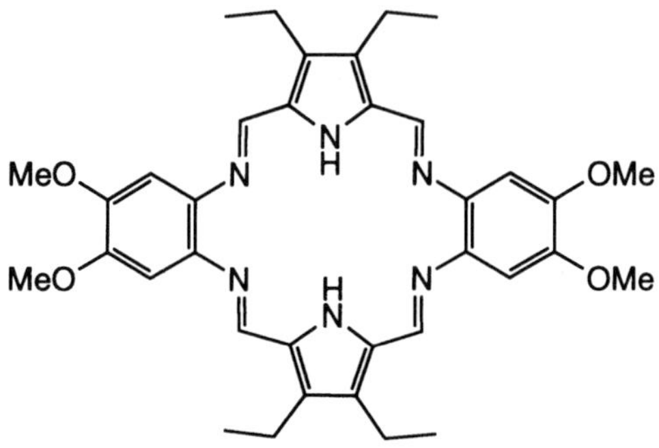

9.28

A further fascinating feature of the above crystal structure is that two molecules of methanol were found to be "tethered" to the core NH groups of the macrocycle *via* hydrogen bonds. This unexpected result led Sessler and coworkers to propose that this system could function as a potential receptor for neutral substrates. In accord with this notion, these workers found that macrocycle **9.26** forms 1:1 complexes in CD_2Cl_2 solution not only with methanol, but also with other neutral molecules such as ethanol, trifluoroethanol, phenol, and catechol. Interestingly, this latter substrate was found to be bound with an affinity constant that exceeds 10^4 M^{-1}. By contrast, neither this substrate nor any of the other neutral species tested was found to be bound appreciably by the smaller "2 + 2" analog **9.28**.[18]

Because macrocycles **9.24–9.27** contain large octaaza macrocyclic cores, it was envisioned that these systems could serve as ditopic receptors for metal cations. In preliminary work involving the dimethoxy-substituted derivative **9.26**, this expectation appears to have been borne out, at least in the case of certain transition metal ions. Specifically, it was found that when **9.26** is treated with divalent cations of Ni, Cu, Mn, Zn, Pd, or Ru, the corresponding dinuclear complexes **9.29–9.33** are formed in good (*ca.* 50–90%) yield (Scheme 9.1.5).[19] A single crystal X-ray diffraction structural analysis carried out on the bis-nickel complex **9.29** revealed that each nickel atom is bound by two imine nitrogen atoms located at one end of the macrocycle (Figure 9.1.5). One nickel ion rests 0.70 Å above and the other 1.10 Å below the mean macrocyclic plane defined by the eight internal nitrogen atoms. The pyrrole nitrogen atoms do not interact with the metal centers directly. Instead they are

Figure 9.1.4   Single Crystal X-Ray Diffraction Structure of the Free-base Macrocycle **9.26**.

This figure was generated using information down-loaded from the Cambridge Crystallographic Data Centre and corresponds to a structure originally reported in reference 18. Two molecules of methanol are held proximate to the macrocycle by hydrogen bonding interactions (not shown) involving the pyrrolic NH groups and the imine nitrogen atoms. Atom labeling scheme: carbon: ○; nitrogen: ●; oxygen: ⊜. Hydrogen atoms have been omitted for clarity

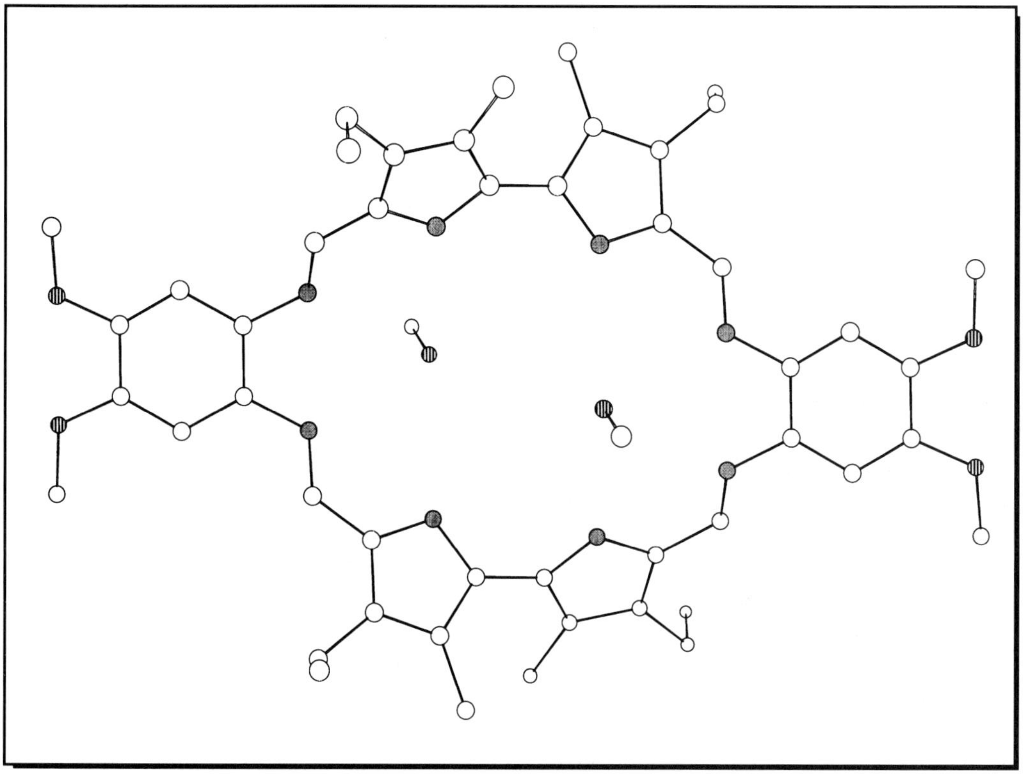

involved in "second sphere" hydrogen-bonding interactions involving the ancillary ligands (acetate and water, not shown) bound to the metal centers.

A final example of a bipyrrole-derived "[2 + 2]" Schiff base expanded porphyrin was reported by Johnson, *et al.* in 1995 in the form of a preliminary abstract.[21] In this instance, macrocycle **9.34** was prepared *via* the condensation of hydrazine with the diformyl bipyrrole **9.23** (Scheme 9.1.6). This macrocycle was reported to exhibit spectral characteristics indicating that it is antiaromatic. Further, treatment with manganese dioxide was said to result in oxidation of **9.34** to an aromatic 22 π-electron system. This latter system proved, however, to be much less stable than its 24 π-electron "parent", and this precluded isolation and characterization. In fact, with full experimental details not yet reported in the literature, the characterization, and structural assignment for **9.34** must also be considered tentative.

Scheme 9.1.5

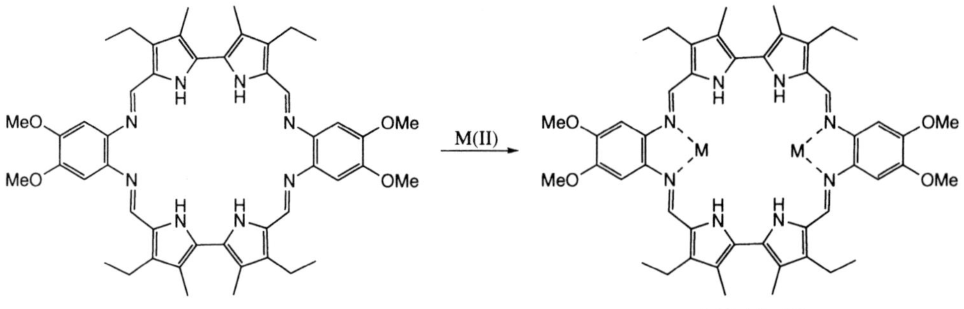

9.26

M(II) →

9.29. M = Ni
9.30. M = Cu
9.31. M = Zn
9.32. M = Pd
9.33. M = Ru

9.1.1.3   Diformylterpyrrole-derived Systems

Sessler and coworkers have also prepared the diformylterpyrrole-derived macrocycles **9.37** and **9.38**.[22] These were prepared using a procedure analogous to that used to prepare the diformylbipyrrole-derived systems **9.24–9.27**. That is, two molecules of diformylterpyrrole (either **9.35** or **9.36**) were condensed with two molecules of dimethoxy *o*-phenylenediamine **9.21** under acidic conditions. This afforded the corresponding macrocycles **9.37** and **9.38** (Scheme 9.1.7). These systems contain a formal 36 π-electron, main conjugation pathway and have ten inward-pointing nitrogen atoms. While these systems are expected to exhibit binding characteristics similar to those of their smaller bipyrrole-containing congeners (*vide supra*), they have yet to be tested in this regard.

9.1.2   The "[1 + 1]" Approach

9.1.2.1   Texaphyrin

In the preceding sections of this chapter, the general applicability of a "[2 + 2]" Schiff base-derived approach to a number of expanded porphyrins was high-lighted. The following section will focus on a "[1 + 1]" Schiff base approach and includes one particularly exciting class of Schiff base-derived macrocycles known as "texaphyrins" (for Texas-size porphyrins; Figure 9.1.6). The texaphyrins have pro-ven to be among the most promising of expanded porphyrins prepared to date in terms of their rich coordination chemistry and their potential utility in a range of biomedical applications (see Chapter 10). The promise that the texaphyrins engender in terms of their potential use has also served to inspire the synthesis of several related macrocycles. Therefore, the synthesis of the texaphyrins and of these conceptually similar macrocycles will be discussed in this section. Various applications of the texaphyrins are reviewed separately in the next chapter.

Figure 9.1.5   Single Crystal X-Ray Diffraction Structure of the Bis[diacetatonickel(II)] Complex **9.29**.

This figure was generated using data provided by Sessler, *et al.*; however, this structure was originally reported in reference 19. The two nickel(II) ions lie in a distorted octahedral coordination geometry and are on opposite "faces" of the macrocycle. One nickel atom sits 0.70 Å above the mean N_8 plane and the other 1.10 Å below. The nickel ions are bound by the imine nitrogen atoms and not the bipyrrolic ones. Each nickel(II) ion is also bound by a water molecular (not shown). The Ni–Ni interatomic separation is 5.37 Å. Atom labeling scheme: nickel: ◉; carbon: ○; nitrogen: ●; oxygen: ⊖. Hydrogen atoms have been omitted for clarity

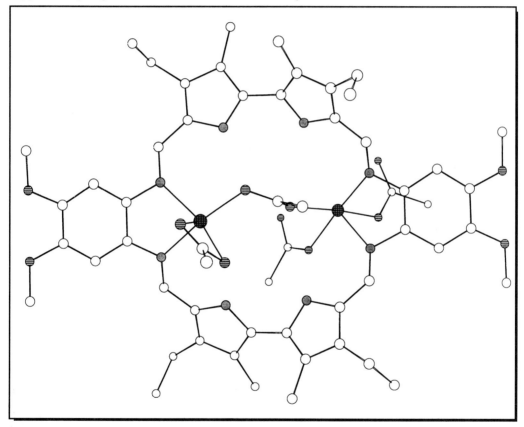

9.1.2.1.1   "Texaphyrinogen"

The general synthetic underpinnings for what would subsequently lead to the synthesis of texaphyrin (*vide infra*) were introduced by Sessler and coworkers in 1987.[23] In that early report, the synthesis of the tripyrrane-containing macrocycle

9.40 was described. This "porphyrinogen-like" system was prepared from the condensation of diformyltripyrrane **9.39** and *o*-phenylenediamine **9.19** effected under conditions of either HCl or Lewis acid (e.g., UO_2^{2+}, Pb^{2+}) catalysis (Scheme 9.1.8). Yields of the nearly colorless macrocycle ranged from 44% when HCl catalysis was employed to 61–69% for reactions catalyzed by UO_2Cl_2 and $Pb(SCN)_2$, respectively. When $Pb(SCN)_2$ and HCl were used together in an attempt to increase the yield of the reaction, a 68% yield of a red solid was obtained that, on the basis of spectroscopic and analytic data, was formulated as being the mono-HSCN salt **9.40b**. This latter conclusion was confirmed by a single crystal X-ray analysis which revealed both a non-planar macrocycle containing a nearly circular core of 5 nitrogen atoms and an SCN^- anion hydrogen bound to a protonated imine moiety (Figure 9.1.7).

Scheme 9.1.6

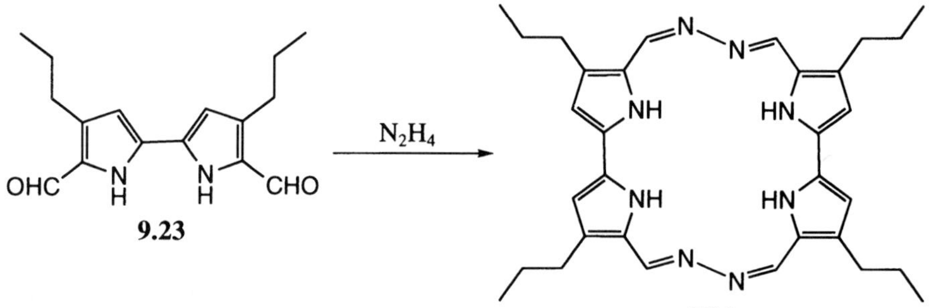

9.23

9.34

Scheme 9.1.7

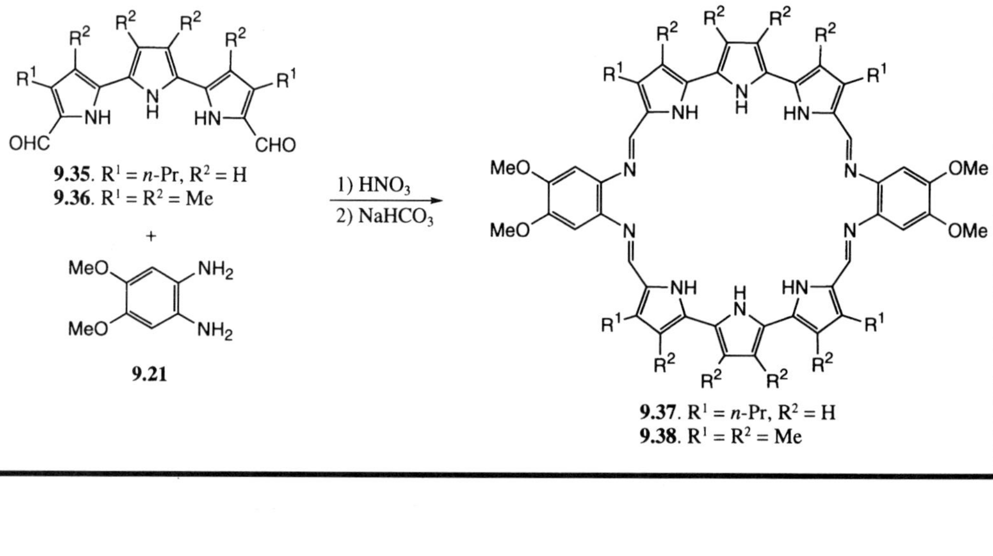

9.35. $R^1 = n$-Pr, $R^2 = H$
9.36. $R^1 = R^2 = Me$

+

9.21

9.37. $R^1 = n$-Pr, $R^2 = H$
9.38. $R^1 = R^2 = Me$

Figure 9.1.6 Structure of Texaphyrin and the "Lone-star" Pentagonal Binding Geometry Provided by its Pentaaza Core

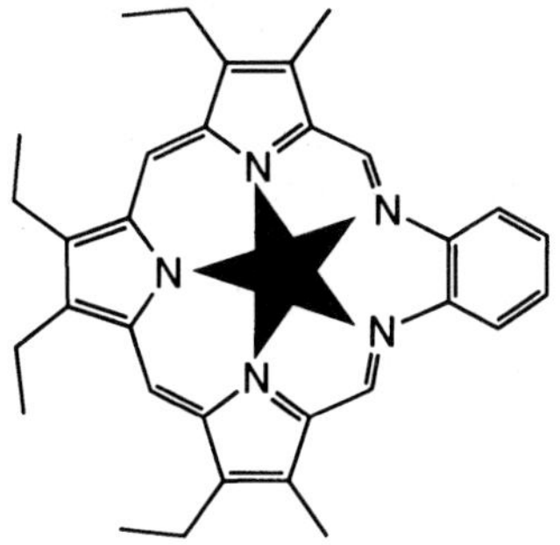

"texaphyrin"
(anionic form)

Scheme 9.1.8

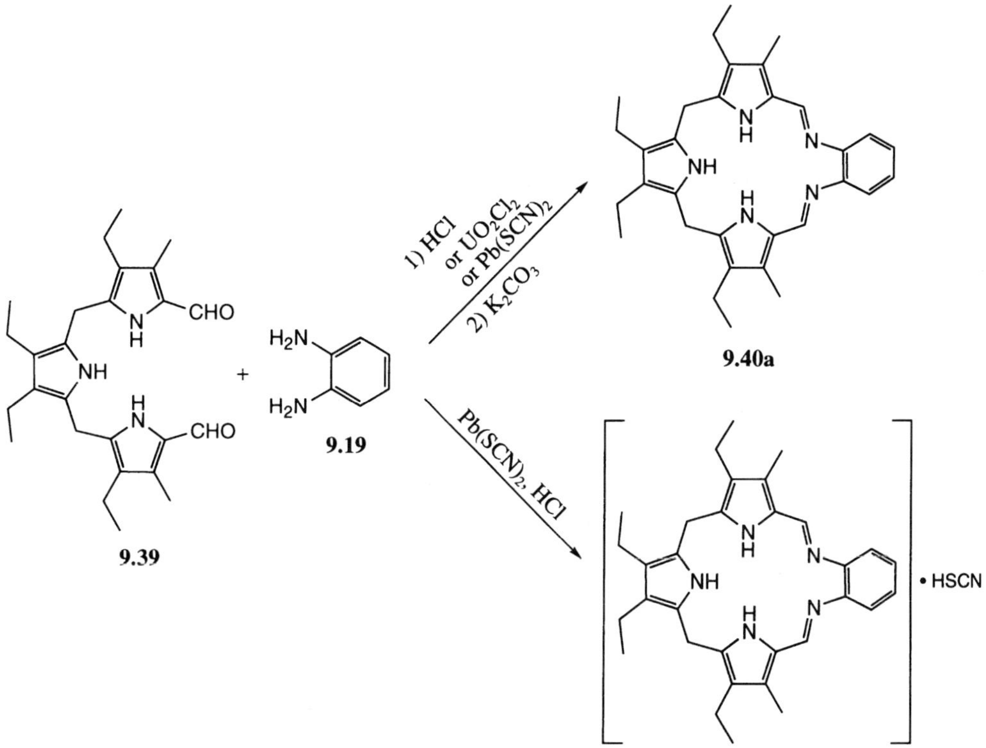

9.39 **9.19** **9.40a** **9.40b**

Figure 9.1.7   Single Crystal X-Ray Diffraction Structure of "Texaphyrinogen" mono-HSCN Salt **9.40**.

This figure was generated using information down-loaded from the Cambridge Crystallographic Data Centre and corresponds to a structure originally reported in reference 23. All three pyrrole rings are tilted toward the capping thiocyanate ion which sits 1.76 Å above the mean N_5 coordination plane. The estimated N_5 core center-to-nitrogen radius is 2.5 Å. The central pyrrole ring of the tripyrrane subunit is rotated away from the adjacent pyrrole rings by angles of 74.6° and 70.0°, respectively. Atom labeling scheme: carbon: ○; nitrogen: ●; sulfur: ◑. Hydrogen atoms have been omitted for clarity

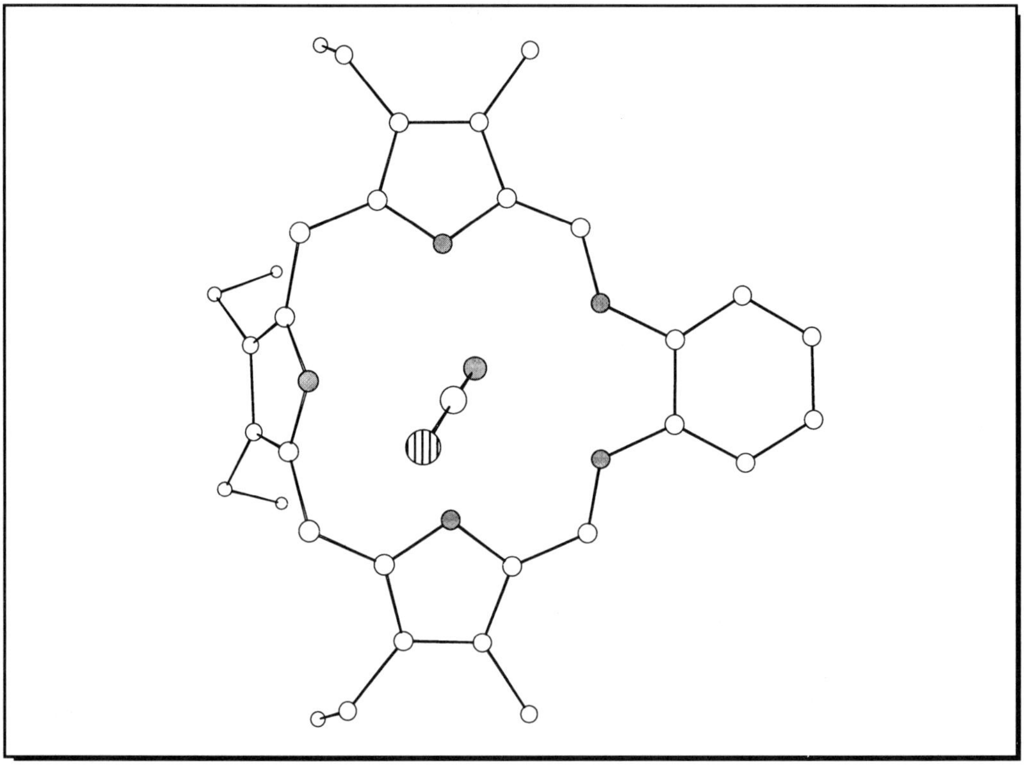

   Subsequent to this initial synthetic report, this same general approach was applied to the synthesis of quite a wide variety of related tripyrrane-containing macrocycles (e.g., **9.41–9.52**; Figure 9.1.8).[23–31] In general, these compounds were made by reacting the appropriate tripyrrane with *o*-phenylenediamine, or a functionalized derivative thereof. Using a similar approach, the naphthalene and phenanthrene-containing derivatives **9.55** and **9.56** were prepared from tripyrrane **9.39**, and the corresponding diaminonaphthalene and diaminophenanthrene derivatives **9.53** and **9.54** (Scheme 9.1.9).[32] Likewise, it proved possible to prepare "benz-free" texaphyrinogen derivative **9.58** from the *cis*-diaminoalkene **9.57** (Scheme 9.1.10).[32]

Figure 9.1.8   Structures **9.41–9.52**

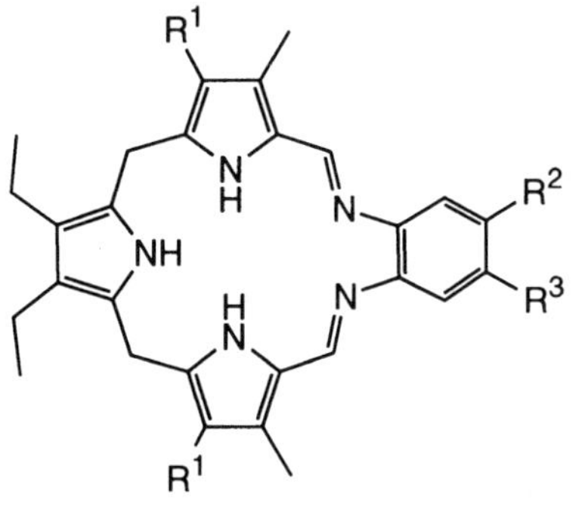

9.41. R^1 = Et, R^2 = R^3 = Me
9.42. R^1 = Et, R^2 = R^3 = OMe
9.43. R^1 = Et, R^2 = R^3 = -O(CH$_2$CH$_2$O)$_4$-
9.44. R^1 = Et, R^2 = R^3 = O(CH$_2$)$_3$OH
9.45. R^1 = Et, R^2 = CO$_2$H, R^3 = H
9.46. R^1 = Et, R^2 = OMe, R^3 = H
9.47. R^1 = Et, R^2 = Me, R^3 = H
9.48. R^1 = Et, R^2 = Cl, R^3 = H
9.49. R^1 = Et, R^2 = NO$_2$, R^3 = H
9.50. R^1 = (CH$_2$)$_3$OH, R^2 = R^3 = H
9.51. R^1 = (CH$_2$)$_3$OH, R^2 = R^3 = O(CH$_2$)$_3$OH
9.52. R^1 = (CH$_2$)$_3$OH, R^2 = OCH$_2$CO$_2$H, R^3 = H

9.1.2.1.2   *Oxidation to aromatic texaphyrins*

The core size and geometry of these non-aromatic, methylene-bridged macro-cycles (as shown by the X-ray structure of **9.40b**; Figure 9.1.7) led to the prediction that they might be well-suited for the complexation of large metal cations provided deprotonation of one or more of the pyrrolic NH protons could be effected. Unfortunately, all efforts to prepare stable, crystalline metal complexes of these methylene-bridged macrocycles have met with failure.[23–25] It was, therefore, consid-ered expeditious to try an alternative strategy. This had as its main conceptual component the idea that effecting a formal four-electron oxidation of these macro-cycles would generate the corresponding conjugated, aromatic species (e.g., **9.59**), which, in turn, would serve as good *monoanionic* ligands. After considerable effort, this desired oxidation was accomplished in the case of **9.40a** by exposing the macro-cycle to O$_2$ in the presence of the non-nucleophilic Brønsted base $N,N,N'N'$-tetra-methyl-1,8-diaminonaphthalene (Scheme 9.1.11).[24] Unfortunately, however, this oxidation afforded macrocycle **9.59** in but low yield ($\leqslant 12\%$), and without reliable reproducibility. Nonetheless, it was observed that this oxidized texaphyrin product proved to be much less prone to decompose than the reduced, porphyrinogen-like precursor from which it was made. This was presumed to be the result of the increased stability afforded by the 18 π-electron and 22 π-electron aromatic conjugation pathways present in macrocycle **9.59**, which were not, of course, available in **9.40a**.

Scheme 9.1.9

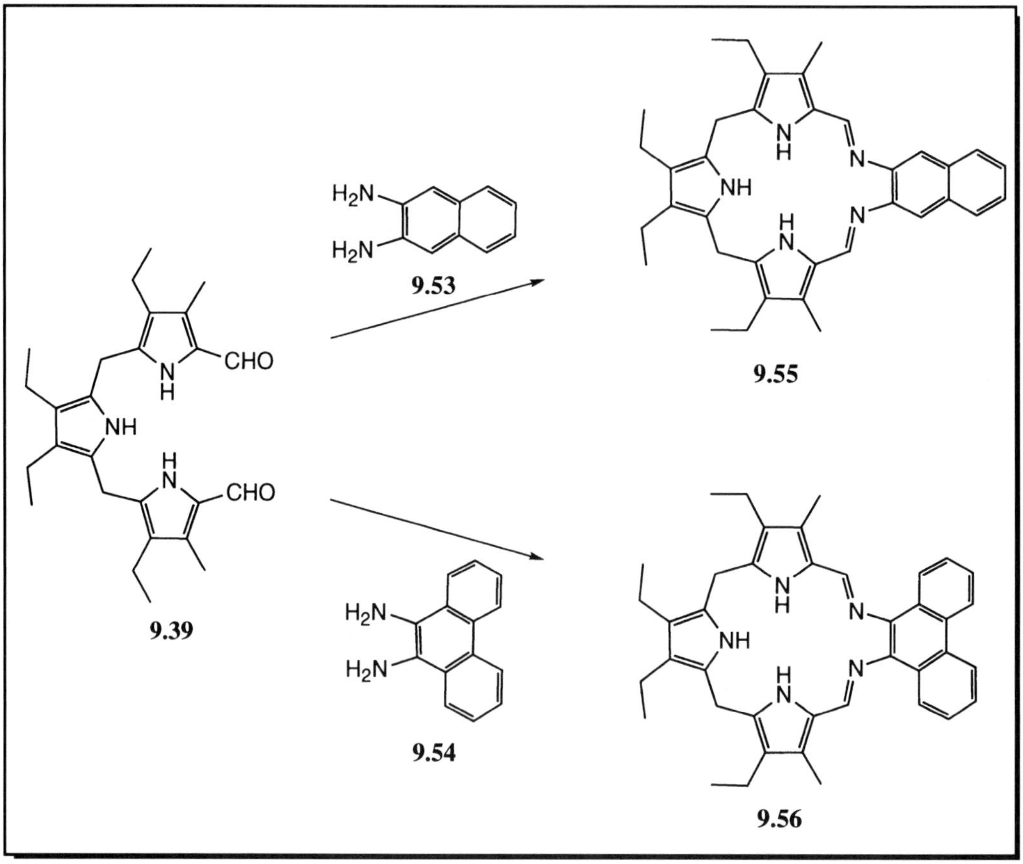

Because of the difficulty encountered in preparing metal-free oxidized texaphyrins, efforts were made to bring about ligand oxidation concurrent with metal insertion. This type of oxidative metal ion insertion was first effected successfully using macrocycle **9.40a** as the texaphyrin precursor, CdCl$_2$ as the metal cation source, and air as the oxidant.[23–25] Specifically, it gave the aromatic cadmium(II) texaphyrin complex **9.60** in *ca.* 25% yield (as the chloride salt). Interestingly, when Cd(NO$_3$)$_2$ was used in place of CdCl$_2$, a metallotexaphyrin derivative (**9.61a**) was isolated in which a molecule of benzimidazole is axially coordinated to the cadmium center.[26] Here, the benzimidazole was thought to derive from side-reactions involving decomposition of the starting texaphyrinogen and oxidative insertion of a formyl group into an *o*-phenylenediamine fragment.

Scheme 9.1.10

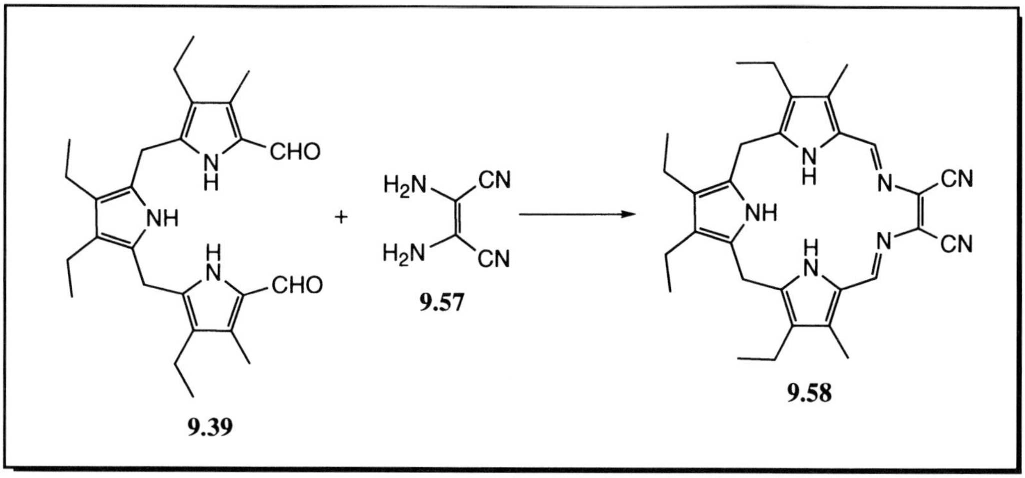

Scheme 9.1.11

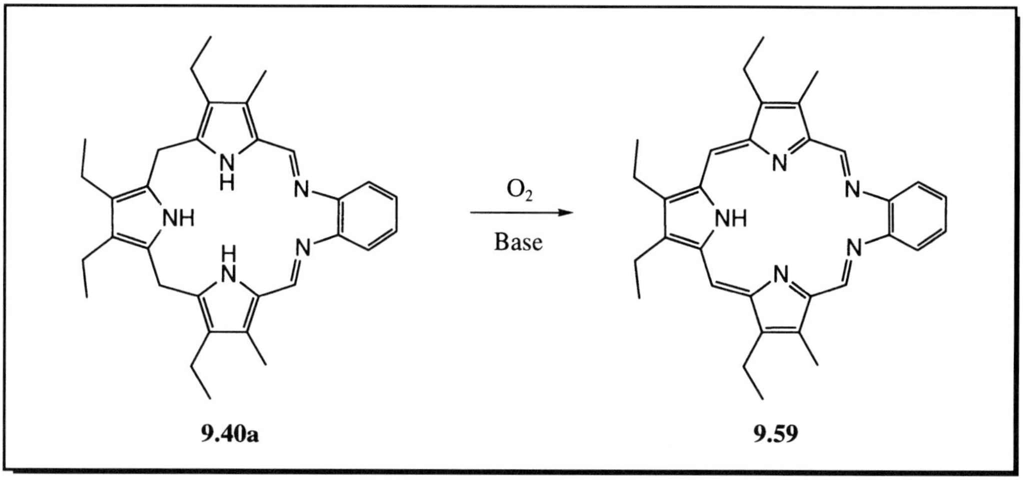

Using an oxidative metalation procedure analogous to that used to prepare **9.60**, the cadmium(II) texaphyrin analogs **9.62**–**9.64** were prepared (Scheme 9.1.13).[32] By generalizing the procedure, it also proved possible to prepare a wide range of metallotexaphyrins (e.g., **9.65**–**9.79**) using a variety of other metal cations.[26–32] As is evident from inspection of Figure 9.1.9, the range of metalated texaphyrins is large. Indeed, it is so impressive that it led Sessler to draw up a "Periodic Table of the Texaphyrins". This table, which illustrates all the stable metal complexes currently known for texaphyrin, is given in Figure 9.1.10.

Of particular interest is the fact that stable 1:1 metallotexaphyrin complexes are known for all of the non-radioactive trivalent lanthanides. This has made the texaphyrins of special interest in the context of potential biomedical applications including MRI imaging, radiation sensitization, and photodynamic therapy.[29,33–43] These aspects of texaphyrin chemistry are discussed in Chapter 10.

Scheme 9.1.12

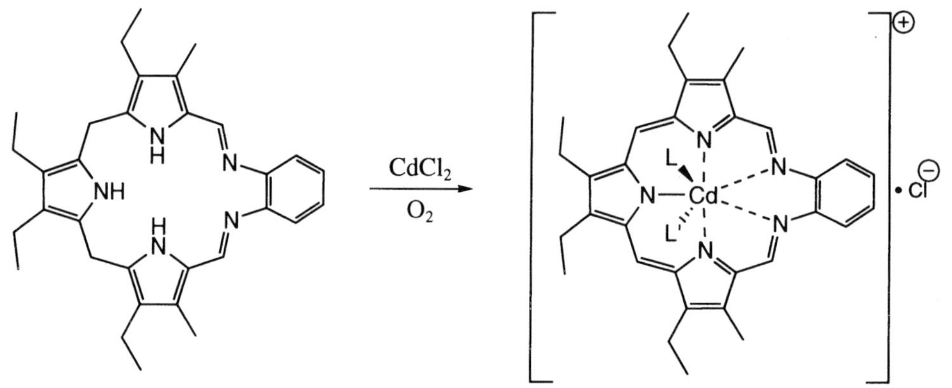

9.40a

9.60. L and L' absent
9.61a. L = benzimidazole, L' absent
9.61b. L and L' = pyridine

The solid-state structures of many metallotexaphyrin complexes have been determined by single crystal X-ray diffraction analysis. The first structure to be elucidated in this way was the cadmium derivative **9.60**, a complex that crystallized in the form of its bis-pyridine adduct **9.61b** (Figure 9.1.11).[24,26,29] Here, inspection of the structure revealed the Cd(II) center to be lying at the center of a near-perfect pentagonal bipyramidal arrangement of donor atoms. In this arrangement, the five equatorial donor atoms are provided by the monoanionic texaphyrin ligand, while the two apical ones derive from the ligated pyridines. The net result is a cadmium atom that is coordinated directly within the mean plane of the macrocycle. This in-plane coordination of cadmium, which is not observed in Cd(II) porphyrins,[44–47] is made possible, presumably, by the larger core size of the texaphyrin ligand. In fact, from this crystallographic study it could be ascertained that the center-to-nitrogen radius of the texaphyrin core is *ca.* 2.4 Å, some 20% larger than that of a typical porphyrin.[26]

As alluded to above, metal complexes of a number of lanthanide and actinide texaphyrin complexes have also been prepared.[27–29,31–43] In the case of Dy(III) texaphyrin **9.74**, as in the case of the bis-pyridine cadmium complex **9.61b**, the metal center sits directly within the mean plane of the macrocycle (Figure 9.1.12).[48] This result stands in direct contrast to the highly labile, typically "sandwich-type" 2:1 or 3:2 complexes observed for porphyrin complexes with these larger metal cations.[49–51]

It thus attests further to the fact that the central ligating core of texaphyrin is larger than that of the porphyrins.[29]

Scheme 9.1.13

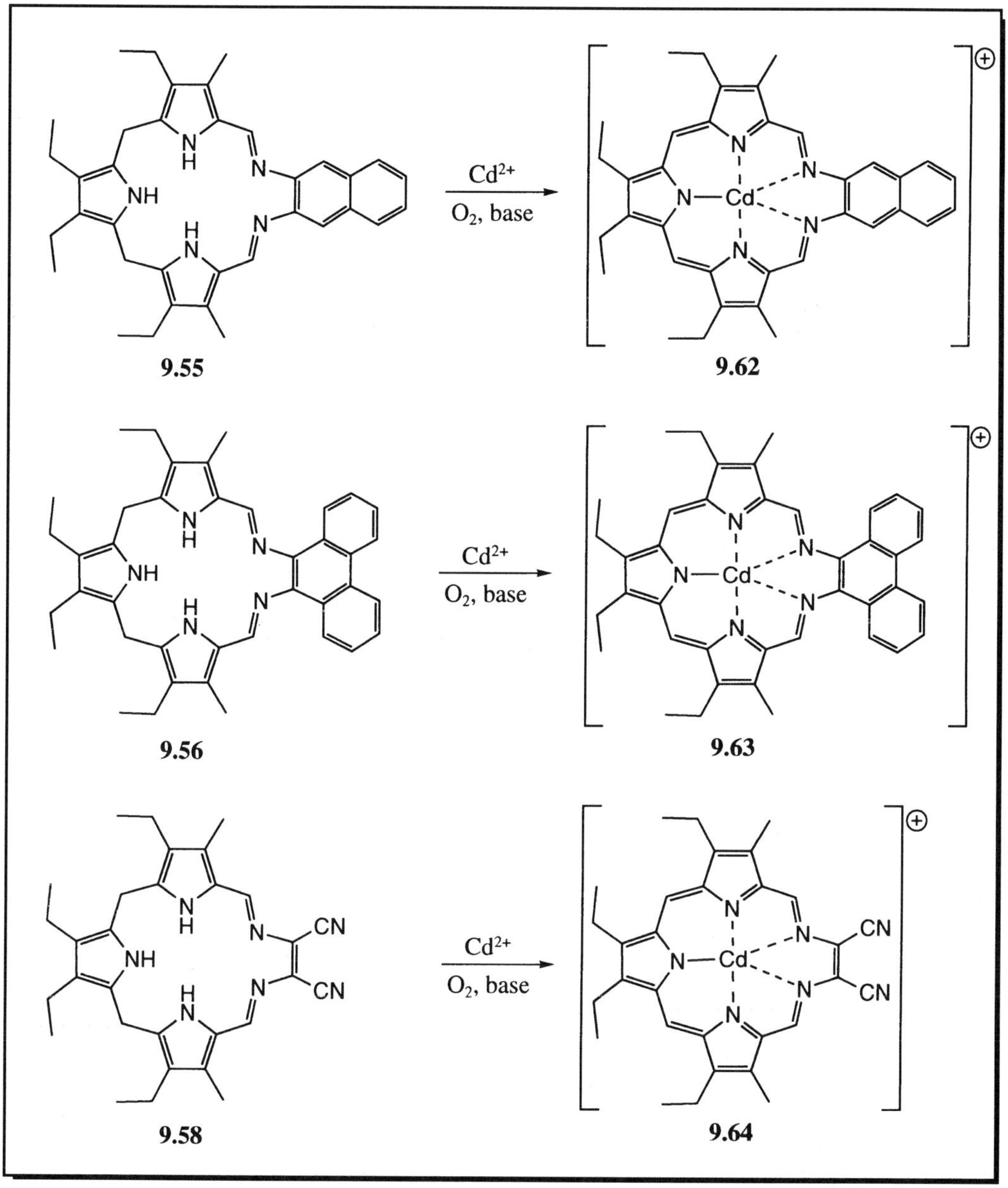

9.55
9.62
9.56
9.63
9.58
9.64
Cd^{2+}
O_2, base

Figure 9.1.9   Structures **9.65–9.79**

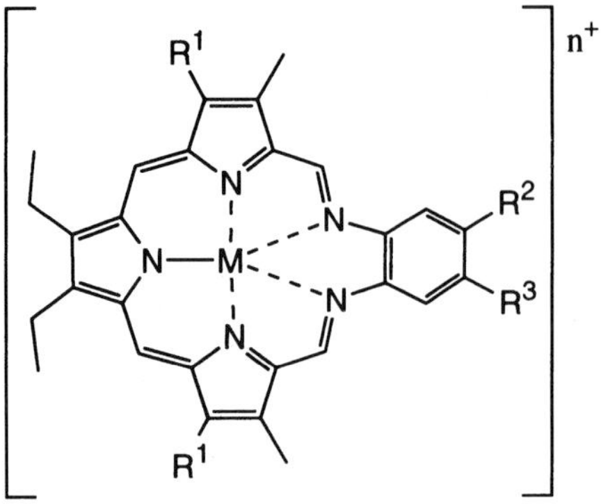

9.65. R^1 = Et, R^2 = R^3 = H, M = Zn, n = 1
9.66. R^1 = Et, R^2 = R^3 = H, M = Mn, n = 1
9.67. R^1 = Et, R^2 = R^3 = H, M = Sm, n = 2
9.68. R^1 = Et, R^2 = R^3 = H, M = Nd, n = 2
9.69. R^1 = Et, R^2 = R^3 = Me, M = Cd, n = 1
9.70. R^1 = Et, R^2 = R^3 = Me, M = Gd, n = 2
9.71. R^1 = Et, R^2 = R^3 = Me, M = Eu, n = 2
9.72. R^1 = Et, R^2 = R^3 = Me, M = Sm, n = 2
9.73. R^1 = Et, R^2 = R^3 = OMe, M = La, n = 2
9.74. R^1 = Et, R^2 = R^3 = OMe, M = Dy, n = 2
9.75. R^1 = Et, R^2 = R^3 = O(CH$_2$)$_3$OH, M = Lu, n = 2
9.76. R^1 = Et, R^2 = CO$_2$H, R^3 = H, M = In, n = 2
9.77. R^1 = Et, R^2 = NO$_2$, R^3 = H, M = Cd, n = 1
9.78. R^1 = (CH$_2$)$_3$OH, R^2 = R^3 = O(CH$_2$)$_3$OH, M = Gd, n = 2
9.79. R^1 = (CH$_2$)$_3$OH, R^2 = OCH$_2$CO$_2$H, R^3 = H, M = Eu, n = 2

A number of lanthanide(III) texaphyrin complexes have been crystallized that possess an unsymmetric ligand environment. In such cases the lanthanide(III) cation is generally found to reside out of the mean texaphyrin plane. The extent of out-of-plane character, however, was found to be a function of metal cation size (and, of course, apical ligand type). This is best illustrated by the series of F^0, F^7, and F^{14} complexes **9.73**, **9.78**, and **9.75** (La, Gd, and LuTx, respectively).[29] In the case of the F^0 La(III) complex **9.73**, the metal is 0.91 Å out of the plane and is ten-coordinate with the coordination being completed by an apical methanol and two bidentate nitrate counter anions (Figure 9.1.13; not shown).[28] These same kinds of ligands complete the coordination sphere in the case of the Gd(III) complex **9.78**.[18] In this instance, the metal is 0.60 Å out of the plane and nine-coordinate. For the F^{14} Lu(III) complex **9.75**, the smaller size of the metal precludes coordination of both neutralizing nitrate counteranions.[18] The result is an eight-coordinate complex with the metal a mere 0.27 Å above the mean texaphyrin plane. Since, in generalized terms, the "other" ligands are the same for **9.73**, **9.78**, and **9.75**, these three complexes, taken together, serve to show how the size of the cation (1.27 Å, 1.11 Å, and 0.98 Å radius for La(III), Gd(III), and Lu(III), respectively[52]) can influence in quite a dramatic way the exact details of binding. This series also provides a nice illustration of the effects of the classic contraction in size that occurs as the trivalent lanthanide series is traversed.

Figure 9.1.10   The Periodic Table of the Texaphyrins.
Shaded elements indicate those for which complexes with texaphyrins are known

The Periodic Table of the Texaphyrins

H																	He
Li	Be											B	C	N	O	F	Ne
Na	Mg											Al	Si	P	S	Cl	Ar
K	Ca	Sc	Ti	V	Cr	**Mn**	Fe	Co	Ni	Cu	**Zn**	Ga	Ge	As	Se	Br	Kr
Rb	Sr	**Y**	Zr	Nb	Mo	Tc	Ru	Rh	Pd	Ag	**Cd**	**In**	Sn	Sb	Te	I	Xe
Cs	Ba	**La**	Hf	Ta	W	Re	Os	Ir	Pt	Au	**Hg**	Tl	Pb	Bi	(Po)	(At)	(Rd)
(Fr)	Ra	Ac															

Ce	**Pr**	**Nd**	(Pm)	**Sm**	**Eu**	**Gd**	**Tb**	**Dy**	**Ho**	**Er**	**Tm**	**Yb**	**Lu**
Th	Pa	U	Np	(Pu)	(Am)	(Cm)	(Bk)	(Cf)	(Es)	(Fm)	(Md)	(No)	(Lr)

9.1.2.1.3   *General properties of the metallotexaphyrins*

Because of their highly delocalized π-electron conjugation pathway, the metallotexaphyrins absorb strongly in the visible region of the electromagnetic spectrum. Like other aromatic pyrrole-containing macrocycles, the metallotexaphyrin absorption spectrum is characterized by strong Soret-like and Q-like transitions. The Soret-like absorbances of these macrocycles generally fall in the 430–450 nm spectral region and are less intense than those of comparable porphyrin complexes. Q-like texaphyrin bands, on the other hand, are both more intense and fall further to the red than those of the porphyrins. A case in point is the Cd(II) texaphyrin chloride complex **9.60**. This species displays a Soret-like band at 427 nm (ε = 72 700 $M^{-1}cm^{-1}$ in $CHCl_3$) and a Q-like absorbance at 765.5 nm (ε = 41 200 $M^{-1}cm^{-1}$).[26,30]

Considerable effort has been devoted to exploring the ground and excited state properties of the metallotexaphyrins.[30,32,33,34] Much of the original motivation for this work was triggered by the observation that these systems absorb strongly in the spectral region where living tissues are relatively transparent (i.e., 700–1000 nm).[53] This fact, along with the finding that diamagnetic metallotexaphyrins form long-lived triplet states in high yield and act as efficient photosensitizers for the formation of singlet oxygen,[54] led to the suggestion that effective photochemotherapeutic agents might be derived from these systems.[1,29,43] Studies along these lines have indeed been carried out, and are discussed in detail in the following chapter.

Figure 9.1.11   Single Crystal X-Ray Diffraction Structure of the Cadmium(II) Texaphyrin Complex **9.61b**.

Two pyridine ligands are bound to the metal center. This figure was generated using information down-loaded from the Cambridge Crystallographic Data Centre and corresponds to a structure originally reported in reference 24. The cadmium(II) ion lies in the plane of the nearly planar macrocycle (the maximum deviation from planarity of the texaphyrin atoms is 0.10 Å). The average Cd–N distance for macrocyclic nitrogen atoms is 2.39 Å. The average Cd–N$_{pyridine}$ bond length is 2.46 Å. Atom labeling scheme: cadmium: ●; carbon: ○; nitrogen: ●. Hydrogen atoms have been omitted for clarity

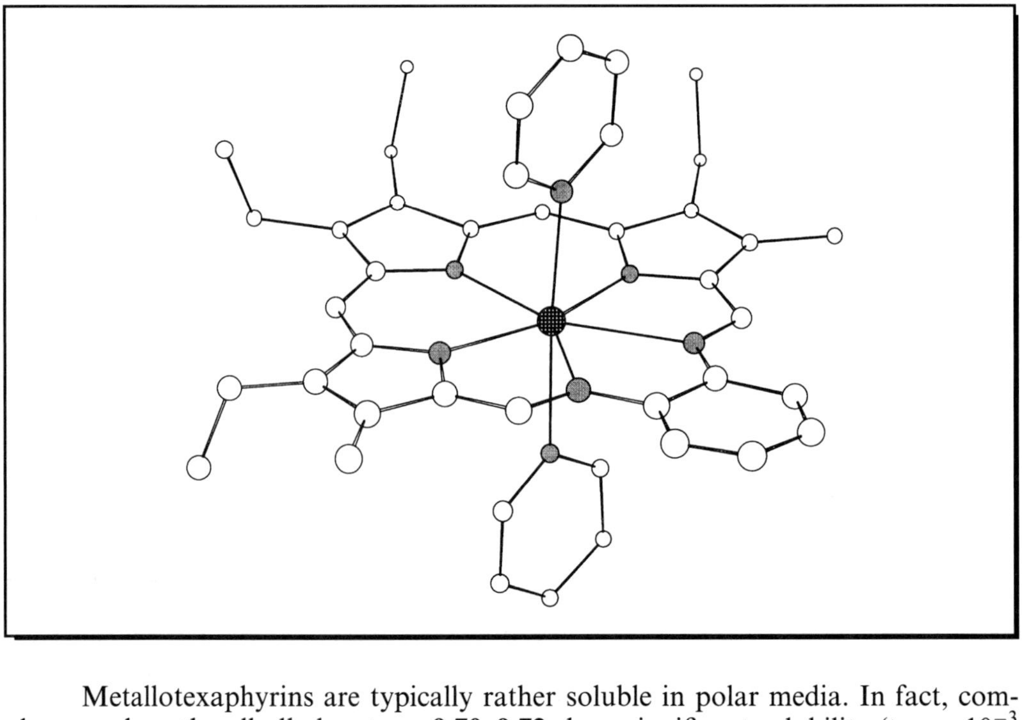

Metallotexaphyrins are typically rather soluble in polar media. In fact, complexes such as the all-alkyl systems **9.70–9.72** show significant solubility (to *ca.* 10^{-3} M concentrations) in 1 : 1 (v/v) methanol/water mixtures.[27] Perhaps more importantly, metallotexaphyrins have been demonstrated to be rather hydrolytically inert. For instance, detailed kinetic studies carried out on a solution of gadolinium(III) texaphyrin **9.70** in a 1 : 1 methanol/water mixture revealed that the half-life for decomplexation/decomposition is $\geq$ 37 days.[27] These facts, taken together, led Sessler and coworkers to design a host of water-soluble monomeric and oligomeric systems, including **9.78** and **9.80–9.89** (Figures 9.1.14–9.1.17). The highly paramagnetic gadolinium(III) complexes **9.78** and **9.81–9.83** were designed with the specific intent of developing suitable systems to act as effective MRI contrast agents.[29,39–41] Lutetium(III) texaphyrin **9.80** is currently being studied in terms of its ability to act as a photosensitizer for photodynamic therapy.[31,42,43] The oligonucleotide-bearing water-soluble lanthanide(III) systems **9.84–9.89**, on the other hand, were designed to explore the potential use of such systems in antisense applications involving RNA hydrolysis and DNA photolysis.[36,38] In all instances, the relevant texaphyrin complexes show "shelf-life" stabilities in aqueous media that are well in excess of one year.

Figure 9.1.12   Single Crystal X-Ray Diffraction Structure of the Dysprosium(III) Texaphyrin Complex **9.74**.

Two phenyl phosphate counter anions are bound to the metal center. This figure was generated using information down-loaded from the Cambridge Crystallographic Data Centre and corresponds to a structure originally reported in reference 48. The dysprosium(III) ion is displaced by 0.07 Å above an essentially planar N_5 mean plane. The average Dy–N distance is 2.40 Å. The average Dy–O bond length is 2.23 Å. Atom labeling scheme: dysprosium: ●; carbon: ○; nitrogen: ●; oxygen: ⊖; phosphorus: ⊕. Hydrogen atoms have been omitted for clarity

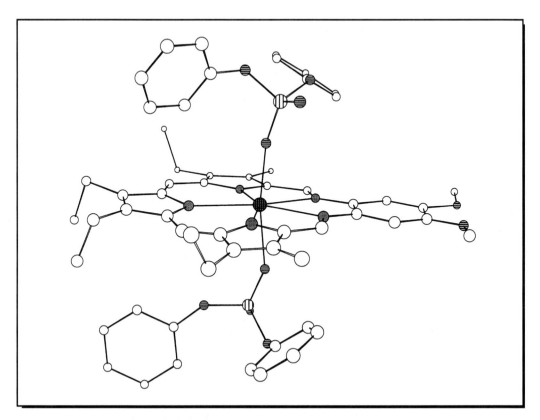

Figure 9.1.13   Views of the Single Crystal X-Ray Diffraction Structures of Lanthanum(III) Texaphyrin **9.73** (top), Gadolinium(III) Texaphyrin **9.78** (middle), and Lutetium(III) Texaphyrin **9.75** (bottom) Showing the Out-of-plane Displacement of the Metal Ions.

This figure was generated using information down-loaded from the Cambridge Crystallographic Data Centre and corresponds to structures originally reported in references 28 and 35. In the lanthanum(III) complex **9.73**, the metal center is 0.91 Å above the mean-square pentaaza plane. The corresponding out-of-plane displacement for the gadolinium(III) and lutetium(III) centers in complexes **9.78** and **9.75** are 0.60 Å and 0.27 Å, respectively. Atom labeling scheme: lanthanum, gadolinium, lutetium: ◉; carbon: ○; nitrogen: ●; oxygen: ⊖. In all cases, the axially coordinated counter anions are not shown. Hydrogen atoms have been omitted for clarity

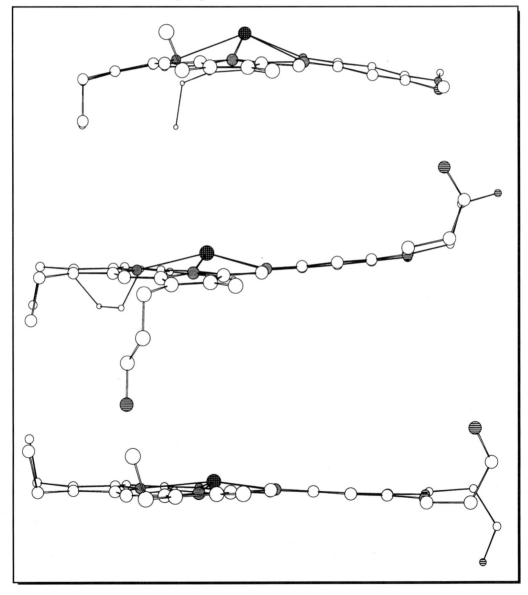

Figure 9.1.14   Structure **9.80**

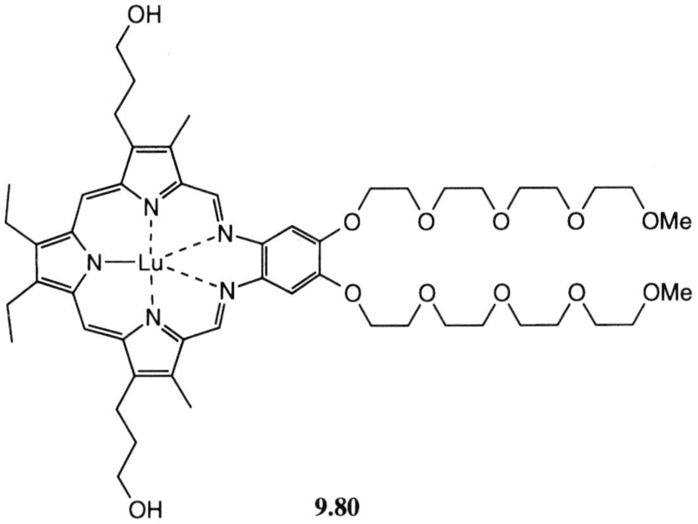

The metallotexaphyrins have also been found to be easy to reduce and capable of "capturing" hydrated electrons in aqueous solution.[41] This has made them of potential interest as X-ray radiation therapy (XRT) enhancement agents. While a discussion of this and other applications of metallotexaphyrins is deferred to Chapter 10 of this book, it is worth mentioning that preliminary results have been obtained that are highly encouraging. This, in turn, is stimulating on-going efforts to prepare additional texaphyrin and texaphyrin-like macrocycles.[39]

9.1.2.2   "Anthraphyrin"

As implied above, the success enjoyed with the texaphyrins has inspired the synthesis of several other "[1 + 1]" Schiff base systems (e.g., **9.92**, **9.93**, and **9.95**; Schemes 9.1.14 and 9.1.15).[23,25,55] One of these, an anthracene-derived "expanded porphyrinogen" (structure **9.95**; Scheme 9.1.15) is considered particularly noteworthy .

Macrocycle **9.95**, termed "anthraphyrin" for short, was prepared from the acid-catalyzed condensation of diformyl tripyrrane **9.39** with 1,8-diaminoanthracene **9.94**.[55] In the presence of HCl and HBF$_4$, single crystals of the protonated form of macrocycle **9.95** (i.e., **9.95a**) suitable for X-ray analysis could be obtained. From the resulting structure of this mixed salt, it was determined that the chloride anion is centrally located within the cavity of the macrocycle, being held there by four hydrogen bonds (Figure 9.1.18). The net result is a 1:1 anion-to-expanded porphyrin adduct that is reminiscent of the tightly organized fluoride anion complex that was

Figure 9.1.15   Structures **9.81** and **9.82**

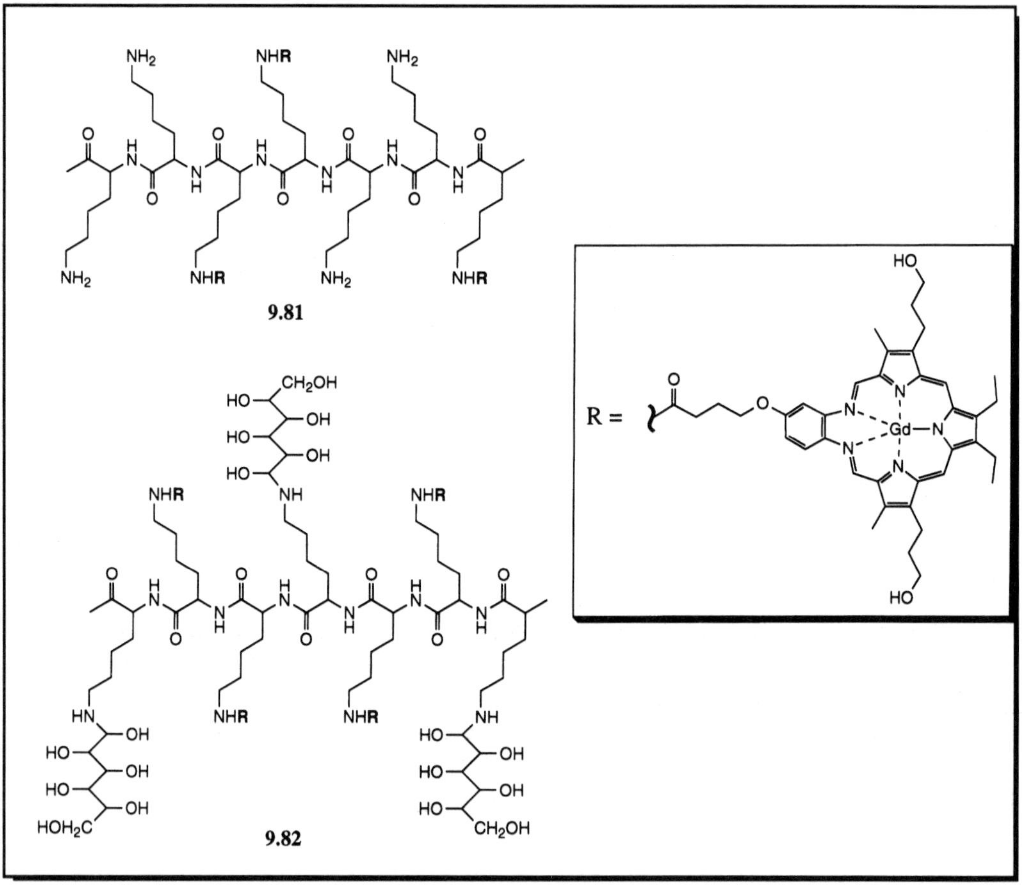

observed in the case of diprotonated sapphyrin (see Chapter 5). However, unlike sapphyrin, macrocycle **9.95** shows a higher affinity for chloride anion, rather than fluoride anion.[55] Presumably, this reflects the fact that anthraphyrin, with a larger core than sapphyrin, provides a better size match-up for this larger anion. These and other anion chelation-related points are discussed further in Chapter 10 (Section 10.5).

9.1.2.3   Expanded Texaphyrinogens

Two other notable examples of expanded porphyrins prepared *via* the "[1 + 1]" Schiff base approach are systems **9.97** and **9.99**.[22,29,39,56] In both cases the macrocycles were prepared by condensing a linear diformyl*tetra*pyrrole (i.e., **9.96** and **9.98**) with *o*-phenylenediamine (**9.19**) (Schemes 9.1.16 and 9.1.17). Formally, both these new macrocycles can be considered as being higher-order homologs of

Figure 9.1.16   Structure **9.83**

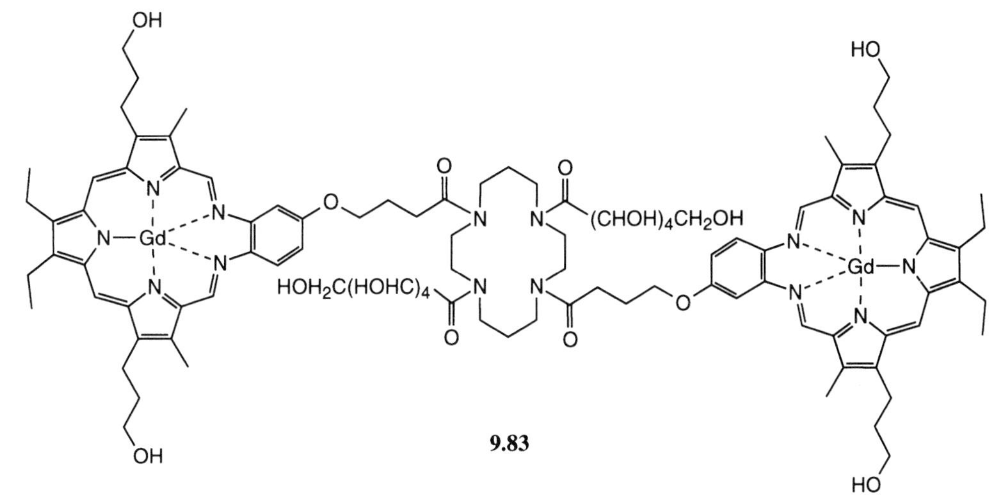

9.83

the tripyrrane-derived "texaphyrinogens" (discussed in Section 9.1.2). Unfortunately, attempts to effect oxidation of macrocycle **9.97**, both in the presence and absence of metal cation, met with failure. Likewise, attempts to effect metalation of the formally conjugated (i.e., 22 π-electron), yet non-aromatic, system **9.99** proved similarly unsuccessful. Thus, direct comparisons to the smaller texaphyrins, which are best characterized in their oxidized, metal-containing forms, cannot be made at present.

9.2   Other Nitrogen-bridged Expanded Porphyrins

9.2.1   *Superphthalocyanines*

Phthalocyanines, of general structure **9.100**, may formally be regarded as being tetrabenzo-tetraazaporphyrins (Figure 9.2.1). Unlike the porphyrins, however, phthalocyanines are generally prepared *via* metal-templated condensations involving phthalonitrile and its derivatives, rather than through directed organic synthesis. The same is true for the known phthalocyanine analogs that are "expanded" and "contracted". Examples of these latter systems, derived from boron-templated cyclizations of phthalonitrile (e.g., subphthalocyanine **9.101**), are presented in Chapter 2. In this section of this chapter, a similar but opposite deviation from "standard" phthalocyanine chemistry is presented. It concerns the larger *pentameric* macrocycle **9.105** and related systems, such as **9.106** and **9.107**, which are obtained *via* the uranyl cation-templated condensation of phthalonitrile and its derivatives (e.g., **9.102**–**9.104**) (Scheme 9.2.1).

Figure 9.1.17   Structures **9.84–9.89**

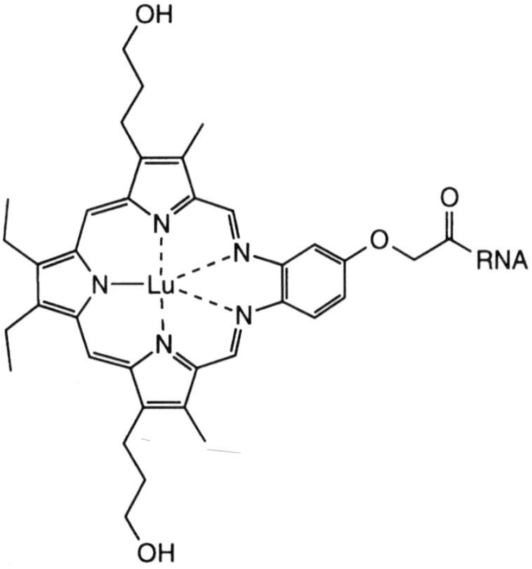

9.84. RNA = NH(CH$_2$)$_6$-PO$_4$-5'-CAU CUG UGA GCC GGG-3' (2'-OMe)
9.85. RNA = NH(CH$_2$)$_6$-PO$_4$-5'-CUC GGC CAU AGC GAA-3' (2'-OMe)

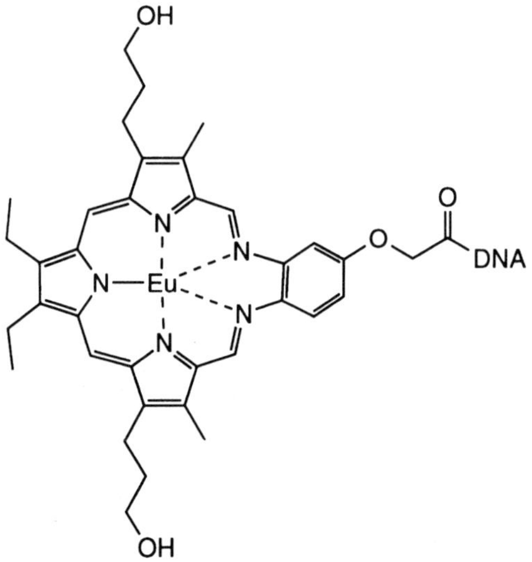

9.86. DNA = NH(CH$_2$)$_6$-NHCO(CH)$_2$-5'-CAT CTG TGA GCC GGG TGT TG-3'
9.87. DNA = NH(CH$_2$)$_6$-NHCO(CH)$_2$-5'-CTC GGC CAT AGC GAA TGT TC-3'
9.88. DNA = NH(CH$_2$)$_6$-PO$_4$-5'-CAT CTG TGA GCC GGG TGT TG-3'
9.89. DNA = NH(CH$_2$)$_6$-PO$_4$-5'-CTC GGC CAT AGC GAA TGT TC-3'

Scheme 9.1.14

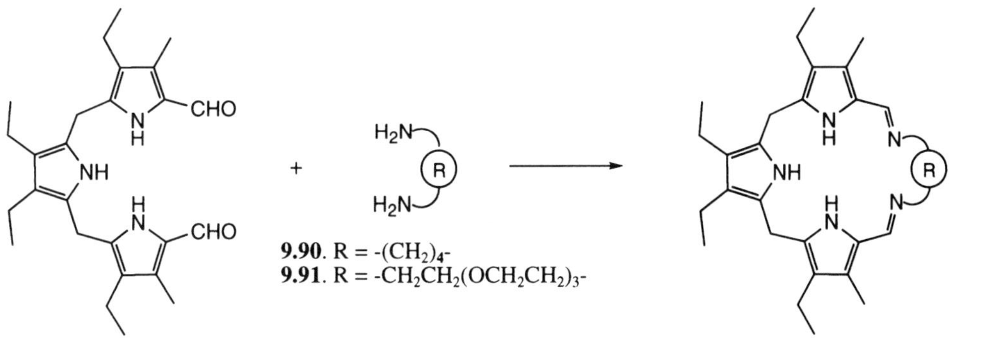

9.90. R = -(CH₂)₄-
9.91. R = -CH₂CH₂(OCH₂CH₂)₃-

9.39

9.92. R = -(CH₂)₄-
9.93. R = -CH₂CH₂(OCH₂CH₂)₃-

Scheme 9.1.15

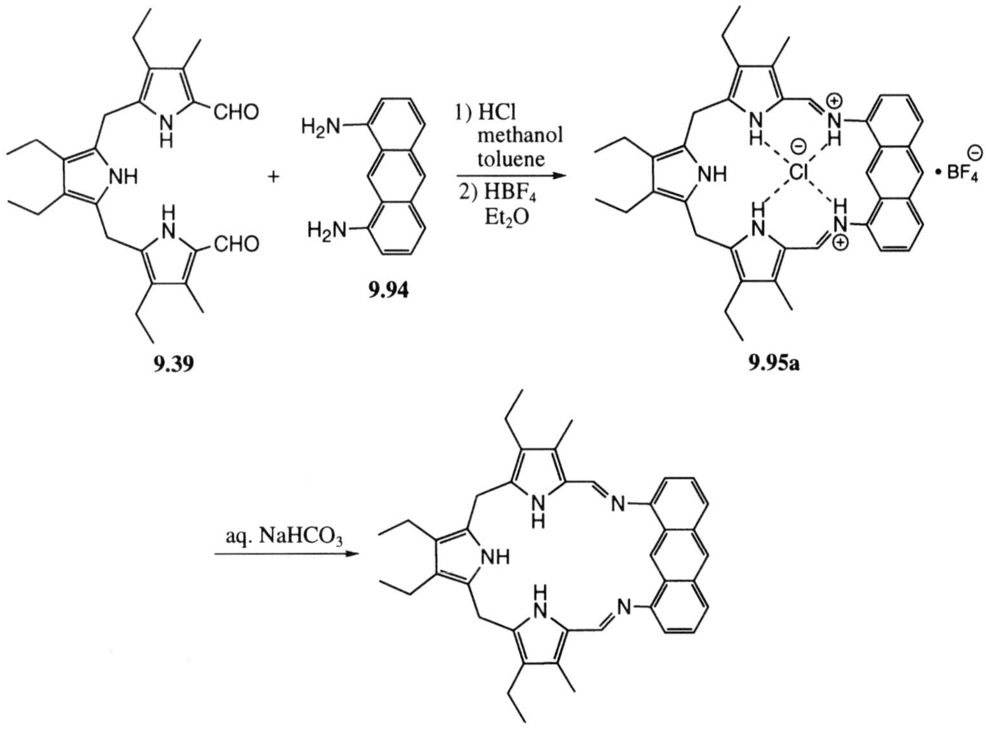

9.39

9.94

1) HCl
methanol
toluene

2) HBF₄
Et₂O

9.95a

• BF₄

aq. NaHCO₃

9.95b

Figure 9.1.18   Single Crystal X-Ray Diffraction Structure of the Mixed HCl and HBF₄ Salt of "Anthraphyrin" **9.95a**.

This figure was generated using information down-loaded from the Cambridge Crystallographic Data Centre and corresponds to a structure originally reported in reference 55. All three pyrrole rings are tilted toward the chloride ion which is in a pseudo square-planar geometry. The Cl⁻ sits roughly 0.80 Å above the planar portion of the macrocycle defined by the anthracene moiety and the end pyrroles of the tripyrrane unit. The central pyrrole ring of the tripyrrane subunit is almost perpendicular to this plane (dihedral angle: 79.3°). Atom labeling scheme: carbon: ○; nitrogen: ●; chlorine: ☺. Not shown is an H-bonded BF₄⁻ counterion that sits directly above the central tripyrrane pyrrole subunit. Hydrogen atoms have been omitted for clarity

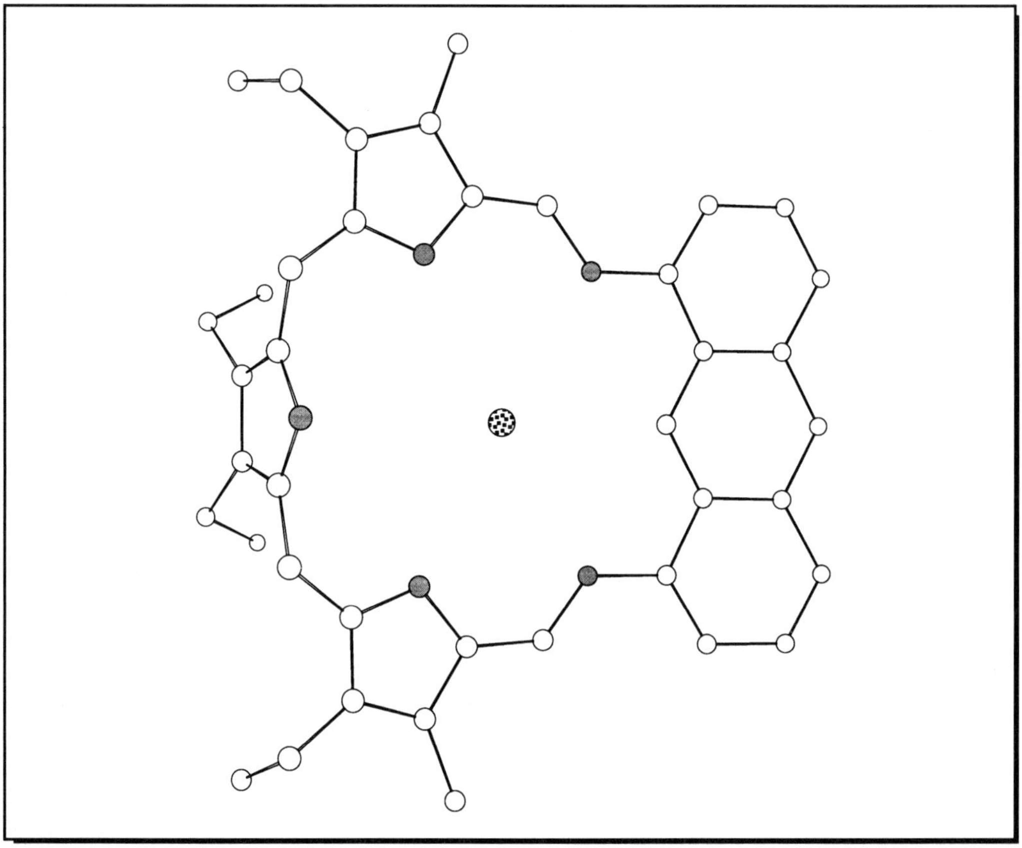

Macrocycle **9.105**, in the form of its pentaligated uranyl complex, was first characterized by Marks and Day.[57–61] While earlier reports had claimed the successful preparation of "normal" UO₂-phthalocyanine from the reaction of UO_2^{2+} and phthalonitrile,[62] the results of Marks and Day revealed that it is the pentameric uranyl superphthalocyanine, contaminated with only small quantities of metal-free phthalocyanine, that is the dominant product obtained as the result of such a process. The findings of Marks and Day were thus consistent with one other earlier report (see reference 57 and references therein) wherein mass spectrometric evidence was put forward in support of a species in which five dicyanobenzene subunits were proposed as being coordinated to the uranyl cation.

Scheme 9.1.16

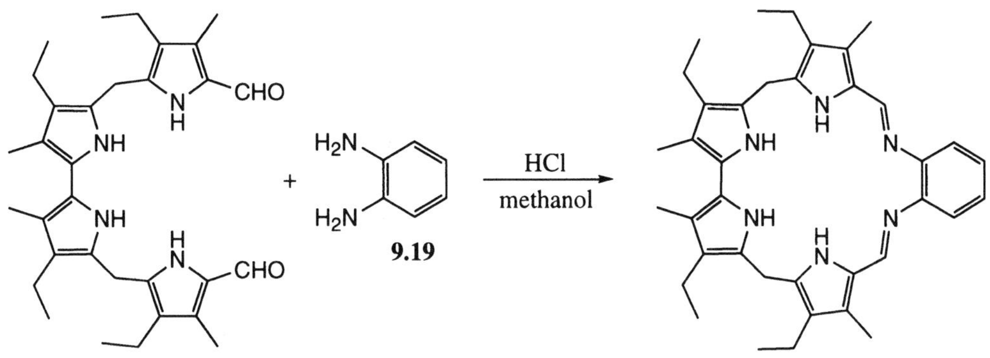

9.96 **9.19** **9.97**

Scheme 9.1.17

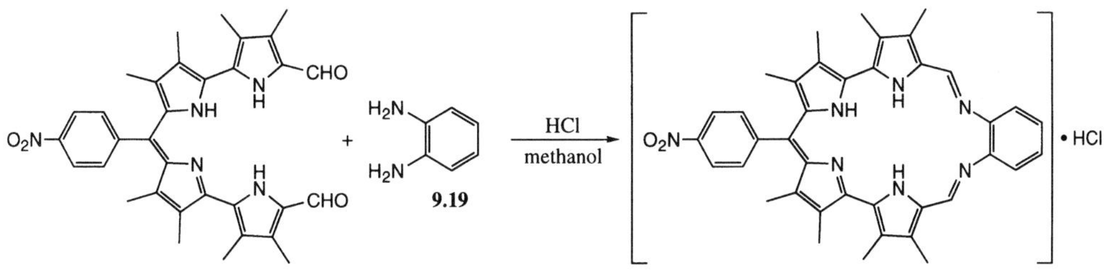

9.98 **9.19** **9.99**

Figure 9.2.1 Structures **9.100** and **9.101**

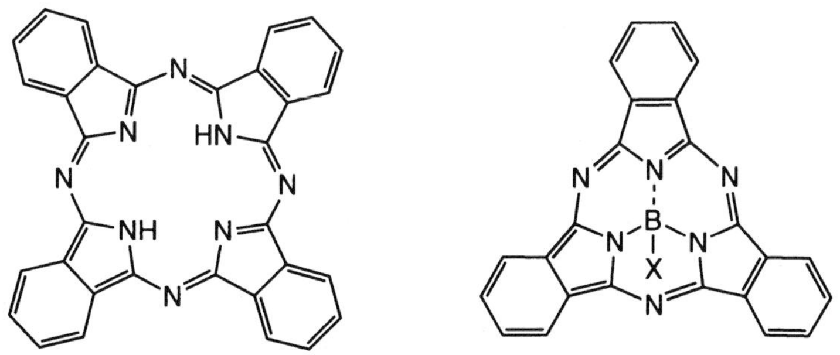

9.100 **9.101**

Scheme 9.2.1

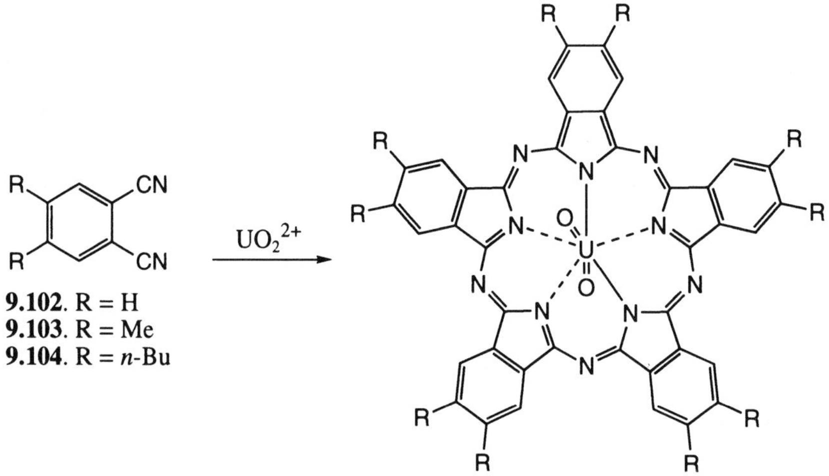

9.102. R = H
9.103. R = Me
9.104. R = n-Bu

9.105. R = H
9.106. R = Me
9.107. R = n-Bu

The definitive evidence put forward by Marks and Day consisted of a single crystal X-ray diffraction structure of **9.105** (Figure 9.2.2).[57] This structure confirmed the close match in size between the superphthalocyanine ligand and the coordinated uranyl cation. While the uranyl cation sits directly within the mean plane of the macrocycle, the macrocycle itself is somewhat buckled, presenting a "wave-like" appearance when viewed sequentially from one benzpyrrole subunit to the next. Nonetheless, this structure does serve to confirm, at least in general terms, the intuitively appealing proposition that superphthalocyanine is formed in preference to phthalocyanine because the large uranyl cation will fit in a pentaaza pentabenzpyrrole macrocyclic core but not a smaller tetraaza tetrabenzpyrrolic one.

The considerable excitement aroused by the synthesis and structural characterization of superphthalocyanine **9.105** inspired attempts to use other large cations to effect this type of macrocyclization. Unfortunately, however, it was found that only uranyl cation would effect cyclization of phthalonitrile and its derivatives to the pentameric superphthalocyanine. Sadly, it was also found that attempted demetalation or transmetalation of uranyl superphthalocyanine led to ring contraction (giving the corresponding phthalocyanine or metallophthalocyanine). Thus, uranyl superphthalocyanine **9.105** remains the only structurally characterized example of this not-very-general class of macrocycles. Nevertheless, the very fact that this system could be made did serve as an important inspiration for Gossauer[63] and Sessler[15,20,64–66] who have succeeded in preparing other uranyl-ligating expanded porphyrins, which are, in some measures, structurally analogous (see Chapters 5, 6, and 8).

Figure 9.2.2   Single Crystal X-Ray Diffraction Structure of Uranyl(VI) Superphthalocyanine **9.105**.
This figure was generated using information down-loaded from the Cambridge Crystallographic Data Centre and corresponds to a structure originally reported in reference 57. The ligand is significantly distorted from planarity (the maximum displacement of the five coordinated nitrogen atoms from the mean N_5 plane is 0.30 Å). The uranium(VI) ion sits in an axially compressed pentagonal bipyramidal coordination geometry. The average U–N and U–O bond lengths are 2.52 Å and 1.74 Å, respectively. Atom labeling scheme: uranium: ◐; carbon: ○; nitrogen: ●; oxygen: ⊖. Hydrogen atoms have been omitted for clarity

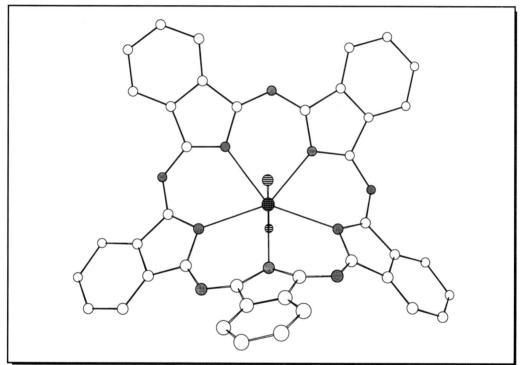

9.2.2   Porphocyanines

Another class of nitrogen-bridged expanded porphyrins was first reported in 1993 by Dolphin and coworkers. Referred to as porphocyanines,[67–71] the first of these tetrapyrrolic expanded porphyrins to be reported was system **9.110**. It was prepared *via* the oxidative dimerization (with loss of ammonia) of the bis(amino-methyl)dipyrrylmethane derivative **9.109**, a precursor derived, in turn, from the bis(-cyano)-substituted dipyrrylmethane **9.108** (Scheme 9.2.2).[67,68,71]

The general procedure of Scheme 9.2.2 has also been applied to the synthesis of *meso*-aryl-substituted porphocyanines. For instance, *meso*-diphenylporphocyanine **9.115** was prepared *via* the reduction and subsequent oxidative dimerization of the *meso*-phenyl-substituted dipyrrylmethane **9.111** (Scheme 9.2.3).[69,71] The *meso*-aryl-substituted porphocyanines **9.116–9.118** were also prepared in an analogous fashion starting from the corresponding aryl-substituted dipyrrylmethanes **9.112–9.114**.[69,71] This procedure has also been applied to the construction of the unsymmetrically substituted porphocyanine **9.119**. Here, a mixed condensation of the LiAlH₄-reduction product of dipyrrylmethanes **9.108** and **9.111** afforded the unsymmetrical system **9.119** in 21% yield, along with the symmetrical species **9.110** and **9.115** in 32% and <3% yield, respectively (Scheme 9.2.4).[71]

Scheme 9.2.2

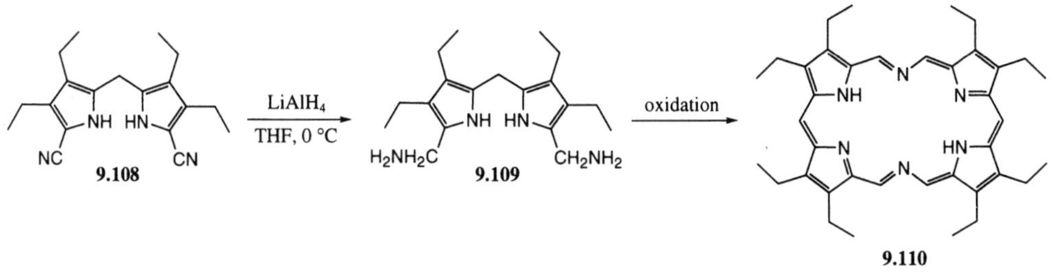

Scheme 9.2.3

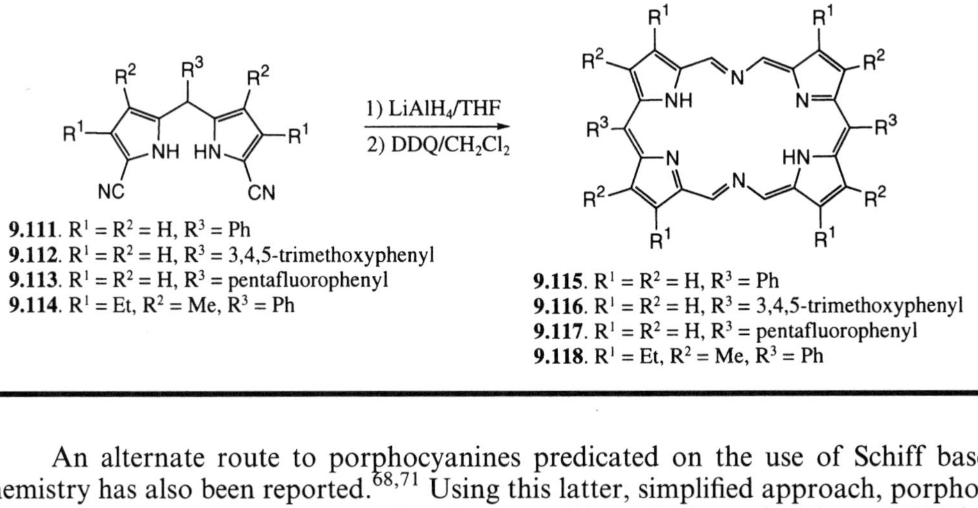

9.111. R¹ = R² = H, R³ = Ph
9.112. R¹ = R² = H, R³ = 3,4,5-trimethoxyphenyl
9.113. R¹ = R² = H, R³ = pentafluorophenyl
9.114. R¹ = Et, R² = Me, R³ = Ph

9.115. R¹ = R² = H, R³ = Ph
9.116. R¹ = R² = H, R³ = 3,4,5-trimethoxyphenyl
9.117. R¹ = R² = H, R³ = pentafluorophenyl
9.118. R¹ = Et, R² = Me, R³ = Ph

An alternate route to porphocyanines predicated on the use of Schiff base chemistry has also been reported.[68,71] Using this latter, simplified approach, porphocyanine **9.110** could be prepared directly from diformyl dipyrrylmethane **9.120** by treating with ammonia-saturated ethanol followed by oxidation with oxygen (Scheme 9.2.5). In a similar way, the *tetrakis*(ethoxycarbonyl)-derived porphocyanine **9.122** was also prepared. As might be expected, this highly functionalized system could not be obtained using the initial synthetic pathway.

The porphocyanines contain six internally oriented nitrogen atoms, of which four are theoretically capable of being protonated in acidic solution. Using a biphasic spectrometric titration technique, pK_a values, corresponding to the deprotonation of doubly protonated octaethylporphocyanine **9.110** have been determined.[70,71] Interestingly, only the monoprotonated and diprotonated forms of porphocyanine were observed in these experiments, for which pK_{a1} values of 4.4 ± 0.1 (diprotonated $\rightarrow$ monoprotonated form) and 6.0 ± 0.1 (monoprotonated $\rightarrow$ neutral form) were recorded. On this basis it was proposed that protonation of free-base porphocyanine occurs at the inner pyrrolinene nitrogen atoms (Scheme 9.2.6), rather than at the *meso*-like imine nitrogen atoms.

Scheme 9.2.4

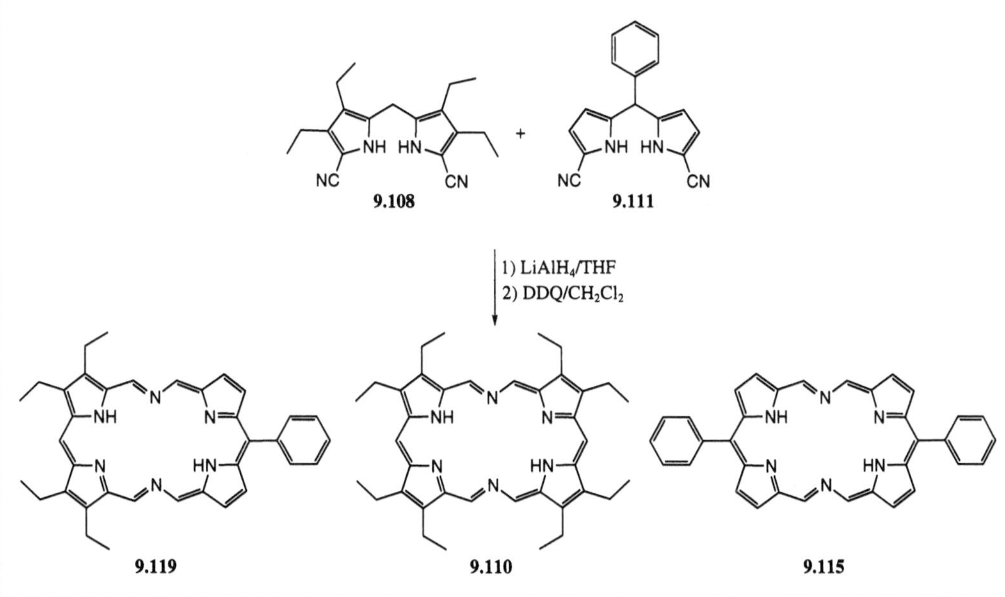

Scheme 9.2.5

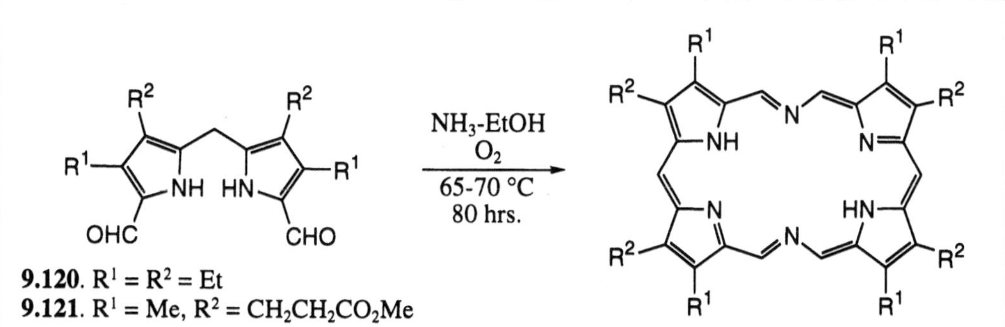

9.120. R^1 = R^2 = Et
9.121. R^1 = Me, R^2 = CH$_2$CH$_2$CO$_2$Me

9.110. R^1 = R^2 = Et
9.122. R^1 = Me, R^2 = CH$_2$CH$_2$CO$_2$Et

Scheme 9.2.6

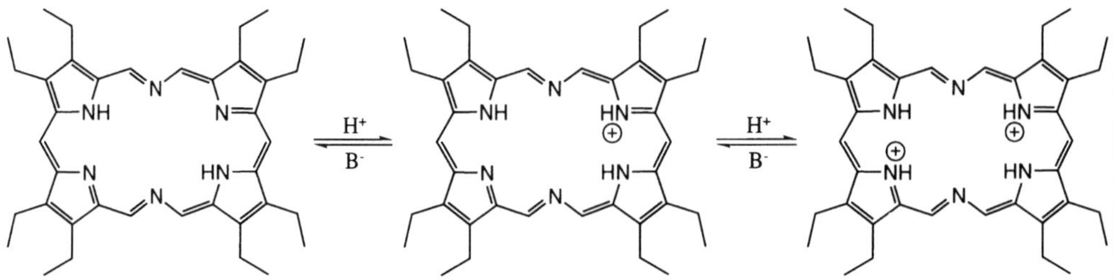

9.110a   9.110b   9.110c

Porphocyanines are nominally 22 π-electron aromatic macrocycles. Certainly, they exhibit visible absorption characteristics consistent with such a formulation. Specifically, they display both an intense Soret-like absorbance and weaker Q-type bands when dissolved in organic solvents such as dichloromethane. For instance, for the free-base form of **9.110**, the Soret band appears at 457 nm (ε = 240 000 $M^{-1} cm^{-1}$), and the Q-like bands are found at 592, 633, 728, and 797 nm (ε = 17 400, 5800, 3200, and 27 000 $M^{-1} cm^{-1}$, respectively). The protonated form of porphocyanine exhibits UV-vis absorption bands that are sharper than those of the free-base form. Interestingly, these bands are also somewhat blue-shifted, for instance, the Soret band of **9.110** in TFA/CH$_2$Cl$_2$ is at 453 nm (ε = 760 000 $M^{-1} cm^{-1}$). When excited at 450 nm, the free-base macrocycle **9.110** fluoresces at 805 nm in THF solution and at 750 nm in a mixed THF/HOAc solution.[67]

Initial investigations into the metal-coordination ability of the porphocyanines led to the isolation of the ZnCl$_2$ adduct **9.123**. This mononuclear complex was prepared by heating a methanol-dichloromethane solution of the free-base porphocyanine **9.110** with ZnCl$_2$ (Scheme 9.2.7). The structure of this complex was determined *via* single crystal X-ray diffraction analysis.[67,71] It revealed a macrocyclic ligand system that is remarkably planar (Figure 9.2.3). This same structure also confirmed that: (1) the complex is mononuclear in nature; and (2) the presence of a zinc atom that is tetrahedrally coordinated. Two of the zinc-coordination sites are occupied by chloride anions and two are occupied by pyrrolic nitrogen atoms "donated" from one end of the macrocyclic core. The net result of this coordination is that the other two pyrrolic nitrogen atoms bear protons. These NH protons are within hydrogen bonding distance of two chloride anions that are ligated to the zinc center. Interestingly, these pyrrolic protons of **9.123** are acidic enough to be removed by treating the complex with strong bases such as Et$_3$N or DBU.

Aside from the zinc complex, the coordination chemistry of the porphocyanines remains largely unexplored. Nonetheless, the pseudohexagonal arrangement of the six potentially donating nitrogen atoms present in porphocyanine has led to speculation that such systems could serve as ligands for larger cations such as those of the lanthanide or actinide series (as illustrated by the hypothetical structure

9.124; Figure 9.2.4).[67] Unfortunately, at present, it is too early to tell whether this promise will be borne out by experimental reality.

Scheme 9.2.7

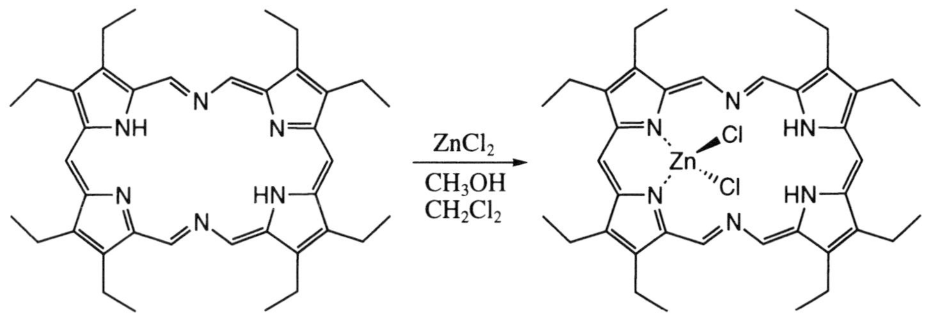

9.110   9.123

9.2.3   *14-Aza[26]porphyrin-(5.1.5.1)*

The nitrogen-bridged expanded porphyrin **9.126** was first prepared by Franck and coworkers in 1993. This material, which is referred to as 14-aza[26]porphyrin-(5.1.5.1),[72] was prepared from the linear tetrapyrrolic species **9.125** by treating it with a solution of ammonia-saturated methanol for 10 minutes at 20 °C, followed by oxidation with DDQ (Scheme 9.2.8).

Interestingly, macrocycle **9.126** is prepared from the same "building block" that was used to prepare the tetravinylogous porphyrin **9.127** (Figure 9.2.5) (see Chapter 4). Thus, compound **9.126** may properly be considered as being an aza-bridged tetravinylogous porphyrin. Nonetheless, this 14-aza[26]porphyrin-(5.1.5.1) species, **9.126**, like its all-carbon-bridged "cousins" (e.g., **9.127**), is a 26 π-electron aromatic macrocycle. This is evident from an inspection of its ^{1}H NMR spectrum. Although ring-current effects observed for **9.126** are smaller than those observed in the case of the parent *meso*-carbon system **9.127**, strong shifts to higher field are seen for the NH protons of **9.126** compared to **9.125**. Further consistent with the aromaticity proposed for **9.126** is the finding that the free-base form of this macrocycle exhibits a very intense Soret-like absorbance and two intense Q-like transitions in its optical spectrum (λ_{max} = 490, 649, and 697 nm (ε = 169 800, 28 200, and 15 100 $M^{-1}cm^{-1}$, respectively) in CH_2Cl_2). The protonated form of **9.126** exhibits a similar electronic spectrum. However, in this instance the single Soret and two Q-like bands are red-shifted relative to what is seen in the case of the free-base form (λ_{max} = 503, 681, and 735 nm (ε = 363 100, 32 400, and 27 500 $M^{-1}cm^{-1}$, respectively) in 1% TFA in CH_2Cl_2).[72]

Figure 9.2.3   Single Crystal X-Ray Diffraction Structure of Dichlorozinc(II) Porphocyanine **9.123**.

This figure was generated using information down-loaded from the Cambridge Crystallographic Data Centre and corresponds to a structure originally reported in reference 67. The macrocyclic framework is essentially planar (the average deviation of the porphocyanine atoms from the mean macrocyclic plane is 0.01 Å). The zinc(II) ion sits directly within the mean macrocyclic plane and is tetrahedrally coordinated to two pyrrolic nitrogen atoms and two chlorine atoms. The average Zn–N and Zn–Cl bond lengths are 2.06 Å and 2.24 Å, respectively. Atom labeling scheme: zinc: ●; carbon: ○; nitrogen: ●; chlorine: ☒. Hydrogen atoms have been omitted for clarity

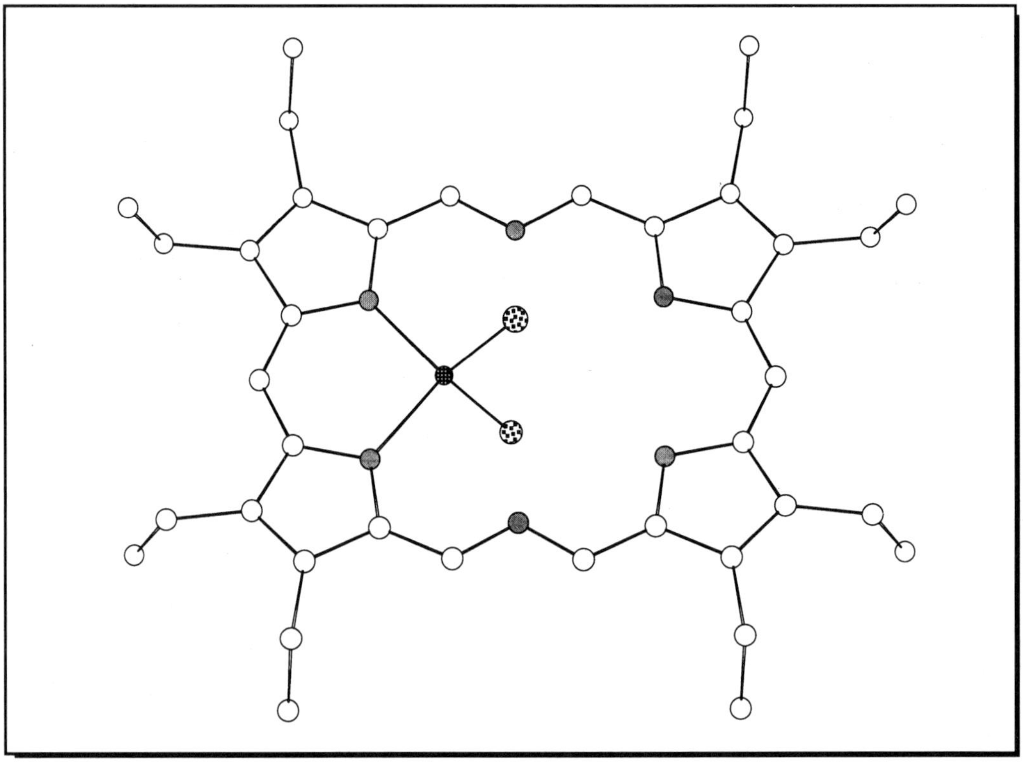

Figure 9.2.4   Structure **9.124**

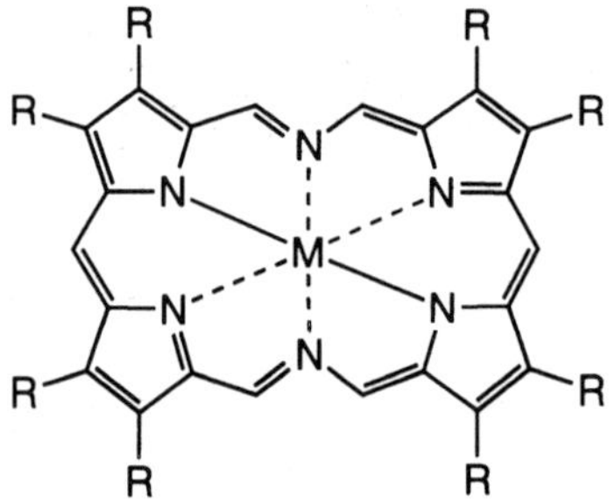

9.124

Scheme 9.2.8

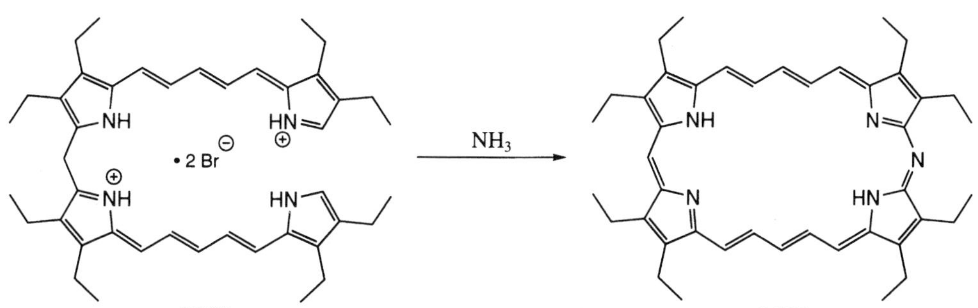

9.125   **9.126**

Figure 9.2.5   Structure **9.127**

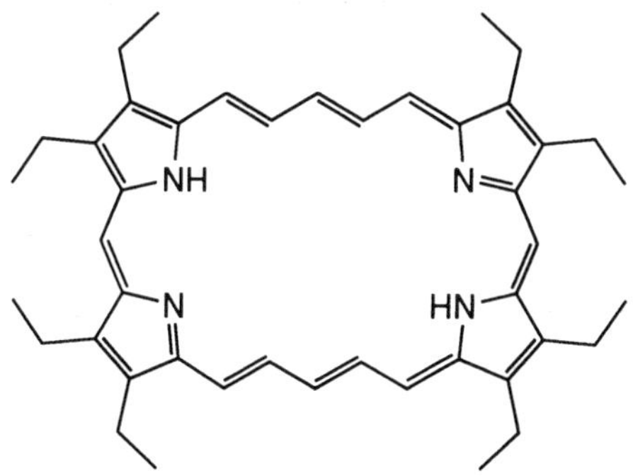

9.127

9.3   Cryptand-like Expanded Porphyrins

A very different type of nitrogen-bridged expanded porphyrin is defined by two "cryptand-like" species recently reported by Sessler and coworkers.[39,73] The first of these, the tripyrrane-strapped porphyrin **9.130**, was prepared *via* the Schiff base-type condensation of one equivalent of diformyltripyrrane **9.128** with the $\alpha,\beta,\alpha,\beta$-atropisomer of tetrakis(2-aminophenyl)porphyrin **9.129** (Scheme 9.2.9). When two equivalents of tripyrrane **9.128** are used, the doubly strapped porphyrin **9.131** can be isolated. Both of these systems can be consistently prepared in 30–45% yields.

The demonstrated ability of nonaromatic expanded porphyrins to act as complexing agents for both neutral and anionic substrates[55,74,75] led Sessler and coworkers to suggest that systems such as **9.130** and **9.131** might exhibit interesting host–guest-type chemistry. That is to say, the anion and neutral substrate-binding

Scheme 9.2.9

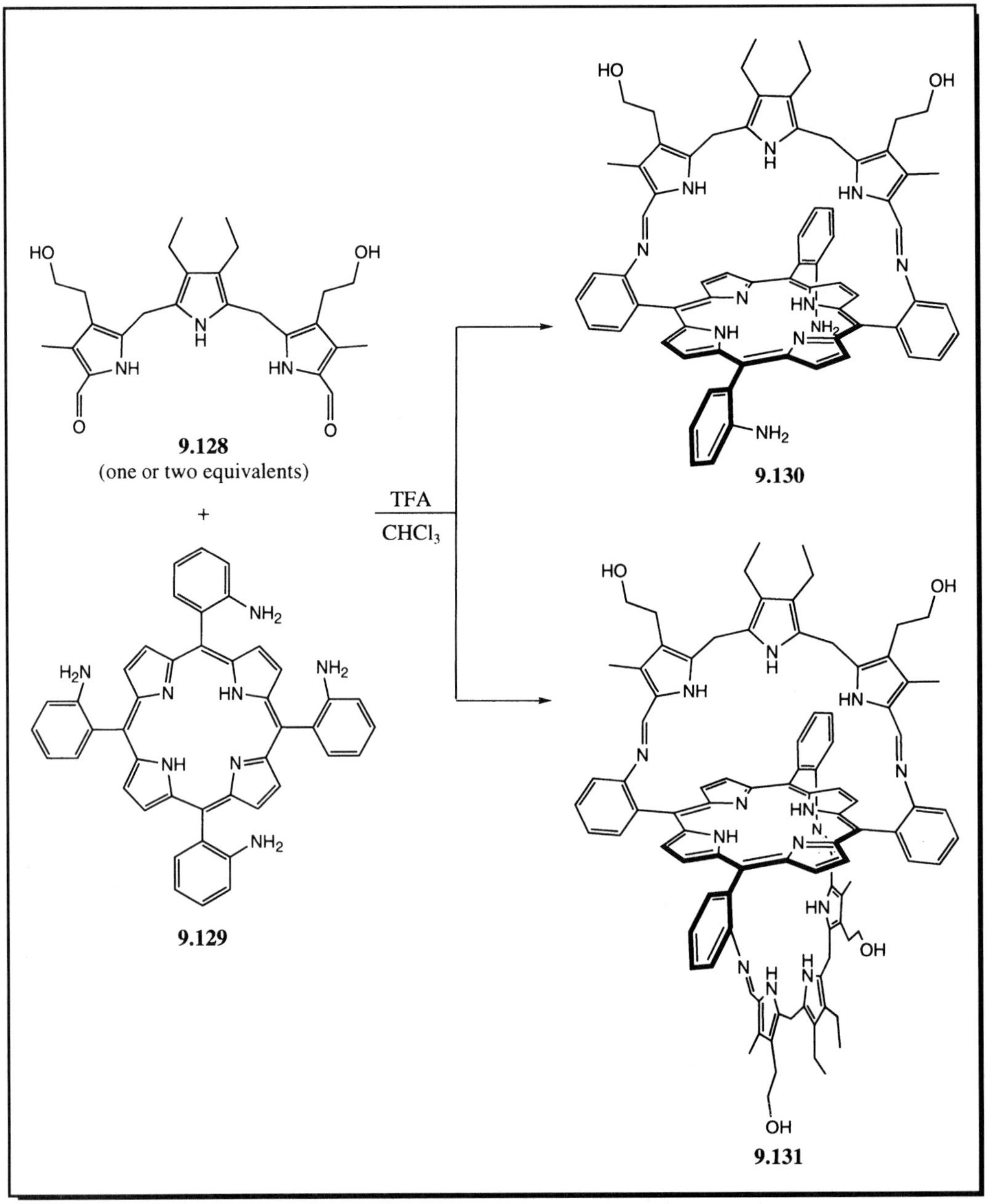

9.128
(one or two equivalents)

+

9.129

TFA

CHCl₃

9.130

9.131

capability of the oligopyrrole strap(s) of such systems could act in concert with the metal-coordinating ability of the porphyrin to provide novel polytopic receptors. Based on mass spectrometric analyses, the ability of **9.130** and **9.131** to bind small molecules such as MeOH and CD_3CN, within the cavity defined by the bridging tripyrrane straps has been confirmed. Additionally, it has proved possible to prepare porphyrin-centered Ni(II) and Cu(II) complexes of the singly strapped system **9.130**. These results, taken together, provide an indication that cryptand-like systems such as **9.130** and **9.131** might be used to "link" within one supramolecular receptor the sometimes disparate areas of coordination chemistry and molecular recognition.

One final example of a cryptand-like expanded porphyrin is the niobium(IV) bicyclophthalocyaninato system **9.132** reported by Gingl and Strähle in 1990.[76] Interestingly, **9.132** was isolated as a by-product of a standard phthalocyanine-forming reaction (i.e., metal-templated cyclocondensation of phthalonitrile **9.102**) in which $NbOCl_3$ was used as the catalyst (Scheme 9.2.10). As such, it represents the fourth type of system to be prepared using this type of procedure (the other three being the parent phthalocyanines, the subphthalocyanines discussed in Chapter 2, and the superphthaolcyanines discussed in Section 9.2.1 of this chapter).

Scheme 9.2.10

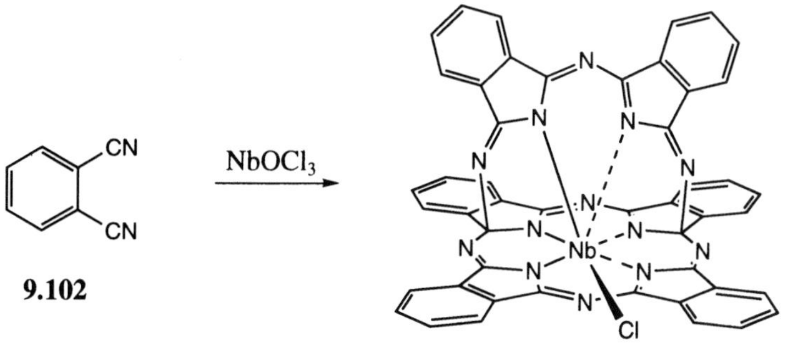

9.132

The formation of **9.132** was confirmed *via* a single crystal X-ray diffraction analysis (Figure 9.2.6). Here, it was shown that **9.132** is composed of six isoindole units and contains one niobium atom centrally bound within the macrocyclic cavity. The metal atom rests in a capped trigonal prism coordination geometry; the internally oriented nitrogen atoms form a trigonal prism-shaped core, and there is one axially coordinating chloride anion. It was presumed that this heptacoordinate arrangement actually serves as the primary driving force in the formation of this system.[76] Interestingly, similar bicyclophthalocyaninato complexes have been proposed as intermediates in the synthesis of cobalt and copper phthalocyanines.[77] However, complex **9.132** provides the first structural evidence in support of this conjecture.

Figure 9.2.6   Single Crystal X-Ray Diffraction Structure of Bicyclophthalocyaninatochloroniobium(IV) **9.132**.

This figure was generated using information down-loaded from the Cambridge Crystallographic Data Centre and corresponds to a structure originally reported in reference 76. The nitrogen atoms of the six isoindole subunits provide a trigonal prism coordination environment for the niobium(IV) ion. The chloride anion completes the coordination sphere forming a monocapped trigonal prism. The average Nb–N bond length is 2.21 Å and the Nb–Cl distance is 2.52 Å. Atom labeling scheme: niobium: ◉; carbon: ○; nitrogen: ●; chlorine: ⊗. Hydrogen atoms have been omitted for clarity

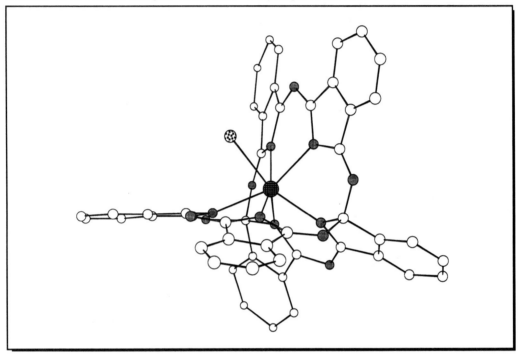

Note in proof

Details pertaining to the synthesis and structure of compound **9.34** have recently appeared.[78]

9.4   References

1.   Sessler, J. L.; Burrell, A. K. *Top. Curr. Chem.* **1991**, *161*, 177–273.
2.   Nelson, S. M.; Esho, F. S.; Drew, M. G. B. *J. Chem. Soc., Chem. Commun.* **1981**, 388–389.
3.   Bailey, N. A.; Eddy, M. M.; Fenton, D. E.; Jones, G.; Moss, S.; Mukhopadhyay, A. *J. Chem. Soc., Chem. Commun.* **1981**, 628–630.

4.   Drew, M. G. B.; Esho, F. S.; Nelson, S. M. *J. Chem. Soc., Chem. Commun.* **1982**, 1347–1348.
5.   Nelson, S. M.; Esho, F.; Lavery, A.; Drew, M. G. B. *J. Am. Chem. Soc.* **1983**, *105*, 5693–5695.
6.   Bailey, N. A.; Eddy, M. M.; Fenton, D. E.; Moss, S.; Mukhopadhyay, A. *J. Chem. Soc., Dalton Trans.* **1984**, 2281–2288.
7.   Abid, K. K.; Fenton, D. E. *Inorg. Chim. Acta* **1984**, *82*, 223–226.
8.   Adams, H.; Bailey, N. A.; Fenton, D. E.; Moss, S. *Inorg. Chim. Acta* **1984**, *83*, L79–L80.
9.   Drew, M. G. B.; Yates, P. C.; Trocha-Grimshaw, J.; McKillop, K. P.; Nelson, S. M. *J. Chem. Soc. Chem. Commun.* **1985**, 262–263.
10.   Adams, H.; Bailey, N. A.; Fenton, D. E.; Moss, S.; Rodriguez de Barbarin, C. O.; Jones, G. *J. Chem. Soc. Dalton Trans.* **1986**, 693–699.
11.   Adams, H.; Bailey, N. A.; Fenton, D. E.; Good, R. J.; Moody, R.; Rodriguez de Barbarin, C. O. *J. Chem. Soc. Dalton Trans.* **1987**, 207–218.
12.   Drew, M. G. B.; Yates, P. C. *J. Chem. Soc. Dalton Trans.* **1987**, 2563–2572.
13.   Fenton, D. E.; Moody, R. *J. Chem. Soc. Dalton Trans.* **1987**, 219–220.
14.   Vigato, P. A.; Fenton, D. E. *Inorg. Chim. Acta* **1987**, *139*, 39–48.
15.   Sessler, J. L.; Mody, T. D.; Dulay, M. T.; Espinoza, R.; Lynch, V. *Inorg. Chim. Acta* **1996**, *246*, 23–30.
16.   Acholla, F. V.; Mertes, K. B. *Tetrahedron Lett.* **1984**, *25*, 3269–3270.
17.   Acholla, F. V.; Takusagawa, F.; Mertes, K. B. *J. Am. Chem. Soc.* **1985**, *107*, 6902–6908.
18.   Sessler, J. L.; Mody, T. D.; Lynch, V. *J. Am. Chem. Soc.* **1993**, *115*, 3346–3347.
19.   Mody, T. D., Ph.D. Dissertation, University of Texas at Austin, Austin, TX, USA, **1993**.
20.   Sessler, J. L.; Mody, T. D.; Lynch, V. *Inorg. Chem.* **1992**, *31*, 529–531.
21.   Johnson, M. R.; Slebodnick, C.; Ibers, J. A. Presented at the 209th ACS National Meeting, Anaheim, CA (Org. Abstr. No. 018), April 2–6, 1995.
22.   Sessler, J. L.; Weghorn, S. J.; Meyer, S. Unpublished results.
23.   Sessler, J. L.; Johnson, M. R.; Lynch, V. *J. Org. Chem.* **1987**, *52*, 4394–4397.
24.   Sessler, J. L.; Murai, T.; Lynch, V.; Cyr, M. *J. Am. Chem. Soc.* **1988**, *110*, 5586–5588.
25.   Sessler, J. L.; Johnson, M. R.; Lynch, V.; Murai, T. *J. Coord. Chem.* **1988**, *18*, 99–104.
26.   Sessler, J. L.; Murai, T.; Lynch, V. *Inorg. Chem.* **1989**, *28*, 1333–1341.
27.   Sessler, J. L.; Murai, T.; Hemmi, G. *Inorg. Chem.* **1989**, *28*, 3390–3393.
28.   Sessler, J. L.; Mody, T. D.; Hemmi, G. W.; Lynch, V. *Inorg. Chem.* **1993**, *32*, 3175–3187.
29.   Sessler, J. L.; Hemmi, G.; Mody, T.; Murai, T.; Burrell, A.; Young, S. W. *Acc. Chem. Res.* **1994**, *27*, 43–50.
30.   Maiya, B. G.; Mallouk, T. E.; Hemmi, G.; Sessler, J. L. *Inorg. Chem.* **1990**, *29*, 3738–3745.
31.   Young, S. W.; Woodburn, K. W.; Wright, M.; Mody, T. D.; Fan, Q.; Sessler, J. L.; Dow, W. C.; Miller, R. A. *Photochem. Photobiol.* **1996**, *63*, 892–897.

32.   Sessler, J. L.; Hemmi, G.; Maiya, B. G.; Harriman, A.; Judy, M. L.; Boriak, R.; Matthews, J. L.; Ehrenberg, B.; Malik, Z.; Nitzan, Y.; Rück, A. *SPIE Proc. Soc. Opt. Eng.* **1991**, *1426*, 318–329.

33.   Maiya, B. G.; Harriman, A.; Sessler, J. L.; Hemmi, G.; Murai, T.; Mallouk, T. E. *J. Phys. Chem.* **1989**, *93*, 8111–8115.

34.   Ehrenberg, B.; Malik, Z.; Nitzan, Y.; Ladan, H.; Johnson, F. M.; Hemmi, G.; Sessler, J. L. *Lasers in Med. Sci.* **1993**, *8*, 197–203.

35.   Sessler, J. L.; Mody, T. D.; Hemmi, G. W.; Lynch, V.; Young, S.; Miller, R. A. *J. Am. Chem. Soc.* **1993**, *115*, 10368–10369.

36.   Magda, D.; Miller, R. A.; Sessler, J. L.; Iverson, B. L. *J. Am. Chem. Soc.* **1994**, *116*, 7439–7440.

37.   Iverson, B. L.; Shreder, K.; Král, V.; Smith, D. A.; Smith, J.; Sessler, J. L. *Pure Appl. Chem.* **1994**, *66*, 845–850.

38.   Magda, D.; Wright, M.; Miller, R. A.; Sessler, J. L.; Sansom, P. I. *J. Am. Chem. Soc.* **1995**, *117*, 3629–3630.

39.   Sessler, J. L.; Král, V.; Hoehner, M.; Chin, K. O. A.; Dávila, R. M. *Pure Appl. Chem.* **1996**, *68*, 1291–1295.

40.   Geraldes, C. F. G. C.; Sherry, A. D.; Vallet, P.; Maton, F.; Muller, R. N.; Mody, T. D.; Hemmi, G.; Sessler, J. L. *J. Magn. Reson. Imaging* **1995**, *5*, 725–729.

41.   Young, S. W.; Qing, F.; Harriman, A.; Sessler, J. L.; Dow, W. C.; Mody, T. D.; Hemmi, G. W.; Hao, Y.; Miller, R. A. *Proc. Natl. Acad. Sci. USA* **1996**, *93*, 6610–6615.

42.   Sessler, J. L.; Dow, W. C.; O'Connor, D.; Harriman, A.; Hemmi, G.; Mody, T. D.; Miller, R. A.; Qing, F.; Springs, S.; Woodburn, K.; Young, S. W. *J. Alloys Comp.* **1996**, in press.

43.   Woodburn, K. W.; Young, S. W.; Fan, Q.; Kessel, D.; Miller, R. A. *Proc. SPIE Int. Opt. Eng.* **1996**, *2675*, 172–178.

44.   Miller, J. R.; Dorough, G. D. *J. Am. Chem. Soc.* **1952**, *74*, 3977–3981.

45.   Kirksey, C. H.; Hambright, P. *Inorg. Chem.* **1970**, *9*, 958–960.

46.   Rodesiler, P. F.; Griffith, E. H.; Ellis, P. D.; Amma, E. L. *J. Chem. Soc. Chem. Commun.* **1980**, 492–493.

47.   Hazell, A. *Acta Crystallogr. Sect. C. Cryst. Struct. Commun.* **1986**, *C42*, 296–299.

48.   Lisowski, J.; Sessler, J. L.; Lynch, V.; Mody, T. D. *J. Am. Chem. Soc.* **1995**, *117*, 2273–2285.

49.   Buchler, J. W.; De Cian, A.; Fischer, J.; Kihn-Botulinski, M.; Paulus, J.; Weiss, R. *J. Am. Chem. Soc.* **1986**, *108*, 3652–3659.

50.   Schaverien, C. J.; Orpen, G. *Inorg. Chem.* **1991**, *30*, 4968–4978.

51.   Buchler, J. W.; Löffler, J.; Wicholas, M. *Inorg. Chem.* **1992**, *31*, 524–526.

52.   Shannon, R. D. *Acta Crystallogr.* **1976**, *A32*, 751–767.

53.   Wan, S.; Parrish, J. A.; Anderson, R. R.; Madden, M. *Photochem. Photobiol.* **1981**, *34*, 679–681.

54.   Harriman, A.; Maiya, B. G.; Murai, T.; Hemmi, G.; Sessler, J. L.; Mallouk, T. E. *J. Chem. Soc. Chem. Commun.* **1989**, 314–316.

55. Sessler, J. L.; Mody, T. D.; Ford, D. A.; Lynch, V. *Angew. Chem., Int. Ed. Eng.* **1992**, *31*, 452–455.

56. Hoehner, M. C., Ph.D. Dissertation, University of Texas at Austin, Austin, TX, USA, **1996**.

57. Day, V. W.; Marks, T. J.; Wachter, W. A. *J. Am. Chem. Soc.* **1975**, *97*, 4519–4527.

58. Marks, T. J.; Stojakovic, D. R. *J. Chem. Soc., Chem. Commun.* **1975**, 28–29.

59. Marks, T. J.; Stojakovic, D. R. *J. Am. Chem. Soc.* **1978**, *100*, 1695–1705.

60. Cuellar, E. A.; Marks, T. J. *Inorg. Synth.* **1980**, *20*, 97–100.

61. Cuellar, E. A.; Marks, T. J. *Inorg. Chem.* **1981**, *20*, 3766–3770.

62. Bloor, J. E.; Schlabitz, J.; Walden, C. C.; Demerdache, A. *Can. J. Chem.* **1964**, *42*, 2201–2208.

63. Gossauer, A. *Chimia* **1984**, *38*, 45–47.

64. Burrell, A. K.; Hemmi, G.; Lynch, V.; Sessler, J. L. *J. Am. Chem. Soc.* **1991**, *113*, 4690–4692.

65. Burrell, A. K.; Cyr, M. J.; Lynch, V.; Sessler, J. L. *J. Chem. Soc. Chem. Commun.* **1991**, 1710–1713.

66. Sessler, J. L.; Weghorn, S. J.; Lynch, V.; Johnson, M. R. *Angew. Chem. Int. Ed. Eng.* **1994**, *33*, 1509–1512.

67. Dolphin, D.; Rettig, S. J.; Tang, H.; Wijesekera, T.; Xie, L. Y. *J. Am. Chem. Soc.* **1993**, *115*, 9301–9302.

68. Xie, L. Y.; Dolphin, D. *J. Chem. Soc. Chem. Commun.* **1994**, 1475–1476.

69. Boyle, R. W.; Xie, L. Y.; Dolphin, D. *Tetrahedron Lett.* **1994**, *35*, 5377–5380.

70. Xie, L. Y.; Dolphin, D. *Can. J. Chem.* **1995**, *73*, 2148–2152.

71. Xie, L. Y.; Boyle, R. W.; Dolphin, D. *J. Am. Chem. Soc.* **1996**, *118*, 4853–4859.

72. Wessel, T.; Franck, B.; Möller, M.; Rodewald, U.; Läge, M. *Angew. Chem. Int. Ed. Eng.* **1993**, *32*, 1148–1151.

73. Sessler, J. L.; Dávila, R. M.; Král, V. *Tetrahedron Lett.* **1996**, *37*, 6469–6472.

74. Sessler, J. L.; Burrell, A. K.; Furuta, H.; Hemmi, G. W.; Iverson, B. L.; Král, V.; Magda, D. J.; Mody, T. D.; Shreder, K.; Smith, D.; Weghorn, S. J. In *Supramolecular Chemistry, NATO ASI Series*; Fabbrizzi, L. and Poggi, A., Eds; Kluwer: Amsterdam, 1994, Series C, Vol. 448, pp. 391–408.

75. Král, V.; Furuta, H.; Shreder, K.; Lynch, V.; Sessler, J. L. *J. Am. Chem. Soc.* **1996**, *118*, 1595–1607.

76. Gingl, F.; Strähle, J. *Acta Crystallogr.* **1990**, *C46*, 1841–1843.

77. Baumann, F.; Bienert, B.; Rösch, G.; Vollmann, H.; Wolf, W. *Angew. Chem.* **1956**, *68*, 133–150.

78. Johnson, M. R.; Sleboduick, C.; Ibers, J. A. *J. Porph. Phthalocy.* **1997**, *1*, 87–92.

10   Introduction

Like the porphyrins, expanded, contracted, and isomeric porphyrins are of interest with regard to a wide range of potential applications. These run the gamut from bioinorganic modeling and pharmaceutical development to sensor applications and materials research. In this chapter, five applications of polypyrrole porphyrin analogs have been chosen for highlight. Two of these, namely magnetic resonance imaging (MRI) contrast agent development and RNA hydrolysis, are predicated on the fact that certain expanded porphyrins coordinate trivalent lanthanide cations in a stable 1:1 way that is reminiscent of how the better-studied porphyrins chelate biologically important cations of the first transition series. These two applications, discussed in the first and fourth sections of this chapter, thus have antecedents, as least intellectually, in the chemistry of the porphyrins.

The same is true for photodynamic therapy (PDT), an application also chosen for highlight in this chapter. Here, as discussed in Section 10.2, the special optical properties of certain isomeric and expanded porphyrins, deriving from their electronic differences *vis à vis* the porphyrins, makes these macrocycles attractive as PDT photosensitizers. The special electronic properties of one expanded porphyrin, viz. texaphyrin, are also allowing it to be developed as a sensitizer for X-ray radiation therapy (XRT). As detailed further in Section 10.3, such an application, which is nearly without precedent in the porphyrin literature, could be wide ranging in terms of its significance. The lack of porphyrin-based precedent is perhaps even more striking in the case of the last application chosen for highlight, anion binding (Section 10.5). This application, or more precisely, this potentially useful phenomenon, was considered to be orthogonal to the porphyrin field until it was discovered to be directly relevant to the chemistry of protonated sapphyrins. Interestingly, it now defines an important research direction in the porphyrin analog area.

10.1   Magnetic Resonance Imaging

10.1.1   Background: Utility of Contrast Agents

Magnetic resonance imaging, or MRI, is now firmly entrenched as a clinical tool of considerable importance. It is a non-invasive, non-radiative method that continues to see increasing application in the diagnosis and staging of many diseases.[1–6] Unfortunately, however, the difference in MRI signal for diseased vs. normal tissues is often small. This has limited the utility of MRI in certain clinical situations, including many associated with cancer and cardiovascular diagnosis. To address this issue, considerable effort is currently being devoted to the preparation of MRI contrast reagents. Here, certain expanded porphyrins, especially the gadolinium(III) texaphyrins, appear to have an important role to play.

The physical basis of current MRI methods has its origin in the fact that, in a strong magnetic field, the nuclear spins of water protons in different tissues relax back to equilibrium at different rates, when subject to perturbation from the resting Boltzmann distribution by the application of a short radio frequency (rf) pulse.[5,7–9] For the most common type of spin-echo imaging, return to equilibrium takes place in accord with equation 1 and is governed by two time constants T_1 and T_2, the longitudinal and transverse relaxation times, respectively.

$$SI = [H]H(v)\{\exp(-T_E/T_2)\}\{1 - \exp(-T_R/T_1)\} \tag{1}$$

Here, SI represents the signal intensity, $[H]$ is the concentration of water protons in some arbitrary volume element (termed a voxel), $H(v)$ a motion factor corresponding to motion (if any) in and out of this volume element, and T_E and T_R are the echo-delay time and the pulse-repetition times, respectively. The various pulse sequences associated with obtaining an MRI image thus correspond to choosing T_E and T_R by setting the times associated with (and between) the excitation and interrogation rf pulses, and determining SI, which, as illustrated above, is a function of the particular T_1 and T_2 values in force. Both T_1 and T_2 are a function of the local (bulk) magnetic environment and, as such, are a function of the particular tissue in which the water proton is situated. Differences in these values (and hence SI) thus allow for image reconstruction. On the other hand, only when these local, tissue-dependent, relaxation differences are large can tissue differentiation be effected.[4–6,8,9]

In practice for biological systems, T_2 values are very short (and T_E and T_R are chosen to accentuate this situation). Thus, it is differences in the longitudinal time constant (T_1) which dominate relaxation effects and relative signal intensity: decreases in T_1 correspond to increasing signal intensity. Any factors, therefore, which will serve to decrease T_1 selectively for a particular tissue or organ, will thus lead to increased intensity for that area and better contrast (signal to noise) relative to surrounding tissue background.[8,9] This is where paramagnetic MRI contrast agents come into play.[4–6,10]

It has been known since the earliest days of magnetic resonance spectroscopy that paramagnetic compounds, containing one or more unpaired spins, enhance the relaxation rates for the water protons in which they are dissolved.[11] The extent of this enhancement, termed relaxivity,[4–6,8,9] is highly dependent on the magnitude of the dipole–dipole interactions between the electron spin on the paramagnetic metal complex and the proton spin on the water molecule in question. These interactions are often quite complex. They can be treated on a formal (theoretical) level by the Solomon–Bloembergen equations.[12,13] On a strictly practical level, however, the development of an effective MRI contrast agent requires finding a highly paramagnetic species, which (1) binds water molecules directly in a so-called inner sphere fashion (such that the effects on the water spins are magnified); and (2) allows fast aquo ligand exchange (so that overall relaxivity is optimized).[4-6] Such a species must, of course, be biocompatible and, ideally, show a high degree of tissue-targeting selectivity.

Currently, trivalent gadolinium is the paramagnetic cation of choice for MRI contrast agents. This cation contains seven unpaired electrons and is the most paramagnetic single cation known. Exchange of water about gadolinium(III) centers is also generally quick. A priori, it thus engenders a high relaxivity.[4-6,14] Unfortunately, however, the aquo ion of Gd^{3+} is too toxic to allow its use *in vivo*. As a consequence, considerable effort has been devoted in recent years to the preparation of Gd(III) complexes that might be less toxic and hence suitable for use in clinical situations. Indeed, several such complexes have now been approved for human use in the United States, including the bis-*N*-methylglucamine salt of Gd(III) diethylenetriaminepentaacetic acid (DTPA) (Magnevist[TM]), the bis-*N*-methylamide of Gd(III) DTPA (Omniscan[TM]), and the Gd(III) complex of the 10-(2-hydroxypropyl) derivative of 1,4,7,10-tetraazacyclododecane-*N*, *N'*,*N''*-triacetic acid (DO3A) (Prohance[TM]). It is important to appreciate, however, that all U.S. Food and Drug Administration (FDA)-approved agents rely on the same general metal-binding strategy, namely polydentate oxyanion-based chelation. This generates agents that are of high water solubility and of low relaxivity. This, in turn, has limited their use in such possibly beneficial MRI contrast applications as atherosclerotic plaque enhancement or small tumor detection that necessarily involve biological loci of a rather hydrophobic nature.

One feature of porphyrins, long appreciated in the area of PDT photosensitizer development (*vide infra*), is that certain ones, such as hematoporphyrin, do in fact localize selectively in neoplastic tissues including sarcomas and carcinomas.[15-19] While the reasons for this selectivity remain obscure, this fact has made porphyrins logical choices for trying to develop tumor-selective MRI contrast agents. Indeed, promising work along these lines is currently being carried out with Mn(III) systems.[20-25] Unfortunately, the porphyrins fail to form stable complexes with gadolinium(III),[26-28] a cation that is more paramagnetic than Mn(III) and hence intrinsically more desirable for MRI.[4]

The above realizations led to the consideration that a larger "porphyrin-like" macrocycle, capable of coordinating gadolinium(III) in a stable, non-labile manner, could prove useful as a cancer-indicating MRI contrast agent, provided it retained the tumor-localizing characteristics of the porphyrins and could be rendered biocompatible (i.e., water-soluble and non-toxic). In fact, it was this basic analog idea that provided much of the early impetus to develop the texaphyrins as lanthanide(III)-ligating agents.[29-34] These particular expanded porphyrins do coordinate gadolinium(III) with seemingly unsurpassed kinetic stability (see Chapter 9)[32] and, as detailed below, show promise as tumor-localizing MRI-detectable radiation sensitizers. Interestingly, at present, the texaphyrins appear to be the only expanded porphyrins being studied in either of these two applications (i.e., MRI and radiation sensitization).

10.1.2   *Gadolinium(III) texaphyrins*

There are several chemical properties that make the gadolinium(III) texaphyrins attractive as potential MRI agents. First, as noted earlier (see above and Chapter 9), they do form stable, 1:1 non-labile complexes with this highly paramagnetic lanthanide(III) cation (as they do with other Ln(III) ions).[32] Second, as judged from single X-ray diffraction analysis (Figure 10.1.1), the texaphyrins coordinate this metal center in such a way (meridianal, nearly in plane) that fast and efficient exchange of 4–5 water molecules might be expected at the apical ligation sites.[32] In accord with this expectation, a first-generation water-soluble texaphyrin **10.1** (Figure 10.1.2) was found to show an exceptionally high relaxivity in aqueous medium (R_1 = 16.9 ± 1.5 mM^{-1} s^{-1}; 50 MHz; room temperature).[32,35,36] At the time it was obtained, this important result was considered to augur well for the *in vivo* usefulness of **10.1**; it meant that high efficacy could be expected at doses considerably lower than those required for more conventional, carboxylate-derived gadolinium(III) chelates.

The *in vivo* efficacy of the first generation system **10.1** was demonstrated by Young, *et al.* using a number of tumor model systems, including a V2 carcinoma implanted into the thigh of a New Zealand white rabbit.[35,37,38] Specifically, these workers found that good tumor-depicting MRI enhancement could be achieved when complex **10.1**, in its bis-acetate form, was administered intravenously at a dose level of 5 µmol kg^{-1} (Figure 10.1.3). Under these conditions, little contrast enhancement of the surrounding muscle tissues was seen. Further, no signs of acute toxicity were observed in the context of these studies, and indeed, no serious toxicity was observed in healthy rats given thrice weekly 20 µmol kg^{-1} doses of **10.1** for a period of 21 days. Using this same agent, Young and coworkers were also able to demonstrate *ex vivo* the MRI enhancement of atheromatous plaques in human aorta[37] as well as acute cerebral ischemia in a rabbit model.[39] Thus, these initial studies served to show that the gadolinium(III) texaphyrin approach to MRI contrast enhancement could prove potentially useful in imaging applications involving a wide range of important diseases.

Support for the above conclusions is coming from studies involving the second-generation gadolinium(III) texaphyrin complex **10.2** (Figure 10.1.2). This product, to which the code name PCI-0120 has been assigned (for the bis-acetate form), is being developed jointly by researchers at Pharmacyclics Inc. in Sunnyvale, California and The University of Texas at Austin as an MRI-detectable radiation sensitizer (see Section 10.3 below).[40] It is currently in Phase Ib/II human clinical trials for such applications.[41] As the result of these studies[41] and predicative preclinical ones,[40,42] it is now clear that this new gadolinium(III) texaphyrin-based agent not only facilitates the MRI-based detection of neoplastic tissues but does so without engendering a toxic response except when administered at dose levels greatly exceeding those needed for effective imaging.[41]

Figure 10.1.1   Single Crystal X-Ray Diffraction Structure of Complex **10.1** (Bis-nitrate Form) Showing the Planar Nature of the Basic Monoanionic Texaphyrin Ligand and the Four Putative Inner Sphere Coordination Sites for Water (Occupied by Two Apical Methanol Molecules and a Bidentate Nitrate Anion in this Structure).

This figure was generated using information down-loaded from the Cambridge Crystallographic Data Centre and corresponds to a structure originally reported in reference 32. Atom labeling scheme: gadolinium: ◑; carbon: ○; nitrogen: ●; oxygen: ◒. Hydrogen atoms have been omitted for clarity

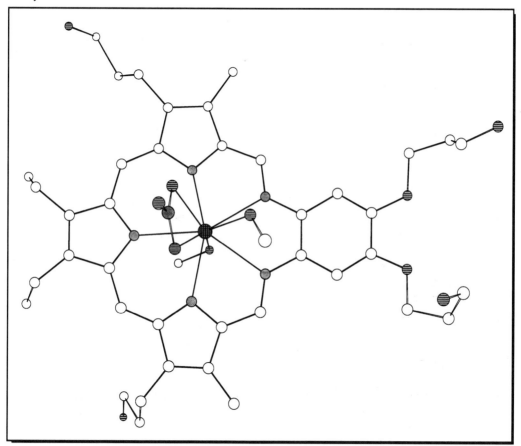

10.2   Photodynamic Therapy and Photodynamic Viral Inactivation

10.2.1   Introduction: Need for Photosensitizers

Photodynamic therapy is one of the more promising new modalities currently being explored for use in the control and treatment of tumors.[43-51] This technique is based on the use of a photosensitizing dye, such as a porphyrin,[†] which localizes at, or near, the tumor site, and when irradiated in the presence of oxygen serves to produce cytotoxic materials, such as singlet oxygen ($O_2(^1\Delta_g)$), from otherwise benign precursors (e.g., ($O_2(^3\Sigma_g^-)$)). Much of the current excitement associated with PDT derives from this property, since a level of control and selectivity may be attained, which is not otherwise possible. Also, PDT may be used for the treatment of solid tumors, such as lung cancer (177 000 cases/year in the USA) for which good curative strategies are often lacking.[52]

Figure 10.1.2   Structures **10.1–10.4**

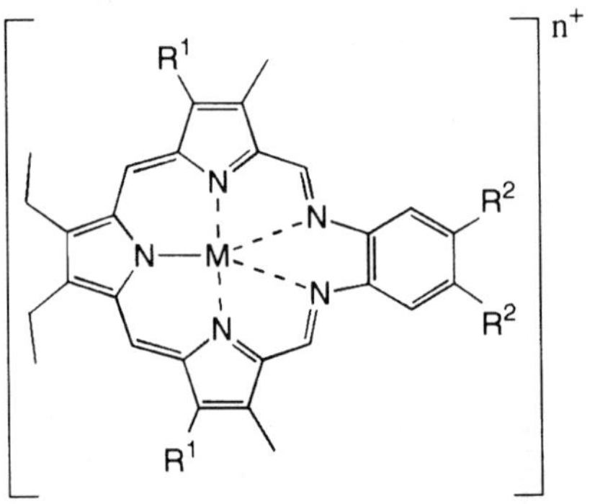

10.1. $R^1 = (CH_2)_3OH$, $R^2 = O(CH_2)_3OH$, M = Gd, n = 2
10.2. $R^1 = (CH_2)_3OH$, $R^2 = O(CH_2CH_2O)_2CH_3$, M = Gd, n = 2
10.3. $R^1 = Et$, $R^2 = H$, M = Cd, n = 1
10.4. $R^1 = (CH_2)_3OH$, $R^2 = O(CH_2CH_2O)_2CH_3$, M = Lu, n = 2

    Included within the overall aegis of photodynamic therapy is photodynamic viral inactivation[53–58] (abbreviated PDI[59] or, alternatively, PDV[60]). Here, the objective is to use a combination of photosensitizing drug and light activation to eradicate photodynamically a virus or other unwanted pathogen from a given targeted medium. As originally conceived by Matthews and coworkers,[53] PDI would see its greatest application in blood-banking operations, where, by passing a mixture of blood and photosensitizer in front of a light source, it could be used to remove vestigial impurities, such as AIDS or hepatitis viruses, not removed by other means. Subsequently, this idea has been generalized to include conceptually the possible *in vitro* sterilization of, e.g., biotechnology products as well as, in the extreme, the *ex vivo* treatment of certain blood-borne diseases. On a more practical level, the demonstration of a PDI effect *in vitro* might be expected to provide an important first "hint" that a given dye could in fact be a good candidate for use as a PDT photosensitizer *in vivo*. Indeed, such *in vitro*, PDI-like screenings are often carried out in the context of the more mainstream, cancer-targeted parts of the PDT field.[43–51]

[†]For early literature describing the ability of porphyrins to localize selectively in cancerous tissues, see references 15–19.

Figure 10.1.3   A Series of Magnetic Resonance Imaging Scans Obtained through the Left Thigh of a New Zealand White Rabbit Bearing a V2 Carcinoma Before (Panel A), Immediately After (Panel B), and 3½ Hours (Panel C) After the Administration of 5 µmol kg^{-1} of PCI-0101 (Complex **10.1**; Acetate Counter Anions).

Noteworthy is the persistent, marked increase in contrast enhancement of the V2 carcinoma compared with surrounding muscle obtained under these conditions (1.5 T; T_1-weighted pulsing sequence; 300/15 repetition time/echo time). This figure was reproduced with permission from reference 37

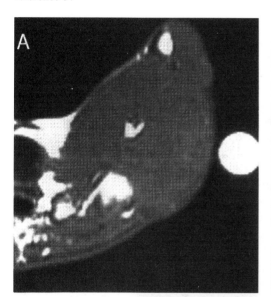
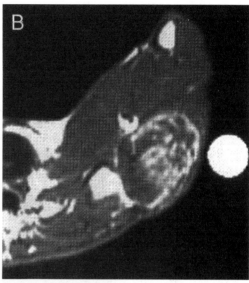
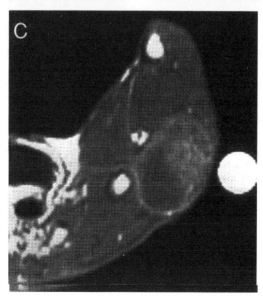

Historically, finding a good PDT photosensitizer has not been easy. Indeed, at present, there is only one PDT-photosensitizing agent approved for clinical use in the USA. This substance, marketed under the trade name Photofrin®, is produced by treating hematoporphyrin dihydrochloride with acetic acid–sulfuric acid followed by dilute base,[61] and consists of the fractions rich in the oligomeric species that have the best tumor-localizing ability.[62] While this tumor-localizing ability has led to important clinical advances, the fact that Photofrin® is a mixture is seen as a drawback. Further, Photofrin® is known to engender cutaneous photosensitivity in patients for weeks after PDT treatment is finished,[50,63,64] perhaps as the result of containing hard-to-catabolize and/or hard-to-excrete oligomers within its overall mixed constitution. Finally, Photofrin®, with an absorption maximum of 630 nm, absorbs but poorly in the red part of the electronic spectrum (i.e., $\lambda \geqslant 700$ nm) where blood and other bodily tissues are most transparent.[65] Taken together, these deficiencies have prompted efforts to develop new, so-called "second-generation" PDT photosensitizers[66] that might better fulfill what are considered to be an idealized set of desiderata (see Figure 10.2.1).[59,60,66] While many leads are currently being pursued,[67–87‡] several expanded porphyrins, including texaphyrins (e.g., **10.3** and **10.4**; Figure 10.1.2), cumulene porphycenes (e.g., **10.5**; Figure 10.2.2), sapphyrins (e.g., **10.6**; Figure 10.2.3), and vinylogous porphyrins (e.g., **10.7** and **10.8**; Figure 10.2.4), look particularly attractive in this regard. The simple porphyrin isomer, porphycene (e.g., **10.9–10.11**; Figure 10.2.5) and related systems (e.g., **10.12–10.14**; Figures 10.2.5 and 10.2.6) also appear to be of interest. Of these expanded porphyrin-based approaches, the lutetium(III) texaphyrin system **10.4** (Figure 10.1.2) appears to be the most thoroughly studied (*vide infra*).

The main reason isomeric and expanded porphyrin systems are attractive as potential PDT photosensitizers is that they generally absorb at wavelengths at or above 700 nm rather than at wavelengths of $\leqslant 630$ nm as is true for Photofrin®. This ability to absorb in the near-infrared is important. It means more of the activating light, administered at a wavelength of $\geqslant 700$ nm, is available for singlet oxygen production with commensurately less being lost to tissue absorption and scatter.[65,88] In fact, as can be appreciated from an inspection of Figure 10.2.7, a switch in activation wavelength from $\leqslant 630$ nm to $\geqslant 700$ nm is expected to increase the viable depth of light penetration by 2–6-fold.[65,88] To the extent this is true, it would make PDT more efficacious for treating both deep-seated tumors and/or highly pigmented ones such as melanoma. This assumes, of course, that the other key criteria of Figure 10.2.1 are appropriately met.

10.2.2   *Cadmium(II) and Lutetium(III) Texaphyrins*

Harriman, Sessler, and Mallouk were, apparently, the first to suggest that expanded porphyrins could be useful as PDT photosensitizers.[30,89‖] This proposal

---

‡Reference 87 gives an example of a promising expanded porphyrin "lead" whose current level of development (i.e., *in vitro* and/or *in vivo* analysis) does not yet warrant specific mention in this chapter.
‖Interestingly, nearly coincident with Sessler and coworkers, Franck proposed that certain vinylogous porphyrins might also be useful as PDT sensitizers (see footnote 3 in reference 90).

was made based on a consideration of the photophysical properties of the first-generation cadmium(II) texaphyrin **10.3**.[89] This species, in its mono-nitrate form, was found to absorb light at a physiologically desirable wavelength (lowest energy $\lambda_{max} = 759$) with good efficiency (log $\varepsilon = 4.59$) in MeOH.[89,91] It was also found to produce singlet oxygen in good quantum yield ($\Phi_\Delta = 0.69 \pm 0.09$ in MeOH).[89] Subsequent *in vitro* studies confirmed that this substance could be used to effect the efficient photodynamic eradication of K562 human leukemia cells[91,92] and both Gram-positive *Staphylococcus aureus*[91,93,94] and Gram-negative *Escherichia coli* bacteria, provided the latter was treated concurrently with polymyxin nonapeptide.[94] However, issues involving administration and out-of-body clearance (**10.3** is not appreciably water-soluble), as well as residual concerns about cadmium toxicity, served to limit follow-up *in vivo* studies of this material.[95]

Figure 10.2.1   Idealized Criteria for PDT or PDI Photosensitizers

1.	Readily available
2.	Chemical pure and stable
3.	Easily subject to synthetic modification
4.	Soluble in water such that administration can be effected without resorting to the use of an ancillary carrier.
5.	Intrinsically non-toxic in the absence of light
6.	Taken up or retained selectively in diseased tissues or other targeted biological loci.
7.	Cleared quickly after administration
8.	Free of lingering photo-toxicity effects, cutaneous or otherwise
9.	Endowed with one or more strong absorption bands in the ≥ 700 nm spectra region
10.	Capable of acting as an efficient singlet oxygen photosensitizer when irradiated with near-infrared light.

In view of the above concerns, recent work in the texaphyrin-based PDT area have focused on the water-solubilized lutetium(III) derivative **10.4** (lowest energy $\lambda_{max} = 732$ nm).[42,96] This compound, code named PCI-0123 in its bis-acetate form, is being developed to promising effect by Pharmacyclics Inc. In fact, a Phase I clinical study involving this substance is now on-going. While the so-called maximum tolerated dose or MTD has not yet been reached, it is already clear from this study that PCI-0123, when irradiated at 732 nm, 75 mW cm^{-2}, and 150 J cm^{-2}, is an effective PDT photosensitizer active against a variety of cancers, including pigmented melanoma (including patients with failed isolated limb perfusion), breast cancer, AIDS-related Kaposi's sarcoma, and basal cell carcinoma.[97]

Figure 10.2.2   Structure **10.5**

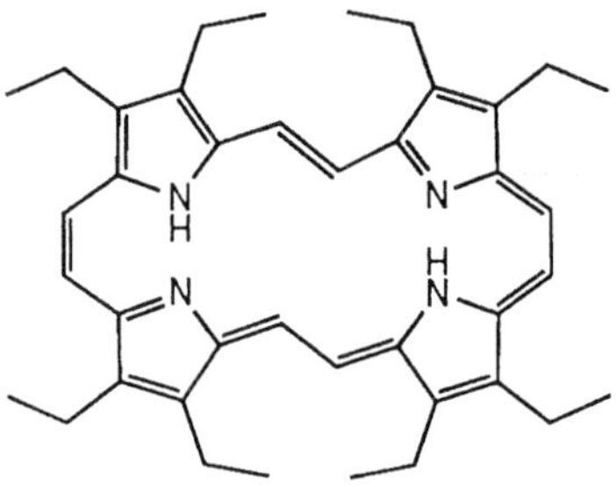

10.5

Figure 10.2.3   Structure **10.6**

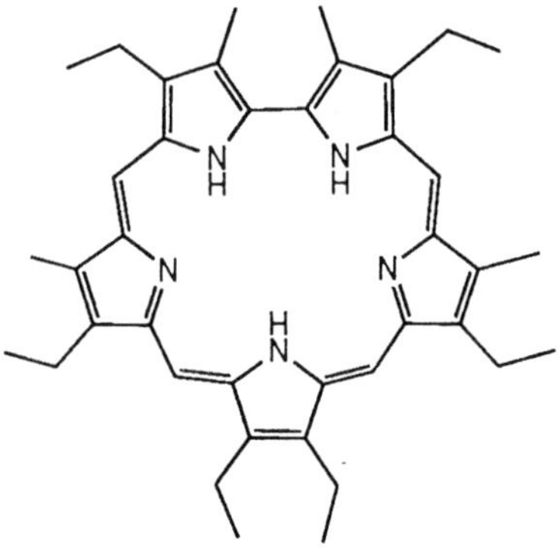

10.6

Not surprisingly, the above Phase I human clinical study is supported by numerous pre-clinical studies. These have served to show, for instance, that complete cures of implanted fast-growing spontaneous mouse mammary tumors (SMT-F) in female DBA/2N mice can be achieved using a proper combination of light fluence, PCI-0123 drug dose, and post-administration irradiation time (see Figure 10.2.8).[42]

Such studies have also shown that PCI-0123 is photo-active against other murine cancer models including the EMT6 mammary sarcoma and the B16 pigmented melanoma.[96,98] Finally, in recent preclinical work, it has been demonstrated that PCI-0123 (complex **10.4**, bis-acetate form) may be used to effect the irradiation-based eradication of atheromatous plaque in diet-induced hypercholesterolemic New Zealand white rabbits.[98,99]

Figure 10.2.4   Structures **10.7** and **10.8**

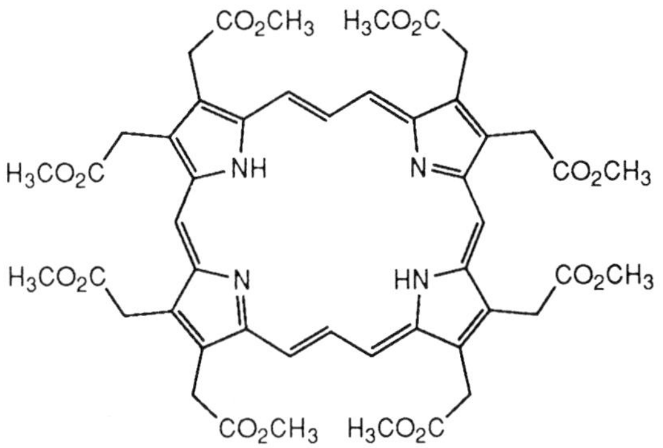

10.7 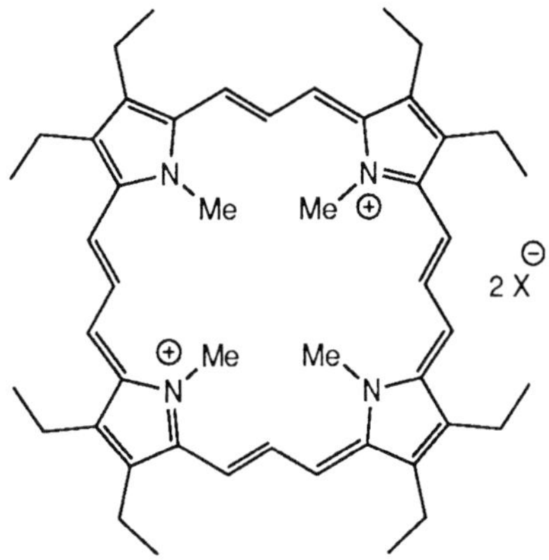   10.8

Figure 10.2.5   Structures **10.9–10.12**

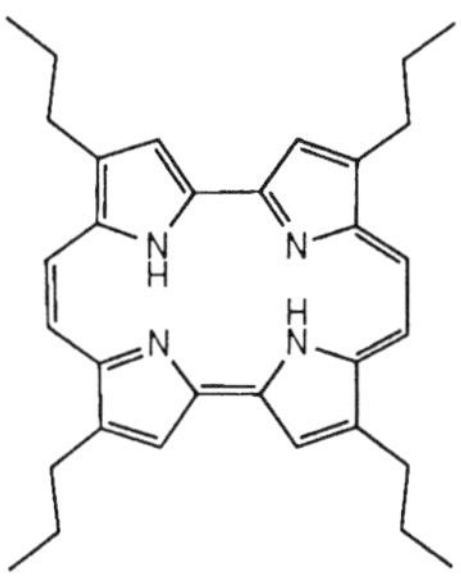

10.9

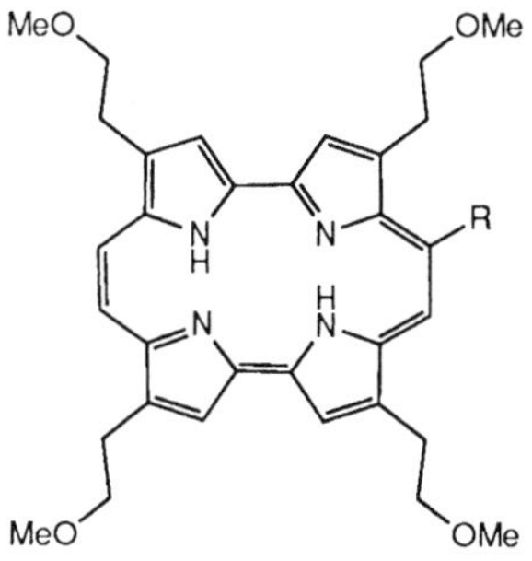

10.10. R = H
10.11. R = OCOMe

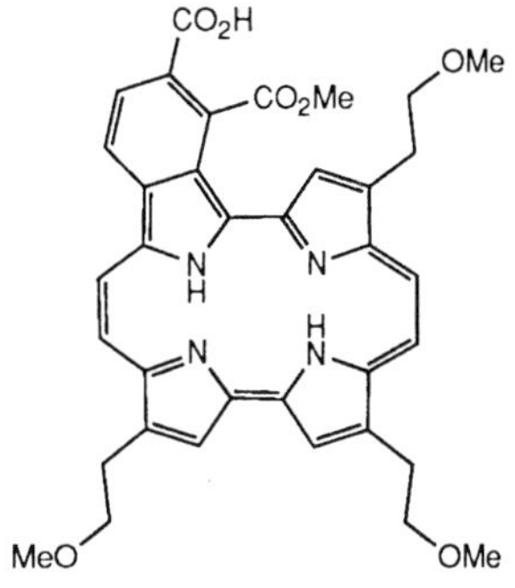

10.12

Figure 10.2.6   Structures **10.13** and **10.14**

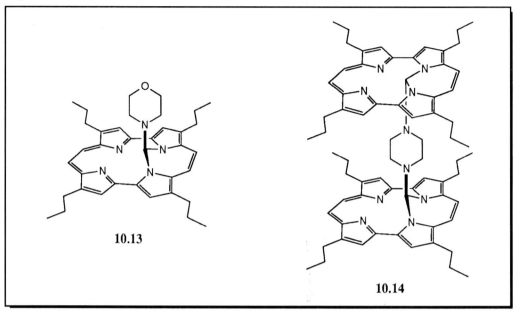

Figure 10.2.7   Spectral Transmittance through Human Chest Wall with Rib Thickness of 22 mm. Redrawn with permission from reference 65.

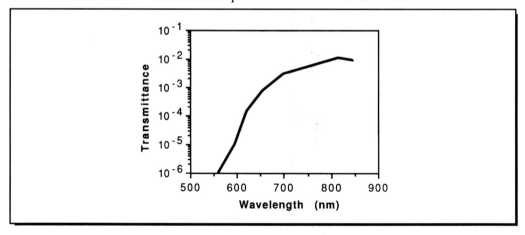

10.2.3   *Porphycenes and Expanded Porphycenes*

Much less is currently known about the actual *in vivo* photosensitizing capabilities of other expanded porphyrins. In fact, only the expanded porphycene **10.5**[100–102] (lowest energy λ_{max} = 795; log ε = 4.78 in benzene[100,101]) appears to have been studied in this regard. It was found by G. Jori and coworkers to be an effective PDT agent when incorporated into unilamellar liposomes, administered intravenously at the 0.35 mg kg^{-1} body weight dose level to mice bearing trans-planted tumors, and activated by subsequent irradiation (720–800 nm; 180 mW cm^{-2}; 450 J cm^{-2}). While tumor necrosis was often substantial and long-term survival was enhanced, complete cures are not in general recorded (e.g., 3/ 10 in the case of pigmented melanoma model).[102] Nonetheless, the encouraging results obtained with this sensitizer are important if only because they match those found[100–102] with the smaller, isomeric porphyrin system **10.9** (tetrapropylporphycene). This latter substance, although lacking a truly red-shifted absorption band (lowest energy λ_{max} = 637 nm; log ε = 4.70 in toluene), shows favorable photophysical properties,[102,103] and is currently being developed as a PDT agent by Cytopharm Inc. of Menlo Park, California.[104] It is also being tested as a photosensitizer for use in light-based dermatological applications by Glaxo-Welcome.[105] Analogous systems, including functionalized monomers (e.g., **10.10**,[106,107] **10.11**,[108] and **10.12**[106,107]) and interesting N-bridged systems such as **10.13** and **10.14**,[109–112] are also being developed; these too show promise for use in PDT.

Figure 10.2.8   Kaplan–Meier Survival Curves for SMT-F-bearing Mice Treated with 10 μmol kg^{-1} PCI-0123 (Complex **10.4**; Acetate Counter Anions).
The mice were irradiated 3 h (n = 9, open squares), 5 h (n = 16, s), 12 h (n = 8, l) and 24 h (n = 7, m) post-injection of PCI-0123 with 150 J cm^{-2} @ 150 mW cm^{-2} at 732 nm. A matched set of control animals received light irradiation alone (n = 10, n). Initial tumor volumes were 70 ± 35 mm^3. All animals at day 40 that were still in the study displayed no evidence of disease at the tumor site. This figure is redrawn with permission from reference 42

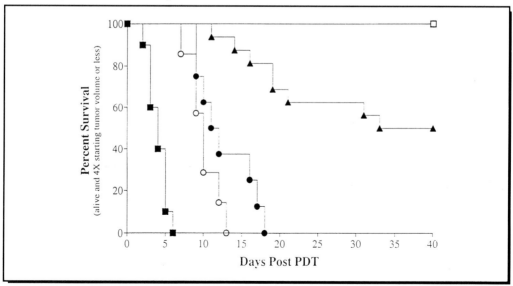

10.2.4   *Sapphyrins and Vinylogous Porphyrins*

While apparently not yet tested *in vivo*, both Sessler's sapphyrin **10.6**[59,113] and Franck's vinylogous porphyrins **10.7** and **10.8**[60,114] have been proposed as possible PDI sensitizers. In both cases this has been done in the context of collaborative working involving Dr. J. L. Matthews and coworkers at the Baylor Research Foundation in Dallas, Texas. From these studies it has emerged that sapphyrin (lowest energy $\lambda_{max} \approx 680$ nm) is a highly effective photosensitizer, allowing for the *in vitro* eradication of both Herpes simplex virus (HSV) and human immunodeficiency virus (HIV).[59,113] At a light fluence of 10 J cm^{-2} ($\lambda_{excit.}$ = 680 nm) and dye concentrations of 34 μmol and 16 μmol sapphyrin **10.6** was found to eradicate 99.999% and $\geqslant$99% of these two viruses, respectively. In the case of the vinylogous porphyrins **10.7** and **10.8**, the corresponding viral inhibition studies are apparently not yet complete. However, it is to be noted that one of these dyes, the bisvinylogous system **10.7**, was found, upon photolysis, to be very effective at preventing the *in vitro* growth of certain leukemic cell lines.[60] This augurs well for the eventual use of this compound and its analogs in both PDI and PDT applications.

10.3   X-ray Radiation Therapy Enhancement

10.3.1   *General Overview: Potential Benefit of Sensitizers*

Another potentially curative application of expanded porphyrins is to use them as sensitizers in X-ray tumor therapy (XRT). While so far only proposed in the context of the gadolinium(III) texaphyrin system **10.2**, such a use is one that carries within it the possible seeds of enormous societal benefit. This is because it would allow one of the most important of all cancer control strategies, namely radiation therapy, to be made more efficacious.

Every year, roughly 1 400 000 new cancer cases are recorded in the USA alone.[52] Of these, roughly one half undergo radiation therapy in one form or another.[115–117] Unfortunately, in all too many instances the hoped-for curative effects are not realized. Thus, there is a constant, on-going search for ways to improve the efficacy of this time-honored approach. One avenue being explored involves the use of radiosensitizers, compounds specifically designed to enhance the effects of localized, tumor-targeted radiation.[118–122] However, in spite of a great deal of effort and a number of promising leads,[123–134] not a single agent has been approved as a radiation sensitizer in the United States. The texaphyrins, by virtue of their special electronic character (*vide infra*), could serve to fill this gap.

10.3.2   *Gadolinium(III) Texaphyrins*

There are two critical features of the texaphyrins that led to a consideration that it could function as a radiation sensitizer. First, as detailed in Sections 10.1 and

10.2 above, they localize with high selectivity in cancerous tissues. Second, they contain a low-lying LUMO; in comparison to porphyrins and most other endogenous species, they are thus very easy to reduce ($E_{1/2} \approx 0.08$ vs. NHE; aqueous, pH 7).[40] Taken together, these two facts led to the proposal that the water-soluble, MRI-detectable gadolinium(III) texaphyrin agent PCI-0120 (complex **10.2**, bis-acetate form) would make a good radiation sensitizer.

The idea that an easy-to-reduce species could function as a radiation sensitizer is not new.[118–122] Indeed, there is considerable precedent for it both in the radiation sensitization literature *per se*[127–132] and in classic mechanistic explanations of radiation-induced cytotoxicity.[135] While the biological effects of radiation therapy are certainly complex, it is well established that exposure of bodily tissues to high-energy gamma- and X-ray radiation (i.e., of the type commonly used clinically[117]) causes the ejection of one or more high-energy electrons from water (the most prevalent available target). These electrons can exert a cytotoxic effect directly by, e.g., breaking the sugar phosphate bonds in DNA.[117] More frequently, however, they exert their effect indirectly *via* the follow-up formation of solvated electrons and hydroxyl radicals (equation 2).

$$H_2O \xrightarrow{X-ray} H_2O^+ + e_{aq}^- \xrightarrow{H_2O} H_3O^+ + HO^{\cdot} + e_{aq}^- \qquad (2)$$

Of these daughter products, it is the hydroxyl radical, $HO^{\cdot}$, that constitutes the dominant cytotoxin.[117,135] Hydroxyl radicals are highly reactive, oxidizing entities, that cause damage (chromosomal or otherwise) within a *ca.* ± 4 nm three-dimensional radius of where they are formed.[117,135] The hydrated electrons, on the other hand, are strongly reducing, rather mobile, and relatively long-lived ($\tau_{1/2} \sim 25$ ns).[117,135] Although these latter hydrated electrons can cause DNA damage themselves (in analogy to the direct, electron-mediated process described above), they are also capable of recombining with hydroxyl radicals.[117,135] Such an $e_{aq}^- + HO^{\cdot}$ deactivation process produces hydroxide anion (HO^-) and thus serves effectively to deactivate both cytotoxic daughter products, namely e_{aq}^- and $HO^{\cdot}$. In the presence of oxygen, on the other hand, the solvated electrons get trapped, at least initially, in the form of superoxide anion.[117,135] This trapping precludes recombinatorial deactivation of the hydroxyl radicals, leaving the latter free to carry out cell killing.

The above chemistry explains why hypoxia is a major contributor to the failure of radiation therapy. Hypoxic cells in solid tumors contain a lower oxygen tension (and correspondingly reduced number of e_{aq}^- traps) than their highly vascularized, rapidly growing "outside neighbors" and, indeed, have been observed to be 2.5–3 times more resistant to the effects of ionizing radiation.[136–138] So-called fractionation, wherein the total chosen radiation dose (generally on the order of 3–5 Gy) is administered in many small fractions helps; it allows healthy tissues time to undergo cell repopulation and permits reoxygenation of hypoxic regions originally distant from functional vasculature.[122,139] Unfortunately, even this approach, which is standard clinical practice, does not provide a complete, satisfactory solution to the problems of hypoxic-region cancer control. Hence the appeal of electron-deficient species, such as PCI-0120 (complex **10.2**, bis-acetate form). These, by virtue of "soaking up" the unwanted e_{aq}^- (i.e, by acting as "oxygen surrogates"), could serve to increase substantially the clinical benefit of XRT.[127–132]

Currently, PCI-0120 is being tested as radiation sensitizer in an on-going Phase Ib/II clinical study.[41] While good patient tolerance and MRI enhancement ability were noted in the context of the predicative Phase I study (*vide supra*), at present only anecdotal efficacy results are available from these Phase Ib/II XRT tests. Nonetheless, a considerable body of preclinical data exist that supports the suggestion that this agent will be a good radiation sensitizer when used clinically. First, it is known to react at near diffusion-controlled rates with hydrated electrons.[40] Second, it acts as an efficient *in vitro* radiation sensitizer, showing a sensitization enhancement ration (SER) of 1.92 against L1210 cancer cells at 2 Gy irradiation and at a texaphyrin concentration of 2.2×10^{-3} M.[40] Finally, under *in vivo* conditions, it was found to improve significantly the long-term survival of DBA/2N mice containing implanted SMT-F tumors under single-dose irradiation conditions (Figure 10.3.1) as well as BALB/c mice bearing EMT6 neoplasms treated in accord with a model fractionation protocol.[40]

While the above results are encouraging in the extreme, it is important to appreciate that they do not necessarily serve to confirm (or refute) the mechanistic analyses that led PCI-0120 (**10.2**; bis-acetate form) to be considered initially as a potential radiation sensitizer. Indeed, many other explanations for the observed sensitization effects, including ones involving inhibition of DNA repair or cell-cycle modification, can be conceived. Still, to the extent that the basic "tumor-localized electron capture" idea is valid, it would lead to the suggestion that other easy-to-reduce porphyrin analogs might be worth testing as possible radiation sensitizers. For the same reason, other (i.e., non-texaphyrin) imine-containing entities, either cyclic or non-cyclic, might also warrant consideration. In any event, the very fact that texaphyrin **10.2** appears to work as a radiation sensitizer (at least as judged from animal studies), serves to highlight what could prove to be an important new application for expanded, contracted, and isomeric porphyrins.

10.4   Antisense Applications: RNA Hydrolysis and DNA Photolysis

10.4.1   Introduction

An additional biomedical area in which the expanded porphyrins are seen as being potentially beneficial is in antisense drug design.[140–162] Antisense therapy, as originally conceived, involves the synthesis of relatively short (< 20 bases) synthetic "antisense" oligodeoxynucleotides (ODNs) that can bind to mRNA and inhibit competitively the process of RNA–protein translation. Subsequently, the antisense concept was generalized to include triple-helix DNA-binding systems designed to prevent initial DNA–RNA transcription.

Figure 10.3.1   Single-dose Radiation Sensitization Study Following i.v. Administration of PCI-0120 (Gadolinium(III) Complex **10.2**, Bis-Acetate Form). Twelve DBA/2N mice were injected intravenously via the tail vein with 10 μmol kg^{-1} of PCI-0120, 4–7 days after intramuscular injection of SMT-F tumor cells into the hind flank of the right rear leg. Two hours later, a single dose of 30 Gy of radiation was administered. A control group of 12 DBA/2N mice with implanted SMT-F tumors were treated with 30 Gy of radiation only. The results show that after 51 days all of the animals given the PCI-0120 as a putative radiation sensitizer are alive and without evidence of disease while 4 of the 12 irradiation-only animals died and, of the remaining 8 mice, only 6 are without evidence of disease. This figure was drawn using data provided by Pharmacyclics Inc. and highlights findings similar to those originally reported in reference 40

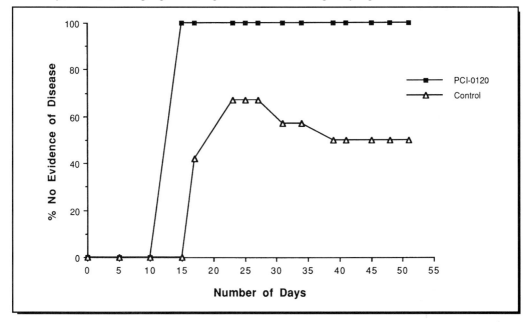

Much of the appeal in antisense stems from the fact that it would allow one, at least in principle, to target any disorder for which the causative gene is known. Indeed, because the antisense ODN–target binding interactions might be expected to be governed by the normal rules of Watson–Crick or Hoogsteen base-pairing, one could imagine designing a specifically tailored drug for any selected disease.[§] This could be particularly advantageous in the treatment of cancer, as there are 150–200 different neoplastic diseases, each correlating with different genetic lesions.[165,166] Further, because these drugs could function at the genotypic level, they would hold out the hope of allowing very subtle distinctions to be made between normal and diseased states. This, in turn, makes antisense particularly attractive in both the cancer and viral treatment areas, since here clinical success often rests on an ability to effect just these kinds of most precise distinctions. It should be noted in this context that there are advantages and disadvantages to targeting nucleus-contained DNA

[§]Limitations on this simplistic view are known. See references 163 and 164.

versus cytoplasmic, messenger RNA. The latter is less stable than duplex DNA and (generally) single-stranded. This, at least in principle, makes it easier to target. However, mRNA is present in multiple copies making it an elusive "prey" in an absolute sense. The DNA targets, on the other hand, are singular but far more stable. This stability, and their duplex nature, also makes them challenging sites in terms of antisense targeting. Additionally, there are concerns associated with the mutagenic or teratogenic effects that are often displayed by agents that modify DNA. Thus, on balance, RNA may represent the more attractive target from a therapeutic stand-point. Such arguments are discussed more fully in several leading reviews.[140–142]

To date, progress in the burgeoning field of antisense has been nothing short of phenomenal.[141,148,159] Nonetheless, at present, much of the promise inherent in anti-sense remains to be attained.[157,158] Many key limitations, such as nuclease-dependent instability, low for-target affinity, and cellular impermeability, have been identified, and these and other problems need to be addressed before clinically useful therapeu-tics are likely to be obtained. One major problem stems from the fact that simple steric blocking (i.e., competitive inhibition) is often insufficient to prevent gene expression.[167] This means that, in contrast to how the field was originally conceived, it is often not sufficient to generate an antisense ODN which simply binds to the targeted nucleic acid strand. Rather, one has to effect the actual "destruction" of the target portion of the antisense-target hybrid. This can be done, at least in principle, using either chemical[146,167–169] or biochemical[170–178] means, including those that are ribozyme and enzyme (e.g., RNase H) based. To date, most progress has been achieved *via* the latter biochemical approach. However, in the end, chemical-based solutions, perhaps predicated on the use of porphyrin analogs, could prove superior; they simply offer more synthetic flexibility for pharmacological optimization and use.

The chemical approach to antisense is based on the tethering of a reactive group, or "artificial active site" to a short (12–20-base pair) site-directing oligonu-cleotide.[146,167,169] The resulting conjugate, if appropriately designed, should then serve to effect an irreversible (or more strictly speaking, non-reparable) chemical modification of the targeted RNA or DNA strand, thereby blocking any further genetic expression. The power of this approach is that it does not rely on endogenous adjuvants (e.g., RNase H[170]) as do many of the biochemically based strategies. Also in contrast to the biochemical approaches, this chemical strategy is, at least in prin-ciple, compatible with a wide range of DNA backbones, including some such as 2'-O alkyl[162] and phosphoramidates[179–181] that generally display very desirable uptake and stability properties.[141,182] Finally, because of its inherent "ability to go at it alone", the chemical approach should allow one to consider the targeting of either RNA or DNA substrates. This would allow for the inhibition of gene expression at either the translational or transcriptional levels.

While considerable effort has been devoted to generating chemical modifiers for DNA and RNA,[143,145,146,162,167,169,183] a consensus opinion[184–188] has emerged in recent years that selective oligonucleotide cleavage may best be accomplished in an antisense (i.e., directed) fashion using either chemically enhanced hydroly-sis[162,183,189–193] or absorption-triggered photolysis.[183,186–188,194–205] As it currently stands, the first of these approaches appears more suitable for applications involving

RNA targets, whereas the latter shows more promise for those involving DNA. In any event, the success or failure of the chemical approach is predicated on having good cleavage-inducing "active sites" at one's disposition. This is where expanded porphyrins have a role to play. Depending on the choice of macrocycle, they can act as the key oligonucleotide-modifying hydrolysis or photolysis agents, or both.

10.4.2   *Photolytic Strategies*

10.4.2.1   Lutetium(III) Texaphyrins

The reasons expanded porphyrins are attractive as potential photolysis agents for use in antisense applications are basically the same ones that make them attractive for PDT and PDI (Section 10.2), namely an ability to absorb light at long wavelengths (i.e., $\geq$ 700 nm) and generate singlet oxygen in good quantum yield. In accord with precedent in the porphyrin area,[186–188,194–204] it would be expected, therefore, that attachment of an appropriate photoactivatable expanded porphyrin to a suitable carrier ODN would generate a conjugate capable of photomodifying a complementary DNA target sequence. In contrast to the porphyrin case, however, the use of a tethered expanded porphyrin should confer an ability to effect photomodification at wavelengths where bodily tissues are most transparent; this would be advantageous in terms of eventual clinical use.

The validity of the above concept was first demonstrated by Magda, *et al.*[183,205] These workers attached a functionalized lutetium(III) texaphyrin (LuTx) center (**10.15**; Figure 10.4.1), formally analogous to the experimental PDT photosensitizer PCI-0123 (complex **10.4**; bis-acetate form), to two 2'-O-methyl RNA 15-mers and two DNA 15-mers to generate conjugate pairs **10.16** and **10.18**, and **10.17** and **10.19**, respectively (Figure 10.4.2). The synthetic DNA 36-mers **10.20** and **10.21**, each complementary in sequence to only one of the paired sets of the lutetium(III) texaphyrin-bearing conjugates, were selected as substrates (Figure 10.4.2). Following incubation of the 5'-^{32}P-labeled forms of these targets with an excess of the 15-mer conjugates, irradiation was effected at 732 nm. It was found that the presence of a complementary LuTx 15-mer conjugate, but not a non-complementary control, led to photomodification of the DNA targets (70–90% chemical yield), as assayed by treatment with piperidine. The cleavage products were found to comigrate exclusively with bands generated by the Maxim–Gilbert sequencing (G) reaction, as would be expected for a mechanism involving singlet oxygen as the actual photogenerated DNA-modifying agent.

Subsequent to this initial work, similar experiments were carried out using RNA targets of identical sequence. In general, analogous results were obtained, namely the complementary targets (only) were seen to be photocleaved. However, the actual chemical yields of cleavage products were found to be lower. This was considered to reflect differences in work-up (necessitated by the change from DNA to RNA), rather than intrinsic differences in per photon reactivity.[183]

Figure 10.4.1   Structures **10.15**, **10.22**, **10.26**, and **10.28**

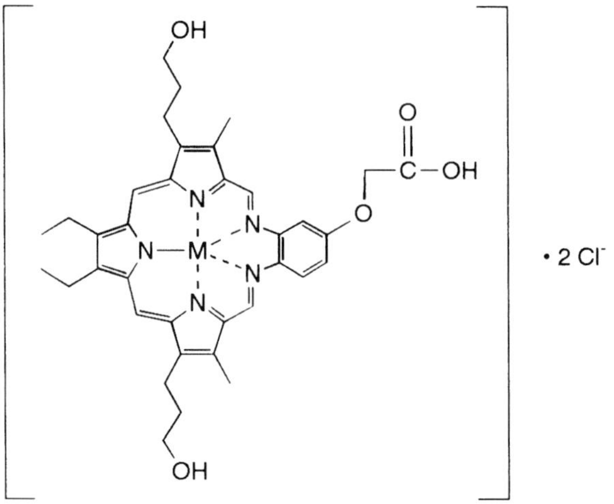

10.15 M = Lu
10.22 M = Y
10.26 M = Eu
10.28 M = Dy

10.4.2.2   Sapphyrins

At present, the lutetium(III) texaphyrins remain the best studied of the many possible expanded porphyrin chromophores that could be used as the basis for generating light-activatable antisense agents. Some analyses of the corresponding Y(III) texaphyrin systems (i.e., those derived from **10.22**) have, however, been made recently in the context of generating an antisense agent capable of effecting both RNA hydrolysis and DNA photolysis (*vide infra*). In addition, a first round of preliminary, proof of concept, experiments involving sapphyrin-derived conjugates had been completed at the time this book was being prepared for publication.

The sapphyrin system tested consisted of the functionalized thymidine 12-mer **10.23** shown in Figure 10.4.3.[206] When irradiated at $\geq$ 620 nm, this conjugate was able to effect the regio-selective photomodification of the complementary target **10.24** as indicated in Figure 10.4.3. A range of control experiments served to confirm the critical importance of the attached sapphyrin, light, oxygen, and complementary hybridization. For instance, no cross-linking or cleavage was observed in the absence of illumination or when the sapphyrin-functionalized 12-mer was replaced by a simple thymidine dodecamer (i.e., $(dT)_{12}$). Likewise, no photomodification was observed when conjugate **10.23** was irradiated in the presence of various non-complementary sequences. Finally, no other residues besides G (i.e., C, A, or T) were found to be significantly cleaved when **10.23** and **10.24** were subject to irradiation.

Taken together, these results are consistent with a photomodification process being mediated by singlet oygen; this is as one would predict in light of the fact that sapphyrin is an effective PDI photosensitizer (*vide supra*).

Figure 10.4.2   Synthetic DNA 36-mers **10.20** and **10.21** Used as Targets for Sequence-specific Photocleavage and Complementary 2'-*O*-Methyl RNA and DNA Conjugates Bearing a Covalently Attached Lutetium(III) Texaphyrin (LuTx) Center Derived from **10.15** (Structures **10.16** and **10.18**, and **10.17** and **10.19**, respectively).
The arrows indicate positions of strong, intermediate, or weak DNA modification as exposed by treatment with piperidine

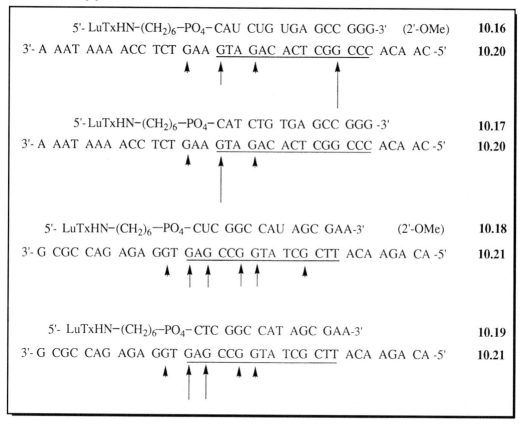

10.4.3   *RNA Hydrolysis: Lanthanide(III) Texaphyrins*

Effecting targeted, site-selective RNA hydrolysis is the other antisense-related area where the expanded porphyrins could prove advantageous. In point of fact, however, it is only the texaphyrins (of all the expanded porphyrin systems known) that have actually seen application in this context.[193] This is because the texaphyrins, as described in Chapter 9, have a "special" ability to form strong, non-labile 1:1 "porphyrin-like" complexes with the trivalent lanthanides. This, in turn, allows these Lewis acidic cations (in the form of their texaphyrin complexes) to be conjugated to RNA-targeting oligonucleotide (ODN) backbones.

Figure 10.4.3 Schematic Representation of the Site-directed Photocleavage of Substrate **10.24** Effected by Conjugate **10.23**.
Hybridized regions are underlined and the arrows show the positions of hydrolysis

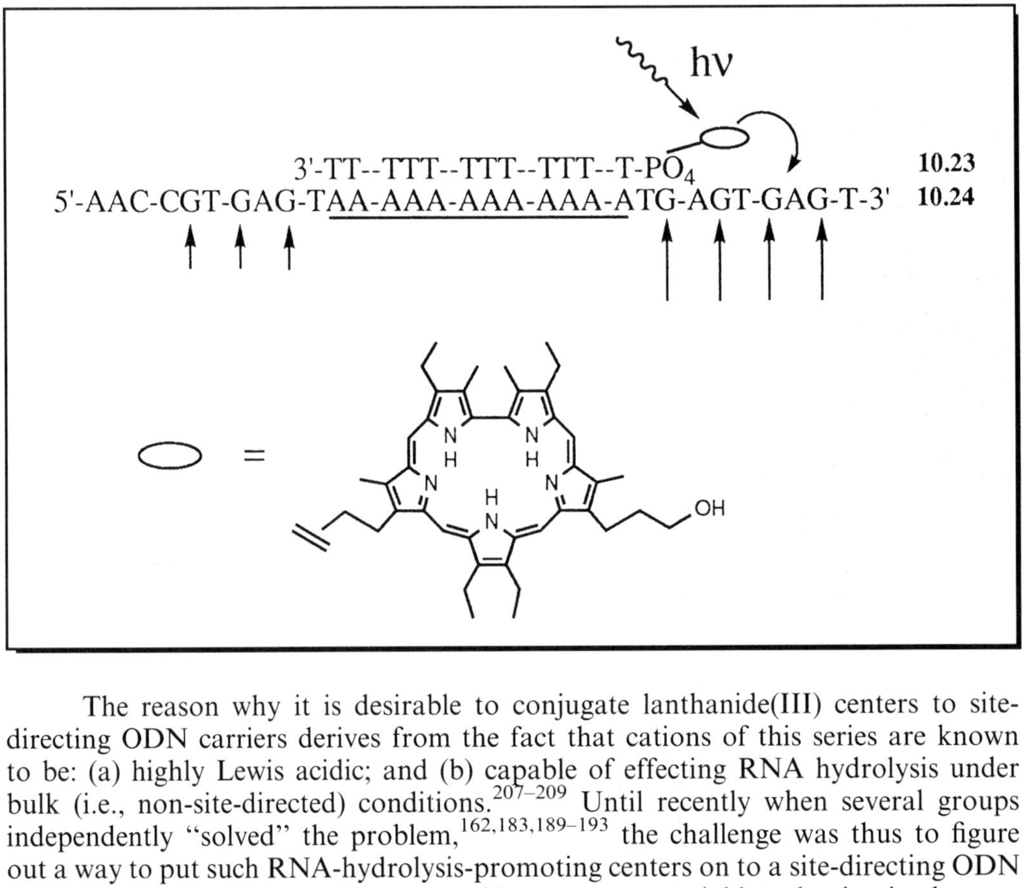

The reason why it is desirable to conjugate lanthanide(III) centers to site-directing ODN carriers derives from the fact that cations of this series are known to be: (a) highly Lewis acidic; and (b) capable of effecting RNA hydrolysis under bulk (i.e., non-site-directed) conditions.[207–209] Until recently when several groups independently "solved" the problem,[162,183,189–193] the challenge was thus to figure out a way to put such RNA-hydrolysis-promoting centers on to a site-directing ODN without suffering either significant metal loss or an appreciable reduction in cleavage activity.

The texaphyrin-based solution to the above problem, illustrated schematically in Figure 10.4.4, is one of considerable appeal. It involves, as first demonstrated by Magda *et al.*,[193] attaching a Lewis acidic lanthanide(III) texaphyrin to the 5'-end of a DNA-type oligomer and using the resulting conjugate to effect hydrolytic cleavage of an RNA target. This idea, which was made further credible by the finding that phosphodiesters coordinate to lanthanide(III) texaphyrins both in solution and in the solid state (see Figure 10.4.5),[210] was tested initially using conjugate **10.25** (Figure 10.4.6).[193] This conjugate, derived from the DNA 15-mer, 5'-CTCGGCCATAGCGAATGTTC-3', and the europium(III) texaphyrin complex **10.26** (Figure 10.4.1) was designed so as to be complementary to the synthetic RNA 36-mer, **10.27**, a substrate selected from a unique site within the gene transcript for multiple drug resistance (MDR). The expectation was thus that this conjugate, but not various non-complementary or texaphyrin-free controls, would be able to effect the hydrolysis of this designated RNA target at more reactive[211] single-stranded sites near the "across-the-duplex" point of europium(III) texaphyrin attachment. As illustrated schematically in Figure 10.4.6, and detailed further in reference 193, this expectation was in fact realized. At a concentration approximately 10^4 times lower than was required for either free europium(III) texaphyrin or any of several non-complementary Eu(III) texaphyrin conjugates prepared as controls, system **10.25** proved able to cleave target **10.27** efficiently, hydrolytically, and site selectively.

Figure 10.4.4   Schematic Representation of the Lanthanide(III) Texaphyrin-based Approach to Antisense-type RNA Hydrolysis

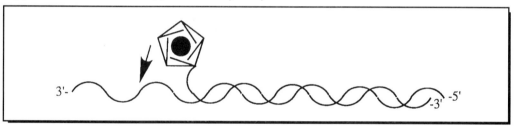

To build further on the above foundation, Magda, *et al.* have looked recently at a variety of ODN carriers, RNA targets, and lanthanide(III) centers as well as modified texaphyrin–ODN linking tethers and different texaphyrin cores. From this work, a partial structure–function picture is beginning to emerge. For instance, within the confines of given texaphyrin ligand system (e.g., **10.15**, **10.22**, **10.26**, and **10.28** of Figure 10.4.1), it has been found that trivalent cations near the middle of lanthanide series are more efficacious (i.e., effect hydrolysis faster) than those at the ends. Specifically, in the case of analogs of conjugate **10.25**, identical except for a change in the ligated metal, the following efficiency order was derived: Nd < < Eu < Gd < Tb ≈ Dy > Ho > Er > Tm > Lu.[183] On the other hand, by keeping the metal the same (i.e., Dy(III)), it has proved possible to show that the optimal number of atoms, linking the 5'-end of the ODN to the texaphyrin macrocycle, should be three, not nine (as originally used in **10.25**).[212] Finally, with such short linkers it has been found that changes in the peripheral texaphyrin substitution pattern (i.e., hydroxyl propyl vs. ethyl) have little effect on hydrolysis efficacy.[212] Thus, from this basic set of studies the parameters are being established whereby this approach to antisense may be optimized.

Figure 10.4.5 Single Crystal Structure of a Dysprosium(III) Texaphyrin Complex Wherein Two Phenyl Phosphate Counter Anions are Found Bound to the Metal Center.

This figure was generated using information down-loaded from the Cambridge Crystallographic Data Centre and corresponds to a structure originally reported in reference 210. Atom labeling scheme: dysprosium: ●; carbon: ○; nitrogen: ◉; oxygen: ⊖; phosphorus: ⦶. Hydrogen atoms have been omitted for clarity

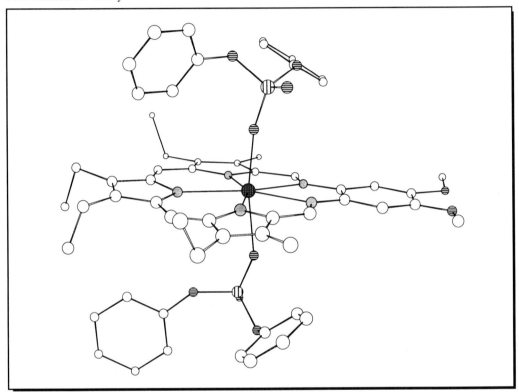

Figure 10.4.6 Synthetic RNA Substrate **10.27** Used as a Target for Sequence-specific Hydrolysis by a Complementary Europium(III) Texaphyrin DNA Conjugate **10.25**.

The arrows show the positions of hydrolysis

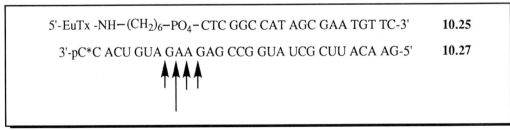

5'-EuTx -NH−(CH$_2$)$_6$−PO$_4$−CTC GGC CAT AGC GAA TGT TC-3'   **10.25**

3'-pC*C ACU GUA GAA GAG CCG GUA UCG CUU ACA AG-5'   **10.27**

10.4.4   Combined Strategies: Yttrium(III) Texaphyrins

One critical finding to emerge from the above work on hydrolysis-mediated cleavage was the discovery that the most paramagnetic lanthanide(III) cations (viz. Dy(III), Tb(III), Gd(III), and Eu(III)) worked the best, at least within the context of a texaphyrin-based approach. Such species, however, are known to be ineffectual in terms of generating singlet oxygen (the excited state simply relaxes too quickly).[89] They cannot, therefore, be used to effect photolytic cleavage of a targeted oligonucleotide. On the other hand, the lutetium(III) texaphyrins, being diamagnetic, produce singlet oxygen in good quantum yield (*vide supra*). Unfortunately, these latter species are poor reagents when it comes to promoting the site-selective hydrolysis of RNA.

The above dichotomy led Magda *et al.* to search for alternative metal centers that might work for both RNA hydrolysis and DNA (and, perhaps, RNA) photolysis.[183] To date, one such metal center has been found—yttrium(III). When attached to the same sequences used to prepare systems such as **10.25**, conjugates are produced that not only demonstrate hydrolytic RNA-cleaving activity approximately equal to that of the original EuTx-based systems (i.e., **10.25** itself), but also show light-induced DNA and RNA cleaving potential as good as the various lutetium(III) texaphyrin-based ODN systems discussed above (see Figure 10.4.2).[183] This important result is considered to augur well in terms of eventual clinical use; situations can easily be envisioned where a combined hydrolytic *and* photolytic approach to antisense would prove advantageous.

10.5   Expanded Porphyrins as Anion-binding Agents

10.5.1   Introduction

The final application of expanded porphyrins to be discussed involves anion binding. This is an area of tremendous ferment, both within the expanded porphyrin field, and the far more generalized disciplines of molecular recognition and biological chemistry. The reasons for this are multi-faceted but derive in large measure from the fact that anions, like cations, are ubiquitous in biology.[213] They play roles in processes as diverse as chemical energy manipulation (e.g., ATP hydrolysis),[214] information storage and processing (e.g., DNA and RNA),[215] and protein conformational modification.[216–218] Many anions, especially phosphates, chloride, and carboxylates (including amino acids), are involved in metabolic cycles[219–222] and are, of course, critical cellular components.[223,224] They can thus be the cause or cure for disease. For instance, cystic fibrosis, a hereditary disorder striking 1 in 2500 Caucasians (live births),[225–229] has its genesis in a single inoperative chloride anion channel,[226,227] while, on the other hand, active antiviral agents are often phosphorylated nucleotide analogues generated *in situ* (i.e., within the cell).[232] Anions can also mediate salubrious or deleterious effects in other ways. They can act as direct toxins (e.g., cyanide or arsenate anion)[233] or promote undesirable consequences, such as river and lake eutrification (e.g., nitrate and phosphate),[234] which are less acute but equally insidious. Conversely, the metal coordinating properties of anions, particularly carboxylates, have made possible a variety of beneficial applications such as the MRI ones discussed in the introduction to Section 10.1.

Not surprisingly in view of their importance, a tremendous effort has been devoted to the generation of species capable of binding, recognizing, or transporting anions.[235–240] Here, a range of receptor strategies have been explored, including those based on the use of amides,[241–244] polyammonium cations,[245–249] guanidinium and guanidinium-like subunits,[250–253] Lewis acidic centers,[254–256] and coordinated metal cations.[257–266] Taken as a whole, this anion binding work is characterized by a diversity, creativity, and elegance that has given it a prominent role in the emerging area of supramolecular science;[235–240,267,268] it is nonetheless notable for one salient fact, namely a general non-reliance on porphyrin-based approaches. While metalloporphyrins will coordinate anions as axial ligands, such one-point binding is not generally useful.[269–273] Further, free-base porphyrins are notoriously tough to protonate and thus do not bear the positive charge needed to function as anion receptors under normal (i.e., near-neutral) laboratory conditions.[††] By contrast, many expanded porphyrins are quite basic, making them capable of binding anions both in solution and in the solid state.

As detailed below, and discussed more fully in a separate review,[237] more is known about the anion binding properties of sapphyrin and its derivatives (e.g., **10.6**, **10.29**–**10.31**; Figure 10.5.1) than any other expanded porphyrin system. In fact, apart from sapphyrin, only two other polypyrrolic macrocycles, namely anthraphyrin (**10.32**; Figure 10.5.2) and rubyrin (**10.33**; Figure 10.5.2), have been studied in solution as potential anion binding agents. Nonetheless, on the basis of solid-state structural studies, a number of other systems, including various heterosapphyrins (e.g., **10.34**–**10.37**; Figure 10.5.3), vinylogous porphyrins of the [22]-(3.1.3.1) and [26]-(5.1.5.1) varieties (e.g., **10.38**, **10.39**, and **10.40**; Figure 10.5.4), rosarin (**10.41**; Figure 10.5.5), amethryin (**10.42**; Figure 10.5.6), orangarin (**10.43**; Figure 10.5.6), turcasarin (**10.44**; Figure 10.5.7), and octaphyrin(2.1.0.1.2.1.0.1) (**10.45**; Figure 10.5.7) look attractive as anion chelators. In light of this, a discussion of solid-state anion-to-protonated expanded porphyrin interactions will be given first (Section 10.5.2). Following this, a summary of the solution-phase anion binding properties of protonated sapphyrin, rubyrin, and anthraphyrin macrocycles (Section 10.5.3) will be given. Finally, this chapter will conclude with a brief overview of a new and intriguing area of expanded porphyrin-based anion recognition that is defined by the chemistry of sapphyrin-oligonucleotide interactions.

10.5.2   *Anion Binding in the Solid State*

The first discovery, made by Sessler and Ibers and coworkers in early 1990,[280] that expanded porphyrins can act as anion-binding agents was a fortuitous one. It resulted from work directed at solving the single crystal X-ray diffraction structure of

---

[††]For a rare example, see reference 274; for examples of the more extensive molecular recognition chemistry of cationic porphyrins, especially with regards to DNA interactions, see references 274–279.

what was thought to be the bis-HPF$_6$ salt of sapphyrin **10.6** (Figure 10.5.1). Here, instead of the expected structure, what was found was a structure with only one PF$_6^-$ counter anion (per sapphyrin) and "extra" electron density at the sapphyrin core. On the basis of scattering parameters, charge considerations, independent synthesis (of the bis-HF salt), and solution-phase ^{19}F NMR spectroscopic studies, this extra density was assigned to a fluoride counter anion held within the cavity by five radially oriented hydrogen bonds of *ca.* 2.7 Å (N-H···F) length (Figure 10.5.8). Apparently, under the crystallization conditions, sapphyrin **10.6** (or its protonated form) is able to "pull" fluoride anion away from a PF$_6^-$ counter anion in accord with the known equilibrium (eq. 3):[281]

$$HF + PF_5 \rightleftharpoons HPF_6 \tag{3}$$

Figures 10.5.1   Structures **10.6, 10.29–10.31**

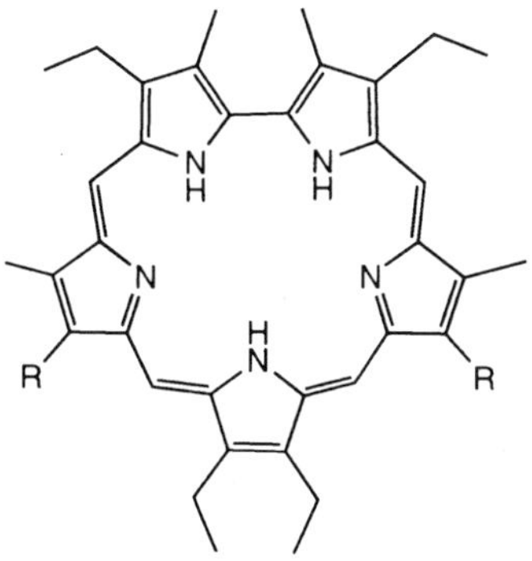

10.6. R = Et
10.29. R = (CH$_2$)$_3$OH
10.30. R = (CH$_2$)$_2$CO$_2$H
10.31. R = (CH$_2$)$_2$CON(CH$_2$CH$_2$OH)$_2$

While it can be debated whether or not a PF$_6^-$ counter anion actually provided the source of adventitious fluoride anion in the original Sessler–Ibers structure (see footnote 17 in reference 280), its very existence nonetheless stands as clear proof that the doubly protonated form of sapphyrin can indeed "coordinate" fluoride anion in the solid state. This seminal result thus led directly to the proposal[280,282–287] that sapphyrins and other expanded porphyrins could constitute an interesting new class of anion-binding agents.

As implied in the introductory paragraphs above (Section 10.5.1), support for this proposal is now extensive, especially in the solid state. In early work, for instance, Franck and coworkers found that their [22]porphyrins-(3.1.3.1) **10.38**[288] and **10.39**[289] crystallized nicely as the bis-HCl and bis-trifluoracetic acid (TFA) salts, respectively. The resulting X-ray diffraction structures (Figures 10.5.9 and 10.5.10) revealed near-planar systems with the counter anions "chelated" above and below the mean macrocycle plane (mean N_4-anion distances are 1.87 Å and 1.42 Å for chloride and trifluoracetate, respectively) by a combination, presumably, of electrostatic interactions and hydrogen bonds. This same group also was able to show that this larger system, **10.40**, could be crystallized as the bis-TFA salt.[290] Here, the resulting X-ray structure revealed the counter anions being bound, respectively, 1.30 Å and 1.42 Å above and below the macrocyclic plane (Figure 10.5.11). While these workers failed (apparently) to appreciate the significance of these structures in terms of defining a new anion recognition motif, they are important in an historical sense in that they provide experimental support for the notion that sapphyrin is not unique among expanded porphyrins in being able to function as an anion binding agent.

Figure 10.5.2   Structures **10.32** and **10.33**

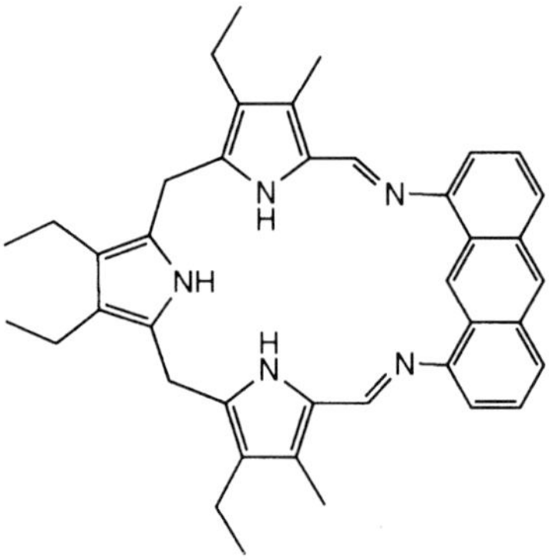

10.32 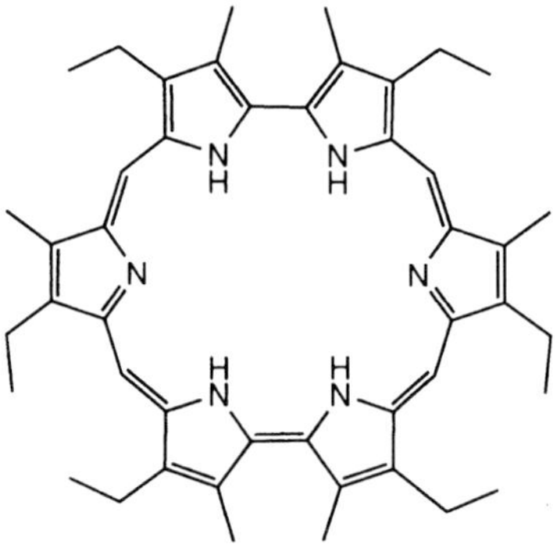   **10.33**

Figure 10.5.3   Structures **10.34–10.37**

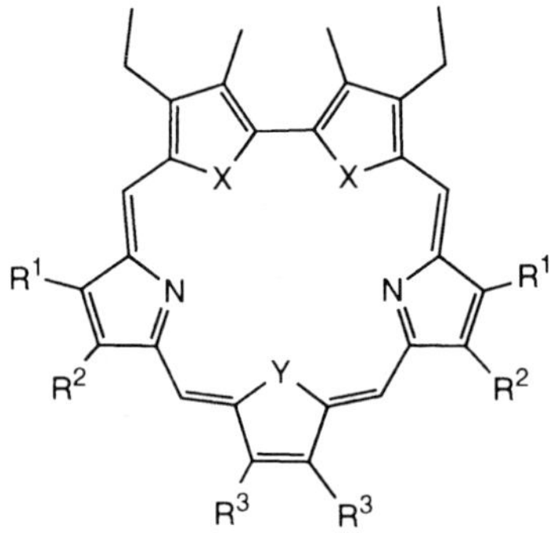

10.34. X = NH, Y = O, R^1 = Et, R^2 = R^3 = Me
10.35. X = O, Y = NH, R^1 = Me, R^2 = R^3 = Et
10.36. X = Y = O, R^1 = Et, R^2 = R^3 = Me
10.37. X = NH, Y = Se, R^1 = Et, R^2 = Me, R^3 = H

Figure 10.5.4   Structures **10.38–10.40**

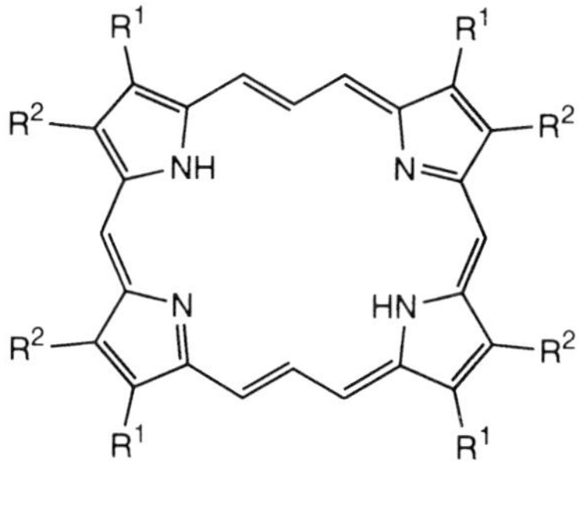

10.38. R^1 = Me, R^2 = CH$_2$CH$_2$CO$_2$Me
10.39. R^1 = Me, R^2 = Et

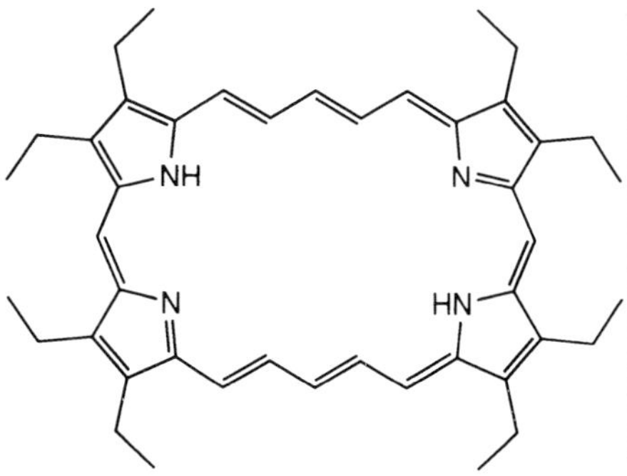

10.40

Figure 10.5.5   Structure **10.41**

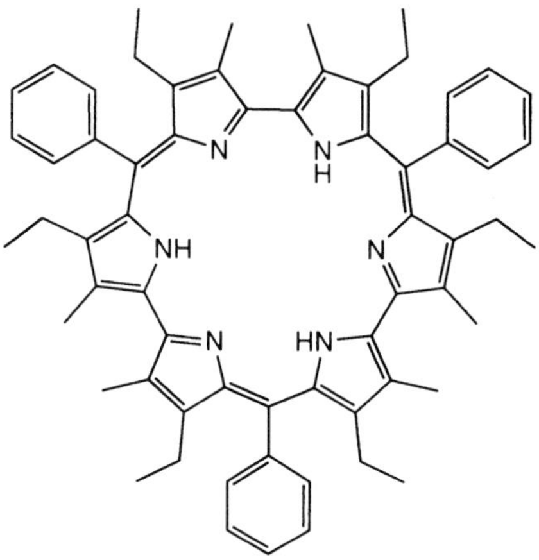

10.41

Out-of-plane anion chelation was also observed in the historically important structure of the sapphyrin bis-hydrochloride salt **10.6**·2HCl (Figure 10.5.12).[282] Here, in analogy to what was seen in Franck's systems, the chloride anions were found to reside above and below the mean N$_5$ plane of the doubly protonated macrocycle (at *ca.* 1.77 Å and 1.88 Å, respectively), being held there by a combination of electrostatic interactions and hydrogen bonds (either 3 or 2, as the case may be). Thus, while showing congruence with what was observed for the [22]porphyrins-(3.1.3.1), this result stands in marked contradistinction to what was observed in the case of the sapphyrin-fluoride structure mentioned above. Apparently, the larger chloride coun-ter anion (ionic radius $\approx$ 1.67 Å for Cl$^-$ vs. $\approx$ 1.19 Å for F$^-$ [291]) is simply too large to fit within the proton-bearing N$_5$ sapphyrin binding core.

Figure 10.5.6   Structures **10.42** and **10.43**

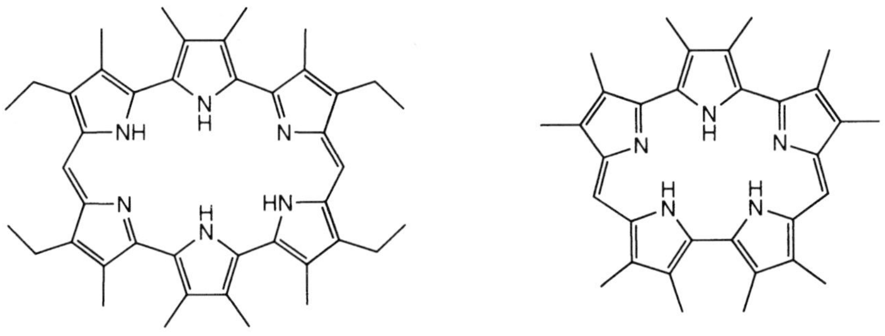

10.42   10.43

Figure 10.5.7   Structures **10.44** and **10.45**

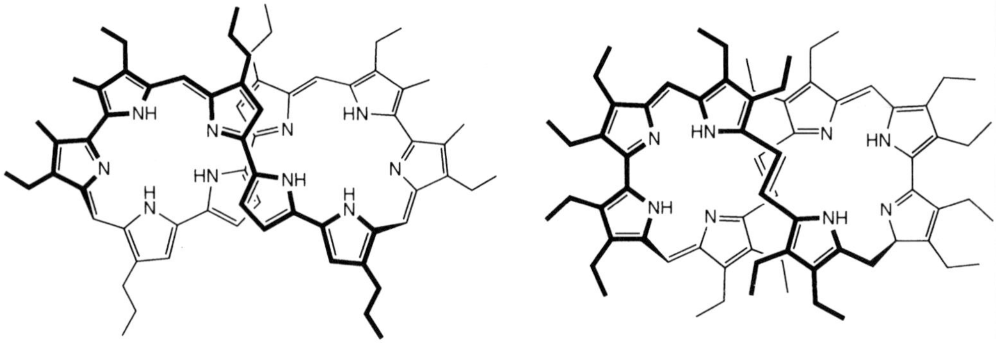

10.44   10.45

    Subsequent to solving the single crystal X-ray structure of the bis-HCl salt of sapphyrin **10.6**, Sessler and coworkers succeeded in obtaining structural information for the corresponding mono-HCl salt (Figure 10.5.13).[286] In this instance, the chloride counter anion is again found to reside out of the mean sapphyrin plane (by *ca*. 1.72 Å), being held in place by four, as opposed to three or two (*vide supra*), hydrogen bonds. A related structure, involving azide anion, was also obtained (Figure 10.5.14).[283] While identical in many respects, in this structure the anionic terminus, bound 1.13 Å above the mean N$_5$ plane, is found to reside closer to sap-phyrin core. This is as would befit the smaller atoms involved (nitrogen vs. chlorine).

Figure 10.5.8   Single Crystal X-Ray Structure of the Mixed HF-HPF$_6$ Salt of Sapphyrin **10.6**.

In this structure the sapphyrin is doubly protonated and the fluoride anion is bound nearly within the mean macrocyclic plane; the PF$_6^-$ counter anion, on the other hand, is not proximate to the sapphyrin skeleton. This figure was generated using information down-loaded from the Cambridge Crystallographic Data Centre and corresponds to a structure originally reported in reference 280. Atom labeling scheme: carbon: ◯; nitrogen: ●; fluorine: ⊗; hydrogen: ◦. Selected hydrogen atoms have been omitted for clarity

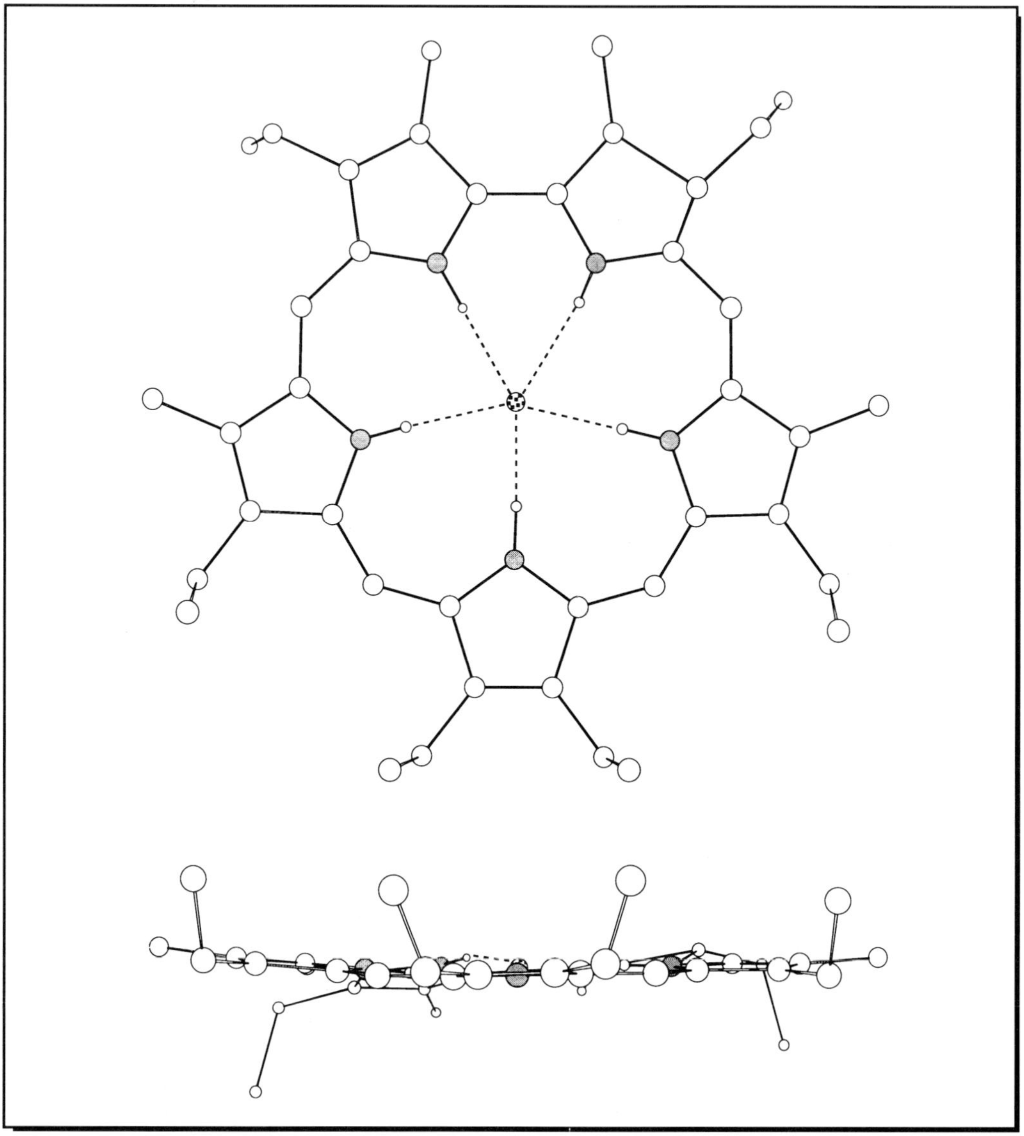

Figure 10.5.9   Single Crystal X-Ray Structure of the Bis-HCl Salt of
[22]Porphyrin-(3.1.3.1) **10.38**.
This figure was generated using information down-loaded from the Cambridge Crystallographic
Data Centre and corresponds to a structure originally reported in reference 288. Atom labeling
scheme: carbon: ○; nitrogen: ●; chlorine: ⊗; hydrogen: ○. Selected hydrogen atoms have been
omitted for clarity

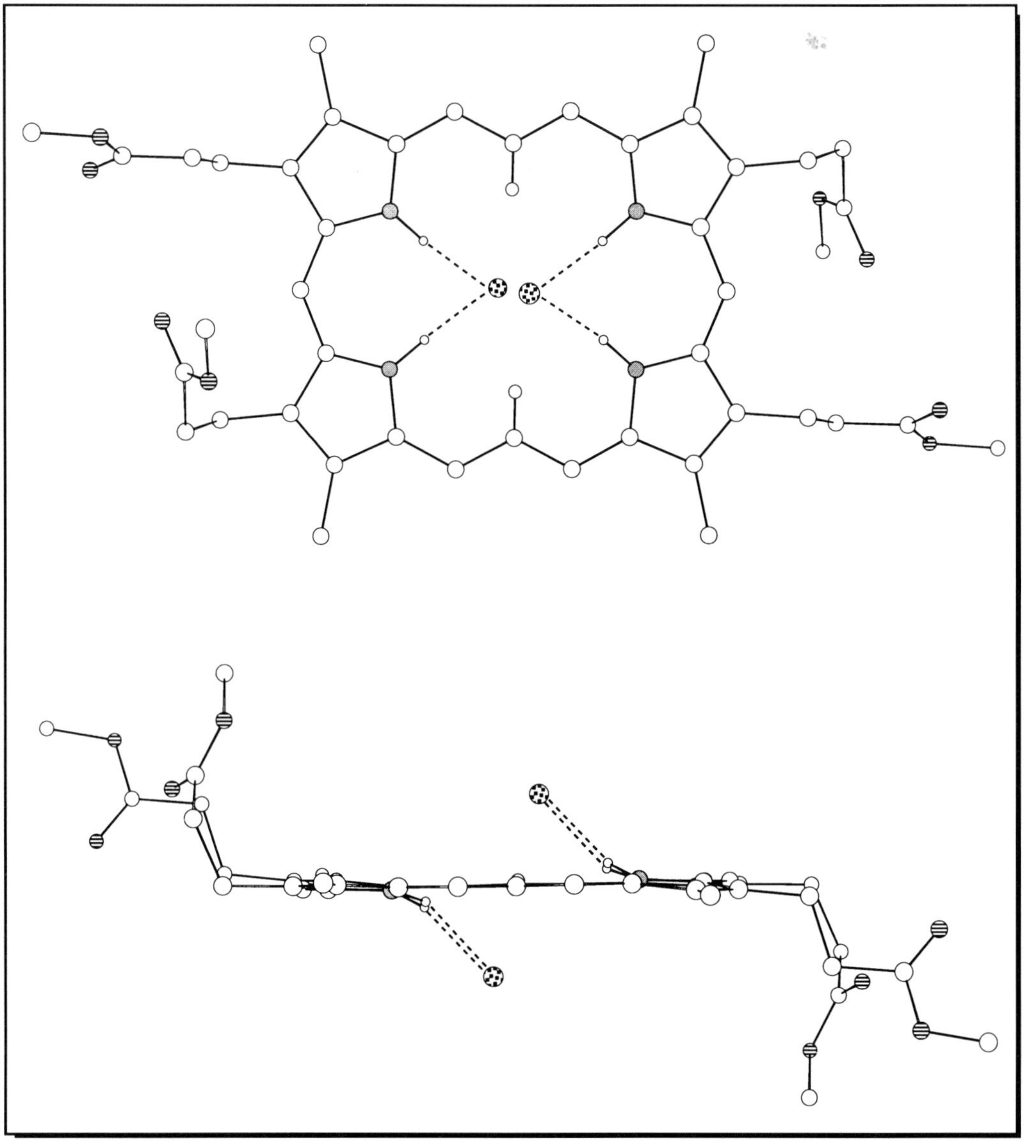

Figure 10.5.10 Single Crystal X-Ray Structure of the Bis-HTFA Salt of
[22]Porphyrin-(3.1.3.1) **10.39**.
This figure was generated using information down-loaded from the Cambridge Crystallographic
Data Centre and corresponds to a structure originally reported in reference 289. Atom labeling
scheme: carbon: ○; nitrogen: ●; oxygen: ⊜; fluorine: ⊗; hydrogen: ○. Selected hydrogen atoms have
been omitted for clarity

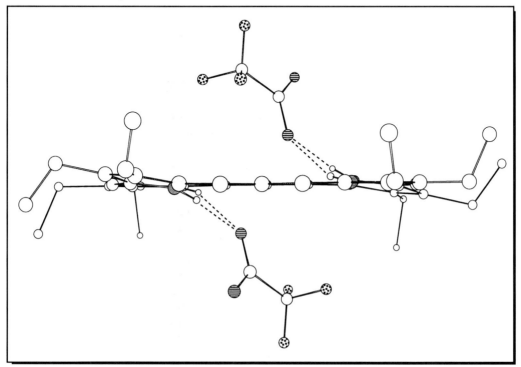

Taken together, these two structures, the only ones so far obtained for monoprotonated sapphyrin derivatives, are important. They demonstrate clearly that, at
least in the solid state, sapphyrins need not be doubly protonated to bind anions.
These results thus provide critical structural support for the notion that sapphyrins
could prove useful as anion receptors in solution. This is because in aqueous media it
is the mono-, not the diprotonated, form of sapphyrin that dominates at or near
neutral pH (*vide infra*).[292]

In order to support the contention that the protonated forms of sapphyrin
have a role to play as anion receptors, particularly for physiologically important
species such as phosphates and carboxylates, efforts were made to crystallize complexes containing these anions. In the case of phosphate, a number of such structures
have in fact been obtained.[292,293] Two are considered particularly informative. The
first involves the 1:1 complex formed between monobasic phosphoric acid and
diprotonated sapphyrin **10.6** (Figure 10.5.15)[292] and the second, the mixed salt of
diprotonated sapphyrin **10.29**, chloride anion, and cyclic-AMP (Figure 10.5.16).[293]

In both cases, the phosphate oxyanion is clearly "chelated" to the diprotonated sapphyrin core by anisotropic hydrogen-bonding interactions as well as by, presumably, charge effects. Taken together, these effects appear to result in strong binding, at least as inferred from bond distances; the phosphate oxyanion is found 1.22 Å and 1.38 Å above the mean N_5 macrocyclic plane in the case of these two complexes, respectively.[292,293]

Figure 10.5.11   Single Crystal X-Ray Structure of the Bis-HTFA Salt of [26]Porphyrin-(5.1.5.1) **10.40**.
This figure was generated using information down-loaded from the Cambridge Crystallographic Data Centre and corresponds to a structure originally reported in reference 290. Atom labeling scheme: carbon: ○; nitrogen: ●; oxygen: ◓; fluorine: ⊗; hydrogen: ○. Selected hydrogen atoms have been omitted for clarity

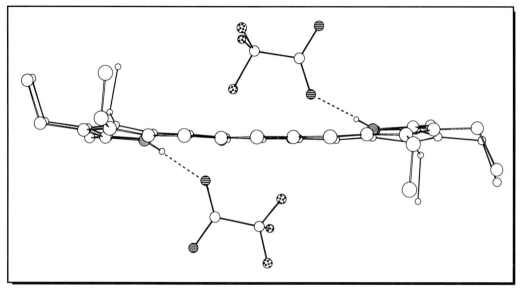

In the case of carboxylates, less is available in terms of structural information. In fact, the only pentaaza-sapphyrin to yield X-ray-quality diffraction single crystals showing carboxylate anion recognition is the doubly substituted system **10.30**.[294] Interestingly, this compound, in the presence of trifluoroacetic acid, crystallizes as a self-assembled dimer (Figure 10.5.17) wherein the "tail" of one functionalized sapphyrin "bites" the protonated center of a second sapphyrin subunit. Trifluoroacetate anions, bound on the "top" and "bottom" faces of the resulting "sandwich", then serve to complete the overall supramolecular ensemble. While not a result that would have necessarily been predicted a priori, this finding is nonetheless important. It shows that sapphyrin–carboxylate anion interactions can be used to assemble large structures of not insignificant complexity.[294]

In addition to the results obtained with "pure", all-aza sapphyrins, structural information is now available for anion complexes of two doubly protonated hetero-

Figure 10.5.12   Single Crystal X-Ray Structure of the Bis-HCl Salt of Sapphyrin
10.6.

This figure was generated using information down-loaded from the Cambridge Crystallographic
Data Centre and corresponds to a structure originally reported in reference 282. Atom labeling
scheme: carbon: ○; nitrogen: ●; chlorine: ⊗; hydrogen: ○. Selected hydrogen atoms have been
omitted for clarity

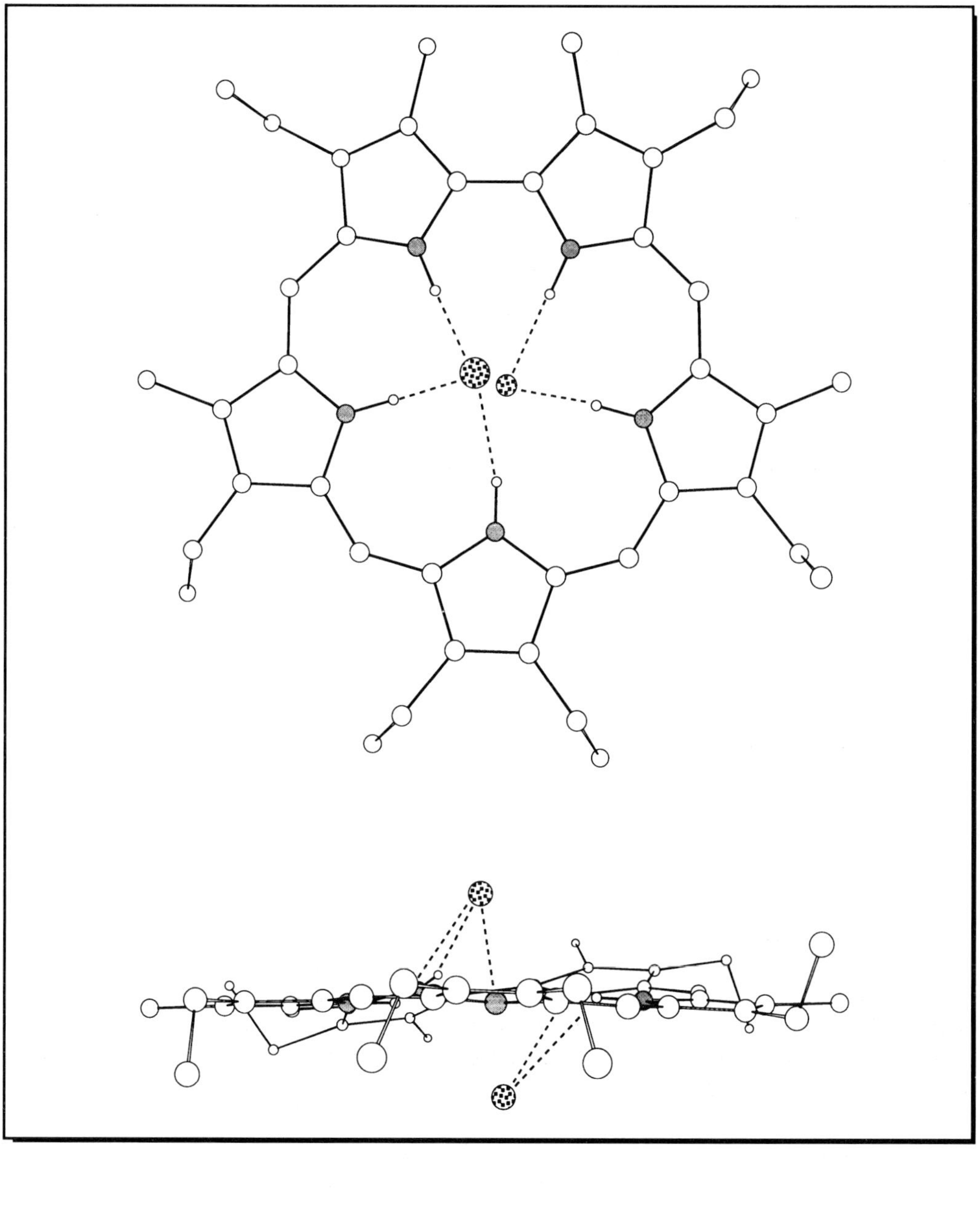

Figure 10.5.13   Single Crystal X-Ray Structure of the Mono-HCl Salt of Sapphyrin **10.6**.

This X-ray structural figure was generated using unpublished data provided by Sessler, *et al.*, but corresponds to a structure originally reported in reference 286. Atom labeling scheme: carbon: ○; nitrogen: ●; chlorine: ☉; hydrogen: ○. Selected hydrogen atoms have been omitted for clarity

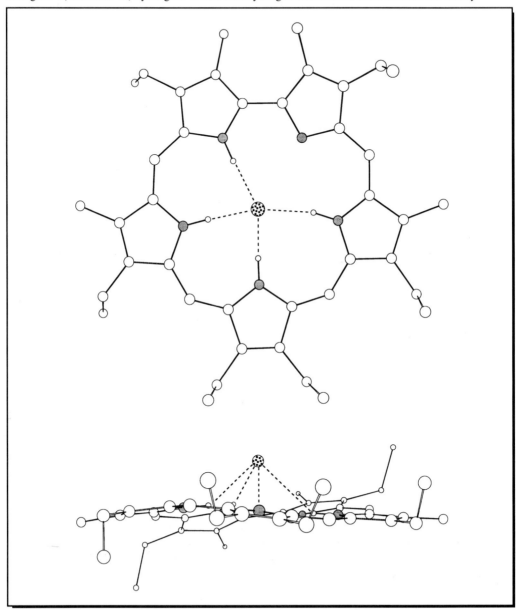

Figure 10.5.14   Single Crystal X-Ray Structure of the Mono-HN$_3$ Salt of Sapphyrin **10.6**.

This X-ray structural figure was generated using data provided by Sessler, *et al.*, but corresponds to a structure originally reported in reference 283. Atom labeling scheme: carbon: ○; nitrogen: ●; hydrogen: ○. Selected hydrogen atoms have been omitted for clarity

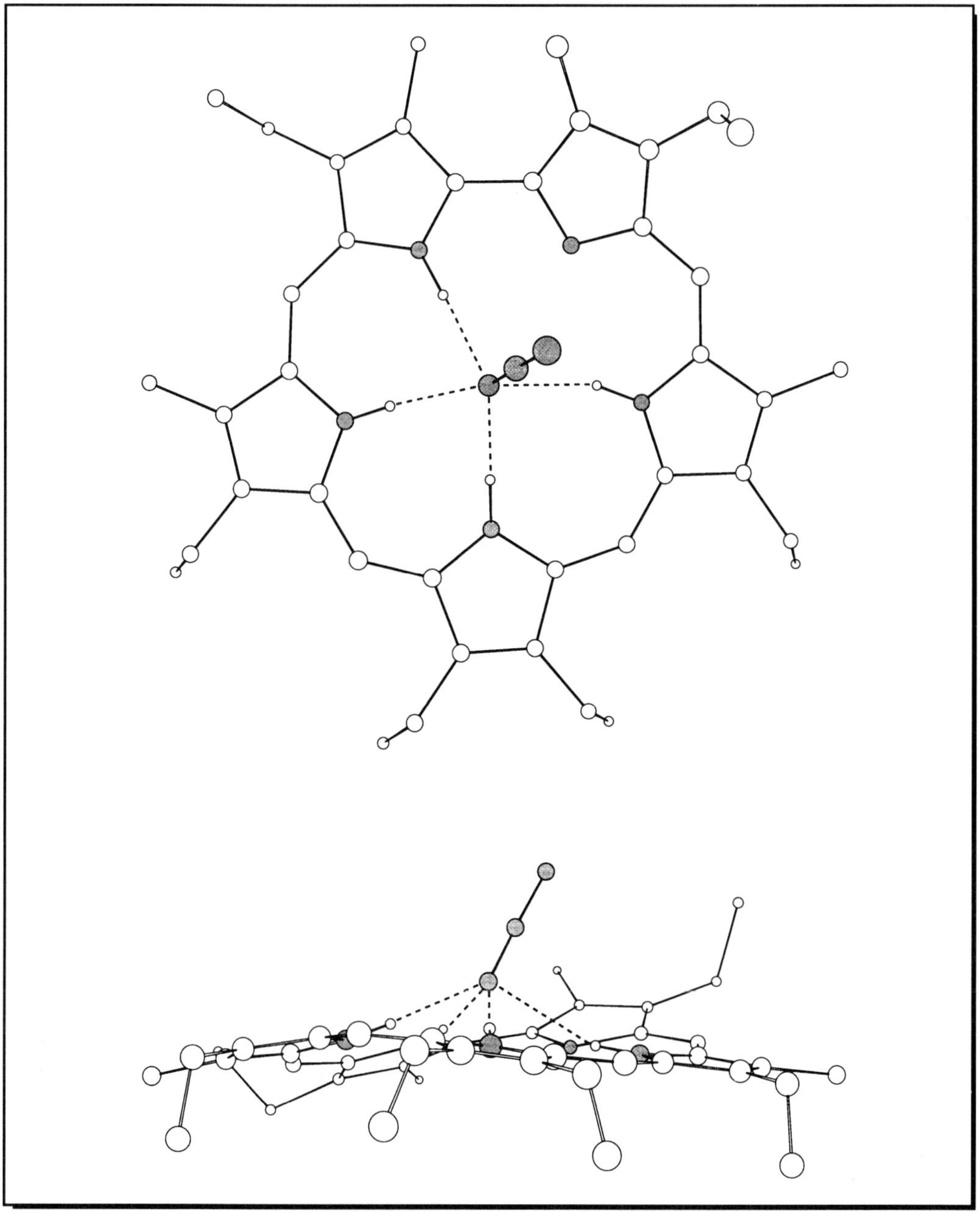

Figure 10.5.15   View of the 1 : 1 Cationic Inner-sphere Complex Formed between Sapphyrin **10.6** and Monobasic Phosphoric Acid.

The actual species crystallized, **10.6**·[H₃PO₄]₃·0.68H₂O contains one molar equivalent of phosphoric acid and dihydrogen phosphate, respectively, per sapphyrin. However, these species are not proximate to the sapphyrin skeleton. This X-ray structural figure was generated using information down-loaded from the Cambridge Crystallographic Data Centre and corresponds to a structure originally reported in reference 292. Atom labeling scheme: carbon: ○; nitrogen: ●; oxygen: ⊖; phosphorus: ⓪; hydrogen: ○. Selected hydrogen atoms have been omitted for clarity

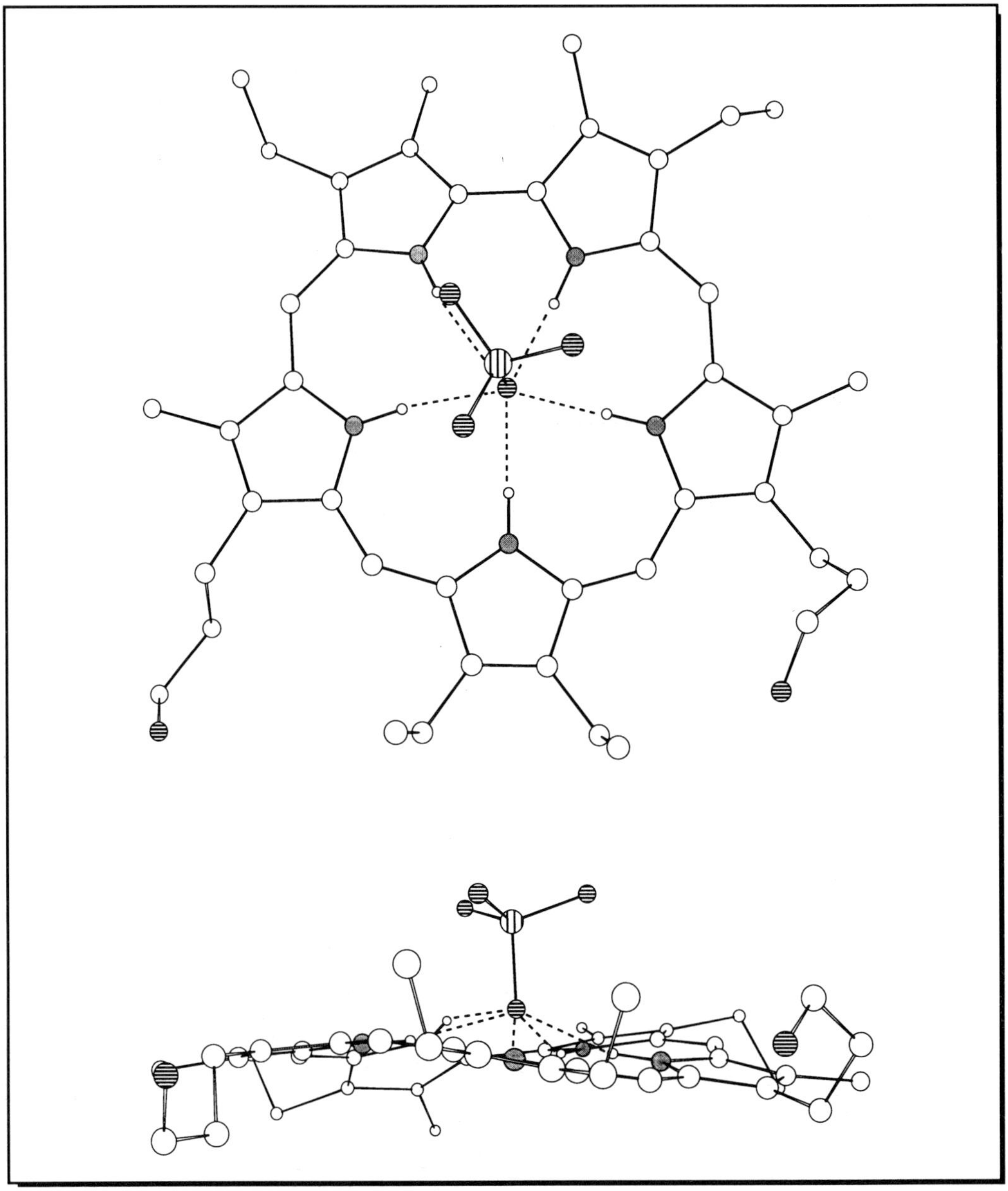

Figure 10.5.16   Single Crystal X-Ray Structure of the Mixed Salt Formed between Sapphyrin **10.29**, HCl, and the Acid Form of *c*-AMP.

This figure was generated using information down-loaded from the Cambridge Crystallographic Data Centre and corresponds to a structure originally reported in reference 293. Atom labeling scheme: carbon: ○; nitrogen: ●; oxygen: ⊖; chlorine: ⊘; phosphorus: ⦶; hydrogen: ○. Selected hydrogen atoms have been omitted for clarity

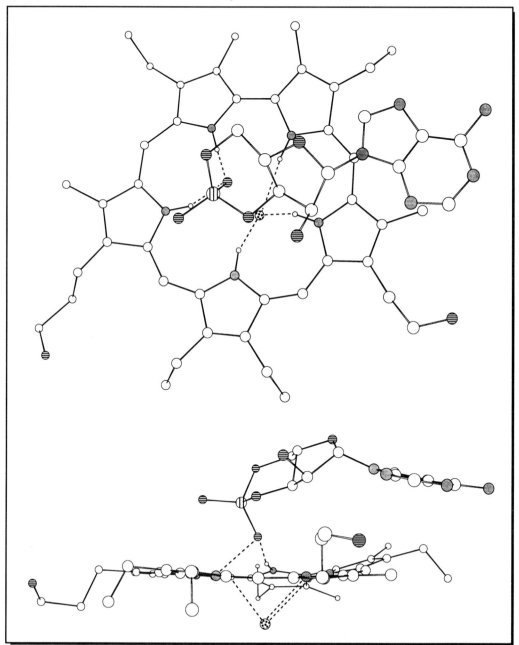

Figure 10.5.17   View of the Self-assembled Dimer Formed from Sapphyrin **10.30** in the Presence of Trifluoroacetic Acid.

This X-ray structural figure was generated using information down-loaded from the Cambridge Crystallographic Data Centre and corresponds to a structure originally reported in reference 294. Atom labeling scheme: carbon: ○; nitrogen: ●; oxygen: ⊜; fluorine: ⊗; hydrogen: ○. Selected hydrogen atoms have been omitted for clarity

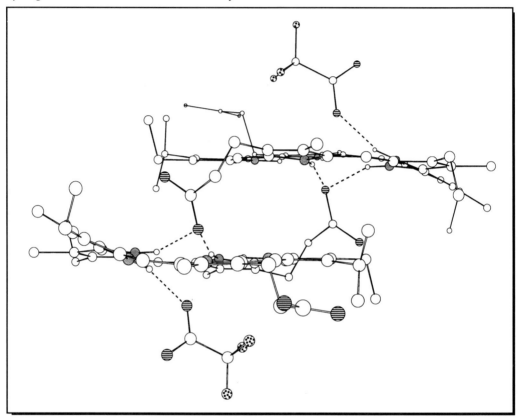

sapphyrin derivatives, namely the bis-HCl and bis-TFA salts of the mono-oxasapphyrin **10.34** and the bis-hydrochloride adduct of monoselenasapphyrin **10.37** (Figures 10.5.18–10.5.20). While these structures bear general resemblance to those obtained in the all-aza series, the total number of hydrogen bonds is reduced from five to four. As a likely consequence of this, the counter anions are found further from the mean macrocycle plane than one might otherwise expect (mean sapphyrin–chloride anion distances of 1.814 Å and 1.936 Å,[295] and *ca.* 2.50 Å[296] were determined for **10.34**·2HCl and **10.37**·2HCl, respectively, vs. corresponding distances of 1.77Å and 1.88 Å for **10.6**·2HCl; *vide supra*). This, in turn, leads to the conclusion that hydrogen-bonding effects, in addition to purely electrostatic ones (likely to be the same for **10.6**·2HCl, **10.34**·2HCl, and **10.37**·2HCl), play a significant role in defining the nature and strength of the observed anion-binding interactions. While

Figure 10.5.18   Single Crystal X-Ray Structure of the Bis-TFA Salt of
Monooxasapphyrin **10.34**.
This X-ray structural figure was generated using unpublished data provided by Sessler, *et al.* and
corresponds to a structure originally reported in reference 295. Atom labeling scheme: carbon: ○;
nitrogen: ◉; oxygen: ⊖; fluorine: ⊘; hydrogen: ○. Selected hydrogen atoms have been omitted for
clarity

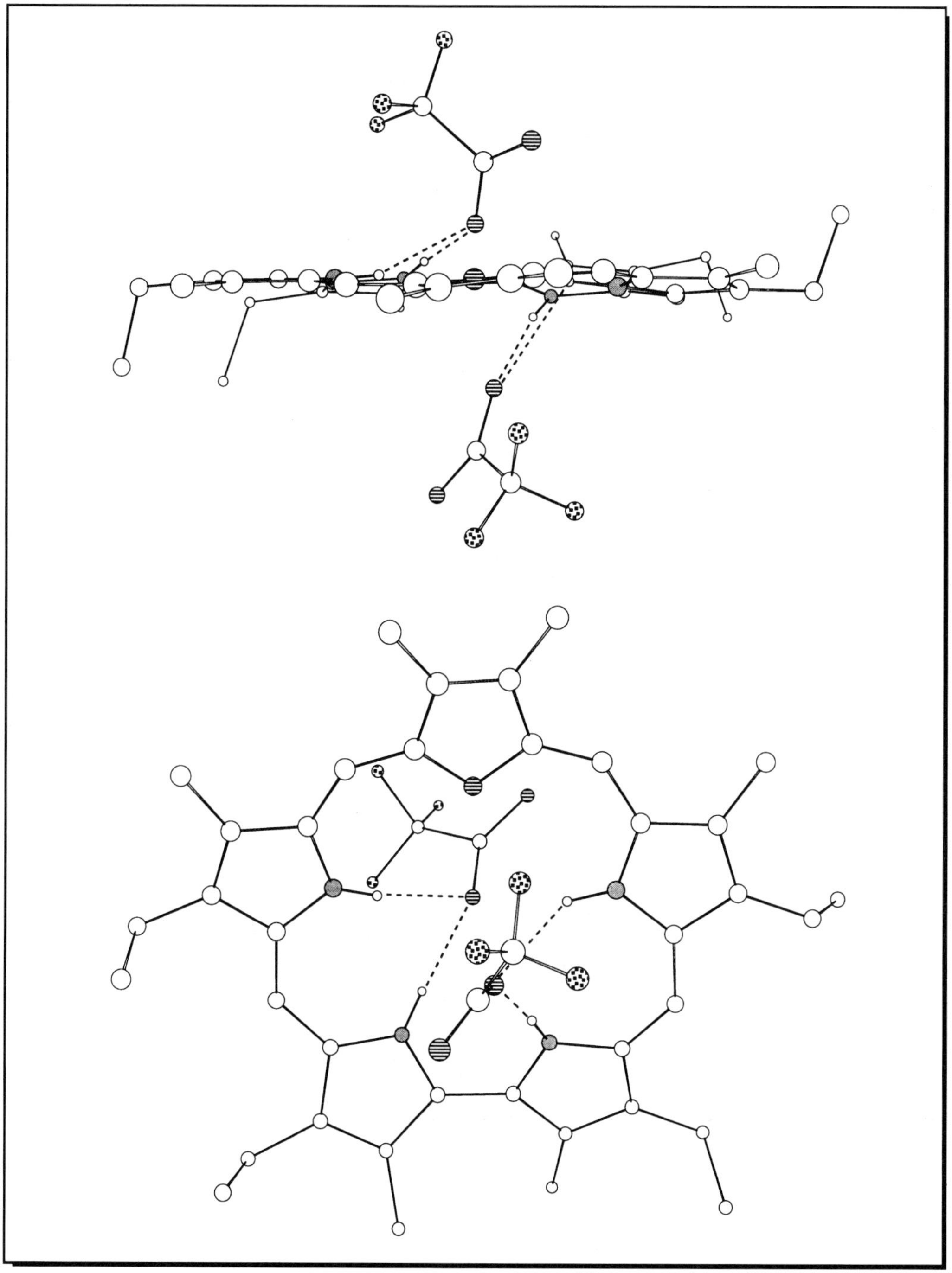

Figure 10.5.19 Single Crystal X-Ray Structure of the Bis-HCl Salt of Monooxasapphyrin **10.34**.
This X-ray structural figure was generated using unpublished data provided by Sessler, *et al.* and corresponds to a structure originally reported in reference 295. Atom labeling scheme: carbon: ○; nitrogen: ●; oxygen: ⊜; chlorine: ◌; hydrogen: ○. Selected hydrogen atoms have been omitted for clarity

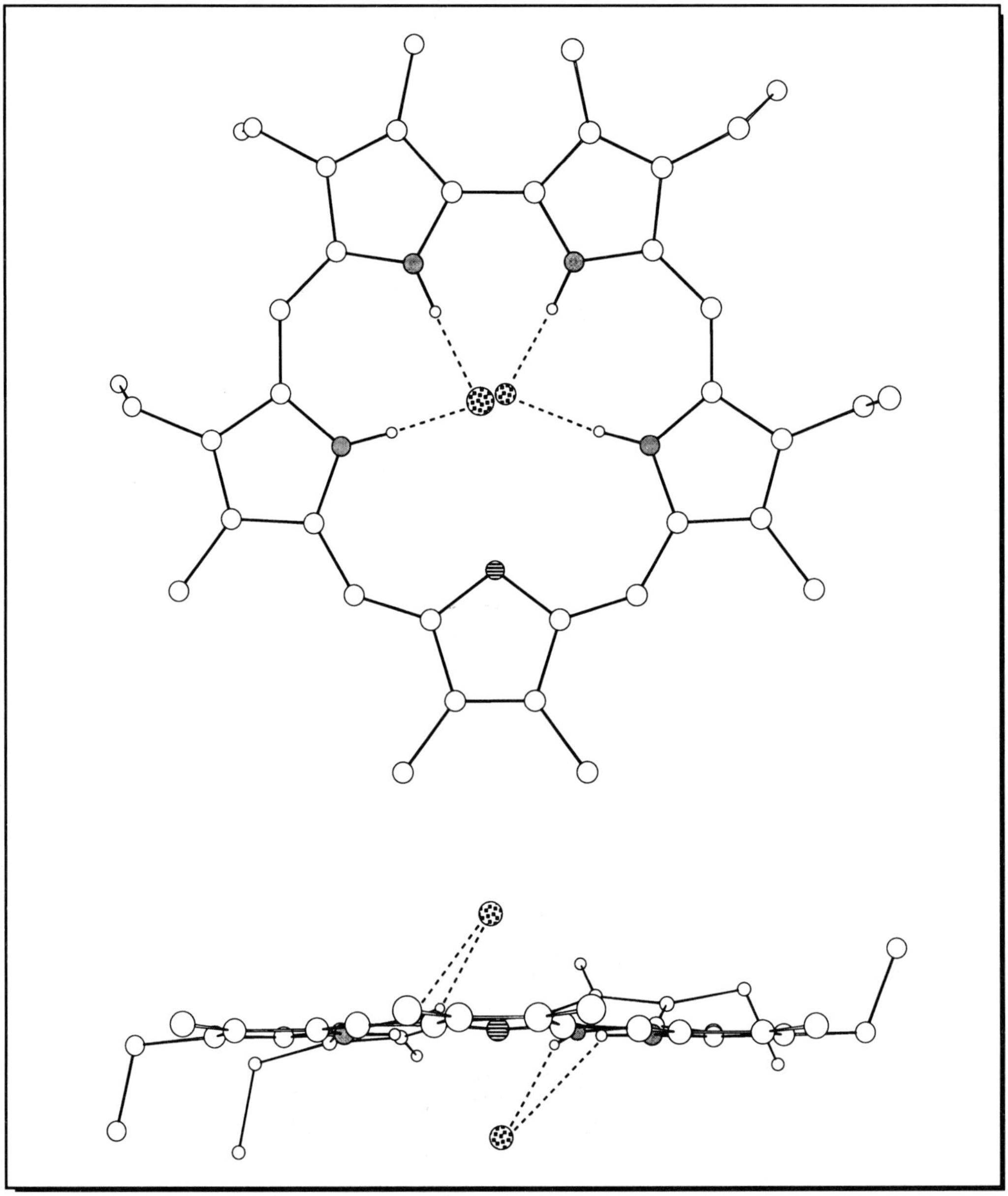

Figure 10.5.20   Single Crystal X-Ray Structure of the Bis-HCl Salt of
Selenasapphyrin **10.37**.

This figure was generated using information down-loaded from the Cambridge Crystallographic
Data Centre and corresponds to a structure originally reported in reference 296. Atom labeling
scheme: carbon: ○; nitrogen: ●; chlorine: ☉; selenium: ⊘; hydrogen: ○. Selected hydrogen atoms
have been omitted for clarity

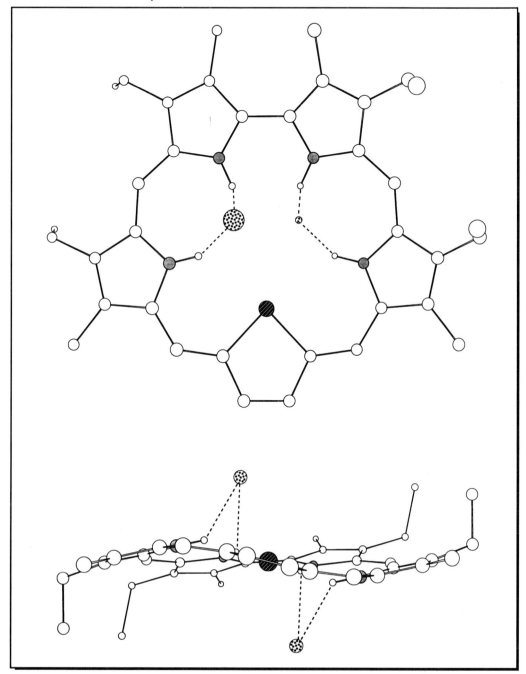

the merits of this conclusion have yet to be tested in terms of detailed solution-phase quantitative binding studies, the series defined by heterosapphyrins **10.34–10.36**[295] might allow this to be done in a rather controlled way. To the extent this proves true, it would allow the relative contributions of hydrogen bonding and electrostatics to be assessed in at least this class of prototypic anion-binding receptors.

Figure 10.5.21   Single Crystal X-Ray Structure of the Mixed Salt Formed between Anthraphyrin **10.32**, HCl, and HBF$_4$.

In this structure the anthraphyrin skeleton is doubly protonated and the chloride anion is bound nearly within the pyrrolic receptor cavity; the BF$_4^-$ counter anion, on the other hand, is not proximate to the macrocylic ring. This figure was generated using information down-loaded from the Cambridge Crystallographic Data Centre and corresponds to a structure originally reported in reference 297. Atom labeling scheme: carbon: ○; nitrogen: ●; chlorine: ⊗; hydrogen: ○. Selected hydrogen atoms have been omitted for clarity

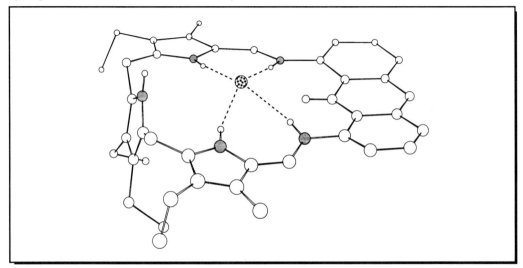

The first non-sapphyrin expanded porphyrin to be recognized as being a bona fide anion binding agent was the so-called anthraphyrin system **10.32**.[297] This macrocycle, when protonated, binds chloride anion effectively in the solid state (Figure 10.5.21). While in purely structural terms the resulting supramolecular complex (characterized structurally in the form of the mixed HCl–HBF$_4$ salt) resembles that observed earlier in the case of the HSCN salt (Figure 10.5.22) of monoprotonated "texaphyrinogen" **10.46** (Figure 10.5.23),[298] it differs significantly in that: (1) the macrocycle is diprotonated; and (2) the chloride counter anion is encapsulated within the pyrrolic receptor cavity. Specifically, in the mixed anthraphyrin salt of Figure 10.5.21, the chloride anion was found to reside only 0.795 Å above the planar portion of the macrocycle.[297] By contrast, in the texaphyrinogen·HSCN salt of Figure 10.5.22 the nitrogen atom of the thiocyanate counter anion rests a full 2.11 Å above the mean plane obtained by considering the four flattest nitrogen atoms.[298]

Figure 10.5.22   Single Crystal X-Ray Structure of the Mono-HSCN Salt of
"Texaphyrinogen" **10.46**.

This figure was generated using information down-loaded from the Cambridge Crystallographic
Data Centre and corresponds to a structure originally reported in reference 298. Atom labeling
scheme: carbon: ○; nitrogen: ◉; sulfur: ◑; hydrogen: ○. Selected hydrogen atoms have been omitted
for clarity

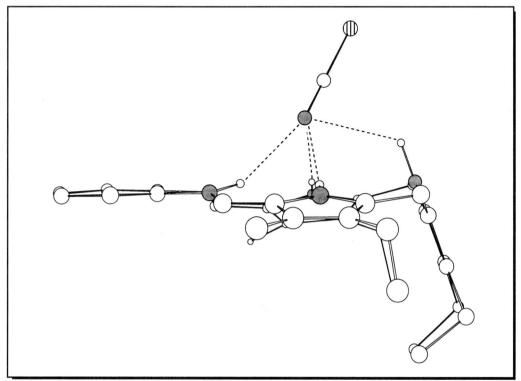

Chronologically, the next protonated expanded porphyrin–anion system to be
characterized in the solid state was the bis-hydrochloride salt of rubyin **10.33**.[299] The
free-base form of this compound may be considered as being a larger 26 π-electron
homolog of sapphyrin. In accord with this notion, the single crystal X-ray structure
of **10.33·2HCl** (Figure 10.5.24) bears considerable resemblance to that of **10.6·2HCl**
(see Figure 10.5.12 above). However, in the case of rubyrin, the chloride counter
anions were found to reside *closer* to the macrocyclic core, being ≈ 1.6 Å above and
below the mean N_6 plane (vs. the corresponding above and below distances of ≈ 1.8
Å for sapphyrin; *vide supra*). This result, which is not surprising on reflection, was
rationalized in terms of the larger rubyrin core being better able to accommodate
concurrently large, negatively charged chloride counter anions.[299]

Although only prepared quite recently,[300] amethryin (**10.42**) and orangarin
(**10.43**) are of interest in that they define smaller analogs of rubyrin and sapphyrin,
respectively. One set of comparisons such congruence could allow is in terms of

anion binding. Specifically, having a range of related compounds could allow such critical features as cavity size, nitrogen basicity, and macrocycle flexibility to be assessed in this capacity. So far, however, this possibility has only been explored in the solid state. Here, as illustrated in Figures 10.5.25 and 10.5.26, it has been determined (from X-ray structural analyses) that both amethryin and orangarin, when doubly protonated, do in fact bind anions in the solid state when crystallized under appropriate conditions. In the case of amethryin, it was the bis-hydrochloride salt that was determined crystallographically; the resulting structure reveals two bound chloride anions held above and below the macrocyclic plane (at a distance of *ca.* 1.80 Å) by a single well-defined hydrogen bond.[300] By contrast, with orangarin, it was the bis-TFA salt that yielded X-ray diffraction crystals.[301] In this instance, the trifluoroacetate counter anions were found to reside *ca.* 1.70 Å and 1.76 Å above and below the mean N_5 plane, being held in place there by three and two hydrogen bonds, respectively.

Figure 10.5.23   Structure **10.46**

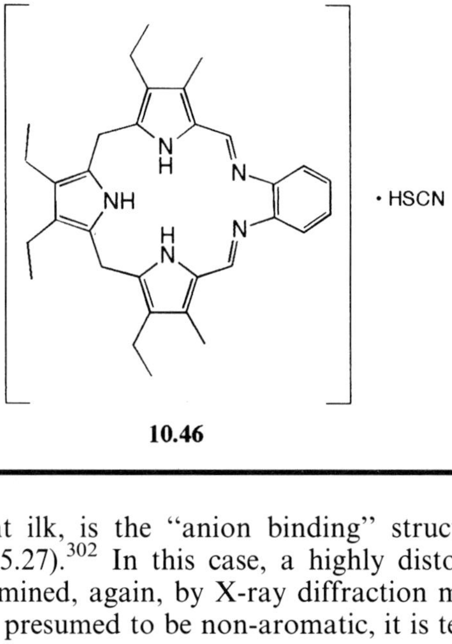

10.46

Of a far different ilk, is the "anion binding" structure of rosarin trishydrochloride (Figure 10.5.27).[302] In this case, a highly distorted system is seen in the solid state (as determined, again, by X-ray diffraction means). While this is as befits a structure that is presumed to be non-aromatic, it is tempting to suggest that the observed anion binding interactions, holding two of the three chloride counter anions within the macrocycle "core", play a role in stabilizing the overall supramolecular geometry.

It is likely that anion–protonated macrocycle interactions also play a critical role in establishing the three-dimensional geometry of turcasarin tetrahydrochloride.[303] The quadruply protonated form of this macrocycle (**10.44**·4HCl) is found to adopt a helix-like "figure-eight" twist both in solution and in the solid state. While the determinants of conformation in solution are potentially difficult to "deconvolute", the single crystal X-ray diffraction structure (Figure 10.5.28) of **10.44**·4HCl provides a strong "hint" that protonated turcasarin–chloride anion interactions likely play an important stabilizing role, at least in the solid state.

Figure 10.5.24   Single Crystal X-Ray Structure of the Bis-HCl Salt of Rubyrin **10.33**.

This figure was generated using information down-loaded from the Cambridge Crystallographic Data Centre and corresponds to a structure originally reported in reference 299. Atom labeling scheme: carbon: ○; nitrogen: ●; chlorine: ◉; hydrogen: ○. Selected hydrogen atoms have been omitted for clarity

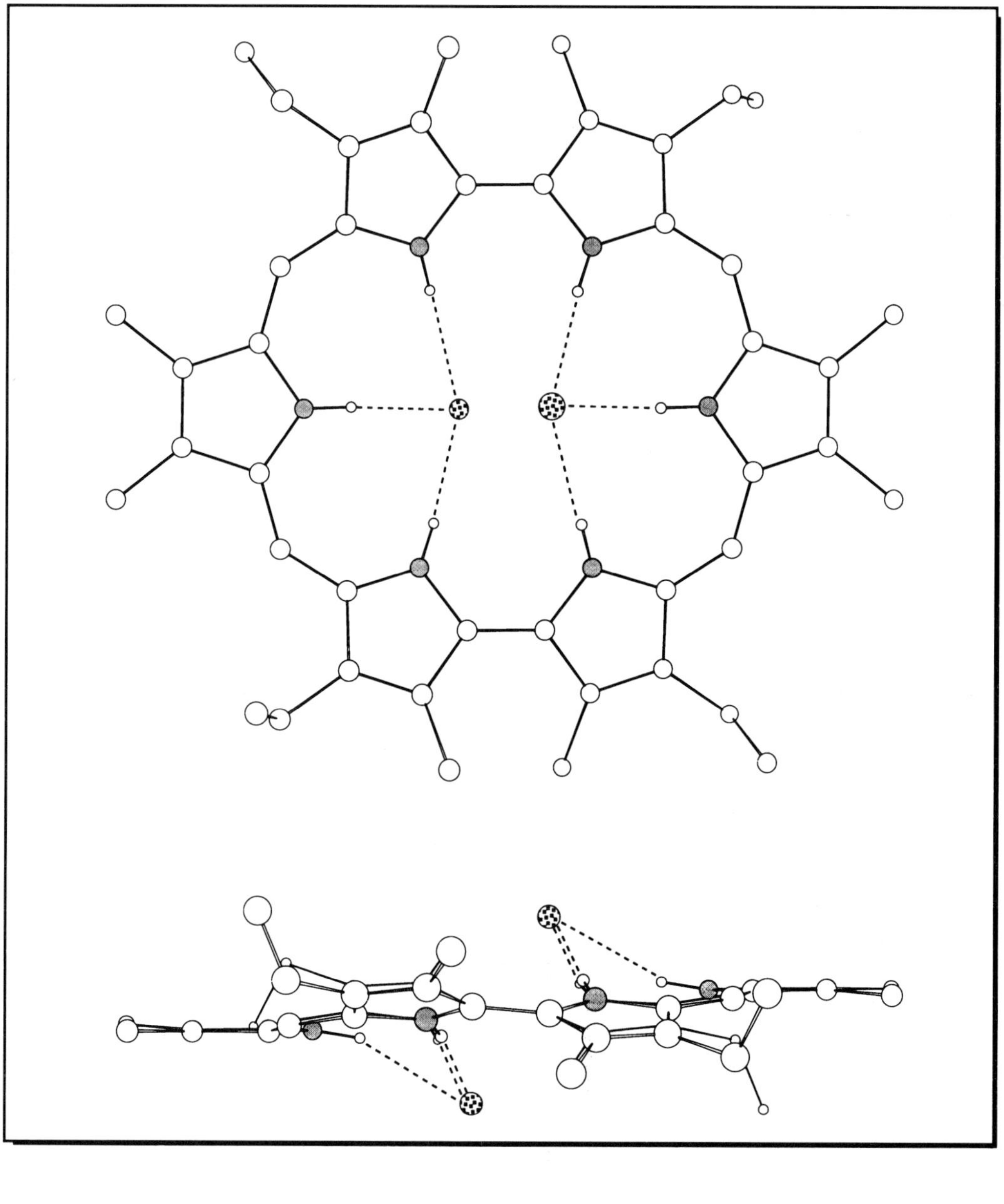

Figure 10.5.25   Single Crystal X-Ray Structure of the Bis-HCl Salt of Amethryin **10.42**.

In this structure, each of the chloride anions is within hydrogen-bonding distance (N-H···Cl = 3.214(6) Å) of only one of the pyrrolic nitrogens, and is found to reside either above or below the mean N_6 plane. This figure was generated using information down-loaded from the Cambridge Crystallographic Data Centre and corresponds to a structure originally reported in reference 300. Atom labeling scheme: carbon: ○; nitrogen: ●; chlorine: ⊙; hydrogen: ○. Selected hydrogen atoms have been omitted for clarity

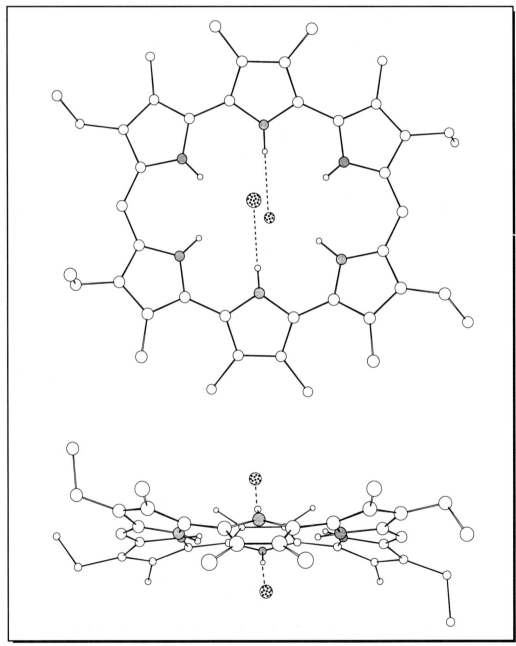

Figure 10.5.26   Single Crystal X-Ray Structure of the Bis-TFA Salt of Orangarin **10.43**.

This figure was generated using unpublished data provided by Sessler, Lynch, and Guba, and corresponds to a structure originally reported in reference 301. Atom labeling scheme: carbon: ○; nitrogen: ◉; oxygen: ⊖; fluorine: ⊗; hydrogen: ○. Selected hydrogen atoms have been omitted for clarity

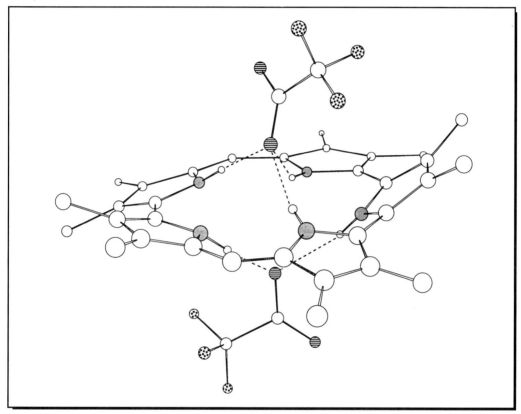

Less clear at present is how important anion-binding effects are in the case of Vogel's octaphyrin-(2.1.0.1.2.1.0.1) system **10.45**.[304] This system, like Sessler's turcasarin, adopts a figure-eight conformation both in solution and in the solid state. However, being smaller and hence more rigid, it is possible that the observed three-dimensional geometry is more intrinsic to the system and less a function of anion-mediated folding effects. Nonetheless, as illustrated by Figure 10.5.29, this macrocycle, when tetraprotonated, can be induced to crystallize under conditions where macrocycle–anion interactions are evident. This, in turn, leads the authors of this monograph to suggest that these new systems, at least formally, should be considered as being anion-binding receptors. On the other hand, the apparent differences between turcasarin and octaphyrin-(2.1.0.1.2.1.0.1) serve to illustrate the complexities and challenges that await synthetic chemists as the chemistry of expanded porphyrins begins to move into three dimensions.

Figure 10.5.27   Single Crystal X-Ray Structure of the Tris-HCl Salt of Rosarin **10.41**.

In this structure, the macrocycle is triply protonated. However, only two of the three chloride anions reside within the macrocyclic "core"; the other is not proximate to rosarin skeleton. This figure was generated using information down-loaded from the Cambridge Crystallographic Data Centre and corresponds to a structure originally reported in reference 302. Atom labeling scheme: carbon: ○; nitrogen: ●; oxygen: ⊖; chlorine: ⊘; hydrogen: ○. Selected hydrogen atoms have been omitted for clarity

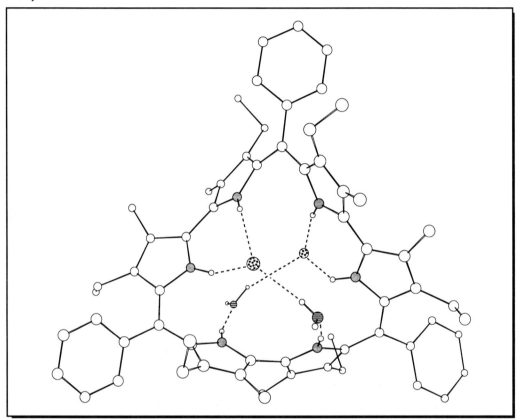

10.5.3   Anion Binding in Solution: Sapphyrins, Rubyrins, Anthraphyrins

In spite of what is now an overwhelming body of data supporting the contention that protonated expanded porphyrins can chelate anions in the solid state, it is important to appreciate that, in most instances, the presumed anion-binding properties have yet to be confirmed *via* what are often more demanding solution-phase analyses. In fact, as mentioned in the introduction to this portion of the chapter (i.e., Section 10.5.1), only three expanded porphyrin systems, sapphyrin, rubyrin and anthraphyrin, have been established as being bona fide anion-binding agents in solution and only one of these, namely sapphyrin, has been studied in detail.

Figure 10.5.28   Single Crystal X-Ray Structure of the Tetrakis-HCl Salt of Turcasarin **10.44**.

In this structure, the macrocycle is quadruply protonated with all four chloride anions residing (two each) within the two "pentapyrrolic pockets". This figure was generated using information down-loaded from the Cambridge Crystallographic Data Centre and corresponds to a structure originally reported in reference 303. Atom labeling scheme: carbon: ○; nitrogen: ●; chlorine: ⊗; hydrogen: ○. Selected hydrogen atoms have been omitted for clarity

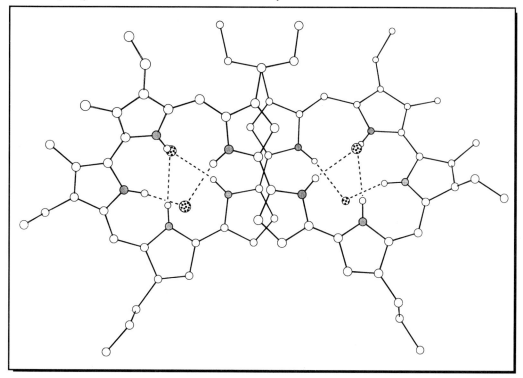

In general, two basic approaches have been used to study the anion-binding properties of expanded porphyrins in solution, namely transport experiments and direct affinity-constant determinations. For sapphyrin, a third approach, involving making elaborated sapphyrin derivatives (including functionalized solid supports), has also been used to good effect.

In the case of expanded porphyrins, the transport-based approach, which is relevant to potential applications involving the regulation of anion flux (e.g., the development of chloride anion carriers for treatment of cystic fibrosis or adjuvants to allow improved uptake into cells of antiviral agents), has for the most part involved the use of a U-tube type[284,285,287,297,305,306] aqueous I–dichloromethane–aqueous II model membrane system (see Figure 10.5.30). Here, the relevant experiments consisted of seeing whether an organic soluble expanded porphyrin, added to the intervening dichloromethane phase, could be used to enhance the rate of diffusion of an organic insoluble anion, such as fluoride, chloride, or phosphate, from the first aqueous phase (Aq. I) to the second (Aq. II). Positive "hits" were then defined as those combinations of anion, carrier, and conditions (i.e., expanded porphyrin concentration, initial pH, etc.) that allowed for enhanced Aq. I to Aq. II anion transport.

Figure 10.5.29   Single Crystal X-Ray Structure of the Tetrakis-HClO$_4$ Salt of Octaphyrin-(2.1.0.1.2.1.0.1) **10.45**.
This figure was generated using information down-loaded from the Cambridge Crystallographic Data Centre and corresponds to a structure originally reported in reference 304. Atom labeling scheme: carbon: ○; nitrogen: ●; oxygen: ◉; chlorine: ⊗; hydrogen: ○. Selected hydrogen atoms have been omitted for clarity

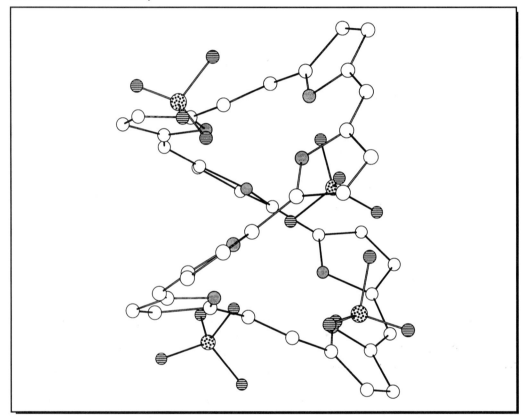

Using the above approach, in early work it was found that sapphyrin **10.6** and anthraphyrin **10.32** both acted as effective carriers for halide anions at neutral (Aq. I) pH.[285,297] Interestingly, in the case of anthraphyrin where a detailed comparison was carried out, it was fluoride anion, rather than chloride anion, that was transported at the faster rate ($\Phi_{F^-}/\Phi_{Cl^-} \approx 3.5$).[297] On the other hand, as was to be expected in view of the solid-state data (see Figure 10.5.21 above), the affinity constant for chloride anion binding in dichloromethane was found to be over a factor of 10 larger than that of fluoride ($K_{a[Cl^-]} \approx 2 \times 10^5 \text{ M}^{-1}$ vs. $K_{a[F^-]} \approx 1.4 \times 10^4 \text{ M}^{-1}$).[297] Accordingly, the enhanced rate for fluoride anion transport was rationalized in terms of anion release at the organic–Aq. II interface being rate limiting.[297]

Figure 10.5.30   Schematic Representation of Standard U-Tube-type Model Membrane System Used to Test the Anion Carrier Capability of a Given Expanded Porphyrin System (Illustrated with a Generalized Nucleotide Serving as the Putative Substrate).
Here, the shaded portion on the left side of the figure corresponds to the initial aqueous phase (Aq. I), the unshaded portion, the intervening dichloromethane "membrane", and the shaded portion on the right the receiving aqueous phase (Aq. II)

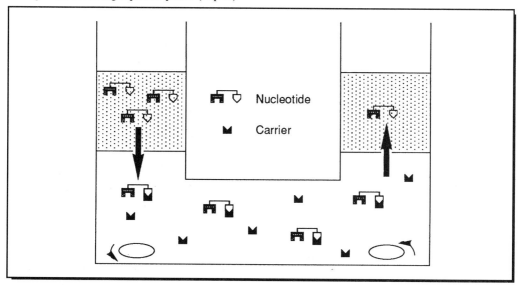

Sapphyrin **10.6**[284,287] and rubyrin **10.33**[306] were also studied as potential phosphate anion carriers. While it was found that sapphyrin could effect the through-model-membrane transport of the monoanionic substrate *c*-AMP at neutral (Aq. I) pH,[287] neither it nor rubyrin was found to function as a viable carrier for GMP under identical neutral Aq. I conditions. This was rationalized in terms of these systems being monoprotonated at neutral pH (pK_a values of 4.8 and 8.8 have recently been recorded for the diprotonated form of sapphyrin **10.31**[292]) and hence unsuited for binding GMP or other phosphorylated substrates that are largely *dianionic* at pH 7. Consistent with this supposition were findings that: (1) at low pH (i.e. $\leqslant$ 3.5) both of these systems did function as GMP carriers;[287] and (2) at near-neutral pH the larger and presumably more basic rubyrin system could be used to effect the through-model-membrane transport of GMP, provided triisopropylsilyl protected cytosine (C-Tips) was added as a cocarrier.[306]

To build on the above promise and to generate systems that would be capable of effecting the through-membrane transport of mononucleotides, systems **10.47** and **10.48** (Figure 10.5.31) were synthesized by Sessler and coworkers.[287,305,307] These contain Watson–Crick hydrogen bonding "complements" built into the molecule.

The thought was thus that a combination of hydrogen bonding and sapphyrin-based phosphate "chelation" interactions would allow for the selective recognition and through membrane transport of the appropriate, complementary nucleotide. As it transpired, this approach, illustrated schematically in Figure 10.5.32, did in fact work; conjugate **10.47** was found to be a selective carrier for GMP whereas **10.48** was found to be selective for CMP.[287,305,307] Taken together, these results were considered to augur well for the eventual use of sapphyrin-based systems in antiviral therapy applications. They might be used to enhance the uptake into cells of phosphorylated nucleotide-type drugs, such as 9-(β-D-xylofuranosyl)guanine,[308] whose cytoplasmic concentration might otherwise be too low to mediate a therapeutic response.[287,305,307]

Figure 10.5.31   Structures **10.47** and **10.48**

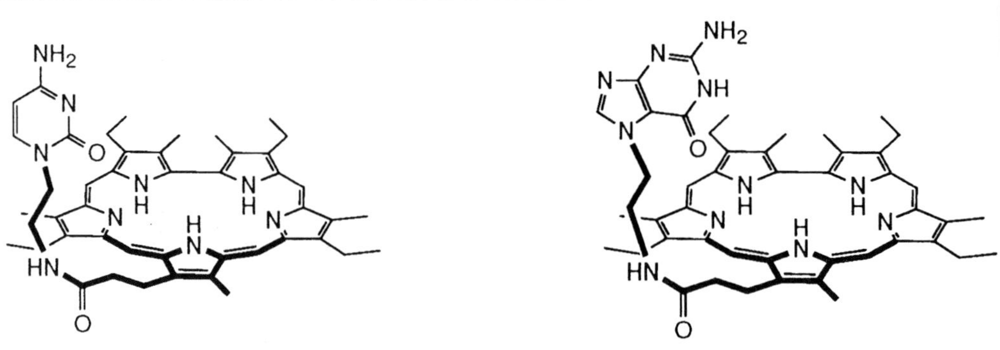

10.47   10.48

Figure 10.5.32   Schematic representation of the proposed supramolecular complex formed between conjugate **10.47** and monobasic GMP

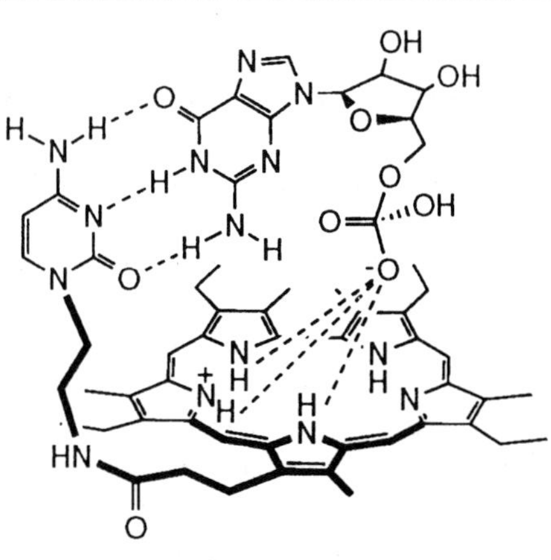

Carboxylate anion binding and transport was also studied qualitatively using a combination of transport experiments and tailored conjugate construction. In this instance, it was quickly appreciated, on the basis of preliminary experiments, that simple monomeric sapphyrins such as **10.6** were ineffectual as carriers for biologically interesting carboxylated substrates such as aromatic amino acids or rigid α,ω-dicarboxylic acids. Thus, conjugates **10.49**[309] and **10.50**[310] (Figure 10.5.33) were synthesized by Sessler and coworkers; as expected these ditopic systems did indeed function as receptors and carriers for these two classes of targets, respectively.[309,310] More elaborate analogs of these latter all-sapphyrin receptors, including the trimeric and tetrameric sapphyrin systems **10.51** and **10.52** (Figure 10.5.34), have also been prepared; these latter were found to function as carriers for nucleotide di- and triphosphates, respectively.[311] This chemistry, as well as that of conjugates **10.49** and **10.50** has been reviewed recently.[237]

Figure 10.5.33   Structures **10.49** and **10.50**

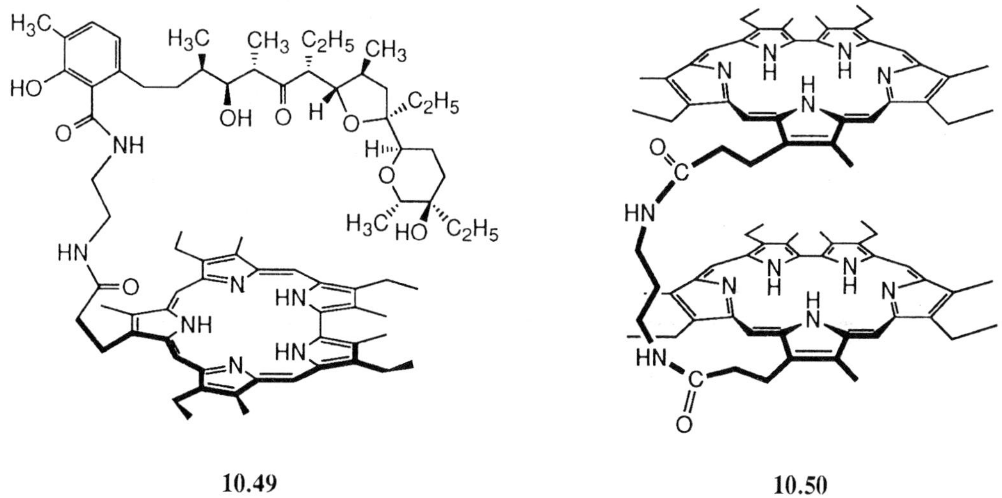

10.49   10.50

As with anthraphyrin (*vide supra*), direct measurements of affinity constants were made as a way of complementing the basic "yes, it binds anions" conclusions obtained from the sapphyrin-predicated through-model-membrane transport studies described above.[282,292,310,312] In general, the requisite measurements were made by monitoring the changes in a given spectroscopic property (e.g., ^{1}H or ^{31}P NMR chemical shift, UV-vis absorbance, fluorescence yield) that were observed to occur as the sapphyrin–anionic substrate ratios were systematically varied. Occasionally, affinity constants were derived by competition. Here, spectroscopic changes were monitored as one anion (of unknown affinity) was added to a solution of sapphyrin containing another anion for which the relevant K_a-binding value was known.

Figure 10.5.34   Structures **10.51** and **10.52**

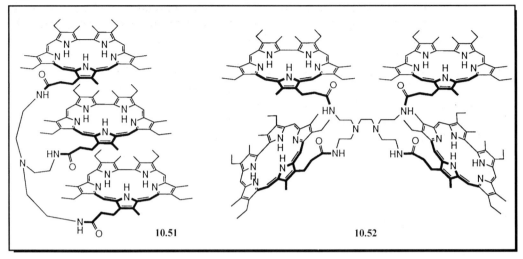

By such methods, it proved possible to determine that fluoride anion is bound to the diprotonated form of sapphyrin with highest affinity ($K_a > 10^8$ M^{-1} in CH_2Cl_2; $K_a \approx 1 \times 10^5$ M^{-1} in CH_3OH^{282}) but that phosphorylated entities, such as $H_2PO_4^-$, are also bound quite well ($K_a \approx 1 \times 10^4$ M^{-1} in CH_3OH; $K_a \approx 1 \times 10^2$ M^{-1} aqueous bis-Tris buffer, pH 6.1[292,293]). By contrast, chloride anion ($K_a \approx 2 \times 10^7$ in CH_2Cl_2; $K_a \approx 10^2$ M^{-1} in CH_3OH^{282}) and bromide anion ($K_a \approx 1.5 \times 10^6$ in CH_2Cl_2; $K_a \leqslant 10^2$ M^{-1} in CH_3OH^{282}) were found to be bound but weakly to simple diprotonated sapphyrin derivatives. Aromatic carboxylate anions, on the other hand, were found to be reasonably well bound by diprotonated sapphyrins (for benzoic acid: $K_a \approx 2.5 \times 10^3$ M^{-1} in CH_3OH^{292}) but not so well by monoprotonated sapphyrins ($K_a \approx 1 \times 10^3$ in CH_2Cl_2; $K_a \leqslant 10$ M^{-1} in $CH_3OH^{310,312}$).

Interestingly, in spite of the fact that carboxylate anions are not bound overly well by monoprotonated sapphyrins, it proved possible to construct the supramolecular porphyrin–sapphyrin ensemble shown in Figure 10.5.35.[312] This pseudo dimer (**10.53**) is of interest because it represents a new kind of non-covalent energy- and electron-transfer model system. It is also of interest because it provides yet further proof of anion binding in solution. This is because irradiation at 573 nm gives rise to optical changes that are consistent with rapid, Förster-type singlet–singlet energy transfer from the excited porphyrin subunit (**10.53a**) to the lower energy sapphyrin "acceptor" (**10.53b**).[312] Since such rapid energy transfer was not observed in various monomeric and non-bound controls, its very occurrence was considered consistent with the key predicative concept that the monoprotonated sapphyrin "mouth" can in fact "bite" the porphyrin carboxylate "tail". Thus, these results support the conclusions reached from studies of the carboxyl-functionalized sapphyrin derivative **10.30** (*vide supra*), namely that sapphyrin–carboxylate interactions can be used to build up supramolecular systems of considerable complexity.

Figure 10.5.35   Schematic representation of the proposed non-covalent energy-transfer ensemble **10.53** formed *via* carboxylate anion recognition[312]

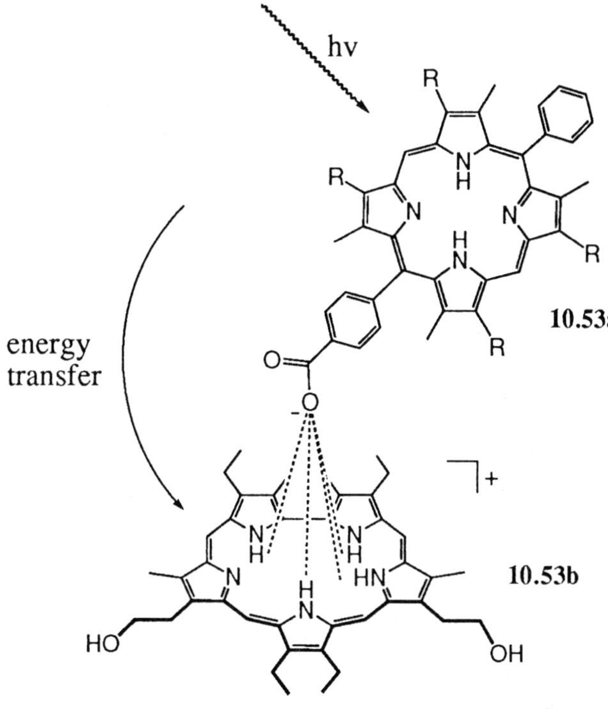

ENSEMBLE 10.53

10.5.4   *Sapphyrin–Oligonucleotide Interactions*

Inspired by the finding that monoprotonated sapphyrin would bind *c*-AMP and other monoanionic phosphate species at neutral pH, Iverson and Sessler proposed that sapphyrins should bind well to various oligonucleotides, including DNA.[33,313,314] Specifically, the proposal put forward was that sapphyrin would "chelate" to the anionic phosphate backbone of these phosphodiester polymers in an analogy to what is seen in the solid-state structure of Figure 10.5.15. To the extent this proposal proved correct, it was recognized that it would define a new mode of small molecule–DNA interaction, which, in terms of potential applications, could complement the three other general molecular recognition strategies currently employed, namely intercalation, groove binding, and simple electrostatic attraction.[315–318]

To date, considerable evidence has been put forward that supports the conclusion that sapphyrins do in fact bind to DNA as proposed above, namely *via* a molecular recognition mode that can best be termed "phosphate chelation".[293,313] For instance, in early work it was found that adding an excess of sapphyrin **10.31** to dsDNA at neutral pH led to an immediate precipitation of the DNA as visible green fibers.[313] The solid-state ^{31}P NMR spectrum of this precipitate revealed that the critical DNA ^{31}P signal had undergone a 3.6 ppm upfield shift as compared to DNA precipitated in the absence of sapphyrin.[293] In the case of a porphyrin control (structure not shown), an upfield shift of only 1.6 ppm was observed under identical experimental conditions. Along these same lines, the complex produced from phosphoric acid and sapphyrin **10.31** (but not various porphyrin controls) yielded similar upfield shifts of 3.8 ppm in the solid-state ^{31}P NMR spectrum.[293] Taken together, these results were considered consistent with the phosphorous atoms of the DNA backbone being "chelated" above the plane of **10.31** and thus subject to sapphyrin-derived aromatic ring-current effects.

Further evidence for the proposed mode of sapphyrin–DNA interaction came from UV-vis spectroscopic studies. Here, in what was important background work, it was discovered that, broadly speaking, a water-soluble sapphyrin such as **10.31** can exist in three limiting (and spectroscopically distinct) aggregation states.[293] These are a purely monomeric form, possessing a λ_{max} value of *ca.* 450 nm, a dimer (with λ_{max} of *ca.* 420 nm), and a highly aggregated, stacked state of sapphyrin that is characterized by an absorption maximum between 400 and 410 nm. Using these predicative findings as a gauge, it was determined that adding anion DNA, either single stranded (ssDNA) or double stranded (dsDNA) leads to a build up of the less aggregated forms of sapphyrin, with the final spectroscopic state, observed at high phosphate–sapphyrin ratios, being characteristic (in both cases) of a phosphate-bound sapphyrin dimer. On this basis, it was concluded that groove binding plays little role in defining sapphyrin–DNA interactions (since ssDNA lacks a well-defined groove) but that so-called "outside stacking" effects[277,278] could not be entirely ignored.[293]

The above studies also provided an important "hint" that intercalation, a well-known mode of porphyrin–DNA interaction,[318] is not a significant contributor to the sapphyrin–DNA molecular recognition process. This is because, if intercalation were occurring, sapphyrin **10.31** would be bound in the form of a monomer rather than a dimer. Further support for this proposed non-intercalation came from both viscometric analyses and topoisomerase I unwinding studies. In both cases sapphyrin **10.31** was found to act in complete contrast to ethidium bromide, a known DNA intercalative agent.[315–318]

In order to support further the basic underlying postulate that monoprotonated sapphyrins will "chelate" oligonucleotides, Iverson, *et al.* prepared the sapphyrin–ethylenediaminetetraacetic acid (EDTA) conjugate **10.54** (Figure 10.5.36).[319] Previous studies of small molecule–EDTA conjugates have shown that these species, in the presence of iron(II), O_2, and a suitable reducing agent, can effect the oxidative cleavage of DNA.[320] This so-called "affinity cleavage", in turn, can be used to probe features of the chosen DNA target as well as to establish the nature of a given kind of small molecule–DNA interaction. In the case of **10.54**, a standard supercoiled plasmid DNA assay[321,322] was employed to test whether the synthetic "addition" of a sapphyrin subunit augmented, diminished, or left unaffected the cleavage efficiency of the appended iron EDTA "active site". As it transpired, the iron chelate of **10.54** (prepared *in situ* by treating **10.54** with ferrous ammonium sulfate) was found, in the presence of O_2 and the reducing agent dithiothreitol (DTT), to effect cleavage of supercoiled plasmid pBR322 DNA at concentrations that were at least an order of magnitude lower than those at which Fe–EDTA alone was found to effect similar levels of cleavage (Figure 10.5.37).[319] On this basis it was concluded that sapphyrin does in fact bind DNA with high avidity.

Figure 10.5.36   Structure **10.54**

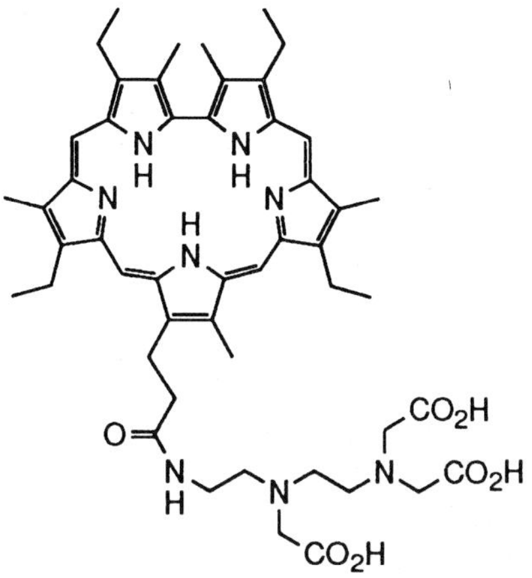

10.54

One final "proof" that monoprotonated sapphyrins bind oligonucleotides under neutral aqueous conditions was put forward by the Iverson–Sessler team. It consisted of showing that high-performance liquid chromatography columns prepared using sapphyrin-functionalized silica gels (as, for instance, shown schematically in Figure 10.5.38) could separate not only mononucleotides (i.e., AMP, from ADP, from ATP) but also oligonucleotides of intermediate size.[323] Specifically, under isochratic conditions using a simple phosphate buffer as the eluent, it proved possible to effect the separation of a poly-adenine 9-mer from the corresponding 8-mer with near-complete resolution (Figure 10.5.39). Such separations, effected on the basis of the net substrate charge, are not only consistent with the basic "phosphate chelation" hypothesis, they are also interesting in their own right. Specifically, they led Iverson and Sessler to propose that columns based on sapphyrin-functionalized silica gels could provide alternatives to other more conventional oligonucleotide separation procedures.[323] Whether this promise is in fact realized using sapphyrin or some other porphyrin analog system remains to be seen. Nonetheless, these chromatographic applications, like the ones involving more direct interactions with DNA, highlight the potentially important role expanded porphyrins have to play in the newer, more biochemically targeted areas of bioorganic chemistry and biological molecular recognition. Here, the fact that many expanded porphyrins, including sapphyrin, are photoactive and hence capable of effecting light-induced cleavage of RNA and DNA (see Section 10.4 above) makes this line of research seem like a fertile one indeed.

Figure 10.5.37 Photograph of a 0.8% Agarose Gel Stained with Ethidium Bromide Showing the Results of a Supercoiled pBR322 DNA Cleavage Assay Using the Sapphyrin–EDTA Conjugate Given in Figure 10.5.35 (Structure **10.54**). All reactions were run for 40 minutes with the following specific reagent concentrations being used: DNA, 15 μg mL^{-1}; dithiothreitol (DTT), 2.0 mM; PIPES buffer, 5 mM, pH 7.0, 23 °C. This figure corresponds to one that originally appeared in reference 319 and has been reproduced with permission

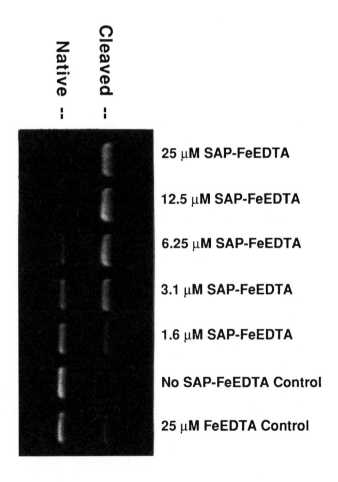

25 μM SAP-FeEDTA

12.5 μM SAP-FeEDTA

6.25 μM SAP-FeEDTA

3.1 μM SAP-FeEDTA

1.6 μM SAP-FeEDTA

No SAP-FeEDTA Control

25 μM FeEDTA Control

Figure 10.5.38   Schematic Representation of the Sapphyrin-Functionalized Silica Gel **10.55** Used to Generate Oligonucleotide-separating HPLC Chromatography Columns

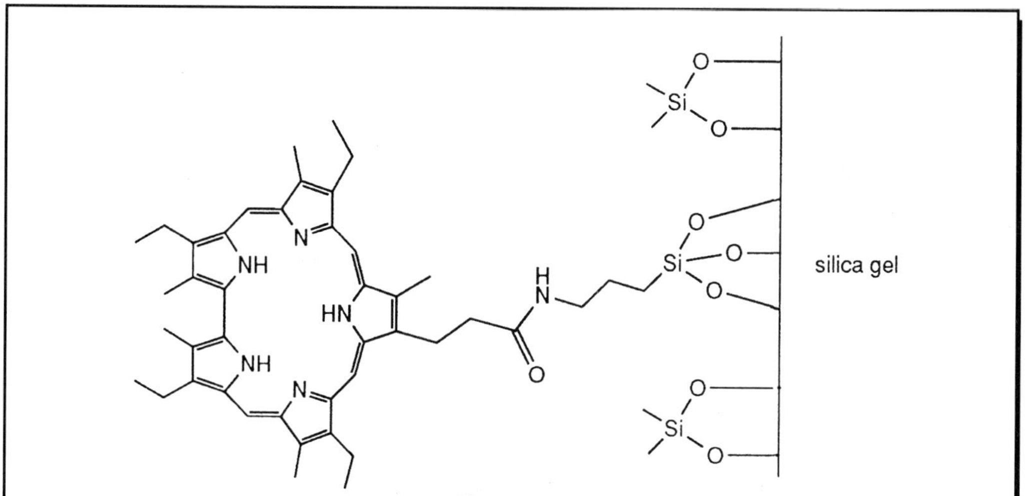

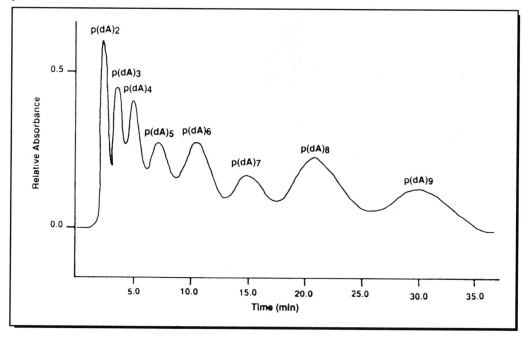

10.55

Figure 10.5.39   HPLC Chromatogram Showing the Separation of 2–9 mers of Polydeoxyadenylic Acids.

The solid support was the silyl-capped, sapphyrin-functionalized silica gel shown schematically as structure **10.55** in Figure 10.5.38. Conditions: Isochratic buffer consisting of 1.0 M ammonium phosphate dibasic, at 1.5 mL min^{-1}, pH 7.0, monitored at 260 nm. This figure was redrawn with permission from reference 323

Note in proof

Subsequent to the submission of this manuscript, Pharmacyclics Inc. announced the results of the Phase I clinical study of Lu(III) texaphyrin **10.4**.[324] A total response (partial and complete) rate of 78% and 50% for skin or subcutaneous breast cancer and melanoma lesions, respectively.

10.6   References

1.   Edelman, R.; Warach, S. *New Engl. J. Med.* **1993**, *328*, 708–716.
2.   Moonen, C. T.; van-Zijil, P. C.; Frank, J. A.; Le-Bihan, D.; Becker, E. D. *Science* **1990**, *250*, 53–61.
3.   Young, S. W. *Magnetic Resonance Imaging: Basic Principles*; Raven Press: New York, 1988.
4.   Lauffer, R. B. *Chem. Rev.* **1987**, *87*, 901–927.
5.   Tweedle, M. F.; Brittain, H. G.; Eckelman,W. C.; Gaughan, G. T.; Hagan, J. J.; Wedeking, P. W.; Runge, V. M. In *Magnetic Resonance Imaging*, Vol. 1, 2nd edn, Partain C. L., Ed.; W.B. Saunders: Philadelphia, 1988; pp. 793–809.
6.   Mann, J. S.; Brasch, R. C. In *Handbook of Metal–Ligand Interactions in Biological Fluids: Bioinorg. Med.*, Vol. 2; Berthon, G., Ed.; Dekker: New York, 1995; pp. 1358–1373.
7.   Morris, P. G. *Nuclear Magnetic Resonance Imaging in Medicine and Biology*, Clarendon Press: Oxford, 1986.
8.   Koenig, S. H.; Brown, R. D., III. *Magn. Res. in Med.* **1984**, *1*, 437–449.
9.   Koenig, S. H.; Brown, R. D. In *Handbook of Metal–Ligand Interactions in Biological Fluids: Bioinorganic Medicine*, Vol. 2; Berthon, G., Ed.; Dekker: New York, 1995; pp. 1093–1108.
10.   Øksendal, A. N.; Hals, F.-A. *J. Magn. Res. Imag.* **1993**, *3*, 157–165.
11.   Bloch, F. *Phys. Rev.* **1946**, *70*, 460–474.
12.   Bloembergen, N; Purcell, E. M.; Pound, E. V. *Phys. Rev.* **1948**, *73*, 679–712.
13.   Solomon, I. *Phys. Rev.* **1955**, *99*, 559–565.
14.   Fossheim, S.; Johansson, C.; Fahlvik, A. K.; Grace, D.; Klaveness, J. *Magn. Res. Med.* **1996**, *35*, 201–206.
15.   Policard, A. *CR Soc. Biol.* **1924**, *91*, 1423–1424.
16.   Auler, H.; Banzer, G. *Krebsforsch.* **1942**, *53*, 65–68.
17.   Figge, F. H. J.; Weiland, G. S.; Manganiello, L. O. J. *Proc. Soc. Exp. Med.* **1948**, *68*, 640–644.
18.   Rasmussen-Taxdall, D. S.; Ward, G. E.; Figge, F. H. *Cancer (Phila.)* **1955**, *8*, 78–81.
19.   Lipson, R. L.; Baldes, E. J.; Olsen, A. M. *J. Natl Cancer Inst.* **1961**, *26*, 1–10.
20.   Chen, C.-W.; Cohen, J. S.; Myers, C. E.; Sohn, M. *FEBS Lett.* **1984**, *168*, 70–74.
21.   Jackson, L. S.; Nelson, J. A.; Case, T. A.; Burnham, B. F. *Invest. Radiol.* **1985**, *2*, 226–229.

22. Patronas, N. J.; Cohen, J. S.; Knop, R. H.; Dwyer, A. J.; Colcher, D.; Lundy, J.; Mornex, F.; Hambright, P.; Sohn, M.; Myers, C. E. *Cancer Treat. Rep.* **1986**, *70*, 391–395.

23. Bohdiewicz, P. J.; Lavallee, D. K.; Fawwaz, R. A.; Newhouse, J. H.; Oluwole, S. F.; Alderson, P. O. *Invest. Radiol.* **1990**, *25*, 765–770.

24. Lyon, R. C.; Faustino, P. J.; Cohen, J. S.; Katz, A.; Mornex, F.; Colcher, D.; Baglin, C.; Koenig, S. H.; Hambright, P. *Magn. Reson. Med.* **1987**, *4*, 24–33.

25. Huang, L. R.; Staubinger, R. M.; Kahl, S. B.; Koo, M.-S.; Alletto, J. J.; Mazurchuk, R.; Chau, R. I.; Thamer, S. L.; Fiel, R. J. *J. Magn. Res. Imaging* **1993**, *3*, 352–356.

26. Hambright, P.; Adams, C.; Vernon, K. *Inorg. Chem.* **1988**, *27*, 1660–1662.

27. Haye, S.; Hambright, P. *J. Chem. Soc., Chem. Commun.* **1988**, 666–668.

28. Lyon, R. C.; Faustino, P. J.; Cohen, J. S.; Katz, A.; Mornex, F.; Colcher, D.; Baglin, C.; Koenig, S. H.; Hambright, P. *Magn. Reson. Med.* **1987**, *4*, 24–33.

29. Sessler, J. L.; Murai, T.; Lynch, V.; Cyr, M. *J. Am. Chem. Soc.* **1988**, *110*, 5586–5588.

30. Stinson, S. *C & E News* **August 8, 1988**, 26–27.

31. Sessler, J. L.; Murai, T.; Hemmi, G. *Inorg. Chem.* **1989**, *28*, 3390–3393.

32. Sessler, J. L.; Mody, T. D.; Hemmi, G. W.; Lynch, V. *Inorg. Chem.* **1993**, *32*, 3175–3187.

33. Sessler, J. L.; Burrell, A. K.; Furuta, H.; Hemmi, G. W.; Iverson, B. L.; Král, V.; Magda, D. J.; Mody, T. D.; Shreder, K.; Smith, D.; Weghorn, S. J. In *Transition Metals in Supramolecular Chemistry (NATO ASI Series C*, Vol. 448), Fabbrizzi, L. and Poggi, A., Eds; Kluwer: Dordrecht, 1994; pp. 391–408.

34. Sessler, J. L.; Hemmi, G.; Mody, T. D.; Murai, T.; Burrell, A.; Young, S. W. *Acc. Chem. Res.* **1994**, *27*, 43–50.

35. Sessler, J. L.; Mody, T. D.; Hemmi, G. W.; Lynch, V.; Young, S. W.; Miller, R. A. *J. Am. Chem. Soc.* **1993**, *115*, 10368–10369.

36. Geraldes, C. F. G. C.; Sherry, A. D.; Vallet, P.; Maton, F.; Muller, R. N.; Mody, T. D.; Hemmi, G.; Sessler, J. L. *J. Magn. Res. Imaging.* **1995**, *5*, 725–729.

37. Young, S. W.; Sidhu, M. K.; Qing, F.; Muller, H. H.; Neuder, M.; Zanassi, G.; Mody, T. D.; Hemmi, G.; Dow, W.; Mutch, J. D.; Sessler, J. L.; Miller, R. A. *Invest. Radiol.* **1994**, *29*, 330–338.

38. Young, S. W.; Qing, F. *Invest. Radiol.* **1996**, *31*, 280–283.

39. Young, S. W.; Qing, F.; Kunis, D. M.; Steinberg, G. K. *Invest. Radiol.* **1996**, *31*, 353–358.

40. Young, S. W.; Qing, F.; Harriman, A.; Sessler, J. L.; Dow, W. C.; Mody, T. D.; Hemmi, G.; Hao, Y.; Miller, R. A. *Proc. Natl Acad. Sci. USA* **1996**, *93*, 6610–6615.

41. Rosenthal, D. I.; Becerra, C. R.; Nurenberg, P.; Young, S. W.; Miller, R. A.; Engel, J. F.; Carbone, D. P.; Frenkel, E. P. *Abstracts of the 1996 American Society for Clinical Oncology, Annual Meeting*, May 18–21, 1996.

42. Young, S. W.; Woodburn, K. W.; Wright, M.; Mody, T. D.; Fan, Q.; Sessler, J. L.; Dow, W. C.; Miller, R. A. *Photochem. Photobiol.* **1996**, *63*, 892–897.

43. Buskard, N. A.; Wilson, B. C. *Sem. Oncol.* **1994**, *21(6)S15*, 1–3.
44. Kessel, D. *The Spectrum* **1990**, *3*, 13–15.
45. Gomer, C. J. *Sem. Hematol.* **1989**, *26*, 27–34.
46. Manyak, M. J.; Russo, A.; Smith, P. D.; Glatstein, E. *J. Clin. Oncol.* **1988**, *6*, 380–391.
47. Brown, S. B.; Truscott, T. G. *Chem. in Brit.* **1993**, 955–958.
48. Pass, H. I. *J. Natl Cancer Inst.* **1993**, *85*, 443–456.
49. Dougherty, T. J. *Photochem. Photobiol.* **1993**, *58*, 895–900.
50. Henderson, B. W.; Dougherty, T. J. *Photochem. Photobiol.* **1992**, *55*, 145–157.
51. Dougherty, T. J. *Adv. Photochem.* **1992**, *17*, 275–311.
52. Parker, S. L.; Tong, T.; Bolden, S.; Wingo, P. A. *CA Cancer J. Clin.* **1996**, *46*, 5–27.
53. Matthews, J. L.; Newman, J. T.; Sogandares-Bernal, F.; Judy, M. M.; Kiles, H.; Leveson, J. E.; Marengo-Rowe, A. J.; Chanh, T. C. *Transfusion* **1988**, *28*, 81–83.
54. Matthews, J. L.; Sogandares-Bernal, F.; Judy, M. M.; Marengo-Rowe, A. J.; Leveson, J.; Skiles, H.; Newman, J.; Chanh, T. C. *Transfusion* **1991**, *231*, 636–641.
55. Ben-Hur, E.; Horowitz, B. *Photochem. Photobiol.* **1995**, *62*, 383–388.
56. Ben-Hur, E.; Geacintov, N. E.; Studamire, B.; Kenney, M. E.; Horowitz, B. *Photochem. Photobiol.* **1995**, *61*, 190–195.
57. Rywkin, S.; Ben-Hur, E.; Malik, Z.; Prince, A. M.; Li, Y.-S.; Kenney, M. E.; Oleinick, N. L.; Horowitz, B. *Photochem. Photobiol* **1994**, *60*, 165–170.
58. Gottlieb, P.; Shen, L. G.; Chimezie, E.; Bahng, S.; Kenney, M. S.; Horowitz, B.; Ben-Hur, E. *Photochem. Photobiol.* **1995**, *62*, 869–874.
59. Sessler, J. L.; Cyr, M.; Maiya, B. G.; Judy, M. L.; Newman, J. T.; Skiles, H.; Boriack, R.; Matthews, J. L.; Chanh, T. C. *Proc. SPIE Int. Opt. Eng.* **1990**, *1203*, 233–245.
60. Franck, B.; Nonn, A. *Angew. Chem. Int. Ed. Eng.* **1995** *34*, 1795–1811.
61. Dougherty, T. J. *Photochem. Photobiol.* **1987**, *46*, 569–573.
62. Dougherty, T. J. *Adv. Photochem.* **1992**, *17*, 275–311.
63. Dougherty, T. J. *Photochem. Photobiol.* **1987**, *45*, 879–889.
64. Christensen, T.; Sandquist, T.; Feren, K.; Waksvik, H.; Moan, J. *Br. J. Cancer* **1983**, *48*, 35–43.
65. Wan, S.; Parrish, J. A.; Anderson, R. R.; Madden, M. *Photochem. Photobiol.* **1981**, *34*, 679–681.
66. Kreimer-Birnbaum, M. *Sem. Hematol.* **1989**, *26*, 157–173.
67. Hua, Z.; Gibson, S. L.; Foster, T. H.; Hilf, R. *Cancer Res.* **1995**, *55*, 1723–1731.
68. Marcus, S. L.; Sobel, R. S.; Golub, A. L.; Carroll, R. L.; Lundahl, S.; Shulman, D. G. *Proc. SPIE Int. Opt. Eng.*, **1996**, *2675*, 32–42.
69. Szeimies, R.-M.; Sassy, T.; Lanthaler, M. *Photochem. Photobiol.* **1994**, *59*, 73–76.
70. Henderson, B. W.; Vaughan, L.; Bellnier, D. A.; van Leengoed, H.; Johnson P. G.; Oseroff, A. R. *Photochem. Photobiol.* **1995**, *62*, 780–789.

71. Morgan, A. R.; Rampersauld, A.; Garbo, G. M. *J. Med. Chem.* **1989**, *32*, 904–908.
72. Kessel, D. *Photochem. Photobiol.* **1989**, *50*, 169–174.
73. Razum, N. J.; Snyder, A. B.; Doiron, D. R. *Proc. SPIE Int. Opt. Eng.*, **1996**, *2675*, 43–46.
74. Gomer, C. J.; Ferrario, A. *Cancer Res.* **1990**, *50*, 3985–3990.
75. Spikes, J. D.; Bommer, J. C. *J. Photochem. Photobiol. B Biol.* **1993**, *17*, 135–143.
76. Spikes, J. D.; Bommer, J. C. *Photochem. Photobiol.* **1993**, *58*, 346–350.
77. Richter, A. M.; Waterfield, E.; Jain, A. K.; Allison, B.; Sternberg, E. D.; Dolphin, D.; Levy, J. G. *Br. J. Cancer* **1991**, *63*, 87–93.
78. Aveline, B.; Hasan, T.; Redmond, R. W. *Photochem. Photobiol.* **1994**, *59*, 328–335.
79. Cincotta, L.; Szeto, D.; Lampros, E.; Hasan, T.; Cincotta, A. H. *Photochem. Photobiol.* **1996**, *63*, 229–237.
80. Waterfield, E. M.; Renke, M. E.; Smits, C. B.; Gervais, M. D.; Bower, R. D.; Stonefield, M. S.; Levy, J. G. *Photochem. Photobiol.* **1994**, *60*, 383–387.
81. Rosenthal, I. *Photochem. Photobiol.* **1991**, *53*, 859–870.
82. Firey, P. A.; Rodgers, M. A. J. *Photochem. Photobiol.* **1987**, *45*, 535–538.
83. Oleinick, N. L.; Antunez, A. R.; Clay, M. E.; Richter, B. D.; Kenney, M. E. *Photochem. Photobiol.* **1993**, *57*, 242–247.
84. Zuk, M. M.; Rihter, B. D.; Kenney, M. E.; Rodgers, M. A. *Photochem. Photobiol.* **1996**, *63*, 132–140.
85. Biolo, R.; Jori, G.; Soncin, M.; Rihter, B.; Kenney, M. E.; Rodgers, M. A. J. *Photochem. Photobiol.* **1996**, *63*, 224–228.
86. Hahn, K. A.; Panjehpour, M.; Lu, X. *Photochem. Photobiol.* **1996**, *63*, 117–122.
87. Xie, L. Y.; Boyle, R. W.; Dolphin, D. *J. Am. Chem. Soc.* **1996**, *118*, 4853–4859.
88. Wilson, B. C. In *Photosensitizing Compounds: Their Chemistry, Biology, and Clinical Use (Ciba Foundation Symposium 146)*, Bock, G. and Harnett, S., Eds; Wiley: Chichester, 1989; pp. 60–77.
89. Harriman, A.; Maiya, B. G.; Murai, T.; Hemmi, G.; Sessler, J. L.; Mallouk, T. E. *J. Chem. Soc., Chem. Commun.* **1989**, 314–316.
90. Knübel, G.; Franck, B. *Angew. Chem. Int. Ed. Eng.* **1988**, *27*, 1170–1172.
91. Sessler, J. L.; Hemmi, G. L.; Maiya, B. G.; Harriman, A.; Judy, M. L.; Boriak, R.; Matthews, J. L.; Ehrenberg, B.; Malik, Z.; Nitzan, Y.; Rück, A. *Proc. SPIE Int. Opt. Eng.* **1991**, *1426*, 318–329.
92. Ehrenberg, B.; Roitman, L.; Lavi, A.; Nitzan, Y.; Sessler, J. L. *Proc. SPIE Int. Opt. Eng.* **1994**, *2325*, 68–79.
93. Ehrenberg, B.; Lavi, A.; Nitzan, Y.; Malik, Z.; Ladan, H.; Johnson, F. M.; Sessler, J. L. *Proc. SPIE Int. Opt. Eng.* **1992**, *1645*, 259–263.
94. Ehrenberg, B.; Malik, Z.; Nitzan, Y.; Ladan, H.; Johnson, F. M.; Hemmi, G.; Sessler, J. L. *Lasers Med. Sci.* **1993**, *8*, 197–203.

95. König, K.; Genze, F.; Miller, K.; Rück, A.; Reich, E.; Repassy, D. *Laser Surg. Med.* **1993**, *13*, 522–527.

96. Woodburn, K. W.; Young, S. W.; Miller, R. A. *Proc. SPIE Int. Opt. Eng.* **1996**, *2675*, 172–178.

97. Wieman, T. J.; Fingar, V.; Taber, S.; Panjehpour, M.; Julius, C.; Panella, T. J.; Yuen, A.; Horning, S.; Woodburn, K.; Engel, J.; Miller, R. A.; Renschler, M. F.; Young, S. W. *Abstracts of 24th Annual Meeting American Society for Photobiology*, Atlanta, GA, June 15–20, 1996.

98. Woodburn, K. W.; Fan, Q.; Kessel, D.; Wright, M.; Mody, T. D.; Hemmi, G.; Magda, D.; Sessler, J. L.; Dow, W. C.; Miller, R. A.; Young, S. W. *J. Clin. Lasers Surg. Med.* **1996**, *14*, 343–348.

99. Woodburn, K. W.; Young, S. W.; Fan, Q.; Kessel, D.; Miller, R. A. *Proc. SPIE Int. Opt. Eng.* **1996**, *2671*, 62–71.

100. Guardiano, M.; Biolo, R.; Jori, G.; Schaffner, K. *Cancer Lett.* **1989**, *44*, 1–6.

101. Milanesi, C.; Biolo, R.; Jori, G.; Schaffner, K. *Lasers. Med. Sci.* **1991**, *6*, 437–442.

102. Schaffner, K.; Vogel, E.; Jori, G. In *Biologic Effects of Light* **1993**, Jung, E. G. and Holick, M. F., Eds; Walter de Gruyter: Berlin, 1994; pp. 312–321.

103. Aramendia, P. F.; Redmond, R. W.; Nonell, S.; Schuster, W.; Braslavsky, S. E.; Schaffner, K.; Vogel, E. *Photochem. Photobiol.* **1986**, *44*, 555–559.

104. Cross, A. D. *Proc. SPIE Int. Opt. Eng.* **1996**, *2625*, 262–264.

105. Private communication from Professor E. Vogel, June 12, 1996.

106. Richert, C.; Wessels, J. M.; Müller, M.; Kisters, M.; Benninghaus, R.; Goetz, A. E. *J. Med. Chem.* **1994**, *37*, 2797–2807.

107. Leunig, M.; Richert, C.; Gamarra, F.; Lumper, W.; Vogel, E.; Jocham, D.; Goetz, A. E. *Br. J. Cancer* **1993**, *68*, 225–234.

108. Szeimies, R.-M.; Karrer, S.; Abels, C.; Steinbach, P.; Fickweiler, S.; Messmann, H.; Baeumler, W.; Landthaler, M. *J. Photochem. Photobiol. B* **1996**, *34*, 67–72.

109. Chang, C. K.; Morrison, I.; Wu, W.; Chern, S.-S.; Peng, S.-M. *J. Chem. Soc., Chem. Commun.* **1995**, 1173–1174.

110. Luo, Y.; Chang, C. K.; Kessel, D. *Photochem. Photobiol.* **1996**, *63*, 528–534.

111. Kessler, D.; Luo, Y.; Woodburn, K.; Chang, C. K.; Henderson, B. W. *SPIE* **1995**, 2392, 122–128.

112. Luo, Y.; Chang, C. K.; Kessel, D. *SPIE*, **1996**, *2675*, 132–137.

113. Judy, M. L.; Matthews, J. L.; Newman, J. T.; Skiles, H.; Boriack, R.; Cyr, M.; Maiya, B. G.; Sessler, J. L. *Photochem. Photobiol.* **1991**, *53*, 101–107.

114. Franck, B.; Schneider, U.; Schröder, D.; Gulliya, K. S.; Matthews, J. L. In *Biologic Effects of Light 1993*, Jung, E. G.; Holick, M. F., Eds; Walter de Gruyter: Berlin, 1994; pp. 289–302.

115. Otto, S. E. *Oncology Nursing,* Mosby-Year Book, Inc.: St. Louis, 1991.

116. Felmeier, J. J. In *Clinical Oncology*, Weiss, G. R., Ed.; Appleton & Lange: Norwalk, 1993; pp. 74–88.

117. Hendrickson, F. R.; Wither, H. R. In *American Cancer Society Textbook of Clinical Oncology*, Holleb, A. I., Fink, D. J. and Murphy, G. P., Eds; American Cancer Society: Washington, D.C., 1991; pp. 35–37.

118. Russo, A.; Mitchell, J.; Kinsella, T.; Morstyn, G.; Glatstein E. *Semin. Oncol.* **1985**, *12*, 332–349.

119. Brada, M.; Ross, G. *Curr. Opin. Oncol.* **1995**, 7, 214–219.

120. Hall, E. J. *Radiobiology for the Radiologist*, 3rd edn; Lippincott: Philadelphia, 1988.

121. Beard, C. J.; Coleman, C. N.; Kinsella, T. J. In *Cancer: Principles and Practice of Oncology*, Devita, V. T. Jr.; Hellman, S. and Rosenberg, S. A., Eds; J. B. Lippincott, Philadelphia, 1993; pp. 2701–2710.

122. Shenoy, M. A.; Singh, B. B. *Cancer Invest.* **1992**, *10, 533–551*.

123. Kinsella, T. J.; Russo, A.; Mitchell, J. B.; Rowland, J.; Jenkins, J.; Schwade, J.; Myers, C. E.; Collins, J. M.; Speyer, J.; Kornbluth, P.; Smith, B.; Kufta, C.; Glatstein, E. *Int. J. Radiat. Oncol. Biol. Phys.* **1984**, *10*, 69–76.

124. Kinsella, T. J.; Russo, A.; Mitchell, J. B.; Collins, J. M.; Rowland, J.; Wright, D.; Glatstein, E. *Int. J. Radiat. Oncol. Biol. Phys.* **1984**, *11*, 1941–1946.

125. O'Connell, M. J.; Martenson, J. A.; Wieand, H. S.; Krook, J. E.; MacDonald, J. S.; Haller, D. G.; Mayer, R. I.; Gunderson, L. L.; Rich, T. A. *New Engl. J. Med.* **1994**, *331*, 502–507.

126. Landoni, F.; Maneo, A.; Zanetta, G.; Colombo, A.; Nava, S.; Placa, F.; Tancini, G.; Mangioni, C. *Gynecol. Oncol.* **1996**, *61*, 321–327.

127. Roberts, J. T.; Bleehen, N. M.; Workman, P.; Walton, M. I. *Int. J. Radiat. Oncol. Biol. Phys.* **1984**, *10*, 1755–1758.

128. Newman, H. F. V.; Ward, R.; Workman, P.; Bleehen, N. M. *Int. J. Radiat. Oncol. Biol. Phys.* **1988**, *15*, 1073–1083.

129. Dische, S.; Saunders, M. I.; Bennett, M. H.; Chir, B.; Dunphy, E. P.; Des Rochers, C.; Stratford, M. R. L.; Minchinton, A. I.; Wardman, P. *Br. J. Radiol.* **1986**, *59*, 911–917.

130. Saunders, M. I.; Anderson, P. J.; Bennett, M. H.; Dische, S.; Minchinton, A.; Stratford, M. R. L.; Tothill, M. *Int. J. Radiat. Oncol. Biol. Phys.* **1984**, *10*, 1759–1763.

131. Coleman, C. N.; Wasserman, T. H.; Urtasun, R. C.; Halsey, J.; Hirst, V. K.; Hancock, S.; Phillips, T. L. *Int. J. Radiat. Oncol. Biol. Phys.* **1986**, *12*, 1105–1108.

132. Coleman, C. N.; Halsey, J.; Cox, R. S.; Hirst, V. K.; Blaschke, T.; Howes, A. E.; Wasserman, T. H.; Urtasun, R. C.; Pajak, T.; Hancock, S.; Phillips, T. L.; Noll, L. *Cancer Res.* **1987**, *47*, 519–522.

133. Fields, A. L.; Anderson, P. S.; Goldberg, G. L.; Walder, S.; Beitler, J.; Sood, B.; Runowicz, C. D. *Gynecol. Oncol.* **1996**, *61*, 416–422.

134. Suzuki, M.; Saga, Y.; Sekiguchi, I.; Sato, I. *Curr. Therap. Res.* **1996**, *57*, 430–437.

135. Mulcahy, R. T.; Sutherland, R. M.; Siemann, D. W. In *Clinical Oncology. A Multidisciplinary Approach for Physicians and Students*, Rubin, P., Ed.; W.B. Saunders: Philadelphia, 1993; pp. 87–90.

136. Tannock, I. F. *Br. J. Radiol.* **1972**, *45*, 515–524.
137. Watson, E. R.; Halnan, K. E.; Dische, S.; Saunders, M. I.; Cade, I. S.; McEwan, J. B.; Wienik, F.; Perrins, D. J. D.; Sutherland, I. *Br. J. Radiol.* **1978**, *51*, 879–887.
138. Brown, J. M.; Lee, W. W. *Radiation Sensitizers: Their Use in the Clinical Management of Cancer*, Masson: New York, 1980; pp. 2–13.
139. Wang, C. C. *Clinical Radiation Oncology: Indications, Techniques and Results*, PSG Publishing: Littleton, MA, 1988.
140. Bischofberger, N.; Shea, R. G. In *Nucleic Acid Targeted Drug Design*, Propst, T. J. and Perun, C. L., Eds; Marcel Dekker: New York, 1992; pp. 579–612.
141. Uhlmann, E.; Peyman, A. *Chem. Rev.* **1990**, *90*, 543–584.
142. *Oligodeoxynucleotides: Antisense Inhibitors of Gene Expression*, Cohen, J. S., Ed., CRC Press, Inc.: Boca Raton, 1989.
143. *Prospects for Antisense Nucleic Acid Therapy of Cancer and AIDS*, Wickstrom, E., Ed. Wiley-Liss: New York, 1991.
144. Hélène, C.; Toulmé, J.-J. *Biochim. Biophys. Acta* **1990**, *1049*, 99–125.
145. Cazenave, C.; Hélène, C. In *Antisense Nucleic Acids and Proteins*, Mol, J. N. M. and van der Krol, A. R., Eds; Marcel Dekker: New York, 1991; pp. 47–94.
146. Thuong, N. T.; Hélène, C. *Angew. Chem. Int. Ed. Eng.* **1993**, *32*, 666–690.
147. Englisch, U.; Gauss, D. H. *Angew. Chem. Int. Ed. Eng.* **1991**, *30*, 613–629.
148. Zon, G. *Pharm. Res.* **1988**, *5*, 539–549.
149. Milligan, J. F.; Matteucci, M. D.; Martin, J. C. *J. Med. Chem.* **1993**, *36*, 1923–1937.
150. Tidd, D.M. *Anticancer Res.* **1990**, *10*, 1169–1182.
151. Stein, C. A. *Cancer Res.*, **1988**, *48*, 2659–2668.
152. Stein, C. A.; Tonkinson, J. L.; Yakubov, L. *Pharmac. Ther.* **1991**, *52*, 365–384.
153. Stein, C. A. In *Cancer: Principles and Practice of Oncology*, 4th Edn; DeVita, V. T., Jr, Hellman, S. and Rosenberg, S. A., Eds; J. B. Lippincott: Philadelphia, 1993; pp. 2646–2655.
154. Stein, C. A.; Cheng, Y.-C. *Science* **1993**, *261*, 1004–1012.
155. Stein, C. A. *Contemp. Oncol.* **1994**, *4*, 31–39.
156. Ts'o, P.O.P.; Miller, P.S.; Aurelian, L.; Murakami, A.; Agris, C.; Blake, K.R.; Lin, S.-B.; Lee, B.L.; Smith, C.C. *Ann. N. Y. Acad. Sci.* **1993**, *14*, 220–241.
157. Gura, T. *Science*, **1995**, *270*, 575–577.
158. Wagner, R. W. *Nature* **1994**, *372*, 333–335.
159. Meunier, B., Ed. *DNA and RNA Cleavers and Chemotherapy of Cancer and Viral Diseases (NATO ASI Series C, Vol. 479)*, Kluwer: Dordrecht, 1996.
160. *Antisense Research and Applications*; Crooke, S. T. and Lebleu, B, Eds; CRC Press: Boca Raton, 1993.
161. Heider, H. K. Bardos, T. J. In *Cancer Chemotherapeutic Agents*; Foye, W. O., Ed.; American Chemical Society: Washington, DC, 1995; pp. 529–576.
162. De Mesmaeker, A.; Häner, R.; Martin, P.; Moser, H. E. *Acc. Chem. Res.* **1995**, *28*, 366–374.
163. Lima, W. F.; Monia, B. P.; Ecker, D. J.; Freier, S. M. *Biochemistry* **1992**, *31*, 12055–12061.

164. Rittner, K.; Burmester, C.; Sczakiel, G. *Nucleic Acids Res.* **1993**, *21*, 1381–1387.

165. Milligan, J. F.; Jones, R. J.; Froehler, B. C.; Matteucci, M. D. In *Ann. N.Y. Acad. Sci.: Gene Therapy for Neoplastic Diseases*; Huber, B. E. and Lazo, J. S., Eds; New York Academy of Sciences: New York, 1994; pp. 228–241.

166. Bishop, J. M. *Science* **1987**, *235*, 305–311.

167. Goodchild, J. *Bioconj. Chem.* **1990**, *1*, 165–187.

168. Pei, D.; Schultz, P. G. In *Nucleases*, Linn, S. M., Lloyd, R. S. and Roberts, R. J., Eds; Cold Spring Harbor Laboratory Press: New York, 1993; pp. 317–340.

169. Knorre, D. G.; Zaryotova, V. F. In *Prospects for Antisense Nucleic Acid Therapy of Cancer and AIDS*, Wickstrom, E., Ed.; Wiley-Liss: New York, 1991; pp. 195–218.

170. Walder, R. Y.; Walder, J. A. *Proc. Natl Acad. Sci. USA* **1988**, *85*, 5011–5115.

171. Shuttleworth, J.; Colman, A. *EMBO J.* **1988**, *7*, 427–434.

172. Sullenger, B. A.; Cech, T. R. *Science* **1993**, *262*, 1566–1569.

173. Sarver, N.; Cantin, E. M.; Chang, P. S.; Zaia, J. A.; Ladne, P. A.; Stephens, D. A.; Rossi, J. J. *Science* **1990**, *247*, 1222–1225.

174. Yu, M.; Ojwang, J.; Yamada, O.; Hampel, A.; Rapapport, J.; Looney, D.; Wong-Staal, F. *Proc. Natl Acad. Sci. USA* **1993**, *90,* 6340–6344.

175. Zuckerman, R. N.; Schultz, P. G. *Proc. Natl Acad. Sci. USA* **1989**, *86*, 1766–1779.

176. Corey, D. R.; Zuckerman, R. N.; Schultz, P. G. In *Frontiers of Bioorganic Chemistry*; H. Dugas, Ed.; Springer-Verlag: Berlin, 1990; pp. 3–31.

177. Uhlenbeck, O.C. In *Antisense Research and Applications*; Crooke, S. T. and Lebleu, B., Eds; CRC Press: Boca Raton, 1993; pp. 83–96.

178. Baringa, M. *Science* **1993**, *262*, 1512–1514.

179. Marshall, W. S.; Caruthers, M. H. *Science* **1993**, *259*, 1564–1570.

180. Graznov, S.; Chen, J.-K. *J. Am. Chem. Soc.* **1994**, *116*, 3143–3144.

181. Graznov, S.; Skorski, T.; Cacco, C.; Skosska, M.; Chiu, C. Y.; Loyd, D.; Chen, J.-K.; Koziolkiewicz, M.; Calabretta, B. *Nucleic Acids Res.* **1996**, *24*, 1508–1514.

182. Nielson, P. E.; Egholm, M.; Buchardt, O. *Bioconj. Chem.* **1994**, *5*, 3–7.

183. Magda, D.; Miller, R. A.; Wright, M.; Rao, J.; Sessler, J. L.; Iverson, B. L.; Sansom, P. I. In *DNA and RNA Cleavers and Chemotherapy of Cancer and Viral Diseases* (*NATO ASI Series C*, Vol. 479); Meunier, B., Ed.; Kluwer: Dordrecht, 1996; pp. 337–353.

184. Modak, A.S.; Gard, J. K.; Merriman, M. C.; Winkeler, K. A.; Bashkin, J. K.; Stern, M. K. *J. Am. Chem. Soc.* **1991**, *113*, 283–291.

185. Morrow, J. R.; Buttrey, L. A.; Shelton, V. M.; Berback, K. A. *J. Am. Chem. Soc.* **1992**, *114*, 1903–1905.

186. Le Doan, T.; Praseuth, D.; Perrouault, L.; Chassignol, M.; Thuong, N. T.; Hélène, C. *Bioconj. Chem.* **1990,** *1*, 108–113.

187. Vlassov, V. V.; Deeva, E. A.; Ivanova, E. M.; Knorre, D. G.; Maltseva, T. V.; Prolova, E. I. *Nucl. & Nucl.* **1991**, *10*, 641–643.

188. Fedorova, O. S.; Savitskii, A. P.; Shoikhet, K. G.; Ponomarev, G. V. *FEBS Lett.* **1990**, *259*, 335–337.

189. Bashkin, J. K.; Frolova, E.I.; Sampath, U. *J. Am. Chem. Soc.* **1994**, *116*, 5981–5982.

190. Hall, J.; Hüsken, D.; Pieles, U.; Moser, H. E.; Häner, R. *Chem. & Biol.* **1994**, *1*, 185–190.

191. Häner, R.; Hall, J.; Hüsken, D.; Moser, H. E. In *DNA and RNA Cleavers and Chemotherapy of Cancer and Viral Diseases* (*NATO ASI Series C*, Vol. 479); Meunier, B., Ed.; Kluwer: Dordrecht, 1996; pp. 307–320.

192. Matsumura, K.; Endo, M.; Komiyama, M. *J. Chem. Soc., Chem. Comm.* **1994**, 2019–2020.

193. Magda, D.; Miller, R. A.; Sessler, J. L.; Iverson, B. L. *J. Am. Chem. Soc.* **1994**, *116*, 7439–7440.

194. Mastruzzo, L.; Woisard, A.; Ma, D. D. F.; Rizzarelli, E.; Favre, A.; Le Doan, T. *Photochem. Photobiol.* **1994**, *60*, 316–322.

195. Le Doan, T. ; Perrouault, L.; Praseuth, D.; Habhoub, N.; Decout, J.-L.; Thuong, N. T.; Lhomme, J.; Hélène, C. *Nucleic Acids Res.* **1987**, *15*, 7749–7760.

196. Levina, A. S.; Berezovskii, M. V.; Venjaminova, A. G.; Dobrikov, M. I.; Repkova, M. N.; Zarytova, V. F. *Biochimie* **1993**, *75*, 25–27.

197. Praseuth, D.; Perrouault, L.; Le Doan, T.; Chassignol, M.; Thuong, N.; Hélène, C. *Proc. Natl Acad. Sci. USA* **1988**, *85*, 1349–1353.

198. Perrouault, L.; Asseline, U.; Rivalle, C.; Thuong, N. T.; Bisagni, E.; Giovannangeli, C.; Le Doan, T.; Hélène, C. *Nature* **1990**, *344*, 358–360.

199. Bhan, P.; Miller, P. S. *Bioconj. Chem.* **1990**, *1*, 82–88.

200. Teare, J.; Wollenzien, P. *Nucleic Acids Res.* **1989**, *17*, 3359–3372.

201. Pieles, U.; Englisch, U. *Nucleic Acids Res.* **1989**, *17*, 285–299.

202. Takasugi, M.; Guendouz, A.; Chassignol, M.; Decout, J. L.; Lhomme, J.; Thuong, N. T.; Hélène, C. *Proc. Natl Acad. Sci. USA* **1991**, *88*, 5602–5606.

203. Praseuth, D.; Le Doan, T.; Chassignol, M.; Decout, J.-L.; Habhoub, N.; Lhomme, J.; Thuong, N. T.; Hélène, C. *Biochemistry* **1988**, *27*, 3031–3038.

204. Boutorine, A. S.; Tokuyama, H.; Takasugi, M.; Isobe, H.; Nakamura, E.; Hélène, C. *Angew. Chem. Int. Ed. Eng.* **1994**, *33*, 2462–2465.

205. Magda, D.; Wright, M.; Miller, R. A.; Sessler, J. L.; Sansom, P. I. *J. Am. Chem. Soc.* **1995**, *117*, 3629–3630.

206. Sessler, J. L.; Sansom, P. I.; Král, V.; O'Connor, D.; Iverson, B. L. *J. Am. Chem. Soc.* **1996**, *118*, 12322–12330.

207. Eichorn, G. L.; Butzow, J. J. *Biopolymers* **1965**, *3*, 79–94.

208. Breslow, R.; Huang, D. L. *Proc. Natl Acad. Sci. USA* **1991**, *88*, 4080–4083.

209. Morrow, J. R.; Buttrey, L. A.; Berback, K. A. *Inorg. Chem.* **1992**, *31*, 16–20.

210. Lisowski, J.; Sessler, J. L.; Lynch, V.; Mody, T. D. *J. Am. Chem. Soc.* **1995**, *117*, 2273–2285.

211. Ksala, K. A.; Morrow, J. R.; Sharma, A. P. *Inorg. Chem.* **1993**, *32*, 3983–3984.

212. Magda, D.; Crofts, S.; Lin, A.; Miles, D.; Wright, M.; Sessler, J. L. *J. Am. Chem. Soc.,* **1997**, *119*, 2293–2294.

213. Voet, D.; Voet, J. G. *Biochemistry*, 2nd Edn, Wiley: New York, 1995.

214. Adams, R. L. P.; Knowler, J. T., Leader, D. P. Eds; *The Biochemistry of Nucleic Acids*, 10th Edn; Chapman and Hall: New York, 1986.

215. Saenger, W. *Principles of Nucleic Acid Structure*, Springer Verlag: New York, 1988.

216. Perutz, M. F.; Shih, D. T.-B.; Williamson, D. *J. Mol. Biol.* **1994**, *239,* 555–560.

217. Bhat, M. K.; Pickersgill, R. W.; Perry, B. N.; Brown, R. A.; Jones, S. T.; Mueller-Harvey, I.; Sumner, I. G.; Goodenough, P. W. *Biochemistry* **1993**, *32,* 12203–12208.

218. Ranganathan, D.; Vaish, N. K.; Shah, K. *J. Am. Chem. Soc.* **1994**, *116,* 6545–6557.

219. Gerenscser, G. Ed.; *Chloride Transport Coupling in Biological Membranes and Epithelia*, Elsevier: Amsterdam, 1984.

220. LaNoue, K.; Mizani, S. M.; Kilingenberg, M. *J. Biol. Chem.* **1978**, *253,* 191–198.

221. Numa, S. Ed.; *Fatty Acid Metabolism and Its Regulation*, Elsevier: Amsterdam, 1984.

222. Meister, A. *Biochemistry of Amino Acids*, 2nd Edn, Vols 1 and 2, Academic Press: New York, 1965.

223. Hall, J. L.; Baker, D. A. *Cell Membranes and Ion Transport,* Longmans-Green: New York, 1977.

224. Pedersen, P. L.; Carafoli, E. *Trends Biochem. Sci.* **1987**, *12,* 146–150.

225. Boat, T. F.; Welsh, M. J.; Beaudet A. L. In *The Metabolic Basis of Inherited Disease*, 6th Edn; Scriver, C. R., Beaudet, A. L., Sly, W. S. and Valle, D., Eds; McGraw-Hill: New York, 1989.

226. Davis, P. B. *New Engl J. Med.* **1991**, *325,* 575–576.

227. Quinton, P. M. *FASEB* **1990**, *4,* 2709.

228. Jentsch, T. J. *Curr. Opin. Neurobiol.* **1994**, *4,* 294–303.

229. Weiss, R. *Science News* **1991**, *139,* 132.

230. Tsui, L.-C., Romeo, G., Greger, R. and Gorini, S., Eds; *The Identification of The CF (Cystic Fibrosis) Gene, Recent Progress and New Research Strategies*; Plenum: New York, 1991.

231. Jentsch, T. J. *Curr. Opin. Cell. Biol.* **1994**, *6,* 600–606.

232. Martin, J. C., Ed.; *Nucleotide Analogs as Antiviral Agents (ACS Symposium Series 401)*; American Chemical Society: Washington, DC, 1989.

233. Stryer, L.; *Biochemistry*, 3rd Edn, W. H. Freeman and Co.: New York, 1988; pp. 386 and 413.

234. Moss, B. *Chem. Ind.* **1996**, *11,* 407–411.

235. Atwood, J. L.; Holman, T.; Steed, J. W. *Chem. Commun.* **1996**, 1401–1407.

236. Beer, P. D.; Smith, D. K. *Prog. Inorg. Chem.* **1997**, *46,* 1–96

237. Sessler, J. L.; Andrievsky, A.; Genge, J. W. in *Advances in Supramolecular Chemistry* Vol. 4 Gokel, G. W., Ed.; JAI Press: Greenwich, CT, in press.

238. Sessler, J. L.; Sansom, P. I.; Andrievsky, A.; Král, V. In *The Supramolecular Chemistry of Anions,* Vol. 4; Bianchi, A.; Bowman-James, K.; Garcia-España, E., Eds.; VCH: Weinheim, in press.

239. Izatt, R. M.; Pawlak, K.; Bradshaw, J. S.; Bruening, R. L. *Chem. Rev.* **1995**, *95*, 2529–2586.

240. Dietrich, B. *Pure Appl. Chem.* **1993**, *65*, 1457–1464.

241. Beer, P. D.; Gale, P. A.; Hesek, D. *Tetrahedron Lett.* **1995**, *36*, 767–770.

242. Scheerder, J.; van Duynhoven, J. P. M.; Engbersen, J. F. J.; Reinhoudt, D. N. *Angew. Chem. Int. Ed. Eng.* **1996**, *35*, 1090–1093.

243. Davis, A. P.; Gilmer, J. F.; Perry, J. J. *Angew. Chem. Int. Ed. Eng.* **1996**, *35*, 1312–1315.

244. Valiyaeettil, S.; Engbersen, J. F. J.; Verboom, W.; Reinhoudt, D. N. *Angew. Chem. Int. Ed. Eng.* **1993**, *32*, 900–901.

245. Lehn, J.-M. *Science,* **1985**, *227*, 849–856.

246. Kimura, E. *Top. Curr. Chem.* **1985**, *128*, 113–141.

247. Schmidtchen, F. P. *Top. Curr. Chem.* **1988**, *132*, 101–133.

248. Li, T.; Krasne, S. J.; Persson, B.; Kaback, H. R.; Diederich, F. *J. Org. Chem.* **1993**, *58*, 380–384.

249. van Arman, S. A.; Czarnik, A. W. *J. Am. Chem. Soc.* **1994**, *116*, 9397–9398.

250. Wilson, H. R.; Williams, R. J. P. *J. Chem. Soc. Faraday Trans. 1* **1987**, *83*, 1885–1892.

251. Hannon, C. L.; Anslyn, E. V. In *Biorganic Chemistry Frontiers*, Vol. 3; Dugas, H., Ed.; Springer-Velag: Berlin, 1993; pp. 193–225.

252. Galán, A.; de Mendoza, J.; Toirin, C.; Bruix, M.; Delongchamps, G.; Rebek, J., Jr. *J. Am. Chem Soc.* **1991**, *113*, 9424–9425.

253. Dixon, R. P.; Geib, S. J.; Hamilton, A. D. *J. Am. Chem. Soc.* **1992**, *114*, 365–366.

254. Beauchamp, A. L.; Olievier, M. J.; Wuest, J. D.; Zacharie, B. *J. Am. Chem. Soc.* **1986**, *108*, 73–77.

255. Newcomb, M.; Horner, J. H.; Blanda, M. T.; Squatritto, P. J. *J. Am. Chem. Soc.* **1989**, *111*, 6294–6301.

256. Yang, X.; Knobler, C. B.; Hawthorne, M. F. *J. Am. Chem. Soc.* **1992**, *114*, 380–382.

257. Beer, P. D.; Drew, M. G. B.; Graydon, A. R.; Smith, D. K.; Stokes, S. E. *J. Chem. Soc. Dalton Trans.* **1995**, 403–408.

258. Beer, P. D.; Drew, M. G. B.; Hesek, D.; Jagessar, R. *J. Chem. Soc. Chem. Commun.* **1995**, 1187–1189.

259. Lanza, S.; Scolaro, L. M.; Rosace, G. *Inorg. Chim. Acta* **1994**, *227*, 63–69.

260. Burrows, A. D.; Mingos, M. P.; White, A. J. P.; Williams, D. J. *J. Chem Soc. Chem. Commun.* **1996**, 97–99.

261. Slone, R. V.; Yoon, D. I.; Calhoun, R. M.; Hupp, J. T. *J. Am. Chem. Soc.* **1995**, *117*, 11813–11814.

262. Katz, H. E. *J. Am. Chem. Soc.* **1985**, *107*, 1420–1421.

263. Beer, P. D. *Chem. Commun.* **1996**, 689–696.

264. Shionoya, M.; Kimura, E.; Shiro, M. *J. Am. Chem. Soc.* **1993**, *115*, 6730–6737.

265. Shionoya, M.; Ikeda, T.; Kimura, E.; Shiro, M. *J. Am. Chem. Soc.* **1994**, *116*, 3848–3859.

266. Koike, T.; Takashige, M.; Kimura, E.; Fujioka, H.; Shiro, M. *Chem. Eur. J.* **1996**, *2*, 617–623.

267. Lehn, J. M. *Supramolecular Chemistry*, VCH: Weinheim, 1995.

268. Bianchi, A., Bowman-James, K. and Garcia-España, E., Eds; *The Supramolecular Chemistry of Anions*; VCH: Weinheim, in press.

269. Kibbey, C. E.; Meyerhoff, M. E. *Anal. Chem.* **1993**, *65*, 2189–2196.

270. Kliza, D. M.; Meyerhoff, M. E. *Electroanalysis* **1992**, *4*, 841–849.

271. Xiao, J.; Savina, R.; Martin, G. B.; Francis, A. H.; Meyerhoff, M. E. *J. Am. Chem. Soc.* **1994**, *116*, 9341–9342.

272. Mizutani, T.; Yagi, S.; Honmaru, A.; Ogoshi, H. *J. Am. Chem. Soc.* **1996**, *118*, 5318–5319.

273. Kuroda, Y.; Kato, Y.; Higashioji, T.; Hasegawa, J.; Kawanami, S.; Takahashi, M.; Shiraishi, N.; Tanabe, K.; Ogoshi, H. *J. Am. Chem. Soc.* **1995**, *117*, 10950–10958.

274. Anderson, H. L.; Sanders, J. K. M. *J. Chem. Soc., Chem. Commun.* **1992**, 946–947.

275. Carvlin, M. J.; Datta-Gupta, N.; Fiel, R. J. *Biochem. Biophys. Res. Commun.* **1982**, *108*, 66–73.

276. Pasternack, R. F.; Giannetto, A. *J. Am. Chem. Soc.* **1991**, *113*, 7799–7800.

277. Pasternack, R. F.; Schaefer, K. F.; Hambright, P. *Inorg. Chem.* **1994**, *33*, 2062–2065.

278. Marzilli, L. G.; Pethö, G.; Lin, M.; Kim, M. S.; Dixon, D. W. *J. Am. Chem. Soc.* **1992**, *114*, 7575–7577.

279. Mukundan, N. E.; Pethö, G.; Dixon, D. W.; Marzilli, L. G. *Inorg. Chem.* **1995**, *34*, 3677–3687.

280. Sessler, J. L.; Cyr, M. J.; Lynch, V.; McGhee, E.; Ibers, J. A. *J. Am. Chem. Soc.* **1990**, *112*, 2810–2813.

281. Cotton, F. A.; Wilkinson, G. *Advanced Inorganic Chemistry*, 4th Edn. Wiley Interscience: New York, 1980; p. 452.

282. Shionoya, M.; Furuta, H.; Lynch, V.; Harriman, A.; Sessler, J. L. *J. Am. Chem. Soc.* **1992**, *114*, 5714–5722.

283. Sessler, J. L.; Cyr, M. J.; Burrell, A. K. *SynLett* **1991**, 127–133.

284. Furuta, H.; Cyr, M. J.; Sessler, J. L. *J. Am. Chem. Soc.* **1991**, *113*, 6677–6678.

285. Sessler, J. L.; Ford, D.; Cyr, M. J.; Furuta, H. *J. Chem. Soc., Chem. Commun.* **1991**, 1733–1735.

286. Sessler, J. L.; Cyr, M.; Furuta, H.; Král, V.; Mody, T.; Morishima, T.; Shionoya, M.; Weghorn, S. *Pure Appl. Chem.* **1993**, *65*, 393–398.

287. Sessler, J. L.; Furuta, H.; Král, V. *Supramolec. Chem.* **1993**, *1*, 209–220.

288. Beckmann, S.; Wessel, T.; Franck, B.; Hönle, W.; Borrmann, H.; von Schnering, H.-G. *Angew. Chem. Int. Ed. Eng.* **1990**, *29*, 1395–1397.

289. König, H.; Eickemeier, C.; Möller, M.; Rodewald, U.; Franck, B. *Angew. Chem. Int. Ed. Eng.* **1990**, *29*, 1393–1395.

290. Wessel, T.; Franck, B.; Möller, M.; Rodewald, U.; Läge, M. *Angew. Chem. Int. Ed. Eng.* **1993**, *32*, 1148–1151.

291. Shannon, R. D. *Acta Crystallogr.* **1976**, *A32*, 751–767.

292. Král, V.; Furuta, H.; Shreder, K.; Lynch, V.; Sessler, J. L. *J. Am. Chem. Soc.* **1996**, *118*, 1595–1607.

293. Iverson, B. L.; Shreder, K.; Král, V.; Sansom, P.; Lynch, V.; Sessler, J. L. *J. Am. Chem. Soc.* **1996**, *118*, 1608–1616.

294. Gale, P.; Andrievsky, A.; Lynch, V.; Sessler, J. L. *Angew. Chem. Intl. Ed. Eng.* **1996**, *35*, 2782–2785.

295. Hoehner, M. C. Ph.D. Dissertation, University of Texas at Austin, 1996.

296. Lisowski, J.; Sessler, J. L.; Lynch, V. *Inorg. Chem.* **1995**, *34*, 3567–3571.

297. Sessler, J. L.; Mody, T. D.; Ford, D. A.; Lynch, V. *Angew. Chem., Int. Ed. Eng.* **1992**, *31*, 452–455.

298. Sessler, J. L.; Johnson, M. R.; Lynch, V. *J. Org. Chem.* **1987**, *52*, 4394–4397.

299. Sessler, J. L.; Morishima, T.; Lynch, V. *Angew. Chem. Int. Ed. Eng.* **1991**, *30*, 977–980.

300. Sessler, J. L.; Weghorn, S. J.; Hiseada, Y.; Lynch, V. *Chem. Eur. J.* **1995**, *1*, 56–67.

301. Guba, A. Diplomarbeit, Universität zu Jena, 1996.

302. Sessler, J. L.; Weghorn, S. J.; Morishima, T.; Rosingana, M.; Lynch, V.; Lee, V. *J. Am. Chem. Soc.* **1992**, *114*, 8306–8307.

303. Sessler, J. L.; Weghorn, S. J.; Lynch, V.; Johnson, M. R. *Angew. Chem. Int. Ed. Eng.* **1994**, *33*, 1509–1512.

304. Vogel, E.; Bröring, M.; Fink, J.; Rosen, D.; Schmickler, H.; Lex, J.; Chan, K. W. K.; Wu, Y.-D.; Plattner, D. A.; Nendel, M.; Houk, K. N. *Angew. Chem. Int. Ed. Eng.* **1995**, *34*, 2511–2514.

305. Král, V.; Sessler, J. L.; Furuta, H. *J. Am. Chem. Soc.* **1992**, *114*, 8704–8705.

306. Furuta, H.; Morishima, T.; Král, V.; Sessler, J. L. *Supramolec. Chem.* **1993**, *3*, 5–8.

307. Sessler, J. L.; Král, V. *Tetrahedron* **1995**, *51*, 539–554.

308. Holý, A.; De Clercq, E.; Votruba, I. In *Nucleotide Analogues as Antiviral Agents*, Harden, M. R., Ed.; VCH: Deerfield Beach, 1985; pp. 51–71.

309. Sessler, J. L.; Anrievsky, A. *Chem. Commun.* **1996**, 1119–1120.

310. Král, V.; Andrievsky, A.; Sessler, J. L. *J. Am. Chem. Soc.* **1995**, *117*, 2953–2954.

311. Král, V.; Andrievsky, A.; Sessler, J. L. *J. Chem. Soc., Chem. Commun.* **1995**, 2349–2351.

312. Král, V.; Springs, S. L.; Sessler, J. L. *J. Am. Chem. Soc.* **1995**, *117*, 8881–8882.

313. Iverson, B. L.; Shreder, K.; Král, V.; Sessler, J. L. *J. Am. Chem. Soc.* **1993**, *115*, 11022–11023.

314. Iverson, B. L.; Shreder, K.; Král, V.; Smith, D. A.; Smith, J.; Sessler, J. L. *Pure Appl. Chem.* **1994**, *66*, 845–850.

315. Long, E. C.; Barton, J. K. *Acc. Chem. Res.* **1990**, *23*, 271–273.

316. Bailly, C.; Henichart. J.-P. *Bioconj. Chem.* **1991**, *2*, 379–393.

317. Schneider, H.-J.; Blatter, T. *Angew. Chem. Int. Ed. Eng.* **1992**, *31*, 1207–1208.

318. Fiel, R. J. *J. Biomol. Struct. Dyn.* **1989**, *6*, 1259–1274.

319. Iverson, B. L.; Shreder, K.; Morishima, T.; Rosingana, M.; Sessler, J. L. *J. Org. Chem.* **1995**, *60*, 6616–6620.

320. Dervan, P. B. *Science* **1986**, *283*, 464–471.
321. Praseuth, D.; Gaudemer, A.; Verlhac, J.-B.; Kraljic, I.; Sissoëff, I.; Guillé, E. *Photochem. Photobiol.* **1986**, *44*, 717–724.
322. Croke, D. T.; Perrouault, L.; Sari, M. A.; Battioni, J.-P.; Mansuy, D.; Hélène, C.; Le Doan, T. *J. Photochem. Photobiol. B Biol.* **1993**, *18*, 41–50.
323. Iverson, B. L.; Thomas, R. E.; Král, V.; Sessler, J. L. *J. Am. Chem. Soc.* **1994**, *116*, 2663–2664.
324. PRNewswire, January 7th, 1997.

Index

absorption spectrum 49, 61, 62, 66, 80, 86, 87, 104, 113, 129, 140, 159, 163, 171, 189, 217, 225, 229, 233, 239, 242, 271, 289, 295, 306, 323, 361, 362, 379, 382, 403

accordion 373, 386, 387

acetic acid (HOAc) 83, 193, 195, 418, 436

acetylene-cumulene macrocycle 216, 217, 223

Acholla, F. V. 386

acid catalyst 24, 174, 259, 334, 389

actinide 2, 272, 318, 400, 418

Adler, A. D. 340, 349

ADP 487

affinity cleavage 486

aggregation 35, 38, 270, 272, 486

AIDS 434, 438

air 54, 56, 60, 62, 84, 135, 162, 175, 193, 205, 224, 257, 279, 322, 347, 353, 398

alkene 311, 352

aluminum, -oxo-di 146

aluminum(III), Al(III) 146, 164

amethryin 454, 473, 474
   meso-unsubstituted 351
   ρ-NO$_2$-phenyl-substituted 351

amethyst 351

amino acid 288, 289, 294, 453, 483

ammonia 26, 28, 286, 415, 416, 419
   ligand 332

AMP 461, 481, 485, 487

anion
   binding 3, 244, 286, 287, 289, 293, 329, 338, 341, 344, 453, 454,

      455, 456, 462, 468, 472, 474, 477, 478, 479, 480, 483, 484
   carriers 286, 479, 480, 481, 482, 483
   chelation 408, 431, 457, 482, 486, 487
   recognition 3, 453, 454, 456, 462, 482, 486
   sensing 3
   template 173, 389
   transport 3, 257, 287, 288, 289, 338, 454, 479, 480, 481, 482, 483

anisotropic effect 306

annulene
   heteroatom-bridged 2, 107, 113, 116, 173
   oxido-bridged 359
   porphyrin-related 361

[30]annulene
   pentasulfide 325
   pentoxide 324

annulenone 116, 117, 118, 120
   epiminodiepoxy 117
   epithiodiepoxy 117
   furan-containing 116
   pyrrole-containing 117
   thiophene-containing 117
   triepoxy 117, 120

[22]annulenoquinone 241

anthraphyrin 407, 408, 454, 472, 478, 480, 483

antiaromatic 185, 235, 245, 304, 382, 391

antiaromaticity 204, 205

antiferromagnetic 355

antimony pentachloride 320

antisense 2, 405, 444, 445, 446, 447, 448, 449, 451, 453
antiviral therapy 339, 482
aorta 432
apical ligation 38, 400, 402, 432
aromaticity 108
arsenate 453
atheromatous plaque 432, 439
ATP 453, 487
azide 386, 458

Ba^{2+} 386
bacteria 437
Badger, G. M. 107, 108, 110
Bartczak, T. J. 13
Barton, D. H. R. 257
Battersby, A. R. 13
Baylor Research Foundation 442
BBr_3 153, 154
Bekaroglu, Ö. 96, 100
Bell, T. W. 241
benzaldehyde 45, 114, 172, 173, 210, 240, 259, 260, 261, 340
benzimidazole 398
ρ-benzoquinone 329
benzoyl peroxide 27
benzpyrrole 414
Berger, R. A. 203, 205
bidentate 402
bifuran 20, 22, 41, 83, 84, 272, 298, 339
    diformyl 20, 136, 265, 266, 295, 298, 344
biladiene 26, 27, 28, 29, 31, 32, 33, 34, 35, 36, 37, 39, 40, 42, 43, 46, 47, 53, 54, 62, 199, 200
    1,19-dihalo 28, 29
    bis 85
    bisvinylogous 206
    metallo 31
    porphyrin conjugate 87
    tetravinylogous 209
bimetallic 87, 274, 370
    hetero 87, 276
binucleating 2, 352

bioinorganic 2, 429
biotechnology 434
bipyrrole 41, 88, 89, 128, 132, 134, 148, 157, 224, 253, 255, 256, 257, 259, 260, 266, 295, 297, 298, 304, 308, 337, 340, 344, 369, 371, 372, 376, 379, 389, 391
    alkyne-stretched 222, 311
    diacid 41
    diacyl 132
    diformyl 41, 88, 128, 134, 148, 157, 255, 256, 259, 266, 297, 304, 308, 336, 344, 363, 369, 371, 372, 376, 379, 389, 391
    pyrrolyl 257, 260, 295, 297, 298
bipyrrolic 29, 41, 130, 223, 256, 271, 337, 339, 352, 371
biscopper 386
bishydrochloride 269, 270, 311, 373, 457, 468, 473, 474
bisnickel 390
bispyridine 48, 400
bistrifluoroacetate 208, 214, 215, 360
bis(trifluoroacetoxy)iodobenzene 25, 214, 233, 360
bithiophene 140, 339, 363
    bis(hydroxymethyl) 339
    diformyl 140
Black, D. St. C. 114
blood 434, 436
blue 4, 253, 311, 316, 336
Boltzman distribution 430
boric acid 154
boron 96, 97, 409
Boschi, T. 32, 35, 37, 51
Br_2 136, 212, 215, 231
breast cancer 438
Broadhurst, M. J. 20, 22
Brønsted base 397
bronzaphyrin 309, 346, 347, 348, 350, 351
    bisoxa 309
    bridged 350
    hexathia 347, 348
bronze 346

C-TIPS 338, 481
Ca²⁺ 386

Ca²⁺ 386

cadmium, cadmium(II), Cd(II), Cd²⁺ 272, 398, 399, 400, 403, 436, 437

calix[3]indole 114, 115

Callot, H. J. 167, 188, 196, 198

camphorsulfonic acid 297

carbene 174, 188, 199

carbodiimide 286

carbonyldiimidazole 286

carboxylate 432, 462, 483, 484

carcinoma 431, 432, 438

cardiovascular 430

Cava, M. P. 110, 140, 235, 236, 239, 346, 347, 363, 364, 380

cavitand 245, 246, 247, 362, 363

cell, hypoxic 443

cerebral ischemia 432

ceric sulfate 27

Chang, C. K. 157, 158

chiral 260, 289, 291, 369, 371, 372

chloranil 239, 261, 315, 351

chlorin 150, 192

m-chloroperoxybenzoic acid (*m*-CPBA) 148

chlorophyll 1

Chmielewski, P. J. 175

chromium(III), Cr(III) 37, 52

chromium(V) 55

CMP 482

cobalt 33, 34, 45, 48, 49, 61, 62, 63, 160, 164, 170, 272, 282, 284

cobalt(II), Co(II), Co²⁺ 22, 35, 41, 45, 49, 50, 62, 73, 77, 87, 93, 141, 164, 193, 272, 273, 282, 284

cobalt(III), Co(III) 34, 35, 41, 42, 43, 44, 45, 48, 49, 50, 60, 61, 62, 63, 64, 73, 77, 85, 86, 94, 146, 188, 319

cobalt(IV) 61

complexing agent
  anionic substrate, for 421
  dinuclear 310
  neutral substrate, for 421

condensation
  2 + 1 + 1 + 1 259
  2 + 2 20, 22, 29, 41, 175
  3 + 2 256, 261, 266, 295, 297, 298, 306
  3 + 3 315, 329
  4 + 1 256, 297
  MacDonald-type 256, 295, 297, 315, 336, 376
  Wittig-type 110, 116, 119, 230, 234, 237, 244, 249, 324, 358, 359
  [1 + 1] 379, 392, 407, 408
  [2 + 2] 369, 376, 379, 386, 389, 391

conformation
  cyclophane-type 134, 136, 151
  figure-eight 369, 370, 371, 372, 373, 378, 379, 474, 477
  near-planar 55, 98, 134, 142, 144, 164, 171, 174, 207, 224, 310, 313, 321, 330, 337, 352, 377, 455
  non-planar 89, 99, 104, 108, 117, 140, 185, 249, 340, 364, 375, 378, 382, 394
  planar 108, 114, 117, 119, 129, 136, 137, 139, 150, 205, 211, 212, 217, 219, 230, 244, 260, 295, 337, 340, 389, 418

conformational flipping 127, 371

contrast agent 2, 405, 429, 430, 431

control pore glass (CPG) 287

coordination
  chemistry 2, 31, 93, 171, 249, 272, 331, 333, 392, 418, 423
  environment 142, 191, 278, 282
  sphere 164, 332, 402

copper 32, 47, 55, 60, 77, 160, 163, 170, 353, 358, 386, 387

copper(I) 353

copper(II), Cu(II) 19, 20, 22, 32, 33, 47, 64, 70, 76, 79, 141, 158,

copper(II), Cu(II) (*contd.*)
  163, 164, 187, 193, 199, 202,
  206, 353, 358, 423
copper(III) 60, 63, 64
corrin 3, 11, 12, 13, 45
Corriu, R. J. P. 240
corrole
 5-hydrotetraoxa 24
 5-oxotetraoxa 24
 bifuran-containing 20, 22, 41, 83, 84
 bithiophene-containing 22
 C-allyl 71
 C-methyl 72
 carbethoxy 67
 deprotonation 33, 64
 dioxa 24
 from bipyrrolic precursors 29
 from tetrapyrrolic precursors 26
 furan-containing 20, 22
 hetero 12, 17, 20, 22, 83
 imino 18, 19
 meso phenyl 42, 43, 45
 metallo 17, 18, 31, 32, 33, 35, 39,
  41, 42, 45, 46, 47, 54, 55, 56,
  60, 62, 63, 64, 70, 71, 76, 78,
  79, 85
 metallo imino 18
 metallo oxa 18
 N-acetyl 64
 N-acyl 67
 N-alkyl 64, 69, 74, 75, 76, 78, 79, 83
 N-aryl 73
 N-ethyl 76, 79
 N-methyl 66, 67, 68, 70, 72, 74, 75,
  76, 82, 83
 N-methylimino 18
 N-substituted 64, 66, 73, 74, 78, 83
 N,N'-dimethyl 68, 69
 numbering 13
 oxa 18
 oxidation 61, 62, 63, 64
 protonation 16, 41, 62, 72, 76, 79,
  80, 82, 83
 reduction 24, 62, 63, 64, 73, 84
 reviews 11, 12, 13
 tetraoxa 24
 thia 19, 20
corrole-to-porphyrin relationship 14,
 186
corrologen 24, 84, 91
 tetraoxa 24
corrphycene 162, 163, 164, 171, 372
 metallo 163, 164
coupling
 exitonic 159
 low-valent titanium-mediated 216,
  218, 223
 McMurry 110, 111, 128, 162, 216,
  219, 220, 225, 228, 233, 236,
  239, 308, 325, 359, 363
 reductive carbonyl 88, 91, 130, 132,
  134, 136, 139, 140, 148, 169
 Wittig 119
Cram, D. J. 245, 246, 362
Cresp, T. M 116, 117, 118
cryptand 421, 423
cumulene 216, 217, 222, 313, 436
cyanide 279, 453
cycloheptatriene 186
cyclooctapyrrole 371, 379
cyclopentadiene 186
cyclopentadienyl 14, 179
cyclophane 134, 136, 151
cyclotetramer 359
cystic fibrosis 453, 479
Cytopharm, Inc. 441
cytosine 286, 338, 481
cytotoxicity 318, 443
cytotoxin 443

Day, V. 412, 414
DBU 154, 418
DDQ 45, 63, 84, 135, 139, 174, 206,
  209, 232, 235, 239, 244, 322,
  338, 339, 340, 351, 419
decaphyrin 369, 380
decaphyrin-(1.0.1.0.0.1.0.1.0.0) 369
decaphyrin-(2.0.0.0.0.2.0.0.0.0) 380
decapyrrolic 369, 380

definitions 3
dehydrocorrin 3, 11, 13
  octa 11
  tetra 11
[22]dehydropentaphyrin-(2.1.0.0.1) 311
[22]dehydrosapphyrin-(2.1.0.0.1) 311
demetalation 19, 154, 163, 167, 193,
    199, 204, 206, 276, 414
deprotonation 33, 64, 282, 342, 397,
    416
di-*t*-butylperoxide 27
diagnosis 429, 430
dianion 140, 141, 213
diastereomer 278
diatropic 120, 140, 202, 208
  dianion 140
  shift 141
dicarboxylate 154, 294
dichloroacetic acid 287
2,3-dichloro-5,6-dicyano-1,4-
    benzoquinone (DDQ) 45, 63,
    84, 135, 139, 174, 206, 210, 232,
    235, 239, 240, 244, 322, 338,
    339, 340, 351, 419
[22]didehydroporphyrin-(2.2.2.2) 217
Diels-Alder reaction 154
dihydroporphycene 148
diiminoisoindoline 99, 100, 101, 103
dimer
  corrole-corrole 85, 86
  corrole-porphyrin 85, 86, 87
  homo-corrole 87
  µ-oxo-bridged 146
  subphthalocyanine 103
dimerization 169, 270
  McMurry-type 137, 157, 226, 230,
    236, 308, 345, 381
  oxidative 415
dipyrrylmethane 20, 22, 29, 41, 42, 43,
    44, 45, 175, 205, 206, 209, 255,
    257, 260, 261, 266, 305, 307,
    315, 376, 386, 415, 416
dipyrrylmethene 91
distorted square pyramidal 53
dithiothreitol (DTT) 487

divalent
  metal 22, 32, 47, 55, 62, 79, 141,
    142, 144, 160, 171, 176, 199,
    278, 390
  metalloporphyrin 33
DNA 288, 443, 444, 445, 446, 447,
    450, 451, 453, 485, 486, 487,
    488
  backbone 446, 486
  double stranded 486
  photolysis 405, 444, 447, 448, 453
  plasmid 486, 487
  single stranded 446, 451, 486
DO3A 431
Dolphin, D. 175, 415
DTPA 431
Dy(III) 400, 451, 453

EDTA 486, 487
electrochemical 62, 63, 140, 195
electron
  14 π- 97
  16 π- 117, 120
  18 π- 14, 24, 33, 119, 120, 128, 136,
    139, 162, 163, 270, 307, 322,
    398
  20 π- 128, 136, 137, 139, 157, 193,
    244, 304
  22 π- 4, 140, 205, 206, 207, 225,
    227, 232, 237, 254, 270, 298,
    307, 309, 311, 316, 322, 327,
    340, 344, 391, 398, 409, 418
  24 π- 308, 322, 340, 341, 344, 352,
    391
  26 π- 209, 210, 213, 214, 223, 322,
    329, 331, 335, 336, 345, 346,
    419
  28 π- 212, 213, 215, 229, 345, 346,
    348
  30 π- 212, 213, 215, 364
  32 π- 379
  34 π- 215, 216, 377
  36 π- 361, 372, 392
  40 π- 369

electron (*contd.*)
   hydrated 407, 443, 444
   solvated 443
Elix, J. A. 110, 230, 324, 358, 359
emerald 297, 308
enantiomer 278, 289, 370, 373, 379
ene-diyne 229
Engel, J. 28
enzyme 2, 446
epr 62, 358
europium(III), Eu(III) 451, 453
eutrification 454
extraction 247

Falk, H. 162
FDA 431
FeCl$_3$ 27, 37, 51, 61, 91, 162, 169, 322
ferromagnetic 355
figure-eight 369, 370, 371, 372, 373,
   378, 379, 474, 477
flexibility 225, 227, 263, 333, 474
Floriani, C. 13
flow chart of porphyrin analogues 6
fluctional 226, 330, 341
formic acid 233, 253, 254
Förster-type singlet-singlet energy
   transfer 484
fractionation 443, 444
Franck, B. 204, 206, 208, 211, 215,
   219, 326, 419, 442, 455
   nomenclature 4, 5, 162, 329
   notation 309, 340
furan
   2,5-bis(triphenylphosphonium
      chloride) 116
   2,5-diacetic acid 116
   bis(pyrrolyl) 346
   dibenzo 362
   diethyl 213
   diformyl 110, 111, 116, 386
Furuta, H. 172, 174

gadolinium 431
gadolinium(III), Gd(III) 402-404, 405,
   430, 431, 432, 442, 443, 453

gallium(III) 146
gamma-ray 443
Gebauer, A. 338
gene 445, 451
   expression 446
genotypic 445
germanium 40
germanium(IV) 146
Ghosh, A. 174
Gingl, F. 423
Glaxo-Welcome 441
GMP 337, 481, 482
gold(III) 342
Gosmann, M. 211
Gossauer, A. 29, 315, 319, 329, 331,
   333, 414
green 253, 268, 297, 307, 308, 311,
   316, 486
Grigg, R. 46, 74, 186, 256, 266, 295,
   295, 297
groove binding 485, 486
guanidinium 454
guanosine 286
guanosine 5′-monophosphate 338
Guilard, R. 162

H$_2$O$_2$ 27
H$_2$SO$_4$ 16, 18, 19, 83, 151, 163, 199,
   201, 244, 347, 389, 436
Hanack, M. 97, 100, 103, 107
Harriman, A. 436
HBF$_4$ 407, 472
HBr 16, 24, 28, 41, 172, 205, 209, 212,
   253, 255, 271, 295, 315, 316,
   329, 364, 389
HCl 24, 39, 54, 109, 116, 118, 193,
   260, 263, 269, 270, 271, 277,
   304, 311, 318, 337, 338, 352,
   369, 370, 373, 374, 389, 394,
   407, 455, 457, 458, 468, 472,
   473, 474, 475
HClO$_4$ 24, 61, 62, 136, 137, 157, 212,
   214, 221, 229, 231, 233, 373
helical twist 369

hemiporphycene 167, 168, 169, 171
hepatitis 434
heptacoordinate 423
herpes simplex virus (HSV) 442
heterodimerization, McMurry-type 347
heterosapphyrin 454
hexamer 321, 324, 325, 334, 358, 359, 360
[24]hexaphyrin-(1.0.1.0.1.0) 340
[26]hexaphyrin-(1.1.0.1.1.0) 336
[26]hexaphyrin-(1.1.1.1.1.1) 4, 329
[30]hexaphyrin-(2.0.2.0.2.0), hexathia 363
[34]hexaphyrin-(2.2.2.2.2.2), hexathia 360
[36]hexaphyrin-(2.2.2.2.2.2)
    hexaoxa 358, 359
    hexathia 358, 359
hexaphyrin 4, 5, 314, 315, 329, 330, 331, 332, 333, 336, 351, 363, 364
    decamethyl 331
    triply-contracted 340
hexaphyrin-(2.0.0.2.0.0) 346
hexaphyrinogen, hexathia 334
hexapyrrin 344
hexapyrrolic macrocycle 4, 260, 315, 329, 336, 337, 340, 346, 349
hexavalent osmium 148
Hg(II) 317
HIV 442
HOAc 83, 418
homoannulenes 118
homoazaporphyrin 186, 187, 188
homoporphycenes 346
homoporphyrin 167, 185, 186, 188, 189, 192, 193, 195, 198, 199, 200
homoporphyrinoid 185
Hoogsteen (hydrogen bonding) 445
HPLC 97, 487
HSCN 394, 472
HSV 442
Hückel (theory) 2, 185, 213, 216, 378

human immunodeficiency virus (HIV) 442
Hush, N. S. 62
hybrid 446
hybridization 189, 448
hybridized
    sp 217, 311
    sp^2 3, 4, 127, 185, 193
    sp^3 120, 185, 186, 278
hydrazine 55, 85, 239, 391
    hydrate 63
hydrogen bonding 227, 260, 338, 352, 390, 418, 462, 468, 472, 482
hydroquinone 26, 340
hydroxyl radical 443

Ibers, J. A. 308, 346, 454, 455
imidazole 180
imine 390, 394, 417, 444
indium(III), In(III) 40, 53, 146
indole
    2-(α-hydroxyamide) 115
    2-hydroxymethyl 114
    7-(α-hydroxyamide) 115
    7-hydroxymethyl 114
intercalation 485, 486
iodine (I_2) 52, 253, 254, 276, 288
iridium 280, 341
iridium(I), Ir(I) 221, 274, 280
iron 38, 45, 55, 60, 61, 171, 322, 487
iron(II), Fe^{2+} 51, 62, 272, 486
iron(III), Fe(III) 37, 38, 39, 51, 52, 54, 55, 62, 64, 144, 146, 164, 199, 220, 229
iron(IV) 54, 55, 60
isocorrole 12, 88, 89, 90, 91, 93, 94
    etio 90, 91, 94
    formyl-substituted 89, 91
    octaalkyl 88
    octaethyl 89, 90, 93
isoporphycene 170
isosmaragdyrin 306, 307
Iverson, B. L. 485, 486, 487

Johnson, A. W. 17, 18, 19, 22, 26, 27, 31, 47, 48, 49, 50, 66, 67, 70, 73, 77, 84, 256, 265, 266, 280, 295, 295
Johnson, M. R. 308, 346, 391
Jori, G. 441

$K_2Fe(CN)_6$ 27
Kaposi's sarcoma 438
Kasuga, K. 100
Kay, I. T. 48, 66
kinetic
    acidity 174
    barrier 134
    control 66, 72
    instability 93
    stability 431
    studies 404
Knübel, G. 215
Kobayashi, N. 97, 99
Kodadek, T. J. 259, 260

lanthanide 2, 272, 399, 400, 402, 403, 405, 418, 429, 431, 432, 449, 450, 451, 453
lanthanum(III), La(III) 402, 403
lasalocid 288
Latos-Grażyński, L. 172, 173, 174, 175, 261, 262
lead, Pb^{2+} 218, 278, 394
LeGoff, E. 205, 206, 209, 210, 364
leukemia 437
Lewis acid 450, 454
    catalyst 84, 259, 394
$LiAlH_4$ 118, 149, 347, 416
Licoccia, S. 38, 42, 45
ligand
    apical 38, 400, 402, 432
    binucleating 2, 352
    dianionic 32, 141, 317
    ditopic 286, 289, 290, 390, 483
    monoanionic 397, 400
    neutral 141, 246, 282, 390

pentadentate 318, 370
pentaligating 280
trianionic 17, 272
ligation 75, 273, 274, 353, 432
    axial 39, 48
    out-of-plane 37, 274
light harvesting arrays 290
Lind, E. 364
Lindlar reduction 218, 311
Liu, B. Y. 174
LUMO 443
lung cancer 434
lutetium(III), Lu(III) 405, 436, 437, 447, 448, 453

MacDonald, S. F. 256, 295, 297, 315, 336, 376
macrocycle
    decaheterocyclic 380, 382
    decapyrrolic 369, 380
    hexameric 358
    hexapyrrolic 4, 260, 315, 329, 336, 336, 340, 346, 351, 352
    metalated 18, 20, 76, 77, 86, 200, 274, 331, 386, 399
    methylated 67, 70, 72, 157, 176, 371
    non-conjugated 239, 326, 372
    octaaza 390
    octapyrrolic 371, 372, 380
    pentameric 315, 321, 324, 326, 409, 412, 414
    pentapyrrolic 4, 253, 254, 255, 261, 295, 303, 306, 308, 311, 326
    tetrameric 359
    [4n + 2]-type 364
Magda, D. 447, 450, 451, 453
magnesium, Mg(II), Mg^{2+} 164, 171, 386
magnetic resonance imaging 2, 429, 454
Magnevist™ 431
Mallouk, T. E. 436
manganese 45
manganese(II), Mn(II), Mn^{2+} 32, 45, 52, 272, 431

manganese(III), Mn(III) 37, 51, 52, 146, 431
manganese dioxide 391
Märkl, G. 230
Marks, T. 412, 414
Mathey, F. 13, 249
Matthews, J. L. 442
Maxim-Gilbert sequencing 447
maximum tolerated dose (MTD) 438
McMurry, J. E. 91, 110, 111, 112, 128, 137, 157, 162, 216, 219
melanoma 436, 438, 439, 441
Meller, A. 95
membrane, U-tube model 289, 338, 479, 481, 482, 483
Mertes, K. B. 386
Merz, A. 140, 346, 348, 363, 364
metal chelating cores 329
metal chelation 180, 185, 272, 282, 341
metal coordination 22, 34, 53, 65, 74, 75, 94, 141, 144, 164, 171, 191, 240, 277, 282, 331, 333, 339, 353, 370, 392, 400, 418, 423
metal
 divalent 22, 32, 46, 47, 55, 57, 62, 79, 94, 141, 142, 144, 160, 171, 176, 199, 278, 390
 higher-valent 141
 ligation 37, 75, 273, 274, 353
 monovalent 22, 46, 74, 141
 tetravalent 17, 46, 61, 64, 146, 171
 trivalent 33, 35, 36, 46, 60, 94, 144, 146, 171, 272, 399, 403, 429, 431, 450, 451
metalloporphycene 141
methanol
 addition 193, 278, 279
 binding 227, 390, 402
methanolysis 195
migration 192
 methyl 161, 162
 proton 193
molybdenum, oxo 55
molybdenum(V) 55
monometalated 148, 273

mononuclear complex 418
monovalent 22, 46
 metal 22, 46, 74, 141
Moreau, J. J. E. 240
Mössbauer spectrum 61
mouse 438
MRI 2, 400, 405, 429, 430, 431, 432, 444, 454
mRNA 444, 446
MTD 438
Murakami, Y. 32, 55, 62
mutant porphyrin 172, 173, 180, 326

N-confused porphyrin 172, 173, 180, 326
N-meso homoporphyrin 186
naphthalene 97, 396
 1-chloro 107
 2,3-dicyano 97
 diamino 396, 397
near-infrared (IR) 361, 436
Neidlein, R. 139, 346, 348, 363, 364
neoplastic 431, 432, 445
neutral pH 286, 289, 339, 461, 481, 485, 486
neutral substrate 227, 253, 390, 423
nickel 2, 32, 55, 60, 63, 87, 142, 160, 171, 176, 191, 195, 196, 390
nickel(II), Ni(II), Ni^{2+} 22, 31, 32, 33, 45, 47, 56, 62, 64, 70, 71, 75, 79, 141, 143, 144, 154, 164, 167, 174, 188, 192, 193, 195, 199, 206, 272, 280, 331, 423
niobium 423
niobium(IV) 423
nitric acid 136, 137, 157, 389
n.O.e. 332
nomenclature 3
non-aromatic 33, 47, 108, 113, 114, 185, 186, 193, 226, 228, 229, 229, 231, 235, 237, 240, 278, 304, 307, 340, 341, 344, 352, 369, 379, 389, 397
non-racemic 260

Nonn, A. 204
1R-(+)-nopinone 260
norsapphyrin 295, 297
  dioxa 295, 296
  hetero 295
  pentaaza 297
nuclease 446
nucleobase 286, 288
nucleoside 286
nucleotide 3, 286, 287, 294, 453, 482, 483

O_2 193, 209, 397, 434, 486, 487
[26]octadehydroporphyrin-(2.4.2.4) 223
[28]octadehydroporphyrin-(6.0.6.0) 229
octamer 372
octaphyrin 371, 376, 378, 379, 454, 477
octaphyrin-(1.0.1.0.1.0.1.0) 379
octaphyrin-(1.1.1.0.1.1.1.0) 376
octaphyrin-(2.1.0.1.2.1.0.1) 371, 376, 454, 477
  tetrahydro 371
ODN 444, 445, 446, 447, 450, 451, 453
oligodeoxynucleotide 444
oligomer 50, 285, 291, 293, 405, 436, 450
oligopyrrole 423
Omniscan^{TM} 431
orangarin 303, 304, 369, 454, 473, 474
orange 316
orange-red 336
Osa, T. 99
osmium 148
osmium(II) 148
osmium(VI) 148
Ossko, A. 95
outside stacking 486
oxidant 139, 347, 351, 361, 398
oxidation 37, 39, 45, 51, 55, 56, 61, 62, 63, 64, 84, 93, 128, 135, 136, 137, 140, 151, 157, 162, 169, 174, 175, 199, 209, 212, 214, 216, 221, 223, 225, 229, 231, 233, 235, 240, 253, 254, 256, 257, 261, 269, 279, 315, 321, 322, 329, 336, 339, 340, 344, 347, 351, 353, 360, 361, 365, 391, 397, 398, 409, 416, 419
  electrochemical 63, 140, 199
oxygen 1, 28, 91, 193, 209, 397, 416, 434, 443, 448, 486, 487
  singlet 404, 434, 436, 437, 447, 453
  surrogate 444
Oz 308
ozaphyrin 308, 309, 310

palladium 47, 141, 160, 332, 372
palladium(II), Pd(II) 19, 47, 72, 75–79, 141, 171, 274, 332
Paolesse, R. 28, 42, 43, 44, 45, 85, 86, 87
paratropic 117, 215, 234
paratropicity 117, 120
PCI-0120 432, 443, 444
PCI-0123 437, 438, 439, 447
PDI 433, 434, 436, 442, 447, 449
PDT 2, 152, 205, 271, 429, 431, 433, 434, 436, 437, 438, 441, 442, 447
PDV 434
pentagonal 310
  bipyramidal 320, 400
  planar 278
pentamer 321, 324, 334, 381
[20]pentaphyrin-(2.1.0.0.1) 303
[22]pentaphyrin-(1.1.1.1.1), pentathia 321
[22]pentaphyrin-(2.0.2.0.0) 309
[22]pentaphyrin-(2.1.0.0.1) 311
[30]pentaphyrin-(2.2.2.2.2)
  pentaoxa 324
  pentathia 325
pentaphyrin 4, 5, 313, 314, 315, 318, 322
  dehydro 311
  hetero 321
  inverted 326

pentaphyrin (*contd.*)
  metallo 318
  pentaaza 315
  pentathia 321
  tetrahydroxy 318
  uranyl 319
  water-soluble 318
pentaphyrinogen
  decavinylogous 326
  inverted 327
  pentathia 321, 322
pentapyrrolic
  expanded porphyrin 303, 326,
    329
  macrocycle 4, 253, 254, 255, 261,
    295, 303, 306, 309, 311, 326
perchloric acid ($HClO_4$) 24, 61, 62,
    136, 137, 157, 212, 214, 220,
    229, 231, 233, 373
Periodic Table of the Texaphyrins 399
pH 268, 480, 481, 484
  neutral 286, 289, 338, 443, 461, 480,
    481, 485, 486
  physiological 338
pharmaceutical 429
Pharmacyclics, Inc. 432, 438
o-phenylenediamine 389, 392, 394,
    396, 398, 408
phosphodiester 288, 450, 485
phosphoramidate 446
phosphoric acid 461, 486
photodynamic therapy 2, 152, 180,
    205, 400, 405, 429, 431, 433
photodynamic viral inactivation 433,
    434
Photofrin® 436
photolysis 405, 442, 444, 446, 447,
    448, 453
photomodification 288, 447, 448, 449
photosensitizer 2, 152, 271, 404, 405,
    431, 433, 434, 436, 438, 441,
    442, 447, 449
photosynthesis 1
phthalocyanine 3, 95, 98, 99, 101, 103,
    409, 412, 414, 423

phthalonitrile 95, 96, 104, 409, 412,
    414, 423
Pigments of Life 1
pink 340
PKa 269, 416, 417, 481
planar coordination 35, 50, 142, 143,
    191, 282, 332, 418
platinum 84, 141
platinum(II), Pt(II) 141
platyrin 4, 205, 209
polyammonium 454
polyether 350
porphocyanine 415, 416, 417, 418
porphycene 127
  2,3-dihydro 148, 150, 151, 157
  9,10-dihydro 151
  benzo 134, 135, 154, 155, 226
  cumulene 436
  dihydro 134, 136, 151
  dihydro-dithia- 139
  dihydro-O_4 137
  dioxa 137
  dithia 139
  expanded 440
  for PDT 152
  hetero 136
  N-methyl 157, 159, 161
  N,N′-bridged 158, 159
  N,N′-dimethyl 157
  N,N′N″-trimethyl 157
  O_4 137
  stretched 216, 221, 224, 226, 230
  tetra-*N*-methyl 157
  tetraoxa 136, 137
  unsymmetrically-functionalized 130
  vinyl 154
porphycenogen
  dioxa 157
  dithia 139
  tetrathia 363
porphycerin 162
porphylene 162
porphyrazines 3
[18]porphyrin-(2.1.0.1) 162
[18]porphyrin-(2.1.1.0) 167

[18]porphyrin-(3.0.1.0) 171
[22]porphyrin-(2.1.2.1), tetrathia 239
[22]porphyrin-(2.2.2.2) 217, 218, 219, 220, 222
  octaethyl 220
  tetraoxa 230, 232, 233
  tetrathia 235, 236
[22]porphyrin-(3.0.3.0), tetraoxa 239
[22]porphyrin-(3.1.3.1) 4, 205, 206, 219
[22]porphyrin-(4.0.4.0) 224
[24]porphyrin-(2.2.2.2) 220
  dioxadithia 234
  tetraoxa 230, 231, 245
  tetrathia 234, 235, 236, 236
[26]porphyrin-(3.3.3.3) 212
  tetraoxa 214, 215
[26]porphyrin-(5.1.5.1) 209
  14-aza 419
[26]porphyrin-(6.0.6.0) 226
[34]porphyrin-(5.5.5.5) 215, 216
porphryin
  aminophenyl 289, 421
  aza 409
  bisvinylogous 205, 206, 208, 240, 241, 442
  capped 284
  dimer 285
  hemato 431, 436
  homologue 329
  inverted 174, 175, 176
  metallo 33, 60, 87, 187, 318, 454
  mutant 170, 173, 178, 326
  N-confused 170, 173, 178, 326
  octaethyl 186, 207, 241
  octavinylogous 201, 215, 326
  oligomer 285, 436
  phenyl-substituted 265
  picket-fence 284
  spiro 13
  tetraaryl 188, 351
  tetraphenyl 172, 188, 192, 198, 261
  tetravinylogous 201, 209, 211, 212, 419
  Texas-size 4, 392

  vinylogous 4, 185, 204, 205, 208, 215, 216, 240, 436, 442, 454
porphyrin-(2.0.2.0) 127
porphyrin-like 11, 129, 130, 180, 199, 204, 240, 322, 331, 333, 353, 374, 390, 431, 450
porphyrinogen 3, 173, 393, 397
  expanded 407
  inverted 173
  tetraoxa 24, 213
  tetraphenyl 45
  uro 45
porphyrinogen-(2.1.2.1), tetrathia 239, 240
porphyrinogen-(2.2.2.2), tetrathia 235
porphyrinogen-(3.0.3.0), tetraoxa 239
porphyrinogen-(3.3.3.3) 211, 212
Prohance^{TM} 431
proton hole 15
pseudohexagonal 418
pyrazole 364
pyridine 19, 34, 35, 38, 41, 47, 48, 49, 51, 54, 55, 62, 73, 77, 94, 110, 164, 295, 332, 400
  -containing macrocycle 240, 244
  -like nitrogen 269, 316
  t-butyl 94
pyrrolinene nitrogen 417

quinone
  [22]annuleno 244
  ρ-benzo 329
  2,3-dichloro-5,6-dicyano-1,4-benzo 45, 135, 206, 339
  hydro 26, 340

rabbit 432, 439
racemic 260
radiation
  dose 443
  sensitization 400
  sensitizer 432, 442, 443, 444
  therapy 2, 407, 429, 442, 443, 444

Rauschnabel, J. 97, 107
receptor
  anion 287, 294, 329, 461, 472, 477
  cation 454
  ditopic 289, 483
  metal 352, 370, 390
  neutral substrate 253, 390
  π–π stacking 362
  polytopic 423
  stereo-differentiating 260
  supramolecular 423
  tritopic 286
rectangular 132, 164
red 4, 109, 340, 394
reduction
  electrochemical 62, 140, 195
  hydrazine hydrate 63
  hydrogenation over platinum
    catalyst 84
  hydroxide-mediated 62
  Lindlar-type 218, 220, 311
  lithium aluminum hydride 118
  potassium metal 212, 215
  sodium borohydride 24, 63, 73, 118,
    215, 322
  sodium film 62
  sodium metal 140, 151
  zinc in acetic acid 195
relaxation time 430
relaxivity 430, 431, 432
resonance hybrid 224
rhenium(I) 147
rhodium 22, 35, 36, 45, 46, 53, 74,
    162, 171, 222, 241, 341
rhodium(I), Rh(I) 22, 37, 75, 222, 225,
    228, 274, 280
rhodium(III), Rh(III) 36, 37, 52, 161
ribozyme 446
Richert, C. 153, 154
ring contraction 45, 88, 89, 92, 167,
    186, 187, 188, 192, 316, 414
ring current
  diamagnetic 98, 120, 129, 141, 163,
    171, 173, 205
  paramagnetic 136

RNA 2, 444, 445, 446, 447, 450, 451,
    453, 488
  hydrolysis 405, 429, 444, 448, 449,
    450, 451, 453
  photolysis 453
RNase H 446
rosarin 340, 341, 342, 454, 474
  tris(phenyl) 340, 344
rosarinogen 344
  dioxa 344
  hetero 344
  meso-unsubstituted 344
  mono-aryl 344
  mono-phenyl 344
Rothemund, P. 172, 340, 351
rubyrin 4, 336, 337, 338, 454, 473, 478,
    481
  hetero 339
  hexathia 339, 340
ruthenium 52, 148
ruthenium(III) 37

samarium(III) 342
sapphyrin
  aggregation 270, 272, 486
  azacrown 290
  capped 285, 289, 290
  carboxylic acid 286, 291
  conjugate 287, 288, 289, 448, 482,
    486
  diamine 291
  dimer 270, 291, 293, 462, 486
  dioxa 266, 269, 271, 280
  diphenyl 259, 260, 261, 269
  hetero 265, 266, 269, 271, 280, 282,
    283, 454, 462, 472
  isomer 309, 311, 312
  mono-oxa 271, 282, 468
  mono-selena 280, 468
  mono-thia 266, 271, 280, 282
  oligomer 291, 293
  oxa 271
  protonated 270, 408, 429, 454, 461,
    462, 484, 485, 486, 487

sapphyrin (*contd.*)
  pseudo dimer 289, 484
  reduction 278
  tetraphenyl 261, 262, 263, 271
  three dimensional 284
sapphyrin-like 265, 278
sarcoma 431, 439
Sargent, M. V. 116, 117, 118
SbCl$_5$ 321
Schiff-base 385, 386, 389, 391, 392,
    407, 408, 416, 421
Schumacher, K.-H. 326
sensitization enhancement ratio 444
sensitizer
  photodynamic therapy 2, 152, 205,
    271, 405, 431, 434, 436, 438,
    441, 442, 447, 449
  radiation 432, 442, 443, 444
    therapy 2, 429, 442
sensor 429
SER 444
serendipitous 167, 199, 253, 265, 369,
    371
Sessler 162, 167, 169, 256, 257, 259,
    260, 261, 273, 274, 278, 279,
    280, 282, 287, 289, 291, 298,
    311, 316, 317, 335, 339, 344,
    348, 353, 369, 371, 389, 390,
    392, 393, 399, 405, 414, 421,
    436, 442, 454, 455, 458, 477,
    481, 483, 485, 487
signal intensity (SI) 430
silica gel 274, 487
sitting-a-top 274, 280
smaragdyrin 295, 297, 298, 303
  dioxa 298
  hetero 295, 298
Smith, K. M. 13, 29
sodium
  borohydride 24, 63, 73, 118, 162,
    215, 322
  metal 62, 140, 151
  sulfide 19
Solomon-Bloembergen equation 430
spiro porphyrin 13

square 137, 164
  planar 35, 50, 132, 142, 143, 191,
    332
  pyramidal 34
Sr^{2+} 386
staging 429
Strähle, J. 423
Strand, A. 234
stretched porphycene 216, 221, 224,
    226, 230
structure, cyclophane-type 134, 136,
    151
Stuart-Briegleb model analysis 119
subnaphthalocyanine 97
subphthalocyanine 95, 98, 99, 100,
    103, 104
  chloro-substituted 98
  dimer 104, 107
  hexakis(alkylthio)-substituted 96,
    100
  nitro-substituted 96
  phenyl-substituted 98
  t-butyl substituted 95, 99
subtriazaporphyrin 107
sulfur-containing 117
sulfuric acid 16, 18, 19, 83, 151, 163,
    199, 204, 244, 347, 389, 436
superarene 209
superoxide 443
superphthalocyanine 409, 412, 414
symmetry
  C_2 230, 257, 372
  C_{2h} 220, 331
  C_{2v} 331
  C_3 340, 341
  centrosymmetric 141, 224
  D_2 373
  D_{2h} 221, 246
  D_{3d} 362
  D_{5h} 321
  loss in 278
synthesis
  3 + 2 256, 257, 261, 266, 295, 297,
    298, 306, 329
  3 + 3 315, 329

synthesis (*contd.*)
  4 + 1 256, 297
  4 + 2 = 5 260
  MacDonald-type 256, 295, 297, 315,
    336, 376
  McMurry-type 93, 110, 111, 112,
    128, 137, 157, 162, 216, 219,
    220, 225, 226, 229, 230, 232,
    233, 235, 236, 308, 325, 346,
    350, 359, 363, 381
  Rothemund-Adler-type 340, 351
  Rothemund-type 172
system, higher order 204, 369, 371,
  381

T_1 430
T_2 430
Tb(III) 453
terpyrrole 304, 306, 311, 351, 369
  diformyl 307, 392
  hexaalkyl 351
  hexamethyl 371
[22]tetradehydroporphyrin-
  (2.2.2.2) 216, 217, 218, 220, 222
tetrahedral 31, 282, 418
tetrahydrochloride 474
tetraligated 273
tetrapyrrole 19, 20, 254, 261, 379
  bisvinylogous 206
  dibromo 163
  diformyl 408
  monoaldehyde 171
  mono-amide-containing 18
tetrapyrrolic
  intermediate 44, 169
  precursor 26, 31, 91, 162, 253, 256
  skeleton 150
texaphyrin 4, 392, 393, 397, 399, 400,
  401, 402, 403, 405, 407, 429,
  432, 437, 444, 450, 451
  metallo- 398, 399, 400, 402, 403,
    404, 405, 407, 432, 436, 437,
    442, 443, 447, 448, 450, 451,
    453

texaphyrinogen 393, 398, 472
  expanded 409
Texas-size 4, 392
TFA 25, 44, 72, 76, 79, 173, 175, 193,
  199, 201, 205, 208, 214, 227,
  287, 291, 306, 307, 349, 351,
  361, 418, 419, 455, 456, 462,
  468, 474
thiacycloheptadecin 118, 119
thiaphlorin 20, 24, 29, 67, 77, 269
thiophene
  2-hydroxymethyl 321, 334
  2,5-bis(triphenylphosphonium
    chloride) 117
  2,5-diacetic acid 108, 118
  3,4-diethyl 339
  bis(pyrrolyl) 346
  diformyl 110, 111, 235, 236, 325,
    359, 386
$TiCl_3/LiAlH_4$ 149, 347
$TiCl_4/Zn$ 149, 347
tin 171
tin(II) 39
tin(IV), Sn(IV) 39, 40, 146, 164
titanium, low-valent 93, 110, 136, 139,
  140, 169, 216, 218, 223, 230,
  235, 308, 347, 348
p-toluenesulfonic acid 114, 334
p-tolylisocyanide 41
topoisomerase I 486
torand 242
Torres, T. 96
toxicity 432, 437
trapezoid 164
triethylamine 36, 195
trifluoroacetate 79, 462
trifluoroacetic acid 25, 44, 72, 76, 79,
  173, 175, 193, 199, 204, 205,
  208, 214, 227, 287, 291, 305,
  307, 318, 351, 360, 361, 418,
  419, 455, 456, 462, 468, 474
  deutero 16
trigonal prism 423
trihydrochloride 340, 474

tripyrrane 174, 256, 261, 266, 267, 315,
    316, 329, 394, 396, 407, 421
  strapped porphyrin 421
Tschamber, T. 188
tumor 318, 431, 432, 433, 434, 436,
    438, 441, 442, 443, 444
turcasarin 369, 370, 454, 474, 475
turquoise 369

uranyl (UO_2^{2+}) 277, 278, 279, 319,
    370, 389, 394, 409, 412, 413,
    414
uroporphyrin 29, 84

van der Waals 132
Vilsmeier's reagent 158
Vilsmeier-type formylation 158
vinylogous porphyrin 4, 185, 204, 205,
    208, 215, 216, 240, 436, 442,
    454
  bis 241, 442
  octa 215
  tetra 209, 211, 212, 419
vinylporphycene 154
virus 434, 442
vitamin B_{12} 11, 13, 253
Vogel, E. 22, 24, 47, 54, 55, 56, 60, 63,
    64, 88, 92, 93, 94, 110, 111,
    112, 127, 128, 129, 130, 132,
    134, 136, 137, 157, 162, 167,
    169, 170, 216, 218, 219, 222,

224, 226, 321, 325, 339, 371,
    376, 379, 477

Watson-Crick hydrogen bonding 445,
    481
Weaver, O. G. 209, 210
Weghorn, S. J. 353
window of transparency 269
Wittig, G. 110, 116, 119, 230, 234,
    237, 244, 249, 324, 358, 359
Wöhrle, D. 100, 102, 103
Woodward, R. B. 253, 255, 256, 272,
    277, 278, 297, 298
Woolsey, I. S. 62

X-ray radiation therapy (XRT) 2, 407,
    429, 442, 444

yellow-green 316
Young, S. W. 432
Yttrium(III), Y(III) 448, 453

Zard, S. Z. 257
zinc 144, 160, 171, 195, 199, 272, 331,
    339, 418
zinc(II), Zn(II), Zn^{2+} 31, 47, 103, 141,
    144, 187, 198, 272, 273, 280,
    319, 331
  acetate dihydrate 102